Birkhäuser

ISNM

International Series of Numerical Mathematics

Volume 153

For further volumes:
www.birkhauser-science.com/series/4819

Tomáš Roubíček

Nonlinear Partial Differential Equations with Applications

Second Edition

Birkhäuser

Tomáš Roubíček
Mathematical Institute
Charles University
Sokolovská 83
186 75 Praha 8
Czech Republic
and
Institute of Thermomechanics of the ASCR
Dolejškova 5
182 00 Praha 8
Czech Republic
and
Institute of Information Theory and Automation of the ASCR
Pod vodárenskou věží 4
182 08 Praha 8
Czech Republic

ISBN 978-3-0348-0512-4 ISBN 978-3-0348-0513-1 (eBook)
DOI 10.1007/978-3-0348-0513-1
Springer Basel Heidelberg New York Dordrecht London

Library of Congress Control Number: 2012956219

Mathematics Subject Classification (2010): Primary: 35Jxx, 35Kxx, 35Qxx, 47Hxx, 47Jxx, 49Jxx;
Secondary: 65Nxx, 74Bxx, 74Fxx, 76Dxx, 80Axx

Printed on acid-free paper

Springer Basel is part of Springer Science+Business Media (www.birkhauser-science.com)

To the memory of professor Jindřich Nečas

Contents

II EVOLUTION PROBLEMS 199

Preface

The theoretical foundations of differential equations have been significantly developed, especially during the 20th century. This growth can be attributed to fast and successful development of supporting mathematical disciplines (such as functional analysis, measure theory, and function spaces) as well as to an ever-growing call for applications especially in engineering, science, and medicine, and ever better possibility to solve more and more complicated problems on computers due to constantly growing hardware efficiency as well as development of more efficient numerical algorithms.

A great number of applications involve distributed-parameter systems (which can be, in particular, described by *partial differential equations*[1]) often involving various nonlinearities. This book focuses on the theory of such equations with the aim of bringing it as fast as possible to a stage applicable to real-world tasks. This competition between rigor and applicability naturally needs many compromises to keep the scope reasonable. As a result (or, conversely, the reason for it) the book is primarily meant for graduate or PhD students in programs such as mathematical modelling or applied mathematics. Although some preliminary knowledge of modern methods in linear partial differential equations is useful, the book is basically self-contained if the reader consults Chapter 1 where auxiliary material is briefly presented without proofs.

The prototype tasks addressed in this book are boundary-value problems for *semilinear*[2] equations of the type

$$-\Delta u + c(u) = g, \qquad \text{or more general} \qquad -\operatorname{div}\big(\kappa(u)\nabla u\big) + c(u) = g, \tag{0.1}$$

or, still more general, for *quasilinear*[3] equations of the type

$$-\operatorname{div}\big(a(u, \nabla u)\big) + c(u, \nabla u) = g, \tag{0.2}$$

[1]The adjective "partial" refers to occurrence of partial derivatives.

[2]In this book the adjective "semilinear" will refer to equations where the highest derivatives stand linearly and the induced mappings on function spaces are weakly continuous.

[3]The adjective "quasilinear" refers to equations where the highest derivatives occur linearly but multiplied by functions containing lower-order derivatives, which means here the form $-\sum_{i,j=1}^{n} a_{ij}(x, u, \nabla u)\partial^2 u/\partial x_i \partial x_j + c(x, u, \nabla u) = g$. After applying the chain rule, one can see that (0.2) is only a special case, namely an equation in the so-called divergence form.

and various generalizations of those equations, in particular variational inequalities. Furthermore, systems of such equations are treated with emphasis on various real-world applications in (thermo)mechanics of solids and fluids, in electrical devices, engineering, chemistry, biology, etc. These applications are contained in Part I.

Part II addresses evolution variants of previously treated boundary-value problems like, in case of (0.2),[4]

$$\frac{\partial u}{\partial t} - \mathrm{div}\big(a(u,\nabla u)\big) + c(u,\nabla u) = g, \tag{0.3}$$

completed naturally by boundary conditions and initial or periodic conditions.

Let us emphasize that our restriction on the quasilinear equations (or inequalities) in the divergence form is not severe from the viewpoint of applications. However, in addition to fully nonlinear equations of the type $a(\Delta u) = g$ or $\frac{\partial}{\partial t}u+a(\Delta u) = g$, topics like problems on unbounded domains, homogenization, detailed qualitative aspects (asymptotic behaviour, attractors, blow-up, multiplicity of solutions, bifurcations, etc.) and, except for a few remarks, hyperbolic equations are omitted.

In particular cases, we aim primarily at formulation of a suitable definition of a solution and methods to prove existence, uniqueness, stability or regularity of the solution.[5] Hence, the book balances the presentation of general methods and concrete problems. This dichotomy results in two levels of discourse interacting with each other throughout the book:

- abstract approach – can be explained systematically and lucidly, has its own interest and beauty, but has only an auxiliary (and not always optimal) character from the viewpoint of partial differential equations themselves,
- targeted concrete partial differential equations – usually requires many technicalities, finely fitted with particular situations and often not lucid.

The addressed methods of general purpose can be sorted as follows:

- ∘ indirect in a broader sense: construction of auxiliary approximate problems easier to solve (e.g. Rothe method, Galerkin method, penalization, regularization), then a-priori estimates and a limit passage;
- ∘ direct in a broader sense: reformulation of the differential equation or inequality into a problem solvable directly by usage of abstract theoretical results, e.g. potential problems, minimization by compactness arguments;
- ∘ iterational: fixed points, e.g. Banach or Schauder's theorems;

---

[4]In fact, a nonlinear term of the type $c(u)\frac{\partial}{\partial t}u$ can easily be considered in (0.3) instead of $\frac{\partial}{\partial t}u$; see p. 277 for a transformation to (0.3) or Sect. 11.2 for a direct treatment. Besides, nonlinearity like $C(\frac{\partial}{\partial t}u)$ will be considered, too; cf. Sect. 11.1.1 or 11.1.2.

[5]To complete the usual mathematical-modelling procedure, this scheme should be preceded by a formulation of the model, and followed by numerical approximations, numerical analysis, with computer implementation and graphic visualization. Such, much broader ambitions are not addressed in this book, however.

We make the general observation that simple problems usually allow several approaches while more difficult problems require sophisticated combination of many methods, and some problems remain even unsolved.

The material in this book is organized in such a way that some material can be skipped without losing consistency. At this point, Table 1 can give a hint:

	steady-state	evolution
basic minimal scenario	Chapters 2,4	Chapter 7, Sect. 8.1–8.8
variational inequalities	Chapter 5	Chapter 10
accretive setting	Chapter 3	Chapter 9
systems of equations	Chapters 6	Chapter 12, Sect. 9.5
some special topics	—	Sect. 8.9–8.10, Chapter 11
auxiliary summary of general tools	Chapter 1	

Table 1. General organization of this book.

Except for the basic minimal scenario, the rest can be combined (or omitted) quite arbitrarily, assuming that the evolution topics will be accompanied by the corresponding steady-state part. Most chapters are equipped with exercises whose solution is mostly sketched in footnotes. Suggestions for further reading as well as some historical comments are in biographical notes at the ends of the chapters.

The book reflects both my experience with graduate classes I taught in the program "Mathematical modelling" at Charles University in Prague during 1996–2005[6] and my own research[7] and computational activity in this area during the past (nearly three) decades, as well as my electrical-engineering background and research contacts with physicists and material scientists. My thanks and deep

[6]In the usual European 2-term organization of an academic year, a natural schedule was Part I (steady-state problems) for one term and Part II (evolution problems) for the other term. Yet, only a selection of about 60% of the material was possible to expose (and partly accompanied by exercises) during a 3-hour load per week for graduate- or PhD-level students. Occasionally, I also organized one-term special "accretive-method" course based on Chapters 3 and 9 only.

[7]It concerns in particular a research under the grants 201/03/0934 (GA ČR), IAA 1075402 (GA AV ČR), and MSM 21620839 (MŠMT ČR) whose support is acknowledged.

gratitude are to a lot of my colleagues, collaborators, or tutors, in particular M. Arndt, M. Beneš, M. Bulíček, M. Feistauer, J. Franců, J. Haslinger, K.-H. Hoffmann, J. Jarušek, J. Kačur, M. Kružík, J. Málek, J. Malý, A. Mielke, J. Nečas, M. Pokorný, D. Pražák, A. Świerczewska, and J. Zeman for numerous discussions or/and reading the manuscript.

Praha, 2005 T.R.

Preface to the 2nd edition

Although the core of the book is identical with the 1st edition, at particular spots this new edition modifies and expands it quite considerably, reflecting partly the further research of my own[8] and my colleagues, as well as a feedback from continuation of classes in the program "Mathematical modelling" at Charles University in Prague during 2005–2012, partly executed also by my colleague, M. Bulíček.

More specifically, the main changes are as follows: On an abstract level, Rothe's method has been improved by using a finer discrete Gronwall inequality and by refining some estimates to work if the governing potential is only semiconvex, as well as various semi-implicit variants have been added. Also the presentation of Galerkin's method in Sect. 8.4 has been simplified. Morever, needless to list, some particular assertions have been strengthened or their proofs simplified.

On the level of concrete partial differential equations, boundary conditions in higher-order equations in Section 2.4.4 are now more elaborated. Interpolation by using Gagliardo-Nirenberg inequality has been applied more systematically; in particular the exponent $p^{\circledast}$ has been defined more meticuously and a new "boundary" exponent $p^{\oplus}$ has been introduced and used in a modified presentation of the parabolic equations in Sect. 8.6. Interpolation has also been exploited in new estimates especially in examples of thermally coupled systems which have been more elaborated or completely rewritten, cf. Sects. 6.2 and 12.1, and even some newer ones have been added, cf. Sects. 12.7–12.9. Other issues concern e.g. newly added singularly-perturbed problems, positivity of solutions (typically temperature in the heat equation), Navier's boundary conditions, etc.

Moreover, some rather local augments have been implemented. Various exercises have been expanded or added and some new applications have been involved. Also the list of references has been expanded accordingly. Of course, various typos or mistakes have been corrected, too.

Last but not least, it should be emphasized that inspiring discussions with M. Bulíček, J. Málek, J. Malý, P. Podio-Guidugli, U. Stefanelli, and G. Tomassetti have thankfully been reflected in this 2nd edition.

Praha, 2012 T.R.

[8]In particular it concerns the GA ČR-projects 201/09/0917, 201/10/0357, and 201/12/0671.

Notational conventions

A	a mapping (=a nonlinear operator), usually $V \to V^*$ or $\mathrm{dom}(A) \to X$,
a.a., a.e.	a.a.=almost all, a.e.=almost each, referring to Lebesgue measure,
$C(\bar{\Omega})$	the space of continuous functions on $\bar{\Omega}$, equipped with the norm $\|u\|_{C(\bar{\Omega})} = \max_{x\in\bar{\Omega}} \|u(x)\|$, sometimes also denoted by $C^0(\bar{\Omega})$,
$C^{0,1}(\Omega)$	the space of the Lipschitz continuous functions on Ω,
$C(\bar{\Omega};\mathbb{R}^n)$	the space of the continuous $\mathbb{R}^n$-valued functions on $\bar{\Omega}$,
$C^k(\bar{\Omega})$	the space of functions whose all derivatives up to k-th order are continuous on $\bar{\Omega}$,
$\mathrm{cl}(\cdot)$	the closure,
curl	the rotation of a vector-valued field on $\mathbb{R}^3$, see p. 181,
$\mathscr{D}(\Omega)$	the space of infinitely smooth functions with a compact support in Ω, see p. 10,
$\mathrm{D}\Phi(u,v)$	the directional derivative of Φ at u in the direction v,
$\mathrm{diam}(S)$	the diameter of a set $S \subset \mathbb{R}^n$; i.e. $\mathrm{diam}(S) := \sup_{x,y\in S} \|x-y\|$,
div	the divergence of a vector field; i.e. $\mathrm{div}(v) = \frac{\partial}{\partial x_1}v_1 + \cdots + \frac{\partial}{\partial x_n}v_n$ for $v = (v_1,\ldots,v_n)$,
div_S	the surface divergence, $\mathrm{div}_\mathrm{S} := \mathrm{Tr}(\nabla_\mathrm{S})$, see p. 57,
$\mathrm{dom}(A)$	the definition domain of the mapping A; in case of a set-valued mapping $A : V_1 \rightrightarrows V_2$ we put $\mathrm{dom}(A) := \{v\in V_1;\ A(v) \neq \emptyset\}$,
$\mathrm{dom}(\Phi)$	the domain of $\Phi : V \to \mathbb{R}\cup\{+\infty\}$; $\mathrm{dom}(\Phi) := \{v\in V;\ \Phi(v) < +\infty\}$,
$e(\cdot)$	the symmetric gradient, cf. 22,
$\mathrm{epi}(\Phi)$	the epigraph of Φ; i.e. $\{(v,a)\in V\times\mathbb{R};\ a \geq \Phi(v)\}$,
I	the time interval $[0,T]$,
$\mathbf{I}$	the identity mapping,
$\mathbb{I}$	the unit matrix,
$\mathrm{int}(\cdot)$	the interior,
J	the duality mapping,
$\mathscr{L}(V_1,V_2)$	the Banach space of linear continuous mappings $A : V_1 \to V_2$ normed by $\|A\|_{\mathscr{L}(V_1,V_2)} = \sup_{\|v\|_{V_1}\leq 1} \|Av\|_{V_2}$,
$L^p(\Omega)$	the Lebesgue space of p-integrable functions on Ω, equipped with the norm $\|u\|_{L^p(\Omega)} = \big(\int_\Omega \|u(x)\|^p \mathrm{d}x\big)^{1/p}$,
$L^p(\Omega;\mathbb{R}^n)$	the Lebesgue space of $\mathbb{R}^n$-valued p-integrable functions on Ω,
$\mathscr{M}(\bar{\Omega})$	the space of regular Borel measures, $\mathscr{M}(\bar{\Omega}) \cong C(\bar{\Omega})^*$, cf. p.10,
$\mathrm{meas}_n(\cdot)$	n-dimensional Lebesgue measure of a set,

n	the spatial dimension,
$\mathbb{N}$	the set of all natural numbers,
$\mathscr{N}_a$	the Nemytskiĭ mapping induced by an integrand a,
$N_K(\cdot)$	the normal cone, cf. p.6,
$\mathscr{O}(\cdot)$	the "great O" symbol: $f(\varepsilon) = \mathscr{O}(\varepsilon^\alpha)$ for $\varepsilon\searrow 0$ means $\limsup_{\varepsilon\searrow 0}\frac{\lvert f(\varepsilon)\rvert}{\varepsilon^\alpha}<\infty$,
$o(\cdot)$	the "small O" symbol: $f(\varepsilon) = o(\varepsilon^\alpha)$ for $\varepsilon\searrow 0$ means $\lim_{\varepsilon\searrow 0} f(\varepsilon)/\varepsilon^\alpha = 0$,
p	the exponent related to the polynomial growth/coercivity of the highest-order term in a differential operator,
$p' = \frac{p}{p-1}$	the conjugate exponent to $p \in [1, +\infty]$, cf. (1.20) on p.12,
p^*	the exponent in the embedding $W^{1,p}(\Omega) \subset L^{p^*}(\Omega)$, see (1.34)on p.16,
p^{**}	the exponent in the embedding $W^{2,p}(\Omega) \subset L^{p^{**}}(\Omega)$, i.e. $p^{**} = (p^*)^*$,
$p^\#$	the exponent in the trace operator $u \mapsto u\vert_\Gamma : W^{1,p}(\Omega) \to L^{p^\#}(\Gamma)$, see (1.37) on p.17; e.g. $p^{\#'}$ or $p^{*\#'}$ mean $(p^\#)'$ or $((p^*)^\#)'$, respectively,
$p^{\circledast}$	the exponent in the continuous embedding $L^p(I; W^{1,p}(\Omega)) \cap L^\infty(I; L^2(\Omega)) \subset L^{p^{\circledast}}(Q)$, see (8.131) on p.253,
$p^{\oplus}$	the exponent in the trace operator $L^p(I; W^{1,p}(\Omega)) \cap L^\infty(I; L^2(\Omega)) \to L^{p^{\oplus}}(\Sigma)$, see (8.136) on p.254,
Q	a space-and-time cylinder, $Q = I \times \Omega$,
$\mathbb{R}, \mathbb{R}^+, \mathbb{R}^-$	the set of all (resp. positive, or negative) reals,
$\bar{\mathbb{R}}$	the set of extended reals $\mathbb{R} \cup \{+\infty, -\infty\}$,
$\mathbb{R}^n$	the Euclidean space with the norm $\lvert s\rvert = \lvert(s_1, \ldots, s_n)\rvert = \big(\sum_{i=1}^n s_i^2\big)^{1/2}$,
sign	the single-valued "signum", i.e. the mapping $\mathbb{R} \to [-1, 1]$, $\mathrm{sign}(0) = 0$, $\mathrm{sign}(\mathbb{R}^+) = 1$, $\mathrm{sign}(\mathbb{R}^-) = -1$, cf. Figure 10a on p.133,
Sign	the set-valued "signum", i.e. the mapping $\mathbb{R} \rightrightarrows [-1, 1]$, $\mathrm{Sign}(0) = [-1, 1]$, $\mathrm{Sign}(\mathbb{R}^+) = \{1\}$, $\mathrm{Sign}(\mathbb{R}^-) = \{-1\}$, cf. Figure 10b on p.133,
$\mathrm{span}(\cdot)$	the linear hull of the specified set,
$\mathrm{supp}(u)$	the support of a function u, i.e. the closure of $\{x \in \Omega;\ u(x) \neq 0\}$,
T	a fixed time horizon, $T > 0$,
V	a separable reflexive Banach space (if not said otherwise), $\lVert\cdot\rVert_V$ (or briefly $\lVert\cdot\rVert$) its norm,
V^*	a topological dual space with $\lVert\cdot\rVert_{V^*}$ (or briefly $\lVert\cdot\rVert_*$) its norm,
$W^{k,p}(\Omega)$	the Sobolev space of functions whose distributional derivatives up to $k^{\rm th}$ order belongs to $L^p(\Omega)$, cf. (1.30) on p.15.
$W_0^{1,p}(\Omega)$	the Sobolev space of functions from $W^{1,p}(\Omega)$ whose traces on Γ vanish,
$W^{-1,p'}(\Omega)$	the dual space to $W_0^{1,p}(\Omega)$,
$W_{\rm loc}^{k,p}(\Omega)$	the set of functions v on Ω whose restrictions $v\vert_O$, with any open O such that $\bar{O} \subset \Omega$, belong to $W^{k,p}(O)$,

$W^{1,p,q}$	the Sobolev space of abstract functions having the time-derivative, see (7.1) on p. 201,
$W^{1,p,\mathscr{M}}$	the Sobolev space of abstract functions whose derivatives are measures, see (7.40) on p. 211,
$W^{2,\infty,p,q}$	the Sobolev space of abstract functions having the second time-derivative, see (7.4) on p. 202,
$W^{1,p}_{0,\mathrm{div}}$	the set of divergence-free functions $v\in W^{1,p}_0(\Omega;\mathbb{R}^n)$, see (6.29) on p. 178,
Γ	the boundary of a domain Ω,
δ_K	the indicator function of a set K; i.e. $\delta_K(\cdot)=0$ on K and $\delta_K(\cdot)=+\infty$ on the complement of K,
δ_x	the Dirac distribution (measure) supported at a point x,
Δ	the Laplace operator: $\Delta u=\mathrm{div}(\nabla u)=\frac{\partial^2}{\partial x_1^2}u+\cdots+\frac{\partial^2}{\partial x_n^2}u$,
Δ_p	the p-Laplace operator: $\Delta_p u=\mathrm{div}(\lvert\nabla u\rvert^{p-2}\nabla u)$ with $p>1$,
ν	the unit outward normal to Γ at $x\in\Gamma$, $\nu=\nu(x)$,
Σ	the side surface of the cylinder Q, i.e. $I\times\Gamma$, or a σ-algebra of sets,
χ_S	the characteristic function of a set S; i.e. $\chi_S(\cdot)=1$ on S and $\chi_S(\cdot)=0$ on the complement of S,
Ω	a bounded, connected, Lipschitz domain, $\Omega\subset\mathbb{R}^n$,
$\bar\Omega$	the closure of Ω,
$\subset$	a subset, or a continuous embedding,
$\Subset$	a compact embedding,
$\int_\Omega\ldots\mathrm{d}x$	integration according to the n-dimensional Lebesgue measure,
$\int_\Gamma\ldots\mathrm{d}S$	integration according to the $(n{-}1)$-dimensional surface measure on Γ,
$\partial\Phi$	the subdifferential of the convex functional $\Phi:V\to\mathbb{R}$,
$(\cdot)^{-1}$	the inverse mapping,
$(\cdot)^*$	the dual space, see p.3, or the adjoint operator, see p.5, or the Legendre-Fenchel conjugate functional, see p.294,
$(\cdot)^+$, $(\cdot)^-$	the positive and the negative parts, respectively, i.e. $u^+=\max(u,0)$ and $u^-=\min(u,0)$,
$(\cdot)'$	the Gâteaux derivative, cf. p.5, or a partial derivative, or the conjugate exponent, see p.12,
$(\cdot)\lvert_S$	the restriction of a mapping or a function on a set S,
$(\cdot)^\top$	the transposition of a matrix,
$\to$	a convergence (in a locally convex space) or a mapping between sets,
$\mapsto$	a mapping of elements into other ones, e.g. $A:u\mapsto f$ where $f=A(u)$,
$\searrow$	convergence on $\mathbb{R}$ from the right; similarly $\nearrow$ means from the left,
$\rightrightarrows$	a set-valued mapping (e.g. $A:X\rightrightarrows Y$ abbreviates $A:X\to 2^Y=$ the set of all subsets of Y),

∇	the spatial gradient: $\nabla u = (\frac{\partial}{\partial x_1}u, \ldots, \frac{\partial}{\partial x_n}u)$,
$\nabla_{\rm S}$	the surface gradient, see p. 57,
$\cdot$	a position of an unspecified variable, or the scalar product of vectors; i.e. $u{\cdot}v := \sum_{i=1}^m u_i v_i$ for $u, v \in \mathbb{R}^m$,
$:$	the scalar product of matrices; i.e. $A{:}B := \sum_{i=1}^n \sum_{j=1}^m A_{ij} B_{ij}$,
$\vdots$	the product of 3rd-order tensors $A \vdots B := \sum_{i=1}^n \sum_{j=1}^m \sum_{k=1}^l A_{ijk} B_{ijk}$,
$:=$	the definition of a left-hand side by a right-hand-side expression,
$\otimes$	the tensorial product of vectors: $[u \otimes v]_{ij} = u_i v_j$,
$\langle \cdot, \cdot \rangle$	the bilinear pairing of spaces in duality, cf. p.3,
$\langle \cdot, \cdot \rangle_{\rm s}$	the semi-inner product in a Banach space, cf. (3.7) on p.97,
$(\cdot, \cdot)$	the inner (i.e. scalar) product in a Hilbert space, cf. (1.4) on p.2,
$\|\cdot\|$	a norm on a Banach space, see p.1,
$|\cdot|$	a seminorm on a Banach space, or an Euclidean norm in $\mathbb{R}^n$.

Chapter 1

Preliminary general material

For the reader's convenience, this chapter summarizes some concepts, definitions and results which are mostly relevant to the undergraduate curriculum and are thus assumed as basically known, or have specific roots in rather distant areas and have rather auxiliary character with respect to the purpose of this book, which is to push pure theory towards possible applications as fast as possible. As such, all assertions in Chapter 1 are made without proofs and the scope has been minimized to only material actually needed in the book. If a reader decides to skip this Chapter, (s)he can easily come back through references in the Index or in the text itself, if in need.

1.1 Functional analysis

The universal framework used in (even nonlinear) differential equations is based on linear functional analysis, and also on convex analysis.

1.1.1 Normed spaces, Banach spaces, locally convex spaces

Considering a (real) linear space V,[1] a non-negative, degree-1 homogeneous, sub-additive functional $\|\cdot\|_V : V \to \mathbb{R}$ is called a *norm* if it vanishes only at 0; often, we will write briefly $\|\cdot\|$ instead of $\|\cdot\|_V$ if V is clear from the context.[2] A linear space equipped with a norm is called a *normed linear space*. If the last property (i.e. $\|u\|_V = 0 \Rightarrow u = 0$) is missing, we call such a functional a *seminorm*; i.e. a

[1]This means V is endowed by a binary operation $(v_1, v_2) \mapsto v_1 + v_2 : V \times V \to V$ which makes it a commutative group, i.e. $v_1 + v_2 = v_2 + v_1$, $v_1 + (v_2 + v_3) = (v_1 + v_2) + v_3$, $\exists\, 0 \in V$: $v + 0 = v$, and $\forall v_1 \in V\ \exists v_2$: $v_1 + v_2 = 0$, and furthermore it is equipped with a multiplication by scalars $(a, x) \mapsto ax : \mathbb{R} \times V \to V$ satisfying $(a_1 + a_2)v = a_1 v + a_2 v$, $a(v_1 + v_2) = av_1 + av_2$, $(a_1 a_2)v = a_1(a_2 v)$, and $1v = v$.

[2]The mentioned properties mean respectively: $\|v\| \geq 0$, $\|av\| = |a|\,\|v\|$, $\|u+v\| \leq \|u\| + \|v\|$ for any $u, v \in V$ and $a \in \mathbb{R}$, and $\|v\| = 0 \Rightarrow v = 0$.

functional $|\cdot|_\xi : V \to \mathbb{R}$ is a seminorm if it satisfies

$$\forall u,v\in V\ \forall a\in\mathbb{R}: \quad 0\le|u+v|_\xi\le|u|_\xi+|v|_\xi \quad\text{and}\quad |au|_\xi=|a|\,|u|_\xi. \tag{1.1}$$

Having V equipped with a collection $\{|\cdot|_\xi\}_{\xi\in\Xi}$ of seminorms $|\cdot|_\xi$ with an arbitrary index set Ξ, we call V a *locally convex space*. A sequence $\{u_k\}_{k\in\mathbb{N}}$ in V is then called a *Cauchy sequence* if

$$\forall\xi\in\Xi\ \ \forall\varepsilon>0\ \ \exists k_0\in\mathbb{N}\ \ \forall k_1,k_2\ge k_0: \quad |u_{k_1}-u_{k_2}|_\xi\le\varepsilon. \tag{1.2}$$

Moreover, a sequence $\{u_k\}_{k\in\mathbb{N}}$ is called *convergent* to some $u\in V$ if

$$\forall\xi\in\Xi\ \ \forall\varepsilon>0\ \ \exists k_0\in\mathbb{N}\ \ \forall k\ge k_0: \quad |u_k-u|_\xi\le\varepsilon. \tag{1.3}$$

In this case, u is called its limit and we will write $u=\lim_{k\to\infty}u_k$ or $u_k\to u$ (or, depending on a particular collection of seminorms, $u_k\rightharpoonup u$). A subset A of a locally convex space V is called *closed* if any limit of any convergent sequence contained in A is itself in A.[3] Moreover, A is called *open* if $V\setminus A$ is closed. The *closure* of A, denoted by $\mathrm{cl}(A)$, is the smallest closed set $B\supset A$, while $\mathrm{int}(A):=A\setminus\mathrm{cl}(V\setminus A)$ is called the *interior* of A. The concrete collection of seminorms will often be specified by various adjectives as "strong" or "weak" or "weak*". If $|u|_\xi=0$ for all $\xi\in\Xi$ implies $u=0$, V is called a *Hausdorff* locally convex space. Then every convergent sequence has a uniquely determined limit. The normed linear space is a Hausdorff locally convex space with its only seminorm being then just the norm. A subset $A\subset V$ is called *bounded* if $\sup_{x\in A}\|x\|<+\infty$, and *dense* (in V) if $\mathrm{cl}(A)=V$. If there is a countable dense subset of V, we say that V is *separable.* If every Cauchy sequence in a normed linear space V converges, we say that this space is *complete* and then V is called a *Banach space.*[4] An example of a Banach space is $\mathbb{R}^n$ endowed by the norm, denoted usually by $|\cdot|$ instead of $\|\cdot\|$, defined by $|s|=(\sum_{i=1}^n s_i^2)^{1/2}$; such a Banach space is called an n-dimensional *Euclidean space.*

If V is a Banach space such that, for any $v\in V$, $V\to\mathbb{R}: u\mapsto\|u+v\|^2-\|u-v\|^2$ is linear, then V is called a *Hilbert space.* In this case, we define the *inner product* (also called *scalar product*) by

$$(u,v):=\frac14\|u+v\|^2-\frac14\|u-v\|^2. \tag{1.4}$$

By the assumption, $(\cdot,\cdot):V\times V\to\mathbb{R}$ is a bilinear form which is obviously symmetric[5] and satisfies $(u,u)=\|u\|^2$. E.g., the Euclidean space $\mathbb{R}^n$ is a Hilbert space.

---

[3] We will always work with Ξ at most countable and therefore, for simplicity, we define "closedness" in terms of convergence of sequences, which would be in general situations called rather "sequential closedness".

[4] This fundamental concept has been introduced in [33] in 1922.

[5] This means both $u\mapsto(u,v)$ and $v\mapsto(u,v)$ are linear functionals on V and $(u,v)=(v,u)$.

Let us call a Banach space V *strictly convex* if the sphere[6] in V does not contain any line segment. The space V is *uniformly convex* if

$$\forall\varepsilon>0\ \exists\delta>0\ \forall u,v\!\in\! V:\quad \left.\begin{array}{l}\|u\|\!=\!\|v\|\!=\!1\\ \|u-v\|\geq\varepsilon\end{array}\right\}\ \Rightarrow\ \left\|\frac{1}{2}u+\frac{1}{2}v\right\|\leq 1-\delta. \tag{1.5}$$

Any uniformly convex Banach space is strictly convex but not vice versa.

1.1.2 Functions and mappings on Banach spaces, dual spaces

Having in mind a specific mode of convergence, a function $f:V\to\mathbb{R}\cup\{\pm\infty\}$ is called *lower* (resp. *upper*) *semicontinuous*[7] if

$$\forall u\!\in\! V,\ u_k\to u:\qquad f(u)\leq\liminf_{k\to\infty}f(u_k)\quad\Big(\text{resp. } f(u)\geq\limsup_{k\to\infty}f(u_k)\Big). \tag{1.6}$$

Having two normed linear spaces V_1 and V_2 and a mapping $A:V_1\to V_2$, we say that A is *continuous* if it maps convergent sequences in V_1 onto convergent ones in V_2, and is a *linear operator* if it satisfies $A(a_1v_1+a_2v_2)=a_1A(v_1)+a_2A(v_2)$ for any $a_1,a_2\in\mathbb{R}$ and $v_1,v_2\in V_1$. Often we will write briefly Av instead of $A(v)$. If $V_1=V_2$, a linear continuous operator $A:V_1\to V_2$ is called a *projector* if $A\circ A=A$. The set of all linear continuous operators $V_1\to V_2$ will be denoted by $\mathscr{L}(V_1,V_2)$, being itself a normed linear space when equipped with the addition and multiplication by scalars defined respectively by $(A_1+A_2)v=A_1v+A_2v$ and $(aA)v=a(Av)$, and with the norm

$$\|A\|_{\mathscr{L}(V_1,V_2)}:=\sup_{\|v\|_{V_1}\leq1}\|Av\|_{V_2}=\sup_{v\neq0}\frac{\|Av\|_{V_2}}{\|v\|_{V_1}}. \tag{1.7}$$

A continuous (possibly even nonlinear) mapping $A:V_1\to V_2$ is called a *homeomorphism* if the inverse $A^{-1}:V_2\to V_1$ does exist and is continuous. Moreover, $A:V_1\to V_2$ is called a *homeomorphical embedding* if $A:V_1\to A(V_1)$ is a homeomorphism and $A(V_1)$ is dense in V_2.

As $\mathbb{R}$ itself is a linear topological space[8], we can consider the linear space $\mathscr{L}(V,\mathbb{R})$, being also denoted by V^* and called the *dual space* to V. The original space V is then called *predual* to V^*. For an operator (=now a functional) $f\in V^*$, we will write $\langle f,v\rangle$ instead of fv. The bilinear form $\langle\cdot,\cdot\rangle_{V^*\times V}:V^*\times V\to\mathbb{R}$ is called a canonical *duality pairing*. Instead of $\langle\cdot,\cdot\rangle_{V^*\times V}$, we will occasionally write briefly $\langle\cdot,\cdot\rangle$. Always, V^* is a Banach space if endowed by the norm (1.7), denoted often briefly $\|\cdot\|_*$ instead of $\|\cdot\|_{V^*}$, i.e. $\|f\|_*=\sup_{\|v\|\leq1}\langle f,v\rangle$. Obviously,

$$\langle f,u\rangle=\|u\|\left\langle f,\frac{u}{\|u\|}\right\rangle\leq\|u\|\sup_{\|v\|\leq1}\langle f,v\rangle=\|f\|_*\,\|u\|. \tag{1.8}$$

---

[6] A "sphere" $\{v\!\in\! V;\ \|v\|=\varrho\}$ (of the radius $\varrho>0$) is a surface of a "ball" $\{v\!\in\! V;\ \|v\|\leq\varrho\}$.
[7] The property (1.6) is often called sequential lower semicontinuity.
[8] The conventional norm on $\mathbb{R}$ is the absolute value $|\cdot|$.

If V is a Hilbert space, then $(u \mapsto (f,u)) \in V^*$ for any $f \in V$, and the mapping $f \mapsto (u \mapsto (f,u))$ identifies V with V^*. Then (1.8) turns into a so-called *Cauchy-Bunyakovskiĭ* inequality $(f,u) \le \|f\|\,\|u\|$.

Having two Banach spaces V_1, V_2, we have

$$\big(V_1 \cap V_2\big)^* \cong V_1^* + V_2^* := \big\{f = f_1 + f_2;\ f_1 \in V_1^*,\ f_2 \in V_2^*\big\} \tag{1.9}$$

if the duality pairing is defined by $\langle f, v\rangle = \langle f_1, v\rangle_{V_1^*\times V_1} + \langle f_2, v\rangle_{V_2^*\times V_2}$ with $f = f_1 + f_2$, and if the norm on $V_1 \cap V_2$ is taken as $\|v\|_{V_1\cap V_2} := \max(\|v\|_{V_1}, \|v\|_{V_2})$ while $\|f\|_{V_1^*+V_2^*} := \inf_{f=f_1+f_2}(\|f_1\|_{V_1^*} + \|f_2\|_{V_2^*})$.

Theorem 1.1 (Banach-Steinhaus principle [36]). *Let $\{A_\alpha\}_{\alpha\in S}$ be a family in $\mathscr{L}(V_1, V_2)$, V_1 a Banach space, V_2 a normed linear space. Then boundedness of $\{A_\alpha(v)\}_{\alpha\in S} \subset V_2$ for any $v \in V_1$ implies boundedness of $\{A_\alpha\}_{\alpha\in S} \subset \mathscr{L}(V_1, V_2)$.*

One can consider a normed linear space V equipped with the collection of seminorms $\{v \mapsto |\langle f, v\rangle|\}_{f\in V^*}$, which makes it a locally convex space. Sequences converging in this locally convex space will be called *weakly convergent.* Any sequence $\{u_k\}_{k\in\mathbb{N}}$ converging in the original norm, i.e. $\lim_{k\to\infty}\|u_k - u\|_V = 0$, will be called norm-convergent or also *strongly convergent* and is automatically weakly convergent.[9]

If the convergence in (1.6) refers to the weak one, the function $f : V \to \mathbb{R}$ will be called *weakly lower* (resp. upper) *semicontinuous.*

Theorem 1.2.[10] *If V is uniformly convex, $u_k \rightharpoonup u$, and $\|u_k\| \to \|u\|$, then $u_k \to u$.*

Likewise, also the Banach space V^* can be endowed by the collection of seminorms $\{f \mapsto |\langle f, v\rangle|\}_{v\in V}$, which makes it a locally convex space. Sequences converging in this locally convex space will be called *weakly* convergent.* Any sequence $\{f_k\}_{k\in\mathbb{N}}$ converging in the original norm, i.e. $\lim_{k\to\infty}\|f_k - f\|_* = 0$, is automatically weakly* convergent. The duality pairing is continuous if $V^* \times V$ is equipped with weak*×norm or norm×weak topology[11] and separately (weak*,weak)-continuous.[12]

Proposition 1.3. *If V^* is separable, then so is also the space V.*

The Banach-Steinhaus Theorem 1.1 has immediately the following

Corollary 1.4. *Every weakly* convergent sequence in V^* must be bounded if V is a Banach space. In particular, every weakly convergent sequence in a reflexive Banach V must be bounded.*

[9] The converse implication holds only if the normed linear space V is finite-dimensional, i.e. isometrically isomorphic to an Euclidean space.

[10] See Fan and Glicksberg [140] for thorough investigation and various modifications.

[11] This means $\lim_{k\to\infty}\langle f_k, u_k\rangle = \langle f, u\rangle$ if either $f_k \to f$ weakly* and $u_k \to u$ strongly or $f_k \to f$ strongly and $u_k \to u$ weakly.

[12] This means both $\lim_{k\to\infty}\lim_{l\to\infty}\langle f_k, u_l\rangle = \langle f, u\rangle$ and $\lim_{l\to\infty}\lim_{k\to\infty}\langle f_k, u_l\rangle = \langle f, u\rangle$ if $f_k \to f$ weakly* and $u_k \to u$ weakly.

Having two locally convex spaces V_1 and V_2 and an operator $A \in \mathscr{L}(V_1, V_2)$, we define the so-called *adjoint operator* $A^* \in \mathscr{L}(V_2^*, V_1^*)$ by the identity $\langle A^* f, v\rangle = \langle f, Av\rangle$ to be valid for any $v \in V_1$ and $f \in V_2^*$. If V_1 and V_2 are normed linear spaces, then $A \mapsto A^*$ realizes an isometrical (i.e. norm-preserving) isomorphism between $\mathscr{L}(V_1, V_2)$ and $\mathscr{L}(V_2^*, V_1^*)$.

Besides, forgetting that V^* has been created by its predual V, one can think about the weak convergence on V^*, induced by the collection of seminorms $\{f \mapsto |\langle \phi, f\rangle|\}_{\phi \in (V^*)^*}$. The space $V^{**} := (V^*)^*$ is called a *bidual* to V and the space V itself is embedded into it by the canonical embedding $i : V \to V^{**}$ defined by $\langle i(v), f\rangle = \langle f, v\rangle$. We will often identify[13] V with its image $i(V)$ in V^{**} so that always $V \subset V^{**}$, and therefore any weakly convergent sequence in V^* is also weakly* convergent (to the same limit). The converse implication holds if and only if $V \equiv i(V) = V^{**}$, at which case the Banach space V is called *reflexive*. The *Milman-Pettis theorem* [285, 335] says that every uniformly convex Banach space must be reflexive. The *Asplund theorem* [23] says that any reflexive Banach space can equivalently be renormed[14] so that it is, together with its dual, strictly convex. By the *Clarkson theorem* [97], also separable Banach spaces can be renormed to be strictly convex.[15]

A mapping $A : V_1 \to V_2$ is called *bounded* if it maps bounded sets in V_1 into bounded sets in V_2. The normed structure of V_1 and V_2 allows us to define a mapping $A : V_1 \to V_2$ to be *Lipschitz continuous* if, for some $\ell \in \mathbb{R}$ and all $u, v \in V$, it holds that $\|A(u) - A(v)\|_{V_2} \leq \ell\|u - v\|_{V_1}$; in this case, ℓ is called a Lipschitz constant. Obviously, any Lipschitz continuous mapping is bounded and *uniformly continuous* in the sense that $\|A(u) - A(v)\|_{V_2} \leq \zeta(\|u - v\|_{V_1})$ with ζ increasing, $\zeta(0) = 0$. If $\ell \leq 1$ (resp. $\ell < 1$), A is called *non-expansive* (resp. a *contraction*).

The linear structure of V_1 and V_2 allows us to investigate *smooth*ness of A. We say that $A : V_1 \to V_2$ has the *directional derivative* at $u \in V$ in the direction $h \in V$, denoted by $\mathrm{D}A(u, h)$, if there is the limit

$$\lim_{\varepsilon \searrow 0} \frac{A(u + \varepsilon h) - A(u)}{\varepsilon} =: \mathrm{D}A(u, h). \tag{1.10}$$

If the mapping $h \mapsto \mathrm{D}A(u, h)$ is linear and continuous, then we say that A has a *Gâteaux* [173] *differential* at $u \in V$, denoted by $A'(u) \in \mathscr{L}(V_1, V_2)$. If the Gâteaux differential exists in any point, A is called Gâteaux differentiable and $A' : V \mapsto \mathscr{L}(V_1, V_2)$. In the special case $V_2 = \mathbb{R}$, a Gâteaux-differentiable functional $\Phi : V_1 \to \mathbb{R}$ has the differential $\Phi' : V_1 \to V_1^*$.

---

[13] This identification is indeed very natural because the mapping $i : V \to V^{**}$ realizes a (weak,weak*)- as well as (norm,norm)-homeomorphical embedding.

[14] Two norms $\|\cdot\|_1$ and $\|\cdot\|_2$ on V are called equivalent to each other if $\exists \varepsilon > 0\ \forall v \in V$: $\varepsilon\|v\|_1 \leq \|v\|_2 \leq \varepsilon^{-1}\|v\|_1$.

[15] These renormalization results can be still improved for so-called locally uniformly convex spaces, i.e. δ in (1.5) depends on u; cf. Troyanski [412].

1.1.3 Convex sets

A set A in a linear space is called *convex* if $\lambda u + (1-\lambda)v \in A$ whenever $u, v \in A$ and $\lambda \in [0,1]$, and it is called a *cone* (with the vertex at the origin 0) if $\lambda v \in A$ whenever $v \in A$ and $\lambda \geq 0$. A function $f : V \to \mathbb{R}$ is called *convex* if $f(\lambda u + (1-\lambda)v) \leq \lambda f(u) + (1-\lambda)f(v)$ for any $u, v \in V$, $\lambda \in [0,1]$. If $u \neq v$ and $\lambda \in (0,1)$ imply a strict inequality, then f is called *strictly convex*. A function $f : V \to \mathbb{R}$ is convex (resp. lower semicontinuous) if and only if its *epigraph* $\mathrm{epi}(f) := \{(x,a) \in V \times \mathbb{R};\ a \geq f(x)\}$ is a convex (resp. closed) subset of $V \times \mathbb{R}$.

Theorem 1.5 (HAHN [196] and BANACH [34]). *Let K be an open convex nonempty subset of a locally convex space V and L be a linear manifold that does not intersect K. Then there is a closed hyperplane $\bar{L}$ such that $L \subset \bar{L}$ and $K \cap \bar{L} = \emptyset$. In other words, there is $f \in V^*$ such that $\langle f, u\rangle > \langle f, v\rangle$ whenever $u \in K$ and $v \in L$.*

Proposition 1.6. *A closed set A is convex if and only if $\frac{1}{2}u + \frac{1}{2}v \in A$ whenever $u, v \in A$. Closed convex sets are weakly closed.*

This is related to the fact that a lower semicontinuous functional f is convex if and only if

$$\frac{1}{2}f(u_1) + \frac{1}{2}f(u_2) \geq f\Big(\frac{u_1 + u_2}{2}\Big), \tag{1.11}$$

cf. Exercise 4.18 below.

Having a convex subset K of a locally convex space V and $u \in K$, we define the *tangent cone* $T_K(u) \subset V$ by

$$T_K(u) := \mathrm{cl}\Big(\bigcup_{a>0} a(K-u)\Big)\,. \tag{1.12}$$

Obviously, $T_K(u)$ is a closed convex cone and $v \in T_K(u)$ means precisely that $u + a_k v_k \in K$ for suitable sequences $\{a_k\}_{k\in\mathbb{N}} \subset \mathbb{R}$ and $\{v_k\}_{k\in\mathbb{N}} \subset V$ such that $\lim_{k\to\infty} v_k = v$. Besides, we define the *normal cone* $N_K(v)$ as

$$N_K(u) := \big\{f \in V^*;\ \ \forall v \in T_K(u) : \ \ \langle f, v\rangle \leq 0\big\}\,. \tag{1.13}$$

Again, the normal cone is always a closed convex cone in V^*.

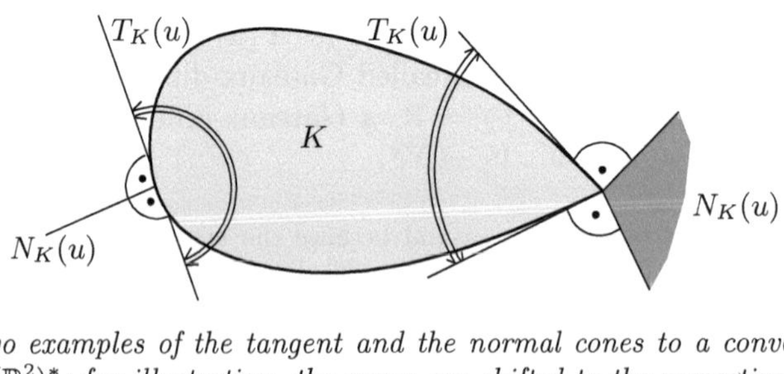

Figure 1. *Two examples of the tangent and the normal cones to a convex set $K \subset \mathbb{R}^2 \equiv (\mathbb{R}^2)^*$; for illustration, the cones are shifted to the respective points u's.*

1.1.4 Compactness

An important notion on which a lot of powerful tools are based is compactness. Let us agree, for a certain simplicity[16], to say that a set V in a locally convex space V is *compact* if every sequence in A contains a convergent subsequence whose limit belongs to A. Having in mind a Banach space V (possibly having a predual) with a structure of the norm (resp. weak or possibly weak*) locally convex space, we will specify compactness by the adjective "norm" (resp. "weak" or possibly "weak*"); if no adjective is mentioned, then the "norm" one will implicitly be understood. A weakening of the notion of compactness is precompactness: we say that a set A is *precompact* in the sense that every sequence in such a set contains a Cauchy subsequence. Another modification is the following: we say that A is *relatively compact* if its closure is compact. The reader can easily guess what e.g. "relatively weakly* compact" or "relatively norm compact" mean. If every Cauchy sequence converges (in particular in a Banach space), the prefix/adjective "pre-" and "relatively" coincide with each other. Thus, in a Banach space, relative norm compactness and norm precompactness are the same.

A mapping $A : V_1 \to V_2$, V_1, V_2 Banach spaces, is called *totally continuous* if it is (weak,norm)-continuous, i.e. it maps weakly convergent sequences to strongly convergent ones. It is called *compact* if it maps bounded sets in V_1 into precompact[17] sets in V_2. If V_1 is reflexive, then any totally continuous mapping is compact[18] but not conversely[19].

Theorem 1.7 (SELECTION PRINCIPLE, BANACH [35, Chap.VIII, Thm.3][20]). *In a Banach space with a separable predual, any bounded sequence contains a weakly* convergent subsequence.*

Often, Theorem 1.7 is also called, not completely exactly however, *Alaoglu-Bourbaki's theorem*[21].

[16] In fact, our "compactness" is in general topology called "sequential compactness" while compactness means that every net (=a generalized sequence) contains a convergent finer net (=a suitable generalization of the notion of "subsequence"), or, equivalently, that every covering of A by open sets contains a finite sub-covering. These two concepts coincide with each other in a lot of important cases, primarily in normed spaces (considering norm compactness). More generally, it happens if the structure of a locally convex spaces can be equally induced by a countable collection of seminorms $\{|\cdot|_\xi\}_{\xi\in\mathbb{N}}$, e.g. it concerns weak (resp. weak*) compactness if V has a separable dual (resp. predual). A far less trivial fact, known as the Eberlein-Šmuljan theorem, is that in a Banach space the relatively weakly compact (in the general-topology sense) sets coincide with the relatively weakly sequentially compact ones.

[17] As V_2 is assumed a Banach space, precompact sets are just those which are relatively compact.

[18] For $B \subset V_1$ bounded, any sequence in $A(B)$, say $\{A(v_k)\}_{k\in\mathbb{N}}$ with $v_k \in B$ contains a subsequence convergent in V_2, e.g. $\{A(v_{k_l})\}_{l\in\mathbb{N}}$ with $\{v_{k_l}\}_{l\in\mathbb{N}}$ weakly convergent in V_1; here both reflexivity of V_1 guaranteeing existence of $\{v_{k_l}\}_{l\in\mathbb{N}}$ and compactness of A guaranteeing convergence of $\{A(v_{k_l})\}_{l\in\mathbb{N}}$ has been used.

[19] See Remark 2.39 and Exercise 2.64 below.

[20] For the proof cf. Exercise 2.51.

[21] In fact, Alaoglu-Bourbaki's theorem says that the polar set to a neighbourhood of the origin in a locally convex space is weakly* compact, see Alaoglu [13] and Bourbaki [63].

Theorem 1.8 (BOLZANO and WEIERSTRASS[22]). *Every lower (resp. upper) semi-continuous function $X \to \mathbb{R}$ on a compact set attains its minimum (resp. maximum) on this set.*

1.1.5 Fixed-point theorems

A point $u \in V$ is called a *fixed point* of a mapping $M : V \to V$ if $M(u) = u$.

Theorem 1.9 (SCHAUDER FIXED-POINT THEOREM [379]). *A continuous compact mapping on a closed, bounded, convex set in a Banach space has a fixed point. Alternatively, a continuous mapping on a compact, convex, set in a Banach space has a fixed point.*

As a special case for $V = \mathbb{R}^n$, one gets a historically older achievement which has been tremendously generalized by the previous Theorem 1.9:

Theorem 1.10 (BROUWER FIXED-POINT THEOREM [72]). *A continuous mapping on a compact convex set in $\mathbb{R}^n$ has a fixed point.*

Further generalization of Theorem 1.9 is very useful in applications. A useful concept of a *set-valued mapping* $M : V \rightrightarrows V$ just means that $M : V \to 2^V$, the set of all subsets of V. Such M is called *upper semicontinuous* if it has a closed graph, i.e. $u_k \to u$, $f_k \to f$, and $f_k \in M(u_k)$ implies $f \in M(u)$.

Theorem 1.11 (KAKUTANI FIXED-POINT THEOREM [224][23]). *An upper semicontinuous set-valued mapping $M : V \rightrightarrows V$ with nonempty closed convex values which maps a nonempty compact convex set K in a locally convex space into itself has a fixed point, i.e. $\exists u \in K: u \in M(u)$.*

A completely different principle is based on metric properties and exploits completeness instead of compactness:

Theorem 1.12 (BANACH FIXED-POINT THEOREM [33][24]). *A contractive mapping on a Banach space has a fixed point which is even unique.*

1.2 Function spaces

We consider the Euclidean space $\mathbb{R}^n$, $n \geq 1$, endowed with standard Euclidean topology and for Ω a subset of $\mathbb{R}^n$ we will define various spaces of functions $\Omega \to$

[22]In fact, this is rather a tremendous generalization of the original assertion by Bolzano [60] who showed, rather intuitively (because completion of irrational numbers by transcendental ones which locally compactifies $\mathbb{R}$ has been discovered only much later) that any real continuous function of a bounded closed interval is bounded. An essence, called the Bolzano-Weierstrass principle, is that every sequence in a compact set has a cluster point, i.e. a point whose each neighbourhood contains infinitely many members of this sequence.

[23]The original version [224] formulated in $\mathbb{R}^n$ has been generalized for locally convex spaces by Fan [139] and Glicksberg [181].

[24]In fact, this theorem works (and was formulated in [33]) in a complete metric space, too.

$\mathbb{R}^m$. If endowed by a pointwise addition and multiplication[25] the linear-space structure of $\mathbb{R}^m$ is inherited by these spaces. Besides, we will endow them by norms, which makes them normed linear (or, mostly even Banach) spaces. Having two such spaces $U \subset V$, we say that the mapping $I : U \to V : u \mapsto u$ is a *continuous embedding* (or, that U is embedded continuously to V) if the linear operator I is continuous (hence bounded). This means that $\|u\|_V \le N\|u\|_U$; for N one can take the norm $\|I\|_{\mathscr{L}(U,V)}$. If I is compact, we speak about a *compact embedding* and use the notation $U \Subset V$. If U is a dense subset in V, we will speak about a *dense embedding*; this property obviously depends on the norm of V but not of U. It follows by a general functional-analysis argument that the adjoint mapping $I^* : V^* \to U^*$ is continuous and injective provided $U \subset V$ continuously and densely[26]; then we can identify V^* as a subset of U^*.

1.2.1 Continuous and smooth functions

The notation $C(\cdot)$, $C^0(\cdot)$, and $C^{0,1}(\cdot)$ will indicate sets of continuous, bounded continuous, and Lipschitz continuous functions, respectively.[27] E.g. $C^0(\Omega;\mathbb{R}^m)$ denote the set of bounded continuous functions $\Omega \to \mathbb{R}^m$. We denote by $\bar\Omega$ the closure of Ω in the Euclidean space $\mathbb{R}^n$. By the Bolzano-Weierstrass Theorem 1.8, $C^0(\Omega;\mathbb{R}^m) = C(\bar\Omega;\mathbb{R}^m)$ if Ω is bounded. Equally, we can understand $C(\bar\Omega;\mathbb{R}^m)$ as composed from functions $\Omega \to \mathbb{R}^m$ having a continuous extension on the closure $\bar\Omega$. For $m = 1$, we will write briefly $C^0(\bar\Omega)$ (resp. $C^{0,1}(\bar\Omega)$). If equipped by a pointwise addition and multiplication and by the norm

$$\|u\|_{C^0(\bar\Omega;\mathbb{R}^m)} := \max_{x\in\bar\Omega} |u(x)|, \qquad \|u\|_{C^{0,1}(\bar\Omega;\mathbb{R}^m)} := \sup_{\substack{x,\xi\in\bar\Omega\\ x\ne\xi}} \left(|u(x)| + \frac{|u(x)-u(\xi)|}{|x-\xi|} \right)$$

with $|\cdot|$ denoting the Euclidean norm in the corresponding spaces, both $C^0(\bar\Omega;\mathbb{R}^m)$ and $C^{0,1}(\bar\Omega;\mathbb{R}^m)$ become Banach spaces.

Furthermore, for $k \ge 1$, we define spaces of *smooth* functions, having derivatives up to k-th order continuous up to the boundary, i.e.

$$C^k(\bar\Omega;\mathbb{R}^m) := \Big\{ u \in C^0(\bar\Omega;\mathbb{R}^m);\ \forall (i_1,\dots,i_n) \in (\mathbb{N}\cup\{0\})^n, \\ \sum_{\alpha=1}^n i_\alpha \le k : \quad \frac{\partial^{i_1+\dots+i_n} u}{\partial^{i_1}x_1 \cdots \partial^{i_n}x_n} \in C^0(\bar\Omega;\mathbb{R}^m) \Big\}. \tag{1.14}$$

If endowed by the linear structure of $C^0(\bar\Omega;\mathbb{R}^m)$ and by the norm $\|u\|_{C^k(\bar\Omega;\mathbb{R}^m)} := \|u\|_{C^0(\bar\Omega;\mathbb{R}^m)} + \sum_{i_1+\dots+i_n\le k} \big\|\frac{\partial^{i_1+\dots+i_n}}{\partial^{i_1}x_1\cdots\partial^{i_n}x_n} u\big\|_{C^0(\bar\Omega;\mathbb{R}^m)}$, they become Banach spaces.

[25] This means, for $u, v : \Omega \to \mathbb{R}^m$ and $\lambda \in \mathbb{R}$, we define $[u+v](x) := u(x) + v(x)$ and $[\lambda u](x) := \lambda u(x)$ for all $x \in \bar\Omega$.

[26] Indeed, I^* is injective (because two different linear continuous functionals on V must have also different traces on any dense subset, in particular on U).

[27] For $\Omega = \mathbb{R}^n$, the adjectives "continuous" and "Lipschitz continuous" can be understood as in Section 1.1 because both $\mathbb{R}^n$ and $\mathbb{R}^m$ are normed spaces, while for a general Ω it is just a restriction of these definitions on $\bar\Omega$.

The continuous embedding $C^k(\bar\Omega;\mathbb{R}^m) \subset C^l(\bar\Omega;\mathbb{R}^m)$ holds for any $k \ge l \ge 0$, i.e. $\|u\|_{C^l(\bar\Omega;\mathbb{R}^m)} \le N\|u\|_{C^k(\bar\Omega;\mathbb{R}^m)}$. In this particular case, $N = 1$. This embedding is even dense and, if $k > l \ge 0$, also compact. As functions with bounded derivatives are Lipschitz continuous, we have also $C^1(\bar\Omega;\mathbb{R}^m) \subset C^{0,1}(\bar\Omega;\mathbb{R}^m)$.

By Riesz's theorem, the dual space $C^0(\bar\Omega)^*$ is isometrically isomorphic with the space $\mathscr{M}(\bar\Omega)$ of regular Borel measures on $\bar\Omega$; a *measure* is a σ-additive set function[28] and a *regular Borel measure* $\mu \in \mathscr{M}(\bar\Omega)$ is a measure on the so-called Borel σ-algebra[29] which is regular[30] and has a finite *variation*[31] $|\mu|$ over $\bar\Omega$, i.e. $|\mu|(\bar\Omega) < +\infty$. The mentioned isomorphism $f \mapsto \mu : C^0(\bar\Omega)^* \to \mathscr{M}(\bar\Omega)$ is by $\langle f, v\rangle = \int_{\bar\Omega} v\,\mu(\mathrm{d}x)$.[32] Then $\|f\|_{C^0(\bar\Omega)^*} = |\mu|(\bar\Omega)$. For $x \in \bar\Omega$, a measure $\delta_x \in \mathscr{M}(\bar\Omega)$ defined by $\langle \delta_x, v\rangle = v(x)$ is called *Dirac's measure* supported at x.

Considering an increasing sequence of compact subsets $K_i \subset \Omega$ such that $\Omega = \bigcup_{i\in\mathbb{N}} K_i$, we put

$$\mathscr{D}(\Omega) := \bigcup_{i\in\mathbb{N}} \bigcap_{k\in\mathbb{N}} C^k_{K_i}(\Omega), \tag{1.15}$$

where $C^k_{K_i}(\Omega)$ denotes the space of all functions $\Omega \to \mathbb{R}$ which are continuous together with all their derivatives up to the order k and which have the support contained in K_i. Each $C^\infty_{K_i} := \bigcap_{k\in\mathbb{N}} C^k_{K_i}(\Omega)$ is endowed by the collection of seminorms $\{|\cdot|_{k,K_i}\}_{k\in\mathbb{N}}$ with $|u|_{k,K_i} := \big\|u|_{K_i}\big\|_{C^k(K_i)}$, which makes it a locally convex space. Then $\mathscr{D}(\Omega) = \bigcup_{i\in\mathbb{N}} C^\infty_{K_i}$ is equipped with the finest topology that makes all the embeddings $C^\infty_{K_i} \to \mathscr{D}(\Omega)$ continuous,[33] which makes it a locally convex space. The elements of the dual space $\mathscr{D}(\Omega)^*$ are called *distributions* on Ω.

1.2.2 Lebesgue integrable functions

The n-dimensional *outer Lebesgue measure* $\mathrm{meas}_n(\cdot)$ on the Euclidean space $\mathbb{R}^n$, $n \ge 1$, is defined as

$$\mathrm{meas}_n(A) := \inf\Big\{\sum_{k=1}^{\infty}\prod_{i=1}^{n} b_i^k - a_i^k \,:\, A \subset \bigcup_{k=1}^{\infty}[a_1^k, b_1^k]\times\cdots\times[a_n^k, b_n^k],\ \ a_i^k \le b_i^k\Big\} \tag{1.16}$$

and then we call a set $A \subset \mathbb{R}^n$ Lebesgue *measurable* if $\mathrm{meas}_n(A) = \mathrm{meas}_n(A\cap S) + \mathrm{meas}_n(A \setminus S)$ for any subset $S \subset \mathbb{R}^n$.[34] The collection Σ of Lebesgue measurable

[28] This means $\mu(\bigcup_{i\in\mathbb{N}} A_i) = \sum_{i\in\mathbb{N}} \mu(A_i)$ for any mutually disjoint A_i from an algebra in question.

[29] A collection Σ of subsets of $\bar\Omega$ is called a σ-algebra if $A_i{\in}\Sigma \Rightarrow \bigcup_{i\in\mathbb{N}} A_i{\in}\Sigma$, $\emptyset{\in}\Sigma$, and $A{\in}\Sigma \Rightarrow \Omega\backslash A{\in}\Sigma$. Borel's σ-algebra is the smallest σ-algebra containing all open subsets of $\bar\Omega$.

[30] A set function μ is regular if $\forall A{\in}\Sigma\ \forall\varepsilon{>}0\ \exists A_1, A_2{\in}\Sigma$: $\mathrm{cl}(A_1){\subset}A{\subset}\mathrm{int}(A_2)$ and $|\mu|(A_2\backslash A_1){\le}\varepsilon$.

[31] For μ additive, we define the variation $|\mu|$ of μ by $|\mu|(A) = \sup_{(A_1,\dots,A_I)\in M(A)} \sum_{i=1}^{I} |\mu(A_i)|$, where $M(A)$ denotes the set of all finite collections $(A_1, \dots, A_I)$ of mutually disjoint $A_i \in \Sigma$ such that $A_i \subset A$ for any $i = 1, \dots, I$.

[32] The integral via μ is defined by limit of simple functions similarly as in Section 1.2.2.

[33] The convergence $f_k \to f$ in $\mathscr{D}(\Omega)$ means that $\exists K \subset \Omega$ compact $\exists k_0{\in}\mathbb{N}\ \forall k \ge k_0$: $\mathrm{supp}(f_k) \subset K$ and $f_k \to f$ in $C^l_K(\Omega)$ for any $l{\in}\mathbb{N}$; cf. e.g. [425, Sect.I.1].

[34] For example, all closed sets are Lebesgue measurable, hence every open set too, as well as their countable union or intersection, etc.

subsets of Ω forms a so-called σ-algebra[35] which, together with the function $\text{meas}_n : \Sigma \to \mathbb{R} \cup \{+\infty\}$, have (and are characterized by) the following properties:

1. A open implies $A \in \Sigma$,
2. $A = [a_1, b_1] \times \cdots \times [a_n, b_n]$ with $a_i \le b_i$ implies $\text{meas}_n(A) = \prod_{i=1}^n (b_i - a_i)$,
3. meas_n is countably additive, i.e. $\text{meas}_n\big(\bigcup_{k\in\mathbb{N}} A_k\big) = \sum_{k\in\mathbb{N}} \text{meas}_n(A_k)$ for any countable collection $\{A_k\}_{k\in\mathbb{N}}$ of pairwise disjoint sets $A_i \in \Sigma$,
4. $A \subset B \in \Sigma$ and $\text{meas}_n(B) = 0$ implies $A \in \Sigma$ and $\text{meas}_n(A) = 0$.

The function $\text{meas}_n : \Sigma \to \mathbb{R} \cup \{+\infty\}$ is called the *Lebesgue measure.* Having a set $\Omega \in \Sigma$, we say that a property holds *almost everywhere* on Ω (in abbreviation a.e. on Ω) if this property holds everywhere on Ω with the possible exception of a set of Lebesgue-measure zero; referring to those x where this property holds, we will also say that it holds at *almost all* $x \in \Omega$ (in abbreviation a.a. $x \in \Omega$).

A function $u : \mathbb{R}^n \to \mathbb{R}^m$ is called (Lebesgue) *measurable* if $u^{-1}(A) := \{x \in \mathbb{R}^n;\ u(x) \in A\}$ is Lebesgue measurable for any $A \in \mathbb{R}^m$ open.

We call $u : \mathbb{R}^n \to \mathbb{R}^m$ *simple* if it takes only a finite number of values $v_i \in \mathbb{R}^m$ and $u^{-1}(v_i) = \{x;\ u(x) = v_i\} \in \Sigma$; then we define the integral $\int_{\mathbb{R}^n} u(x)\,\mathrm{d}x$ naturally as $\sum_{\text{finite}} \text{meas}_n(A_i) v_i$. Furthermore, a measurable u is called *integrable* if there is a sequence of simple functions $\{u_k\}_{k\in\mathbb{N}}$ such that $\lim_{k\to\infty} u_k(x) = u(x)$ for a.a. $x \in \Omega$ and $\lim_{k\to\infty} \int_{\mathbb{R}^n} u_k(x)\,\mathrm{d}x$ does exist in $\mathbb{R}$. Then, this limit will be denoted by $\int_{\mathbb{R}^n} u(x)\,\mathrm{d}x$ and we call it the (Lebesgue) *integral* of u. It is then independent of the particular choice of the sequence $\{u_k\}_{k\in\mathbb{N}}$.

We will now consider $\Omega \subset \mathbb{R}^n$ measurable with $\text{meas}_n(\Omega) < +\infty$. The notions of measurability and the integral of functions $\Omega \to \mathbb{R}^m$ can be understood as before provided all these functions are extended on $\mathbb{R}^n \setminus \Omega$ by 0. By $L^p(\Omega;\mathbb{R}^m)$ we will denote the set of all measurable functions[36] $u : \Omega \to \mathbb{R}^m$ such that $\|u\|_{L^p(\Omega;\mathbb{R}^m)} < +\infty$, where[37]

$$\|u\|_{L^p(\Omega;\mathbb{R}^m)} := \begin{cases} \sqrt[p]{\int_\Omega |u(x)|^p \mathrm{d}x} & \text{for } 1 \le p < +\infty\,, \\ \operatorname*{ess\,sup}_{x\in\Omega} |u(x)| & \text{for } p = +\infty \end{cases} \tag{1.17}$$

and $|\cdot|$ is the Euclidean norm on $\mathbb{R}^m$. The set $L^p(\Omega;\mathbb{R}^m)$, endowed by a pointwise addition and scalar multiplication, is a linear space. Besides, $\|\cdot\|_{L^p(\Omega;\mathbb{R}^m)}$ is a norm

[35] We call Σ an algebra if $\emptyset \in \Sigma$, $A \in \Sigma \Rightarrow \Omega \setminus A \in \Sigma$, and $A_1, A_2 \in \Sigma \Rightarrow A_1 \cup A_2 \in \Sigma$. If also $A_i \in \Sigma \Rightarrow \bigcup_{i\in\mathbb{N}} A_i \in \Sigma$, then Σ will be called a σ-algebra. An example is the so-called Borel σ-algebra: the smallest σ-algebra containing all open subsets of Ω. In fact, Σ is the so-called Lebesgue extension of the Borel σ-algebra, created by adding all subsets of sets having the measure zero.

[36] As usual, we will not distinguish between functions that are equal to each other a.e., so that, strictly speaking, $L^p(\Omega;\mathbb{R}^m)$ contains classes of equivalence of such functions.

[37] The "essential supremum" is defined as

$$\operatorname*{ess\,sup}_{x\in\Omega} f(x) = \inf_{\substack{N\subset\Omega \\ \text{meas}_n(N)=0}} \ \sup_{x\in\Omega\setminus N} f(x).$$

on $L^p(\Omega;\mathbb{R}^m)$ which makes it a Banach space, called a *Lebesgue space.*

If a measure $\mu \in \mathscr{M}(\bar\Omega)$ possesses a density $d_\mu \in L^1(\Omega)$, which means $\mu(A) = \int_A d_\mu(x)\,\mathrm{d}x$ for any measurable $A \subset \Omega$, then μ has a certain special property, namely it is *absolutely continuous* with respect to the Lebesgue measure[38] and also the converse assertion is true: every absolutely continuous measure possesses a density belonging to $L^1(\Omega)$. This fact is known as the Radon-Nikodým theorem.

An important question is how to characterize concretely the dual spaces. The "natural" duality pairing comes from the inner product in L^2-spaces, which means $\langle u,v\rangle := \int_\Omega u(x)\cdot v(x)\,\mathrm{d}x$, where $u\cdot v := \sum_{i=1}^m u_i v_i$ is the inner product in $\mathbb{R}^m$. If $1<p<+\infty$, then $L^p(\Omega;\mathbb{R}^m)$ is reflexive. From the algebraic Young inequality

$$ab \le \frac{1}{p}a^p + \frac{1}{p'}b^{p'} \tag{1.18}$$

one gets[39] the *Hölder* [208] (or, for p=2, the Hölder-Bunyakovskiĭ [85]) *inequality:*[40]

$$\int_\Omega \big|u(x)\cdot v(x)\big|\,\mathrm{d}x \le \sqrt[p]{\textstyle\int_\Omega |u(x)|^p \mathrm{d}x}\;\sqrt[p']{\textstyle\int_\Omega |v(x)|^{p'}\mathrm{d}x}, \tag{1.19}$$

where p' is the so-called *conjugate exponent* defined as

$$p' := \begin{cases} p/(p-1) & \text{for } 1<p<+\infty,\\ \quad 1 & \text{for } p=+\infty,\\ \;+\infty & \text{for } p=1.\end{cases} \tag{1.20}$$

In fact, the modification of (1.19) for $p=1$ or $p=+\infty$ looks trivially as $\int_\Omega |u\cdot v|\mathrm{d}x \le \|u\|_{L^\infty(\Omega;\mathbb{R}^n)}\|u\|_{L^1(\Omega;\mathbb{R}^n)}$. From (1.19), it can be shown that the dual space is isometrically isomorphic with $L^{p'}(\Omega;\mathbb{R}^m)$ if $1\le p<+\infty$. On the other hand, the dual space to $L^\infty(\Omega;\mathbb{R}^m)$ is substantially larger than $L^1(\Omega;\mathbb{R}^m)$.[41] Applying the algebraic Young inequality (1.18) to $|u(x)\cdot v(x)| \le |u(x)|\,|v(x)|$ and integrating it over Ω, we obtain another important inequality, the integral *Young inequality*

$$\int_\Omega \big|u(x)\cdot v(x)\big|\,\mathrm{d}x \le \frac{1}{p}\int_\Omega |u(x)|^p\,\mathrm{d}x + \frac{1}{p'}\int_\Omega |v(x)|^{p'}\,\mathrm{d}x. \tag{1.21}$$

Instead of (1.18), we will often apply the modified Young inequality

$$\forall\varepsilon>0:\quad ab \le \varepsilon a^p + C_\varepsilon b^{p'}, \qquad \text{where } C_\varepsilon := \frac{\sqrt[p-1]{\varepsilon}\;\sqrt[p']{p}}{p-1}. \tag{1.22}$$

Moreover, for $1<p<+\infty$, the space $L^p(\Omega;\mathbb{R}^m)$ is *uniformly convex.*[42]

[38] This means that $\forall\varepsilon>0\ \exists\delta>0\ \forall A\subset\Omega$ measurable: $\mathrm{meas}_n(A)\le\delta \Longrightarrow |\mu(A)|\le\varepsilon$.

[39] Hint: $\|u\|_{L^p(\Omega)}^{-1}\|v\|_{L^{p'}(\Omega)}^{-1}\int_\Omega |u\cdot v|\mathrm{d}x \le \frac{1}{p}\|u\|_{L^p(\Omega)}^{-p}\int_\Omega |u|^p\mathrm{d}x + \frac{1}{p'}\|v\|_{L^{p'}(\Omega)}^{-p'}\int_\Omega |u|^{p'}\mathrm{d}x = 1$.

[40] Originally, Hölder [208] states this in a less symmetrical form for sums in place of integrals.

[41] The elements of $L^\infty(\Omega;\mathbb{R}^m)^*$ are indeed very abstract objects and can be identified with finitely-additive measures vanishing on zero-measure sets, see Yosida and Hewitt [426].

[42] This result is due to Clarkson [97], see also e.g. Adams [3, Corollary 2.29] or Kufner at al. [245, Remark 2.17.8].

If $\text{meas}_n(\Omega) < +\infty$ and $1 \le q \le p \le +\infty$, the embeddings $C^0(\bar\Omega;\mathbb{R}^m) \subset L^p(\Omega;\mathbb{R}^m) \subset L^q(\Omega;\mathbb{R}^m)$ are continuous[43]. Moreover, for $p < +\infty$ these embeddings are dense and then, as $C^0(\bar\Omega;\mathbb{R}^m)$ is separable, $L^p(\Omega;\mathbb{R}^m)$ is separable, too. On the other hand, $L^\infty(\Omega)$ is not separable.[44]

Hölder's inequality also allows for an *interpolation* between $L^{p_1}(\Omega)$ and $L^{p_2}(\Omega)$: for $p_1, p_2, p \in [1,+\infty]$, $\lambda \in [0,1]$, it holds that[45]

$$\frac{1}{p} = \frac{\lambda}{p_1} + \frac{1-\lambda}{p_2} \qquad \Rightarrow \qquad \|v\|_{L^p(\Omega)} \le \|v\|^{\lambda}_{L^{p_1}(\Omega)} \|v\|^{1-\lambda}_{L^{p_2}(\Omega)}. \tag{1.23}$$

Moreover, if p is as in (1.23), $p_1 \le p_2$, and similarly $q^{-1} = \lambda q_1^{-1} + (1-\lambda)q_2^{-1}$, $q_1 \le q_2$, and A is a bounded linear operator $L^{p_1}(\Omega) \to L^{q_1}(\Omega)$ whose restriction on $L^{p_2}(\Omega)$ belongs to $\mathscr{L}(L^{p_2}(\Omega), L^{q_2}(\Omega))$, then

$$\|A\|_{\mathscr{L}(L^p(\Omega),L^q(\Omega))} \le C\|A\|^{\lambda}_{\mathscr{L}(L^{p_1}(\Omega),L^{q_1}(\Omega))} \|A\|^{1-\lambda}_{\mathscr{L}(L^{p_2}(\Omega),L^{q_2}(\Omega))} \tag{1.24}$$

for some constant $C = C(p_1, p_2, q_1, q_2, \lambda)$.

We say that a sequence $u_k : \Omega \to \mathbb{R}^m$ *converges in measure* to u if

$$\forall \varepsilon > 0: \qquad \lim_{k\to\infty} \text{meas}_n\big(\{x \in \Omega;\ |u_k(x) - u(x)| \ge \varepsilon\}\big) = 0. \tag{1.25}$$

Naturally, the convergence a.e. means that $u_k(x) \to u(x)$ for a.a. $x \in \Omega$. Let us emphasize that convergence in measure does not imply convergence a.e.[46]. Anyhow:

Proposition 1.13 (Various modes of convergences).

(i) *Any sequence converging a.e. converges also in measure.*

(ii) *Any sequence converging in measure admits a subsequence converging a.e.*

(iii) *Any sequence converging in $L^1(\Omega)$ converges in measure.*

Theorem 1.14 (Lebesgue [254]). *Let $\{u_k\}_{k\in\mathbb{N}} \subset L^1(\Omega)$ be a sequence converging a.e. to some u and $|u_k(x)| \le v(x)$ for some $v \in L^1(\Omega)$. Then u lives in $L^1(\Omega)$ and $\lim_{k\to\infty} \int_A u_k(x)\,\mathrm{d}x = \int_A u(x)\,\mathrm{d}x$ for any $A \subset \Omega$ measurable.*

Theorem 1.15 (Fatou [143]). *Let $\{u_k\}_{k\in\mathbb{N}} \subset L^1(\Omega)$ be a sequence of non-negative functions[47] such that $\liminf_{k\to\infty} \int_\Omega u_k(x)\,\mathrm{d}x < +\infty$. Then the function $x \mapsto \liminf_{k\to\infty} u_k(x)$ is integrable and*

$$\liminf_{k\to\infty} \int_\Omega u_k(x)\,\mathrm{d}x \ge \int_\Omega \Big(\liminf_{k\to\infty} u_k(x)\Big)\,\mathrm{d}x. \tag{1.26}$$

---

[43] Cf. Exercise 2.68 below.

[44] To see it, consider $\Omega = (0,1)$ and a collection $\{\chi_{(0,a)}\}_{a\in(0,1)}$ of characteristic functions of an interval $(0,a)$, i.e. $\chi_{(0,a)}(x) = 1$ for $x \in (0,a)$ and $\chi_{(0,a)}(x) = 0$ for $x \in [a,1)$. This collection is an uncountable subset of the unit sphere in $L^\infty(0,1)$ and $\|\chi_{(0,a)} - \chi_{(0,b)}\|_{L^\infty(0,1)} = 1$ for $a \ne b$, hence $L^\infty(0,1)$ cannot be separable.

[45] Cf. Exercise 2.59 below. See e.g. [268, p.26].

[46] An example for a sequence converging in the measure on $[0,1]$ to 0 but not a.e. is $u_{(l,m)}$ a characteristic function of the interval of the type $[(m-1)2^{-l}, m2^{-l}]$ for $1 \le m \le 2^l$, arranged lexicographically as $(l,m) = (1,1), (1,2), (2,1), (2,2), (2,3), (2,4), (3,1), \ldots$. Selection of a subsequence converging a.e. to 0 can be done, e.g., by taking $m = 1$ only.

[47] Obviously, existence of a common integrable minorant can weaken this assumption.

In particular, Theorem 1.14 says that the set $\{u_k;\ k \in \mathbb{N}\}$ is relatively weakly compact[48] in $L^1(\Omega)$. This property is related to both equi-absolute continuity and uniform integrability:

Theorem 1.16 (DUNFORD and PETTIS [128]). *Let $M \subset L^1(\Omega;\mathbb{R}^m)$ be bounded. Then the following statements are equivalent to each other:*
(i) *M is relatively weakly compact in $L^1(\Omega;\mathbb{R}^m)$,*
(ii) *the set M is uniformly integrable, which means:*

$$\forall \varepsilon > 0 \ \ \exists K \in \mathbb{R}^+ : \qquad \sup_{u\in M} \int_{\{x\in\Omega;|u(x)|\ge K\}} |u(x)|\,\mathrm{d}x \le \varepsilon\,, \tag{1.27}$$

(iii) *the set M is equi-absolutely-continuous, which means:*

$$\forall \varepsilon > 0 \ \ \exists \delta > 0 : \qquad \sup_{u\in M} \sup_{|A|\le\delta} \int_A |u(x)|\,\mathrm{d}x \le \varepsilon\,. \tag{1.28}$$

Useful generalizations of the Lebesgue dominated-convergence Theorem 1.14 and the Fatou Theorem 1.15 are:

Theorem 1.17 (VITALI [421][49]). *Let $\{u_k\}_{k\in\mathbb{N}} \subset L^1(\Omega)$ be a sequence converging a.e. to some u. Then $u \in L^p(\Omega)$ and $u_k \to u$ in $L^p(\Omega)$ if and only if $\{|u_k|^p\}_{k\in\mathbb{N}}$ is uniformly integrable.*

Theorem 1.18 (FATOU, GENERALIZED[50]). *The conclusion of Theorem* 1.15 *holds if $u_k \ge 0$ is replaced by $u_k \ge v_k$ with $\{v_k\}_{k\in\mathbb{N}}$ being uniformly integrable.*

Theorem 1.19 (FUBINI [160]). *Considering two Lebesgue measurable sets $\Omega_1 \subset \mathbb{R}^{n_1}$ and $\Omega_2 \subset \mathbb{R}^{n_2}$, the following identity holds provided $g \in L^1(\Omega_1\times\Omega_2)$ (in particular, each of the following double-integrals does exist and is finite):*

$$\int_{\Omega_1\times\Omega_2} g(x_1,x_2)\,\mathrm{d}(x_1,x_2) = \int_{\Omega_1}\Big(\int_{\Omega_2} g(x_1,x_2)\,\mathrm{d}x_2\Big)\mathrm{d}x_1 = \int_{\Omega_2}\Big(\int_{\Omega_1} g(x_1,x_2)\,\mathrm{d}x_1\Big)\mathrm{d}x_2.$$

1.2.3 Sobolev spaces

Modern theory of differential equations is based on spaces of functions whose derivatives exist in a generalized sense and enjoy a suitable integrability. Those spaces, developed since the works by Beppo Levi [258], Leray [256], Sobolev [389],

[48] Note that the linear hull of all characteristic functions χ_A with $A \subset \Omega$ measurable is dense in $L^\infty(\Omega) \cong L^1(\Omega)^*$, so that the sequence $\{u_k\}$, being bounded in $L^1(\Omega)$, converges weakly in $L^1(\Omega)$ and, as such, it is relatively sequentially weakly compact, hence by the Eberlein-Šmuljan theorem relatively weakly compact, too.

[49] More precisely, in [421], the integration of summable series is investigated rather than mere sequences.

[50] See Ash [21, Thm.7.5.2], or also Klei and Miyara [233] or Saadoune and Valadier [377].

and Tonelli [406], have thus an absolute importance in this context and their theory is presently very broad. We present here only a rather minimal extent; for more detailed exposition we refer to Adams [3], Adams, Fournier [4], Giusti [180], Kufner, Fučík, John [245], Maz'ya [277], or Ziemer [430].

Having a function $u \in L^p(\Omega)$, we define its *distributional derivative* $\partial^k u/\partial x_1^{k_1}\cdots\partial x_n^{k_n}$ with $k_1+\cdots+k_n = k$ and $k_i \ge 0$ for any $i = 1,\dots,n$ as a distribution such that

$$\forall g\in\mathscr{D}(\Omega): \qquad \Big\langle\frac{\partial^k u}{\partial x_1^{k_1}\cdots\partial x_n^{k_n}}, g\Big\rangle = (-1)^k\int_\Omega u\,\frac{\partial^k g}{\partial x_1^{k_1}\cdots\partial x_n^{k_n}}\,\mathrm{d}x. \tag{1.29}$$

The n-tuple of the first-order distributional derivatives $\big(\frac{\partial}{\partial x_1}u,\dots,\frac{\partial}{\partial x_n}u\big)$ is denoted by ∇u and called a *gradient* of u. We will now consider $\Omega\subset\mathbb{R}^n$ open with $\mathrm{meas}_n(\Omega) < +\infty$; such a set will be called a *domain.* For $p < +\infty$, we define a *Sobolev space* [390]

$$W^{1,p}(\Omega):=\big\{u\in L^p(\Omega);\ \nabla u\in L^p(\Omega;\mathbb{R}^n)\big\},\ \ \text{equipped with the norm} \tag{1.30a}$$

$$\|u\|_{W^{1,p}(\Omega)} := \begin{cases} \sqrt[p]{\|u\|^p_{L^p(\Omega)}+\|\nabla u\|^p_{L^p(\Omega;\mathbb{R}^n)}} \\ \qquad = \sqrt[p]{\int_\Omega |u(x)|^p+|\nabla u(x)|^p\mathrm{d}x} & \text{if } p<+\infty, \\ \max\big(\|u\|_{L^\infty(\Omega)}, \|\nabla u\|_{L^\infty(\Omega;\mathbb{R}^n)}\big) \\ \qquad = \mathrm{ess\,sup}_{x\in\Omega}\max\big(|u(x)|,|\nabla u(x)|\big) & \text{if } p=+\infty. \end{cases} \tag{1.30b}$$

As, by Rademacher's theorem, Lipschitz functions are a.e. differentiable, it holds that $W^{1,\infty}(\Omega) = C^{0,1}(\Omega)$.

Analogously, for $k > 1$ integer, we define

$$W^{k,p}(\Omega) := \big\{u\in L^p(\Omega;\mathbb{R}^m);\ \nabla^k u\in L^p(\Omega;\mathbb{R}^{n^k})\big\}, \tag{1.31}$$

where $\nabla^k u$ denotes the set of all k-th order partial derivatives of u understood in the distributional sense. The standard norm on $W^{k,p}(\Omega)$ is $\|u\|_{W^{k,p}(\Omega)} = \big(\|u\|^p_{L^p(\Omega)}+\|\nabla^k u\|^p_{L^p(\Omega;\mathbb{R}^{n^k})}\big)^{1/p}$, which makes it a Banach space. Likewise for Lebesgue spaces, for $1\le p<+\infty$ the Sobolev spaces $W^{k,p}(\Omega;\mathbb{R}^m)$ are separable and, if $1<p<+\infty$, they are uniformly convex,[51] hence by Milman-Pettis' theorem also reflexive.

The spaces of $\mathbb{R}^m$-valued functions, $W^{k,p}(\Omega;\mathbb{R}^m)$, are defined analogously.[52]

To give a good sense to traces on the *boundary* $\Gamma := \partial\Omega := \bar\Omega\setminus\Omega$, we must qualify Ω suitably. We say that Ω is a *Lipschitz domain* if there is a finite number of overlapping parts Γ_i of the boundary of Γ and corresponding coordinate systems (i.e. transformation unitary matrices A_i and open sets $G_i\in\mathbb{R}^{n-1}$) such that each

[51] See Adams [3, Theorem 3.5].

[52] This means $W^{k,p}(\Omega;\mathbb{R}^m) := \{(u_1,\dots,u_m);\ u_i\in W^{k,p}(\Omega)\}$.

Γ_i can be expressed as a graph of a Lipschitz function $g_i \in C^{0,1}(\mathbb{R}^{n-1})$ in the sense that

$$\Gamma_i = \Big\{A_i\xi;\ \ \xi\in\mathbb{R}^n,\ (\xi_1,\ldots,\xi_{n-1})\in G_i,\ \ \xi_n = g_i(\xi_1,\ldots,\xi_{n-1})\Big\} \tag{1.32}$$

and Ω lies on one side of Γ in the sense that $\big\{A_i\xi;\ \xi \in \mathbb{R}^n,\ (\xi_1,\ldots,\xi_{n-1}) \in G_i,\ g_i(\xi_1,\ldots,\xi_{n-1}){-}\varepsilon < \xi_n < g_i(\xi_1,\ldots,\xi_{n-1})\big\} \subset \Omega$ and simultaneously $\big\{A_i\xi;\ \xi\in\mathbb{R}^n,\ (\xi_1,\ldots,\xi_{n-1})\in G_i,\ g_i(\xi_1,\ldots,\xi_{n-1}) < \xi_n < g_i(\xi_1,\ldots,\xi_{n-1}){+}\varepsilon\big\} \subset \mathbb{R}^n\setminus\bar{\Omega}$ for some $\varepsilon > 0$. For an example of a Lipschitz domain $\Omega \subset \mathbb{R}^2$, see Figure 2a where three indicated coordinate systems are sufficient to cover Γ, and Figure 2b where this condition fails in a variety of ways.

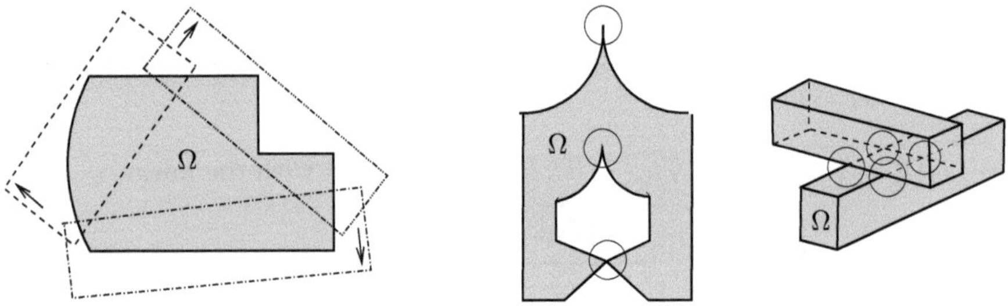

Fig. 2. a) Example of a Lipschitz domain whose boundary can be covered by three charts. *b) Examples of a 2D and a 3D domains which are not Lipschitz because of 3 and 4 spots, respectively (marked by circles).*

If the mappings g_i belong to $C^{k,\alpha}(\mathbb{R}^{n-1})$, we say that the domain Ω is *of $C^{k,\alpha}$-class.* Lipschitz domains are then naturally called of the $C^{0,1}$-class. We will always assume Ω to be Lipschitz and bounded.

Theorem 1.20 (SOBOLEV EMBEDDING [389]). *The continuous embedding*

$$W^{1,p}(\Omega) \subset L^{p^*}(\Omega) \tag{1.33}$$

holds provided the exponent p^* is defined as

$$p^* := \begin{cases} \dfrac{np}{n-p} & \text{for } p < n\,, \\ \text{an arbitrarily large real} & \text{for } p = n\,, \\ +\infty & \text{for } p > n\,. \end{cases} \tag{1.34}$$

Theorem 1.21 (RELLICH, KONDRACHOV [348, 236][53]). *The compact embedding*

$$W^{1,p}(\Omega) \Subset L^{p^*-\epsilon}(\Omega)\,, \qquad \epsilon \in (0, p^*{-}1], \tag{1.35}$$

holds for p^* from (1.34).

[53]The pioneering Rellich's work deals with $p = 2$ only.

For higher-order Sobolev spaces, one gets, e.g., $W^{2,p}(\Omega) \subset W^{1,np/(n-p)}(\Omega)$ by applying Theorem 1.20 on first derivatives, and applying Theorem 1.20 once again for $np/(n-p)$ instead of p one comes to $W^{1,np/(n-p)}(\Omega) \subset L^{np/(n-2p)}(\Omega)$ provided $2p < n$. It identifies $p^{**} := (p^*)^*$ as $np/(n-2p)$ if $2p < n$. Proceeding further by induction, one comes to:

Corollary 1.22 (HIGHER-ORDER SOBOLEV EMBEDDING[54]).

(i) *If $kp < n$, the continuous embedding $W^{k,p}(\Omega) \subset L^{np/(n-kp)}(\Omega)$ and the compact embedding $W^{k,p}(\Omega) \Subset L^{np/(n-kp)-\epsilon}(\Omega)$ hold for any $\epsilon \in (0, \frac{np}{n-kp} - 1]$.*

(ii) *For $kp = n$, it holds that $W^{k,p}(\Omega) \Subset L^q(\Omega)$ for any $q < +\infty$.*

(iii) *For $kp > n$, it holds that $W^{k,p}(\Omega) \Subset C(\bar{\Omega})$.*

Theorem 1.23 (TRACE OPERATOR[55]). *There is exactly one linear continuous operator $T : W^{1,p}(\Omega) \to L^1(\Gamma)$ such that, for any $u \in C^1(\bar{\Omega})$, it holds that $Tu = u|_\Gamma$ (=the restriction of u on Γ). Moreover, T remains continuous (resp. is compact) as the mappings*

$$u \mapsto u|_\Gamma : W^{1,p}(\Omega) \to L^{p^\#}(\Gamma), \quad \textit{resp.,} \tag{1.36a}$$

$$u \mapsto u|_\Gamma : W^{1,p}(\Omega) \to L^{p^\#-\epsilon}(\Gamma)\ , \qquad \epsilon \in (0, p^\#{-}1], \tag{1.36b}$$

provided the exponent $p^\#$ is defined as

$$p^\# := \begin{cases} \dfrac{np-p}{n-p} & \textit{for } p < n\ , \\ \textit{an arbitrarily large real} & \textit{for } p = n\ , \\ +\infty & \textit{for } p > n\ . \end{cases} \tag{1.37}$$

We will call the operator T from Theorem 1.23 the *trace operator*, and write simply $u|_\Gamma$ instead of Tu even if $u \in W^{1,p}(\Omega) \setminus C^1(\bar{\Omega})$. Then we define $W_0^{1,p}(\Omega) := \{v \in W^{1,p}(\Omega);\ v|_\Gamma = 0\}$. For $k > 1$, we define similarly $W_0^{k,p}(\Omega) := \{v \in W^{k,p}(\Omega);\ \nabla^i v \in W_0^{k-i,p}(\Omega; \mathbb{R}^{n^i}),\ i = 0, \ldots, k-1\}$.

Another useful result allows for a certain *interpolation* between the Sobolev space $W^{k,p}(\Omega)$ and the Lebesgue space $L^q(\Omega)$:

Theorem 1.24 (GAGLIARDO-NIRENBERG INEQUALITY [161, 312, 313]). *Let $\beta = \beta_1 + \cdots + \beta_n$, $\beta_1, \ldots, \beta_n \in \mathbb{N} \cup \{0\}$, $k \in \mathbb{N}$, r, q, and p satisfy*

$$\frac{1}{r} = \frac{\beta}{n} + \lambda\Big(\frac{1}{p} - \frac{k}{n}\Big) + (1-\lambda)\frac{1}{q}, \quad \frac{\beta}{k} \le \lambda \le 1, \quad 0 \le \beta \le k-1, \tag{1.38}$$

[54] In fact, the point (iii) can be improved. E.g. for $k = 1$ and Ω smooth, one has $W^{1,p}(\Omega) \subset C^{0,\alpha}(\bar{\Omega})$ with $\alpha = 1 - n/p$, which is referred to a Morrey inequality. In general, $W^{k,p}(\Omega) \subset C^{k-[n/p]-1,\alpha}(\bar{\Omega})$ with $\alpha < 1$ if $n/p \in \mathbb{N}$ or $\alpha = [n/p] + 1 - n/p$ if $n/p \notin \mathbb{N}$.

[55] In fact, $u \mapsto u|_\Gamma : W^{1,p}(\Omega) \to W^{1-1/p,p}(\Gamma)$, where the Sobolev-Slobodeckiĭ space $W^{1-1/p,p}(\Gamma)$ is defined as in (1.42) but on an $(n{-}1)$-dimensional manifold Γ instead of n-dimensional domain Ω. Then, similarly as in Theorem 1.20, we have the embedding $W^{1-1/p,p}(\Gamma) \subset L^{p^\#}(\Gamma)$, resp. $W^{1-1/p,p}(\Gamma) \Subset L^{p^\#-\epsilon}(\Gamma)$. For the rather less standard notation of the exponent $p^\#$, see e.g. Ciarlet [93, Thm 6.1-7].

then it holds

$$\left\| \frac{\partial^\beta v}{\partial x_1^{\beta_1} \cdots \partial x_1^{\beta_n}} \right\|_{L^r(\Omega)} \le C_{\text{GN}} \big\| v \big\|_{W^{k,p}(\Omega)}^{\lambda} \big\| v \big\|_{L^q(\Omega)}^{1-\lambda}, \tag{1.39}$$

provided $k - \beta - n/p$ is not a negative integer (otherwise it holds for $\lambda = |\beta|/k$).

For example, if $n = 2$, then

$$\|v\|_{L^4(\Omega)} \le C_{\text{GN}} \sqrt{\|v\|_{L^2(\Omega)} \big(\|v\|_{L^2(\Omega)} + \|\nabla v\|_{L^2(\Omega;\mathbb{R}^2)} \big)}. \tag{1.40}$$

There is another definition of Sobolev spaces, not so explicit as that one we used. Namely, $\mathbb{W}^{k,p}(\Omega) := \text{cl}\big(C^\infty(\bar{\Omega})\big)$ where "cl" stands for the closure in, say, $L^p(\Omega)$ with respect to the $W^{k,p}$-norm introduced above; for $k = 1$ see (1.30b). Similarly, $\mathbb{W}_0^{k,p}(\Omega) := \text{cl}\big(C_0^\infty(\Omega)\big)$ where $C_0^\infty(\Omega)$ stands for the set of infinitely smooth functions with a compact support in Ω. For the following important assertion an important (here permanently accepted) assumption is that Ω has a Lipschitz boundary.

Theorem 1.25 (Density of smooth functions). *It holds that*

$$W^{k,p}(\Omega) = \mathbb{W}^{k,p}(\Omega), \qquad W_0^{k,p}(\Omega) = \mathbb{W}_0^{k,p}(\Omega). \tag{1.41}$$

A generalization of Sobolev spaces for $k \ge 0$ a non-integer is often useful for various finer investigations:

$$W^{k,p}(\Omega) := \left\{ u \in W^{[k],p}(\Omega);\ \int_{\Omega\times\Omega} \frac{|\nabla^{[k]} u(x) - \nabla^{[k]} u(\xi)|^p}{|x-\xi|^{n+p(k-[k])}} \,\mathrm{d}x\mathrm{d}\xi < +\infty \right\}, \tag{1.42}$$

where $[k]$ denotes the integer part of k. For k non-integer, $W^{k,p}(\Omega)$ is called the *Sobolev-Slobodeckiĭ space*. They are Banach spaces if normed by the norm

$$\|u\|_{W^{k,p}(\Omega)} := \left(\|u\|_{W^{[k],p}(\Omega)}^p + \int_{\Omega\times\Omega} \frac{|\nabla^{[k]} u(x) - \nabla^{[k]} u(\xi)|^p}{|x-\xi|^{n+p(k-[k])}} \,\mathrm{d}x\mathrm{d}\xi \right)^{1/p}. \tag{1.43}$$

In fact, Corollary 1.22 holds for $k \ge 0$ non-integer, too. A complete *interpolation* theory holds for the Hilbertian case $p = 2$ for general k, k_1, and k_2 non-negative:

$$k = \lambda k_1 + (1-\lambda) k_2 \quad \Longrightarrow \quad \big\|u\big\|_{W^{k,p}(\Omega)} \le C \big\|u\big\|_{W^{k_1,p}(\Omega)}^{\lambda} \big\|u\big\|_{W^{k_2,p}(\Omega)}^{1-\lambda} \tag{1.44}$$

with some $C = C(k_1, k_2, \lambda)$, while for $p \ne 2$, the inequality (1.44) holds provided k is not integer and $k_1, k_2 \in \mathbb{N}$.[56] Like in (1.24),

$$\big\|A\big\|_{\mathscr{L}(W^{k,p}(\Omega), W^{l,p}(\Omega))} \le C \big\|A\big\|_{\mathscr{L}(W^{k_1,p}(\Omega), W^{l_1,p}(\Omega))}^{\lambda} \big\|A\big\|_{\mathscr{L}(W^{k_2,p}(\Omega), W^{l_1,p}(\Omega))}^{1-\lambda} \tag{1.45}$$

[56] For $k \in \mathbb{N}$ the Sobolev space $W^{k,p}(\Omega)$ is to be replaced by a so-called Besov space.

for some constant $C = C(k_1, k_2, l_1, l_2, \lambda)$ provided $A \in \mathscr{L}(W^{k_1,2}(\Omega), W^{l_1,2}(\Omega)) \cap \mathscr{L}(W^{k_2,2}(\Omega), W^{l_2,2}(\Omega))$, k is as in (1.44), and $l^{-1} = \lambda l_1^{-1} + (1-\lambda) l_2^{-1}$.

The Sobolev-Slobodeckiĭ spaces are sometimes well fitted with some sophisticated nonlinear estimates, e.g.:

Theorem 1.26 (Málek, Pražák, Steinhauer [269][57]). *If $p \geq 2$ and $\epsilon > 0$, then there is $c > 0$ such that*

$$c\|u\|_{W^{2/p-\epsilon,p}(\Omega)} \leq \Big(\|u\|^p_{L^p(\Omega)} + \int_\Omega |u|^{p-2}|\nabla u|^2\,\mathrm{d}x\Big)^{1/p}. \tag{1.46}$$

1.3 Nemytskiĭ mappings

Considering integers $j, m_0, m_1, \ldots, m_j$, we say that a mapping $a : \Omega \times \mathbb{R}^{m_1} \times \cdots \times \mathbb{R}^{m_j} \to \mathbb{R}^{m_0}$ is a *Carathéodory mapping* if $a(\cdot, r_1, \ldots, r_j) : \Omega \to \mathbb{R}^{m_0}$ is measurable for all $(r_1, \ldots, r_j) \in \mathbb{R}^{m_1} \times \cdots \times \mathbb{R}^{m_j}$ and $a(x, \cdot) : \mathbb{R}^{m_1} \times \cdots \times \mathbb{R}^{m_j} \to \mathbb{R}^{m_0}$ is continuous for a.a. $x \in \Omega$. Then the so-called *Nemytskiĭ mappings* $\mathscr{N}_a$ map functions $u_i : \Omega \to \mathbb{R}^{m_i}$, $i = 1, \ldots, j$, to a function $\mathscr{N}_a(u_1, \ldots, u_j) : \Omega \to \mathbb{R}^{m_0}$ defined by

$$\big[\mathscr{N}_a(u_1, \ldots, u_j)\big](x) = a\big(x, u_1(x), \ldots, u_j(x)\big). \tag{1.47}$$

Theorem 1.27 (Nemytskiĭ mappings in Lebesgue spaces[58]). *If $a : \Omega \times \mathbb{R}^{m_1} \times \cdots \times \mathbb{R}^{m_j} \to \mathbb{R}^{m_0}$ is a Carathéodory mapping and the functions $u_i : \Omega \to \mathbb{R}^{m_i}$, $i = 1, \ldots, j$, are measurable, then $\mathscr{N}_a(u_1, \ldots, u_j)$ is measurable. Moreover, if a satisfies the growth condition*

$$\big|a(x, r_1, \ldots, r_j)\big| \leq \gamma(x) + C \sum_{i=1}^{j} \big|r_i\big|^{p_i/p_0} \quad \text{for some } \gamma \in L^{p_0}(\Omega), \tag{1.48}$$

with $1 \leq p_i < +\infty$, $1 \leq p_0 < +\infty$, then $\mathscr{N}_a$ is a bounded continuous mapping $L^{p_1}(\Omega; \mathbb{R}^{m_1}) \times \cdots \times L^{p_j}(\Omega; \mathbb{R}^{m_j}) \to L^{p_0}(\Omega; \mathbb{R}^{m_0})$. If some $p_i = +\infty$, $i = 1, \ldots, j$, the same holds if the respective term $|\cdot|^{p_i/p_0}$ is replaced by any continuous function.

The case $p_0 = +\infty$ has to be excluded, cf. Exercise 2.61. Also, it should be emphasized that $\mathscr{N}_a$ cannot be weakly continuous unless $a(x, \cdot)$ is affine for a.a. $x \in \Omega$; cf. Exercise 2.62. A "canonical" example is a sequence $u_k(x) := \mathrm{sign}(\sin(k\pi x))$, $\Omega = (0,1) \subset \mathbb{R}^1$, which converges weakly (or weakly*) to 0 in any $L^p(0,1)$, but for $a(x,r) = |r|$ one obviously gets (cf. Figure 3) that $\operatorname*{w-lim}_{k\to\infty} a(u_k) = \lim_{k\to\infty} 1 = 1 \neq 0 = a(0) = a\big(\operatorname*{w-lim}_{k\to\infty} u_k\big)$.

[57] In fact, [269] works with periodic functions having zero means. The proof relies on the so-called Nikolskiĭ spaces. Here, we present an obvious modification involving the L^p-norm in (1.46).

[58] Cf. Exercise 2.60 below.

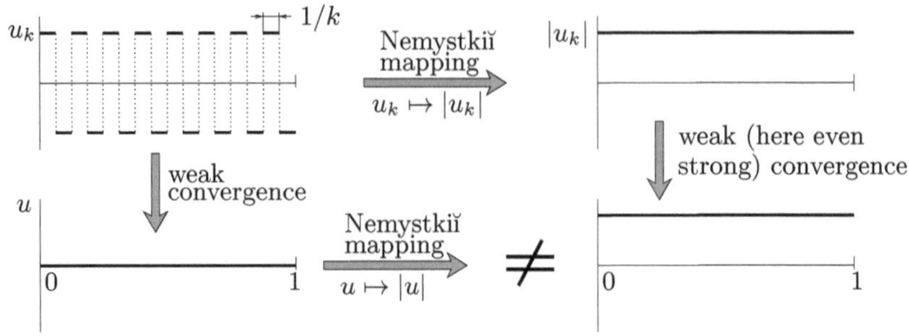

Fig. 3. A counterexample for weak continuity of nonlinear Nemytskiĭ mappings.

Sometimes, it is useful to consider Nemytskiĭ mappings in a Sobolev space. Here we confine ourselves to a special case of a one-argument Nemytskiĭ mapping with an integrand independent of x; in fact, it is then a usual superposition.

Proposition 1.28 (SUPERPOSITION OPERATOR IN SOBOLEV SPACES[59]). *If $a : \mathbb{R} \to \mathbb{R}$ is Lipschitz continuous, then $a(u) \in W^{1,p}(\Omega)$ for any $u \in W^{1,p}(\Omega)$, $p \in [1, +\infty]$, and moreover it holds that*

$$\nabla a(u) = a'(u)\nabla u \qquad \text{(a.e. on } \Omega). \tag{1.49}$$

Denoting $u^+ = \max(u, 0)$ and $u^- = \min(u, 0)$, Proposition 1.28 yields[60]

$$\nabla(u^+) = \begin{cases} \nabla u & \text{if } u > 0, \\ 0 & \text{if } u \leq 0, \end{cases} \qquad \nabla(u^-) = \begin{cases} 0 & \text{if } u \geq 0, \\ \nabla u & \text{if } u < 0, \end{cases} \qquad \text{(a.e. on } \Omega). \tag{1.50}$$

Another useful assertion addresses the change of the order of an integration and a differentiation, and is based on the Lebesgue dominated convergence Theorem 1.14.

Theorem 1.29 (CHANGE OF INTEGRATION AND DIFFERENTIATION). *Let $(x, r) \mapsto \varphi : \Omega \times \mathbb{R} \to \mathbb{R}$ and its partial derivative $\frac{\partial}{\partial r}\varphi : \Omega \times \mathbb{R} \to \mathbb{R}$ be Carathéodory functions, $|\varphi(\cdot, 0)| \in L^1(\Omega)$ and the collection $\{\frac{\partial}{\partial r}\varphi(\cdot, r)\}_{|r| \leq \varepsilon}$ have a common integrable majorant. Then $\Phi : r \mapsto \int_\Omega \varphi(x, r)\,\mathrm{d}x$ is differentiable at 0 and $\frac{\mathrm{d}}{\mathrm{d}t}\Phi(0) = \int_\Omega \frac{\partial}{\partial r}\varphi(x, 0)\,\mathrm{d}x$.*

1.4 Green formula and some inequalities

Let us define the *unit outward normal* $\nu = \nu(x) \in \mathbb{R}^n$ to the boundary Γ at a point $x \in \Gamma$ as the vector

$$\nu(x) = \frac{A_i\left(\dfrac{\partial g_i}{\partial \xi_1}, \ldots, \dfrac{\partial g_i}{\partial \xi_{n-1}}, 1\right)}{\sqrt{\left|\dfrac{\partial g_i}{\partial \xi_1}\right|^2 + \cdots + \left|\dfrac{\partial g_i}{\partial \xi_{n-1}}\right|^2 + 1}} \tag{1.51}$$

[59] See, e.g., monographs by Cazenave and Haraux [90, Prop. 1.3.5] or Ziemer [430, Theorem 2.1.11]. Original results are, in particular, due to Marcus and Mizel [273, 274].

[60] See, e.g., Cazenave and Haraux [90, Corollary 1.3.6] or Ziemer [430, Corollary 2.1.8].

where $x = A_i\xi = A_i\big(\xi_1,\dots,\xi_{n-1},g_i(\xi_1,\dots,\xi_{n-1})\big)$ as in (1.32). It is again natural to assume Ω a Lipschitz domain. Then the functions g_i are Lipschitz continuous and, by the so-called *Rademacher theorem* [342], they possess derivatives almost everywhere on $\Gamma_i \subset \mathbb{R}^{n-1}$, cf. (1.32), hence ν is defined almost everywhere on Γ and does not depend on the concrete covering of Γ by the coordinate systems (G_i, A_i). For a function $f:\Gamma\to\mathbb{R}$, we can define the *surface integral* $\int_\Gamma f(x)\,\mathrm{d}S$ through $(n-1)$-dimensional Lebesgue measure as

$$\int_\Gamma f(x)\,\mathrm{d}S = \sum_i \int_{\tilde G_i} f\big(A_i(\xi_1,\dots,\xi_{n-1},g_i(\xi_1,\dots,\xi_{n-1})\big) \\ \times\sqrt{\Big|\frac{\partial g_i}{\partial\xi_1}\Big|^2+\dots+\Big|\frac{\partial g_i}{\partial\xi_{n-1}}\Big|^2+1}\;\;\mathrm{d}(\xi_1,\dots,\xi_{n-1}) \tag{1.52}$$

where $\tilde G_i \subset G_i$ are chosen suitably to realize, instead of the overlapping covering as on Figure 2, a disjoint covering of Γ. Again, the value $\int_\Gamma f(x)\,\mathrm{d}S$ does not depend on the particular coordinate systems $(\tilde G_i, A_i)$.

Theorem 1.30 (MULTIDIMENSIONAL BY-PART INTEGRATION). *If* $v\in W^{1,p}(\Omega)$ *and* $z\in W^{1,p'}(\Omega)$, *then*

$$\forall i=1,\dots,n: \qquad \int_\Omega \Big(v\frac{\partial z}{\partial x_i}+\frac{\partial v}{\partial x_i}z\Big)\,\mathrm{d}x = \int_\Gamma vz\,\nu_i\,\mathrm{d}S. \tag{1.53}$$

Considering $z=(z_1,\dots,z_n)$, writing (1.53) for z_i instead of z, and eventually summing it over $i=1,\dots,n$ with abbreviating $\operatorname{div} z := \sum_{i=1}^n \frac{\partial}{\partial x_i} z_i$ the *divergence* of the vector field z, we arrive atthe formula which we will often use:

Theorem 1.31 (GREEN'S FORMULA [188][61]). *For any* $v\in W^{1,p}(\Omega)$ *and* $z\in W^{1,p'}(\Omega;\mathbb{R}^n)$, *the following formula holds:*

$$\int_\Omega \big(v\,(\operatorname{div} z)+z\cdot\nabla v\big)\,\mathrm{d}x = \int_\Gamma v\,(z\cdot\nu)\,\mathrm{d}S. \tag{1.54}$$

Theorem 1.32 (POINCARÉ-TYPE INEQUALITIES [339]). *Let* $1\le q\le p^*$. *Then there is* $C_{\rm P}<+\infty$ *such that*

$$\|u\|_{W^{1,p}(\Omega)} \le C_{\rm P}\big(\|\nabla u\|_{L^p(\Omega;\mathbb{R}^n)}+\|u\|_{L^q(\Omega)}\big). \tag{1.55}$$

Let $1\le q\le p^\#$, *let* Ω *be connected*[62], *and let* $\Gamma_{\rm D},\Gamma_{\rm N}\subset\Gamma$ *be such that* $\mathrm{meas}_{n-1}(\Gamma_{\rm D})>0$ *and* $\mathrm{meas}_{n-1}(\Gamma_{\rm N})>0$. *Then there is* $C_{\rm P}<+\infty$ *such that*

$$\|u\|_{W^{1,p}(\Omega)} \le C_{\rm P}\big(\|\nabla u\|_{L^p(\Omega;\mathbb{R}^n)}+\big\|u|_{\Gamma_{\rm N}}\big\|_{L^q(\Gamma_{\rm N})}\big) \tag{1.56}$$

[61] Putting $z=\nabla u$ into (1.54), we get $\int_\Omega v\Delta u+\nabla v\cdot\nabla u\,\mathrm{d}x=\int_\Gamma v\frac{\partial}{\partial\nu}u\,\mathrm{d}S$ derived in [188]. In fact, (1.54) holds, by continuous extension, even under weaker assumptions, cf. Nečas [302].

[62] This means that, for any $x_0,x_1\in\Omega$, there is $z:[0,1]\to\Omega$ continuous with $z(0)=x_0$ and $z(1)=x_1$. It has the consequence that functions with zero gradient must be constant on Ω.

and

$$u|_{\Gamma_\mathrm{D}}=0 \quad\Rightarrow\quad \|u\|_{W^{1,p}(\Omega)}\le C_\mathrm{P}\|\nabla u\|_{L^p(\Omega;\mathbb{R}^n)}, \tag{1.57}$$

and also

$$\|u\|_{W^{1,p}(\Omega)}\le C_\mathrm{P}\Big(\|\nabla u\|_{L^p(\Omega;\mathbb{R}^n)}+\Big|\int_\Omega u\,\mathrm{d}x\Big|\Big). \tag{1.58}$$

A special case of (1.56) with $\Gamma_\mathrm{D}=\Gamma$ and $p=q=2$, is sometimes also called *Friedrichs' inequality* [157].

Theorem 1.33 (Korn inequality [237][63]). *Let $1<p<+\infty$. There is a constant $C_\mathrm{K}=C_\mathrm{K}(p,\Omega)$ such that for any $v\in W_0^{1,p}(\Omega)$ it holds that*

$$\|v\|_{W_0^{1,p}(\Omega;\mathbb{R}^n)}\le C_\mathrm{K}\|e(v)\|_{L^p(\Omega;\mathbb{R}^{n\times n})} \qquad \text{where} \quad e(v)=\frac{\nabla v+(\nabla v)^\top}{2}. \tag{1.59}$$

1.5 Bochner spaces

We will now define spaces of abstract functions on a bounded interval $I\subset\mathbb{R}$ valued in a Banach space V, invented by Bochner [59], which is a basic tool on the abstract level for Part II. We say that $u:I\to V$ is *simple* if it takes only finite number of values $v_i\in V$ and $A_i:=u^{-1}(v_i)$ is Lebesgue measurable; then $\int_0^T u(t)\,\mathrm{d}t:=\sum_\mathrm{finite}\mathrm{meas}_1(A_i)v_i$. We say that $u:I\to V$ is *Bochner measurable* if it is a point-wise limit (in the strong topology) of V of a sequence $\{u_k\}_{k\in\mathbb{N}}$ of simple functions; i.e. $u_k(t)\to u(t)$ for a.a. $t\in I$. We say that $u:I\to V$ is *absolutely continuous*[64] if, for each $\varepsilon>0$, there is $\delta>0$ such that $\sum_{k=1}^K\|u(t_k)-u(s_k)\|_V\le\varepsilon$ whenever $t_{k-1}\le s_k\le t_k\le T$ for $k=1,\dots,K\in\mathbb{N}$, $t_0=0$, and $\sum_{k=1}^K t_k-s_k\le\delta$. A point $t\in I$ is called a *Lebesgue point* of $u:I\to V$ if $\lim_{h\searrow0}\frac1h\int_{-h/2}^{h/2}\|u(t+\vartheta)-u(t)\|\mathrm{d}\vartheta=0$. Analogously, a right Lebesgue point $t\in I$ means $\lim_{h\searrow0}\frac1h\int_0^h\|u(t+\vartheta)-u(t)\|\mathrm{d}\vartheta=0$.

Theorem 1.34 (Pettis [334][65]). *If V is separable, then u is Bochner measurable if and only if it is weakly measurable in the sense: $\langle v^*,u(\cdot)\rangle$ is Lebesgue measurable for any $v^*\in V^*$.*

Considering simple functions $\{u_k\}_{k\in\mathbb{N}}$ as above, we call $u:I\to V$ *Bochner integrable* if $\lim_{k\to\infty}\int_0^T\|u(t)-u_k(t)\|_V\,\mathrm{d}t=0$. Then $\int_0^T u(t)\,\mathrm{d}t:=\lim_{k\to\infty}\int_0^T u_k(t)\,\mathrm{d}t$; this limit exists and is independent of the particular choice

[63] See Nečas [307] or monographs [130, 308, 427]. A counterexample for p=1 is by Ornstein [318].

[64] If $V=\mathbb{R}^1$, this definition naturally coincides with the absolute-continuity with respect to the Lebesgue measure on I used on p. 12.

[65] In fact, [334] works with a general Banach space, showing equivalence of the Bochner measurable mappings with a.e. separably valued weakly measurable ones. See Example 1.42 for a weakly* continuous function u valued in $V=L^\infty(0,1)$ which is not Bochner measurable.

of the sequence $\{u_k\}_{k\in\mathbb{N}}$. Moreover, if V is separable, then a Bochner measurable function u is Bochner integrable if and only if $\|u(\cdot)\|_V$ is Lebesgue integrable. Then also $\|\int_0^T u(t)\,\mathrm{d}t\|_V \le \int_0^T \|u(t)\|_V\mathrm{d}t$. From this, we can see that $\lim_{h\searrow 0}\frac{1}{h}\int_{-h/2}^{h/2} u(t+\vartheta)\mathrm{d}\vartheta \to u(t)$ at each Lebesgue point $t\in I$.[66]

Theorem 1.35. *If $u\in L^1(I;V)$, then a.e. $t\in I$ is a Lebesgue point for u.*

An analogous assertion holds for right Lebesgue points.

An example for Bochner integrable functions is any $u \in C(I;(V,\text{weak}))$.[67] However, some classical "scalar-valued" results need not hold in vectorial cases[68].

For $1\le p<\infty$, a *Bochner space* $L^p(I;V)$ is the linear space (of classes with respect to equivalence a.e.) of Bochner integrable functions $u: I\to V$ satisfying $\int_0^T \|u(t)\|_V^p\mathrm{d}t<+\infty$. This space is a Banach space if endowed with the norm

$$\|u\|_{L^p(I;V)} := \left(\int_0^T \|u(t)\|_V^p\,\mathrm{d}t\right)^{1/p}. \tag{1.60}$$

For $p=\infty$, we modify it standardly, i.e. $\|u\|_{L^\infty(I;V)} := \text{ess sup}_{t\in I}\|u(t)\|_V$. Later, we will often use partition of I to subintervals of the length $\tau := 2^{-K}T$, $K\in\mathbb{N}$.

Proposition 1.36 (Density of piecewise constant functions). *If $1\le p<+\infty$, then the set $\{v: I\to V;\ \exists K\in\mathbb{N}:\ \forall 1\le k\le 2^K:\ v|_{((k-1)\tau,k\tau)}$ is constant, $\tau=2^{-k}T\}$ is dense in $L^p(I;V)$. In particular, if $p\in[1,+\infty)$ and V is separable, $L^p(I;V)$ is separable too.*

Proposition 1.37 (Uniform convexity). *If V is uniformly convex and $1<p<+\infty$, then $L^p(I;V)$ is uniformly convex, too.*

Proposition 1.38 (Dual space). *If $p\in[1,+\infty)$, the dual space to $L^p(I;V)$ always contains $L^{p'}(I;V^*)$ and the equality holds if V^* is separable, the duality pairing being given by the formula*

$$\langle f,u\rangle_{L^{p'}(I;V^*)\times L^p(I;V)} := \int_0^T \langle f(t),u(t)\rangle_{V^*\times V}\,\mathrm{d}t\,. \tag{1.61}$$

Thus, if $p\in(1,+\infty)$ and V is reflexive and separable, then $L^p(I;V)$ is reflexive.

---

[66] This follows simply from $\|u(t)-\frac{1}{h}\int_{-h/2}^{h/2}u(t+\vartheta)\mathrm{d}\vartheta\|_V = \|\frac{1}{h}\int_{-h/2}^{h/2}u(t+\vartheta)-u(t)\mathrm{d}\vartheta\|_V \le \frac{1}{h}\int_{-h/2}^{h/2}\|u(t+\vartheta)-u(t)\|_V\mathrm{d}\vartheta$.

[67] The weak* continuity would not be sufficient, however, as Example 1.42 shows.

[68] E.g., $u\in C(I;(V,\text{weak*}))$ need not be Bochner integrable, as Example 1.42 shows. Also, a σ-additive absolutely continuous mapping $u: I\to V$ need not be represented by an integrable function v in the sense $u(t)=\int_0^t v(\vartheta)\mathrm{d}\vartheta$; for the counterexample see Yosida [425, Sect.V.5].

Theorem 1.39 (KOMURA [235]). *If V is reflexive and $u : I \to V$ is absolutely continuous, then u is (strongly) differentiable a.e.*[69] *and $u(t) = u(0) + \int_0^t u'$, $u' \in L^1(I;V)$.*[70]

Convention 1.40. Often, we will use $V = L^q(\Omega;\mathbb{R}^m)$. Considering $p,q \in [1,+\infty]$, it is then natural to identify $L^p(I;L^q(\Omega;\mathbb{R}^m))$ with

$$\Big\{\tilde{u} : I\times\Omega \to \mathbb{R}^m;\quad \int_0^T \Big(\int_\Omega \big|\tilde{u}(t,x)\big|^q \mathrm{d}x\Big)^{p/q} \mathrm{d}t < +\infty\Big\} \tag{1.62}$$

through the natural isomorphism $\tilde{u} \mapsto u$: $[u(t)](x) := \tilde{u}(t,x)$. Likewise, $L^p(I;W^{1,q}(\Omega))$ is identified via this isomorphism with functions $\tilde{u} : I\times\Omega \to \mathbb{R}$ for which $\int_0^T \big(\int_\Omega |\tilde{u}(t,x)|^q + |\nabla\tilde{u}(t,x)|^q \mathrm{d}x\big)^{p/q} \mathrm{d}t < +\infty$.

Proposition 1.41 (INTERPOLATION OF $L^{p_1}(I;L^{q_1}(\Omega))$ AND $L^{p_2}(I;L^{q_2}(\Omega))$ [71]). *Let $p_1,p_2,q_1,q_2 \in [1,+\infty]$, $\lambda \in [0,1]$, and $v \in L^{p_1}(I;L^{q_1}(\Omega)) \cap L^{p_2}(I;L^{q_2}(\Omega))$. Then*

$$\frac{1}{p} = \frac{\lambda}{p_1} + \frac{1-\lambda}{p_2} \quad \textit{and} \quad \frac{1}{q} = \frac{\lambda}{q_1} + \frac{1-\lambda}{q_2}$$
$$\Longrightarrow \quad \big\|v\big\|_{L^p(I;L^q(\Omega))} \le \big\|v\big\|^{\lambda}_{L^{p_1}(I;L^{q_1}(\Omega))} \big\|v\big\|^{1-\lambda}_{L^{p_2}(I;L^{q_2}(\Omega))}. \tag{1.63}$$

Example 1.42. (*The space $L^\infty(I;L^\infty(\Omega))$ is not $L^\infty(Q)$.*[72]) Using the isomorphism from Convention 1.40, we can identify abstract functions from $L^\infty(I;L^\infty(\Omega))$ with functions on $Q := I\times\Omega$. However, $L^\infty(I;L^\infty(\Omega)) \neq L^\infty(Q)$. For $\Omega = I$, the function $u(t) = \chi_{[0,t]}$ induces a function $\tilde{u}(t,x) = 1$ if $x \le t$ and $= 0$ if $x > t$ which obviously belongs to $L^\infty(Q)$. Even u is weakly* continuous but it is not Bochner measurable, hence $u \notin L^\infty(I;L^\infty(\Omega))$.

Considering Banach spaces V_0, V_1, ..., V_k and $a : I\times V_1\times\cdots\times V_k \to V_0$, let us still define the *Nemytskiĭ mappings* $\mathscr{N}_a$ again by the formula (1.47). The following generalization of Theorem 1.27 holds:

Theorem 1.43 (NEMYTSKIĬ MAPPINGS IN BOCHNER SPACES [265][73]). *Let V_0, V_1, ..., V_k be separable Banach spaces, $a : I\times V_1\times\cdots\times V_k \to V_0$ a Carathéodory mapping*[74] *and the growth condition*

$$\big\|a(t,r_1,\ldots,r_k)\big\|_{V_0} \le \gamma(t) + C\sum_{i=1}^k \big\|r_i\big\|_{V_i}^{p_i/p_0} \quad \textit{for some } \gamma \in L^{p_0}(I), \tag{1.64}$$

[69] This means $\lim_{\varepsilon\to 0} \frac{1}{\varepsilon}u(t+\varepsilon) - \frac{1}{\varepsilon}u(t)$ does exist for a.a. $t \in I$. The reflexivity of X is indeed necessary, e.g. $u : I \to X := L^\infty(0,1)$ defined by $[u(t)](x) := x\sin(t/x)$ is Lipschitz continuous but nowhere differentiable.

[70] If V is a uniformly convex space, this formula has been proved already by Clarkson [97].

[71] Cf. Exercise 8.60 on p.266.

[72] This observation is due to Fattorini [144, Example 5.0.10].

[73] Besides the original paper [265] by Lucchetti and Patrone, we refer also, e.g., to Hu and Papageorgiou [209, Part I, Sect.9.1].

[74] Here, it means that, for a.a. $t \in I$, $a(t,\cdot) : V_1\times\cdots\times V_k \to V_0$ is to be (norm,norm)-continuous and, as in Section 1.3, $a(\cdot,r_1,\ldots,r_k) : I \to V_0$ is to be Bochner measurable.

hold with $p_0, p_1, \dots, p_k$ as for (1.48). *Then $\mathscr{N}_a$ maps $L^{p_1}(I;V_1) \times \cdots \times L^{p_k}(I;V_k)$ continuously into $L^{p_0}(I;V_0)$.*

1.6 Some ordinary differential equations

As an auxiliary problem, we will often use the initial-value problem for the system of k ordinary differential equations in the form:

$$\frac{\mathrm{d}u}{\mathrm{d}t} = f\big(t, u(t)\big) \quad \text{for a.a. } t \in I, \qquad u(0) = u_0, \tag{1.65}$$

where $f : (0, +\infty) \times \mathbb{R}^k \to \mathbb{R}^k$ is a Carathéodory mapping. By a solution on a time interval $[0, T]$ we will understand an absolutely continuous mapping $u : [0, T] \to \mathbb{R}^k$ such that the equation (1.65) holds a.e. on $[0, T]$ and $u(0) = u_0$ holds, too.

Theorem 1.44 (LOCAL-IN-TIME EXISTENCE). *Let $f : (0, +\infty) \times \mathbb{R}^k \to \mathbb{R}^k$ be a Carathéodory mapping. Then there is $T > 0$ (not given a-priori) such that the initial-value problem has a solution on the interval $[0, T]$.*

The main ingredient for estimation of evolution systems in general is the so-called *Gronwall inequality*,[75] which we will also often use. In the general form, this inequality says that, for all $t \geq 0$,

$$y(t) \leq \Big(C + \int_0^t b(\theta) \mathrm{e}^{-\int_0^\theta a(\vartheta)\mathrm{d}\vartheta} \mathrm{d}\theta\Big) \mathrm{e}^{\int_0^t a(\theta)\mathrm{d}\theta} \tag{1.66}$$

whenever we know that $y(t) \leq C + \int_0^t (a(\vartheta)y(\vartheta) + b(\vartheta))\mathrm{d}\vartheta$ for some $a, b \geq 0$ integrable.[76] For $a \geq 0$ constant, (1.66) simplifies to $y(t) \leq \mathrm{e}^{at}(C + \int_0^t b(\vartheta)\mathrm{e}^{-a\vartheta}\mathrm{d}\vartheta) \leq (C + \int_0^t b(\vartheta)\mathrm{d}\vartheta)\mathrm{e}^{aT}$ for $t \in I$.

Theorem 1.45 (EXISTENCE AND UNIQUENESS). *Let T be fixed and $f : I \times \mathbb{R}^k \to \mathbb{R}^k$ be a Carathéodory mapping satisfying the growth condition $|f(t, r)| \leq \gamma(t) + C|r|$ with some $\gamma \in L^1(I)$. Then:*

(i) *The initial-value problem* (1.65) *has a solution $u \in W^{1,1}(I; \mathbb{R}^k)$ on the interval $I = [0, T]$.*

(ii) *If $f(t, \cdot)$ is also Lipschitz continuous in the sense $|f(t, r_1) - f(t, r_2)| \leq \ell(t)|r_1 - r_2|$ with some $\ell \in L^1(I)$, then the solution is unique.*

---

[75] In the general form presented here, which can be found, e.g., in Ioffe and Tihomirov [211, Sect. 9.1, Lemma 3], it is also called the Bellman-Gronwall inequality according to the original works [193] (for $C = 0$, a, b constant) and [41] (for $C \geq 0$, $b = 0$, $a \in L^1(0, T)$).

[76] A generalization for slightly superlinear growth in y, namely $y \ln(y)$ or $y \ln(y) \ln(\ln(y))$, is possible, too.

Applying the so-called *bootstrap argument*, i.e. knowing that $u \in W^{1,1}(I;\mathbb{R}^k)$ and therefore $f(t,u) \in W^{1,1}(I;\mathbb{R}^k)$ provided f is smooth in the sense

$$\forall \rho\in\mathbb{R}^+ \quad \exists \gamma_\rho\in L^1(I),\ C_\rho<+\infty \quad \forall t\in I,\ |r|\le\rho: \\ \Big|\frac{\partial f}{\partial t}(t,r)\Big| \le \gamma_\rho(t) \ \text{ and } \ |f(t,r)|\le C_\rho, \tag{1.67}$$

we know immediately that $\frac{\mathrm{d}}{\mathrm{d}t}u = f(t,u) \in W^{1,1}(I;\mathbb{R}^k)$.[77] Hence:

Theorem 1.46 (REGULARITY). *If f satisfies all assumptions in Theorem 1.45 and is also smooth in the sense* (1.67), *then $u \in W^{2,1}(I;\mathbb{R}^k)$.*

The following *discrete* version of the *Gronwall inequality* will be often used:[78]

$$y_l \le \Big(C + \tau\sum_{k=1}^{l-1} b_k\Big) \mathrm{e}^{\tau\sum_{k=1}^{l-1} a_k} \tag{1.68}$$

provided $y_l \le C + \tau\sum_{k=1}^{l-1}(a_k y_k + b_k)$ for any $l \ge 0$ (of course, for $l=0$ it means $y_l \le C$). We will often use $a_k \equiv a$ constant, and the condition

$$y_l \le C + \tau\sum_{k=1}^{l}(a y_k + b_k), \tag{1.69}$$

from which we can easily derive $y_l \le (1-a\tau)^{-1}(C+\tau b_1 + \tau\sum_{k=1}^{l-1}(a y_k + b_{k+1}))$, so that (1.68) gives

$$y_l \le \frac{\mathrm{e}^{\tau l a/(1-a\tau)}}{1-a\tau}\Big(C + \tau\sum_{k=1}^{l} b_k\Big) \qquad \text{if } \tau < \frac{1}{a}. \tag{1.70}$$

There is a variant of the discrete Gronwall inequality by Nochetto, Savaré and Verdi [314] designed for fine estimates for variable time steps and even better fit to discretisation of some evolutionary problems. While the condition (1.69) is an analog of $\frac{\mathrm{d}}{\mathrm{d}t}y \le ay + b$, the modified version is based rather on the continuous condition[79] $y\frac{\mathrm{d}}{\mathrm{d}t}y \le ay + b + cy^2$: If non-negative sequences $\{y_k\}_{k=0}^K$ and $\{(z_k, a_k, b_k)\}_{k=1}^K$ satisfy[80]

$$y_k\frac{y_k - y_{k-1}}{\tau} + z_k \le a_k y_k + b_k + c y_k^2 \qquad \text{for } k = 1,\dots,K \tag{1.71a}$$

[77] We use $\frac{\mathrm{d}}{\mathrm{d}t}f(t,u) = \frac{\partial}{\partial t}f(t,u) + \frac{\partial}{\partial r}f(t,u)\frac{\mathrm{d}}{\mathrm{d}t}u \in L^1(I;\mathbb{R}^k)$ because $\frac{\partial}{\partial t}f(t,u) \in L^1(I;\mathbb{R}^k)$ while $|\frac{\partial}{\partial r}f(t,u)| \le \ell$ and $\frac{\mathrm{d}}{\mathrm{d}t}u = f(t,u) \in L^\infty(I;\mathbb{R}^k)$.

[78] See Lees [255], Krejčí and Roche [241, Lemma A.1], Quarteroni and Valli [341, Lemma 1.4.2] or Thomée [405].

[79] Note that $\frac{1}{2}\frac{\mathrm{d}}{\mathrm{d}t}y^2 = y\frac{\mathrm{d}}{\mathrm{d}t}y \le ay + b + cy^2 \le (\frac{1}{2}a + c)y^2 + \frac{1}{2}a + b$, from which the boundedness of y follows by the classical Gronwall inequality (1.66) provided a, b, and c are integrable.

[80] More specifically, this is (up to some scaling) a special case of [314, Lemma 4.12] for c independent of k.

with a constant $c \geq 0$, then one has both $\max_{k=1,\dots,K} \mathrm{e}^{-kc\tau/(1-c\tau)} y_k$ and $(2\tau \sum_{k=1}^{K} \mathrm{e}^{-((2k-1)c\tau)/(1-c\tau)} z_k)^{1/2}$ estimated from above by

$$(y_0^2 + 2\tau \sum_{k=1}^{K} \mathrm{e}^{-(2k-1)c\tau} b_k)^{1/2} + \sqrt{2}\tau \sum_{k=1}^{K} \mathrm{e}^{-(k-1)c\tau} a_k$$

provided $0 < \tau < 1/c$. A particular but well applicable case can be obtained for $K = T/\tau$: assuming (1.71a) and also

$$0 < \tau \leq \tau_0 \quad \text{with} \quad \tau_0 < \frac{1}{c}, \tag{1.71b}$$

one has, with some constant C_T independent of τ,

$$\max_{k=1,\dots,K} y_k + \Big(\tau \sum_{k=1}^{T/\tau} z_k\Big)^{1/2} \leq C_T \bigg(y_0 + \Big(\tau \sum_{k=1}^{T/\tau} b_k\Big)^{1/2} + \tau \sum_{k=1}^{T/\tau} a_k \bigg). \tag{1.72}$$

Part I

STEADY-STATE PROBLEMS

Chapter 2

Pseudomonotone or weakly continuous mappings

The basic modern approach to boundary-value problems in differential equations of the type (0.1)–(0.2) is the so-called *energy-method* technique which took the name after a-priori estimates having sometimes physical analogies as bounds of an energy.[1] This technique originated from modern theory of linear partial differential equations where, however, other approaches are efficient, too. On the abstract level, this method relies on relative weak compactness of bounded sets in reflexive Banach spaces, and either pseudomonotonicity or weak continuity of differential operators which are understood as bounded from one Banach space to another (necessarily different) Banach space. On the concrete-problem level, the main tool is a weak formulation of boundary-value problems in question, Poincaré and Hölder inequalities, and fine issues from the theory of Sobolev spaces.

2.1 Abstract theory, basic definitions, Galerkin method

Throughout this chapter (and most of the others), V will be a separable reflexive Banach space and V^* its dual space, with $\|\cdot\|$ and $\|\cdot\|_*$ denoting briefly their norms instead of $\|\cdot\|_V$ and $\|\cdot\|_{V^*}$, respectively.

Definition 2.1 (Monotonicity modes). Let A be a mapping $V \to V^*$.

(i) $A : V \to V^*$ is *monotone* iff $\forall u, v \in V : \langle A(u) - A(v), u - v\rangle \geq 0$.

(ii) If A is monotone and $u \neq v$ implies $\langle A(u) - A(v), u - v\rangle > 0$, then A is *strictly monotone.*

[1] Cf. Example 6.7 or e.g. also (11.120) or (12.11).

(iii) Considering an increasing function $d : \mathbb{R}^+ \to \mathbb{R}$, we say that $A : V \to V^*$ is *d-monotone* with respect to a seminorm $|\cdot|$,

$$\big\langle A(u) - A(v), u - v\big\rangle \geq \Big(d\big(|u|\big) - d\big(|v|\big)\Big)\big(|u| - |v|\big). \tag{2.1}$$

If $|\cdot|$ is the norm $\|\cdot\|$ on V, we say simply that A is d-monotone. Moreover, A is called *uniformly monotone* if

$$\big\langle A(u) - A(v), u - v\big\rangle \geq \zeta\big(\|u - v\|\big)\|u - v\| \tag{2.2}$$

for some increasing continuous function $\zeta : \mathbb{R}^+ \to \mathbb{R}^+$. If $\zeta(r) = \delta r$ for some $\delta > 0$, then A is called *strongly monotone.*

(iv) The mapping $A : V \to V^*$ is called *pseudomonotone* iff

$$A \text{ is bounded, and} \tag{2.3a}$$

$$\left.\begin{array}{r} u_k \rightharpoonup u \\ \limsup\limits_{k\to\infty} \big\langle A(u_k), u_k - u\big\rangle \leq 0 \end{array}\right\} \Rightarrow \begin{array}{l} \forall v \in V: \\ \big\langle A(u), u - v\big\rangle \leq \liminf\limits_{k\to\infty} \big\langle A(u_k), u_k - v\big\rangle. \end{array} \tag{2.3b}$$

Remark 2.2. Let us emphasize that the monotonicity due to Definition 2.1(i) has no direct relation with monotonicity of mappings with respect to an ordering. E.g., if $V^* = V$, the composition of monotone operators has a good sense but need not be monotone. Definition 2.1(iv) represents a suitable extent[2] of generalization of the monotonicity concept from the viewpoint of quasilinear differential equations of the type (0.2).

Definition 2.3 (Continuity modes).

(i) $A : V \to V^*$ is *hemicontinuous* iff $\forall u, v, w \in V$ the function $t \mapsto \langle A(u+tv), w\rangle$ is continuous, i.e. A is directionally weakly continuous.

(ii) If it holds only for $v = w$, i.e. $\forall u, v \in V : t \mapsto \langle A(u+tv), v\rangle$ is continuous, then A is called *radially continuous.*

(iii) $A : V \to V^*$ is *demicontinuous* iff $\forall w \in V$ the functional $u \mapsto \langle A(u), w\rangle$ is continuous; i.e. A is continuous as a mapping $(V, \text{norm}) \to (V^*, \text{weak})$.

(iv) $A : V \to V^*$ is *weakly continuous* iff $\forall w \in V$ the functional $u \mapsto \langle A(u), w\rangle$ is weakly continuous; i.e. A is continuous as a mapping $(V, \text{weak}) \to (V^*, \text{weak})$.

(v) $A : V \to V^*$ is *totally continuous* if it is continuous as a mapping $(V, \text{weak}) \to (V^*, \text{norm})$.

Lemma 2.4. *Any pseudomonotone mapping A is demicontinuous.*

Proof. Suppose $u_k \to u$. By (2.3a), the sequence $\{A(u_k)\}_{k\in\mathbb{N}}$ is bounded in a reflexive space V^*. Then, as V is assumed separable, by the Banach Theorem 1.7 after taking a subsequence (denoted, for simplicity, by the same indices) we have

[2] In the sense that the premise of (2.3b) still can be proved under reasonable assumptions and the conclusion of (2.3b) still suffices to prove convergence of various approximate solutions.

$A(u_k) \rightharpoonup f$ for some $f \in V^*$. Then $\lim_{k\to\infty}\langle A(u_k), u_k - u\rangle = \langle f, u - u\rangle = 0$ and therefore, by (2.3b),

$$\big\langle A(u), u - v\big\rangle \leq \liminf_{k\to\infty} \big\langle A(u_k), u_k - v\big\rangle = \langle f, u - v\rangle \tag{2.4}$$

for any $v \in V$. From this we get $A(u) = f$. In particular, f is determined uniquely, and thus even the whole sequence (not only the selected subsequence) must converge. □

Definition 2.5 (Coercivity). $A : V \to V^*$ is *coercive* iff $\exists \zeta : \mathbb{R}^+ \to \mathbb{R}^+$: $\lim_{s\to+\infty} \zeta(s) = +\infty$ and $\langle A(u), u\rangle \geq \zeta(\|u\|)\|u\|$. In other words, A coercive means

$$\lim_{\|u\|\to\infty} \frac{\big\langle A(u), u\big\rangle}{\|u\|} = +\infty. \tag{2.5}$$

Theorem 2.6 (Brézis [64]). *Any A pseudomonotone and coercive is surjective; this means, for any $f \in V^*$, there is at least one solution to the equation*

$$A(u) = f. \tag{2.6}$$

Proof. Let us divide the proof into four particular steps.

Step 1: (*Abstract Galerkin approximation.*) As V is supposed separable, we can take a sequence of finite-dimensional subspaces

$$\forall k \in \mathbb{N}: \quad V_k \subset V_{k+1} \subset V \quad \text{and} \quad \bigcup_{k\in\mathbb{N}} V_k \ \text{ is dense in } V. \tag{2.7}$$

Then we define a *Galerkin approximation* $u_k \in V_k$ by the identity:

$$\forall v \in V_k: \qquad \big\langle A(u_k), v\big\rangle = \langle f, v\rangle. \tag{2.8}$$

Step 2: (*Existence of approximate solutions u_k.*) In other words, we seek $u_k \in V_k$ solving $I_k^*(A(u_k) - f) = 0$ where $I_k : V_k \to V$ is the canonical inclusion so that the adjoint operator $I_k^* : V^* \to V_k^*$ represents the restriction $I_k^* f = f|_{V_k}$. Besides, as V_k is finite-dimensional, we will identify $V_k \cong V_k^*$ by a linear homeomorphism $J_k : V_k \to V_k^*$ such that $\langle J_k u, u\rangle = \|u\|_{V_k}^2$, $\|J_k u\|_{V_k^*} = \|u\|_{V_k}$, and $\langle J_k u, J_k^{-1} f\rangle = \langle f, u\rangle$.[3]

As A is coercive, for ϱ sufficiently large we have

$$\|u\|_{V_k} = \varrho \quad \Longrightarrow \quad \big\langle A(u) - f, u\big\rangle \geq \big\langle A(u), u\big\rangle - \|f\|_*\|u\| > 0. \tag{2.9}$$

[3] If necessary, we can re-norm the finite-dimensional V_k to impose a Hilbert structure (i.e. V_k is then homeomorphic with a Euclidean space). Then J_k can be taken as in (3.1) below; note that, by Lemma 3.2, J_k is a homeomorphism. Also note that (2.5) restricted on V_k holds in the new, equivalent norm, as well; possibly, the function ζ in Definition 2.5 is changed by this renormalization.

Suppose, for a moment, that $I_k^*A(u) \neq I_k^*f$ for any $u \in V_k$ with $\|u\|_{V_k} \leq \varrho$. Then the mapping

$$u \mapsto \varrho \frac{J_k^{-1}I_k^*(f - A(u))}{\big\|I_k^*(f - A(u))\big\|_{V_k^*}} \tag{2.10}$$

maps the convex compact set $\{u \in V_k;\ \|u\| \leq \varrho\}$ into itself because $\|J_k^{-1}\| = 1$; note that $\|J_k^{-1}f\|_{V_k} = \|f\|_{V_k^*}$. Also, by Lemma 2.4, the mapping $u \mapsto \langle A(u), v\rangle : V_k \to \mathbb{R}$ is continuous for any v so that also $u \mapsto I_k^*A(u) : V_k \to V_k^*$ is continuous. By the Brouwer fixed-point Theorem 1.10, the mapping (2.10) has a fixed point u, this means

$$u = \varrho \frac{J_k^{-1}I_k^*(f - A(u))}{\big\|I_k^*(f - A(u))\big\|_{V_k^*}}. \tag{2.11}$$

As $\|J_k^{-1}f\|_{V_k} = \|f\|_{V_k^*}$, (2.11) implies $\|u\|_{V_k} = \varrho$. Testing (2.11) by $J_k u \|I_k^*(f - A(u))\|_{V_k^*}$, one gets

$$\begin{aligned}\varrho^2\big\|I_k^*(f - A(u))\big\|_{V_k^*} &= \langle J_k u, u\rangle\big\|I_k^*(f - A(u))\big\|_{V_k^*}\\ &= \varrho\big\langle J_k u, J_k^{-1}I_k^*(f - A(u))\big\rangle = \varrho\big\langle I_k^*(f - A(u)), u\big\rangle\\ &= \varrho\big\langle f - A(u), I_k u\big\rangle = \varrho\big\langle f - A(u), u\big\rangle\end{aligned} \tag{2.12}$$

which yields $\langle A(u) - f, u\rangle = -\varrho\|I_k^*(A(u) - f)\|_{V_k^*} \leq 0$, a contradiction with (2.9).

STEP 3: (*An a-priori estimate.*) Moreover, putting $v := u_k$ into (2.8), we can estimate[4]

$$\zeta\big(\|u_k\|\big)\|u_k\| \leq \big\langle A(u_k), u_k\big\rangle = \langle f, u_k\rangle \leq \|f\|_*\|u_k\| \tag{2.13}$$

with a suitable increasing function $\zeta : \mathbb{R}^+ \to \mathbb{R}^+$ such that $\lim_{\xi\to\infty}\zeta(\xi) = +\infty$, cf. the coercivity (2.5) of A. Then $\|u_k\| \leq \zeta^{-1}(\|f\|_*) < +\infty$, so that u_k is bounded in V independently of k. This holds even for any solution to (2.8).

STEP 4: (*Limit passage.*) Since $\{u_k\}_{k\in\mathbb{N}}$ is bounded and V is reflexive and separable, by the Banach Theorem 1.7 together with Proposition 1.3, there is a subsequence and $u \in V$ such that $u_k \rightharpoonup u$. From (2.8), we have also

$$\big\langle A(u_k), v_m - u_k\big\rangle = \langle f, v_m - u_k\rangle \tag{2.14}$$

for any $k \geq m$ and $v_m \in V_m \subset V_k$. By density of $\bigcup_{k\in\mathbb{N}} V_k$ in V, we can take $v_k \to u$. Then, by (2.14), one gets

$$\begin{aligned}\limsup_{k\to\infty}\,\langle A(u_k), u_k - u\rangle &= \limsup_{k\to\infty}\Big(\langle A(u_k), u_k - v_k\rangle + \langle A(u_k), v_k - u\rangle\Big)\\ &\leq \lim_{k\to\infty}\Big(\langle f, u_k - v_k\rangle + \|A(u_k)\|_*\|v_k - u\|\Big) = 0.\end{aligned} \tag{2.15}$$

[4]Here we forget possible renormalization of the finite-dimensional subspaces V_k and come back to the original norm on V.

Note that the sequence $\{u_k\}_{k\in\mathbb{N}}$ has been proved bounded so $\{\|A(u_k)\|_*\}_{k\in\mathbb{N}}$ is bounded by (2.3a) and that, in fact, even an equality holds in (2.15) and "limsup" is a limit. By pseudomonotonicity (2.3b) of A, we get

$$\forall v\in V: \quad \liminf_{k\to\infty} \big\langle A(u_k), u_k - v\big\rangle \geq \big\langle A(u), u - v\big\rangle. \tag{2.16}$$

On the other hand, from (2.14) we also have

$$\forall v\in \bigcup_{m\in\mathbb{N}} V_m: \quad \liminf_{k\to\infty} \big\langle A(u_k), u_k - v\big\rangle = \lim_{k\to\infty} \langle f, u_k - v\rangle = \langle f, u - v\rangle. \tag{2.17}$$

Combining (2.16) and (2.17), one gets $\langle A(u), u - v\rangle \leq \langle f, u - v\rangle$ for any v ranging over a dense subset of V, namely $\bigcup_{m\in\mathbb{N}} V_m$, which shows that $A(u) = f$. □

Remark 2.7 (Nonconstructivity). Let us emphasize three aspects of high nonconstructivity of the above proof:
✓ usage of Brouwer's fixed-point theorem,
✓ a contradiction argument, and
✓ a selection of a convergent subsequence by a compactness argument.

Remark 2.8 (Necessity of approximation). The approximation (Step 1) is necessary in the above proof, otherwise one would have to think about usage of Schauder's type fixed point Theorem 1.9 instead of the Brouwer one. This would need additional assumptions about weak continuity of A and the Hilbert structure of V, cf. Exercise 2.56, which is not fitted with general quasilinear differential equations, cf. Sect. 2.5 where omitting the approximation would also hurt for not allowing for a weaker concept of A as $V \to Z^*$ with $Z \subsetneqq V$.

2.2 Some facts about pseudomonotone mappings

Brézis' Theorem 2.6 showed the importance of the class of pseudomonotone mappings. It is therefore worth knowing some more specific cases leading to such mappings.

Lemma 2.9 (Brézis [64]). *Radially continuous monotone mappings satisfy* (2.3b). *In particular, bounded radially continuous monotone mappings are pseudomonotone.*

Proof. Take $u_k \rightharpoonup u$ and assume $\limsup_{k\to\infty}\langle A(u_k), u_k - u\rangle \leq 0$. Since A is monotone, $\langle A(u_k), u_k - u\rangle \geq \langle A(u), u_k - u\rangle \to 0$ so that $\liminf_{k\to\infty}\langle A(u_k), u_k - u\rangle \geq 0$ and therefore altogether

$$\lim_{k\to\infty} \langle A(u_k), u_k - u\rangle = 0. \tag{2.18}$$

We take $u_\varepsilon = (1-\varepsilon)u + \varepsilon v$, $\varepsilon > 0$, and write the monotonicity condition of A between u_k and u_ε:

$$0 \leq \langle A(u_k) - A(u_\varepsilon), u_k - u_\varepsilon\rangle = \langle A(u_k) - A(u_\varepsilon), \varepsilon(u - v) + u_k - u\rangle \tag{2.19}$$

and, by a simple algebraic manipulation, we obtain

$$\varepsilon\langle A(u_k), u-v\rangle \geq \langle A(u_\varepsilon), u_k-u\rangle - \langle A(u_k), u_k-u\rangle + \varepsilon\langle A(u_\varepsilon), u-v\rangle. \tag{2.20}$$

Therefore, fixing $\varepsilon > 0$ and passing with k to infinity, by (2.18) we get

$$\varepsilon \liminf_{k\to\infty}\langle A(u_k), u-v\rangle \geq \varepsilon\langle A(u_\varepsilon), u-v\rangle. \tag{2.21}$$

Then divide it by ε, which gives $\liminf_{k\to\infty}\langle A(u_k), u-v\rangle \geq \langle A(u_\varepsilon), u-v\rangle = \langle A(u+\varepsilon(v-u)), u-v\rangle$. Passing with $\varepsilon \to 0$ and using the radial continuity of A, we eventually get

$$\liminf_{k\to\infty}\langle A(u_k), u-v\rangle \geq \lim_{\varepsilon\searrow 0}\langle A(u+\varepsilon(v-u)), u-v\rangle = \langle A(u), u-v\rangle. \tag{2.22}$$

The pseudomonotonicity of A then follows by using (2.22) with (2.18):

$$\liminf_{k\to\infty}\langle A(u_k), u_k-v\rangle = \lim_{k\to\infty}\langle A(u_k), u_k-u\rangle + \liminf_{k\to\infty}\langle A(u_k), u-v\rangle \geq \langle A(u), u-v\rangle.$$

□

Lemma 2.10. *Any bounded demicontinuous mapping $A : V \to V^*$ satisfying*

$$\Big(u_k \rightharpoonup u \quad \& \quad \limsup_{k\to\infty}\big\langle A(u_k)-A(u), u_k-u\big\rangle \leq 0\Big) \quad \Rightarrow \quad u_k \to u \tag{2.23}$$

is pseudomonotone.

Proof. The premise of (2.3b), i.e. $u_k \rightharpoonup u$ and $\limsup_{k\to\infty}\langle A(u_k), u_k-u\rangle \leq 0$, yields

$$\limsup_{k\to\infty}\big\langle A(u_k)-A(u), u_k-u\big\rangle = \limsup_{k\to\infty}\big\langle A(u_k), u_k-u\big\rangle - \lim_{k\to\infty}\big\langle A(u), u_k-u\big\rangle \leq 0,$$

so that by (2.23) we have $u_k \to u$, and by demicontinuity of A also $A(u_k) \rightharpoonup A(u)$, and eventually $\lim_{k\to\infty}\langle A(u_k), u_k-v\rangle = \langle A(u), u-v\rangle$ for any $v \in V$. □

Lemma 2.11.

(i) *The sum of any pseudomonotone mappings remains pseudomonotone, i.e. A_1 and A_2 pseudomonotone implies $u \mapsto A_1(u) + A_2(u)$ pseudomonotone.*

(ii) *A shift of a pseudomonotone mapping remains pseudomonotone, i.e. A pseudomonotone implies $u \mapsto A(u+w)$ pseudomonotone for any $w \in V$.*

Proof. The boundedness (2.3a) of $A_1 + A_2$ and $A(\cdot + w)$ is obvious hence we need to show only (2.3b).

To prove (i), let A_1, A_2 be pseudomonotone, $u_k \rightharpoonup u$ and $\limsup_{k\to\infty}\langle [A_1 + A_2](u_k), u_k - u\rangle \leq 0$. Let us verify that

$$\limsup_{k\to\infty}\big\langle A_1(u_k), u_k-u\big\rangle \leq 0 \quad \text{and} \quad \limsup_{k\to\infty}\big\langle A_2(u_k), u_k-u\big\rangle \leq 0. \tag{2.24}$$

Suppose, for a moment, that $\limsup_{k\to\infty}\langle A_2(u_k), u_k - u\rangle = \varepsilon > 0$. Taking a subsequence, we can suppose that $\lim_{k\to\infty}\langle A_2(u_k), u_k - u\rangle = \varepsilon > 0$ and therefore

$$\limsup_{k\to\infty} \big\langle A_1(u_k), u_k - u\big\rangle \le -\varepsilon < 0. \tag{2.25}$$

As A_1 is pseudomonotone, we get $\liminf_{k\to\infty}\langle A_1(u_k), u_k - v\rangle \ge \langle A_1(u), u - v\rangle$ for any $v \in V$. In particular, for $v = u$ we get $\liminf_{k\to\infty}\langle A_1(u_k), u_k - u\rangle \ge 0$, which contradicts (2.25). Thus (2.24) holds. By the pseudomonotonicity both for A_1 and for A_2, we get

$$\begin{aligned}\liminf_{k\to\infty} \big\langle [A_1{+}A_2](u_k), u_k{-}v\big\rangle &\ge \liminf_{k\to\infty} \big\langle A_1(u_k), u_k{-}v\big\rangle + \liminf_{k\to\infty} \big\langle A_2(u_k), u_k{-}v\big\rangle\\ &\ge \big\langle A_1(u), u{-}v\big\rangle + \big\langle A_2(u), u{-}v\big\rangle \ge \big\langle [A_1{+}A_2](u), u{-}v\big\rangle.\end{aligned}$$

As to (ii), let $u_k \rightharpoonup u$ and $\limsup_{k\to 0}\langle A(u_k{+}w), u_k{-}u\rangle \le 0$. Then obviously $u_k{+}w \rightharpoonup u{+}w$ and $\limsup_{k\to 0}\langle A(u_k{+}w), (u_k{+}w) - (u{+}w)\rangle \le 0$. If A is pseudomonotone, then $\liminf_{k\to 0}\langle A(u_k{+}w), u_k{-}v\rangle = \liminf_{k\to 0}\langle A(u_k{+}w), (u_k{+}w) - (v{+}w)\rangle \ge \langle A(u{+}w), (u{+}w) - (v{+}w)\rangle = \langle A(u{+}w), u{-}v\rangle$, hence $A(\cdot{+}w)$ is pseudomonotone. □

Corollary 2.12. *A perturbation of a pseudomonotone mapping by a totally continuous mapping is pseudomonotone.*

Proof. Realize that any totally continuous mapping is pseudomonotone; indeed, it is bounded (which can be easily proved by contradiction) and, if $u_k \rightharpoonup u$, then $A(u_k) \to A(u)$ and thus $\lim_{k\to\infty}\langle A(u_k), u_k - v\rangle = \langle A(u), u - v\rangle$ so that (2.3b) is trivial. □

2.3 Equations with monotone mappings

Monotone mappings (with boundedness and radial continuity properties) are a special class of pseudomonotone mappings, cf. Lemma 2.9, and, as such, they allow special treatment with a bit stronger results than a general "pseudomonotone theory" can yield, cf. Theorem 2.14 vs. Proposition 2.17.

Lemma 2.13 (MINTY'S TRICK [286]). *Let $A : V \to V^*$ be radially continuous and let $\langle f{-}A(v), u{-}v\rangle \ge 0$ for any $v{\in}V$. Then $f = A(u)$.*

Proof. Replace v with $u + \varepsilon w$ with $w \in V$ arbitrary. This gives

$$\big\langle f - A(u{+}\varepsilon w), -\varepsilon w\big\rangle \ge 0. \tag{2.26}$$

Divide it by $\varepsilon > 0$ and pass to the limit with ε by using radial continuity of A:

$$0 \ge \big\langle f - A(u{+}\varepsilon w), w\big\rangle \to \big\langle f - A(u), w\big\rangle. \tag{2.27}$$

As w is arbitrary, one gets $A(u) = f$. □

Theorem 2.14. *Let A be bounded,*[5] *radially continuous, monotone, coercive. Then:*

(i) *A is surjective; this means, for any $f \in V^*$, there is u solving (2.6). Moreover, the set of solutions to (2.6) is closed and convex.*

(ii) *If, in addition, A is strictly monotone, then $A^{-1} : V^* \to V$ does exist, is strictly monotone, bounded, and demicontinuous. If A is also d-monotone and V uniformly convex, then $A^{-1} : V^* \to V$ is continuous.*

(iii) *If, in addition, A is uniformly (resp. strongly) monotone, then $A^{-1} : V^* \to V$ is uniformly (resp. Lipschitz) continuous.*

Proof. By Lemma 2.9, A is pseudomonotone. As A is supposed also coercive, the surjectivity of A follows from Theorem 2.6. By Lemma 2.4, A is demicontinuous, hence the set of solutions to (2.6) is closed in the norm topology of V. Hence, to prove convexity of this set, it suffices to show that $u = \frac{1}{2}u_1 + \frac{1}{2}u_2$ solves (2.6) provided u_1 and u_2 do so, cf. Proposition 1.6. Thus we have

$$\begin{aligned}\langle f-A(v), u-v\rangle &= \frac{1}{2}\langle f - A(v), u_1 - v\rangle + \frac{1}{2}\langle f - A(v), u_2 - v\rangle \\ &= \frac{1}{2}\big\langle A(u_1)-A(v), u_1-v\big\rangle + \frac{1}{2}\big\langle A(u_2)-A(v), u_2-v\big\rangle \geq 0 \qquad (2.28)\end{aligned}$$

because of $A(u_1) = f = A(u_2)$ and of monotonicity of A. Then, by Lemma 2.13, one gets $A(u) = f$.

Let us go on to (ii). If A is strictly monotone, we have $\langle A(u_1) - A(u_2), u_1 - u_2\rangle = \langle f - f, u_1 - u_2\rangle = 0$ which is possible only if $u_1 = u_2$. In other words, the equation (2.6) has a unique solution so that the inverse A^{-1} does exist.

The mapping A^{-1} is strictly monotone: For $f_1, f_2 \in V^*$, $f_1 \neq f_2$, put $u_i = A^{-1}(f_i)$. Then also $u_1 \neq u_2$. As A is strictly monotone, one has

$$\big\langle f_1 - f_2, A^{-1}(f_1) - A^{-1}(f_2)\big\rangle = \big\langle A(u_1) - A(u_2), u_1 - u_2\big\rangle > 0. \qquad (2.29)$$

The mapping A^{-1} is bounded: by the coercivity of A, there is $\zeta : \mathbb{R}^+ \to \mathbb{R}$ such that $\lim_{\xi\to\infty} \zeta(\xi) = +\infty$ and $\langle A(u), u\rangle \geq \|u\|\zeta(\|u\|)$. Therefore

$$\zeta\big(\|u\|\big)\|u\| \leq \big\langle A(u), u\big\rangle = \big\langle f, u\big\rangle \leq \|f\|_*\|u\| \qquad (2.30)$$

so that $\zeta(\|A^{-1}(f)\|) = \zeta(\|u\|) \leq \|f\|_*$. Thus A^{-1} maps bounded sets in V^* into bounded sets in V.

The mapping A^{-1} is demicontinuous, i.e. (norm,weak)-continuous: take $f_k \to f$ in V^*. As A^{-1} was shown to be bounded, $\{A^{-1}(f_k)\}_{k\in\mathbb{N}}$ is bounded and (possibly up to a subsequence) $u_k = A^{-1}(f_k) \rightharpoonup u$ in V by Banach's Theorem 1.7. It remains to show $A(u) = f$. By the monotonicity of A, for any $v \in V$:

$$0 \leq \big\langle A(u_k) - A(v), u_k - v\big\rangle = \big\langle f_k - A(v), u_k - v\big\rangle. \qquad (2.31)$$

[5] If proved directly, i.e. without passing through pseudomonotone mappings, the boundedness assumption can be omitted; cf. Theorem 2.18 below.

Therefore, by (norm×weak)-continuity of the duality pairing, passing to the limit with $k \to \infty$ yields

$$0 \le \lim_{k\to\infty} \langle f_k - A(v), u_k - v\rangle = \langle f - A(v), u - v\rangle. \tag{2.32}$$

Then we apply again the Minty-trick Lemma 2.13, which gives $A(u) = f$. Thus even the whole sequence $\{u_k\}_{k\in\mathbb{N}}$ converges weakly.

If A is d-monotone, we can refine (2.31) used for $v := u$ as follows:

$$\begin{aligned}\big(d(\|u_k\|) - d(\|u\|)\big)(\|u_k\| - \|u\|) &\le \big\langle A(u_k) - A(u), u_k - u\big\rangle \\ &= \big\langle f_k - A(u), u_k - u\big\rangle \to \big\langle f - A(u), u - u\big\rangle = 0, \end{aligned}\tag{2.33}$$

which gives $\|u_k\| \to \|u\|$ because $d : \mathbb{R} \to \mathbb{R}$ is increasing. Hence $u_k \to u$ by Theorem 1.2. In other words, A^{-1} is continuous.

The point (iii): By (2.2) one has for any $A(u_1) = f_1$ and $A(u_2) = f_2$ the estimate

$$\begin{aligned}\zeta(\|u_1 - u_2\|)\|u_1 - u_2\| &\le \big\langle A(u_1) - A(u_2), u_1 - u_2\big\rangle \\ &= \langle f_1 - f_2, u_1 - u_2\rangle \le \|f_1 - f_2\|_*\|u_1 - u_2\| \end{aligned}\tag{2.34}$$

so that $\zeta(\|u_1 - u_2\|) \le \|f_1 - f_2\|_*$. By the assumed properties of ζ, the inverse mapping A^{-1} is uniformly continuous. The case of strong monotonicity is obvious. $\square$

Lemma 2.15. *Any monotone mapping $A : V \to V^*$ is locally bounded in the sense:*

$$\forall u\in V\ \exists\varepsilon > 0\ \exists M\in\mathbb{R}^+\ \forall v\in V:\quad \|v-u\| \le \varepsilon \ \Rightarrow\ \|A(v)\|_* \le M. \tag{2.35}$$

Proof. Suppose the contrary, i.e. (2.35) does not hold at some $u \in V$. Without loss of generality, assume $u = 0$. This means that there is a sequence $\{v_k\}$, $v_k \to 0$, such that $\|A(v_k)\|_* \to \infty$. Putting $c_k := 1 + \|A(v_k)\|_*\|v_k\|$, we can estimate by monotonicity of A that

$$\begin{aligned}\Big\langle \frac{A(v_k)}{c_k}, v\Big\rangle &\le \frac{\big\langle A(v_k), v_k\big\rangle + \big\langle A(v), v - v_k\big\rangle}{c_k} \\ &\le 1 + \big\|A(v)\big\|_*\big(\|v\| + \|v_k\|\big) \to 1 + \|A(v)\|_*\|v\|. \end{aligned}\tag{2.36}$$

Replacing v by $-v$, we can conclude that $\limsup_{k\to\infty}\big|\langle c_k^{-1}A(v_k), v\rangle\big| < +\infty$ for any $v \in V$. By Banach-Steinhaus' Theorem 1.1, $c_k^{-1}\|A(v_k)\|_* \le M$. This means $\|A(v_k)\|_* \le Mc_k = M(1+\|A(v_k)\|_*\|v_k\|)$, and then also $\|A(v_k)\|_* \le M/(1-M\|v_k\|) \to M$, which contradicts the fact that $\|A(v_k)\|_* \to \infty$. $\square$

Lemma 2.16. *Radially continuous monotone mappings are also demicontinuous.*

Proof. Take a sequence $\{u_k\}_{k\in\mathbb{N}}$ convergent to some $u \in V$. By Lemma 2.15, $\{A(u_k)\}_{k\in\mathbb{N}}$ is bounded in V^* and, by Banach Theorem 1.7, we can select a subsequence $\{A(u_{k_l})\}_{l\in\mathbb{N}}$ converging weakly to some $f \in V^*$. Then, by the monotonicity of A, $0 \le \lim_{l\to\infty}\langle A(u_{k_l}) - A(v), u_{k_l} - v\rangle = \langle f - A(v), u - v\rangle$. As v is arbitrary and we assume radial continuity of A, the Minty-trick Lemma 2.13 yields $f = A(u)$. As f is thus determined uniquely, even the whole sequence $\{A(u_k)\}_{k\in\mathbb{N}}$ must converge to it weakly. □

Proposition 2.17. *Let $A = A_1 + A_2 : V \to V^*$ be coercive, and A_1 be radially continuous and monotone and A_2 be totally continuous. Then A is surjective.*

Proof. As in the proof of Brézis' Theorem 2.6, consider $u_k \in V_k$ the Galerkin approximations (2.8), i.e. here

$$\big\langle A_1(u_k) + A_2(u_k), v\big\rangle = \langle f, v\rangle \qquad \forall v\in V_k, \tag{2.37}$$

and the a-priori estimate (2.13), and choose a weakly convergent subsequence $\{u_{k_i}\}_{i\in\mathbb{N}}$ with a limit $u \in V$. Use monotonicity of A_1 to write

$$0 \le \big\langle A_1(v_l) - A_1(u_{k_i}), v_l - u_{k_i}\big\rangle = \big\langle A_1(v_l), v_l - u_{k_i}\big\rangle + \big\langle A_2(u_{k_i}) - f, v_l - u_{k_i}\big\rangle \tag{2.38}$$

for any $v_l \in V_l$ with $l \le k$. Passing to the limit with $i \to \infty$, it gives

$$0 \le \big\langle A_1(v_l), v_l - u\big\rangle + \big\langle A_2(u) - f, v_l - u\big\rangle. \tag{2.39}$$

Then, by density of $\bigcup_{k\in\mathbb{N}} V_k$ in V, consider $v_l \to v$ for $v\in V$ arbitrary, use demicontinuity of A_1 (cf. Lemma 2.16), and pass to the limit with $l \to \infty$ to get:

$$0 \le \big\langle A_1(v), v - u\big\rangle + \big\langle A_2(u) - f, v - u\big\rangle. \tag{2.40}$$

Finally, replace v by $u + \varepsilon w$ with $w \in V$ arbitrary and use Minty's trick as in (2.26)–(2.27) to show that $A_1(u) + A_2(u) = f$. □

In principle, if A_1 is also bounded, one could use Lemma 2.9 and Corollary 2.12 to see that A from Proposition 2.17 is surjective; realize that A_2, being totally continuous, is certainly bounded. The above direct proof allowed us to avoid the boundedness assumption of A_1. In particular, for $A_2 = 0$, one thus obtains the celebrated assertion:

Theorem 2.18 (Browder [73] and Minty [286]). *Any monotone, radially continuous, and coercive $A : V \to V^*$ is surjective.*

As a very special case, one gets another celebrated result:

Theorem 2.19 (Lax and Milgram [253][6]). *Let V be a Hilbert space, $A : V \to V^*$ be a linear continuous operator which is positive definite in the sense $\langle Av, v\rangle \ge \varepsilon\|v\|^2$ for some $\varepsilon > 0$. Then A has a bounded inverse.*

[6]A usual formulation uses a bounded, positive definite, bilinear form $a : V \times V \to \mathbb{R}$. This form then determines $A : V \to V^*$ through the identity $\langle Au, v\rangle = a(u, v)$.

Sometimes, the following modification of Proposition 2.17 can be advantageously applied, obtaining also the strong convergence of Galerkin's approximate solutions.

Proposition 2.20. *Let $A = A_1 + A_2 : V \to V^*$ be coercive, and A_1 be monotone radially continuous and satisfy (2.23), and A_2 be demicontinuous and compact.*[7] *Then A is surjective.*

Proof. We have the Galerkin identity (2.37) and a subsequence $u_k \rightharpoonup u$, and write

$$\begin{aligned}\big\langle A_1(u_k)-A_1(u), u_k-u\big\rangle &= \big\langle A_1(u_k)-A_1(u), v_k-u\big\rangle + \big\langle A_1(u_k)-A_1(u), u_k-v_k\big\rangle \\ &= \big\langle A_1(u_k)-A_1(u), v_k-u\big\rangle + \big\langle f-A_2(u_k)-A_1(u), u_k-v_k\big\rangle =: I_k^{(1)} + I_k^{(2)}. \qquad (2.41)\end{aligned}$$

As A_1 is monotone, for any $\varepsilon > 0$, then

$$\begin{aligned}\big\|A_1(u_k)\big\|_* &= \frac{1}{\varepsilon}\sup_{\|v\|\le\varepsilon}\big\langle A_1(u_k), v\big\rangle \le \frac{1}{\varepsilon}\sup_{\|v\|\le\varepsilon}\Big(\big\langle A_1(u_k), v\big\rangle \\ &+ \big\langle A_1(u_k)-A_1(v), u_k-v\big\rangle\Big) = \frac{1}{\varepsilon}\sup_{\|v\|\le\varepsilon}\Big(\big\langle A_1(u_k), u_k\big\rangle + \big\langle A_1(v), v-u_k\big\rangle\Big). \qquad (2.42)\end{aligned}$$

Now we use that $\{\langle A_1(u_k), u_k\rangle\}_{k\in\mathbb{N}}$ is bounded because $\langle A_1(u_k), u_k\rangle = \langle f - A_2(u_k), u_k\rangle$ and the compact mapping A_2 is certainly bounded, and also $\{\langle A_1(v), v - u_k\rangle; \|v\| \le \varepsilon\}$ is bounded if $\varepsilon > 0$ is small enough because A_1 is locally bounded around the origin due to Lemma 2.15. Thus (2.42) shows that $\|A_1(u_k) - A_1(u)\|_*$ is bounded, and, choosing $v_k \to u$ in V, we obtain $\lim_{k\to\infty} I_k^{(1)} = 0$ in (2.41).

Taking a subsequence such that also $A_2(u_k)$ converges to some $\chi \in V^*$ (as we can because A_2 is compact), we get $I_k^{(2)} = \langle f - A_2(u_k) - A_1(u), u_k - v_k\rangle \to \langle f - \chi - A_1(u), u - u\rangle = 0$. As this limit is determined uniquely, even the whole sequence $\{I_k^{(2)}\}_{k\in\mathbb{N}}$ converges to 0.

Using (2.41), by (2.23) we get $u_k \to u$. By Lemma 2.16, $A_1(u_k) \rightharpoonup A_1(u)$. By demicontinuity of A_2, also $A_2(u_k) \rightharpoonup A_2(u)$. It allows us to pass to the limit in (2.37), obtaining $\langle A_1(u) + A_2(u) - f, v\rangle = 0$ for any $v \in \bigcup_{k\in\mathbb{N}} V_k$, hence $A(u) = f$. $\square$

Remark 2.21 (d-monotone A on a uniformly convex V)**.** Any d-monotone $A : V \to V^*$ satisfies (2.23) if V is uniformly convex. Indeed, the premise of (2.23) with $\langle A(u_k) - A(u), u_k - u\rangle \ge (d(\|u_k\|) - d(\|u\|))(\|u_k\| - \|u\|)$, cf. (2.1), yields $\|u_k\| \to \|u\|$. Then, by uniform convexity of V and Theorem 1.2, we get immediately $u_k \to u$.

The nonconstructivity of Brézis' Theorem 2.6 pointed out in Remark 2.7 can be avoided in special situations by using Banach's fixed-point Theorem 1.12 for the iterative process

$$u_k = T_\varepsilon(u_{k-1}) := u_{k-1} - \varepsilon J^{-1}\big(A(u_{k-1}) - f\big)\,, \quad k\in\mathbb{N}, \quad u_0\in V, \qquad (2.43)$$

[7] In fact, any demicontinuous and compact A_2 is automatically continuous.

if V is a Hilbert space and the linear operator $J : V \to V^*$ is defined by $\langle Ju, v\rangle := (u, v)$ with $(\cdot,\cdot)$ denoting here the inner product in V, cf. Remark 3.10. For weakening of the assumptions by further (constructive) approximation see Example 2.95.

Proposition 2.22 (Banach fixed-point technique). *Let V be a Hilbert space, $A : V \to V^*$ be strongly monotone, i.e. $\zeta(r) = \delta r$ from (2.2) with $\delta > 0$, and also A be Lipschitz continuous, i.e. $\|A(u) - A(v)\|_* \le \ell\|u - v\|$. Then the nonlinear mapping T_ε defined by (2.43) is contractive for any $\varepsilon > 0$ satisfying*

$$\varepsilon < 2\delta/\ell^2 \tag{2.44}$$

and the fixed point of T_ε, i.e. $T_\varepsilon(u) = u$, does exist and obviously solves $A(u) = f$.

Proof. It holds that[8] $\langle f, J^{-1}f\rangle = \|f\|_*^2$, so that one has

$$\begin{aligned}\|T_\varepsilon(u) - T_\varepsilon(v)\|^2 &= \big\langle J(u - v) - \varepsilon(A(u) - A(v)), u - v - \varepsilon J^{-1}(A(u) - A(v))\big\rangle\\ &= \|u - v\|^2 - 2\varepsilon\big(u{-}v, J^{-1}(A(u) - A(v))\big) + \varepsilon^2\|J^{-1}A(u) - J^{-1}A(v)\|^2\\ &= \|u - v\|^2 - 2\varepsilon\langle A(u) - A(v), u - v\rangle + \varepsilon^2\|A(u) - A(v)\|_*^2\\ &\le \|u - v\|^2 - 2\varepsilon\delta\|u - v\|^2 + \varepsilon^2\ell^2\|u - v\|^2.\end{aligned}$$

The condition (2.44) just guarantees the Lipschitz constant $\sqrt{1 - 2\varepsilon\delta + \varepsilon^2\ell^2}$ of T_ε to be less than 1. □

2.4 Quasilinear elliptic equations

We will illustrate the above abstract theory on boundary-value problems for the quasilinear 2nd-order partial differential equation

$$-\mathrm{div}\big(a(x, u, \nabla u)\big) + c(x, u, \nabla u) = g \tag{2.45}$$

considered on a bounded connected Lipschitz domain $\Omega \subset \mathbb{R}^n$. Here $a : \Omega \times \mathbb{R} \times \mathbb{R}^n \to \mathbb{R}^n$ and $c : \Omega \times \mathbb{R} \times \mathbb{R}^n \to \mathbb{R}$; for more qualification see (2.54) and (2.55a,c) below. Recall that $\nabla u := \big(\frac{\partial}{\partial x_1}u, \ldots, \frac{\partial}{\partial x_n}u\big)$ denotes the gradient of u. More in detail, (2.45) means

$$-\sum_{i=1}^{n}\frac{\partial}{\partial x_i}a_i\big(x, u(x), \nabla u(x)\big) + c\big(x, u(x), \nabla u(x)\big) = g(x) \tag{2.46}$$

for $x \in \Omega$ but we will rather use the abbreviated form (2.45) in what follows. For some systems of the 2nd-order equations see Sect. 6.1 below while higher-order equations will be briefly mentioned in Sect. 2.4.4. Besides, we will confine ourselves to data with polynomial-growth; $p \in (1, +\infty)$ will denote the growth of the leading

[8] Realize that, for $v = J^{-1}f$, one has $\langle f, J^{-1}f\rangle = \langle f, v\rangle = \langle Jv, v\rangle = (v, v) = \|v\|^2 = \|f\|_*^2$.

nonlinearity $a(x, u, \cdot)$ which essentially determines the setting and the other data qualification. Also, $a(x, u, \cdot)$ will be assumed to behave monotonically, cf. (2.65), which is related to the adjective *elliptic*. For the linear case $a(x, r, s) = \mathbb{A}s$, the monotonicity (2.65) and coercivity (2.92a) below implies the matrix $\mathbb{A}$ is positive definite, which is what is conventionally called "elliptic", contrary to $\mathbb{A}$ indefinite (resp. semidefinite) which is addressed as *hyperbolic* (resp. *parabolic*).

Convention 2.23 (Omitting x-variable). For brevity, we will often write $a(x, u, \nabla u)$ instead of $a(x, u(x), \nabla u(x))$ (as we already did in (2.45)) or sometimes even $a(u, \nabla u)$ if the dependence on x is automatic; hence, in fact, $\mathscr{N}_a(u, \nabla u) = a(u, \nabla u)$. Thus, e.g. $\int_\Omega c(u, \nabla u) v \, dx$ will mean $\int_\Omega c(x, u(x), \nabla u(x)) v(x) \, dx$.

2.4.1 Boundary-value problems for 2nd-order equations

The equation (2.45) may admit very many solutions, which indicates some missing requirements. This is usually overcome by a boundary condition to be prescribed for the solution on the boundary $\Gamma := \partial\Omega$ of the domain Ω.

One option is to prescribe simply the trace $u|_\Gamma$ of u on the boundary, i.e.

$$u|_\Gamma = u_{\mathrm{D}} \quad \text{on } \Gamma \tag{2.47}$$

with u_{D} a fixed function on Γ. This condition is referred to as a *Dirichlet boundary condition*.

Having in mind the equation (2.45), the alternative natural possibility is to prescribe a local equation for the "boundary flux" $\nu \cdot a$, i.e.

$$\nu \cdot a(x, u, \nabla u) + b(x, u) = h \quad \text{on } \Gamma \tag{2.48}$$

where $\nu = (\nu_1, \ldots, \nu_n)$ denotes the unit outward normal to Γ and $h : \Gamma \to \mathbb{R}$ and $b : \Gamma \times \mathbb{R} \to \mathbb{R}$ are given functions qualified later. More in detail, (2.48) means $\sum_{i=1}^n \nu_i(x) \, a_i(x, u(x), \nabla u(x)) + b(x, u(x)) = h(x)$ for $x \in \Gamma$. This condition is referred to as a (nonlinear) *Newton boundary condition* or sometimes also a *Robin condition*. If $b = 0$, it is called a *Neumann boundary condition*.

One can still think about a combination of (2.47) and (2.48) on various parts of Γ. For this, let us divide (up to a zero-measure set) the boundary Γ on two disjoint open parts Γ_{D} and Γ_{N} such that $\mathrm{meas}_{n-1}\big(\Gamma \setminus (\Gamma_{\mathrm{D}} \cup \Gamma_{\mathrm{N}})\big) = 0$, and then consider so-called *mixed boundary conditions*

$$u|_\Gamma = u_{\mathrm{D}} \qquad \text{on } \Gamma_{\mathrm{D}}, \tag{2.49a}$$

$$\nu \cdot a(x, u, \nabla u) + b(x, u) = h \qquad \text{on } \Gamma_{\mathrm{N}}. \tag{2.49b}$$

As either Γ_{D} or Γ_{N} may be empty, (2.49) covers also (2.48) and (2.47), respectively.

Completing the equation (2.45) with the boundary conditions (2.47) (resp. (2.48) or (2.49)), we will speak about a Dirichlet (resp. Newton or mixed) *boundary-value problem*. One can have an idea to seek a so-called *classical solution* u of it, i.e. such $u \in C^2(\bar{\Omega})$ satisfying the involved equalities everywhere on

Ω and Γ. This requires, however, very strong data qualifications both for a, b, and c and for Ω itself. Therefore, modern theories rely on a natural generalization of the notion of the solution. In this context, ultimate requirements on every sensible definition are[9]:

1. *Consistency*: Any classical solution to the boundary-value problem in question is the generalized solution.
2. *Selectivity*: If all data are smooth and if the generalized solution belongs to $C^2(\bar{\Omega})$, then it is the classical solution. Moreover, speaking a bit vaguely, in qualified cases the generalized solution is unique.

2.4.2 Weak formulation

Here, the generalized solution will arise from a so-called weak formulation of the boundary-value problem, which is the most frequently used concept and which just fits to the pseudomonotonicity approach. Later, we will present some other concepts, too.

For the full generality, we will treat the mixed boundary conditions (2.49). The *weak formulation* of (2.45) with (2.49) arises as follows:

Step 1: Multiply the differential equation, i.e. here (2.45), by a test function v.

Step 2: Integrate it over Ω.

Step 3: Use Green's formula (1.54), here with $z = a(x, u, \nabla u)$.

Step 4: Substitute the Newton boundary condition, i.e. here (2.49b), into the boundary integral, i.e. here $\int_{\Gamma_\mathrm{N}} v(z \cdot \nu)\,\mathrm{d}S = \int_{\Gamma_\mathrm{N}} (\nu \cdot a(x, u, \nabla u))v\,\mathrm{d}S$ in (1.54), while by considering $v|_{\Gamma_\mathrm{D}} = 0$, the integral over Γ_D simply vanishes.

This procedure looks here as

$$\int_\Omega \Big(- \mathrm{div}\big(a(x, u, \nabla u)\big) + c(x, u, \nabla u)\Big) v\,\mathrm{d}x$$

$$\overset{\text{Green's formula}}{=} \int_\Omega a(x, u, \nabla u) \cdot \nabla v + c(x, u, \nabla u)v\,\mathrm{d}x - \int_\Gamma \big(\nu \cdot a(x, u, \nabla u)\big)v\,\mathrm{d}S$$

$$\overset{\text{boundary conditions}}{=} \int_\Omega a(x, u, \nabla u) \cdot \nabla v + c(x, u, \nabla u)v\,\mathrm{d}x + \int_\Gamma \big(b(x, u) - h(x)\big)v\,\mathrm{d}S. \tag{2.50}$$

Realizing still that the left-hand side in (2.50) is just $\int_\Omega gv\,\mathrm{d}x$, we come to the integral identity

$$\int_\Omega a(x, u, \nabla u) \cdot \nabla v + c(x, u, \nabla u)v\,\mathrm{d}x + \int_{\Gamma_\mathrm{N}} b(x, u)v\,\mathrm{d}S = \int_\Omega gv\,\mathrm{d}x + \int_{\Gamma_\mathrm{N}} hv\,\mathrm{d}S. \tag{2.51}$$

[9]See [360, Remark 5.3.8] or [370] for some examples of unsuitable concepts of so-called "measure-valued" solutions, cf. also DiPerna [124] or Illner and Wick [210].

As declared, we confine ourselves to a p-polynomial growth, cf. (2.55a) below, and then it is natural to seek the weak solution in the Sobolev space $W^{1,p}(\Omega)$. It leads to the following definition:

Definition 2.24. We call $u \in W^{1,p}(\Omega)$ a *weak solution* to the mixed boundary-value problem (2.45) and (2.49) if $u|_{\Gamma_D} = u_D$ and if the integral identity (2.51) holds for any $v \in W^{1,p}(\Omega)$ with $v|_{\Gamma_D} = 0$.

The above 4-step procedure to derive (2.51) guarantees automatically its consistency. On the other hand, its selectivity is related to the important fact that the space V of test-functions v's, i.e.

$$V = \{v \in W^{1,p}(\Omega);\ v|_{\Gamma_D} = 0\}, \tag{2.52}$$

is sufficiently rich, the restriction of v on Γ_D being compensated by direct involvement of the boundary condition (2.49a) in Definition 2.24:

Proposition 2.25 (Selectivity of the weak-solution definition). *Let* $a \in C^1(\bar\Omega \times \mathbb{R} \times \mathbb{R}^n; \mathbb{R}^n)$, $c \in C^0(\bar\Omega \times \mathbb{R} \times \mathbb{R}^n)$, *and* $b \in C^0(\bar\Gamma_N \times \mathbb{R})$, $g \in C(\bar\Omega)$, *and* $h \in C(\Gamma_N)$. *Then any weak solution* $u \in C^2(\bar\Omega)$ *is the classical solution.*

Proof. Put $v \in V$ into (2.51) and use Green's formula (1.54). One gets

$$\int_\Omega \Big(\mathrm{div}\, a(x,u,\nabla u) - c(x,u,\nabla u) + g\Big) v \,\mathrm{d}x \\ + \int_{\Gamma_N} \Big(h - b(x,u) - \nu\cdot a(x,u,\nabla u)\Big) v \,\mathrm{d}S = 0. \tag{2.53}$$

Considering $v|_\Gamma = 0$, the boundary integral in (2.53) vanishes. As v is otherwise arbitrary, one deduces that (2.45) holds a.e., and hence even everywhere in Ω due to the assumed smoothness of a and c.[10] Hence, the first integral in (2.53) vanishes. Then, putting a general $v \in V$ into (2.53) shows the latter boundary condition in (2.49) valid,[11] while the former one is directly involved in Definition 2.24. □

The important issue now is to set up basic data qualification to give a sense to all integrals in (2.51). Recall that we keep the permanent assumption Ω to be a bounded Lipschitz domain (so that, in particular, ν is defined a.e. on Γ) and Γ_D and Γ_N are open in Γ (hence, in particular, measurable). To ensure measurability of integrands on the left-hand side of (2.51) we must assume:

$$a_i, c : \Omega \times \mathbb{R} \times \mathbb{R}^n \to \mathbb{R}\,, \quad b : \Gamma \times \mathbb{R} \to \mathbb{R} \quad \text{are Carathéodory functions,} \tag{2.54}$$

for $i = 1, \dots, n$; this means measurability in x and continuity in the other variables. The further ultimate requirement is integrability of all integrands on the left-hand

[10] Here we use the fact that the set of test functions is sufficiently rich, namely that $W_0^{1,p}(\Omega)$ is dense in $L^1(\Omega)$; cf. Theorem 1.25 and the well-known fact that $C_0^\infty(\Omega)$ is dense in $L^1(\Omega)$.

[11] Here the important fact is that the set $\{v|_{\Gamma_N};\ v \in V\}$ is dense in $L^1(\Gamma_N)$. This is guaranteed by the assumption that Γ_N is open in Γ.

side of (2.51). This, and some continuity requirements needed further, lead us to assume the growth conditions on the nonlinearities a, b, and c:

$$|a(x,r,s)| \le \gamma(x) + C|r|^{(p^*-\epsilon)/p'} + C|s|^{p-1} \quad \text{for some } \gamma \in L^{p'}(\Omega)\,, \tag{2.55a}$$

$$|\,b(x,r)\,| \;\le\; \gamma(x) + C|r|^{p^\#-\epsilon-1} \quad \text{for some } \gamma \in L^{p^{\#\prime}}(\Gamma), \tag{2.55b}$$

$$|c(x,r,s)| \le \gamma(x) + C|r|^{p^*-\epsilon-1} + C|s|^{p/p^{*\prime}} \quad \text{for some } \gamma \in L^{p^{*\prime}}(\Omega). \tag{2.55c}$$

Let us recall the notation of the prime denoting the conjugate exponents (i.e., e.g., $p' = p/(p-1)$, cf. (1.20)) and the continuous (resp. compact) embedding $W^{1,p}(\Omega) \subset L^{p^*}(\Omega)$ (resp. $W^{1,p}(\Omega) \Subset L^{p^*-\epsilon}(\Omega)$ with $\epsilon>0$), cf. Theorem 1.20. Moreover, the trace operator $u \mapsto u|_\Gamma$ maps $W^{1,p}(\Omega)$ into $L^{p^\#}(\Gamma)$ continuously and into $L^{p^\#-\epsilon}(\Gamma)$ compactly, cf. Theorem 1.23. For p^* and $p^\#$ see (1.34) and (1.37).

Convention 2.26. For $p > n$, the terms $|r|^{+\infty}$ occurring in (2.55) are to be understood such that $|a(x,\cdot,s)|$, $|b(x,\cdot)|$, and $|c(x,\cdot,s)|$ may have arbitrary fast growth if $|r| \to \infty$.

In view of Theorem 1.27, the growth conditions (2.55) are designed so that respectively

$$\mathscr{N}_a : W^{1,p}(\Omega) \times L^p(\Omega;\mathbb{R}^n) \to L^{p'}(\Omega;\mathbb{R}^n) \text{ is (weak}\times\text{norm,norm)-continuous,} \tag{2.56a}$$

$$u \mapsto \mathscr{N}_b(u|_\Gamma) : W^{1,p}(\Omega) \to L^{p^{\#\prime}}(\Gamma) \quad \text{is (weak,norm)-continuous,} \tag{2.56b}$$

$$\mathscr{N}_c : W^{1,p}(\Omega) \times L^p(\Omega;\mathbb{R}^n) \to L^{p^{*\prime}}(\Omega) \text{ is (weak}\times\text{norm,norm)-continuous.} \tag{2.56c}$$

In particular, for $u, v \in W^{1,p}(\Omega)$, the integrands $a(x,u,\nabla u)\cdot\nabla v$ and $c(x,u,\nabla u)v$ occurring in (2.51) belong to $L^1(\Omega)$ while $b(x,u|_\Gamma)v|_\Gamma$ belongs to $L^1(\Gamma)$.

Furthermore, we will also suppose the right-hand side qualification:

$$g \in L^{p^{*\prime}}(\Omega), \qquad h \in L^{p^{\#\prime}}(\Gamma). \tag{2.57}$$

Note that (2.57) ensures $gv \in L^1(\Omega)$ and $hv|_\Gamma \in L^1(\Gamma)$ for $v \in W^{1,p}(\Omega)$, hence (2.51) has a good sense. Moreover, we must qualify $u_{\rm D}$ occurring in the Dirichlet boundary condition (2.49a). The simplest way is to assume

$$\exists w \in W^{1,p}(\Omega) :\; u_{\rm D} = w|_\Gamma. \tag{2.58}$$

Then, considering V from (2.52) equipped by the norm (1.30b) denoted simply by $\|\cdot\|$, we define $A : W^{1,p}(\Omega) \to V^*$ and $f \in V^*$ simply by

$$\langle A(u), v\rangle := \text{left-hand side of (2.51)}, \tag{2.59}$$

$$\langle f, v\rangle \;:=\; \text{right-hand side of (2.51)}. \tag{2.60}$$

Moreover, referring to (2.58), let us define $A_0 : V \to V^*$ by

$$A_0(u) = A(u+w). \tag{2.61}$$

Note that A_0 has again the form of A from (2.51) but the nonlinearities a, b, and c are respectively replaced by a_0, b_0, and c_0 given by $a_0(x,r,s) := a(x,r+w(x),s+\nabla w(x))$, $b_0(x,r) := b(x,r+w(x))$, and $c_0(x,r,s) := c(x,r+w(x),s+\nabla w(x))$, and these nonlinearities satisfy (2.54)–(2.55) if $w \in W^{1,p}(\Omega)$ and if the original nonlinearities a, b, and c satisfy (2.54)–(2.55). Note also that for zero (or none) Dirichlet boundary conditions, one can assume $w=0$ in (2.58) and then $A_0 \equiv A|_V$ (or simply $A_0 \equiv A$).

Note that, indeed, $f \in V^*$ because of the obvious estimate

$$\|f\|_* = \sup_{\|v\|\le 1}\Big(\int_\Omega gv\,\mathrm{d}x + \int_{\Gamma_\mathrm{N}} hv\,\mathrm{d}S\Big) \le \sup_{\|v\|\le 1}\Big(\|g\|_{L^{p^{*\prime}}(\Omega)}\|v\|_{L^{p^*}(\Omega)} + \|h\|_{L^{p^{\#\prime}}(\Gamma_\mathrm{N})}\|v\|_{L^{p^\#}(\Gamma_\mathrm{N})}\Big) \le N_1\|g\|_{L^{p^{*\prime}}(\Omega)} + N_2\|h\|_{L^{p^{\#\prime}}(\Gamma_\mathrm{N})} \tag{2.62}$$

where N_1 is the norm of the embedding operator $W^{1,p}(\Omega) \to L^{p^*}(\Omega)$ and N_2 is the norm of the trace operator $v \mapsto v|_{\Gamma_\mathrm{N}} : W^{1,p}(\Omega) \to L^{p^\#}(\Gamma_\mathrm{N})$. By similar arguments, (2.54) and (2.55) ensures $A(u) \in V^*$, cf. Lemma 2.31 below.

Proposition 2.27 (Shift for non-zero Dirichlet condition). *The abstract equation* (2.6) *for A_0 has a solution $u_0 \in V$, i.e. $A_0(u_0) = f$, if and only if $u = u_0 + w \in W^{1,p}(\Omega)$ is the weak solution to the boundary-value problem* (2.45) *and* (2.49) *in accord to Definition* 2.24.

Proof. We obviously have $f = A_0(u_0) = A_0(u-w) = A(u-w+w) = A(u)$, hence the assertion immediately follows by the definition (2.59)–(2.60). □

Remark 2.28 (Why both u and v are from V). In principle, Definition 2.24 could work with $v \in Z := W^{1,\infty}(\Omega)$, or even with v's smoother; the selectivity Proposition 2.25 would hold as far as density of Z in V would be preserved, as used in Section 2.5 below. The choice of v's from the same space where the solution u is supposed to live, i.e. here V, is related to the setting $A : V \to Z^*$ which is fitted with the pseudomonotone-mapping concept only for $Z = V$.

Remark 2.29 (Why both Γ_D and Γ_N are assumed open). In principle, Definition 2.24 as well as the existence Theorem 2.36 below could work for Γ_D and Γ_N only measurable. However, we would lose the connection to the original problem, cf. Proposition 2.25: indeed, one can imagine Γ_D measurable dense in Γ and Γ_N of a positive measure. Then, for $p > n$, $v|_\Gamma \in C(\Gamma)$ and the condition $v|_{\Gamma_\mathrm{D}} = 0$ would imply $v|_\Gamma = 0$, so that the Γ_N-integrals in (2.51) vanish and the Newton boundary condition on Γ_N in (2.49b) would be completely eliminated.

Remark 2.30 (Integral balance). The equation (2.45) is a differential alternative to the integral balance

$$\int_O c(x,u,\nabla u) - g(x)\,\mathrm{d}x = \int_{\partial O} a(x,u,\nabla u)\cdot\nu\,\mathrm{d}S \tag{2.63}$$

for any test volume $O \subset \Omega$ with $\bar{O} \subset \Omega$ and a smooth boundary ∂O with the normal $\nu = \nu(x)$. Obviously, one is to identify c as the balanced quantity (depending on u and ∇u) while a as a flux of this quantity[12], and then (2.63) just says that the overall production of this quantity over the arbitrary test volume O is balanced by the overall flux through the boundary ∂O, cf. Figure 4. The philosophy that integral form (2.63) of physical laws is more natural than their differential form (2.45) was pronounced already by David Hilbert[13]. The weak formulation (2.51) implicitly includes, besides information about the boundary conditions, also (2.63). Indeed, it suffices to take v in (2.51) as some approximation of the characteristic function χ_O (which itself does not belong to $W^{1,p}(\Omega)$, however), e.g. v_ε with $v_\varepsilon(x) := (1 - \text{dist}(x, O)/\varepsilon)^+$, and then to pass $\varepsilon \searrow 0$. This limit passage is, however, legal only if $x \mapsto a(x, u, \nabla u)$ is sufficiently regular near ∂O or, in a general case, it holds only in some "generic" sense; cf. e.g. Exercise 2.63.

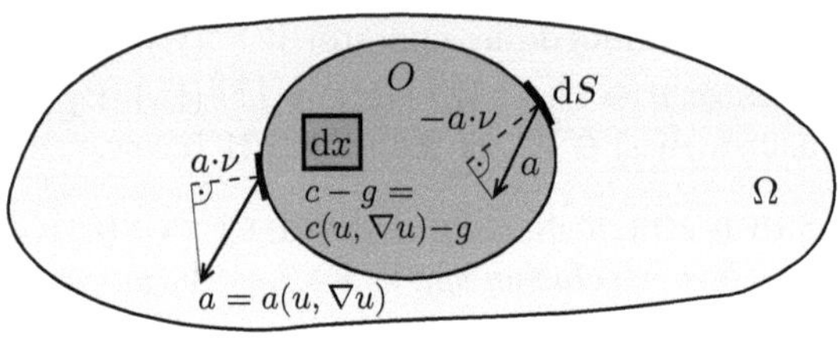

Figure 4. Illustration to balancing the normal flux $a{\cdot}\nu$ through the boundary of a test volume O and the production c inside this volume.

2.4.3 Pseudomonotonicity, coercivity, existence of solutions

In view of Theorem 2.6 with Proposition 2.27, we are to show pseudomonotonicity of $A_0 : V \to V^*$. For simplicity, we can prove it for A as $W^{1,p}(\Omega) \to W^{1,p}(\Omega)^*$, which, by Lemma 2.11(ii), implies pseudomonotonicity of $A_0 : W^{1,p}(\Omega) \to W^{1,p}(\Omega)^*$, and then obviously also of $A_0 : V \to V^*$. Let us prove (2.3a) and (2.3b) respectively in the following lemmas.

Lemma 2.31 (BOUNDEDNESS OF A). *The assumptions* (2.54) *and* (2.55) *ensure* (2.3a), *i.e.* $A : W^{1,p}(\Omega) \to W^{1,p}(\Omega)^*$ *bounded.*

Proof. We prove $A(\{u \in W^{1,p}(\Omega);\ \|u\| \le \rho\})$ bounded in $W^{1,p}(\Omega)^*$ for any $\rho > 0$. Here, $\|\cdot\|$ and $\|\cdot\|_*$ will denote the norms in $W^{1,p}(\Omega)$ and $W^{1,p}(\Omega)^*$, respectively. Indeed, we can estimate

---

[12] In concrete situations, the dependence of a on ∇u may result from a (nonlinear) Fick's, Fourier's, or Darcy's law.

[13] Explicitly, it can be found in his famous Mathematical problems [202, 19th problem]: "Has not every ... variational problem a solution, provided ... if need be that the notion of a solution shall be suitably extended?"

$$\sup_{\|u\|\le\rho}\|A(u)\|_* = \sup_{\|u\|\le\rho}\sup_{\|v\|\le1}\langle A(u),v\rangle$$
$$= \sup_{\|u\|\le\rho}\sup_{\|v\|\le1}\int_\Omega a(u,\nabla u)\cdot\nabla v + c(u,\nabla u)v\,\mathrm{d}x + \int_{\Gamma_\mathrm{N}} b(u)v\,\mathrm{d}S$$
$$\le \sup_{\|u\|\le\rho}\sup_{\|v\|\le1}\big\|a(u,\nabla u)\big\|_{L^{p'}(\Omega;\mathbb{R}^n)}\big\|\nabla v\big\|_{L^p(\Omega;\mathbb{R}^n)}$$
$$+\big\|c(u,\nabla u)\big\|_{L^{p^*\prime}(\Omega)}\big\|v\big\|_{L^{p^*}(\Omega)}+\big\|b(u)\big\|_{L^{p^\#\prime}(\Gamma_\mathrm{N})}\big\|v\big\|_{L^{p^\#}(\Gamma_\mathrm{N})}$$
$$\le \sup_{\|u\|\le\rho}\big\|a(u,\nabla u)\big\|_{L^{p'}(\Omega;\mathbb{R}^n)}+N_1\big\|c(u,\nabla u)\big\|_{L^{p^*\prime}(\Omega)}+N_2\big\|b(u)\big\|_{L^{p^\#\prime}(\Gamma_\mathrm{N})} \tag{2.64}$$

where N_1 and N_2 are as in (2.62). In view of (2.55), it is bounded uniformly for u ranging over a bounded set in $W^{1,p}(\Omega)$. □

Further, we still have to strengthen our data qualification. The crucial assumption we must make for pseudomonotonicity of A is the so-called *monotonicity in the main part*:

$$\forall(\text{a.a.})\ x\in\Omega\ \ \forall r\in\mathbb{R}\ \ \forall s,\tilde{s}\in\mathbb{R}^n: \qquad \big(a(x,r,s)-a(x,r,\tilde{s})\big)\cdot(s-\tilde{s})\ge 0. \tag{2.65}$$

To cover as many situations as possible, we distinguish three cases in accordance with whether $c(x,r,\cdot)$ is constant, linear, or nonlinear, respectively.

Lemma 2.32 (THE IMPLICATION (2.3B)). *Let the assumptions* (2.54) *and* (2.55) *be valid, let a satisfy* (2.65), *and let one of the following three cases hold: c is independent of s, i.e. for some $\widetilde{c}:\Omega\times\mathbb{R}\to\mathbb{R}$,*

$$c(x,r,s)=\widetilde{c}(x,r)\,, \tag{2.66}$$

or c is linearly dependent on s, i.e. for some $\overline{c}:\Omega\times\mathbb{R}\to\mathbb{R}^n$,

$$c(x,r,s)=\overline{c}(x,r)\cdot s, \tag{2.67}$$

or c is generally dependent on s but the strict monotonicity "in the main part" and coercivity of $a(x,r,\cdot)$ hold and the growth of $c(x,\cdot,\cdot)$ is further restricted:

$$\big(a(x,r,s)-a(x,r,\tilde{s})\big)\cdot(s-\tilde{s})=0 \implies s=\tilde{s}, \tag{2.68a}$$
$$\forall s_0\in\mathbb{R}^n:\quad \lim_{|s|\to\infty}\frac{a(x,r,s)\cdot(s-s_0)}{|s|}=+\infty \text{ uniformly for } r \text{ bounded,} \tag{2.68b}$$
$$\exists\gamma\in L^{p^{*\prime}+\epsilon}(\Omega)\,\exists C\in\mathbb{R}:\ |c(x,r,s)|\le\gamma(x)+C|r|^{p^*-\epsilon-1}+C|s|^{(p-\epsilon)/p^{*\prime}} \tag{2.68c}$$

with Convention 2.26 *in mind. Then $A:W^{1,p}(\Omega)\to W^{1,p}(\Omega)^*$ satisfies* (2.3b).

Remark 2.33. Obviously, (2.66) together with the growth condition (2.55c) imply $|\widetilde{c}(x,r)|\le\gamma(x)+C|r|^{p^*-\epsilon-1}$ with γ as in (2.55c). A bit more difficult is to realize that (2.67) together with the growth condition (2.55c) imply that $\overline{c}:\Omega\times\mathbb{R}\to\mathbb{R}^n$ has to satisfy

$$|\overline{c}(x,r)|\leq\gamma(x)+C|r|^{p^*/q-\epsilon_1}\quad\text{with }\ \gamma\in L^{q+\epsilon_1}(\Omega)\ \text{ and some }\ \epsilon_1>0,\qquad\text{where }\ q=\begin{cases}\dfrac{np}{np-2n+p} & \text{if } p<n,\\ p' & \text{if } p\geq n.\end{cases}\tag{2.69}$$

This condition together with the structural condition (2.67) now guarantees

$$\mathscr{N}_c\!:W^{1,p}(\Omega)\times L^p(\Omega;\mathbb{R}^n)\to L^{(p^*-\epsilon)'}(\Omega)\ \text{is (weak}\times\text{weak,weak)-continuous.}\tag{2.70}$$

Eventually, note that the growth condition (2.68c) strengthens (2.55c) and is designed so that, for some $\epsilon>0$ (depending on ϵ used in (2.68c)),

$$\mathscr{N}_c:W^{1,p}(\Omega)\times L^p(\Omega;\mathbb{R}^n)\to L^{p^{*\prime}+\epsilon}(\Omega)\ \text{ is (weak}\times\text{norm,norm)-continuous.}\tag{2.71}$$

Proof of Lemma 2.32. Let us take $u_k\rightharpoonup u$ in $W^{1,p}(\Omega)$ and assume that

$$\limsup_{k\to\infty}\langle A(u_k),u_k-u\rangle\leq 0.\tag{2.72}$$

We are to show that $\liminf_{k\to\infty}\langle A(u_k),u_k-v\rangle\geq\langle A(u),u-v\rangle$ for any $v\in W^{1,p}(\Omega)$. To distinguish between the highest and the lower-order terms, we define $B(w,u)\in W^{1,p}(\Omega)^*$ by

$$\big\langle B(w,u),v\big\rangle:=\int_\Omega a(x,w,\nabla u)\cdot\nabla v+c(x,w,\nabla w)v\,\mathrm{d}x+\int_{\Gamma_\mathrm{N}}b(x,w)v\,\mathrm{d}S\tag{2.73}$$

for $u,w\in W^{1,p}(\Omega)$; recall the Convention 2.23. Obviously, $A(u)=B(u,u)$.

Let us put $u_\varepsilon=(1-\varepsilon)u+\varepsilon v$, $\varepsilon\in[0,1]$. Monotonicity (2.65) implies $\langle B(u_k,u_k)-B(u_k,u_\varepsilon),u_k-u_\varepsilon\rangle\geq 0$. Then, just by simple algebra,

$$\begin{aligned}\varepsilon\big\langle A(u_k),u-v\big\rangle\ &\geq\ -\big\langle A(u_k),u_k-u\big\rangle\\&\quad+\big\langle B(u_k,u_\varepsilon),u_k-u\big\rangle+\varepsilon\big\langle B(u_k,u_\varepsilon),u-v\big\rangle.\end{aligned}\tag{2.74}$$

Let us assume, for a moment, that we have proved

$$\lim_{k\to\infty}\ \big\langle B(u_k,v),u_k-u\big\rangle=0,\tag{2.75}$$

$$\operatorname*{w-lim}_{k\to\infty}\ B(u_k,v)=B(u,v)\qquad\text{(the weak limit in }W^{1,p}(\Omega)^*\text{)},\tag{2.76}$$

and use them here for $v=u_\varepsilon$ to pass successively to the limit in the right-hand-side terms of (2.74). Using also (2.72), we thus obtain

$$\begin{aligned}\varepsilon\liminf_{k\to\infty}\big\langle A(u_k),u-v\big\rangle\ &\geq\ -\limsup_{k\to\infty}\big\langle A(u_k),u_k-u\big\rangle+\lim_{k\to\infty}\big\langle B(u_k,u_\varepsilon),u_k-u\big\rangle\\&\quad+\ \varepsilon\lim_{k\to\infty}\big\langle B(u_k,u_\varepsilon),u-v\big\rangle\ \geq\ \varepsilon\big\langle B(u,u_\varepsilon),u-v\big\rangle.\end{aligned}$$

Divide it by $\varepsilon > 0$. Then the limit passage $\varepsilon \to 0$ gives $u_\varepsilon \to u$ strongly so that we get $B(u, u_\varepsilon) \to B(u, u)$ even strongly[14], which results in $\liminf_{k\to\infty}\langle A(u_k), u - v\rangle \geq \langle B(u, u), u - v\rangle = \langle A(u), u - v\rangle$.

Then, by using the monotonicity in the main part (2.65) once again, now as $\langle B(u_k, u_k) - B(u_k, u), u_k - u\rangle \geq 0$, and by using also (2.75) now with $v = u$, we can claim that

$$\begin{aligned}\liminf_{k\to\infty}\big\langle A(u_k), u_k{-}v\big\rangle &\geq \liminf_{k\to\infty}\big\langle A(u_k), u_k{-}u\big\rangle + \liminf_{k\to\infty}\big\langle A(u_k), u{-}v\big\rangle \\ &= \lim_{k\to\infty}\big\langle B(u_k, u), u_k{-}u\big\rangle + \liminf_{k\to\infty}\big\langle B(u_k, u_k){-}B(u_k, u), u_k{-}u\big\rangle \\ &\quad + \liminf_{k\to\infty}\big\langle A(u_k), u - v\big\rangle \geq \langle A(u), u - v\rangle,\end{aligned} \tag{2.77}$$

which is just the conclusion of (2.3b).

Thus it remains to prove (2.75) and (2.76). Since $u_k \rightharpoonup u$ in $W^{1,p}(\Omega) \Subset L^{p^*-\epsilon}(\Omega)$, we have $u_k \to u$ in $L^{p^*-\epsilon}(\Omega)$. Similarly, $u_k|_\Gamma \to u|_\Gamma$ in $L^{p^\#-\epsilon}(\Gamma)$. Then, by the continuity of the Nemytskiĭ mappings induced by $a(\cdot, \nabla v)$ and b, we get $a(u_k, \nabla v) \to a(u, \nabla v)$ in $L^{p'}(\Omega;\mathbb{R}^n)$, cf. (2.56a), and $b(u_k) \to b(u)$ in $L^{p^{\#\prime}}(\Gamma)$; cf. (2.56b) together with (1.36b); recall again Convention 2.23. Hence, realizing that $\nabla(u_k{-}u) \rightharpoonup 0$ in $L^p(\Omega;\mathbb{R}^n)$ and $(u_k{-}u)|_\Gamma \rightharpoonup 0$ in $L^{p^\#}(\Gamma_{\rm N})$, one gets

$$\int_\Omega a(u_k, \nabla v)\cdot\nabla(u_k - u)\,\mathrm{d}x + \int_{\Gamma_{\rm N}} b(u_k)(u_k - u)\,\mathrm{d}S \to 0. \tag{2.78}$$

By the same reasons, for any $z \in W^{1,p}(\Omega)$, we have also

$$\int_\Omega a(u_k, \nabla v){\cdot}\nabla z\,\mathrm{d}x + \int_{\Gamma_{\rm N}} b(u_k)z\,\mathrm{d}S \to \int_\Omega a(u, \nabla v)z\,\mathrm{d}x + \int_{\Gamma_{\rm N}} b(u)z\,\mathrm{d}S. \tag{2.79}$$

As to the term c, we will distinguish the above suggested three cases.

The case (2.66): By the continuity of the Nemytskiĭ mappings induced by $\tilde{c}$, one has $\tilde{c}(u_k) \to \tilde{c}(u)$ in $L^{p^{*\prime}}(\Omega)$. Therefore, realizing that $u_k - u \rightharpoonup 0$ in $L^{p^*}(\Omega)$, one gets $\int_\Omega \tilde{c}(u_k)(u_k - u)\,\mathrm{d}x \to 0$. Adding it with (2.78), one gets

$$\begin{aligned}\langle B(u_k, v), u_k - u\rangle := \int_\Omega \Big(&a(u_k, \nabla v)\cdot\nabla(u_k - u) \\ &+ \tilde{c}(u_k)(u_k - u)\Big)\,\mathrm{d}x + \int_{\Gamma_{\rm N}} b(u_k)(u_k - u)\,\mathrm{d}S \to 0,\end{aligned} \tag{2.80}$$

which proves (2.75). Similarly, $\int_\Omega \tilde{c}(u_k)z\mathrm{d}x \to \int_\Omega \tilde{c}(u_k)z\mathrm{d}x$, which, together with (2.79), gives just (2.76).

The case (2.67): Here we have a certain reserve in the growth, cf. (2.70), and can thus exploit the compactness of the embedding $W^{1,p}(\Omega) \Subset L^{p^*-\epsilon}(\Omega)$ to use $u_k \to u$

---

[14]Here we use the continuity of the Nemytskiĭ mapping $\mathscr{N}_{a\circ u}$ with $a\circ u : (x, s) \mapsto a(x, u(x), s)$.

strongly in $L^{p^*-\epsilon}(\Omega)$. Also, we can use $\overline{c}(u_k) \to \overline{c}(u)$ in $L^{q+\epsilon_1}(\Omega)$ with q from (2.69) and some $\epsilon_1 > 0$ (depending on ϵ); note that $(q+\epsilon_1)^{-1}+p^{-1}+(p^*-\epsilon)^{-1} \leq 1$ if ϵ is small enough depending on the chosen ϵ_1. As $\nabla u_k \rightharpoonup \nabla u$ in $L^p(\Omega;\mathbb{R}^n)$, we can pass to the limit in the c-term:

$$\int_\Omega \overline{c}(u_k)\cdot\nabla u_k(u_k-u)\,\mathrm{d}x \to 0. \tag{2.81}$$

Adding it with (2.78), one gets (2.75). Similarly, $\int_\Omega \overline{c}(u_k)\cdot\nabla u_k z\,\mathrm{d}x \to \int_\Omega \overline{c}(u_k)\cdot \nabla u z\,\mathrm{d}x$, which, together with (2.79), gives just (2.76).

The case (2.68): We already showed that $u_k \to u$ in $L^{p^*-\epsilon}(\Omega)$. In view of the boundedness (2.71) of $\{c(u_k,\nabla u_k)\}_{k\in\mathbb{N}}$ in $L^{p^{*\prime}+\epsilon}(\Omega)$, we obviously have

$$\int_\Omega c(u_k,\nabla u_k)(u_k-u)\,\mathrm{d}x \;\to\; 0\,. \tag{2.82}$$

Adding it with (2.78), one gets (2.75).

To prove (2.76), we need to show a convergence of ∇u_k to ∇u in a better mode than the weak one only. Let us denote

$$\mathfrak{a}_k(x) := \big(a(x,u_k(x),\nabla u_k(x)) - a(x,u_k(x),\nabla u(x))\big)\cdot\nabla\big(u_k(x)-u(x)\big). \tag{2.83}$$

By the monotonicity (2.65), it holds

$$\begin{aligned} 0 \;\leq\; & \limsup_{k\to\infty}\int_\Omega \mathfrak{a}_k(x)\,\mathrm{d}x = \limsup_{k\to\infty}\big\langle B(u_k,u_k)-B(u_k,u),u_k-u\big\rangle \\ =\; & \limsup_{k\to\infty}\big\langle A(u_k),u_k-u\big\rangle - \lim_{k\to\infty}\big\langle B(u_k,u),u_k-u\big\rangle \leq 0; \end{aligned} \tag{2.84}$$

note that the last limit superior is non-positive by assumption while the last limit equals zero by (2.75) with $v := u$. This implies that $\mathfrak{a}_k \to 0$ in the measure so that we can select a subsequence such that

$$\mathfrak{a}_k(x) \to 0 \tag{2.85}$$

for a.a. $x\in\Omega$. As $u_k \to u$ strongly in $L^{p^*-\epsilon}(\Omega)$, by Proposition 1.13(ii)–(iii) we can further select a subsequence that also

$$u_k(x) \to u(x) \tag{2.86}$$

for a.a. $x\in\Omega$. Take $x\in\Omega$ such that both (2.85) and (2.86) hold and also $\nabla u(x)$, $\nabla u_k(x)$, $k\in\mathbb{N}$, and $\gamma(x)$ from (2.55a) are finite, and $a(x,\cdot,\cdot)$ is continuous. If the sequence $\{\nabla u_k(x)\}_{k\in\mathbb{N}}$ would be unbounded, then the coercivity (2.68b) used for $s_0 = \nabla u(x)$ would yield $\limsup_{k\to\infty}(a(x,u_k(x),\nabla u_k(x)) - a(x,u_k(x),s_0))\cdot(\nabla u_k(x)-s_0) = +\infty$, which would contradict (2.85). Therefore, we can take a suitable $s\in\mathbb{R}^n$ and a (for a moment sub-) sequence such that $\nabla u_k(x)\to s$ in $\mathbb{R}^n$.

By (2.85) and (2.86) and the continuity of $a(x,\cdot,\cdot)$, cf. (2.54), we can pass to the limit in (2.83), which yields

$$\big(a(x,u(x),s)-a(x,u(x),\nabla u(x))\big)\cdot\big(s-\nabla u(x)\big)=0. \tag{2.87}$$

By the strict monotonicity (2.68a), we get $s=\nabla u(x)$. As s is determined uniquely, even the whole sequence $\{\nabla u_k(x)\}_{k\in\mathbb{N}}$ converges to s.[15] Then

$$c(u_k,\nabla u_k)\to c(u,\nabla u)\quad\text{a.e. in }\Omega. \tag{2.88}$$

By Hölder's inequality, for any measurable $S\subset\Omega$, we can estimate

$$\int_S\big|c(u_k,\nabla u_k)-c(u,\nabla u)\big|^{p^{*\prime}}\mathrm{d}x\le\big\|c(u_k,\nabla u_k)-c(u,\nabla u)\big\|_{L^{p^{*\prime}+\epsilon}(\Omega)}\mathrm{meas}_d(S)^{1+p^{*\prime}/\epsilon}. \tag{2.89}$$

Further, we realize that the sequence $\{c(u_k,\nabla u_k)-c(u,\nabla u)\}_{k\in\mathbb{N}}$ is bounded in $L^{p^{*\prime}+\epsilon}(\Omega)$ thanks to the assumption (2.68c). Thus (2.89) verifies the equi-absolute-continuity of the collection $\{|c(u_k,\nabla u_k)-c(u,\nabla u)|^{p^{*\prime}}\}_{k\in\mathbb{N}}$, cf. (1.28). By Dunford-Pettis' theorem 1.16(ii), $\{|c(u_k,\nabla u_k)-c(u,\nabla u)|^{p^{*\prime}}\}_{k\in\mathbb{N}}$ is also uniformly integrable and, since it converges to 0 a.e. due to (2.88), by Vitali's theorem 1.17, $|c(u_k,\nabla u_k)-c(u,\nabla u)|^{p^{*\prime}}\to 0$ in $L^1(\Omega)$, i.e.

$$c(u_k,\nabla u_k)\to c(u,\nabla u)\qquad\text{in }L^{p^{*\prime}}(\Omega). \tag{2.90}$$

As the limit $c(u,\nabla u)$ is determined uniquely, even the whole sequence (not only that one selected for (2.85)–(2.86)) must converge. Then (2.76) follows by joining (2.90) with (2.79). □

Note that, as always $p^{*\prime}+\epsilon>1$, (2.90) also proves that

$$c(u_k,\nabla u_k)\rightharpoonup c(u,\nabla u)\qquad\text{in }L^{p^{*\prime}+\epsilon}(\Omega). \tag{2.91}$$

By the same technique one can also prove $a(u_k,\nabla u_k)\rightharpoonup a(u,\nabla u)$ weakly in $L^{p'}(\Omega;\mathbb{R}^n)$. We however did not need this fact in the above proof.

Remark 2.34 (Critical growth in lower-order terms). The above theorem and its proof permits various modifications: If $b(x,\cdot)$ is monotone, then the splitting (2.73) can involve $b(u)$ instead of $b(w)$, which allows for borderline growth of b, i.e. (2.55b) with $\epsilon=0$. Similarly, if $c=\tilde{c}(x,r)$ as in (2.66) but with $\tilde{c}(x,\cdot)$ is monotone, then (2.73) can involve $c(u)$ instead of $c(w,\nabla w)$, and (2.55c) with $\epsilon=0$ suffices. Modification of the basic space V in these cases would allow for even a super-critical growth, cf. (2.128). The growth restriction can also be eliminated if a maximum principle, guaranting L^∞-estimates, is at our disposal, which unfortunately can be expected only in special cases like the equation $\Delta u=|\nabla u|^2$ or (6.73) below.

[15] The fact that we do not need to select a subsequence at every x in question is important because the set of all such x's should have the full measure in Ω and thus cannot be countable.

Lemma 2.35 (THE COERCIVITY (2.5)). *Let the following coercivity hold:*

$$\exists\varepsilon_1,\varepsilon_2>0,\ k_1\in L^1(\Omega):\ a(x,r,s)\cdot s+c(x,r,s)r\ge\varepsilon_1|s|^p+\varepsilon_2|r|^q-k_1(x),\quad (2.92a)$$
$$\exists c_1<+\infty\ \exists k_2\in L^1(\Gamma):\qquad b(x,r)r\ge -c_1|r|^{q_1}-k_2(x),\quad (2.92b)$$

for some $1<q_1<q\le p$. *Then* $A:W^{1,p}(\Omega)\to W^{1,p}(\Omega)^*$ *is coercive.*

Proof. We use the Poincaré inequality in the form (1.55), i.e. $\|u\|_{W^{1,p}(\Omega)}\le C_{\rm P}(\|\nabla u\|_{L^p(\Omega;\mathbb{R}^n)}+\|u\|_{L^q(\Omega)})$, which implies

$$\begin{aligned}\|u\|^q_{W^{1,p}(\Omega)} &\le 2^{q-1}C_{\rm P}^q\big(\|\nabla u\|^q_{L^p(\Omega;\mathbb{R}^n)}+\|u\|^q_{L^q(\Omega)}\big)\\ &\le C_{p,q}\big(1+\|\nabla u\|^p_{L^p(\Omega;\mathbb{R}^n)}+\|u\|^q_{L^q(\Omega)}\big).\end{aligned}\quad (2.93)$$

Also, by Young's inequality and boundedness of the trace operator[16] $u\mapsto u|_\Gamma: W^{1,p}(\Omega)\to L^q(\Gamma)$ (let N denote its norm), we use the estimate

$$\|u\|^{q_1}_{L^{q_1}(\Gamma)}=\int_\Gamma|u|^{q_1}\mathrm{d}S\le\int_\Gamma\varepsilon|u|^q+C_\varepsilon\mathrm{d}S\le\varepsilon N^q\|u\|^q_{W^{1,p}(\Omega)}+C_\varepsilon\mathrm{meas}_{n-1}(\Gamma)\quad (2.94)$$

with $\varepsilon>0$ arbitrarily small and $C_\varepsilon<+\infty$ chosen accordingly; cf. (1.22) with $q/q_1>1$ in place of p. Then (2.92) implies the estimate

$$\begin{aligned}\langle A(u),u\rangle &\ge \int_\Omega\big(\varepsilon_1|\nabla u|^p+\varepsilon_2|u|^q-k_1\big)\,\mathrm{d}x-\int_\Gamma\big(c_1|u|^{q_1}+k_2\big)\,\mathrm{d}S\\ &\ge \min(\varepsilon_1,\varepsilon_2)\bigg(\frac{\|u\|^q_{W^{1,p}(\Omega)}}{C_{p,q}}-1\bigg)-\big\|k_1\big\|_{L^1(\Omega)}\\ &\quad-\varepsilon N^q\big\|u\big\|^q_{W^{1,p}(\Omega)}-C_\varepsilon\mathrm{meas}_{n-1}(\Gamma)-\big\|k_2\big\|_{L^1(\Gamma)}.\end{aligned}\quad (2.95)$$

When one chooses $\varepsilon<\min(\varepsilon_1,\varepsilon_2)/(C_{p,q}N^q)$ and realizes that $q>1$, the coercivity (2.5) of A, i.e. $\lim_{\|u\|_{W^{1,p}(\Omega)}\to\infty}\langle A(u),u\rangle=+\infty$, is shown. □

Theorem 2.36 (LERAY-LIONS [257]). *Let* (2.54), (2.55), (2.57), (2.58), (2.65), *and* (2.92) *be valid and at least one of the conditions* (2.66) *or* (2.67) *or* (2.68) *be valid, then the boundary-value problem* (2.45)–(2.49) *has a weak solution.*

Proof. Lemmas 2.31, 2.32, and 2.35 proved $A:W^{1,p}(\Omega)\to W^{1,p}(\Omega)^*$ pseudomonotone and coercive. These properties are inherited by $A_0:V\to V^*$, cf. also Lemma 2.11(ii). Then we use Theorem 2.6 with Proposition 2.27. □

Remark 2.37 (Coercivity (2.68b)). Note that the coercivity (2.92a) together with (2.55a) and (2.68c) imply the coercivity (2.68b) because

$$a(x,r,s)\cdot(s-s_0)\ge\varepsilon_1|s|^p+\varepsilon_2|r|^q-k_1(x)-c(x,r,s)r-a(x,r,s)\cdot s_0\quad (2.96)$$

[16] Note that always $q\le p<p^\#$.

for such $x \in \Omega$ that $k_1(x)$ is finite. Realizing that $s \mapsto -c(x,r,s)r$ has a maximal decay as $-|s|^{(p-\epsilon)/p^{*\prime}}$ due to (2.68c) and $s \mapsto -a(x,r,s) \cdot s_0$ maximal decay as $-|s|^{p-1}$ due to (2.55a), the estimate (2.96) shows that $s \mapsto a(x,r,s) \cdot (s - s_0)$ has the p-growth uniformly with respect to r bounded because $\epsilon > 0$ and $p^{*\prime} \geq 1$.

Remark 2.38 (Necessity of monotonicity of $a(x,r,\cdot)$)**.** Boccardo and Dacorogna [55] showed that monotonicity of $a(x,r,\cdot)$ is necessary for pseudomonotonicity of the mapping $A(u) = -\mathrm{div}\, a(x,u,\nabla u)$.

Remark 2.39 (Necessity of Leray-Lions' condition (2.65), (2.68))**.** If a lower-order term c is present, the necessity of strict monotonicity of $a(x,r,\cdot)$ for the pseudomonotonicity was shown by Gossez and Mustonen [184].[17] It is worth observing that, for $c(x,r,\cdot)$ not affine, the mapping $u \mapsto c(u,\nabla u) : W^{1,p}(\Omega) \to W^{1,p}(\Omega)^*$, although representing a lower-order term, is neither totally continuous[18] nor pseudomonotone but it is still compact, cf. Exercise 2.64, and, when added to $u \mapsto -\mathrm{div}\, a(u,\nabla u)$, it may result in a pseudomonotone mapping.

Remark 2.40 (General right-hand sides)**.** The functional $f : v \mapsto \int_\Omega gv\,\mathrm{d}x + \int_{\Gamma_\mathrm{N}} hv\,\mathrm{d}S$ we considered, cf. (2.60), is not the general form of a functional $f \in W^{1,p}(\Omega)^*$. In fact, $W^{1,p}(\Omega)^*$ would allow g and h to be certain distributions on Ω and Γ, respectively. For example, if $p > n$, we have a dense and continuous embedding $W^{1,p}(\Omega) \subset C(\bar{\Omega})$ (resp. the trace operator $W^{1,p}(\Omega) \to C(\Gamma)$), henceforth the functional $f : v \mapsto \int_{\bar{\Omega}} v\,\mu(\mathrm{d}x) + \int_\Gamma v\,\eta(\mathrm{d}S)$ with *measures* $\mu \in \mathscr{M}(\bar{\Omega})$ and $\eta \in \mathscr{M}(\Gamma)$ still belongs to $W^{1,p}(\Omega)^*$. Since, in the case $p > n$, it holds that $p^* = p^\# = +\infty$, we, for the sake of simplicity, have considered (and will consider) only those measures μ and η which are absolutely continuous[19] in the presented text, except Sect. 3.2.5 below.

Convention 2.41 (Coercivity and a-priori estimates)**.** The coercivity estimate (2.95) is just the so-called basic a-priori estimate, obtained by the test by solution u itself. Contrary to (2.95), it is routine to organize the terms having a positive sign in the left-hand side (and to estimate them from below typically by Poincaré-type inequalities) while the other terms are put on the right-hand side (and to estimate them from above, e.g., by Hölder and Young inequalities).

---

[17]Indeed, the mere monotonicity in the main part, i.e. (2.54), and (2.55), (2.57), (2.58), and (2.65), cannot be sufficient for the pseudomonotonicity of A. The counterexample is as follows: take $c(x,r,s) \equiv c(s)$ with some $c : \mathbb{R}^n \to \mathbb{R}$ nonlinear, i.e. $\exists s_1, s_2 \in \mathbb{R}^n : \frac{1}{2}c(s_1) + \frac{1}{2}c(s_2) \neq c(\frac{1}{2}s_1 + \frac{1}{2}s_2)$ and take $a = 0$ at least on the line segment $[s_1, s_2]$. Then take a sequence $\{u_k\}_{k\in\mathrm{N}}$ such that ∇u_k is faster and faster oscillating between s_1 and s_2 (cf. Figure 3) on p.20 and $u_k(x) \to (\frac{1}{2}s_1 + \frac{1}{2}s_2) \cdot x$.

[18]Indeed, the mapping $u \mapsto c(u,\nabla u)$ is not totally continuous because it need not map weakly convergent sequences on strongly convergent ones. An example is as follows: take u_k with an oscillating gradient if k odd and affine $u_k(x) = (\frac{1}{2}s_1 + \frac{1}{2}s_2) \cdot x$ if k even, so that again $\{u_k\}_{k\in\mathrm{N}}$ converges weakly to this affine function but $\{c(\nabla u_k)\}_{k\in\mathrm{N}}$ does not converge at all if, e.g., $c(s) = |2s - s_1 - s_2|$.

[19]Those measures are known to have densities $g \in L^1(\Omega)$ and $h \in L^1(\Gamma)$, respectively.

2.4.4 Higher-order equations

The generalization of the 2nd-order equation to equations involving $2k$-order derivatives, $k \geq 2$, is often desirable. The corresponding boundary-value problems then involve k-boundary conditions, called either the Dirichlet one if they involve only derivatives up to $(k-1)$-order or the Neumann or the Newton one if they involve also derivatives of the order between k and $2k-1$. We present here briefly only quasilinear equations of the 4th order in a special[20] divergence form

$$\mathrm{div}\Big(\mathrm{div}\big(a(x,u,\nabla u,\nabla^2 u)\big)\Big) + c(x,u,\nabla u,\nabla^2 u) = g \tag{2.97}$$

in Ω, with $a : \Omega \times \mathbb{R} \times \mathbb{R}^n \times \mathbb{R}^{n\times n} \to \mathbb{R}^{n\times n}$ and $c : \Omega \times \mathbb{R} \times \mathbb{R}^n \times \mathbb{R}^{n\times n} \to \mathbb{R}$. Here $\nabla^2 u := \big[\frac{\partial^2}{\partial x_i \partial x_j} u\big]_{i,j=1}^n$. More in detail, (2.97) means

$$\sum_{i,j=1}^n \frac{\partial^2}{\partial x_i \partial x_j} a_{ij}\big(x,u,\nabla u,\nabla^2 u\big) + c\big(x,u,\nabla u,\nabla^2 u\big) = g. \tag{2.98}$$

Formulation of natural boundary conditions is more difficult than for the 2nd-order case. The weak formulation is created by multiplying (2.97) by a test function v, by integration over Ω, and by using Green's formula twice. Like in (2.50), this gives

$$\begin{aligned}\int_\Omega & a(x,u,\nabla u,\nabla^2 u) : \nabla^2 v + \big(c(x,u,\nabla u,\nabla^2 u) - g\big) v \,\mathrm{d}x \\ &= \int_\Gamma a(x,u,\nabla u,\nabla^2 u) : (\nu \otimes \nabla v) - \mathrm{div}\big(a(x,u,\nabla u,\nabla^2 u)\big)\cdot \nu\, v \,\mathrm{d}S. \end{aligned}\tag{2.99}$$

From this we can see that we must now cope with two boundary terms. In view of this, the *Dirichlet boundary conditions* look as

$$u|_\Gamma = u_{\mathrm{D}} \qquad \text{and} \qquad \frac{\partial u}{\partial \nu}\Big|_\Gamma = u'_{\mathrm{D}} \quad \text{on } \Gamma \tag{2.100}$$

with u_{D} and u'_{D} given. The weak formulation then naturally works with $v \in V := W_0^{2,p}(\Omega) = \{v \in W^{2,p}(\Omega);\ v|_\Gamma = \frac{\partial v}{\partial \nu}|_\Gamma = 0\}$ with $p > 1$ an exponent related to qualification of the highest-order nonlinearity $a(x,r,s,\cdot)$. This choice makes both boundary terms in (2.99) zero; note that $v|_\Gamma = 0$ makes also the tangential derivative of v zero at a.a. $x \in \Gamma$ hence $\frac{\partial v}{\partial \nu}|_\Gamma = 0$ yields $\nabla v(x) = 0$ on Γ.

By this argument, $v|_\Gamma = 0$ makes $\nabla v = \frac{\partial v}{\partial \nu}\nu$ on Γ and allows us to write

$$a(x,u,\nabla u,\nabla^2 u):(\nu\otimes\nabla v) = a(x,u,\nabla u,\nabla^2 u):\Big(\nu\otimes\frac{\partial v}{\partial \nu}\nu\Big) = \big(\nu^\top a(x,u,\nabla u,\nabla^2 u)\nu\big)\frac{\partial v}{\partial \nu}$$

and suggests that we formulate *Dirichlet/Newton boundary conditions* as

$$u|_\Gamma = u_{\mathrm{D}} \qquad \text{and} \qquad \nu^\top a(x,u,\nabla u,\nabla^2 u)\,\nu + b(x,u,\nabla u) = h \quad \text{on } \Gamma \tag{2.101}$$

[20] See Exercises 2.98 and 4.32 for a more general case.

with $u_{\rm D}$ and h given and $b:\Gamma\times\mathbb{R}\times\mathbb{R}^n\to\mathbb{R}$. This choice with $v|_\Gamma=0$ converts the boundary terms in (2.99) to $\int_\Gamma(h-b(x,u,\nabla u))\frac{\partial v}{\partial\nu}{\rm d}S$, which turns (2.99) just into the integral identity

$$\int_\Omega a(x,u,\nabla u,\nabla^2 u):\nabla^2 v+c(x,u,\nabla u,\nabla^2 u)v\,{\rm d}x \\ +\int_\Gamma b(x,u,\nabla u)\frac{\partial v}{\partial\nu}{\rm d}S=\int_\Omega gv\,{\rm d}x+\int_\Gamma h\frac{\partial v}{\partial\nu}{\rm d}S \tag{2.102}$$

forming the weak formulation provided the test-function space V is taken as $\{v\in W^{2,p}(\Omega);\ v|_\Gamma=0\}$.

If $v|_\Gamma$ is not fixed to zero, one must use a general decomposition $\nabla v=\frac{\partial v}{\partial\nu}\nu+\nabla_{\rm S}v$ on Γ with $\nabla_{\rm S}v=\nabla v-\frac{\partial v}{\partial\nu}\nu$ being the *tangential gradient* of v. On a smooth boundary Γ, one can use another (now $(n-1)$-dimensional) Green-type formula along the tangential spaces:[21]

$$\int_\Gamma a(x,u,\nabla u,\nabla^2 u):(\nu\otimes\nabla v)\,{\rm d}S \\ =\int_\Gamma\Big(\nu^\top a(x,u,\nabla u,\nabla^2 u)\nu\Big)\frac{\partial v}{\partial\nu}+a(x,u,\nabla u,\nabla^2 u):(\nu\otimes\nabla_{\rm S}v)\,{\rm d}S \\ =\int_\Gamma\Big(\nu^\top a(x,u,\nabla u,\nabla^2 u)\nu\Big)\frac{\partial v}{\partial\nu}-{\rm div}_{\rm S}\big(a(x,u,\nabla u,\nabla^2 u)\nu\big)v \\ +\big({\rm div}_{\rm S}\nu\big)\big(\nu^\top a(x,u,\nabla u,\nabla^2 u)\nu\big)v\,{\rm d}S \tag{2.103}$$

where ${\rm div}_{\rm S}:={\rm Tr}(\nabla_{\rm S})$ with ${\rm Tr}(\cdot)$ being the trace of a $(n-1)\times(n-1)$-matrix denotes the $(n-1)$-dimensional *surface divergence* so that ${\rm div}_{\rm S}\nu$ is (up to a factor $-\frac12$) the mean curvature of the surface Γ. Substituting it into (2.99), one obtains

$$\int_\Omega a(x,u,\nabla u,\nabla^2 u):\nabla^2 v+\big(c(x,u,\nabla u,\nabla^2 u)-g\big)v\,{\rm d}x \\ =\int_\Gamma\Big(\nu^\top a(x,u,\nabla u,\nabla^2 u)\nu\Big)\frac{\partial v}{\partial\nu}-\Big({\rm div}\big(a(x,u,\nabla u,\nabla^2 u)\big)\cdot\nu \\ +{\rm div}_{\rm S}\big(a(x,u,\nabla u,\nabla^2 u)\nu\big)-\big({\rm div}_{\rm S}\nu\big)\big(\nu^\top a(x,u,\nabla u,\nabla^2 u)\nu\big)\Big)v\,{\rm d}S. \tag{2.104}$$

This allows us to cast a natural *higher-order Dirichlet/Newton boundary condition:*

$$\frac{\partial u}{\partial\nu}\Big|_\Gamma=u'_{\rm D}\qquad\text{and} \tag{2.105a}$$

$${\rm div}\big(a(x,u,\nabla u,\nabla^2 u)\big)\cdot\nu+{\rm div}_{\rm S}\big(a(x,u,\nabla u,\nabla^2 u)\nu\big) \\ -\big({\rm div}_{\rm S}\nu\big)\big(\nu^\top a(x,u,\nabla u,\nabla^2 u)\nu\big)+b(x,u,\nabla u)=h\qquad\text{on }\Gamma. \tag{2.105b}$$

[21]This "surface" Green-type formula reads as $\int_\Gamma w:((\nabla_{\rm S}v)\otimes\nu)\,{\rm d}S=\int_\Gamma({\rm div}_{\rm S}\nu)(w:(\nu\otimes\nu))v-{\rm div}_{\rm S}(w\cdot\nu)v\,{\rm d}S$. In the vectorial variant, this is used in mechanics of complex (also called nonsimple) continua, cf. [153, 337, 407]. For even $2k$-order problems with $k>2$ see also [244].

The underlying Banach space is then considered as $V = \{v{\in}W^{2,p}(\Omega);\ \frac{\partial v}{\partial \nu} = 0\}$ and the weak formulation is based on the integral identity:

$$\int_\Omega a(x,u,\nabla u,\nabla^2 u):\nabla^2 v + c(x,u,\nabla u,\nabla^2 u)v\,\mathrm{d}x \\ + \int_\Gamma b(x,u,\nabla u)v\,\mathrm{d}S = \int_\Omega gv\,\mathrm{d}x + \int_\Gamma hv\,\mathrm{d}S. \qquad (2.106)$$

Eventually, the formula (2.104) reveals also the natural form of *Newton*-type *boundary conditions*:

$$\mathrm{div}\big(a(x,u,\nabla u,\nabla^2 u)\big)\cdot\nu + \mathrm{div}_\mathrm{S}\big(a(x,u,\nabla u,\nabla^2 u)\nu\big) \\ - \big(\mathrm{div}_\mathrm{S}\nu\big)\big(\nu^\top a(x,u,\nabla u,\nabla^2 u)\nu\big) + b_0(x,u,\nabla u) = h_0 \quad \text{and} \qquad (2.107a)$$

$$\nu^\top a(x,u,\nabla u,\nabla^2 u)\,\nu + b_1(x,u,\nabla u) = h_1 \quad \text{on } \Gamma. \qquad (2.107b)$$

The underlying Banach space can then be considered as $V = W^{2,p}(\Omega)$. The resulting weak formulation of the boundary-value problem (2.97)–(2.107) then employs the integral identity:

$$\int_\Omega a(x,u,\nabla u,\nabla^2 u):\nabla^2 v + c(x,u,\nabla u,\nabla^2 u)v\,\mathrm{d}x \\ + \int_\Gamma b_0(x,u,\nabla u)v + b_1(x,u,\nabla u)\frac{\partial v}{\partial \nu}\,\mathrm{d}S = \int_\Omega gv\,\mathrm{d}x + \int_\Gamma h_0 v + h_1\frac{\partial v}{\partial \nu}\,\mathrm{d}S. \quad (2.108)$$

For nonsmooth boundaries, these arguments based on formula (2.104) are no longer valid however and additional boundary terms can be seen; cf. [338] for boundaries with edges.

We will modify the Leray-Lions' Theorem 2.36 for the case of the Dirichlet conditions (2.100). Let us write naturally[22] $p^{**} := (p^*)^*$ and $p^{*\#} := (p^*)^\#$. For simplicity, the assumptions are not the most general in the following assertion, whose proof, paraphrasing that of Theorem 2.36, is omitted here.

Proposition 2.42 (Existence for Dirichlet problem). *Let $a(x,r,s,\cdot) : \mathbb{R}^{n\times n} \to \mathbb{R}^{n\times n}$ be strictly monotone,*

$$\exists k{\in}L^1(\Omega),\ 1{<}q{\le}p:\ a(x,r,s,S){:}S{+}c(x,r,s,S)r \ge \varepsilon|S|^p{+}\varepsilon|r|^q{-}k(x), \quad (2.109a)$$

$$\exists\gamma\in L^{p'}(\Omega):\qquad |a(x,r,s,S)| \le \gamma(x) + C|r|^{(p^{**}-\epsilon)/p'} \\ + C|s|^{(p^*-\epsilon)/p'} + C|S|^{p-1}, \qquad (2.109b)$$

$$\exists\gamma\in L^{p^{**\prime}+\epsilon}(\Omega):\quad |c(x,r,s,S)| \le \gamma(x) + C|r|^{p^{**}-\epsilon-1} \\ + C|s|^{(p^*-\epsilon)/p^{**\prime}} + C|S|^{(p-\epsilon)/p^{**\prime}}, \qquad (2.109c)$$

---

[22]This means $p^{**} = np/(n{-}2p)$ if $p < n/2$ or $p^{**} < +\infty$ if $p = n/2$ or $p^{**} = +\infty$ if $p > n/2$, cf. Corollary 1.22 for $k = 2$. For $p^{*\#} = (np{-}p)/(n{-}2p)$ if $p < n/2$, cf. Exercise 2.70.

with some $C\in\mathbb{R}^+$ *and* $\varepsilon,\epsilon>0$ *and again the Convention* 2.26 *(now concerning* $p^{**}=+\infty$ *for* $p>n/2$*), and let* $u_{\rm D}=v|_\Gamma$ *and* $u'_{\rm D}=\frac{\partial v}{\partial\nu}$ *for some* $v\in W^{2,p}(\Omega)$*, and* $g\in L^{p^{**\prime}}(\Omega)$*. Then the boundary-value problem* (2.97) *with* (2.100) *has a weak solution, i.e.* (2.99) *holds for all* $v\in W^{2,p}_0(\Omega)$ *together with the boundary conditions* (2.97).

For the Newton boundary conditions (2.107), the analog of the existence assertion looks as follows:

Proposition 2.43 (Existence for Newton problem). *Let* a, c, *and* g *be as in Proposition 2.42 and satisfy (2.109), and let* b_0 *and* b_1 *satisfy*

$$\exists k\in L^1(\Gamma):\qquad b_0(x,r,s)r+b_1(x,r,s)(s\cdot\nu(x))\ge -k(x),\tag{2.110a}$$
$$\exists\gamma\in L^{p^{*\#\prime}}(\Gamma):\qquad \big|b_0(x,r,s)\big|\le\gamma(x)+C|r|^{p^{*\#}-\epsilon-1}+C|s|^{(p^{\#}-\epsilon)/p^{*\#}},\tag{2.110b}$$
$$\exists\gamma\in L^{p^{\#\prime}}(\Gamma):\qquad \big|b_1(x,r,s)\big|\le\gamma(x)+C|r|^{(p^{*\#}-\epsilon)/p^{\#\prime}}+C|s|^{p^{\#}-\epsilon-1}\tag{2.110c}$$

with some $C\in\mathbb{R}^+$ *and* $\epsilon>0$*, and let* $h_0\in L^{p^{*\#\prime}}(\Gamma)$ *and* $h_1\in L^{p^{\#\prime}}(\Gamma)$*. Then the boundary-value problem* (2.97) *with* (2.107) *has a weak solution, i.e.* (2.108) *holds for all* $v\in W^{2,p}_0(\Omega)$.

The modification for other boundary conditions (2.101) or (2.105) can easily be cast and is left as an exercise.

As pointed out before, one should care about consistency and selectivity of the definitions of weak solutions. Consistency is guaranteed by the derivation of the weak formulation itself. Let us illustrate the selectivity, i.e. an analog of Proposition 2.25, on the most complicated case of the Newton boundary-value problem:

Proposition 2.44 (Selectivity of the weak-solution definition). *Let* Γ *be smooth,* $a\in C^2(\bar\Omega\times\mathbb{R}\times\mathbb{R}^n\times\mathbb{R}^{n\times n};\mathbb{R}^{n\times n})$, $c\in C^0(\bar\Omega\times\mathbb{R}\times\mathbb{R}^n\times\mathbb{R}^{n\times n})$*, and* $b_0,b_1\in C^0(\bar\Gamma\times\mathbb{R}\times\mathbb{R}^n)$, $g\in C(\bar\Omega)$*, and* $h_0,h_1\in C(\Gamma)$*. Then any weak solution* $u\in C^4(\bar\Omega)$ *of the boundary-value problem* (2.97) *with* (2.107) *is the also classical solution.*

Proof. Put $v\in V=W^{2,p}(\Omega)$ into (2.108) and use Green's formula (1.54) twice, as well as the surface Green formula (2.103). One gets

$$\begin{aligned}&\int_\Omega\Big({\rm div}^2a(x,u,\nabla u)+c(x,u,\nabla u)-g\Big)v\,{\rm d}x\\&\quad+\int_\Gamma\Big({\rm div}\big(a(x,u,\nabla u,\nabla^2u)\big)\cdot\nu+{\rm div}_{\rm S}\big(a(x,u,\nabla u,\nabla^2u)\nu\big)\\&\qquad-({\rm div}_{\rm S}\nu)\big(\nu^\top a(x,u,\nabla u,\nabla^2u)\nu\big)+b_0(x,u,\nabla u)-h_0\Big)v\,{\rm d}S\\&\quad+\int_\Gamma\Big(\nu^\top a(x,u,\nabla u,\nabla^2u)\,\nu+b_1(x,u,\nabla u)-h_1\Big)\frac{\partial v}{\partial\nu}\,{\rm d}S=0.\end{aligned}\tag{2.111}$$

Considering v with a compact support in Ω, one has $v|_\Gamma = 0 = \frac{\partial v}{\partial\nu}$ and both boundary integrals in (2.111) vanish. As v is otherwise arbitrary, one deduces that (2.97) holds a.e., and hence even everywhere in Ω due to the assumed smoothness of a and c. Hence, the first integral in (2.111) vanishes. Then, put a more general $v\in V$ into (2.111) but still such that $\frac{\partial v}{\partial\nu} = 0$. Thus the second boundary integral in (2.111) vanishes. From the first boundary integral, we recover the boundary condition (2.107a).[23] Due to the assumed smoothness of a and continuity of b_0 and h_0, (2.107a) holds pointwise. Finally, we can take $v \in V$ fully general. Knowing already that the first and the second integral in (2.111) vanish, from the last integral we can recover the remaining boundary condition (2.107b).[24] □

Remark 2.45 (Other boundary conditions). The above four combinations of boundary conditions still do not represent the whole class of variationally consistent boundary conditions for equation (2.97). For $\alpha_0, \alpha_1 \in L^\infty(\Gamma)$, one can consider a combined condition composed from (2.101) and (2.105), namely

$$\alpha_1\frac{\partial u}{\partial\nu} + \alpha_0 u = u_{\rm D} \qquad \text{and} \tag{2.112a}$$

$$\begin{aligned}&\alpha_1\Big(\mathrm{div}\big(a(x,u,\nabla u,\nabla^2 u)\big){\cdot}\nu + \mathrm{div}_{\rm S}\big(a(x,u,\nabla u,\nabla^2 u)\nu\big)\\ &\qquad - \big(\mathrm{div}_{\rm S}\nu\big)\big(\nu^\top a(x,u,\nabla u,\nabla^2 u)\nu\big)\Big)\\ &\qquad + \alpha_0\nu^\top a(x,u,\nabla u,\nabla^2 u)\,\nu + b(x,u,\nabla u) = h \qquad \text{on } \Gamma.\end{aligned} \tag{2.112b}$$

The underlying Banach space is then $V = \{v \in W^{2,p}(\Omega);\ \alpha_1\frac{\partial u}{\partial\nu} + \alpha_0 u = 0\}$ and the weak formulation is again (2.106) with $b = b_2/\alpha_1$ and $h = h_2/\alpha_1$ provided $\alpha_1 \neq 0$. Alternatively, for $\alpha_0 \neq 0$ one can rather pursue the weak formulation based on (2.102).

Example 2.46 (p-biharmonic operator). A concrete choice of a from (2.97)

$$a_{ij}(x,r,s,S) := \begin{cases} \big|\sum_{k=1}^n S_{kk}\big|^{p-2}\sum_{k=1}^n S_{kk} & \text{for } i=j,\\ 0 & \text{for } i\neq j,\end{cases} \tag{2.113}$$

converts $\mathrm{div}\,\mathrm{div}\,a(x,u,\nabla u,\nabla^2 u)$ into the so-called p-biharmonic operator $\Delta(|\Delta u|^{p-2}\Delta u)$. Applying Green's formula twice to this operator tested by v yields the identity

[23] Here the important fact is that the set $\{v|_\Gamma;\ v \in V,\ \frac{\partial v}{\partial\nu} = 0\}$ is still dense in $L^1(\Gamma)$. Indeed, any $v \in W^{1,2}(\Omega)$ can be modified to u_ε so that $(v_\varepsilon - v)|_\Gamma$ is small but $\frac{\partial}{\partial\nu}v_\varepsilon = 0$ on Γ. To outline this procedure, first we rectify Γ locally so that we can consider a half-space, cf. Fig. 8 on p. 91 below, then extend v by reflection of v with respect to Γ, and eventually mollify the extended v.

[24] Here the important fact is that the set $\{\frac{\partial v}{\partial\nu};\ v \in V\}$ is dense in $L^1(\Gamma)$, which can be seen by a local rectification of Γ and by an explicit construction of v in the vicinity of Γ with a given smooth $\frac{\partial v}{\partial\nu}$ and, e.g., zero trace on Γ.

$$
\begin{aligned}
&\int_\Omega \Delta(|\Delta u|^{p-2}\Delta u)v\,\mathrm{d}x \\
&\quad = -\int_\Omega \nabla(|\Delta u|^{p-2}\Delta u)\cdot\nabla v\,\mathrm{d}x + \int_\Gamma \frac{\partial}{\partial\nu}(|\Delta u|^{p-2}\Delta u)\,v\,\mathrm{d}S \\
&\quad = \int_\Omega |\Delta u|^{p-2}\Delta u\Delta v\,\mathrm{d}x + \int_\Gamma \frac{\partial}{\partial\nu}(|\Delta u|^{p-2}\Delta u)\,v - |\Delta u|^{p-2}\Delta u\frac{\partial v}{\partial\nu}\,\mathrm{d}S, \qquad (2.114)
\end{aligned}
$$

from which, besides the Dirichlet conditions (2.100), one can pose naturally also Dirichlet/Newton conditions (2.101) now in the form

$$
u|_\Gamma = u_\mathrm{D} \qquad \text{and} \qquad |\Delta u|^{p-2}\Delta u + b(x,u,\nabla u) = h, \qquad (2.115)
$$

or the higher Dirichlet/Newton conditions (2.105) now in the simpler form

$$
\frac{\partial u}{\partial\nu}\Big|_\Gamma = u'_\mathrm{D} \qquad \text{and} \qquad \frac{\partial}{\partial\nu}(|\Delta u|^{p-2}\Delta u) + b(x,u,\nabla u) = h, \qquad (2.116)
$$

or also the Newton condition (2.107) now in the simpler form

$$
\frac{\partial}{\partial\nu}(|\Delta u|^{p-2}\Delta u) + b_0(x,u,\nabla u) = h_0, \quad |\Delta u|^{p-2}\Delta u + b_1(x,u,\nabla u) = h_1. \qquad (2.117)
$$

Note that (2.116) and (2.117) do not contain the div_S-terms because, instead of $\nu\otimes\nabla v$ in (2.99), one has $\nu\cdot\nabla v = \frac{\partial v}{\partial\nu}$ in (2.114). The pointwise coercivity (2.109a) cannot be satisfied for (2.113), however, and the coercivity of A on V must rely on a delicate interplay with the boundary conditions. E.g., for Dirichlet conditions (2.100) with $u_\mathrm{D} = 0 = u'_\mathrm{D}$ and for $p = 2$, one has by using Green's formula twice $\langle A(u),u\rangle = \int_\Omega |\Delta u|^2\mathrm{d}x = -\int_\Omega \nabla u\cdot\nabla\Delta u\mathrm{d}x = -\int_\Omega \nabla u\cdot\mathrm{div}(\nabla^2 u)\,\mathrm{d}x = \int_\Omega |\nabla^2 u|^2\mathrm{d}x$, which thus controls $\nabla^2 u$ in $L^2(\Omega;\mathbb{R}^{n\times n})$. Another example is the Newton's condition (2.117) with $b_0(x,r,s) = \beta_0(x)r$, $b_1(x,r,s) = -\beta_1(x)(s\cdot\nu)$, and $p = 2$, one has $\langle A(u),u\rangle = \int_\Omega |\Delta u|^2\mathrm{d}x + \int_\Gamma \beta_0 u^2 + \beta_1(\frac{\partial}{\partial\nu}u)^2\mathrm{d}S$. This is a continuous quadratic form on $W^{2,2}(\Omega)$ and for the Poincaré-like inequality $\langle A(u),u\rangle \ge C_\mathrm{P}\|u\|^2_{W^{2,2}(\Omega)}$ it suffices to guarantee that $\langle A(u),u\rangle = 0$ implies $u = 0$. This can be done by assuming $\beta_0,\beta_1 \ge 0$, and β_0 or β_1 positive on a "sufficiently large part" of Γ.[25]

2.5 Weakly continuous mappings, semilinear equations

In case that A is coercive and, instead of being pseudomonotone, is weakly continuous, we can prove existence of a solution to $A(u) = f$ much more easily. Although the assumption of the weak continuity is restrictive, such mappings enjoy still a considerably large application area. Here, we can even advantageously generalize the concept for mappings $A : V \to Z^*$ for some Banach space $Z \subset V$ densely so that $Z^* \supset V^*$. If $V_k \subset Z$ for any $k \in \mathbb{N}$, we can modify (2.5) and then Theorem 2.6:

[25] Here, a certain caution is advisable: e.g. for Ω a square $[0,1]^2$, it is not sufficient if $\beta_0(\cdot) = 1$ on the sides with $x_1 = 0$ and $x_2 = 0$ and otherwise β_0 and β_1 vanishes because of existence of a non-vanishing function $u(x) = x_1x_2$ for which $\langle A(u),u\rangle = 0$.

Proposition 2.47 (Existence). *If a weakly continuous mapping $A : V \to Z^*$ is coercive in the modified sense*

$$\lim_{\substack{\|v\|_V \to \infty \\ v \in Z}} \frac{\langle A(v), v\rangle_{Z^* \times Z}}{\|v\|_V} = +\infty, \tag{2.118}$$

and if $f \in V^*$, then the equation $A(u) = f$ has a solution.

Proof. The technique of the proof of Theorem 2.6 allows for a very simple modification: instead of (2.14), we consider the Galerkin identity (2.8) as $\langle A(u_k) - f, v_k\rangle_{Z^* \times Z} = 0$ for $v_k \in V_k$ such that $v_k \to v$ in Z, and make a direct limit passage. Note that (2.13) looks now as

$$\zeta(\|u_k\|_V)\|u_k\|_V \leq \langle A(u_k), u_k\rangle_{Z^* \times Z} = \langle f, u_k\rangle_{Z^* \times Z} = \langle f, u_k\rangle_{V^* \times V} \leq \|f\|_{V^*}\|u_k\|_V$$

and again yields $\{u_k\}_{k \in \mathbb{N}}$ bounded in V because $f \in V^*$. □

Confining ourselves again to the 2nd-order problems as in Sections 2.4.1–2.4.3, we can easily use this concept for the special case when $a(x, r, \cdot) : \mathbb{R}^n \to \mathbb{R}^n$ and $c(x, r, \cdot) : \mathbb{R}^n \to \mathbb{R}$ are affine, we will call such problems as *semilinear* although sometimes this adjective needs still $a(x, \cdot, s)$ constant as in (0.1). So, here

$$a_i(x, r, s) := \sum_{j=1}^{n} a_{ij}(x, r)s_j + a_{i0}(x, r), \quad i = 1, \ldots, n, \tag{2.119a}$$

$$c(x, r, s) := \sum_{j=1}^{n} c_j(x, r)s_j + c_0(x, r), \tag{2.119b}$$

with $a_{ij}, c_j : \Omega \times R \to \mathbb{R}$ Carathéodory mappings whose growth is now to be designed to induce the Nemytskiĭ mappings $\mathscr{N}_{(a_{i1},\ldots,a_{in})}, \mathscr{N}_{(c_1,\ldots,c_n)} : L^{2^*-\epsilon}(\Omega) \to L^2(\Omega; \mathbb{R}^n)$ and $\mathscr{N}_{a_{i0}}, \mathscr{N}_{c_0} : L^{2^*-\epsilon}(\Omega) \to L^1(\Omega)$ with $\epsilon > 0$. Besides, the boundary nonlinearity $b : \Gamma \times \mathbb{R} \to \mathbb{R}$ is now to induce the Nemytskiĭ mapping $\mathscr{N}_b : L^{2^\# - \epsilon}(\Gamma) \to L^1(\Gamma)$. This means, for $i, j = 1, \ldots, n$,

$$\exists \gamma_1 \in L^2(\Omega),\ C \in \mathbb{R}: \quad |a_{ij}(x, r)| \leq \gamma_1(x) + C|r|^{(2^*-\epsilon)/2},$$
$$|c_j(x, r)| \leq \gamma_1(x) + C|r|^{(2^*-\epsilon)/2}, \tag{2.120a}$$

$$\exists \gamma_2 \in L^1(\Omega),\ C \in \mathbb{R}: \quad |a_{i0}(x, r)| \leq \gamma_2(x) + C|r|^{2^*-\epsilon},$$
$$|c_0(x, r)| \leq \gamma_2(x) + C|r|^{2^*-\epsilon}, \tag{2.120b}$$

$$\exists \gamma_3 \in L^1(\Gamma),\ C \in \mathbb{R}: \quad |b(x, r)| \leq \gamma_3(x) + C|r|^{2^\#-\epsilon}. \tag{2.120c}$$

The exponent $p = 2$ is natural because $a(x, r, \cdot)$ has now a linear growth. Note that these requirements just guarantee that all integrals in (2.51) have a good sense if $v \in W^{1,\infty}(\Omega) =: Z$. Again, Convention 2.26 on p. 46 is considered.

Lemma 2.48 (Weak continuity of A). *Let* (2.119)–(2.120) *hold. Then A is weakly* continuous as a mapping $W^{1,2}(\Omega) \to W^{1,\infty}(\Omega)^*$.*

Proof. Having a weakly convergent sequence $\{u_k\}_{k\in\mathbb{N}}$ in $W^{1,2}(\Omega)$, this sequence converges strongly in $L^{2^*-\epsilon}(\Omega)$. Then, by the continuity of the Nemytskiĭ mappings $\mathscr{N}_{(a_{i1},\dots,a_{in})}, \mathscr{N}_{(c_1,\dots,c_n)} : L^{2^*-\epsilon}(\Omega) \to L^2(\Omega;\mathbb{R}^n)$ and $\mathscr{N}_{a_{i0}}, \mathscr{N}_{c_0} : L^{2^*-\epsilon}(\Omega) \to L^1(\Omega)$, it holds that

$$\lim_{k\to\infty} \int_\Omega \sum_{i=1}^n \Big(\sum_{j=1}^n a_{ij}(u_k)\frac{\partial u_k}{\partial x_j} + a_{i0}(u_k)\Big)\frac{\partial v}{\partial x_i} + \Big(\sum_{j=1}^n c_j(u_k)\frac{\partial u_k}{\partial x_j} + c_0(u_k)\Big)v\,\mathrm{d}x$$
$$= \int_\Omega \sum_{i=1}^n \Big(\sum_{j=1}^n a_{ij}(u)\frac{\partial u}{\partial x_j} + a_{i0}(u)\Big)\frac{\partial v}{\partial x_i} + \Big(\sum_{j=1}^n c_j(u)\frac{\partial u}{\partial x_j} + c_0(u_k)\Big)v\,\mathrm{d}x$$

for $k \to \infty$ and any $v \in W^{1,\infty}(\Omega)$. Also $u_k|_\Gamma \to u|_\Gamma$ in $L^{2^\#-\epsilon}(\Gamma)$, and, by (2.120c), we have convergence in the boundary term $\int_\Gamma b(u_k)v\,\mathrm{d}S \to \int_\Gamma b(u)v\,\mathrm{d}S$. □

Proposition 2.49 (Existence of weak solutions). *Let* (2.119)–(2.120) *hold, $g \in L^{2^{*\prime}}(\Omega)$, $h \in L^{2^{\#\prime}}(\Gamma)$, and, for some $\varepsilon > 0$, $\gamma_1 \in L^2(\Omega)$, $\gamma_2 \in L^1(\Omega)$, $\gamma_3 \in L^1(\Gamma)$, and for a.a. $x \in \Omega$ (resp. $x \in \Gamma$ for* (2.121b)*) and all $(r,s) \in \mathbb{R}^{1+n}$, it holds that*

$$\sum_{i=1}^n \Big(\sum_{j=1}^n a_{ij}(x,r)s_j + a_{i0}(x,r)\Big)s_i + \Big(\sum_{j=1}^n c_j(x,r)s_j + c_0(x,r)\Big)r$$
$$\geq \varepsilon|s|^2 + \varepsilon|r|^2 - \gamma_1(x)|s| - \gamma_2(x), \qquad (2.121a)$$
$$b(x,r)r \geq -\gamma_3(x). \qquad (2.121b)$$

Then the boundary-value problem (2.45) *with* (2.49) *has a weak solution in the sense of Definition* 2.24 *using now $v \in W^{1,\infty}(\Omega)$.*

Proof. We can use the abstract Proposition 2.47 now with $V := W^{1,2}(\Omega)$, $Z := W^{1,\infty}(\Omega)$, and V_k some finite-dimensional subspaces of $W^{1,\infty}(\Omega)$ satisfying (2.7).[26] The coercivity (2.118) is implied by (2.121) by routine calculations.[27] Then we use Lemma 2.48 and Proposition 2.47. □

Remark 2.50 (Conventional weak solutions). Let, in addition to the assumptions of Proposition 2.49, also the growth condition (2.55) with $p = 2$ hold. Then the solution obtained in Proposition 2.49 allows for $v \in W^{1,2}(\Omega)$ in Definition 2.24.

---

[26] Such subspaces always exists, e.g. one can imagine subspaces as in Example 2.67.

[27] We have $\langle A(v), v\rangle = \int_\Omega \sum_{i=1}^n \big(\sum_{j=1}^n a_{ij}(v)\frac{\partial}{\partial x_j}v + a_{i0}(v)\big)\frac{\partial}{\partial x_i}v + \big(\sum_{j=1}^n c_j(v)\frac{\partial}{\partial x_j}v + c_0(v)\big)v\,\mathrm{d}x + \int_\Gamma b(v)v\,\mathrm{d}S \geq \varepsilon\|\nabla v\|^2_{L^2(\Omega;\mathbb{R}^n)} - \frac{1}{2\varepsilon}\|\gamma_1\|^2_{L^2(\Omega)} - \frac{\varepsilon}{2}\|\nabla v\|^2_{L^2(\Omega;\mathbb{R}^n)} - \|\gamma_2\|_{L^1(\Omega)} - \gamma_3\|v\|^1_{L^1(\Gamma)}$.

2.6 Examples and exercises

This section contains both exercises to make the above presented theory more complete and some examples of analysis of concrete semi- and quasi-linear equations. The exercises will mostly be accompanied by brief hints in the footnotes.

2.6.1 General tools

Exercise 2.51 (*Banach's selection principle)*. Assuming the sequential compactness of closed bounded intervals in $\mathbb{R}$ is known, prove Banach's Theorem 1.7 by a suitable diagonalization procedure.[28]

Exercise 2.52 (*Uniform convexity of Hilbert spaces)*. For V being a Hilbert space, prove the assertion of Theorem 1.2 directly.[29] Using (1.4), prove that any Hilbert space is uniformly convex.[30]

Exercise 2.53 (*Pseudomonotonicity)*. Assuming (2.3a), show that (2.3b) is equivalent to[31]

$$\left.\begin{array}{r} u_k \rightharpoonup u, \\ \limsup\limits_{k\to\infty}\big\langle A(u_k), u_k{-}u\big\rangle \le 0 \end{array}\right\} \quad\Rightarrow\quad \left\{\begin{array}{l} \operatorname*{w-lim}\limits_{k\to\infty} A(u_k) = A(u), \\ \lim\limits_{k\to\infty}\big\langle A(u_k), u_k\big\rangle = \big\langle A(u), u\big\rangle. \end{array}\right. \tag{2.122}$$

Exercise 2.54 (Weakening of pseudomonotonicity). Modify the proof of Brézis Theorem 2.6 for A coercive, bounded, demicontinuous, and satisfying[32]

$$\left.\begin{array}{l} u_k \rightharpoonup u \quad \& \quad A(u_k) \rightharpoonup f \\ \limsup\limits_{k\to\infty}\big\langle A(u_k), u_k\big\rangle \le \langle f, u\rangle \end{array}\right\} \quad\Rightarrow\quad f = A(u). \tag{2.123}$$

---

[28]Hint: Consider a sequence $\{f_k\}_{k\in\mathbb{N}}$ bounded in V^* and a countable dense subset $\{v_k\}_{k\in\mathbb{N}}$ in V, take v_1 and select an infinite subset $A_1 \subset \mathbb{N}$ such that the sequence of real numbers $\{\langle f_k, v_1\rangle\}_{k\in A_1}$ converges in $\mathbb{R}$ to some $f(v_1)$, then take v_2 and select an infinite subset $A_2 \subset A_1$ such that $\{\langle f_k, v_2\rangle\}_{k\in A_2}$ converges to some $f(v_2)$, etc. for v_3, v_4, ... Then make a diagonalization procedure by taking l_k the first number in A_k which is greater than k. Then $\{\langle f_{l_k}, v_i\rangle\}_{k\in\mathbb{N}}$ converge to $f(v_i)$ for all $i \in \mathbb{N}$. Show that f is linear on span($\{v_i\}_{i\in\mathbb{N}}$) and bounded because $|f(v_i)| \le \lim_{k\to\infty} |\langle f_{l_k}, v_i\rangle| \le \limsup_{k\to\infty} ||f_k||_*||v_i||$, and finally extend f on the whole V^* just by continuity.

[29]Hint: $\|u_k\|\to\|u\|$ and $u_k\rightharpoonup u$ imply $\|u_k-u\|^2 = \|u_k\|^2 + (u-2u_k, u) \to \|u\|^2 + (u-2u, u) = 0$.

[30]Hint: Realize that $\|u\| = 1 = \|v\|$ and $\|u - v\| \ge \varepsilon$ in (1.5) imply $\frac{1}{2}\|u + v\| = \sqrt{(u, v) - \|u-v\|^2/4} \le \sqrt{1 - \varepsilon^2/4} \le 1 - \delta$ provided $0 < \delta \le 1$ solves $\delta^2 - 2\delta + \varepsilon^2/4 = 0$. Such δ exists if $0 < \varepsilon \le 2$, while for $\varepsilon > 2$ the implication (1.5) is trivial.

[31]Hint: (2.122)⇒(2.3b) is trivial. The converse implication: by (2.3a), assume $A(u_k) \rightharpoonup f$ (a subsequence), then $0 \ge \limsup_{k\to\infty}\langle A(u_k), u_k{-}u\rangle = \limsup_{k\to\infty}\langle A(u_k), u_k\rangle - \langle f, u\rangle$ implies $\langle A(u), u - v\rangle \le \liminf_{k\to\infty}\langle A(u_k), u_k{-}v\rangle \le \limsup_{k\to\infty}\langle A(u_k), u_k\rangle - \langle f, v\rangle \le \langle f, u{-}v\rangle$, from which $A(u) = f$, hence $A(u_k) \rightharpoonup f$ (the whole sequence), and eventually (2.3b) for $v = 0$ yields

$$\langle A(u), u\rangle \le \liminf_{k\to\infty}\langle A(u_k), u_k\rangle \le \limsup_{k\to\infty}\langle A(u_k), u_k\rangle \le \lim_{k\to\infty}\langle A(u_k), u\rangle = \langle A(u), u\rangle.$$

[32]Hint: Modify Step 4 of the proof of Theorem 2.6: as both $\{u_k\}_{k\in N}$ and A are bounded, $A(u_k) \rightharpoonup \chi$ (as a subsequence) and, from (2.8), $\chi = f$, hence $A(u_k) \rightharpoonup f$ (the whole sequence) and, again by (2.8), $\langle A(u_k), u_k\rangle = \langle f, u_k\rangle \to \langle f, u\rangle$. Then by (2.123) $f = A(u)$.

Show that any pseudomonotone A satisfies (2.123).[33]

Exercise 2.55 (Tikhonov-type modification[34] of Schauder's Theorem 1.9). Assuming a reflexive separable Banach space $V \Subset V_1$, show that a weakly continuous mapping $M : V \to V$ which maps a ball B in V into itself has a fixed point.[35]

Exercise 2.56 (Direct method for A weakly continuous). Assume $A : V \to V^*$ weakly continuous, V Hilbert, and modify the Brézis Theorem 2.6 by using directly Schauder fixed-point Theorem 1.9 without approximating the problem.[36]

Exercise 2.57. Try to make a limit passage in (2.38)–(2.39) simultaneously in i and l by considering $i = l$. Realize why it was necessary to make the double limit $\lim_{l\to\infty} \lim_{i\to\infty}$ instead of $\lim_{l=i\to\infty}$ in the proof of Proposition 2.17.

Exercise 2.58. Assuming $1 \le q \le p < +\infty$, evaluate the norms of the continuous embeddings $L^\infty(\Omega) \subset L^p(\Omega) \subset L^q(\Omega)$.[37]

Exercise 2.59 (Interpolation of Lebesgue spaces). Prove (1.23) by using Hölder's inequality.[38]

Exercise 2.60 (Continuity of Nemytskiĭ mappings). Show that the Nemytskiĭ mapping $\mathscr{N}_a$ with a satisfying (1.48) is a bounded continuous mapping $L^{p_1}(\Omega) \times$

[33]Hint: The premise of (2.123) and the pseudomonotonicity implies $\limsup_{k\to\infty}\langle A(u_k), u_k - u\rangle = \limsup_{k\to\infty}\langle A(u_k), u_k\rangle - \lim_{k\to\infty}\langle A(u_k), u\rangle \le \langle f, u\rangle - \langle f, u\rangle = 0$ so that, by (2.3b), $\langle A(u), u - v\rangle \le \liminf_{k\to\infty}\langle A(u_k), u_k - v\rangle \le \limsup_{k\to\infty}\langle A(u_k), u_k - v\rangle \le \langle f, u - v\rangle$ for any $v \in V$, from which $f = A(u)$ indeed follows.

[34]Tikhonov [413] proved a bit more general assertion, known now as Tikhonov's theorem: a continuous mapping from a compact subset of a locally convex space into itself has a fixed point.

[35]Hint: Consider B endowed with a weak topology, realize that $u_k \to u$ in V_1 and $u_k \in B$ implies $u_k \rightharpoonup u$ in B, hence $M(u_k) \rightharpoonup M(u)$ in V and then also $M(u_k) \to M(u)$ in V_1, and then use Schauder's Theorem 1.9.

[36]Hint: Repeat Step 2 of the proof of Brézis Theorem 2.6 directly for V instead of V_k. Use the weak topology on $\{v \in V;\ \|v\| \le \varrho\}$, and realize that I_k is to be omitted while J_k^{-1} is to be weakly continuous (which really is due to its demicontinuity, cf. Corollary 3.3 below, and its linearity, cf. Remark 3.10). Also use Exercise 2.55.

[37]Hint: Estimate

$$\|u\|_{L^p(\Omega)} = \sqrt[p]{\int_\Omega |u|^p \mathrm{d}x} \le \sqrt[p]{\int_\Omega \operatorname*{ess\,sup}_{\xi\in\Omega} |u(\xi)|^p \mathrm{d}x} = \sqrt[p]{\|u\|^p_{L^\infty(\Omega)} \int_\Omega 1\mathrm{d}x} = N\|u\|_{L^\infty(\Omega)}$$

with $N = \left(\mathrm{meas}_n(\Omega)\right)^{1/p}$ being the norm of the embedding $L^\infty(\Omega) \subset L^p(\Omega)$. Likewise, by Hölder's inequality,

$$\|u\|^q_{L^q(\Omega)} = \int_\Omega 1 \cdot |u|^q \mathrm{d}x \le \sqrt[(p/q)']{\int_\Omega 1\mathrm{d}x}\ \sqrt[p/q]{\int_\Omega |u|^p \mathrm{d}x} = N^q \|u\|^q_{L^p(\Omega)}$$

with $N = \left(\mathrm{meas}_n(\Omega)\right)^{(p-q)/(pq)}$.

[38]Hint: Use Hölder's inequality for

$$\int_\Omega |v|^p \mathrm{d}x = \int_\Omega |v|^{\lambda p}|v|^{(1-\lambda)p} \mathrm{d}x \le \Big(\int_\Omega |v|^{\lambda p\alpha} \mathrm{d}x\Big)^{1/\alpha} \Big(\int_\Omega |v|^{(1-\lambda)p\beta} \mathrm{d}x\Big)^{1/\beta}$$

with a suitable $\alpha = p_1/(\lambda p)$ and $\beta = p_2/((1-\lambda)p)$, namely $\alpha^{-1} + \beta^{-1} = 1$ which just means that p satisfies the premise in (1.23).

$L^{p_2}(\Omega;\mathbb{R}^n) \to L^{p_0}(\Omega)$.[39]

Exercise 2.61. Show that p_3 in Theorem 1.27 indeed cannot be $+\infty$: find some a satisfying (1.48) for $p_1, p_2 < +\infty$ and $p_3 = +\infty$ such that $\mathscr{N}_a$ is not continuous.[40]

Exercise 2.62. Show that, for any $c : \mathbb{R}^{m_1} \to \mathbb{R}^{m_2}$ not affine, the Nemytskiĭ mapping $\mathscr{N}_c : u \mapsto c(u)$ is not weakly continuous; modify Figure 3 on p.20.[41]

Exercise 2.63 (Integral balance (2.63))**.** Consider the test volume in the integral balance (2.63) as a ball $O = \{x;\ |x - x_0| \le \varrho\}$ and derive (2.63) for a.a. ϱ by a limit passage in the weak formulation (2.45) tested by $v = v_\varepsilon$ with $v_\varepsilon(x) := (1 - \mathrm{dist}(x,O)/\varepsilon)^+$ provided the basic data qualification (2.55a,c) is fulfilled.[42]

Exercise 2.64. Show that the mapping $u \mapsto c(u,\nabla u)$ is compact, i.e. it maps bounded sets in $W^{1,p}(\Omega)$ into relatively compact sets in $W^{1,p}(\Omega)^*$, cf. Remark 2.39. For this, specify a growth assumption on c.[43]

Exercise 2.65. By using (2.56), show that $A : W^{1,p}(\Omega) \to W^{1,p}(\Omega)^*$ defined by (2.59) is demicontinuous. Note that no monotonicity of this A is needed, contrary to an abstract case addressed in Lemma 2.16.

---

[39]Hint: Take $u_k \to u$ in $L^{p_1}(\Omega)$ and $y_k \to y$ in $L^{p_2}(\Omega;\mathbb{R}^n)$, then take subsequences converging a.e. on Ω. Then, by continuity of $a(x,\cdot,\cdot)$ for a.a. $x \in \Omega$, $\mathscr{N}_a(u_k,y_k) \to \mathscr{N}_a(u,y)$ a.e., and by Proposition 1.13(i), in measure, too. Due to the obvious estimate

$$\big|a(x,u_k,y_k) - a(x,u,y)\big|^{p_0} \le 6^{p_0-1}\Big(\gamma^{p_0}(x) + C\big|u_k(x)\big|^{p_1} + C\big|u(x)\big|^{p_1} + C\big|y_k(x)\big|^{p_2} + C\big|y(x)\big|^{p_2}\Big)$$

for a.a. $x \in \Omega$, show that $\{|a(x,u_k,y_k) - a(x,u,y)|^{p_0}\}_{k\in\mathbb{N}}$ is equi-absolutely continuous since strongly convergent sequences are; use e.g. Theorem 1.16(i)⇒(iii). Eventually combine these two facts to get $\int_\Omega |a(x,u_k,y_k) - a(x,u,y)|^{p_0}\mathrm{d}x \to 0$ and realize that, as the limit $\mathscr{N}_a(u,y)$ is determined uniquely, eventually the whole sequence converges.

[40]Hint: For example, $a(x,r,s) = r/(1+|r|)$ and $u_k = \chi_{A_k}$, a characteristic function of a set A_k, $\mathrm{meas}_n(A_k) > 0$, $\lim_{k\to\infty}\mathrm{meas}_n(A_k) = 0$, and realize that $\|u_k\|_{L^p(\Omega)} = (\mathrm{meas}_n(A_k))^{1/p} \to 0$ but $\|\mathscr{N}_a(u_k)\|_{L^\infty(\Omega)} = 1/2 \not\to 0 = \|\mathscr{N}_a(0)\|_{L^\infty(\Omega)}$.

[41]Hint: Take $r_1, r_2 \in \mathbb{R}^{m_1}$ such that $c(\frac12 r_1 + \frac12 r_2) \ne \frac12 c(r_1) + \frac12 c(r_1)$ and a sequence of functions oscillating faster and faster between r_1 and r_2 (instead of 1 and -1 as used on Figure 3).

[42]Hint: Putting $x_0 = 0$ without any loss of generality, realizing that $\nabla v_\varepsilon(x) = -\varepsilon^{-1}x/|x|$ if $\varrho < |x| < \varrho + \varepsilon$ otherwise $\nabla v_\varepsilon(x) = 0$ a.e. and that $\nu(x) = x/|x|$, the limit passage

$$\begin{aligned} 0 \ =\ &\int_{|x|\le\varrho} c(x,u,\nabla u) - g(x)\,\mathrm{d}x + \int_{\varrho\le|x|\le\varrho+\varepsilon}\bigg(\big(c(x,u,\nabla u) - g(x)\big)\Big(1 - \frac{|x|-\varrho}{\varepsilon}\Big) \\ &- \frac{1}{\varepsilon}a(x,u,\nabla u)\cdot\frac{x}{|x|}\bigg)\,\mathrm{d}x \to \int_{|x|\le\varrho} c(x,u,\nabla u) - g(x)\,\mathrm{d}x - \int_{|x|=\varrho} a(x,u,\nabla u)\cdot\nu\,\mathrm{d}S \end{aligned}$$

holds at every right Lebesgue point of the function $f : \varrho \mapsto \varrho^{-1}\int_{|x|=\varrho} a(x,u,\nabla u)\cdot x\,\mathrm{d}S$, i.e. at every ϱ such that $f(\varrho) = \lim_{\varepsilon\searrow 0}\frac{1}{\varepsilon}\int_\varrho^{\varrho+\varepsilon} f(\xi)\mathrm{d}\xi$. As f is locally integrable thanks to the growth conditions (2.55a,c), it is known that, for a.a. ϱ, it enjoys this property.

[43]Hint: It suffices to design the growth condition so that $\mathscr{N}_c$ maps $L^{p^*}(\Omega) \times L^p(\Omega;\mathbb{R}^n)$ into $L^{(p^*-\epsilon)'}(\Omega)$ which is compactly embedded into $W^{1,p}(\Omega)^*$.

Exercise 2.66 (V-coercivity). Consider, instead of (2.92),

$$\exists \varepsilon_1, k_0{>}0,\ k_1{\in} L^1(\Omega):\ a(x,r,s){\cdot}s{+}c(x,r,s)r \geq \varepsilon_1|s|^p{-}k_0|r|^{q_1}{-}k_1(x), \tag{2.124a}$$
$$\exists \varepsilon_2 > 0,\quad k_2{\in} L^1(\Gamma):\qquad b(x,r)r \geq\ \varepsilon_2\chi_{\Gamma_{\rm N}}(x)|r|^q{-}k_2(x), \tag{2.124b}$$

for some $1 < q_1 < q \leq p$ and $\mathrm{meas}_{n-1}(\Gamma_{\rm N}) > 0$, and prove Lemma 2.35 by using the Poincaré inequality in the form (1.56). Likewise, formulate similar conditions for the case of mixed Dirichlet/Newton conditions (2.49) and use (1.57) to show coercivity of the shifted operator $A_0 = A(\cdot{+}w)$ with $w|_{\Gamma_{\rm D}} = u_{\rm D}$ on V from (2.52).

Example 2.67 (Finite-element method). As an example for the finite-dimensional space V_k used in *Galerkin's method* in the concrete case $V = W^{1,p}(\Omega)$, the reader can think of $\mathscr{T}_k$ as a simplicial partition of a polyhedral domain $\Omega \subset \mathbb{R}^n$, i.e. $\mathscr{T}_k$ is a collection of n-dimensional simplexes having mutually disjoint interiors and covering $\bar{\Omega}$; if $n = 2$ or 3, it means a triangulation or a "tetrahedralization" as on Figure 5a or 5b, respectively. Then, one can consider $V_k := \{v \in W^{1,p}(\Omega);\ \forall S \in \mathscr{T}_k : v|_S \text{ is affine}\}$. A canonical base of V_k is formed by "hat" functions vanishing at all mesh points except one; cf. Figure 5a.[44] Nested triangulations, i.e. each triangulation $\mathscr{T}_{k+1}$ is a refinement of $\mathscr{T}_k$, obviously imply $V_k \subset V_{k+1}$ which we have used in (2.7).

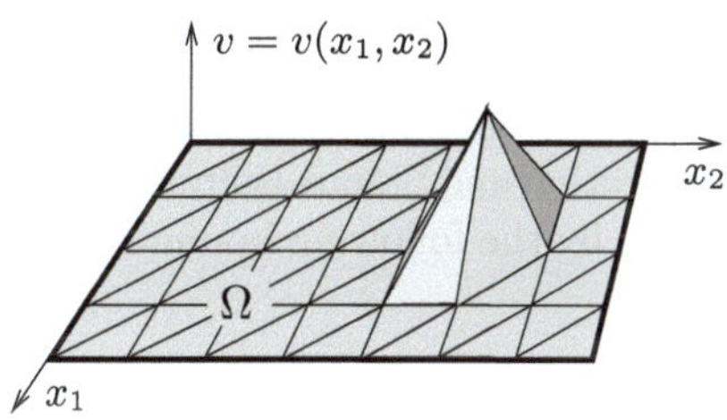

Figure 5a. Triangulation of a polygonal domain $\Omega \subset \mathbb{R}^2$ and one of the piece-wise affine 'hat-shaped' base functions.

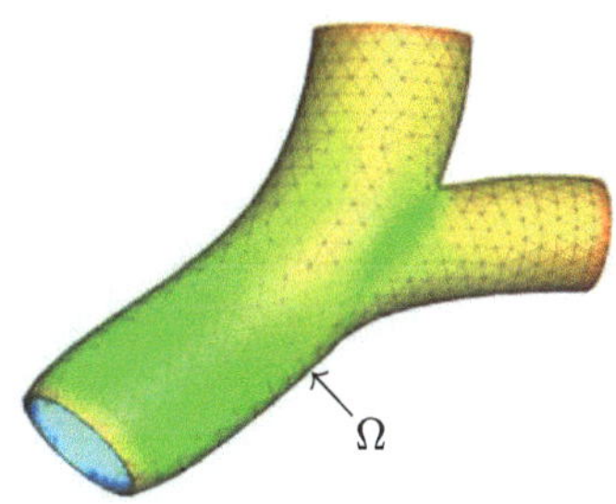

Figure 5b. A fine 3-dimensional tetrahedral mesh on a complicated (but still simply connected) Lipschitz domain $\Omega{\subset}\mathbb{R}^3$; courtesy of M.Mádlík.

This is the so-called P1-finite-element method. Often, higher-order polynomials are used for the base functions, sometimes in combination with non-simplectic meshes. For non-polyhedral domains, one can use a rectification of the curved boundary by a certain homeomorphism as on Figure 8 on p. 91. Efficient software packages based on finite-element methods are commercially available, including routines for automatic mesh generation on complicated domains, as illustrated on Figure 5b.

Exercise 2.68. Assuming $n = 1$ and $\lim_{k\to\infty} \max_{S\in\mathscr{T}_k} \mathrm{diam}(S) = 0$, prove density of $\bigcup_{k\in\mathbb{N}} V_k$ in $W^{1,p}(\Omega)$, cf. (2.7), for V_k from Example 2.67.[45]

[44]In such a base, the local character of differential operators is reflected in the Galerkin scheme that, e.g., linear differential operators result in matrices which are sparse.

[45]Hint: By density Theorem 1.25, take $v \in W^{2,\infty}(\Omega)$ and $v_k \in V_k$ such that $v_k(x) = v(x)$

Remark 2.69. To ensure (2.7) if $n \geq 2$, a qualification of the triangulation is necessary; usually, for some $\varepsilon > 0$, one requires that always $\mathrm{diam}(S)/\varrho_S \geq \varepsilon$ with denoting ϱ_S the radius of a ball contained in S.

Exercise 2.70 (Traces of higher-order Sobolev spaces). Generalize the trace Theorem 1.23 for $W^{2,p}(\Omega)$, and identify the integrability exponent for traces of functions from $W^{2,p}(\Omega)$, namely $p^{*\#} := (p^*)^\#$, as $(np-p)/(n-2p)$ if $2p < n$, otherwise, its integrability is arbitrarily large if $2p = n$ or in $L^\infty(\Gamma)$ if $2p > n$. Continue by induction for $W^{k,p}(\Omega)$, $k \geq 3$.[46]

2.6.2 Semilinear heat equation of type $-\mathrm{div}(\mathbb{A}(x,u)\nabla u) = g$

Here we focus on a heat equation where, from physical reasons, the heat-transfer coefficients depend typically on temperature but not on its gradient, giving rise to a semilinear equation as investigated in Section 2.5. Moreover, we speak about a *critical growth* of the particular nonlinearity when (2.55) would be fulfilled only if $\epsilon = 0$. Here we will meet the situation when even $\epsilon = -1$ in (2.55b) is needed (and by replacing the conventional Sobolev space $W^{1,p}(\Omega)$ by (2.128) eventually allowed) for $b(x,\cdot)$; this is reported as a *super-critical growth.*

Example 2.71 (*Nonlinear heat equation*). The steady-state heat transfer in a nonhomogeneous anisotropic nonlinear[47] medium with a boundary condition controlling the heat flux through two mechanisms, convection and Stefan-Boltzmann-type radiation[48] as outlined on Figure 6a, is described by the following boundary-value problem

$$\left.\begin{aligned} -\mathrm{div}\big(\mathbb{A}(x,u)\nabla u\big) &= g(x) && \text{on } \Omega, \\ \nu^\top \mathbb{A}(x,u)\nabla u &= \underbrace{b_1(x)(\theta - u)}_{\substack{\text{convective}\\ \text{heat flux}}} + \underbrace{b_2(x)(\theta^4 - |u|^3 u)}_{\substack{\text{radiative}\\ \text{heat flux}}} && \text{on } \Gamma, \end{aligned}\right\} \tag{2.125}$$

at every $x \in \bar\Omega$ which is a mesh point of the partition $\mathscr{T}_k$, and $\|\nabla v_k - \nabla v\|_{L^\infty(\Omega;\mathbb{R}^n)} \leq \mathrm{diam}(S)\|\nabla^2 v\|_{L^\infty(\Omega;\mathbb{R}^{n\times n})}$; as $n = 1$, each S is an interval here.

[46]Hint: For $W^{2,p}(\Omega)$ with $2p < n$, consider $W^{2,p}(\Omega) \subset W^{1,p^*}(\Omega)$ and apply Theorem 1.23 for $p^* = np/(n-p)$ instead of p. By induction, $u \mapsto u|_\Gamma : W^{k,p}(\Omega) \to L^{(np-p)/(n-kp)}(\Gamma)$ if $kp < n$.

[47]The adjective "nonhomogeneous" refers to spatial dependence of the material properties, here $\mathbb{A}$. The adjective "anisotropic" means that $\mathbb{A} \neq \mathbb{I}$ in general, i.e. the heat flux is not necessarily parallel with the temperature gradient and applies typically in single-crystals or in materials with a certain ordered structure, e.g. laminates. The adjective "nonlinear" is related here to a temperature dependence of $\mathbb{A}$ which applies especially when the temperature range is large. E.g. heat conductivity in conventional steel varies by tens of percents when temperature ranges hundreds degrees; cf. [358].

[48]Recall the Stefan-Boltzmann radiation law: the heat flux is proportional to $u^4 - \theta^4$ where θ is the absolute temperature of the outer space. In room temperature, the convective heat transfer, proportional to $u - \theta$ through the coefficient b_1, is usually dominant. Yet, for example, in steel-manufacturing processes the radiative heat flux becomes quickly dominant when temperature rises, say, above 1000 K and definitely cannot be neglected; cf. [358].

with

u = temperature in a thermally conductive body occupying Ω,
θ = temperature of the environment,
$\mathbb{A} = [a_{ij}]_{i,j=1}^n$ = a symmetric heat-conductivity matrix, $\mathbb{A}:\Omega\times\mathbb{R}\to\mathbb{R}^{n\times n}$, i.e.
$a_i(x,r,s) = \sum_{j=1}^n a_{ij}(x,r)s_j$,
$\mathbb{A}(x,u)\nabla u = \big(\sum_{j=1}^n a_{ij}(u)\frac{\partial}{\partial x_j}u\big)_{i=1}^n$ =the heat flux,
$\nu^\top\mathbb{A}(x,u)\nabla u = \sum_{i,j=1}^n a_{ij}(u)\nu_i\frac{\partial}{\partial x_j}u$ =the heat flux through the boundary,
b_1, b_2 = coefficients of convective and radiative heat transfer through Γ,
g = volume heat source.

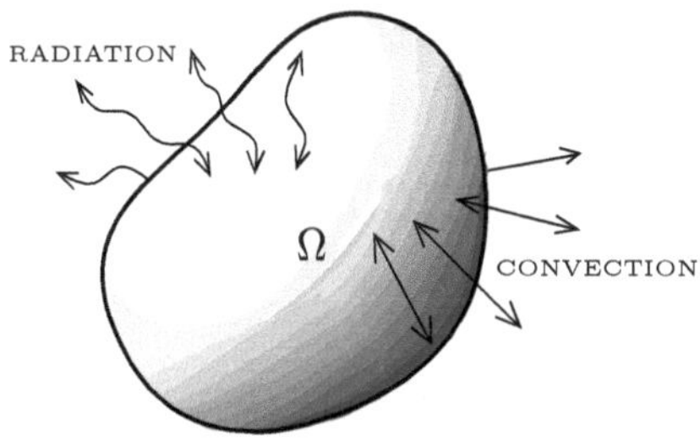

Figure 6a. Illustration of a heat-transfer problem in a 3-dimensional body $\Omega\subset\mathbb{R}^3$.

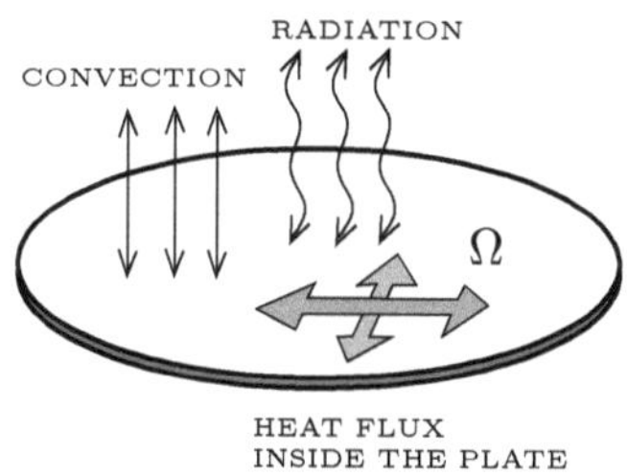

Figure 6b. Illustration of a heat-transfer problem in a 2-dimensional plate $\Omega\subset\mathbb{R}^2$.

In the setting (2.48), $b(x,r) = b_1(x)r + b_2(x)|r|^3r$ and $h(x) = [b_1\theta + b_2\theta^4](x)$. We assume $\theta\ge 0$, $b_1(x)\ge \underline{b}_1 > 0$ and $b_2(x)\ge\underline{b}_2 > 0$, $b_1\in L^{5/3}(\Gamma)$ and $b_2\in L^\infty(\Gamma)$, and the operator A is defined by

$$\langle A(u),v\rangle := \int_\Omega(\nabla v)^\top\mathbb{A}(x,u)\nabla u\,\mathrm{d}x + \int_\Gamma\big(b_1(x)u + b_2(x)|u|^3u\big)v\,\mathrm{d}S. \qquad (2.126)$$

It should be emphasized that no monotonicity of A with respect to the L^2-inner product can be expected if $\mathbb{A}(x,\cdot)$ is not constant.[49]

Exercise 2.72 (Pseudomonotone-operator approach)**.** Check the assumptions (2.55) and (2.65) in Section 2.4[50] as well as the coercivity (2.124)[51]. Realize, in

[49]This means $\int_\Omega(\nabla u_1-\nabla u_2)^\top(\mathbb{A}(x,u_1)\nabla u_1 - \mathbb{A}(x,u_2)\nabla u_2)\,\mathrm{d}x < 0$ may occur.

[50]Hint: The assumption (2.55a) reads here as $|\sum_{j=1}^n a_{ij}(x,r)s_j| \le \gamma(x) + C|r|^{(p^*-\epsilon)/p'} + C|s|^{p-1}$ with $\gamma\in L^{p'}(\Omega)$. This requires $p\ge 2$. The assumption of monotonicity in the main part (2.65) just requires that $\mathbb{A}(x,r) = [a_{ij}(x,r)]$ is positive semi-definite for all r and a.a. $x\in\Omega$, i.e. $s^\top\mathbb{A}(x,r)s\ge 0$. The assumption (2.55b) for the "physical" dimension $n=3$ and for $p=2$ yields $p^\# = (np-p)/(n-p) = 4$, cf. (1.37). This just agrees with the 4-power growth of the Stefan-Boltzmann law at least in the sense that the traces $|u|^3u$ are in $L^1(\Gamma)$ if $u\in W^{1,2}(\Omega)$. Yet (2.55b) admits only $(3-\epsilon)$-power growth of $b(x,\cdot)$ which does not fit with the 4-power growth of Stefan-Boltzmann law.

[51]Hint: The coercivity assumption (2.124a) requires here $s^\top\mathbb{A}(x,r)s\ge\varepsilon_1|s|^p - k_1$, which requires, besides uniform positive definiteness of $\mathbb{A}$, also $p\le 2$. Altogether, $p=2$ is ultimately needed. Note that $p=2$ and (2.55a) need $\mathbb{A}(x,r)$ bounded, i.e. $|a_{ij}(x,r)|\le C$ for any $i,j = 1,\dots,n$. The condition (2.124b) holds trivially with $k_2 = 0$.

particular, that $p=2$ is needed and the qualification (2.57) of (g,h) means

$$g\in\begin{cases} L^1(\Omega), \\ L^{1+\epsilon}(\Omega), \\ L^{2n/(n+2)}(\Omega), \end{cases} \qquad h\in\begin{cases} L^1(\Gamma) & \text{if } n=1, \\ L^{1+\epsilon}(\Gamma) & \text{if } n=2, \\ L^{2-2/n}(\Gamma) & \text{if } n\geq 3. \end{cases} \tag{2.127}$$

In view of this, the pseudo-monotone approach has disadvantages:

✓ direct usage of Leray-Lions Theorem 2.36 on the conventional Sobolev space $W^{1,2}(\Omega)$ is limited to $n\leq 2$ or to $b_2\equiv 0$,

✓ if $n>1$, an artificial integrability of g and h is needed, contrary to the physically natural requirement of a finite energy of heat sources, i.e. $g\in L^1(\Omega)$, $h\in L^1(\Gamma)$,

✓ $\mathbb{A}$ must be bounded.

Modify the setting of Section 2.4.3 by replacing $W^{1,p}(\Omega)$ by

$$V=\big\{v\in W^{1,2}(\Omega);\ v|_\Gamma\in L^5(\Gamma)\big\} \tag{2.128}$$

and show that V becomes a reflexive Banach space densely containing $C^\infty(\bar\Omega)$ if equipped with the norm $\|v\|:=\|v\|_{W^{1,2}(\Omega)}+\big\|v|_\Gamma\big\|_{L^5(\Gamma)}$.[52] Show that $A:V\to V^*$ defined by (2.126) is bounded and coercive. Make a limit passage though the monotone boundary term by Minty's trick instead of the compactness.

Exercise 2.73 (Weak-continuity approach). Use V from (2.128) and $Z=W^{1,\infty}(\Omega)$, assume

$$\exists\varepsilon_1>0: \qquad s^\top\mathbb{A}(x,r)s\geq\varepsilon_1|s|^2, \tag{2.129a}$$

$$\exists\gamma\in L^2(\Omega),\ \epsilon>0: \qquad |\mathbb{A}(x,r)|\leq\gamma(x)+C|r|^{(2^*-\epsilon)/2}, \tag{2.129b}$$

and show that $A:V\to Z^*$ defined by (2.126) is weakly continuous; use the fact that $L^4(\Gamma)$ is an interpolant between $L^2(\Gamma)$ and $L^5(\Gamma)$.[53]

Exercise 2.74 (*Galerkin method*). Consider V_k a finite-dimensional subspace of $W^{1,\infty}(\Omega)$ nested for $k\to\infty$ with a dense union in $W^{1,2}(\Omega)$ and traces dense in

[52]Hint: For $n\leq 2$, simply $V=W^{1,2}(\Omega)$. For $n\geq 3$, any Cauchy sequence $\{v_k\}_{k\in\mathbb{N}}$ in V has a limit v in $W^{1,2}(\Omega)$ and $\{v_k|_\Gamma\}_{k\in\mathbb{N}}$ converges in $L^{p^\#}(\Gamma)$ to $v|_\Gamma$, and simultaneously has some limit w in $L^5(\Gamma)$ but necessarily $v|_\Gamma=w$. As V is (isometrically isomorphic to) a closed subspace in a reflexive Banach space $W^{1,2}(\Omega)\times L^5(\Gamma)$, it is itself reflexive. Density of smooth functions can be proved by standard mollifying procedure.

[53]Hint: Take u_k such that $u_k\rightharpoonup u$ in $W^{1,2}(\Omega)$ and $u_k|_\Gamma\rightharpoonup u|_\Gamma$ in $L^5(\Gamma)$. Use $W^{1,2}(\Omega)\Subset L^{2^*-\epsilon}(\Omega)$ and then $\mathbb{A}(u_k)\to\mathbb{A}(u)$ in $L^2(\Omega;\mathbb{R}^{n\times n})$ and $\nabla u_k\rightharpoonup\nabla u$ weakly in $L^2(\Omega;\mathbb{R}^n)$, and, for $v\in W^{1,\infty}(\Omega)=:Z$, pass to the limit $\int_\Omega(\nabla u_k)^\top\mathbb{A}(x,u_k)\nabla v\,dx\to\int_\Omega(\nabla u)^\top\mathbb{A}(x,u)\nabla v\,dx$. By compactness of the trace operator, $u_k|_\Gamma\to u|_\Gamma$ in $L^{p^\#-\epsilon}(\Gamma)\subset L^2(\Gamma)$, realize that $u_k|_\Gamma\to u|_\Gamma$ in $L^4(\Gamma)$ because

$$\big\|u_k|_\Gamma-u|_\Gamma\big\|_{L^4(\Gamma)}\leq\big\|u_k|_\Gamma-u|_\Gamma\big\|_{L^5(\Gamma)}^{5/6}\big\|u_k|_\Gamma-u|_\Gamma\big\|_{L^2(\Gamma)}^{1/6}\to 0.$$

Then $|u_k|^3u_k|_\Gamma\to|u|^3u|_\Gamma$ in $L^1(\Gamma)$, and $\int_\Gamma|u_k|^3u_kv\,dS\to\int_\Gamma|u|^3uv\,dS$ for any $v\in L^\infty(\Gamma)$.

$L^5(\Gamma)$, and thus in V from (2.128), too. Then the Galerkin method for (2.125) is defined by:

$$\int_\Omega (\nabla u_k)^\top \mathbb{A}(x,u_k)\nabla v - gv\,\mathrm{d}x + \int_\Gamma \Big((b_1 + b_2|u_k|^3)u_k - h\Big)v\,\mathrm{d}S = 0 \tag{2.130}$$

for all $v \in V_k$. Assume $g \in L^{2^{*\prime}}(\Omega)$, $h = b_1\theta + b_2\theta^4$ with $\theta \in L^5(\Gamma)$, see Example 2.71, and assuming existence of u_k, show the a-priori estimate in V from (2.128) by putting $v := u_k$ into (2.130).[54] Then, using the linearity of $s \mapsto a(x,r,s) = \mathbb{A}(x,r)s$, make the limit passage directly in the Galerkin identity (2.130) by using the weak continuity as in Exercise 2.73.

Exercise 2.75 (Strong convergence). Assume again $\mathbb{A}$ bounded as in Exercise 2.72 and, despite the lack of d-monotonicity of $u \mapsto \mathrm{div}\big(\mathbb{A}(x,u)\nabla u\big)$, use strong monotonicity of $u \mapsto \mathrm{div}\big(\mathbb{A}(x,v)\nabla u\big)$ for v fixed, and show $u_k \to u$ in $W^{1,2}(\Omega)$; make the limit passage in the boundary term by compactness[55] if $n \le 2$, or treat it by

---

[54] Hint: By Hölder's and Young's inequalities, this yields the estimate

$$\begin{aligned}
\varepsilon_1 \int_\Omega |\nabla u_k|^2 \mathrm{d}x + \underline{b}_1 \int_\Gamma |u_k|^2 \mathrm{d}S + \underline{b}_2 \int_\Gamma |u_k|^5 \mathrm{d}S \le \int_\Omega (\nabla u_k)^\top \mathbb{A}(x,u_k)\nabla u_k \mathrm{d}x \\
+ \int_\Gamma b_1|u_k|^2 + b_2|u_k|^5 \mathrm{d}S = \int_\Omega g u_k \mathrm{d}x + \int_\Gamma (b_1\theta + b_2\theta^4)u_k \mathrm{d}S \\
\le \|g\|_{L^{p^{*\prime}}(\Omega)}\|u_k\|_{L^{p^*}(\Omega)} + \Big(\|b_1\|_{L^{5/3}(\Gamma)}\|\theta\|_{L^5(\Gamma)} + \|b_2\|_{L^\infty(\Gamma)}\|\,|\theta|^4\|^{5/4}_{L^{5/4}(\Gamma)}\Big)\|u_k\|_{L^5(\Gamma)} \\
< \frac{1}{4\varepsilon}N^2\|g\|^2_{L^{p^{*\prime}}(\Omega)} + \varepsilon\|u_k\|^2_{W^{1,2}(\Omega)} \\
+ C\Big(\|b_1\|_{L^{5/3}(\Gamma)}\|\theta\|_{L^5(\Gamma)} + \|b_2\|_{L^\infty(\Gamma)}\|\theta\|^4_{L^5(\Gamma)}\Big)^{5/4} + \frac{1}{2}\underline{b}_2\|u_k\|^5_{L^5(\Gamma)}
\end{aligned}$$

where N is the norm of the embedding $W^{1,2}(\Omega) \subset L^{2^*}(\Omega)$ and C is a sufficiently large constant, namely $C = 2^9/(5^5\underline{b}_2)$, and $\varepsilon < \min(\varepsilon_1, \underline{b}_1)/C_\mathrm{P}$ with C_P the constant from the Poincaré inequality (1.56) with $p = 2 = q$. Then use (1.56) for the estimate of the left-hand side from below and absorb the right-hand-side terms with u_k in the left-hand side.

[55] Hint: Abbreviate $b(u) = (b_1 + b_2|u|^3)u$ and $\mathfrak{a}_k := \nabla(u_k - u)^\top \mathbb{A}(u_k)\nabla(u_k - u)$, cf. (2.83). Then, use the Galerkin identity (2.130), i.e. $\int_\Omega \nabla(u_k - v_k)^\top \mathbb{A}(u_k)\nabla u_k \mathrm{d}x = \int_\Gamma b(u_k)(v_k - u_k)\,\mathrm{d}S$, to get

$$\begin{aligned}
\int_\Omega \mathfrak{a}_k \,\mathrm{d}x &= \int_\Omega \nabla(u_k - u)^\top\big(\mathbb{A}(u_k)\nabla u_k - \mathbb{A}(u)\nabla u\big) - \nabla(u_k - u)^\top\big(\mathbb{A}(u_k) - \mathbb{A}(u)\big)\nabla u \,\mathrm{d}x \\
&= \int_\Omega \nabla(u_k - v_k)^\top\big(\mathbb{A}(u_k)\nabla u_k - \mathbb{A}(u)\nabla u\big) + \nabla(v_k - u)^\top\big(\mathbb{A}(u_k)\nabla u_k - \mathbb{A}(u)\nabla u\big) \\
&\quad - \nabla(u_k - u)^\top\big(\mathbb{A}(u_k) - \mathbb{A}(u)\big)\nabla u \,\mathrm{d}x = \int_\Gamma \big(b(u) - b(u_k)\big)(u_k - v_k)\,\mathrm{d}S \\
&\quad + \int_\Omega \nabla(v_k - u)^\top\big(\mathbb{A}(u_k)\nabla u_k - \mathbb{A}(u)\nabla u\big) - \nabla(u_k - u)^\top\big(\mathbb{A}(u_k) - \mathbb{A}(u)\big)\nabla u \,\mathrm{d}x
\end{aligned}$$

for any $v_k \in V_k$. In particular, take $v_k \to u$ in $W^{1,2}(\Omega)$. For $n \le 2$, use compactness of the trace operator $W^{1,2}(\Omega) \to L^{p^\# - \epsilon}(\Gamma) \subset L^5(\Gamma)$ and push the first right-hand-side term to zero. Furthermore, use $\nabla v_k \to \nabla u$ in $L^2(\Omega;\mathbb{R}^n)$ and $\mathbb{A}(u_k)\nabla u_k - \mathbb{A}(u)\nabla u$ bounded in $L^2(\Omega;\mathbb{R}^n)$ to push the second term to zero. Finally, push the last expression to zero when using $\big(\mathbb{A}(u_k) - \mathbb{A}(u)\big)\nabla u \to 0$ in $L^2(\Omega;\mathbb{R}^n)$ (note that one cannot rely on $\mathbb{A}(u_k) \to \mathbb{A}(u)$ in $L^\infty(\Omega;\mathbb{R}^{n\times n})$, monotonicity if $n = 3$ when this term has the *super-critical growth*.[56] Note that, for $n = 3$, the super-critical growth of the boundary term is such that, although being formally a lower-order term, it behaves like a highest-order term and must be treated by its monotonicity.[57]

Exercise 2.76 (*Comparison principle*). Put $v := u^- = \min(u, 0)$ into the integral identity (2.51) for the case of (2.125). Show that non-negativity of heat sources, i.e. $h = b_1\theta + b_2\theta^4 \geq 0$ and $g \geq 0$, implies the non-negativity of temperature, i.e. $u \geq 0$.[58] Assume $g = 0$ and $0 \leq \theta(\cdot) \leq \theta_{\max}$ for a constant $\theta_{\max} > 0$ and use $v := (u - \theta_{\max})^+$ in (2.51) to show that $u(\cdot) \leq \theta_{\max}$ almost everywhere.

Exercise 2.77 (Mixed boundary conditions). Perform the analysis by the Galerkin method of the mixed Dirichlet/Newton boundary-value problem[59]

---

however) and when assuming $v_k \to u$ in $W^{1,2}(\Omega)$. Then $\varepsilon_1\|\nabla u_k - \nabla u\|^2_{L^2(\Omega;\mathbb{R}^d)} \leq \int_\Omega \mathfrak{a}_k \mathrm{d}x \to 0$ and, as $u_k \to u$ in $L^2(\Omega)$ by Rellich-Kondrachov's theorem 1.21, $u_k \to u$ in $W^{1,2}(\Omega)$.

[56] Hint: Use identity (2.130) and the previous notation of $\mathfrak{a}_k$ and b to write

$$\begin{aligned}
&\int_\Omega \mathfrak{a}_k \,\mathrm{d}x + \int_\Gamma \big(b(u_k) - b(u)\big)(u_k - u)\,\mathrm{d}S \\
&= \int_\Omega \nabla(u_k - v_k)^\top\big(\mathbb{A}(u_k)\nabla u_k - \mathbb{A}(u)\nabla u\big) + \nabla(v_k - u)^\top\big(\mathbb{A}(u_k)\nabla u_k - \mathbb{A}(u)\nabla u\big) \\
&\quad - \nabla(u_k - u)^\top\big(\mathbb{A}(u_k) - \mathbb{A}(u)\big)\nabla u\,\mathrm{d}x + \int_\Gamma \big(b(u_k) - b(u)\big)(u_k - v_k) + \big(b(u_k) - b(u)\big)(v_k - u)\,\mathrm{d}S \\
&= \int_\Gamma -b(u)(u_k - v_k) + (b(u_k) - b(u))(v_k - u)\,\mathrm{d}S + \int_\Omega -\nabla(u_k - v_k)^\top\mathbb{A}(u)\nabla u \\
&\quad + \nabla(v_k - u)^\top\big(\mathbb{A}(u_k)\nabla u_k - \mathbb{A}(u)\nabla u\big) - \nabla(u_k - u)^\top\big(\mathbb{A}(u_k) - \mathbb{A}(u)\big)\nabla u\,\mathrm{d}x \\
&= I_k^{(1)} + I_k^{(2)} + I_k^{(3)} + I_k^{(4)} + I_k^{(5)}.
\end{aligned}$$

Assume $v_k \to u$ in $W^{1,2}(\Omega)$ and $v_k|_\Gamma \to u|_\Gamma$ in $L^5(\Gamma)$. Use $b(u) \in L^{5/4}(\Gamma)$ and $u_k - v_k \rightharpoonup 0$ in $L^5(\Gamma)$ to show $I_k^{(1)} \to 0$. Use $\{b(u_k)\}_{k\in\mathbb{N}}$ bounded in $L^{5/4}(\Gamma)$ and $v_k - u \to 0$ in $L^5(\Gamma)$ to show $I_k^{(2)} := \int_\Gamma (b(u_k) - b(u))(v_k - u)\,\mathrm{d}S \to 0$. Push the remaining terms $I_k^{(3)}$, $I_k^{(4)}$, and $I_k^{(5)}$ as before. Altogether, conclude $u_k \to u$ in $W^{1,2}(\Omega)$. Moreover, conclude also $\|u_k - u\|_{L^5(\Gamma)} \to 0$.

[57] Hint: Realize the difficulties in pushing $\int_\Gamma (b(u) - b(u_k))(u_k - v_k)\,\mathrm{d}S$ to zero if $n \geq 3$ because we have $\{u_k\}_{k\in\mathbb{N}}$ and $\{b(u_k)\}_{k\in\mathbb{N}}$ only bounded in $L^5(\Gamma)$ and $L^{5/4}(\Gamma)$, respectively, but no strong convergence can be assumed in these spaces.

[58] Hint: Note that $u^- \in W^{1,2}(\Omega)$ if $u \in W^{1,2}(\Omega)$ so $v := u^-$ is a legal test, cf. Proposition 1.28, and then $\int_\Omega (\nabla u)^\top \mathbb{A}(u)\nabla u^-\,\mathrm{d}x = \int_\Omega (\nabla u^-)^\top \mathbb{A}(u)\nabla u^-\,\mathrm{d}x$ due to (1.50). By this way, come to the estimate

$$\begin{aligned}
\varepsilon_1 \int_\Omega |\nabla u^-|^2 \mathrm{d}x + \underline{b}_1 \int_\Gamma (u^-)^2 \mathrm{d}S \leq \int_\Omega (\nabla u)^\top \mathbb{A}(u)\nabla u^-\,\mathrm{d}x \\
+ \int_\Gamma b_1(u^-)^2 + b_2|u^-|^5 \mathrm{d}S = \int_\Omega g u^-\,\mathrm{d}x + \int_\Gamma h u^-\,\mathrm{d}S \leq 0.
\end{aligned}$$

By the Poincaré inequality (1.56), we get $\|u^-\|_{W^{1,2}(\Omega)} = 0$, hence $u^- = 0$ a.e. in Ω.

[59] Hint: Instead of (2.128), use $V = \{v \in W^{1,2}(\Omega);\ v|_{\Gamma_\mathrm{N}} \in L^5(\Gamma_\mathrm{N}),\ v|_{\Gamma_\mathrm{D}} = 0\}$, define Galerkin's approximate solution u_k with approximate Dirichlet conditions $u_k|_{\Gamma_\mathrm{D}} = u_\mathrm{D}^k$, and derive the a-priori estimate by a test $v := u_k - w_k$ where $w_k \in V_k$, a finite-dimensional subspace of V, is chosen so that $w_k|_{\Gamma_\mathrm{N}} = u_\mathrm{D}^k \to u_\mathrm{D}$ in $L^{2^\#}(\Gamma_\mathrm{D})$ and the sequence $\{w_k\}_{k\in\mathbb{N}}$ is bounded in V.

$$\left.\begin{aligned} -\mathrm{div}\big(\mathbb{A}(x,u)\nabla u\big) &= g(x) && \text{in } \Omega, \\ \nu^\top \mathbb{A}(x,u)\nabla u &= b_1(x)(\theta - u) + b_2(x)(\theta^4 - |u|^3 u) && \text{on } \Gamma_{\mathrm{N}}, \\ u|_{\Gamma_{\mathrm{D}}} &= u_{\mathrm{D}} && \text{on } \Gamma_{\mathrm{D}}. \end{aligned}\right\} \tag{2.131}$$

Exercise 2.78 (Heat-conductive plate). Perform the analysis by Galerkin's method of the problem

$$\left.\begin{aligned} -\mathrm{div}\big(\mathbb{A}(x,u)\nabla u\big) &= c_1(x)(\theta - u) + c_2(x)(\theta^4 - |u|^3 u) && \text{in } \Omega, \\ u|_\Gamma &= u_{\mathrm{D}} && \text{on } \Gamma. \end{aligned}\right\} \tag{2.132}$$

In the case $n = 2$, this problem has an interpretation of a plate conducting heat in tangential direction with normal-direction temperature variations neglected, and being cooled/heated by convection and radiation and with fixed temperature on the boundary as outlined on Figure 6b. Consider $n \leq 3$, use the conventional Sobolev space $W_0^{1,2}(\Omega)$, define Galerkin's approximate solution u_k with approximate Dirichlet conditions $u_k|_\Gamma = u_\Gamma^k$, and derive the a-priori estimate by a test by $v := u_k - w_k$ with w_k as in Exercise 2.77.

Example 2.79 (*Special nonlinear media*). Let us consider again the nonlinear heat-transfer problem (2.125) with $\mathbb{A}(x,r) = [a_{ij}(x,r)]$ in the special form

$$a_{ij}(x,r) = b_{ij}(x)\kappa(r) \tag{2.133}$$

with $\mathbb{B} = [b_{ij}] : \Omega \to \mathbb{R}^{n\times n}$ and $\kappa : \mathbb{R} \to \mathbb{R}^+$. Then the so-called *Kirchhoff transformation* employs the primitive function $\widehat{\kappa} : \mathbb{R} \to \mathbb{R}$ to κ, i.e. defined by

$$\widehat{\kappa}(r) := \int_0^r \kappa(\varrho)\mathrm{d}\varrho\,, \tag{2.134}$$

and transforms the nonlinearity of (2.125) inside Ω to the (already nonlinear) boundary conditions. Indeed, $\mathbb{B}(x)\nabla\widehat{\kappa}(u) = \mathbb{B}(x)\kappa(u)\nabla u = \mathbb{A}(x,u)\nabla u$ and $\mathbb{B}(x)\frac{\partial}{\partial\nu}\widehat{\kappa}(u) = \mathbb{B}(x)\kappa(u)\frac{\partial}{\partial\nu}u = \mathbb{A}(x,u)\frac{\partial}{\partial\nu}u$ and, by a substitution $w = \widehat{\kappa}(u)$, one transfers the nonlinearity from the equation on Ω to the boundary conditions which has been nonlinear even originally anyhow due to the Stefan-Boltzmann radiation term. Thus one gets the following semilinear equation for w:

$$\left.\begin{aligned} -\mathrm{div}\big(\mathbb{B}(x)\nabla w\big) &= g && \text{in } \Omega, \\ \mathbb{B}(x)\frac{\partial w}{\partial\nu} + \Big(b_1 + b_2|\widehat{\kappa}^{-1}(w)|^3\Big)\widehat{\kappa}^{-1}(w) &= h && \text{on } \Gamma. \end{aligned}\right\} \tag{2.135}$$

We assume $\mathbb{B} : \Omega \to \mathbb{R}^{n\times n}$ measurable, bounded, and $\mathbb{B}(\cdot)$ uniformly positive definite in the sense $\xi^\top\mathbb{B}(x)\xi \geq \beta|\xi|^2$ for all $\xi \in \mathbb{R}^n$ and some $\beta > 0$. Further, we assume $\kappa(\cdot) \geq \varepsilon > 0$ measurable and bounded; note that this implies $\widehat{\kappa}$ to be continuous and increasing, and one-to-one with $\widehat{\kappa}^{-1}$ Lipschitz continuous, in particular having a linear growth. Furthermore, g and h satisfy (2.127). Again, ultimately $p = 2$, and one can show the coercivity. As the function $\mathbb{R} \to \mathbb{R} : r \mapsto$

$b(x,r) := \big(b_1(x) + b_2(x)|\widehat{\kappa}^{-1}(r)|^3\big)\widehat{\kappa}^{-1}(r)$ is monotone for a.a. $x \in \Gamma$, we can use the monotonicity technique. Then there is just one weak solution $w \in W^{1,2}(\Omega)$. By Proposition 1.28, $u = \widehat{\kappa}^{-1}(w) \in W^{1,2}(\Omega)$ and this u solves the original problem in the weak sense. Moreover, $(g,h) \mapsto u : L^{p^{*\prime}}(\Omega) \times L^{p^{\#\prime}}(\Gamma) \to W^{1,2}(\Omega)$ is (norm×norm,norm)-continuous. Note that the heat-conductivity coefficient κ need not be assumed continuous.[60]

Example 2.80 (*Heat transfer with advection*). The heat equation in moving homogeneous isotropic media, i.e. with advection by a prescribed velocity, say $\vec{v} = \vec{v}(x)$, is

$$-\mathrm{div}\big(\kappa(u)\nabla u\big) + \mathfrak{c}(u)\vec{v}\cdot\nabla u = g, \tag{2.136}$$

where $\mathfrak{c}$ is the heat capacity dependent on temperature. Let us consider, for simplicity, constant Dirichlet boundary conditions and use the Kirchhoff transformation (2.134), i.e. put $w = \widehat{\kappa}(u)$ and using $\nabla\widehat{\kappa}^{-1}(w) = \nabla w/\kappa(\widehat{\kappa}^{-1}(w))$, to arrive at

$$\left.\begin{aligned} -\Delta w + \frac{C(w)\vec{v}\cdot\nabla w}{K(w)} &= g &&\text{in } \Omega,\\ w &= 0 &&\text{on } \Gamma \end{aligned}\right\} \tag{2.137}$$

where we abbreviated $\kappa(\widehat{\kappa}^{-1}(w)) =: K(w)$ and $\mathfrak{c}(\widehat{\kappa}^{-1}(w)) =: C(w)$; note that we can shift $\widehat{\kappa}$ by a constant so that $w = 0$ can be considered on Γ. Note that the pointwise coercivity (2.92a) for $p = 2 \geq q > 1$ is violated if $c(x,r,s) = C(r)\vec{v}(x)\cdot s/K(r)$. Assume the velocity field $\vec{v} \in C^1(\bar{\Omega};\mathbb{R}^n)$ as divergence free, which corresponds to a motion of an incompressible medium, cf. also the equations (6.26c), (12.44c), or (12.95b) below, one can consider an alternative setting with $\mathfrak{c}(u)\vec{v}\cdot\nabla u = \mathrm{div}(\vec{v}\,\widehat{\mathfrak{c}}(u)) = \mathrm{div}(\vec{v}\,\widehat{\mathfrak{c}}(\kappa^{-1}(w)))$ with $\widehat{\mathfrak{c}}$ the primitive function of $\mathfrak{c}$. This leads to $a(x,r,s) = s + \vec{v}(x)\widehat{\mathfrak{c}}(\kappa^{-1}(r))$ which again need not satisfy (2.92a).

Exercise 2.81 (*Uniqueness*). Show uniqueness of the weak solution w to (2.135), and thus of u, as well. Try to show uniqueness in the general case (2.125) and realize the difficulties if smallness of $\|u\|_{W^{1,\infty}(\Omega)}$ is not guaranteed.[61]

Exercise 2.82. Assume $\mathrm{div}\,\vec{v} \leq 0$ in Example 2.80 and show the coercivity of the respective A (in spite of this lack of any pointwise coercivity pointed out in Example 2.80) by derivation of an a-priori estimate again by a test by w.[62]

---

[60] A discontinuity of κ can indeed occur during various phase transformations, cf. [358] for a discontinuity in the heat-conductivity coefficient κ within a recrystallization in steel.

[61] The uniqueness holds even for a general case (2.125) but the proof is rather technical, cf. [243].

[62] Hint: For N the norm of the embedding $W^{1,2}(\Omega) \subset L^{2^*}(\Omega)$, use Green's Theorem 1.31 to estimate

$$\begin{aligned}\int_\Omega |\nabla w|^2\,\mathrm{d}x &\leq \int_\Omega |\nabla w|^2 - (\mathrm{div}\,\vec{v})\Big(\int_0^{w(x)}\frac{C(\xi)}{K(\xi)}\mathrm{d}\xi\Big)\mathrm{d}x = \int_\Omega |\nabla w|^2 + \vec{v}\cdot\nabla\Big(\int_0^{w(x)}\frac{C(\xi)}{K(\xi)}\mathrm{d}\xi\Big)\mathrm{d}x\\ &= \int_\Omega |\nabla w|^2 + \frac{(\vec{v}\cdot\nabla w)C(w)}{K(w)}\,\mathrm{d}x = \int_\Omega gw\,\mathrm{d}x \leq N\|w\|_{W^{1,2}(\Omega)}\|g\|_{L^{2^{*\prime}}(\Omega)}.\end{aligned}$$

Furthermore, assuming Lipschitz continuity of κ, show uniqueness of a solution to (2.137) if $\vec{v}$ is small enough in the L^∞-norm.[63]

2.6.3 Quasilinear equations of type $-\mathrm{div}\big(|\nabla u|^{p-2}\nabla u\big)+c(u,\nabla u)=g$

Here we will address quasilinear equations (2.45) with $a(x,r,\cdot)$ or $c(x,r,\cdot)$ nonlinear so that a limit passage in approximate solutions cannot be made by using mere weak convergence in ∇u and compactness in lower-order terms, unlike in semilinear equations scrutinized in Section 2.6.2. As a "training" quasilinear differential operator in the divergence form, we will frequently use

$$\Delta_p u := \mathrm{div}\big(|\nabla u|^{p-2}\nabla u\big) \tag{2.138}$$

called the *p-Laplacean*; hence the usual Laplacean is what is called here 2-Laplacean. For $p > 2$ one gets a degenerate nonlinearity, while for $p < 2$ a singular one, cf. Figure 9 on p.128 below. Note that, by using the formula $\mathrm{div}(vw) = v\,\mathrm{div}\,w + \nabla v\cdot w$, (2.138) can equally be written in the form

$$\mathrm{div}\big(|\nabla u|^{p-2}\nabla u\big) = |\nabla u|^{p-2}\Delta u + (p{-}2)|\nabla u|^{p-4}(\nabla u)^\top\nabla^2 u\,\nabla u. \tag{2.139}$$

Example 2.83 (d-monotonicity of p-Laplacean). To be more specific, $A = -\Delta_p$ will be understood here as a mapping $W^{1,p}(\Omega) \to W^{1,p}(\Omega)^*$ corresponding to a Neumann-boundary-value problem, i.e.

$$\langle A(u), v\rangle = \int_\Omega |\nabla u|^{p-2}\nabla u\cdot\nabla v\,\mathrm{d}x \tag{2.140}$$

for any $v \in W^{1,p}(\Omega)$. For $p > 1$, the p-Laplacean is always *d-monotone* in the sense (2.1) with respect to the seminorm $|u| := \|\nabla u\|_{L^p(\Omega;\mathbb{R}^n)}$, i.e.

$$\int_\Omega \big(|\nabla u|^{p-2}\nabla u{-}|\nabla v|^{p-2}\nabla v\big)\cdot(\nabla u{-}\nabla v)\,\mathrm{d}x \geq \big(d(|u|){-}d(|v|)\big)\big(|u|{-}|v|\big)$$

[63]Hint: Realizing that also $[C/K](\cdot)$ is Lipschitz continuous, with ℓ denoting the Lipschitz constant, we have

$$\begin{aligned}
&\int_\Omega \vec{v}\cdot\Big(\frac{C(w_1)\nabla w_1}{K(w_1)} - \frac{C(w_2)\nabla w_2}{K(w_2)}\Big)(w_1-w_2)\,\mathrm{d}x\\
&\quad= \int_\Omega \vec{v}\Big(\frac{C(w_1)}{K(w_1)} - \frac{C(w_2)}{K(w_2)}\Big)\cdot\nabla w_1(w_1{-}w_2)\,\mathrm{d}x + \int_\Omega \frac{C(w_2)\vec{v}\cdot\nabla(w_1{-}w_2)}{K(w_2)}(w_1{-}w_2)\,\mathrm{d}x\\
&\quad\leq \|\vec{v}\|_{L^\infty(\Omega;\mathbb{R}^n)}\Big\|\frac{C(w_1)}{K(w_1)} - \frac{C(w_2)}{K(w_2)}\Big\|_{L^4(\Omega)}\|\nabla w_1\|_{L^2(\Omega;\mathbb{R}^n)}\|w_1-w_2\|_{L^4(\Omega)}\\
&\qquad+ \|\vec{v}\|_{L^\infty(\Omega;\mathbb{R}^n)}\Big\|\frac{C(w_2)}{K(w_2)}\Big\|_{L^4(\Omega)}\|\nabla w_1 - \nabla w_2\|_{L^2(\Omega;\mathbb{R}^n)}\|w_1-w_2\|_{L^4(\Omega)}\\
&\quad\leq \|\vec{v}\|_{L^\infty(\Omega;\mathbb{R}^n)}\Big(\|\nabla w_1\|_{L^2(\Omega;\mathbb{R}^n)}\ell N^2 + N\frac{\max c(\cdot)}{\min\kappa(\cdot)}\mathrm{meas}_n(\Omega)^{1/4}\Big)\|w_1-w_2\|^2_{W^{1,2}(\Omega)},
\end{aligned}$$

where N is the norm of the embedding $W^{1,2}(\Omega)\subset L^4(\Omega)$ valid for $n\leq 3$. For $\|\vec{v}\|_{L^\infty(\Omega;\mathbb{R}^n)}$ small enough, conclude that $w_1 = w_2$.

with $d(\xi) = \xi^{p-1}$, which can be proved simply by Hölder's inequality as follows:

$$\begin{aligned}
&\int_\Omega \big(|y|^{p-2}y - |z|^{p-2}z\big)\cdot(y-z)\,\mathrm{d}x \\
&\quad = \|y\|^p_{L^p(\Omega;\mathbb{R}^n)} - \int_\Omega \Big(|y|^{p-2}y\cdot z + |z|^{p-2}z\cdot y\Big)\mathrm{d}x + \|z\|^p_{L^p(\Omega;\mathbb{R}^n)} \\
&\quad \geq \|y\|^p_{L^p(\Omega;\mathbb{R}^n)} - \big\||y|^{p-2}y\big\|_{L^{p'}(\Omega;\mathbb{R}^n)}\|z\|_{L^p(\Omega;\mathbb{R}^n)} \\
&\qquad\qquad - \big\||z|^{p-2}z\big\|_{L^{p'}(\Omega;\mathbb{R}^n)}\big\|y\big\|_{L^p(\Omega;\mathbb{R}^n)} + \big\|z\big\|^p_{L^p(\Omega;\mathbb{R}^n)} \\
&\quad = \big\|y\big\|^p_{L^p(\Omega;\mathbb{R}^n)} - \big\|y\big\|^{p-1}_{L^p(\Omega;\mathbb{R}^n)}\|z\|_{L^p(\Omega;\mathbb{R}^n)} \\
&\qquad\qquad - \big\|z\big\|^{p-1}_{L^p(\Omega;\mathbb{R}^n)}\big\|y\big\|_{L^p(\Omega;\mathbb{R}^n)} + \big\|z\big\|^p_{L^p(\Omega;\mathbb{R}^n)} \\
&\quad = \Big(\big\|y\big\|^{p-1}_{L^p(\Omega;\mathbb{R}^n)} - \big\|z\big\|^{p-1}_{L^p(\Omega;\mathbb{R}^n)}\Big)\Big(\big\|y\big\|_{L^p(\Omega;\mathbb{R}^n)} - \big\|z\big\|_{L^p(\Omega;\mathbb{R}^n)}\Big). \qquad (2.141)
\end{aligned}$$

For $p \geq 2$, from the algebraic inequality[64]

$$\big(|s|^{p-2}s - |\tilde{s}|^{p-2}\tilde{s}\big)\cdot\big(s-\tilde{s}\big) \geq c(n,p)|s-\tilde{s}|^p \qquad (2.142)$$

with some $c(n,p) > 0$, we obtain a uniform monotonicity on $W^{1,p}_0(\Omega)$ in the sense (2.2) with $\zeta(z) = z^{p-1}$ (or with respect to the seminorm $\|\nabla\cdot\|_{L^p(\Omega;\mathbb{R}^n)}$ on $W^{1,p}(\Omega)$):

$$\begin{aligned}
\big\langle A(u)-A(v), u-v\big\rangle &= \int_\Omega \big(|\nabla u|^{p-2}\nabla u - |\nabla v|^{p-2}\nabla v\big)\cdot\nabla(u-v)\,\mathrm{d}x \\
&\geq c(n,p)\int_\Omega |\nabla u - \nabla v|^p \mathrm{d}x.
\end{aligned}$$

It should be emphasized that, for $p < 2$, one has only $\langle A(u)-A(v), u-v\rangle \geq (p-1)\int_\Omega \max(1+|\nabla u|, 1+|\nabla v|)^{p-2}|\nabla u - \nabla v|^2 \mathrm{d}x$.[65]

Exercise 2.84 (Monotonicity of p-Laplacean). Realize that (2.138) corresponds to $a_i(x,r,s) = |s|^{p-2}s_i$ and verify the strict monotonicity (2.65) and (2.68a).[66]

Exercise 2.85 (Strong convergence in $c(\nabla u)$). Consider the Dirichlet boundary-value problem

$$\left.\begin{aligned}
-\mathrm{div}\big(|\nabla u|^{p-2}\nabla u\big) + c(x,\nabla u) &= g && \text{in } \Omega, \\
u &= 0 && \text{on } \Gamma.
\end{aligned}\right\} \qquad (2.143)$$

For some $\epsilon > 0$, assume the growth condition

$$\exists \gamma \in L^{(p^*-\epsilon)'}(\Omega)\ C\in\mathbb{R}\ \forall (\text{a.a.}) x\in\Omega\ \forall s\in\mathbb{R}^n: \ \big|c(x,s)\big| \leq \gamma(x) + C|s|^{p-1-\epsilon}. \qquad (2.144)$$

[64] See DiBenedetto [120, Sect.I.4] or Hu and Papageorgiou [209, Part.I,Sect.3.1].

[65] See Málek et al. [268, Sect.5.1.2].

[66] Hint: Like (2.141), $(|s|^{p-2}s - |\tilde{s}|^{p-2}\tilde{s})\cdot(s-\tilde{s}) = |s|^p - |s|^{p-2}s\cdot\tilde{s} - |\tilde{s}|^{p-2}\tilde{s}\cdot s + |\tilde{s}|^p \geq |s|^p - |s|^{p-1}|\tilde{s}| - |\tilde{s}|^{p-1}|s| + |\tilde{s}|^p = (|s|^{p-1} - |\tilde{s}|^{p-1})(|s| - |\tilde{s}|)$, hence (2.65) holds. If $(|s|^{p-2}s - |\tilde{s}|^{p-2}\tilde{s})\cdot(s-\tilde{s}) = 0$, then $|s| = |\tilde{s}|$, and if $s \neq \tilde{s}$, then $|s|^2 > s\cdot\tilde{s}$ hence $|s|^p - |s|^{p-2}s\cdot\tilde{s} > 0$, and similarly $|\tilde{s}|^p - |\tilde{s}|^{p-2}\tilde{s}\cdot s > 0$, hence $(|s|^{p-2}s - |\tilde{s}|^{p-2}\tilde{s})\cdot(s-\tilde{s}) > 0$, a contradiction, proving (2.68a).

Formulate Galerkin's approximation[67] and prove the a-priori estimate in $W_0^{1,p}(\Omega)$ by testing the Galerkin identity by $v = u_k$[68] and prove strong convergence of $\{u_k\}$ in $W_0^{1,p}(\Omega)$ by using d-monotonicity of $-\Delta_p$, following the scheme of Proposition 2.20 with Remark 2.21 simplified by having boundedness guaranteed explicitly through Lemma 2.31 instead of the Banach-Steinhaus principle through (2.36) and (2.42).[69] Further, considering $p = 2$, formulate a Lipschitz-continuity condition like (2.157) in Exercise 2.90 that would guarantee (uniform) monotonicity of the underlying mapping A.

Exercise 2.86 (Weak convergence in $c(\nabla u)$)**.** Consider the boundary-value problem (2.143) in a more general form:

$$\left.\begin{aligned} -\mathrm{div}\, a(x,\nabla u) + c(x,\nabla u) &= g && \text{in } \Omega, \\ u &= 0 && \text{on } \Gamma, \end{aligned}\right\} \tag{2.145}$$

with $a(x,\cdot) : \mathbb{R}^n \to \mathbb{R}^n$ strictly monotone. The Galerkin approximation looks as

$$\int_\Omega a(\nabla u_k)\cdot\nabla v + \big(c(\nabla u_k) - g\big)v\,\mathrm{d}x = 0 \qquad \forall v \in V_k. \tag{2.146}$$

Assuming coercivity $a(x,s)\cdot s \ge \varepsilon_a|s|^p$ and the growth (2.144), prove the a-priori

---

[67] See (2.146) below for $a(x,s) = |s|^{p-2}s$.

[68] Hint: Use Hölder's inequality between $L^{p/(p-1-\epsilon)}(\Omega)$ and $L^q(\Omega)$ with $q = p/(1+\epsilon)$ to estimate

$$\begin{aligned} \|u_k\|^p_{W_0^{1,p}(\Omega)} &= \|\nabla u_k\|^p_{L^p(\Omega;\mathbb{R}^n)} = \int_\Omega (g - c(\nabla u_k))u_k\,\mathrm{d}x \le \int_\Omega (|g| + \gamma + C|\nabla u_k|^{p-1-\epsilon})|u_k|\,\mathrm{d}x \\ &\le \||g| + \gamma\|_{L^{p^*\prime}(\Omega)}\|u_k\|_{L^{p^*}(\Omega)} + C\|\nabla u_k\|^{p-1-\epsilon}_{L^p(\Omega)}\|u_k\|_{L^q(\Omega)} \\ &\le N_{p^*}\||g| + \gamma\|_{L^{p^*\prime}(\Omega)}\|u_k\|_{W_0^{1,p}(\Omega)} + CN_q\|u_k\|^{p-\epsilon}_{W_0^{1,p}(\Omega)} \end{aligned}$$

with N_q the norm of the embedding $W^{1,p}(\Omega) \subset L^q(\Omega)$, and N_{p^*} with an analogous meaning.

[69] Hint: Take a subsequence $u_k \rightharpoonup u$ in $W_0^{1,p}(\Omega)$. Use the norm $\|v\|_{W_0^{1,p}(\Omega)} := \|\nabla v\|_{L^p(\Omega;\mathbb{R}^n)}$ and, by (2.141) and using still the abbreviation $a(\nabla v) = |\nabla v|^{p-2}\nabla v$, estimate

$$\begin{aligned} \Big(\|u_k\|^{p-1}_{W_0^{1,p}(\Omega)} - \|v\|^{p-1}_{W_0^{1,p}(\Omega)}\Big)\Big(\|u_k\|_{W_0^{1,p}(\Omega)} - \|v\|_{W_0^{1,p}(\Omega)}\Big) &\le \int_\Omega \big(a(\nabla u_k) - a(\nabla v)\big)\cdot\nabla(u_k - v)\,\mathrm{d}x \\ &= \int_\Omega a(\nabla u_k)\cdot\nabla(u_k - v_k) + a(\nabla u_k)\cdot\nabla(v_k - v) - a(\nabla v)\cdot\nabla(u_k - v)\,\mathrm{d}x \\ &= \int_\Omega \big(g - c(\nabla u_k)\big)(u_k - v_k) + a(\nabla u_k)\cdot\nabla(v_k - v) - a(\nabla v)\cdot\nabla(u_k - v)\,\mathrm{d}x \end{aligned}$$

with $v_k \in V_k$. Assume $v_k \to v$. For $v = u$, $u_k - v_k \to u - u = 0$ in $L^{p^*-\epsilon}(\Omega)$ because of the compact embedding $W_0^{1,p}(\Omega) \Subset L^{p^*-\epsilon}(\Omega)$, and then $\int_\Omega c(\nabla u_k)(u_k - v_k)\,\mathrm{d}x \to 0$; note that $\{c(\nabla u_k)\}_{k\in\mathbb{N}}$ is bounded in $L^{(p^*-\epsilon)'}(\Omega)$. Push the other terms to zero, too. Conclude that $u_k \to u$ in $W_0^{1,p}(\Omega)$. Then, having got the strong convergence $\nabla u_k \to \nabla u$, pass to the limit directly in the Galerkin identity.

estimate by testing (2.146) by $v = u_k$.[70] Then prove weak convergence of the Galerkin method as in (2.84).[71]

Exercise 2.87. Modify Exercises 2.85 and 2.86 for non-zero Dirichlet or Newton boundary conditions.

Exercise 2.88 (Monotone case I). Consider the boundary-value problem (2.45)–(2.49) in the special case $a_i(x,r,s) := a_i(x,s)$ and $c(x,r,s) := c(x,r)$, i.e.

$$\left.\begin{aligned} -\mathrm{div}\, a(\nabla u) + c(u) &= g && \text{on } \Omega,\\ \nu\cdot a(\nabla u) + b(u) &= h && \text{on } \Gamma_{\mathrm{N}},\\ u|_{\Gamma_{\mathrm{D}}} &= u_{\mathrm{D}} && \text{on } \Gamma_{\mathrm{D}}. \end{aligned}\right\} \tag{2.147}$$

Assume that $a(x,\cdot)$, $b(x,\cdot)$, and $c(x,\cdot)$ are monotone, coercive (say $a(x,s)\cdot s \ge |s|^p$, $b(x,0) = 0$, $c(x,0) = 0$, and $\mathrm{meas}_{n-1}(\Gamma_{\mathrm{D}}) > 0$) with basic growth conditions, i.e.

$$\begin{aligned} &\big(a(x,s) - a(x,\tilde{s})\big)\cdot(s-\tilde{s}) \ge 0,\\ &\exists \gamma_a \in L^{p'}(\Omega),\ C_a \in \mathbb{R}: \quad |a(x,s)| \le \gamma_a(x) + C_a|s|^{p-1}, \end{aligned} \tag{2.148a}$$

$$\begin{aligned} &\big(b(x,r) - b(x,\tilde{r})\big)(r-\tilde{r}) \ \ge\ 0,\\ &\exists \gamma_b \in L^{p^{\#\prime}}(\Gamma),\ C_b \in \mathbb{R}: \quad |b(x,r)| \le \gamma_b(x) + C_b|r|^{p^{\#}-1}, \end{aligned} \tag{2.148b}$$

$$\begin{aligned} &\big(c(x,r) - c(x,\tilde{r})\big)(r-\tilde{r}) \ \ge\ 0,\\ &\exists \gamma_c \in L^{p^{*\prime}}(\Omega),\ C_c \in \mathbb{R}: \quad |c(x,r)| \le \gamma_c(x) + C_c|r|^{p^{*}-1}, \end{aligned} \tag{2.148c}$$

and prove a-priori estimates[72] and the convergence of Galerkin's approximations by Minty's trick.[73]

---

[70] Hint: Estimate

$$\varepsilon_a\|u_k\|^p_{W_0^{1,p}(\Omega)} = \varepsilon_a\|\nabla u_k\|^p_{L^p(\Omega;\mathbb{R}^n)} \le \int_\Omega a(\nabla u_k)\cdot\nabla u_k \mathrm{d}x \le \int_\Omega (g - c(\nabla u_k))u_k \mathrm{d}x$$

and finish it as in Exercise 2.85.

[71] Hint: Prove $\lim_{k\to\infty}\int_\Omega (a(\nabla u_k) - a(\nabla u))\cdot\nabla(u_k - u)\,\mathrm{d}x = 0$ as in Exercise 2.85. Then, for a selected subsequence, deduce $c(\nabla u_k) \to c(\nabla u)$ a.e. in Ω by the same way as done in (2.88), and similarly also $a(\nabla u_k) \to a(\nabla u)$ a.e. in Ω. Then prove $a(\nabla u_k) \rightharpoonup a(\nabla u)$ in $L^{p'}(\Omega)$ and $c(\nabla u_k) \rightharpoonup c(\nabla u)$ in $L^{p^{*\prime}+\epsilon}(\Omega)$ and pass to the limit directly in (2.146) for any $v \in \bigcup_{h>0} V_k$ without using Minty's trick. Finally, extend the resulted identity by continuity for any $v \in W^{1,p}(\Omega)$.

[72] Hint: Denoting $\bar{u}_{\mathrm{D}} \in W^{1,p}(\Omega)$ an extension of u_{D} test the Galerkin identity determining $u_k \in V_k$ by $v := u_k - \bar{u}_k$ where $\bar{u}_k|_\Gamma \to u_{\mathrm{D}}$ in $W^{1,p}(\Omega)|_\Gamma$ for $k \to \infty$, $\{\bar{u}_k\}_{k\in\mathbb{N}}$ bounded in $W^{1,p}(\Omega)$, $\bar{u}_k \in V_k$, V_k a finite-dimensional subspace of $W^{1,p}(\Omega)$. Arrive to

$$\begin{aligned} \int_\Omega |\nabla u_k|^p \mathrm{d}x &\le \int_\Omega a(\nabla u_k)\cdot\nabla u_k + c(u_k)u_k \mathrm{d}x + \int_{\Gamma_{\mathrm{N}}} b(u_k)u_k \mathrm{d}S\\ &= \int_\Omega a(\nabla u_k)\cdot\nabla\bar{u}_k + c(u_k)\bar{u}_k + g(u_k - \bar{u}_k)\,\mathrm{d}x + \int_{\Gamma_{\mathrm{N}}} b(u_k)\bar{u}_k + h(u_k - \bar{u}_k)\,\mathrm{d}S \end{aligned}$$

and then get u_k estimated in $W^{1,p}(\Omega)$ by Hölder's inequality and Poincaré's inequality (1.57). Alternatively, use the a-priori shift as in Proposition 2.27.

[73] Hint: For $v \in W^{1,p}(\Omega)$, use $v_k \to v$ in $W^{1,p}(\Omega)$, $v_k \in V_k$, $u_k|_{\Gamma_{\mathrm{D}}} = v_k|_{\Gamma_{\mathrm{D}}}$, the monotonicity

Exercise 2.89 (Monotone case II). Consider $A{:}W^{1,\max(2,p)}(\Omega) \to W^{1,\max(2,p)}(\Omega)^*$ given by

$$\langle A(u), v\rangle = \int_\Omega (1+|\nabla u|^{p-2})\nabla u\cdot\nabla v + c(u)v\,\mathrm{d}x + \int_\Gamma b(u)v\,\mathrm{d}S \tag{2.149}$$

so that the equation $A(u) = f$ with f from (2.60) corresponds to the boundary-value problem for the *regularized p-Laplacean*:

$$\left.\begin{array}{ll} -\mathrm{div}\big((1+|\nabla u|^{p-2})\nabla u\big) + c(x,u) = g & \text{in } \Omega,\\ (1+|\nabla u|^{p-2})\dfrac{\partial u}{\partial\nu} + b(x,u) = h & \text{on } \Gamma. \end{array}\right\} \tag{2.150}$$

Assume $c(x,\cdot)$ strongly monotone and $b(x,\cdot)$ either increasing or, if decreasing at a given point r, then being locally Lipschitz continuous with a constant ℓ_b^-:

$$\big(c(x,r) - c(x,\tilde r)\big)(r-\tilde r) \ge \varepsilon_c(r-\tilde r)^2, \tag{2.151}$$

$$\big(b(x,r) - b(x,\tilde r)\big)(r-\tilde r) \ge -\ell_b^-(r-\tilde r)^2. \tag{2.152}$$

Show that A can be monotone even if $b(x,\cdot)$ is not monotone; assume that[74]

$$\ell_b^- \le N^{-2}\min(1,\varepsilon_c). \tag{2.153}$$

Show further strong monotonicity of A with respect to the $W^{1,2}$-norm if (2.153) holds as a strict inequality.

Exercise 2.90 (Monotone case III). Let $A : W^{1,\max(2,p)}(\Omega) \to W^{1,\max(2,p)}(\Omega)^*$ be given by

$$\langle A(u), v\rangle = \int_\Omega (1+|\nabla u|^{p-2})\nabla u\cdot\nabla v + c(\nabla u)v\,\mathrm{d}x + \int_\Gamma b(u)v\,\mathrm{d}S. \tag{2.154}$$

---

and Galerkin's identity

$$0 \le \int_\Omega \big(a(\nabla u_k) - a(\nabla v_k)\big)\cdot\nabla(u_k{-}v_k) + \big(c(u_k)-c(v_k)\big)(u_k{-}v_k)\,\mathrm{d}x + \int_{\Gamma_\mathrm{N}} \big(b(u_k)-b(v_k)\big)$$
$$\times(u_k{-}v_k)\,\mathrm{d}S = \int_\Omega \big(g-c(v_k)\big)(u_k{-}v_k) - a(\nabla v_k)\cdot\nabla(u_k{-}v_k)\,\mathrm{d}x + \int_{\Gamma_\mathrm{N}} \big(h-b(v_k)\big)(u_k{-}v_k)\,\mathrm{d}S$$
$$\to \int_\Omega \big(g-c(v)\big)(u{-}v) - a(\nabla v)\cdot\nabla(u{-}v)\,\mathrm{d}x + \int_{\Gamma_\mathrm{N}} \big(h-b(v)\big)(u{-}v)\,\mathrm{d}S$$

and then put $v := u \pm \varepsilon w$, divide it by $\varepsilon > 0$, and pass $\varepsilon \to 0$.

[74]Hint: Indeed,

$$\langle A(u)-A(v), u-v\rangle = \int_\Omega |\nabla u - \nabla v|^2 + \big(|\nabla u|^{p-2}\nabla u - |\nabla v|^{p-2}\nabla v\big)\cdot(\nabla u - \nabla v)$$
$$+ \big(c(u)-c(v)\big)(u-v)\,\mathrm{d}x + \int_\Gamma \big(b(u)-b(v)\big)(u-v)\,\mathrm{d}S$$
$$\ge \int_\Omega |\nabla u - \nabla v|^2 + \varepsilon_c(u-v)^2\mathrm{d}x - \int_\Gamma \ell_b^-(u-v)^2\mathrm{d}S$$
$$\ge \min(1,\varepsilon_c)\|u{-}v\|^2_{W^{1,2}(\Omega)} - \ell_b^-\|u{-}v\|^2_{L^2(\Gamma)} \ge \big(\min(1,\varepsilon_c) - \ell_b^- N^2\big)\|u{-}v\|^2_{W^{1,2}(\Omega)}.$$

Note that the equation $A(u) = f$ with f from (2.60) corresponds to the boundary-value problem

$$\left.\begin{aligned} -\mathrm{div}\big((1+|\nabla u|^{p-2})\nabla u\big) + c(x,\nabla u) &= g \quad \text{for } x\in\Omega,\\ (1+|\nabla u|^{p-2})\frac{\partial u}{\partial\nu} + b(x,u) &= h \quad \text{for } x\in\Gamma. \end{aligned}\right\} \tag{2.155}$$

Assume $b(x,\cdot)$ strongly monotone and $c(x,\cdot)$ Lipschitz continuous, i.e.

$$\big(b(x,r)-b(x,\tilde r)\big)(r-\tilde r) \ge \varepsilon_b|r-r|^2, \tag{2.156}$$

$$\big|c(x,s)-c(x,\tilde s)\big| \le \ell_c|s-\tilde s|, \tag{2.157}$$

and show monotonicity of A if ℓ_c is sufficiently small, despite that $u\mapsto c(\nabla u)$ alone would not allow for any monotone structure.[75] In particular, if ℓ_c is small enough, realize that A is strictly monotone and uniqueness of the solution follows.

Exercise 2.91 (Monotone case IV: *advection*)**.** Consider a special case of (2.155) with $c(x,s) := \vec v(x)\cdot s$ with $\vec v : \Omega\to\mathbb{R}^n$ being a prescribed velocity field. Assume $\mathrm{div}\,\vec v \le 0$ (as in Exercise 2.82) and $\vec v|_\Gamma\cdot\nu \ge 0$, and show that A enjoys the monotonicity[76] even if there is no point-wise monotonicity.

Exercise 2.92. Consider the following boundary-value problem:

$$\left.\begin{aligned} -\mathrm{div}\Big(|\nabla u|^{p-2}\nabla u + a_0(x,u)\Big) &= g \quad \text{in } \Omega,\\ |\nabla u|^{p-2}\frac{\partial u}{\partial\nu} + b_0(x,u) &= h \quad \text{on } \Gamma. \end{aligned}\right\} \tag{2.158}$$

[75] Hint: Estimate

$$\begin{aligned} \big\langle A(u)-A(v),u-v\big\rangle &= \int_\Omega \Big((1+|\nabla u|^{p-2})\nabla u-(1+|\nabla v|^{p-2})\nabla v\Big)\cdot\nabla(u-v)\\ &\qquad + \big(c(\nabla u)-c(\nabla v)\big)(u-v)\,\mathrm{d}x + \int_\Gamma \big(b(u)-b(v)\big)(u-v)\,\mathrm{d}S\\ &\ge \|\nabla u-\nabla v\|^2_{L^2(\Omega;\mathbb{R}^n)} - \|c(\nabla u)-c(\nabla v)\|_{L^2(\Omega)}\|u-v\|_{L^2(\Omega)} + \varepsilon_b\|u-v\|^2_{L^2(\Gamma)}\\ &\ge \|\nabla u-\nabla v\|^2_{L^2(\Omega;\mathbb{R}^n)} - \ell_c\|\nabla u-\nabla v\|_{L^2(\Omega;\mathbb{R}^n)}\|u-v\|_{L^2(\Omega)} + \varepsilon_b\|u-v\|^2_{L^2(\Gamma)}\\ &\ge \Big(1-\frac{\ell_c^2}{2\delta}\Big)\|\nabla u-\nabla v\|^2_{L^2(\Omega;\mathbb{R}^n)} + \varepsilon_b\|u-v\|^2_{L^2(\Gamma)} - \frac{\delta}{2}\|u-v\|^2_{L^2(\Omega)}\\ &\ge C_{\mathrm P}^{-1}\min\Big(1-\frac{\ell_c^2}{2\delta},\varepsilon_b\Big)\|u-v\|^2_{W^{1,2}(\Omega)} - \frac{\delta}{2}N^2\|u-v\|^2_{W^{1,2}(\Omega)}, \end{aligned}$$

with N the norm of the embedding $W^{1,2}(\Omega)\subset L^2(\Omega)$ and $C_{\mathrm P}$ the constant from the Poincaré inequality (1.56) with $p=2=q$ and $\Gamma_{\mathrm N}=\Gamma$. If ℓ_c is so small that there is some $\delta>0$ such that

$$\min\Big(1-\frac{\ell_c^2}{2\delta},\varepsilon_b\Big) \ge \frac{\delta}{2}N^2C_{\mathrm P},$$

the monotonicity of A follows.

[76] Hint: By using Green's formula, the monotonicity of this linear term is based on the estimate:

$$\int_\Omega(\vec v\cdot\nabla u)u\,\mathrm{d}x = \frac12\int_\Omega \vec v\cdot\nabla u^2\mathrm{d}x = \frac12\int_\Gamma(\vec v\cdot\nu)u^2\mathrm{d}S - \frac12\int_\Omega(\mathrm{div}\,\vec v)\,u^2\mathrm{d}x \ge 0.$$

Assume the basic growth condition: $|a_0(x,r)| \leq \gamma(x) + C|r|^{p^*/p'}$ for some $\gamma \in L^{p'}(\Omega)$ and formulate a definition of the weak solution; denote: $b(x,r) := b_0(x,r) - a_0(x,r)\cdot\nu(x)$. Prove that $u \mapsto \mathrm{div}(a_0(x,u)) : W^{1,p}(\Omega) \to W^{1,p}(\Omega)^*$ is a totally continuous mapping (which allows us to use Theorem 2.6 with Corollary 2.12 to get the existence of a weak solution). Further prove the a-priori estimate by testing by u.[77] Prove convergence of the Galerkin approximation via Minty's trick, and alternatively strong convergence and direct limit passage without Minty's trick. Show uniqueness of the weak solution for Lipschitz continuous $a_0(x,\cdot)$ with a small Lipschitz constant. Make the modification for the Dirichlet boundary condition.[78]

Example 2.93 (Banach fixed-point technique). Consider the boundary-value problem (2.147)) and assume the strong monotonicity of $a(x,\cdot)$ and, e.g., of $c(x,\cdot)$ but no monotonicity of $b(x,\cdot)$, i.e.

$$\big(a(x,s) - a(x,\tilde{s})\big)\cdot(s - \tilde{s}) \geq \varepsilon_a|s - \tilde{s}|^2, \tag{2.159a}$$
$$\big(c(x,r) - c(x,\tilde{r})\big)(r - \tilde{r}) \geq \varepsilon_c(r - \tilde{r})^2, \tag{2.159b}$$

and the Lipschitz continuity

$$|a(x,s) - a(x,\tilde{s})| \leq \ell_a|s - \tilde{s}|, \tag{2.160a}$$
$$-\ell_b^-(r - \tilde{r})^2 \leq \big(b(x,r) - b(x,\tilde{r})\big)(r - \tilde{r}) \leq \ell_b^+(r - \tilde{r})^2, \tag{2.160b}$$
$$|c(x,r) - c(x,\tilde{r})| \leq \ell_c|r - \tilde{r}|, \tag{2.160c}$$

with some $\ell_b^+ \geq \ell_b^- \geq 0$; note that $b(x,\cdot)$ is Lipschitz continuous with the constant ℓ_b^+. Then one can use the Banach fixed-point Theorem 1.12 technique based on the contractiveness of the mapping T_ε from (2.43) where the Lipschitz constant ℓ of A can be estimated as:[79]

---

[77] Hint: Realize that

$$\int_\Omega |\nabla u|^p \mathrm{d}x + \int_\Gamma b_0(u)u\,\mathrm{d}S = \int_\Omega -a_0(u)\cdot\nabla u + gu\,\mathrm{d}x + \int_\Gamma \big(a_0(u)\cdot\nu + h\big)\,u\,\mathrm{d}S.$$

Assume $b_0(x,r)r \geq |r|^p$, and estimate u by assuming further $|a_0(x,r)| \leq \gamma(x) + C|r|^{p-1-\epsilon}$.

[78] Hint: Denoting $\widehat{a}(x,r) = \big(\widehat{a}_1(x,r),\dots,\widehat{a}_n(x,r)\big)$ the component-wise primitive functions to $a_0(x,r) = \big(a_1(x,r),\dots,a_n(x,r)\big)$ and realizing that now $u|_\Gamma = u_\mathrm{D}$, by Green's Theorem 1.31, one gets

$$\int_\Omega a_0(x,u)\cdot\nabla u\,\mathrm{d}x = \int_\Omega \sum_{i=1}^n \frac{\partial}{\partial x_i}\widehat{a}_i(x,u)\,\mathrm{d}x = \int_\Gamma \sum_{i=1}^n \widehat{a}_i(x,u)\nu_i(x)\,\mathrm{d}S = \int_\Gamma \widehat{a}(u_\mathrm{D})\cdot\nu\,\mathrm{d}S = \mathrm{const.}$$

[79] Cf. also (4.17) below.

$$
\begin{aligned}
\|A(u)-A(v)\|_{W^{1,2}(\Omega)^*} &= \sup_{\|z\|_{W^{1,2}(\Omega)}\le 1} \langle A(u)-A(v),z\rangle \\
&= \sup_{\|z\|_{W^{1,2}(\Omega)}\le 1} \int_\Omega \big(a(\nabla u)-a(\nabla v)\big)\cdot\nabla z + \big(c(u)-c(v)\big)z\,\mathrm{d}x + \int_{\Gamma_\mathrm{N}} \big(b(u)-b(v)\big)z\,\mathrm{d}S \\
&\le \sup_{\|z\|_{W^{1,2}(\Omega)}\le 1} \Big(\ell_a\|\nabla u-\nabla v\|_{L^2(\Omega;\mathbb{R}^n)}\|\nabla z\|_{L^2(\Omega;\mathbb{R}^n)} \\
&\qquad + \ell_c\|u-v\|_{L^2(\Omega)}\|z\|_{L^2(\Omega)} + \ell_b^+\|u-v\|_{L^2(\Gamma_\mathrm{N})}\|z\|_{L^2(\Gamma_\mathrm{N})}\Big) \\
&\le \big(\sqrt{2}\max(\ell_a,\ell_c)+N^2\ell_b^+\big)\|u-v\|_{W^{1,2}(\Omega)} =: \ell\|u-v\|_{W^{1,2}(\Omega)}
\end{aligned}
$$

while the constant δ in the strong monotonicity of A can be estimated as $\langle A(u)-A(v),u-v\rangle \ge \big(\min(\varepsilon_c,\varepsilon_a)-N^2\ell_b^-\big)\|u-v\|^2_{W^{1,2}(\Omega)} =: \delta\|u-v\|^2_{W^{1,2}(\Omega)}$; cf. Exercise 2.89. Then, by Proposition 2.22, T_ε from (2.43) with $J: W^{1,2}(\Omega)\to W^{1,2}(\Omega)^*$ defined by[80]

$$
\langle J(u),v\rangle = \int_\Omega \nabla u\cdot\nabla v + uv\,\mathrm{d}x \tag{2.161}
$$

is a contraction provided $\varepsilon>0$ satisfies[81]

$$
\varepsilon < 2\frac{\min(\varepsilon_c,\varepsilon_a)-N^2\ell_b^-}{\big(\sqrt{2}\max(\ell_a,\ell_c)+N^2\ell_b^+\big)^2}. \tag{2.162}
$$

Exercise 2.94. Modify the above Example 2.93 for Dirichlet boundary conditions[82] and/or the term $c(\nabla u)$ instead of $c(u)$[83].

Example 2.95 (Limit passage in coefficients)**.** Consider the problem from Example 2.93 modified, for simplicity, as in Exercise 2.94 with zero Dirichlet boundary conditions. Assume $s\mapsto a(x,s)$ and $r\mapsto c(x,r)$ monotone, $a(x,s)\cdot s+c(x,r)\cdot r\ge \varepsilon_0|s|^p-C$, $|a(x,s)|\le\gamma(x)+C|s|^{p-1}$ with $\gamma\in L^{p'}(\Omega)$ and $1<p\le 2$. Such a problem does not satisfy (2.159) and (2.160a,c). Therefore, we approximate a and c respectively by some a_ε and c_ε which will satisfy both (2.159) and (2.160a,c) and such that $a_\varepsilon(x,\cdot)\to a(x,\cdot)$ uniformly on bounded sets in $\mathbb{R}^n$, and $c_\varepsilon(x,\cdot)\to c(x,\cdot)$ uniformly on bounded sets in $\mathbb{R}$, and such that the collection $\{(a_\varepsilon,c_\varepsilon)\}_{\varepsilon>0}$ is uniformly coercive in the sense

$$
\exists\delta>0\ \forall\varepsilon>0: \qquad a_\varepsilon(x,s)\cdot s+c_\varepsilon(x,r)\cdot r\ge\delta|s|^p-1/\delta. \tag{2.163}
$$

[80]Note that $\langle J(u),u\rangle = \|u\|^2_{W^{1,2}(\Omega)}$ and also $\|u\|_{W^{1,2}(\Omega)} = \|J(u)\|_{W^{1,2}(\Omega)^*}$ if one considers the standard norm $\|u\|_{W^{1,2}(\Omega)} = \sqrt{\|\nabla u\|^2_{L^2(\Omega;\mathbb{R}^n)}+\|u\|^2_{L^2(\Omega)}}$; cf. Remark 3.15.

[81]Cf. (2.44) on p. 42.

[82]Hint: Instead of (2.161) use $\langle J(u),v\rangle = \int_\Omega \nabla u\cdot\nabla v\,\mathrm{d}x$, cf. Proposition 3.14.

[83]Hint: In case of Newton boundary conditions, $b(x,\cdot)$ has to be strongly monotone as in Exercise 2.90.

E.g. one can put $a_\varepsilon(x,\cdot) := \mathscr{Y}^{\rm M}_{n,\varepsilon}(a(x,\cdot))$ and $c_\varepsilon(x,\cdot) := \mathscr{Y}^{\rm M}_{1,\varepsilon}(c(x,\cdot))$ where $\mathscr{Y}^{\rm M}_{n,\varepsilon}$: $\mathbb{R}^n \to \mathbb{R}^n$ denotes a suitable *modification of Yosida's approximation* $\mathscr{Y}_{n,\varepsilon}:\mathbb{R}^n \to \mathbb{R}^n$ defined by

$$\big[\mathscr{Y}^{\rm M}_{n,\varepsilon}(f)\big](s) := \Big[\mathscr{Y}_{n,\varepsilon}\Big(f + \frac{\varepsilon^2}{1-\varepsilon}\mathbf{I}_n\Big)\Big](s) \quad \text{with} \tag{2.164a}$$

$$\big[\mathscr{Y}_{n,\varepsilon}(f)\big](s) := \frac{s - \big(\mathbf{I}_n{+}\varepsilon f\big)^{-1}(s)}{\varepsilon} \tag{2.164b}$$

and $\mathbf{I}_n$ the identity on $\mathbb{R}^n$; cf. also Remark 5.18 below. Unlike the mere Yosida approximation $\mathscr{Y}_{n,\varepsilon}$, the regularization (2.164) turns monotonicity to strong monotonicity; note also that $\mathscr{Y}^{\rm M}_{n,\varepsilon}(\mathbf{I}_n) = \mathbf{I}_n$.

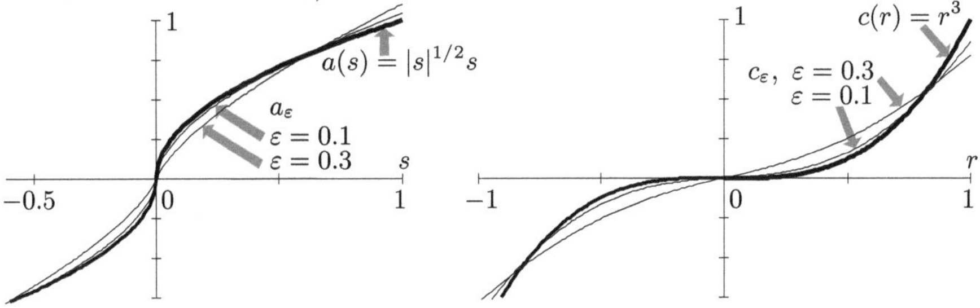

Figure 7. A regularization of the nonlinearities $a(x,s) = |s|^{1/2}s$ and $c(x,r) = r^3$ that makes them both strongly monotone and Lipschitz continuous.

Then we can obtain the weak solution $u_\varepsilon \in W^{1,2}(\Omega)$ of the approximate problem

$$\left.\begin{aligned} -{\rm div}\, a_\varepsilon(\nabla u) + c_\varepsilon(u) &= g \quad \text{in } \Omega, \\ u &= 0 \quad \text{on } \Gamma \end{aligned}\right\} \tag{2.165}$$

constructively by Example 2.93 (modified as in Exercise 2.94). The convergence of $u_\varepsilon \in W^{1,2}_0(\Omega)$ for $\varepsilon \to 0$ relies on an a-priori estimate in $W^{1,p}_0(\Omega)$ which is uniform with respect to $\varepsilon > 0$ due to (2.163), and then a selection of a subsequence $u_\varepsilon \rightharpoonup u$ in $W^{1,p}(\Omega)$. Note that, as $p \le 2$, we have $W^{1,p}_0(\Omega) \supset W^{1,2}_0(\Omega)$. Taking $\widetilde v \in W^{1,\infty}_0(\Omega)$ and using monotonicity, we obtain

$$\begin{aligned} 0 &\le \int_\Omega \big(a_\varepsilon(\nabla u_\varepsilon){-}a_\varepsilon(\nabla\widetilde v)\big){\cdot}(\nabla u_\varepsilon{-}\nabla\widetilde v) + \big(c_\varepsilon(u_\varepsilon){-}c_\varepsilon(\widetilde v)\big)(u_\varepsilon{-}\widetilde v)\,{\rm d}x \\ &= \int_\Omega \big(g - c_\varepsilon(\widetilde v)\big)(u_\varepsilon - \widetilde v) - a_\varepsilon(\nabla\widetilde v){\cdot}(\nabla u_\varepsilon - \nabla\widetilde v)\,{\rm d}x \\ &\to \int_\Omega \big(g - c(\widetilde v)\big)(u - \widetilde v) - a(\nabla\widetilde v){\cdot}(\nabla u - \nabla\widetilde v)\,{\rm d}x \end{aligned} \tag{2.166}$$

for $\varepsilon \to 0$, where we used $a_\varepsilon(\nabla\widetilde v) \to a(\nabla\widetilde v)$ in $L^\infty(\Omega;\mathbb{R}^n)$. Then we can pass $\widetilde v$ to $v \in W^{1,p}_0(\Omega)$; by density of $W^{1,\infty}_0(\Omega)$ in $W^{1,p}_0(\Omega)$, cf. Theorem 1.25, v can be considered arbitrary. By continuity of the Nemytskiĭ mappings $\mathscr{N}_a : L^p(\Omega;\mathbb{R}^n) \to L^{p'}(\Omega;\mathbb{R}^n)$ and $\mathscr{N}_c : L^{p^*}(\Omega) \to L^{p^{*\prime}}(\Omega)$, from (2.166) we get $\int_\Omega (g - c(v))(u - v) - a(\nabla v)\cdot(\nabla u - \nabla v)\,\mathrm{d}x \geq 0$. Eventually, by Minty's trick, we conclude that u solves (2.165); cf. Lemma 2.13.

Remark 2.96 (Constructivity). Let us still point out that, by combining the Banach fixed-point iterations as in Example 2.93 with some coefficient approximation as in Example 2.95, one can solve problems as (2.147) under quite weak assumptions rather constructively, without any Brouwer's fixed-point argument, cf. Remark 2.7. In case of strict monotonicity in (2.147), the whole sequence of approximate solutions converges.

Exercise 2.97. Modify Example 2.95 for the case of Newton boundary conditions.

Exercise 2.98. Add a term $\mathrm{div}\big(b(x,u,\nabla u,\nabla^2 u)\big)$ here with $b:\Omega\times\mathbb{R}\times\mathbb{R}^n\times\mathbb{R}^{n\times n}\to\mathbb{R}^n$ into (2.97) and modify (2.99) and Propositions 2.42 and 2.43.

Exercise 2.99. Realizing that only four out of all six combinations of derivatives up to 3rd-order on the boundary have been used in (2.100), (2.101), (2.105), and (2.107), identify the remaining two combinations and explain why they are not compatible with a consistent and selective weak formulation.[84]

Exercise 2.100 (*Singular higher-order perturbations*). Consider the weak solution $u_\varepsilon \in W^{2,2}(\Omega)\cap W^{1,p}(\Omega)$ of the problem

$$\left.\begin{aligned} \mathrm{div}\big(\varepsilon\mathrm{div}\nabla^2 u - |\nabla u|^{p-2}\nabla u\big) &= g &&\text{in } \Omega,\\ \tfrac{\partial u}{\partial\nu} = u &= 0 &&\text{on } \Gamma. \end{aligned}\right\} \tag{2.167}$$

Prove the a-priori estimates

$$\big\|u_\varepsilon\big\|_{W^{1,p}(\Omega)} \leq C, \qquad \big\|u_\varepsilon\big\|_{W^{2,2}(\Omega)} \leq C/\sqrt{\varepsilon}. \tag{2.168}$$

By using Minty's trick based on the monotonicity of the mapping $\varepsilon\mathrm{div}^2\nabla^2 - \Delta_p$, prove the weak convergence $u_\varepsilon \rightharpoonup u$ in $W_0^{1,p}(\Omega)$ to the solution of the boundary-value problem $\mathrm{div}(|\nabla u|^{p-2}\nabla u) + g = 0$ and $u = 0$ on Γ.[85] Alternatively, make

[84] Hint: These two wrong options would exactly over-determine either the first or the second boundary term in (2.104).

[85] Hint: Taking into account the identity $\int_\Omega |\nabla u_\varepsilon|^{p-2}\nabla u_\varepsilon\cdot\nabla v + \varepsilon\nabla^2 u_\varepsilon{:}\nabla^2 v - gv\mathrm{d}x = 0$, use $\|\varepsilon\nabla^2 u_\varepsilon\|_{L^2(\Omega;\mathbb{R}^{n\times n})} = \mathscr{O}(\sqrt{\varepsilon})$ and, for any $v \in W_0^{2,2}(\Omega)\cap W^{1,p}(\Omega)$, show

$$\begin{aligned} 0 &\leq \int_\Omega \big(|\nabla u_\varepsilon|^{p-2}\nabla u_\varepsilon - |\nabla v|^{p-2}\nabla v\big)\cdot\nabla(u_\varepsilon - v) + \varepsilon\big|\nabla^2 u_\varepsilon - \nabla^2 v\big|^2\,\mathrm{d}x\\ &= \int_\Omega g(u_\varepsilon - v) - |\nabla v|^{p-2}\nabla v\cdot\nabla(u_\varepsilon - v) - \varepsilon\nabla^2 v{:}\nabla^2(u_\varepsilon - v)\,\mathrm{d}x \to \int_\Omega g(u-v) - |\nabla v|^{p-2}\nabla v\cdot\nabla(u-v)\,\mathrm{d}x. \end{aligned}$$

Then extend the limit identity by continuity for all $v \in W_0^{1,p}(\Omega)$, and use $v := u \pm \epsilon z$ and accomplish it by Minty's trick.

it by Minty's trick based on the monotonicity of the mapping $-\Delta_p$.[86] Show also the strong convergence $u_\varepsilon \to u$ in $W_0^{1,p}(\Omega)$ by using d-monotonicity of $-\Delta_p$.[87] Modify it by considering also term $c(\nabla u)$ as in Example 2.85 and/or Newton-type boundary conditions, or a quasilinear regularizing term as in Example 2.46.

2.7 Excursion to regularity for semilinear equations

By *regularity* we understand, in general, that the weak solution has some additional differentiability properties as a consequence of some additional qualification of data, i.e. in case of the boundary-value problem (2.45)–(2.49) a certain differentiability of a, b, c, g, and h, and a qualification of Ω as smoothness or restrictions on angles of possible corners. This represents usually a difficult task and there are examples showing that, in case of higher-order equations or systems of equations, any smoothness of the data need not imply an additional smoothness of weak solutions. Regularity theory is a broad and still developing area which determines a lot of investigations in particular in systems of nonlinear equations and in numerical analysis, and the exposition presented below is to be understood as only an absolutely minimal excursion into this area.

We will confine ourselves to $W^{k,2}$-type regularity for semilinear equations and we start with a so-called interior regularity[88] for the linear equation

[86] Hint: Again first for any $v \in W_0^{2,2}(\Omega) \cap W^{1,p}(\Omega)$, calculate

$$\begin{aligned}0 &\le \int_\Omega \big(|\nabla u_\varepsilon|^{p-2}\nabla u_\varepsilon - |\nabla v|^{p-2}\nabla v\big)\cdot\nabla(u_\varepsilon - v)\,dx\\ &= \int_\Omega g(u_\varepsilon - v) - |\nabla v|^{p-2}\nabla v\cdot\nabla(u_\varepsilon - v) - \varepsilon\nabla^2 u_\varepsilon{:}\nabla^2(u_\varepsilon - v)\,dx\\ &\le \int_\Omega g(u_\varepsilon - v) - |\nabla v|^{p-2}\nabla v\cdot\nabla(u_\varepsilon - v) + \varepsilon\nabla^2 u_\varepsilon{:}\nabla^2 v\,dx \to \int_\Omega g(u-v) - |\nabla v|^{p-2}\nabla v\cdot\nabla(u-v)\,dx.\end{aligned}$$

[87] Hint: Using Example 2.83, for any $v \in W_0^{2,2}(\Omega) \cap W^{1,p}(\Omega)$, show

$$\begin{aligned}&\Big(\|\nabla u_\varepsilon\|^{p-1}_{L^p(\Omega;\mathbb{R}^n)} - \|\nabla u\|^{p-1}_{L^p(\Omega;\mathbb{R}^n)}\Big)\Big(\|\nabla u_\varepsilon\|_{L^p(\Omega;\mathbb{R}^n)} - \|\nabla u\|_{L^p(\Omega;\mathbb{R}^n)}\Big)\\ &\quad\le \int_\Omega \big(|\nabla u_\varepsilon|^{p-2}\nabla u_\varepsilon - |\nabla u|^{p-2}\nabla u\big)\cdot\nabla(u_\varepsilon - u)\,dx\\ &\quad= \int_\Omega g(u_\varepsilon - v) - \varepsilon\nabla^2 u_\varepsilon{:}\nabla^2(u_\varepsilon - v) - |\nabla u|^{p-2}\nabla u\cdot\nabla(u_\varepsilon - u) + |\nabla u_\varepsilon|^{p-2}\nabla u_\varepsilon\cdot\nabla(v-u)\,dx\\ &\quad\le \int_\Omega g(u_\varepsilon - v) + \varepsilon\nabla^2 u_\varepsilon{:}\nabla^2 v - |\nabla u|^{p-2}\nabla u\cdot\nabla(u_\varepsilon - u) + |\nabla u_\varepsilon|^{p-2}\nabla u_\varepsilon\cdot\nabla(v-u)\,dx\\ &\quad\to \int_\Omega g(u-v) + \xi\cdot\nabla(v-u)\,dx\end{aligned}$$

with some $\xi \in L^{p'}(\Omega;\mathbb{R}^n)$ being a weak limit of (a subsequence) of $|\nabla u_\varepsilon|^{p-2}\nabla u_\varepsilon$. Pushing $v \to u$ in $W^{1,p}(\Omega)$ makes the last expression arbitrarily close to zero, which shows $\|\nabla u_\varepsilon\|_{L^p(\Omega;\mathbb{R}^n)} \to \|\nabla u\|_{L^p(\Omega;\mathbb{R}^n)}$, hence the strong convergence $u_\varepsilon \to u$.

[88] This means we get estimates only in subdomains of Ω having a positive distance from Γ.

$$\sum_{i,j=1}^{n} \frac{\partial}{\partial x_i}\Big(a_{ij}(x)\frac{\partial u}{\partial x_j}\Big) = g(x) \qquad \text{on } \Omega \tag{2.169}$$

with nonspecified boundary conditions. By a weak solution to (2.169) we will naturally understand $u \in W^{1,2}(\Omega)$ such that $\int_\Omega (\nabla u)^\top \mathbb{A}\nabla v - gv\,\mathrm{d}x = 0$ for all $v \in W_0^{1,2}(\Omega)$ where $\mathbb{A} : \Omega \mapsto \mathbb{R}^{n\times n} : x \mapsto \mathbb{A}(x) = [a_{ij}(x)]_{i,j=1}^n$.

Proposition 2.101 (Interior $W^{2,2}$-regularity). *Let $\mathbb{A} \in C^1(\Omega;\mathbb{R}^{n\times n})$ satisfy*

$$\exists \delta > 0 \ \ \forall \zeta \in \mathbb{R}^n \ \ \forall (\text{a.a.})\, x \in \Omega : \qquad \zeta^\top \mathbb{A}(x)\,\zeta \geq \delta|\zeta|^2, \tag{2.170}$$

$g \in L^2_{\rm loc}(\Omega)$, and let u be a weak solution to (2.169). Then $u \in W^{2,2}_{\rm loc}(\Omega)$. Moreover, for any open sets $O, O_2 \subset \mathbb{R}^n$ satisfying $\bar{O} \subset O_2$ and $\bar{O}_2 \subset \Omega$, it holds that

$$\|u\|_{W^{2,2}(O)} \leq C\big(\|g\|_{L^2(O_2)} + \|u\|_{L^2(\Omega)}\big) \tag{2.171}$$

with $C = C\big(O, O_2, \|\mathbb{A}\|_{C^1(\Omega;\mathbb{R}^{n\times n})}\big)$.

As the rigorous proof is very technical and not easy to observe, we begin with a heuristic one. Take still an open set O_1 such that $\bar{O} \subset O_1$ and $\bar{O}_1 \subset O_2$, and a smooth "cut-off function" $\zeta : \Omega \to [0,1]$ such that $\chi_O \leq \zeta \leq \chi_{O_1}$. Then, for a test function

$$v := \frac{\partial}{\partial x_k}\Big(\zeta^2 \frac{\partial u}{\partial x_k}\Big) \tag{2.172}$$

with $k = 1, \ldots, n$, by using Green's Theorem 1.30, we have formally the identity

$$\begin{aligned}
&\int_{O_1} \sum_{i,j=1}^{n} a_{ij}\frac{\partial u}{\partial x_j}\frac{\partial}{\partial x_i}\Big(\frac{\partial}{\partial x_k}\Big(\zeta^2\frac{\partial u}{\partial x_k}\Big)\Big)\mathrm{d}x \\
&= -\int_{O_1} \sum_{i,j=1}^{n} \frac{\partial}{\partial x_k}\Big(a_{ij}\frac{\partial u}{\partial x_j}\Big)\frac{\partial}{\partial x_i}\Big(\zeta^2\frac{\partial u}{\partial x_k}\Big)\mathrm{d}x \\
&= -\int_{O_1} \sum_{i,j=1}^{n} \Big(\frac{\partial a_{ij}}{\partial x_k}\frac{\partial u}{\partial x_j} + a_{ij}\frac{\partial^2 u}{\partial x_j \partial x_k}\Big)\Big(\zeta^2\frac{\partial^2 u}{\partial x_i \partial x_k} + 2\zeta\frac{\partial \zeta}{\partial x_i}\frac{\partial u}{\partial x_k}\Big)\mathrm{d}x.
\end{aligned} \tag{2.173}$$

The identity (2.173) leads to the estimate

$$\begin{aligned}
\delta\Big\|\zeta\nabla\frac{\partial u}{\partial x_k}\Big\|^2_{L^2(O_1;\mathbb{R}^n)} &= \delta\int_{O_1}\sum_{i=1}^{n}\zeta^2\Big|\frac{\partial^2 u}{\partial x_i\partial x_k}\Big|^2\mathrm{d}x \\
&\leq \int_{O_1}\sum_{i,j=1}^{n}\zeta^2 a_{ij}\frac{\partial^2 u}{\partial x_i \partial x_k}\frac{\partial^2 u}{\partial x_j\partial x_k}\mathrm{d}x = -\int_{O_1}\sum_{i,j=1}^{n}\zeta^2\frac{\partial a_{ij}}{\partial x_k}\frac{\partial u}{\partial x_j}\frac{\partial^2 u}{\partial x_i\partial x_k} \\
&\qquad + 2\zeta\frac{\partial\zeta}{\partial x_i}\frac{\partial u}{\partial x_k}\Big(\frac{\partial a_{ij}}{\partial x_k}\frac{\partial u}{\partial x_j} + a_{ij}\frac{\partial^2 u}{\partial x_j\partial x_k}\Big) + g\Big(\frac{\partial}{\partial x_k}\Big(\zeta^2\frac{\partial u}{\partial x_k}\Big)\Big)\mathrm{d}x
\end{aligned}$$

$$
\begin{aligned}
&\leq \int_{O_1} \sum_{i,j=1}^{n} \|a_{ij}\|_{C^1(\Omega)} \bigg(\zeta^2 \Big| \frac{\partial u}{\partial x_j} \Big| \Big| \frac{\partial^2 u}{\partial x_i \partial x_k} \Big| \\
&\qquad + 2\zeta \|\zeta\|_{C^1(\Omega)} \Big| \frac{\partial u}{\partial x_k} \Big| \Big(\Big| \frac{\partial u}{\partial x_j} \Big| + \Big| \frac{\partial^2 u}{\partial x_j \partial x_k} \Big| \Big) \bigg) \mathrm{d}x \\
&\qquad + \|g\|_{L^2(O_1)} \Big\| 2\zeta \nabla \zeta \frac{\partial u}{\partial x_k} + \zeta^2 \nabla \frac{\partial u}{\partial x_k} \Big\|_{L^2(O_1;\mathbb{R}^n)} \\
&\leq C_1 \big(\|\nabla u\|_{L^2(O_1;\mathbb{R}^n)} + \|g\|_{L^2(O_1)} \big) \bigg(\Big\| \zeta \nabla \frac{\partial u}{\partial x_k} \Big\|_{L^2(O_1;\mathbb{R}^n)} + \Big\| \zeta \frac{\partial u}{\partial x_k} \Big\|_{L^2(O_1)} \bigg) \\
&\leq \frac{\delta}{2} \Big\| \zeta \nabla \frac{\partial u}{\partial x_k} \Big\|^2_{L^2(O_1;\mathbb{R}^n)} + \Big(\frac{C_1^2}{\delta} + \frac{3}{2} \Big) \big(\|\nabla u\|^2_{L^2(O_1;\mathbb{R}^n)} + \|g\|^2_{L^2(O_1)} \big) \qquad (2.174)
\end{aligned}
$$

with C_1 depending on $\|[a_{ij}]_{i,j=1}^n\|_{C^1(\Omega;\mathbb{R}^{n\times n})}$ and $\|\zeta\|_{C^1(\Omega)}$. Then, letting k range over $1,..,n$, we obtain

$$
\|u\|_{W^{2,2}(O)} \leq C_2 \big(\|g\|_{L^2(O_1)} + \|u\|_{W^{1,2}(O_1)} \big). \qquad (2.175)
$$

Finally, using a smooth "cut-off function" $\eta : \Omega \to [0,1]$ such that $\chi_{O_1} \leq \eta \leq \chi_{O_2}$ and the test-function $v = \eta u$, we get $\delta \|\nabla u\|^2_{L^2(O_1;\mathbb{R}^n)} \leq \delta \int_\Omega \eta |\nabla u|^2 \mathrm{d}x \leq \int_\Omega \eta g u \, \mathrm{d}x \leq \frac{1}{2}\|g\|^2_{L^2(O_2)} + \frac{1}{2}\|u\|^2_{L^2(\Omega)}$, which eventually leads to (2.171). The rigorous proof is, however, more complicated because (2.172) is not a legal test function unless we know that $u \in W^{2,2}_{\mathrm{loc}}(\Omega)$, which is just what we want to prove.

Sketch of the proof of Proposition 2.101. We introduce the difference operator D_k^ε defined by

$$
\big[\mathrm{D}_k^\varepsilon u\big](x) := \frac{u(x+\varepsilon e_k) - u(x)}{\varepsilon}, \qquad \varepsilon \neq 0, \ \ [e_k]_i := \begin{cases} 1 & \text{if } i = k, \\ 0 & \text{if } i \neq k, \end{cases} \qquad (2.176)
$$

and use the test function

$$
v := \mathrm{D}_k^{-\varepsilon} \big(\zeta^2 \mathrm{D}_k^\varepsilon u \big) \qquad (2.177)
$$

with $k = 1, \ldots, n$. Note that, contrary to (2.172), now $v \in W_0^{1,2}(\Omega)$ is a legal test function. The analog of Green's Theorem 1.30 is now

$$
\begin{aligned}
\int_\Omega v \mathrm{D}_k^{-\varepsilon} w \, \mathrm{d}x &= \int_\Omega v(x) \frac{w(x - \varepsilon e_k) - w(x)}{\varepsilon} \mathrm{d}x \\
&= \frac{1}{\varepsilon} \int_\Omega v(x) w(x - \varepsilon e_k) \, \mathrm{d}x - \frac{1}{\varepsilon} \int_\Omega v(x) w(x) \, \mathrm{d}x \\
&= \frac{1}{\varepsilon} \int_\Omega v(x + \varepsilon e_k) w(x) \, \mathrm{d}x - \frac{1}{\varepsilon} \int_\Omega v(x) w(x) \, \mathrm{d}x = -\int_\Omega w \mathrm{D}_k^\varepsilon v \, \mathrm{d}x \qquad (2.178)
\end{aligned}
$$

if $|\varepsilon|$ is smaller than the distance ε_0 of Γ from O_1; note that v vanishes on $\Omega \setminus O_1$. Moreover, by simple algebra, we have the formula

$$
\mathrm{D}_k^\varepsilon (vw) = \mathrm{S}_k^\varepsilon v \, \mathrm{D}_k^\varepsilon w + w \, \mathrm{D}_k^\varepsilon v \qquad (2.179)
$$

with the "shift" operator S_k^ε defined by $\big[\mathrm{S}_k^\varepsilon v\big](x) := v(x+\varepsilon e_k)$. The analog of (2.173) now reads as

$$\begin{aligned}\int_{O_1}\sum_{i,j=1}^n a_{ij}\frac{\partial u}{\partial x_j}\frac{\partial}{\partial x_i}\Big(\mathrm{D}_k^{-\varepsilon}(\zeta^2\mathrm{D}_k^\varepsilon u)\Big)\mathrm{d}x\\ = -\int_{O_1}\sum_{i,j=1}^n \mathrm{D}_k^\varepsilon\Big(a_{ij}\frac{\partial u}{\partial x_j}\Big)\frac{\partial}{\partial x_i}\Big(\zeta^2\mathrm{D}_k^\varepsilon u\Big)\mathrm{d}x\\ = -\int_{O_1}\sum_{i,j=1}^n\Big(\mathrm{D}_k^\varepsilon a_{ij}\frac{\partial u}{\partial x_j}+\mathrm{S}_k^\varepsilon a_{ij}\mathrm{D}_k^\varepsilon\frac{\partial u}{\partial x_j}\Big)\Big(\zeta^2\mathrm{D}_k^\varepsilon\frac{\partial u}{\partial x_i}+2\zeta\frac{\partial\zeta}{\partial x_i}\mathrm{D}_k^\varepsilon u\Big)\mathrm{d}x.\end{aligned} \tag{2.180}$$

We also use that $\|\mathrm{D}_k^\varepsilon v\|_{L^2(\Omega_1)}\le\|\nabla v\|_{L^2(\Omega)}$ if $|\varepsilon|\le\varepsilon_0:=\mathrm{dist}(O_1,\Gamma)$.[89] Then the analog of (2.174) reads as

$$\begin{aligned}\delta\big\|\zeta\,\mathrm{D}_k^\varepsilon\nabla u\big\|^2_{L^2(O_1;\mathbb{R}^n)} \le \frac{\delta}{2}\big\|\zeta\,\mathrm{D}_k^\varepsilon\nabla u\big\|^2_{L^2(O_1;\mathbb{R}^n)}\\ +\Big(\frac{C_1^2}{\delta}+\frac{3}{2}\Big)\big(\big\|\nabla u\big\|^2_{L^2(O_1;\mathbb{R}^n)}+\|g\|^2_{L^2(O_1)}\big).\end{aligned} \tag{2.181}$$

Hence the sequence (selected from) $\{\zeta\,\mathrm{D}_k^\varepsilon\nabla u\}_{0<\varepsilon\le\varepsilon_0}$ is bounded in $L^2(O_1;\mathbb{R}^n)$ and converges, possibly as a subsequence, weakly to some w in $L^2(O_1;\mathbb{R}^n)$. In the sense of distributions, it must hold that $w=\zeta\frac{\partial}{\partial x_k}\nabla u$.[90] In particular, $\frac{\partial}{\partial x_k}\nabla u\in L^2(O;\mathbb{R}^n)$ and, if considering $k=1,\ldots,n$, we have obtained (2.175). Then (2.171) follows as outlined in the heuristics. □

Proposition 2.102 (Interior $W^{3,2}$-regularity). *Let $\mathbb{A}\in C^1(\Omega;\mathbb{R}^{n\times n})\cap W^{2,q}(\Omega;\mathbb{R}^{n\times n})$ with $q=2^*2/(2^*-2)$ with 2^* from (1.34) satisfy (2.170), and let $g\in W^{1,2}_{\mathrm{loc}}(\Omega)$, and let u be a weak solution to (2.169). Then $u\in W^{3,2}_{\mathrm{loc}}(\Omega)$. Moreover, for any open sets $O,O_2\subset\mathbb{R}^n$ satisfying $\bar{O}\subset O_2$ and $\bar{O}_2\subset\Omega$, it holds that*

$$\|u\|_{W^{3,2}(O)}\le C\big(\|g\|_{W^{1,2}(O_2)}+\|u\|_{L^2(\Omega)}\big) \tag{2.182}$$

with $C=C\big(O,\|\mathbb{A}\|_{C^1(\Omega;\mathbb{R}^{n\times n})\cap W^{2,q}(\Omega;\mathbb{R}^{n\times n})}\big)$.

Proof. Applying $\frac{\partial}{\partial x_k}$ to (2.169), we obtain

$$\sum_{i,j=1}^n\frac{\partial}{\partial x_i}\Big(a_{ij}\frac{\partial^2 u}{\partial x_j\partial x_k}\Big)=\frac{\partial g}{\partial x_k}-\sum_{i,j=1}^n\Big(\frac{\partial^2 a_{ij}}{\partial x_i\partial x_k}\frac{\partial u}{\partial x_j}+\frac{\partial a_{ij}}{\partial x_k}\frac{\partial^2 u}{\partial x_j\partial x_i}\Big) \tag{2.183}$$

[89] It holds that $[\mathrm{D}_k^\varepsilon v](x)=\int_0^1\frac{\partial}{\partial x_k}(v+\tau\varepsilon e_k)\mathrm{d}\tau$ so that, by Hölder inequality, we obtain $\|\mathrm{D}_k^\varepsilon v\|^2_{L^2(\Omega_1)}=\int_{\Omega_1}\big|\int_0^1\frac{\partial}{\partial x_k}(v+\tau\varepsilon e_k)\mathrm{d}\tau\big|^2\mathrm{d}x\le\int_\Omega|\nabla v|^2\mathrm{d}x$.

[90] For any $v\in\mathscr{D}(O)$ it holds that $\lim_{\varepsilon\to0}\int_{\Omega_1}(\zeta\,\mathrm{D}_k^\varepsilon\nabla u)v\,\mathrm{d}x=\lim_{\varepsilon\to0}-\int_\Omega\mathrm{D}_k^{-\varepsilon}(\zeta v)\nabla u\,\mathrm{d}x=-\int_\Omega\frac{\partial}{\partial x_k}(\zeta v)\nabla u\,\mathrm{d}x=-\int_O\frac{\partial}{\partial x_k}v\nabla u\,\mathrm{d}x$.

in Ω. Note that, by Proposition 2.101, $u \in W^{2,2}_{\rm loc}(\Omega)$ and therefore (2.183) has indeed a good "weak" sense: $z := \frac{\partial}{\partial x_k}u$ is a weak solution to (2.169) with z instead of u and with $\frac{\partial}{\partial x_k}g - {\rm div}((\frac{\partial}{\partial x_k}\mathbb{A})\nabla u) - (\frac{\partial}{\partial x_k}\mathbb{A})\nabla^2 u \in L^2_{\rm loc}(\Omega)$ instead of g. Hence $\frac{\partial}{\partial x_k}u \in W^{2,2}_{\rm loc}(\Omega)$. □

For linear equations as (2.169) the process suggested in (2.183) can be iterated for $k = 4, \ldots$ to obtain $W^{k,2}_{\rm loc}$-regularity under the assumption that $\mathbb{A} \in C^{k-2}(\Omega;\mathbb{R}^{n\times n}) \cap W^{k-1,q}(\Omega;\mathbb{R}^{n\times n})$ and $g \in W^{k-2,2}(\Omega)$. This differs from nonlinear equations where the regularity has usually a natural bound. Here, we confine ourselves to semilinear equations where results for linear equations can directly be exploited. To be more specific, we will handle the equation

$$\sum_{i,j=1}^{n} \frac{\partial}{\partial x_i}\Big(a_{ij}(x)\frac{\partial u}{\partial x_j} + a_{0i}(u)\Big) + c_0(\nabla u) + |u|^{q-2}u = g(x) \qquad \text{on } \Omega \tag{2.184}$$

again with unspecified boundary conditions. By a weak solution to (2.184) we will naturally understand $u \in W^{1,2}(\Omega)$ such that $\int_\Omega \big((\nabla u)^\top\mathbb{A}+a_0(u)\big)\cdot\nabla v+\big(c_0(\nabla u)+|u|^{q-2}u - g\big)v\,{\rm d}x = 0$ for all $v \in W^{1,2}_0(\Omega)$.

Proposition 2.103 (Regularity for semilinear equations).

(i) *Let $\mathbb{A} \in C^1(\Omega;\mathbb{R}^{n\times n})$ satisfy (2.170), let $1 < q \le (2n-2)/(n-2)$ for $n \ge 3$ (or $q > 1$ arbitrary if $n \le 2$), $a_0 : \mathbb{R} \to \mathbb{R}^n$ be Lipschitz continuous, c_0 have at most linear growth, and $g \in L^2_{\rm loc}(\Omega)$. Then any weak solution $u \in W^{1,2}(\Omega)$ to (2.184) satisfies also $u \in W^{2,2}_{\rm loc}(\Omega)$.*

(ii) *Moreover, let, in addition, $\mathbb{A} \in W^{2,\max(2,n+\epsilon)}(\Omega;\mathbb{R}^{n\times n})$ with $\epsilon > 0$ if $n = 2$ (otherwise $\epsilon = 0$ is allowed), and let also $q \ge 2$, $a_0 \in C^2(\mathbb{R};\mathbb{R}^n)$ with*

$$a_0'' : \mathbb{R} \to \mathbb{R}^n \begin{cases} \textit{having arbitrary growth} & \textit{if } n \le 3, \\ \textit{being bounded} & \textit{if } n = 4, \\ \equiv 0 & \textit{if } n \ge 5, \end{cases} \tag{2.185}$$

$c_0 : \mathbb{R}^n \to \mathbb{R}$ be Lipschitz continuous, and $g \in W^{1,2}_{\rm loc}(\Omega)$. Then any weak solution $u \in W^{1,2}(\Omega)$ to (2.184) belongs also to $W^{3,2}_{\rm loc}(\Omega)$.

Proof. Note that $u \in W^{1,2}(\Omega)$ implies ${\rm div}(a_0(u)) = a_0'(u)\nabla u = \sum_{i=1}^n a_{0i}'(u)\frac{\partial}{\partial x_i}u \in L^2(\Omega)$ if $a_0 \in W^{1,\infty}(\mathbb{R};\mathbb{R}^n)$ as assumed, cf. Proposition 1.28. Also, $c_0(\nabla u) \in L^2(\Omega)$ because of the linear growth of c_0, and eventually $|u|^{q-2}u \in L^{2^*/(q-1)}(\Omega) \subset L^2(\Omega)$ if $1 < q \le (2n-2)/(n-2)$ (or $q > 1$ arbitrary if $n \le 2$). Noting also that the exponent $2^*2/(2^*-2)$ equals $\max(2,n)$ if $n \ne 2$, or is greater than 2 if $n = 2$, we can use simply Proposition 2.101 with g being now $g_1 := g - {\rm div}(a_0(u)) - c_0(\nabla u) - |u|^{q-2}u \in L^2(\Omega)$. The point (i) is thus proved.

Assuming the additional data qualification as specified in the point (ii), we want to show that $g_1 \in W^{1,2}_{\rm loc}(\Omega)$. For $i = 1, \dots, n$, we have

$$\frac{\partial g_1}{\partial x_i} = \frac{\partial g}{\partial x_i} - \sum_{j=1}^{n} \Big(a'_{0j}(u) \frac{\partial^2 u}{\partial x_i \partial x_j} + a''_{0j}(u) \frac{\partial u}{\partial x_i} \frac{\partial u}{\partial x_j} + \frac{\partial c_0}{\partial s_i}(\nabla u) \frac{\partial^2 u}{\partial x_i \partial x_j} \Big) - (q-1)|u|^{q-2} \frac{\partial u}{\partial x_i}. \tag{2.186}$$

For $u \in W^{1,2}(\Omega)$, we have $|u|^{q-2} \in L^{2^*/(q-2)}(\Omega)$ so that, in general, we do not have $|u|^{q-2}\nabla u \in L^2(\Omega)$ guaranteed. Likewise, the a_0- and c_0-terms also do not live in $L^2(\Omega)$ in general if we do not have some additional information about $u \in W^{1,2}(\Omega)$. However, we can use the already proved assertion (i), i.e. $u \in W^{2,2}_{\rm loc}(\Omega)$; this trick is called a *bootstrap*[91]. Then it is easy to show that $g_1 \in W^{1,2}_{\rm loc}(\Omega)$ hence we can use simply Proposition 2.102 with g being now g_1. □

Having the data qualification $\mathbb{A} \in C^1(\Omega; \mathbb{R}^{n\times n})$ and $a_0 \in W^{1,\infty}(\mathbb{R}; \mathbb{R}^n)$ assumed and the $W^{2,2}_{\rm loc}(\Omega)$-regularity at our disposal, it is then straightforward to check that (2.184) holds not only in the weak sense but even a.e. in Ω. Such a mode of a solution to a differential equation is called a *Carathéodory solution.*

Let us now briefly outline how regularity up to the boundary can be obtained. We will confine ourselves to $W^{2,2}$-regularity and the Newton boundary conditions (2.48) and begin with (2.169). Thus (2.48) reads as

$$\sum_{j=1}^{n} \nu_i a_{ij}(x) \frac{\partial u}{\partial x_j} + b(x,u) = h(x) \qquad \text{on } \Gamma. \tag{2.187}$$

Proposition 2.104 ($W^{2,2}$-REGULARITY UP TO BOUNDARY). *Let Ω be of C^2-class, $\mathbb{A} \in C^1(\Omega; \mathbb{R}^{n\times n})$ satisfy (2.170), $b \in C^1(\mathbb{R}^n \times \mathbb{R})$ satisfy, for some $b_0 > 0$ and $C \in \mathbb{R}$,*

$$\forall (\text{a.a.}) x \in \Gamma \ \ \forall r_1, r_2 \in \mathbb{R}: \qquad \big(b(x,r_1) - b(x,r_2)\big)(r_1 - r_2) \geq b_0 \big|r_1 - r_2\big|^2, \tag{2.188a}$$

$$\exists \gamma \in L^2(\Gamma) \ \ \forall (\text{a.a.}) x \in \Gamma \ \ \forall r \in \mathbb{R}: \qquad \Big|\frac{\partial b}{\partial x}(x,r)\Big| \leq \gamma(x) + C|r|^{2^\#/2}, \tag{2.188b}$$

$g \in L^2(\Omega)$, $h \in W^{1,2}(\Gamma)$,[92] *and let $u \in W^{1,2}(\Omega)$ be the unique weak solution to the boundary-value problem (2.169)–(2.187). Then $u \in W^{2,2}(\Omega)$. Moreover, if $b(x,r) = b_1(x)r$ with $b_1 \in W^{1,2^\#2/(2^\#-2)}(\Gamma)$, then*

$$\|u\|_{W^{2,2}(\Omega)} \leq C\big(\|g\|_{L^2(\Omega)} + \|h\|_{W^{1,2}(\Gamma)}\big) \tag{2.189}$$

with $C = C\big(\Omega, \|\mathbb{A}\|_{C^1(\Omega;\mathbb{R}^{n\times n})}, \|b_1\|_{W^{1,2n-2+\epsilon}(\Gamma)}\big)$.

[91] Often, bootstrap is used not only in the order of differentiation but rather in the integrability, which is not possible here because we present the Hilbertian theory only.

[92] The notation $W^{1,2}(\Gamma)$ for Γ smooth means that, after a local rectification like on Figure 8, the transformed and "smoothly cut" functions belong to $W^{1,2}(\mathbb{R}^{n-1})$. Also $\frac{\partial}{\partial x} b$ in (2.188b) refers to the derivatives in the tangential directions only.

Sketch of the proof. First, as Ω is bounded, Γ is a compact set in $\mathbb{R}^n$, and can be covered by a finite number of open sets which are C^2-diffeomorphical images of the unit ball $B = \{\xi \in \mathbb{R}^n;\ |\xi| \leq 1\}$ such that the respective part of Γ is an image of $\{\xi = (\xi_1, \ldots, \xi_n) \in B;\ \xi_1 = 0\}$. Thus we rectified locally the boundary Γ, cf. Figure 8.

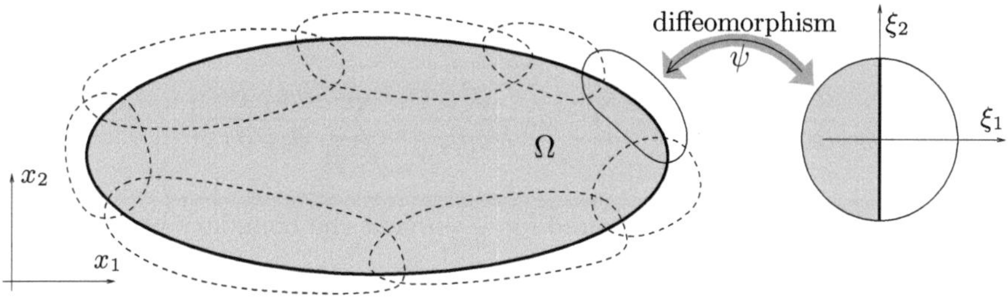

Figure 8. Illustration of finite coverage of $\Gamma \subset \mathbb{R}^2$ and one diffeomorphism rectifying locally a part of Γ.

It is a technical calculation showing that $\tilde{u} \in W^{1,2}(B_0)$ defined by $\tilde{u}(\xi) = u(\psi(\xi))$, where $\psi : B_0 := \{\xi \in B;\ x_1 \leq 0\} \to \bar{\Omega}$ is the homeomorphism in question, is a weak solution to an equation like (2.169) but with the coefficients $\mathbb{A}$ transformed but again being continuously differentiable and satisfying (2.170)[93], and the boundary condition (2.187) transforms to a similar condition for $\tilde{u}|_{\xi_1=0}$. Hence, in fact, it suffices to obtain an estimate like (2.189) only for $\tilde{u} \in W^{1,2}(B_0)$. For simplicity, we will use the original notation.

We again use the test function (2.177) but now only for $k = 2, \ldots, n$, i.e. we use shifts only in the tangential direction, so that we still have (2.178) at our disposal. Now the cut-off function $\zeta : B_0 \to [0, 1]$ can be taken as 1 in a semi-ball $\{\xi \in \mathbb{R}^n;\ |\xi| \leq 1 - \varepsilon_0,\ x_1 \leq 0\}$ and vanishing on $\{\xi \in \mathbb{R}^n;\ |\xi| \geq 1 - \frac{1}{2}\varepsilon_0,\ x_1 \leq 0\}$ with some ε_0. The heuristical estimate (2.173)–(2.174) now involves also the boundary term $\int_\Gamma (b(x, u) - h) \frac{\partial}{\partial x_k}(\zeta^2 \frac{\partial}{\partial x_k} u)\, \mathrm{d}S$ which, in the difference variant, reads and, for $|\varepsilon| \leq \frac{1}{2}\varepsilon_0$, can be estimated as

$$
\begin{aligned}
\int_\Gamma \big(b(x,u) - h\big) \mathrm{D}_k^{-\varepsilon}(\zeta^2 \mathrm{D}_k^\varepsilon u)\, \mathrm{d}S &= -\int_\Gamma \zeta^2 \mathrm{D}_k^\varepsilon \big(b(x,u) - h\big) \mathrm{D}_k^\varepsilon u \,\mathrm{d}S \\
= -\int_\Gamma \zeta^2 \bigg(\frac{b\big(x{+}\varepsilon e_k, u(x{+}\varepsilon e_k)\big) - b\big(x{+}\varepsilon e_k, u(x)\big)}{\varepsilon} &\\
-\mathrm{D}_k^\varepsilon h + \frac{1}{\varepsilon} \int_0^\varepsilon \frac{\partial b}{\partial r}\big(x{+}\tau e_k, u(x)\big) \mathrm{d}\tau \bigg) \mathrm{D}_k^\varepsilon u \,\mathrm{d}S &\\
\leq -b_0 \big\| \zeta \mathrm{D}_k^\varepsilon u \big\|_{L^2(\Gamma)}^2 + \Big(\big\| h \big\|_{W^{1,2}(\Gamma)} + \big\| \gamma{+}C|u|^{2^\#/2} \big\|_{L^2(\Gamma)} \Big) \big\| \zeta \mathrm{D}_k^\varepsilon u \big\|_{L^2(\Gamma)} &
\end{aligned}
$$

[93] To be more specific, $\tilde{u}$ satisfies $\sum_{i,j=1}^n \partial(\tilde{a}_{ij} \partial \tilde{u}/\partial x_j)\partial x_i = \tilde{g}$ with the transformed coefficients $\tilde{a}_{ij}(\xi) = \sum_{k,l=1}^n [a_{kl} \frac{\partial}{\partial x_k} \psi^{-1} \frac{\partial}{\partial x_l} \psi^{-1}](\psi(\xi))$ and $\tilde{g}(\xi) = g(\psi(\xi))$. The boundary conditions are transformed accordingly, i.e. $\tilde{b}(\xi, r) = b(\psi(\xi), r)$ and $\tilde{h}(\xi) = h(\psi(\xi))$.

$$\leq \|h\|^2_{W^{1,2}(\Gamma)} + \frac{1}{b_0}\Big(\|\gamma\|^2_{L^2(\Gamma)} + C^2\|u\|^{2^\#}_{L^{2^\#}(\Gamma)}\Big). \tag{2.190}$$

In this way, we get the local estimates for $\frac{\partial^2}{\partial x_i \partial x_j}u$ for all (i,j) except $i = 1 = j$ meant in the locally rectified coordinate system, cf. Figure 8(right).

The estimate of the normal derivative follows just from the equation itself which has been shown to hold a.e. in Ω. Thus

$$\frac{\partial^2 u}{\partial x_1^2} = \frac{1}{a_{11}}\Big(g - \sum_{i+j>2} a_{ij}\frac{\partial^2 u}{\partial x_i \partial x_j} - \sum_{i,j=1}^{n} \frac{\partial a_{ij}}{\partial x_i}\frac{\partial u}{\partial x_j}\Big) \tag{2.191}$$

from which we get the local L^2-bound for $\frac{\partial^2}{\partial x_1^2}u$ near the boundary because $a_{11}^{-1} \in L^\infty(\Omega)$ due to the uniform ellipticity of $\mathbb{A}$.

For the special case $b(x,r) = b_1(x)r$, the estimate (2.190) can be finalized by $\|[\frac{\partial}{\partial x}b](x,u)\|_{L^2(\Gamma)} \leq \|\frac{\partial}{\partial x}b_1\|_{L^{2^\# 2/(2^\# - 2)}(\Gamma)}\|u\|_{L^{2^\#}(\Gamma)}$. This eventually allows us to derive the a-priori estimate (2.189) by summing the (finite number of) the local estimates on the boundary with one estimate on an open set O from Proposition 2.101 and by using the conventional energy estimate $\|u\|_{W^{1,2}(\Omega)}$ and thus also $\|u\|_{L^{2^\#}(\Gamma)}$ in terms of g and h. $\Box$

Corollary 2.105 ($W^{2,2}$-REGULARITY FOR SEMILINEAR EQUATION). *Let the assumptions of Propositions* 2.103(i) *and* 2.104 *be satisfied. Then any weak solution* u *to the equation* (2.184) *with the boundary conditions*

$$\sum_{j=1}^{n} \nu_i\Big(a_{ij}(x)\frac{\partial u}{\partial x_j} + a_{0i}(u)\Big) + b(x,u) = h(x) \qquad \text{on } \Gamma \tag{2.192}$$

is a Carathéodory solution and belongs also to $W^{2,2}(\Omega)$.

Remark 2.106 (Dirichlet boundary conditions). Alternatively, instead of (2.192), one can think about prescribing $u|_\Gamma = u_{\rm D}$ with $u_{\rm D} = w|_\Gamma$ for some $w \in W^{2,2}(\Omega)$. After a shift by w, cf. Proposition 2.27, one gets a problem for $u_0 = u - w$ with zero Dirichlet condition and a contribution to the right-hand side which is again in $L^2(\Omega)$. The proof of Proposition 2.104 is even simpler because (2.190) simply vanishes.

2.8 Bibliographical remarks

Pseudomonotone mappings have been introduced by Brézis [64].[94] A further reading can involve the books by Nečas [305], Pascali and Sburlan [325], Renardy and

[94] In fact, [64] allows for $A : V \to V_2$ with V_2 "in duality" with V but not necessarily $V_2 = V^*$, and also requires $u \mapsto \langle A(u), u - v\rangle$ to be lower bounded on each compact set in V and for each $v \in V$, which is weaker than (2.3a). In literature, "pseudomonotone" sometimes omit (2.3a) completely, cf. [427, Definition 27.5].

Rogers [349], Růžička [376], and Zeidler [427, Chap.27]. Mere monotone mappings can be found there, too, and also in a lot of further monographs, say [95, 168, 414, 424]. Historically, theory of monotone mappings arises by the works by Browder [73], Minty [286], and Vishik [416].

The mappings weakly continuous when restricted to finite-dimensional subspaces and satisfying (2.123) are called *mappings of the type (M)*, having been invented by Brézis [64], and further generalized e.g. in [201, 227]. This class involves both the pseudomonotone and the weakly continuous mappings[95] but, contrary to those two classes, it is not closed under addition. Mappings of type (M) do not inherit some other nice properties of pseudomonotone mappings, too.[96] As to the weakly continuous mappings, their importance in the context of semilinear equations has been pointed out by Franců [149]. The setting $A : V \to Z^* \supset V^*$ with $V_k \subset Z$ we used in Section 2.5 was used by Hess [201] in the context of the mappings of the type (M), see also [325, Ch.IV, Sect.3.1] or [427, Sect.27.7]. The mappings satisfying (2.23) are called *mappings of the type* (S_+); this notion has been invented by Browder [76, p.279].

The fruitful Galerkin method originated at the beginning of 20th century [171], being motivated by engineering applications.

Concrete quasilinear partial differential equations in the divergence form has been scrutinized, e.g., by Chen and Wu [92, Chap.5], Fučík and Kufner [159], Gilbarg and Trudinger [178, Chap.11], Ladyzhenskaya and Uraltseva [250, Chap.4], Lions [261, Sect.2.2], Nečas [305], Taylor [402, Chap.14], and Zeidler [427, Chap.27]. For semilinear equations see Pao [324]. Quasilinear equations in a non-divergence form (not mentioned in here) can be found, e.g., in Ladyzhenskaya and Uraltseva [250, Chap.6] or Gilbarg and Trudinger [178, Chap.12]. *Fully nonlinear equations* of the type $a(\Delta u) = g$ (also not mentioned in here) are, e.g., in Chen, Wu [92, Chap.7], Caffarelli, Chabré [86], Dong [126, Chap.9,10], Gilbarg and Trudinger [178, Chap.17].

Regularity results in Sect. 2.7 can easily be generalized for the strongly monotone quasilinear equation of the type (2.147) satisfying (2.159a). More general regularity theory for elliptic equations is exposed, e.g., in the monographs by Bensoussan, Frehse [50], Evans [138], Giaquinta [175], Gilbarg, Trudinger [178], Grisvard [190], Lions, Magenes [262], Ladyzhenskaya, Uraltseva [250], Nečas [302, 305], Renardy, Rogers [349], Skrypnik [387], and Taylor [402]. Besides, this active research area is recorded in thousands of papers; e.g. Agmon, Douglis, and Nirenberg [6] and Nečas [303].

[95] For the implication "pseudomonotone ⇒ type-(M)" see Exercise 2.54 while the implication "weakly continuous ⇒ type-(M)" is obvious – note that even $\limsup_{k\to\infty} \langle A(u_k), u_k\rangle \le \langle f, u\rangle$ occurring in (2.123) does not need to have a sense if $A(u_k) \in Z^* \setminus V^*$ or $f \in Z^* \setminus V^*$.

[96] E.g., Φ' of type (M) does not yield weak lower-semicontinuity of Φ, unlike pseudomonotonicity, cf. Theorem 4.4(ii); e.g. $\Phi(u) = -\|u\|^2$ if V is an infinite-dimensional Hilbert space.

Chapter 3

Accretive mappings

Besides bounded mappings from a Sobolev space to its dual, there is an alternative understanding of differential operators as unbounded operators from a (typically dense) subset of a function space to itself. This calls for a generalization of a monotonicity concept for mappings $D \to X$, with X a Banach space and D its subset. Moreover, X need not be reflexive because the weak-compactness arguments will be replaced by metric properties and completeness. The main benefit from this approach will be achieved for evolution problems in Chapter 9 but the method is of some interest in steady-state problems themselves.

3.1 Abstract theory

For brevity, let us agree to write $\|\cdot\|$ and $\|\cdot\|_*$ instead of $\|\cdot\|_X$ and $\|\cdot\|_{X^*}$, respectively.

Definition 3.1. A *duality mapping* (in general set-valued) $J : X \rightrightarrows X^*$ is defined by:

$$J(u) := \big\{f \in X^*;\ \ \langle f, u\rangle = \|u\|^2 = \|f\|_*^2\big\}. \tag{3.1}$$

Lemma 3.2. *Let X be a separable*[1] *Banach space.*

(i) *$J(u)$ is nonempty, closed, and convex, and J is (norm,weak*)-upper semicontinuous.*

(ii) *If X^* is strictly convex, then J is single-valued, demicontinuous (i.e. here (norm,weak*)-continuous), and d-monotone with $d : \mathbb{R} \to \mathbb{R}$ linear.*

(iii) *If X^* is uniformly convex, then J is continuous.*

(iv) *If X is strictly convex, then J is also strictly monotone.*

[1] In fact, if general-topology tools and Alaoglu-Bourbaki's theorem would be used instead of Banach's Theorem 1.7, non-separable spaces can be considered, as well.

Proof. (i) Closedness of the set $J(u)$ is obvious while its convexity follows from the chain of estimates:

$$\|u\|^2 = \Big\langle \frac{1}{2}f_1 + \frac{1}{2}f_2, u\Big\rangle \le \Big\|\frac{1}{2}f_1 + \frac{1}{2}f_2\Big\|_* \|u\| \le \Big(\frac{1}{2}\|f_1\|_* + \frac{1}{2}\|f_2\|_*\Big)\|u\| = \|u\|^2.$$

Nonemptyness of $J(u)$ is a consequence of the Hahn-Banach Theorem 1.5, allowing us to separate any $v \in X$, $\|v\| = 1$, from the interior of the unit ball, i.e. there is $g \in X^*$ such that $\|g\|_* = \sup_{\|\tilde{u}\|=1}\langle g, \tilde{u}\rangle = 1$ and $\langle g, v\rangle = 1$. For arbitrary $0 \ne u \in X$, we then have $f = g\|u\| \in J(u)$ with g selected as previously for $v := u/\|u\|$ because obviously $\langle f, u\rangle = \|u\|\langle g, u\rangle = \|u\|^2\langle g, u/\|u\|\rangle = \|u\|^2\langle g, v\rangle = \|u\|^2$ and $\|f\|_* = \|g\|_*\|u\| = \|u\|$. If $u = 0$, then obviously $J(u) \ni 0$.

To show the (norm,weak*)-upper semicontinuity of J, take $u_k \to u$, $f_k \overset{*}{\rightharpoonup} f$, and $f_k \in J(u_k)$. Then

$$\langle f, u\rangle \leftarrow \langle f_k, u_k\rangle = \|u_k\|^2 \to \|u\|^2 \tag{3.2}$$

and therefore $\langle f, u\rangle = \|u\|^2$. Since $\|\cdot\|_*$ is convex and continuous, it is weakly* lower semicontinuous and thus

$$\|f\|_* \le \liminf_{k\to\infty} \|f_k\|_* = \liminf_{k\to\infty} \|u_k\| = \lim_{k\to\infty} \|u_k\| = \|u\| = \frac{\langle f, u\rangle}{\|u\|} \le \|f\|_*, \tag{3.3}$$

where $\langle f, u\rangle = \|u\|^2$ has been used. Altogether, $\|f\|_* = \|u\|$ and thus $f \in J(u)$.

(ii) Strict convexity of X^* and convexity of $J(u)$ implies that $J(u)$ is a singleton because $J(u)$ always belongs to a sphere in X^* of the radius $\|u\|$. The demicontinuity in the sense of the (norm,weak*)-continuity of J then follows from the point (i).

The d-monotonicity of J follows from the estimate

$$\begin{aligned}\langle J(u) - J(v), u - v\rangle &= \|J(u)\|_*\|u\| + \|J(v)\|_*\|v\| - \langle J(u), v\rangle - \langle J(v), u\rangle \\ &\ge \|J(u)\|_*\|u\| + \|J(v)\|_*\|v\| - \|J(u)\|_*\|v\| - \|J(v)\|_*\|u\| \\ &= \big(\|J(u)\|_* - \|J(v)\|_*\big)\big(\|u\| - \|v\|\big) = \big(\|u\| - \|v\|\big)^2. \end{aligned} \tag{3.4}$$

(iii) Besides $J(u_k) \overset{*}{\rightharpoonup} J(u)$ for $u_k \to u$, we have also $\|J(u_k)\|_* = \|u_k\| \to \|u\| = \|J(u)\|_*$ so that Theorem 1.2 yields $J(u_k) \to J(u)$.

(iv) Suppose $\langle J(u) - J(v), u - v\rangle = 0$. From (3.4) immediately follows $\|u\| = \|v\|$. Suppose, for a moment, that $u \ne v$. Then also $u/\|u\| \ne v/\|v\|$ and thus $\|J(u)\|_* = \langle J(u), u/\|u\|\rangle > \langle J(u), v/\|v\|\rangle$ because the supremum in $\|J(u)\|_* = \sup_{\|z\|\le 1}\langle J(u), z\rangle$ can be attained in at most one point because X is strictly convex; this point is $z = u/\|u\|$. Therefore $\langle J(u), u\rangle > \langle J(u), v\rangle$. Similarly, also $\langle J(v), v\rangle > \langle J(v), u\rangle$. Thus

$$\begin{aligned}\langle J(u) - J(v), u - v\rangle &= \langle J(u), u\rangle + \langle J(v), v\rangle - \langle J(u), v\rangle - \langle J(v), u\rangle \\ &> \langle J(u), v\rangle + \langle J(v), u\rangle - \langle J(u), v\rangle - \langle J(v), u\rangle = 0, \end{aligned} \tag{3.5}$$

a contradiction to $\langle J(u) - J(v), u - v\rangle = 0$. $\square$

Corollary 3.3. *Let X be reflexive and both X and X^* be strictly convex.*[2] *Then J^{-1} exists, and is the duality mapping $X^*{\to}X$. In particular, J^{-1} is demicontinuous.*

Proof. We use Browder-Minty's theorem 2.18. Thus it remains to show the coercivity of J: obviously $\langle J(u),u\rangle/\|u\| = \|u\|^2/\|u\| = \|u\| \to +\infty$ for $\|u\| \to \infty$. By symmetry of the definition (3.1), $J^{-1} : X^* \to X$ is then the duality mapping, and by Lemma 3.2(ii) it is demicontinuous. □

Definition 3.4. The mapping $A : \mathrm{dom}(A){\to}X$, $\mathrm{dom}(A){\subset}X$, is called *accretive* iff

$$\forall u,v\in\mathrm{dom}(A) \quad \exists f\in J(u-v): \qquad \langle f, A(u)-A(v)\rangle \geq 0. \tag{3.6}$$

If, in addition, $\mathbf{I}+A$ is surjective, A is called *m-accretive.*[3]

Remark 3.5. Introducing the notation of a so-called *semi-inner product* $\langle\cdot,\cdot\rangle_{\mathrm{s}}$ by

$$\langle u,v\rangle_{\mathrm{s}} := \sup\big\langle u, J(v)\big\rangle, \tag{3.7}$$

the definition (3.6) can equivalently be written[4] as $\langle A(u)-A(v),u-v\rangle_{\mathrm{s}} \geq 0$.

Remark 3.6. If $-A$ is accretive (resp. m-accretive), A is called *dissipative* (resp. m-dissipative).

Lemma 3.7 (Metric properties). *The mapping A is accretive if and only if $(\mathbf{I}+\lambda A)^{-1}$, defined on* Range$(\mathbf{I}+\lambda A)$, *is non-expansive for any $\lambda>0$, i.e.*

$$\|u-v\| \leq \|u+\lambda A(u)-v-\lambda A(v)\|. \tag{3.8}$$

Proof. The "only if" part: Let $u,v\in\mathrm{dom}(A)$ and $\langle f, A(u)-A(v)\rangle \geq 0$ for some $f\in J(u-v)$. Then

$$\begin{aligned}\|u-v\|^2 &= \langle f,u-v\rangle \leq \big\langle f,u-v+\lambda(A(u)-A(v))\big\rangle\\ &\leq \|f\|_*\big\|u{-}v+\lambda(A(u){-}A(v))\big\| = \|u{-}v\|\big\|u{-}v+\lambda(A(u){-}A(v))\big\| \end{aligned}\tag{3.9}$$

from which (3.8) follows for any $\lambda>0$.

The "if" part: Conversely, suppose that (3.8) holds. Let $f_\lambda \in J(u-v+\lambda(A(u)-A(v)))$. The case $u=v$ is trivial because $f=0\in J(0)=J(u-v)$ satisfies $\langle f, A(u)-A(v)\rangle \geq 0$. Hence we may assume $u\neq v$. Then $f_\lambda\neq 0$ because, by (3.1) and (3.8), $\|f_\lambda\|_* = \|u+\lambda A(u)-v-\lambda A(v)\| \geq \|u-v\| > 0$ and thus we

[2] In fact, the strict convexity is not restrictive in this case because, by Asplund's theorem, every reflexive space can be suitably re-normed so that the new norm is equivalent with the original one and both X and X^* are strictly convex.

[3] Sometimes, this is called "hypermaximal accretive" or "hyperaccretive", cf. Browder [76] Crandal and Pazy [109] or Deimling [118, Sect.13]. For "hyperdissipative" see Yosida [425, Sect.XIV.6].

[4] This equivalence is due to the weak* compactness of $J(v)$, cf. Lemma 3.2(i) and we realize that $J(v)$ is certainly bounded, so that the supremum in (3.7) is certainly attained.

can put $g_\lambda = f_\lambda/\|f_\lambda\|_*$. Then, up to a subsequence, $g_\lambda \rightharpoonup g$ weakly* for $\lambda \to 0$. Also, by (3.8), for all $\lambda > 0$,

$$\begin{aligned}\|u-v\| &\le \|u-v+\lambda A(u)-\lambda A(v)\| = \langle g_\lambda, u-v+\lambda A(u)-\lambda A(v)\rangle \\ &\le \|g_\lambda\|_*\|u-v\| + \lambda\langle g_\lambda, A(u)-A(v)\rangle = \|u-v\| + \lambda\langle g_\lambda, A(u)-A(v)\rangle, \quad (3.10)\end{aligned}$$

from which it follows that $\langle g_\lambda, A(u)-A(v)\rangle \ge 0$ for all $\lambda > 0$. For $\lambda \to 0$, one gets $\langle g, A(u)-A(v)\rangle \ge 0$. By the first part of (3.10), we have also

$$\|u-v\| \le \langle g_\lambda, u-v+\lambda A(u)-\lambda A(v)\rangle \to \langle g, u-v\rangle \le \|g\|_*\|u-v\|. \qquad (3.11)$$

In particular, (3.11) implies $\|g\|_* \ge 1$. Since $\|g\|_* \le 1$ (because $\|\cdot\|_*$ is weakly* lower semicontinuous), we get $\|g\|_* = 1$ and then, again from (3.11), we get $\|u - v\| = \langle g, u-v\rangle$. This shows that $f := g\|u-v\| \in J(u-v)$ from which (3.6) follows with $f = g\|u-v\|$. □

Proposition 3.8 (M-ACCRETIVITY). *If A is m-accretive, then $\mathbf{I} + \lambda A$ is surjective for all $\lambda > 0$.*

Proof. Let $f \in X$ and $\lambda > 0$. Then $u + \lambda A(u) = f$ means just

$$u = (\mathbf{I}+A)^{-1}\Big(\frac{1}{\lambda}f + \big(1-\frac{1}{\lambda}\big)u\Big) =: B_{\lambda,A}(u) \qquad (3.12)$$

because Range$(\mathbf{I} + A) = X$, cf. also Exercise 3.33. Moreover, $(\mathbf{I} + A)^{-1}$ is non-expansive on X (see Lemma 3.7) thus $B_{\lambda,A}$ is Lipschitz continuous with the constant $|1-\frac{1}{\lambda}|$. By the Banach fixed-point Theorem 1.12, (3.12) has a solution provided $\lambda > \frac{1}{2}$. Since f was arbitrary, we may conclude that Range$(\mathbf{I}+\lambda A) = X$ for $\lambda > \frac{1}{2}$. For $\lambda \le \frac{1}{2}$ we just iterate this procedure $[-\log_2(\lambda)]$-times, with $[\cdot]$ denoting here the integer part, relying on the fact that u should also be a fixed point of $B_{\lambda_1,\lambda_2 A}$ with $\lambda = \lambda_1\lambda_2$ and that, if $\lambda_1 > \frac{1}{2}$, $B_{\lambda_1,\lambda_2 A}$ is non-expansive provided we already have proved the surjectivity of $\mathbf{I} + \lambda_2 A$ and non-expansivity of its inverse. □

Remark 3.9 (Maximal accretivity). If A is m-accretive then it is maximal accretive with respect to the ordering of graphs by inclusion.[5]

Remark 3.10 (Special case dom$(A) = X \equiv X^*$ Hilbert). If X is a Hilbert space, then J is linear.[6] If also $X \equiv X^*$, then simply $J(u) = u$. If also dom$(A) = X$,

[5] Cf. Exercise 3.37 below. The opposite implication holds if X is a Hilbert space. In general, it does not hold, as shown by Crandall and Liggett [108]. The latter property equivalently means that, if A is accretive and if, for all $v \in$ dom(A), $\langle f - A(v), u-v\rangle_{\rm s} \ge 0$, then $u \in$ dom(A) and $A(u) = f$.

[6] In this case, one can define the linear operator $J : V \to V^*$ by $\langle Ju, v\rangle := (u, v)$ with $(\cdot,\cdot)$ denoting the inner product in V. Then $\langle Ju, u\rangle = (u, u) = \|u\|^2$ and $\|Ju\|_* = \sup_{\|v\|=1}\langle Ju, v\rangle = \sup_{\|v\|=1}(u, v) = \|u\|$, which obviously coincides with the definition (3.1).

then monotonicity just coincides with accretiveness and, moreover, any accretive radially-continuous A is m-accretive.[7]

Remark 3.11 (Generalized solutions of $u + \lambda A(u) = f$). If V is a normed linear space such that $V \subset X$ densely[8] and A, defined on V, is m-accretive, by (3.8) we can extend the uniformly continuous mapping $(\mathbf{I} + \lambda A)^{-1} : V \to V$ on X. This gives a generalized solution to $u + \lambda A(u) = f$ for $f \in X$, cf. Remark 3.18 below. Sometimes, this equation can be suitably interpreted, cf. (3.33) below.

Remark 3.12 (Solutions of $A(u) = f$). If A is m-accretive, by quite sophisticated arguments, it can be shown that not only $A + \mathbf{I}/\lambda$ but A itself is surjective provided also X and X^* are uniformly convex and A is coercive in the sense $\lim_{\|v\| \to \infty} \|A(v)\| = \infty$; cf. Deimling [118, Thm.13.4] or Hu and Papageorgiou [209, Part I, Thm.III.7.48].

3.2 Applications to boundary-value problems

This section, although having an interest of its own, is rather preparatory for Chapter 9 and will mainly be exploited there.

3.2.1 Duality mappings in Lebesgue and Sobolev spaces

Let us emphasize that we assume $\mathbb{R}^m$ to be endowed by the Euclidean norm $|\cdot|$ so that $r \cdot r = |r|^2$ for $r \in \mathbb{R}^m$. For another choice, see Exercise 3.39 below.

Proposition 3.13 (DUALITY MAPPING FOR LEBESGUE SPACES). *Let* $X = L^p(\Omega; \mathbb{R}^m)$. *If* $1 < p < +\infty$, *then* $J : L^p(\Omega; \mathbb{R}^m) \to L^{p'}(\Omega; \mathbb{R}^m)$ *is given by*

$$J(u)(x) = \frac{u(x)|u(x)|^{p-2}}{\|u\|_{L^p(\Omega;\mathbb{R}^m)}^{p-2}} \tag{3.13}$$

for a.a. $x \in \Omega$. *For* $p = 1$, J *is a set-valued mapping given by*

$$\begin{aligned} J(u) &= \|u\|_{L^1(\Omega;\mathbb{R}^m)} \mathrm{Dir}(u) \qquad \textit{where} \\ & \mathrm{Dir}(u) := \big\{ f \in L^\infty(\Omega; \mathbb{R}^m);\ f(x) \in \mathrm{dir}\big(u(x)\big) \ \textit{ for a.a. } x \in \Omega \big\}, \end{aligned} \tag{3.14}$$

where "dir" denotes the direction of the vector indicated, i.e.

$$\mathrm{dir}(r) := \begin{cases} r/|r| & \textit{if } r \neq 0, \\ \big\{ \tilde{r} \in \mathbb{R}^m;\ |\tilde{r}| \leq 1 \big\} & \textit{if } r = 0. \end{cases} \tag{3.15}$$

[7] Indeed, $\mathbf{I} : X \to X \equiv X^*$ is monotone, bounded, and radially continuous and also coercive, hence so is $\mathbf{I} + A$; the coercivity of $\mathbf{I} + A$ follows from $\frac{\langle u + A(u), u \rangle}{\|u\|} = \|u\| + \frac{\langle A(u), u \rangle}{\|u\|} \to \infty$ for $\|u\| \to \infty$ because, by monotonicity, the term $\langle A(u), u \rangle$ has at most linear decay: $\langle A(u), u \rangle = \langle A(u) - A(0), u - 0 \rangle + \langle A(0), u \rangle \geq -\|A(0)\|_* \|u\|$. Then, by Browder-Minty's Theorem 2.18, A is surjective.

[8] Then X is a so-called completion of V (with respect to a uniformity induced by the norm).

Proof. Recalling that $L^{p'}(\Omega;\mathbb{R}^m)$ is uniformly convex, see Section 1.2.2, by Lemma 3.2(ii), $J(u)$ has just one element, and we are to verify (3.13). Indeed, we obviously have

$$\langle J(u), u\rangle = \frac{\int_\Omega u(x)|u(x)|^{p-2}\cdot u(x)\,\mathrm{d}x}{\|u\|^{p-2}_{L^p(\Omega;\mathbb{R}^m)}} = \|u\|^{2-p}_{L^p(\Omega;\mathbb{R}^m)}\int_\Omega |u|^p\mathrm{d}x = \|u\|^2_{L^p(\Omega;\mathbb{R}^m)}$$

and also

$$\begin{aligned}\|J(u)\|^{p'}_{L^{p'}(\Omega;\mathbb{R}^m)} &= \int_\Omega \Big|u(x)|u(x)|^{p-2}\|u\|^{2-p}_{L^p(\Omega;\mathbb{R}^m)}\Big|^{p'}\,\mathrm{d}x\\ &= \int_\Omega |u(x)|^p\|u\|^{(2-p)p/(p-1)}_{L^p(\Omega;\mathbb{R}^m)}\,\mathrm{d}x = \|u\|^p_{L^p(\Omega;\mathbb{R}^m)}\|u\|^{(2-p)p/(p-1)}_{L^p(\Omega;\mathbb{R}^m)} = \|u\|^{p'}_{L^p(\Omega;\mathbb{R}^m)}.\end{aligned}$$

As to the case $p = 1$, we obviously have

$$\langle J(u), u\rangle = \|u\|_{L^1(\Omega;\mathbb{R}^m)}\int_\Omega f(x)\cdot u(x)\,\mathrm{d}x = \|u\|^2_{L^1(\Omega;\mathbb{R}^m)} \tag{3.16}$$

for any $f \in \mathrm{Dir}(u)$; note that always $u(x)\cdot\mathrm{dir}(u(x)) = |u(x)|$. Also, if $u \neq 0$,

$$\|J(u)\|_{L^\infty(\Omega;\mathbb{R}^m)} = \|u\|_{L^1(\Omega;\mathbb{R}^m)}\ \operatorname*{ess\,sup}_{x\in\Omega}\big|\mathrm{dir}\big(u(x)\big)\big| = \|u\|_{L^1(\Omega;\mathbb{R}^m)}. \tag{3.17}$$

If $u = 0$, then the desired equality $\|J(u)\|_{L^\infty(\Omega;\mathbb{R}^m)} = 0 = \|u\|_{L^1(\Omega;\mathbb{R}^m)}$ holds, too. This proved the inclusion "$\supset$" in (3.14). The opposite inclusion follows by a more detailed analysis of the above formulae. □

Proposition 3.14 (Duality mapping for Sobolev spaces). *Let $X = W^{1,p}_0(\Omega)$, $1 < p < +\infty$, normed by $\|u\|_{W^{1,p}_0(\Omega)} = \|\nabla u\|_{L^p(\Omega;\mathbb{R}^n)}$. Then*

$$J(u) = -\frac{\Delta_p u}{\|u\|^{p-2}_{W^{1,p}_0(\Omega)}}. \tag{3.18}$$

Proof. Uniform convexity of $L^{p'}(\Omega;\mathbb{R}^m)$ makes also $W^{1,p}_0(\Omega)^*$ uniformly convex. Hence, by Lemma 3.2(ii) $J(u)$ has just one element, and we are to verify (3.18). By Green's formula (1.54), we indeed have

$$\langle J(u), u\rangle = \frac{\langle -\Delta_p u, u\rangle}{\|u\|^{p-2}_{W^{1,p}_0(\Omega)}} = \frac{\int_\Omega |\nabla u|^{p-2}\nabla u\cdot\nabla u\;\mathrm{d}x}{\|u\|^{p-2}_{W^{1,p}_0(\Omega)}} = \|u\|^2_{W^{1,p}_0(\Omega)} \tag{3.19}$$

and, using again Green's formula (1.54) and Hölder's inequality, also

$$
\begin{aligned}
\|J(u)\|_{W_0^{1,p}(\Omega)^*} &= \sup_{\|v\|_{W_0^{1,p}(\Omega)}\le 1} \frac{\langle \Delta_p u, v\rangle}{\|u\|^{p-2}_{W_0^{1,p}(\Omega)}} \\
&= \sup_{\|v\|_{W_0^{1,p}(\Omega)}\le 1} \frac{\int_\Omega |\nabla u|^{p-2}\nabla u\cdot\nabla v\,\mathrm{d}x}{\|u\|^{p-2}_{W_0^{1,p}(\Omega)}} \\
&\le \sup_{\|v\|_{W_0^{1,p}(\Omega)}\le 1} \frac{\|\nabla u\|^{p-1}_{L^p(\Omega;\mathbb{R}^n)}\|\nabla v\|_{L^p(\Omega;\mathbb{R}^n)}}{\|u\|^{p-2}_{W_0^{1,p}(\Omega)}} = \|u\|_{W_0^{1,p}(\Omega)}
\end{aligned}
\tag{3.20}
$$

and the supremum (and thus equality) is attained for $v = u/\|u\|_{W_0^{1,p}(\Omega)}$. $\square$

Remark 3.15. For $X = W^{1,p}(\Omega)$ normed by $\|u\|_{W^{1,p}(\Omega)} = (\|\nabla u\|^p_{L^p(\Omega;\mathbb{R}^n)} + \|u\|^p_{L^p(\Omega)})^{1/p}$, we have

$$
J(u) = \frac{|u|^{p-2}u - \Delta_p u}{\|u\|^{p-2}_{W^{1,p}(\Omega)}}. \tag{3.21}
$$

Note that we used (3.21) already for $p = 2$ in (2.161).

3.2.2 Accretivity of monotone quasilinear mappings

Let us consider A corresponding to (2.147), i.e. $A(u) = -\mathrm{div}\, a(x,\nabla u) + c(x,u)$, and investigate its accretivity in $X = L^q(\Omega)$; for this we define[9]

$$
\begin{aligned}
\mathrm{dom}(A) := \Big\{ & u\in W^{1,p}(\Omega);\ \ \mathrm{div}\, a(x,\nabla u) - c(x,u)\in L^q(\Omega),\ \ u|_{\Gamma_\mathrm{D}} = u_\mathrm{D}, \\
& \nu\cdot a(x,\nabla u) + b(x,u) = h(x) \text{ on } \Gamma_\mathrm{N} \text{ in the ``weak sense''}\Big\}
\end{aligned}
\tag{3.22}
$$

with $h\in L^{p^{\#\prime}}(\Gamma_\mathrm{N})$ and u_D satisfying (2.58) fixed. As in Exercise 2.88, we assume here (2.148), i.e. in particular that $a(x,\cdot)$, $b(x,\cdot)$, and $c(x,\cdot)$ are monotone.

Proposition 3.16 (Accretivity). *Let $1 \le q \le p^*$, $q < +\infty$, and let (2.148), (2.57) concerning h, and (2.58) hold. Then $A(u) = -\mathrm{div}\, a(x,\nabla u) + c(x,u)$ posed as (3.22) is accretive on $L^q(\Omega)$.*

Proof. First, note that $\mathrm{dom}(A) \subset W^{1,p}(\Omega) \subset L^q(\Omega) =: X$ provided $q \le p^*$.

If $q = 1$, then "dir" (which is just "sign" for $m = 1$) in (3.15) is not continuous at 0, and we must use its regularization $\mathrm{sign}_\varepsilon : \mathbb{R} \to [-1,1]$ defined, e.g., by

$$
\mathrm{sign}_\varepsilon(r) = \begin{cases} r/|r| & \text{if } |r| \ge \varepsilon, \\ \varepsilon^{-1}r & \text{otherwise.} \end{cases} \tag{3.23}
$$

[9] For $q \ge p^{*\prime}$, (3.22) means that we take all weak solutions $u \in W^{1,p}(\Omega)$ for the boundary-value problem (2.147) with some $g \in L^q(\Omega)$ and with the boundary condition specified.

Then, for $u, v \in \mathrm{dom}(A)$ and for $f = \mathrm{sign}(u-v) \in J(u-v)/\|u-v\|_{L^1(\Omega)}$, we have

$$
\begin{aligned}
\langle f, A(u)-A(v)\rangle &= \int_\Omega \mathrm{sign}(u-v)\Big(c(u)-c(v) - \mathrm{div}\big(a(\nabla u)-a(\nabla v)\big)\Big)\mathrm{d}x \\
&= \lim_{\varepsilon\to 0}\int_\Omega \mathrm{sign}_\varepsilon(u-v)\Big(c(u)-c(v) - \mathrm{div}\big(a(\nabla u)-a(\nabla v)\big)\Big)\mathrm{d}x \\
&= \lim_{\varepsilon\to 0}\int_\Omega \big(a(\nabla u)-a(\nabla v)\big)\cdot\nabla \mathrm{sign}_\varepsilon(u-v) + \big(c(u)-c(v)\big)\mathrm{sign}_\varepsilon(u-v)\,\mathrm{d}x \\
&\quad + \int_{\Gamma_\mathrm{N}} \big(b(u)-b(v)\big)\mathrm{sign}_\varepsilon(u-v)\,\mathrm{d}S \ge 0
\end{aligned}
\tag{3.24}
$$

where we used, beside Green's formula (1.54), the convergence

$$
\lim_{\varepsilon\to 0}\mathrm{sign}_\varepsilon(u(x)-v(x)) = \mathrm{sign}(u(x)-v(x)) = \begin{cases} 1 & \text{if } u(x) > v(x), \\ 0 & \text{if } u(x) = v(x), \\ -1 & \text{if } u(x) < v(x), \end{cases}
\tag{3.25}
$$

for a.a. $x \in \Omega$ and then we used the Lebesgue Theorem 1.14 with the integrable majorant $|\mathrm{div}(a(\nabla u) - a(\nabla v)) - c(u) + c(v)| \in L^1(\Omega)$. The inequality in (3.24) is because $(a(\nabla u) - a(\nabla v))\cdot\nabla\mathrm{sign}_\varepsilon(u-v) = (a(\nabla u) - a(\nabla v))(\nabla u - \nabla v)\mathrm{sign}'_\varepsilon(u-v) \ge 0$, cf. Proposition 1.28, and also $(c(u) - c(v))\mathrm{sign}_\varepsilon(u-v) \ge 0$, and similarly $(b(u) - b(v))\mathrm{sign}_\varepsilon(u-v) \ge 0$.

For $q > 1$, we use J from (3.13). As the case $u = v$ is trivial, let us take $u, v \in \mathrm{dom}(A)$, $u \ne v$, and define $\omega_q(r) := r|r|^{q-2}/\|u-v\|^{q-2}_{L^q(\Omega)}$. Besides, let us consider a Lipschitz continuous regularization $\omega_{q,\varepsilon}$ of ω_q such that $\lim_{\varepsilon\to 0}\omega_{q,\varepsilon}(r) = \omega_q(r)$ for all r and $\omega_{q,\varepsilon}(r) \le \omega_q(r)$ for $r \ge 0$ and $\omega_{q,\varepsilon}(r) \ge \omega_q(r)$ for $r \le 0$. By using Lebesgue's Theorem 1.14 and Green's formula (1.54), we can calculate[10]

$$
\begin{aligned}
\langle J(u-v), A(u)-A(v)\rangle &= \int_\Omega \omega_q(u-v)\Big(c(u)-c(v) - \mathrm{div}\big(a(\nabla u)-a(\nabla v)\big)\Big)\mathrm{d}x \\
&= \lim_{\varepsilon\to 0}\int_\Omega \omega_{q,\varepsilon}(u-v)\Big(c(u)-c(v) - \mathrm{div}\big(a(\nabla u)-a(\nabla v)\big)\Big)\mathrm{d}x \\
&= \lim_{\varepsilon\to 0}\bigg(\int_\Omega \big(a(\nabla u)-a(\nabla v)\big)\cdot\nabla\omega_{q,\varepsilon}(u-v)\,\mathrm{d}x \\
&\quad + \int_\Omega \big(c(u)-c(v)\big)\omega_{q,\varepsilon}(u-v)\,\mathrm{d}x + \int_{\Gamma_\mathrm{N}} \big(b(u)-b(v)\big)\omega_{q,\varepsilon}(u-v)\,\mathrm{d}S\bigg) \\
&=: \lim_{\varepsilon\to 0}\big(I_{1,\varepsilon} + I_{2,\varepsilon} + I_{3,\varepsilon}\big).
\end{aligned}
\tag{3.26}
$$

[10] Note that $(u-v)|u-v|^{q-2} \in L^{q'}(\Omega)$ if $q \le p^*$ while $\mathrm{div}\big(a(\nabla u)-a(\nabla v)\big)+c(u)-c(v) \in L^q(\Omega)$ so that the product is indeed integrable and, up to a factor $\|u-v\|^{2-q}_{L^q(\Omega)}$, it also forms the integrable majorant for the collection $\{\omega_{q,\varepsilon}(u-v)(c(u)-c(v) - \mathrm{div}\big(a(\nabla u)-a(\nabla v)\big))\}_{\varepsilon>0}$ needed for the limit passage by Lebesgue's Theorem 1.14.

The first integral $I_{1,\varepsilon}$ can be estimated as

$$\begin{aligned} I_{1,\varepsilon} &= \int_\Omega \big(a(\nabla u) - a(\nabla v)\big)\cdot\nabla\omega_{q,\varepsilon}(u-v)\,\mathrm{d}x \\ &= \int_\Omega \big(a(\nabla u) - a(\nabla v)\big)\cdot\nabla(u-v)\,\omega'_{q,\varepsilon}(u-v)\,\mathrm{d}x \;\geq\; 0; \end{aligned} \tag{3.27}$$

note that $\omega_{q,\varepsilon}:\mathbb{R}\to\mathbb{R}$ is monotone and $\nabla\omega_{q,\varepsilon}(u-v) = \omega'_{q,\varepsilon}(u-v)\nabla(u-v)$; cf. Proposition 1.28. Of course, we used the monotonicity of $a(x,\cdot)$ and that $\omega'_{q,\varepsilon}$ is bounded. The monotonicity of $c(x,\cdot)$ and of $b(x,\cdot)$ obviously gives $I_{2,\varepsilon}\geq 0$ and $I_{3,\varepsilon}\geq 0$, respectively; the at most linear growth of $\omega_{q,\varepsilon}$ gives a good sense to both $I_{2,\varepsilon}$ and $I_{3,\varepsilon}$. $\square$

Proposition 3.17 (M-ACCRETIVITY). *Let, beside the assumptions of Proposition 3.16, also $q \geq p^{*\prime}$. Then $A(u) = -\mathrm{div}\; a(x,\nabla u) + c(x,u)$ posed as (3.22) is m-accretive on $L^q(\Omega)$.*

Proof. We are to show that the equation $u - \mathrm{div}\; a(\nabla u) + c(u) = g$ has a solution $u\in\mathrm{dom}(A)$ for any $g\in X = L^q(\Omega)$. As we assume $q\geq p^{*\prime}$, we have $L^q(\Omega)\subset L^{p^{*\prime}}(\Omega)$, and there is $u\in W^{1,p}(\Omega)$ solving the boundary-value problem (2.147) in the weak sense. Then it suffices to show $u\in\mathrm{dom}(A)$. Indeed, in the sense of distributions it holds that

$$\mathrm{div}\; a(\nabla u) - c(u) = u - g \;\in L^q(\Omega) \tag{3.28}$$

because $u\in W^{1,p}(\Omega)\subset L^{p^*}(\Omega)\subset L^q(\Omega)$ provided $p^*\geq q$. $\square$

Remark 3.18 (Generalization for $q < p^{*\prime}$). The restriction $p^{*\prime}\leq p^*$ we implicitly made in Propositions 3.16-3.17 requires $p\geq 2n/(n+2)$ and calls for some sort of extension if $q\notin[p^{*\prime},p^*]$ even if this interval is nonempty. For $q\in[1,p^{*\prime})$, one can still perform the estimate $\langle A(u)-A(v),u-v\rangle_{\mathrm{s}}\geq 0$ as in (3.24) or (3.26), and then by (3.9) one can prove the uniform continuity of the mapping $(\mathbf{I}+A)^{-1}$ in the $L^q(\Omega)$-norm. One can then make an extension by continuity of this mapping to get a generalized solution to $\mathrm{div}\,a(\nabla u) - (c(u)+u) = g\in L^q(\Omega)$ with the above boundary conditions, cf. Remark 3.11. Unlike the distributional solution (3.33) below, more concrete interpretation is not entirely obvious unless one gets additional information about ∇u.[11]

Remark 3.19 (The case $q = +\infty$). Investigation of $q > p^*$ would require to show an additional regularity of solutions to the boundary-value problem (2.147) to show $\mathrm{dom}(A)\subset L^q(\Omega)$. The case $q=+\infty$, which we avoided in Propositions 3.16-3.17 anyhow not to speak about $L^\infty(\Omega)^*$, is exceptional and can be treated by a special

[11] A concept of the renormalized and the entropy solutions has been developed for it; see Bénilan et al. [47] and references therein.

comparison technique. Instead of (3.22), let us set[12]

$$\mathrm{dom}(A) := \Big\{u\in W^{1,p}(\Omega)\cap L^\infty(\Omega);\ \mathrm{div}\, a(x,\nabla u)-c(x,u)\in L^\infty(\Omega),\\ u|_{\Gamma_\mathrm{D}}=u_\mathrm{D},\quad \nu\cdot a(x,\nabla u)+b(x,u)=h \text{ on } \Gamma_\mathrm{N}\Big\}. \tag{3.29}$$

Let us assume, in addition, that $a(x,s)\cdot s\ge 0$, $c(x,r)r\ge 0$, $h\in L^\infty(\Gamma)$, and $b(x,\cdot)^{-1}$ does exist and $b^{-1}(h)\in L^\infty(\Gamma)$.

As the dual space to $X=L^\infty(\Omega)$ is a very abstract object, we avoid specifying the duality mapping $J: L^\infty(\Omega)\to L^\infty(\Omega)^*$ and will rather rely on the formula (3.8). For any $g\in L^\infty(\Omega)$ and $\lambda>0$, there is a weak solution $u\in W^{1,p}(\Omega)$ to the boundary-value problem

$$\left.\begin{aligned} u-\lambda\mathrm{div}\, a(\nabla u)+\lambda c(u) &= g && \text{on } \Omega,\\ \nu\cdot a(\nabla u)+b(u) &= h && \text{on } \Gamma_\mathrm{N},\\ u|_{\Gamma_\mathrm{D}} &= u_\mathrm{D} && \text{on } \Gamma_\mathrm{D}. \end{aligned}\right\} \tag{3.30}$$

Putting $G:=\max\big(\|u_\mathrm{D}\|_{L^\infty(\Gamma_\mathrm{D})},\|g\|_{L^\infty(\Omega)},\|b^{-1}(h)\|_{L^\infty(\Omega)}\big)$ and testing the weak formulation of (3.30)[13] by $v=(u-G)^+$, we can see that $u\le G$ a.e. in Ω; cf. Exercise 3.36. Likewise, the test by $v=(u+G)^-$ yields $u\ge -G$ a.e. in Ω. Thus $u\in L^\infty(\Omega)$ and $\mathrm{div}\, a(\nabla u)-c(u)=(u-g)/\lambda\in L^\infty(\Omega)$, so that $u\in\mathrm{dom}(A)$ from (3.29).

Now, take $g_1,g_2\in L^\infty(\Omega)$ and the corresponding $u_1,u_2\in\mathrm{dom}(A)$, subtract the corresponding problems (3.30) from each other, and moreover subtract the constant $G=\|g_1-g_2\|_{L^\infty(\Omega)}$ from both sides, i.e. $(u_1-u_2-G)-\lambda\big(\mathrm{div}\big(a(\nabla u_1)-a(\nabla u_2)\big)-c(u_1)+c(u_2)\big)=g_1-g_2-G$ with the boundary condition $\nu\cdot\big(a(\nabla u_1)-a(\nabla u_2)\big)+b(u_1)-b(u_2)=0$ on Γ_N. Test the weak formulation with $v=(u_1-u_2-G)^+$. Using $\nabla v=\chi_{\{x\in\Omega;\ u_1(x)-u_2(x)>G\}}\nabla(u_1-u_2)$, cf. Proposition 1.28, this gives

$$\int_\Omega\Big((u_1-u_2-G)^+\Big)^2+\lambda\big(c(u_1)-c(u_2)\big)(u_1-u_2-G)^+\mathrm{d}x\\ +\int_{\{x\in\Omega;\ u_1(x)-u_2(x)>G\}}\lambda\Big(a(\nabla u_1)-a(\nabla u_2)\Big)\cdot\nabla(u_1-u_2)\,\mathrm{d}x\\ +\int_{\Gamma_\mathrm{N}}\lambda\big(b(u_1)-b(u_2)\big)(u_1-u_2-G)^+\mathrm{d}S=\int_\Omega(g_1-g_2-G)(u_1-u_2-G)^+\mathrm{d}x\le 0.$$

The left-hand side can be lower-bounded by $\|(u_1-u_2-G)^+\|^2_{L^2(\Omega)}$, which shows $u_1-u_2\le G$ a.e. on Ω. Likewise, testing by $v=(u_1-u_2+G)^-$ yields $u_1-u_2\ge -G$ a.e. in Ω. Altogether, $\|u_1-u_2\|_{L^\infty(\Omega)}\le G=\|g_1-g_2\|_{L^\infty(\Omega)}$ so that the mapping $g\mapsto u$ is a contraction on $X=L^\infty(\Omega)$.

---

[12] In fact, (3.29) coincides with (3.22) for $q=+\infty$ if $p>n$ because then automatically $W^{1,p}(\Omega)\subset L^\infty(\Omega)$.

[13] Here we must assume $p\ge 2n/(n+2)$, so that $p^*\ge 2$ to satisfy $|r+\lambda c(r)|\le C(1+|r|^{p^*-1})$ like in (2.148c).

Remark 3.20 (Alternative setting). We can define $\mathrm{dom}(A)$ more explicitly than (3.22) if a regularity result is employed. E.g., assuming $a(x,s) := \mathbb{A}(x)s$ and a smooth data Ω, $\mathbb{A}$, b, and c as in Corollary 2.105, $\Gamma_{\rm N} := \Gamma$, we can define $\mathrm{dom}(A) = \{u \in W^{2,2}(\Omega);\ \nu^\top\mathbb{A}\nabla u + b(u) = h \text{ a.e. on } \Gamma\}$. The m-accretivity of $A : v \mapsto c(v) - \mathrm{div}(\mathbb{A}\nabla v)$ on $L^2(\Omega)$ then follows from Corollary 2.105.

Example 3.21 (*Advection* term[14]). One can modify A from (3.22) by considering $c(x,r,s) := \vec{v}(x)\cdot s$ with a vector field $\vec{v} : \Omega \to \mathbb{R}^n$ such that $\mathrm{div}\,\vec{v} \le 0$ and $(\vec{v}|_\Gamma)\cdot\nu \ge 0$ as in Exercise 2.91. Again, let $q \in [p^{*\prime}, p^*]$. Then, abbreviating $\omega_q(r) := r\,|r|^{q-2}$ and using Green's formula, the accretivity follows from:

$$\int_\Omega (u{-}v)|u{-}v|^{q-2}\vec{v}\cdot\nabla(u{-}v)\,\mathrm{d}x = \int_\Omega \omega_q(u{-}v)\vec{v}\cdot\nabla(u{-}v)\,\mathrm{d}x$$
$$= \int_\Omega \vec{v}\cdot\nabla\widehat{\omega}_q(u{-}v)\,\mathrm{d}x = \int_\Gamma (\vec{v}{\cdot}\nu)\,\widehat{\omega}_q(u{-}v)\,\mathrm{d}S - \int_\Omega (\mathrm{div}\,\vec{v})\,\widehat{\omega}_q(u{-}v)\,\mathrm{d}x \ge 0 \tag{3.31}$$

where $\widehat{\omega}_q$ is a primitive function of ω_q.

3.2.3 Accretivity of heat equation

We will demonstrate the L^1-accretive structure of the semilinear heat operator in isotropic media, i.e. $u \mapsto -\mathrm{div}(\kappa(u)\nabla u)$, cf. Example 2.79 with $\mathbb{B} = \mathbb{I}$.[15] Except $n = 1$, the previous approaches do not allow treatment of a heat source of finite energy, i.e. bounded only in $L^1(\Omega)$; cf. also (2.127). Here we put off this restriction, i.e. we take $X = L^1(\Omega)$. For simplicity, let c be monotone with at most linear growth, and let $b = 0$. Then $A(u) := -\Delta\widehat{\kappa}(u) + c(u)$, and considering Neumann boundary conditions we put

$$\mathrm{dom}(A) := \Big\{u \in L^1(\Omega);\ \ \Delta\widehat{\kappa}(u) \in L^1(\Omega),\ \ \frac{\partial}{\partial\nu}\widehat{\kappa}(u) = h \text{ on } \Gamma\Big\} \tag{3.32}$$

where $\Delta\widehat{\kappa}(u)$ is understood in the sense of distributions and $\frac{\partial}{\partial\nu}\widehat{\kappa}(u) = h$ is understood in the weak sense. Then $u + A(u) = g$ for $u \in \mathrm{dom}(A)$ means precisely that $u \in L^1(\Omega)$ with $\Delta\widehat{\kappa}(u) \in L^1(\Omega)$ in the sense of distributions and that, by using Green's formula (1.54) twice,

$$\int_\Omega uv - \widehat{\kappa}(u)\,\Delta v + c(u)v\,\mathrm{d}x = \int_\Omega gv\,\mathrm{d}x$$
$$+ \int_\Gamma \Big(\frac{\partial\widehat{\kappa}(u)}{\partial\nu}v - \widehat{\kappa}(u)\frac{\partial v}{\partial\nu}\Big)\,\mathrm{d}S = \int_\Omega gv\,\mathrm{d}x + \int_\Gamma hv\,\mathrm{d}S \tag{3.33}$$

for any $v \in C^\infty(\bar{\Omega})$ such that $\frac{\partial}{\partial\nu}v = 0$ on Γ; note that we used $u \in \mathrm{dom}(A)$ to substitute $\frac{\partial}{\partial\nu}\widehat{\kappa}(u) = h$ into the integral on Γ. The important fact is that

[14] Cf. also Rulla [375].

[15] See Brézis and Strauss [69], or also Barbu [37, Chap. III, Sect. 3.3], Magenes, Verdi, Visintin [267], Showalter [383, Theorem 9.2].

this set of test functions has sufficiently rich traces on Γ.[16] The integral identity (3.33) defines a so-called *distributional solution*, sometimes also called a *very weak solution*.

Proposition 3.22 (M-ACCRETIVITY). *Let $c : \mathbb{R} \to \mathbb{R}$ be monotone with at most linear growth, $0 < \operatorname{ess\,inf} \kappa(\cdot) \le \operatorname{ess\,sup} \kappa(\cdot) < +\infty$, and $h \in L^{2^{\#\prime}}(\Gamma)$. Then $A := -\Delta \widehat{\kappa}(u) + c(u)$ with $\mathrm{dom}(A)$ from (3.32) is m-accretive on $L^1(\Omega)$.*

Proof. For clarity, we divide the proof into four steps.

Step 1: Considering the weak-solution concept, $u + A(u) = g$ has a solution $u \in \mathrm{dom}(A)$ for any $g \in L^{2^{*\prime}}(\Omega)$; see Example 2.79 which gives $u \in W^{1,2}(\Omega)$ satisfying

$$\forall v \in W^{1,2}(\Omega): \quad \int_\Omega \nabla \widehat{\kappa}(u) \cdot \nabla v + uv + c(u)v - gv \,\mathrm{d}x = \int_\Gamma hv \,\mathrm{d}S \tag{3.34}$$

and realize that additionally $\Delta \widehat{\kappa}(u) = u + c(u) - g \in L^{2^{*\prime}}(\Omega) \subset L^1(\Omega)$ and (3.34) implies (3.33) for $v \in C^\infty(\bar{\Omega})$ with $\frac{\partial}{\partial \nu} v = 0$.

Step 2: We will show that the mapping $g \mapsto u$ is non-expansive in the L^1-norm. We consider $g := g_1$ and g_2 and the corresponding u_1 and u_2, write (3.34) for u_1 and u_2, then subtract; note that $\widehat{\kappa}(u_1)$ and $\widehat{\kappa}(u_2)$ live in $W^{1,2}(\Omega)$, cf. Proposition 1.28. The resulting identity holds not only for $v \in \mathscr{D}(\Omega)$ but even for $v \in W^{1,2}(\Omega)$. Thus we can put $v = \mathrm{sign}_\varepsilon(\widehat{\kappa}(u_1) - \widehat{\kappa}(u_2)) \in W^{1,2}(\Omega)$ where $\mathrm{sign}_\varepsilon : \mathbb{R} \to [-1,1]$ was defined by (3.23). This results in

$$\begin{aligned}\int_\Omega &\big(u_1 - u_2 + c(u_1) - c(u_2)\big)\,\mathrm{sign}_\varepsilon\big(\widehat{\kappa}(u_1) - \widehat{\kappa}(u_2)\big)\\ &\quad + \nabla\big(\widehat{\kappa}(u_1) - \widehat{\kappa}(u_2)\big)\cdot\nabla \mathrm{sign}_\varepsilon\big(\widehat{\kappa}(u_1) - \widehat{\kappa}(u_2)\big)\,\mathrm{d}x\\ &\quad = \int_\Omega (g_1 - g_2)\,\mathrm{sign}_\varepsilon\big(\widehat{\kappa}(u_1) - \widehat{\kappa}(u_2)\big)\,\mathrm{d}x \le \int_\Omega |g_1 - g_2|\,\mathrm{d}x.\end{aligned} \tag{3.35}$$

Because of the monotonicity of $\mathrm{sign}_\varepsilon$, c, and $\widehat{\kappa}$, the term

$$\begin{aligned}&\nabla\big(\widehat{\kappa}(u_1) - \widehat{\kappa}(u_2)\big)\cdot\nabla \mathrm{sign}_\varepsilon\big(\widehat{\kappa}(u_1) - \widehat{\kappa}(u_2)\big)\\ &\quad = \big|\nabla\widehat{\kappa}(u_1) - \nabla\widehat{\kappa}(u_2)\big|^2 \mathrm{sign}_\varepsilon'\big(\widehat{\kappa}(u_1) - \widehat{\kappa}(u_2)\big)\end{aligned}$$

is non-negative and also $\big(c(u_1) - c(u_2)\big)\,\mathrm{sign}_\varepsilon\big(\widehat{\kappa}(u_1) - \widehat{\kappa}(u_2)\big) \ge 0$ a.e.; we thus get

$$\int_\Omega (u_1 - u_2)\,\mathrm{sign}_\varepsilon\big(\widehat{\kappa}(u_1) - \widehat{\kappa}(u_2)\big)\,\mathrm{d}x \le \big\|g_1 - g_2\big\|_{L^1(\Omega)}. \tag{3.36}$$

Realizing that $\{(u_1 - u_2)\mathrm{sign}_\varepsilon(\widehat{\kappa}(u_1) - \widehat{\kappa}(u_2))\}_{\varepsilon>0}$ has an integrable majorant, namely $|u_1 - u_2|$, and that this sequence converges to it a.e., we get

$$\lim_{\varepsilon\to 0} \int_\Omega (u_1 - u_2)\,\mathrm{sign}_\varepsilon\big(\widehat{\kappa}(u_1) - \widehat{\kappa}(u_2)\big)\,\mathrm{d}x = \int_\Omega |u_1 - u_2|\,\mathrm{d}x = \big\|u_1 - u_2\big\|_{L^1(\Omega)} \tag{3.37}$$

[16] Here we use the same arguments as in the proof of Proposition 2.44.

thanks to Lebesgue's dominated-convergence Theorem 1.14. Joining (3.36) with (3.37) proves $g \mapsto u$ to be non-expansive in $L^1(\Omega)$.

Step 3. The limit passage: Take $g_k \in L^{2^{*\prime}}(\Omega)$, $g \in L^1(\Omega)$, $g_k \to g$ in $L^1(\Omega)$, $u_k \in \mathrm{dom}(A)$ such that $u_k + A(u_k) = g_k$. By the Step 2, $u_k \to u$ in $L^1(\Omega)$. As $\widehat{\kappa}$ is Lipschitz, also $\widehat{\kappa}(u_k) \to \widehat{\kappa}(u)$ in $L^1(\Omega)$. As c has at most linear growth, we have also $c(u_k) \to c(u)$ in $L^1(\Omega)$. By Green's Theorem 1.31 applied to (3.34), we know that

$$\int_\Omega u_k v - \widehat{\kappa}(u_k)\Delta v + c(u_k)v \,\mathrm{d}x = \int_\Omega g_k v \,\mathrm{d}x + \int_\Gamma hv \,\mathrm{d}S \tag{3.38}$$

which gives in the limit $\int_\Omega uv - \widehat{\kappa}(u)\Delta v + c(u)v \,\mathrm{d}x = \int_\Omega gv \,\mathrm{d}x + \int_\Gamma hv \,\mathrm{d}S$ for any $v \in C^\infty(\bar{\Omega})$; $\frac{\partial}{\partial\nu}v = 0$, i.e. we get (3.33), thus u solves $u + A(u) = g$ with the boundary condition $\frac{\partial}{\partial\nu}\widehat{\kappa}(u) = h$ on Γ in the weak sense. Besides, (3.38) implies, in the sense of distributions, $\Delta\widehat{\kappa}(u) = u + c(u) - g$, so that $\Delta\widehat{\kappa}(u) \in L^1(\Omega)$. Altogether, $u \in \mathrm{dom}(A)$.

Step 4: The accretivity: By the extension of the estimate in Step 2, we get that $(\mathbf{I} + A)^{-1} : g \mapsto u : L^1(\Omega) \to \mathrm{dom}(A)$ is non-expansive. By the same technique, it can be proved that also $(\mathbf{I} + \lambda A)^{-1}$ is non-expansive for any $\lambda > 0$. From this, A is accretive, cf. Lemma 3.7. □

Remark 3.23 (Very weak solution to steady-state heat problem)**.** The previous considerations can immediately give a very weak solution for the heat equation $-\mathrm{div}\big(\kappa(u)\nabla u\big) + c(x,u) = g$ with $g \in L^1(\Omega)$ and with c strongly monotone $(c(x,r_1) - c(x,r_2))(r_1 - r_2) \geq \varepsilon(r_1 - r_2)^2$ so that $u - \varepsilon^{-1}\big(\mathrm{div}\big(\kappa(u)\nabla u\big) + c_0(u)\big) = g$ with $c_0(x,r) := c(x,r) - \varepsilon r$ is still monotone. In case $n = 2$, this describes the heat-conductive plate with the convection coefficient $c_1(x) \geq \varepsilon > 0$, cf. (2.132) and Figure 6b. In the case $c = 0$, which would correspond rather to Figure 6a, the situation is more difficult and Remark 3.12 can apply. Moreover, varying also h and modifying (3.35) appropriately, one gets $\|u_1 - u_2\|_{L^1(\Omega)} \leq \|g_1 - g_2\|_{L^1(\Omega)} + \|h_1 - h_2\|_{L^1(\Gamma)}$, which allows for extension for $h \in L^1(\Gamma)$.

Remark 3.24 (Other boundary conditions)**.** The modification for the Dirichlet boundary condition $u|_\Gamma = u_{\mathrm{D}}$ is quite technical. In (3.32), instead of the Neumann condition, one should involve $\widehat{\kappa}(u)|_{\Gamma_{\mathrm{D}}} = \widehat{u}_{\mathrm{D}}$ with $\widehat{u}_{\mathrm{D}} = \widehat{\kappa}(u_{\mathrm{D}})$, and then (3.33) with $h = 0$ should hold just for $v \in \mathscr{D}(\Omega)$. For the limit passage in Step 3 of the above proof, we need also to show that $\widehat{\kappa}(u_k)|_\Gamma \to \widehat{\kappa}(u)|_\Gamma$ at least weakly in $L^1(\Gamma)$. For this, we use boundedness of $\Delta\widehat{\kappa}(u_k)$ in $L^1(\Omega)$, and the deep results of Boccardo and Gallouët [56, 57], showing that $\widehat{\kappa}(u_k)$ is bounded also in $W^{1,q}(\Omega)$ with $q < n/(n-1)$, so that $\widehat{\kappa}(u_k)|_\Gamma$ is bounded in $L^{q^\#}(\Gamma)$ with, due to (1.37), $q^\# < (n-1)/(n-2)$. More precisely, [56, 57] uses zero boundary condition, hence we must first shift the mapping as in Proposition 2.27 for which we must qualify $\widehat{u}_{\mathrm{D}}$ as being the trace of some $w \in W^{1,q}(\Omega)$ with $\Delta w \in L^1(\Omega)$.

For Newton boundary conditions we refer to Benilan, Crandall, Sacks [49].

Remark 3.25 (Heat equation with *advection*). The mapping $A(u) := \mathfrak{c}(u)\vec{v}\cdot\nabla u - \mathrm{div}(\kappa(u)\nabla u)$, cf. (2.136), allows for L^1-accretivity after a so-called *enthalpy transformation* by introducing the new variable $w := \widehat{\mathfrak{c}}(u)$ where $\widehat{\mathfrak{c}}(r) := \int_0^r \mathfrak{c}(\varrho)\,\mathrm{d}\varrho$ is a primitive function to $\mathfrak{c}$. Then obviously $A(u) = \vec{v}\cdot\nabla\widehat{\mathfrak{c}}(u) - \Delta(\widehat{\kappa}(u)) = \vec{v}\cdot\nabla w - \Delta\beta(w)$ with $\beta = \widehat{\kappa}\circ[\widehat{\mathfrak{c}}]^{-1}$. Then, assuming $\vec{v}\in W^{1,\infty}(\Omega;\mathbb{R}^n)$ such that $\mathrm{div}\,\vec{v}\leq 0$ and $\vec{v}|_\Gamma\cdot\nu = 0$, we can merge the calculations (3.31) with $\mathrm{sign}_\varepsilon$ in place of ω_q with the arguments in Proposition 3.22 with $\widehat{\kappa}\circ[\widehat{\mathfrak{c}}]^{-1}$ in place of $\widehat{\kappa}$. The identity (3.33) augments by the term $u(\vec{v}\cdot\nabla v) - u(\mathrm{div}\,\vec{v})v$ which allows easily for a limit passage as in (3.38); note that the boundary term $u(\vec{v}\cdot\nu)$ is assumed zero otherwise the limit passage would be doubtful.

3.2.4 Accretivity of some other boundary-value problems

An accretive structure may arise also in a so-called *conservation law* posed on a one-dimensional domain $\Omega := (0,1)$ as

$$A(u) = \frac{\mathrm{d}}{\mathrm{d}x}F(u), \qquad X := L^1(0,1), \tag{3.39a}$$

$$\mathrm{dom}(A) := \Big\{u\in W^{1,1}(0,1);\ \ u(0) = u_{\mathrm{D}},\ \ \frac{\mathrm{d}}{\mathrm{d}x}F(u)\in L^1(0,1)\Big\}. \tag{3.39b}$$

Proposition 3.26 (M-ACCRETIVITY). *Let $F:\mathbb{R}\to\mathbb{R}$ be continuous and strongly monotone. Then $A:\mathrm{dom}(A)\to X$ defined by (3.39) is m-accretive.*

Proof. For the accretivity, we choose $f = \mathrm{sign}(u{-}v) \in J(u{-}v)/\|u{-}v\|_{L^1(0,1)}$. Then:

$$\begin{aligned}\big\langle f, A(u){-}A(v)\big\rangle &= \int_0^1 \mathrm{sign}(u-v)\frac{\mathrm{d}}{\mathrm{d}x}\big(F(u){-}F(v)\big)\,\mathrm{d}x\\ &= \int_0^1 \frac{\mathrm{d}}{\mathrm{d}x}\big|F(u){-}F(v)\big|\,\mathrm{d}x = \big|F(u(1)){-}F(v(1))\big| \geq 0,\end{aligned} \tag{3.40}$$

where we used also that $\mathrm{sign}(u-v) = \mathrm{sign}(F(u)-F(v))$ by strict monotonicity of F, and that $u(0) = u_{\mathrm{D}} = v(0)$, and for $g = F(u) - F(v)$ we used also the identity $\int_0^1 \mathrm{sign}(g)\frac{\mathrm{d}}{\mathrm{d}x}g\,\mathrm{d}x = \int_0^1 \frac{\mathrm{d}}{\mathrm{d}x}|g|\,\mathrm{d}x$, which follows by a regularization technique[17]. As to m-accretivity, we are to show that the ordinary differential equation

$$\frac{\mathrm{d}w}{\mathrm{d}x} + G(w(x)) = f(x)\,, \qquad w(0) = F(u_{\mathrm{D}}), \tag{3.41}$$

[17]Using $\mathrm{sign}_\varepsilon:\mathbb{R}\to[-1,1]$ defined for $\varepsilon>0$ by (3.23), by Lebesgue's dominated convergence theorem, it holds that

$$|g(1)|_\varepsilon = \int_0^1 \frac{\mathrm{d}}{\mathrm{d}x}|g|_\varepsilon\,\mathrm{d}x = \int_0^1 \mathrm{sign}_\varepsilon(g) \to \int_0^1 \mathrm{sign}(g)\frac{\mathrm{d}}{\mathrm{d}x}g\,\mathrm{d}x$$

because $\mathrm{sign}_\varepsilon(g)\to\mathrm{sign}(g)$ a.e. and has an L^∞-majorant while $\frac{\mathrm{d}}{\mathrm{d}x}g$ lives in $L^1(0,1)$. On the other hand, also $|g(1)|_\varepsilon \to |g(1)| = \int_0^1 \frac{\mathrm{d}}{\mathrm{d}x}|g|\,\mathrm{d}x$. Note that $\mathrm{sign}_\varepsilon$ has a convex potential which we denote by $|\cdot|_\varepsilon$, i.e. $\mathrm{sign}_\varepsilon(r) = (|r|_\varepsilon)'$. Moreover, we can suppose $|0|_\varepsilon = 0$.

with $G = F^{-1}$ has a solution. As F has growth at least linear, G has growth at most linear, and then existence of w solving (3.41) follows by the standard arguments; cf. Theorem 1.45. For such w, $u = G(w)$ solves $u + A(u) = f$. Moreover, for $f \in L^1(0,1)$ we have $w \in W^{1,1}(0,1)$. Then $\frac{\mathrm{d}}{\mathrm{d}x}u = \frac{\mathrm{d}}{\mathrm{d}x}G(w) = G'(w)\frac{\mathrm{d}}{\mathrm{d}x}w \in L^1(0,1)$ provided $G'(w) \in L^\infty(0,1)$; this requires G to be Lipschitz for bounded arguments, i.e. F strongly monotone. Then $u \in W^{1,1}(0,1)$ and also $\frac{\mathrm{d}}{\mathrm{d}x}F(u) = f - u \in L^1(0,1)$ and therefore $u \in \mathrm{dom}(A)$. □

Remark 3.27 (Scalar conservation law on $\mathbb{R}^n$). Assuming $F \in C^1(\mathbb{R};\mathbb{R}^n)$ and $\limsup_{|u|\to 0} |F(u)|/|u| < +\infty$, the mapping A, defined as the closure in $L^1(\mathbb{R}^n) \times L^1(\mathbb{R}^n)$ of the mapping $u \mapsto \mathrm{div}(F(u)) : C_0^1(\mathbb{R}^n) \to C_0(\mathbb{R}^n)$, has been shown to be m-accretive on $L^1(\Omega)$ in Barbu [38, Section 2.3, Proposition 3.11].

Remark 3.28 (*Hamilton-Jacobi equation*). For $F{:}\mathbb{R}{\to}\mathbb{R}$ increasing and $\Omega = (0,1)$, the m-accretivity of the so-called (one-dimensional) Hamilton-Jacobi operator

$$A(u) = F\Big(\frac{\mathrm{d}u}{\mathrm{d}x}\Big), \qquad X := \big\{v{\in}C([0,1]);\ v(0) = 0 = v(1)\big\}, \tag{3.42a}$$

$$\mathrm{dom}(A) := \Big\{u{\in}C^1([0,1]);\ \ u{\in}X,\ \frac{\mathrm{d}u}{\mathrm{d}x}{\in}X\Big\} \tag{3.42b}$$

has been shown in Deimling [118, Example 23.5].

3.2.5 Excursion to equations with measures in right-hand sides

We saw that the accretivity approach allows us sometimes to solve equations with integrable functions on the right-hand side even if $L^1(\Omega)$ was not contained in $W^{1,p}(\Omega)^*$ and the usual a-priori estimates on $W^{1,p}(\Omega)$ fail. Functions from $L^1(\Omega)$ are special Radon measures on $\bar{\Omega}$, namely those which are absolutely continuous, and a natural question arises whether one can get rid of this absolute continuity and thus consider general *measures in the right-hand sides.* Such generalization is often needed in optimal-control theory of elliptic problems.[18]

It is indeed possible in special cases but one must always employ quite sophisticated techniques not necessarily related to the accretivity approach, and often even a negative answer is known. Here we demonstrate only a rather simple so-called *transposition method* combined with regularity results, applicable to some semilinear equations. We will demonstrate it on the Newton-boundary-value problem:

$$\left.\begin{aligned} -\mathrm{div}\big(\mathbb{A}(x)\nabla u\big) + c(x,u) &= \mu \quad \text{in } \Omega,\\ \nu^\top\mathbb{A}(x)\nabla u + b_1(x)u &= \eta \quad \text{on } \Gamma, \end{aligned}\right\} \tag{3.43}$$

with $\mu \in \mathscr{M}(\bar{\Omega})$ and $\eta \in \mathscr{M}(\Gamma)$. We assume $n \le 3$, $\Omega \subset \mathbb{R}^n$ a domain of C^2-class, $\mathbb{A}{\in}C^1(\bar{\Omega};\mathbb{R}^{n\times n})$ being uniformly positive definite in the sense of (2.170), b_1

[18]Measures typically occur either in the so-called adjoint systems as Lagrange multipliers on state constraints or in the controlled systems as a result of concentration phenomena if an optimal-control problem has only an L^1-coercivity but not coercivity in L^p for $p > 1$.

qualified as in Proposition 2.104, i.e. $b_1(x) \geq b_0 > 0$ and $b_1 \in W^{1,2^\# 2/(2^\# - 2)}(\Gamma)$, and c a Carathéodory function satisfying

$$\exists \gamma_0, \gamma_1 \in L^1(\Omega),\ C \in \mathbb{R}^+,\ \ q < \frac{n}{n-2},\ \ \varepsilon_c > 0\ \ \forall (\text{a.a.})\, x \in \Omega\ \forall r \in \mathbb{R}:$$
$$|c(x,r)| \leq \gamma_0(x) + C|r|^q, \qquad c(x,r)r \geq \big(\varepsilon_c |c(x,r)| - \gamma_1(x)\big)\, |r| \tag{3.44}$$

of course, we mean $q < +\infty$ for $n \leq 2$ (while $q < 3$ for $n = 3$). We call $u \in L^q(\Omega)$ a *distributional solution* to (3.43) if the integral identity obtained like (2.51) but using Green's formula twice, i.e.

$$\int_\Omega c(x,u)v - u \operatorname{div}\big(\mathbb{A}^\top(x)\nabla v\big)\mathrm{d}x = \int_{\bar\Omega} v\, \mu(\mathrm{d}x) + \int_\Gamma v\, \eta(\mathrm{d}S), \tag{3.45}$$

is valid for all

$$v \in W^{2,2}(\Omega): \qquad \nu^\top \mathbb{A}^\top(x)\nabla v + b_1(x)v = 0 \quad \text{ on } \Gamma. \tag{3.46}$$

Lemma 3.29 (The case $c = 0$). *For any $\mu \in \mathscr{M}(\bar\Omega)$ and $\eta \in \mathscr{M}(\Gamma)$, the equation $-\operatorname{div}(\mathbb{A}(x)\nabla u) = \mu$ with the boundary condition $\nu^\top \mathbb{A}(x)\nabla u + b_1(x)u = \eta$ has a unique distributional solution and the a-priori estimate $\|u\|_{W^{\lambda,2}(\Omega)} \leq C\big(\|\mu\|_{\mathscr{M}(\bar\Omega)} + \|\eta\|_{\mathscr{M}(\Gamma)}\big)$ holds for $\lambda < 2-n/2$ and some $C < +\infty$.*

Proof. Let us consider the auxiliary linear problem

$$\left.\begin{aligned} -\operatorname{div}\big(\mathbb{A}^\top(x)\nabla v\big) &= g \quad \text{ in } \Omega, \\ \nu^\top \mathbb{A}^\top(x)\nabla v + b_1(x)v &= 0 \quad \text{ on } \Gamma. \end{aligned}\right\} \tag{3.47}$$

The existence of the weak solution $v \in W^{1,2}(\Omega)$ to (3.47) can be proved by the standard energy method by testing (3.47) by v itself, and we have the estimate $\|v\|_{W^{1,2}(\Omega)} \leq K_1 \|f\|_{W^{1,2}(\Omega)^*}$ with f determined by the pair $(g, 0)$ due to (2.60). As $b_1 > 0$, the solution to (3.47) is unique, and thus defines a linear operator $B : g \mapsto v$. Then we use Proposition 2.104 to claim the $W^{2,2}$-regularity for (3.47), i.e. $\|v\|_{W^{2,2}(\Omega)} \leq K_2 \|g\|_{L^2(\Omega)}$; cf. (2.189).

The interpolation between the linear mappings $B : W^{1,2}(\Omega)^* \to W^{1,2}(\Omega)$ and $B : L^2(\Omega) \to W^{2,2}(\Omega)$ gives a mapping $B : W^{\lambda,2}(\Omega)^* \to W^{2-\lambda,2}(\Omega)$ and an estimate $\|v\|_{W^{2-\lambda,2}(\Omega)} \leq K\|g\|_{W^{\lambda,2}(\Omega)^*}$ for any $\lambda \in [0,1]$ and some K depending on K_1, K_2, and λ, cf. (1.45).

Let us rewrite the integral identity (3.45)–(3.46) with $c \equiv 0$, which defines the distributional solution to the problem considered here, into the form $\langle g, u\rangle = \langle F, Bg\rangle$ for any $g = \operatorname{div}(\mathbb{A}^\top \nabla v) \in L^2(\Omega)$ with F defined by

$$\langle F, v\rangle = \int_{\bar\Omega} v\, \mu(\mathrm{d}x) + \int_\Gamma v\, \eta(\mathrm{d}S). \tag{3.48}$$

We choose $0 \leq \lambda \leq 1$ so small that $W^{2-\lambda,2}(\Omega) \subset C(\bar\Omega)$, i.e. $\lambda < (4-n)/2$, cf. Corollary 1.22(iii) which holds for λ non-integer in place of k, as already mentioned in Section 1.2.3. Then (3.48) indeed defines $F \in W^{2-\lambda,2}(\Omega)^*$; note that

$\mathfrak{F} : (\mu, \eta) \mapsto F : \mathscr{M}(\bar{\Omega}) \times \mathscr{M}(\Gamma) \to W^{2-\lambda,2}(\Omega)^*$ defined by (3.48) is the adjoint mapping to $v \mapsto (v, v|_\Gamma) : W^{2-\lambda,2}(\Omega) \to C(\bar{\Omega}) \times C(\Gamma)$, so we have $F = \mathfrak{F}(\mu, \eta)$. In this notation, $u = B^*F = F \circ B \in W^{\lambda,2}(\Omega)^{**} \cong W^{\lambda,2}(\Omega)$ is a solution to $\langle g, u\rangle = \langle F, Bg\rangle$. Moreover, because g is arbitrary, this solution must be unique. Also, we have the estimate

$$\|u\|_{W^{\lambda,2}(\Omega)} \leq K\|F\|_{W^{2-\lambda,2}(\Omega)^*} \leq NK\|(\mu,\eta)\|_{\mathscr{M}(\bar{\Omega})\times\mathscr{M}(\Gamma)} \tag{3.49}$$

with N the norm of the embedding $W^{2-\lambda,2}(\Omega) \subset C(\bar{\Omega})$. □

Let us realize the embedding $W^{\lambda,2}(\Omega) \subset L^q(\Omega)$ with q from (3.44), cf. Corollary 1.22(i) for λ non-integer in place of k; more precisely, for any $q < n/(n-2)$ we can choose $\lambda < (4-n)/2$. Let us also note that $W^{1,n/(n-1)+\epsilon}(\Omega)$ mentioned in Remark 3.24 is embedded into it, too, i.e. $(n/(n-1)+\epsilon)^* = q < n/(n-2)$ if $\epsilon > 0$ is taken small.

Lemma 3.30 ($W^{\lambda,2}$-ESTIMATE). *If $c : \Omega \times \mathbb{R} \to \mathbb{R}$ satisfies (3.44), $g \in L^{2^{*\prime}}(\Omega)$ and $h \in L^{2^{\#\prime}}(\Gamma)$, then the conventional weak solution $u \in W^{1,2}(\Omega)$ of the equation $-\mathrm{div}(\mathbb{A}(x)\nabla u)+c(x,u) = g$ with the boundary conditions $\nu^\top\mathbb{A}(x)\nabla u+b_1(x)u = h$ satisfies, for any $\lambda < 2 - n/2$, also the a-priori estimate*

$$\|u\|_{W^{\lambda,2}(\Omega)} \leq C\big(\|g\|_{L^1(\Omega)} + \|h\|_{L^1(\Gamma)} + \|\gamma_0\|_{L^1(\Omega)} + \|\gamma_1\|_{L^1(\Omega)}\big). \tag{3.50}$$

Proof. Use the test $v := \mathrm{sign}_\varepsilon(u)$, $\varepsilon > 0$, see (3.23) for the regularized signum function. Passing $\varepsilon \to 0$, we get $\varepsilon_c \int_\Omega |c(u)|\mathrm{d}x + b_0 \int_\Gamma |u| \leq \|g\|_{L^1(\Omega)} + \|h\|_{L^1(\Gamma)} + \|\gamma_1\|_{L^1(\Omega)}$; realize that $c(x,r)\mathrm{sign}(r) \geq \varepsilon_c|c(x,r)| - \gamma_1(x)$ and cf. Step 2 in the proof of Proposition 3.22. In particular, $\|g-c(u)\|_{L^1(\Omega)} \leq \|g\|_{L^1(\Omega)}+\|c(u)\|_{L^1(\Omega)} \leq (1+\varepsilon_c^{-1})\|g\|_{L^1(\Omega)} + \varepsilon_c^{-1}\|h\|_{L^1(\Gamma)} + \varepsilon_c^{-1}\|\gamma_1\|_{L^1(\Omega)}$. Then one can use Lemma 3.29 for $\mu := g - c(u)$ and $\eta := h$. □

Proposition 3.31 (EXISTENCE AND STABILITY). *Let Ω be a C^2-domain, $\mathbb{A} \in C^1(\Omega;\mathbb{R}^{n\times n})$ satisfy* (2.170), *$b_1 \in W^{1,2^\#2/(2^\#-2)}(\Gamma)$, $b_1(x) \geq b_0 > 0$, and c satisfy* (3.44). *Then the problem* (3.43) *has a distributional solution. Moreover, for any sequences $\{g_k\}_{k\in\mathbb{N}} \subset L^{2^{*\prime}}(\Omega)$ and $\{h_k\}_{k\in\mathbb{N}} \subset L^{2^{\#\prime}}(\Gamma)$ converging respectively to the measures μ and η weakly*, the sequence $\{u_k\}_{k\in\mathbb{N}} \subset W^{1,2}(\Omega)$ of the corresponding weak solutions contains a subsequence converging weakly in $W^{\lambda,2}(\Omega)$ for any $\lambda < 2 - n/2$ to some u and any u obtained by this way is a distributional solution to* (3.43).

Proof. It suffices to select a subsequence converging weakly in $W^{\lambda,2}(\Omega)$ and to make a limit passage in the integral identity $\int_\Omega (c(x,u_k) - g_k)v - u_k\mathrm{div}(\mathbb{A}^\top(x)\nabla v)\,\mathrm{d}x = \int_\Gamma h_k v\,\mathrm{d}S$, which just gives (3.45); realize the compact embedding $W^{\lambda,2}(\Omega) \subset L^{n/(n-2)-\varepsilon}(\Omega)$ for any $\varepsilon > 0$ and $\lambda < 2 - n/2$ large enough with respect to this ε, and the continuity of the Nemytskiĭ mapping $\mathscr{N}_c : L^{n/(n-2)-\varepsilon}(\Omega) \to L^1(\Omega)$ provided $\varepsilon := n/(n-2) - q$. □

Remark 3.32 (The case μ and η absolutely continuous). If μ and η are absolutely continuous (and g and h are the respective densities), and $r \mapsto c(x,r) - \varepsilon r$ nondecreasing for some $\varepsilon > 0$, then the distributional solution is the "accretive" solution. Moreover, (3.50) yields an additional estimate of u.

3.3 Exercises

Exercise 3.33. Show that $u + A(u) = f$ has a unique solution if A is m-accretive.[19]

Exercise 3.34. Show what the mapping $(u,v) \mapsto \sup \langle u, J(v)\rangle =: \langle u, v\rangle_{\mathrm{s}}$ is upper semicontinuous with respect to the norm topology on $X \times X$.[20]

Exercise 3.35. Prove the formula (3.21), assuming the uniform convexity of $W^{1,p}(\Omega)$ known.[21]

Exercise 3.36 (The comparison technique). Prove the estimate $u \le G$ for u solving (3.30) in the weak sense by testing $v = (u - G)^+$ with G as suggested in Remark 3.19.[22]

Exercise 3.37 (Maximal accretivity). Show that m-accretive mappings are maximal accretive.[23]

---

[19] Hint: Take u_1 and u_2 two solutions, subtract the corresponding equations, test it by $J(u_1 - u_2)$, and use (3.8).

[20] Hint: Take $u_k \to u$ and $v_k \to v$, and $j_k^* \in J(v_k)$ such that $\langle j_k^*, u_k\rangle = \sup\langle J(v_k), u_k\rangle$ (such j_k^* does exist because $J(v_k)$ is weakly* compact), then take a subsequence $j_k^* \rightharpoonup j^*$ weakly* in X^* (such a subsequence exists because $\{J(v_k)\}$ is bounded), and use Lemma 3.2(i), to show $j^* \in J(v)$. Then also $\sup\langle J(v_k), u_k\rangle = \langle j_k^*, u_k\rangle \to \langle j^*, u\rangle \le \sup\langle J(v), u\rangle$. As this holds for an arbitrary cluster point of $\{j_k^*\}$, we proved the desired upper semicontinuity $\limsup_{k\to\infty} \sup\langle J(v_k), u_k\rangle \le \sup\langle J(v), u\rangle$.

[21] Hint: The modification of (3.19) is routine, while (3.20) needs additionally the estimate

$$\begin{aligned}\left\langle |u|^{p-2}u - \Delta_p u, v\right\rangle &= \int_\Omega |\nabla u|^{p-2}\nabla u \cdot \nabla v + |u|^{p-2}uv \,\mathrm{d}x \\ &\le \|\nabla u\|_{L^p(\Omega;\mathbb{R}^m)}^{p-1}\|\nabla v\|_{L^p(\Omega;\mathbb{R}^m)} + \|u\|_{L^p(\Omega)}^{p-1}\|v\|_{L^p(\Omega)} \\ &\le \left(\|\nabla u\|_{L^p(\Omega;\mathbb{R}^m)}^p + \|u\|_{L^p(\Omega)}^p\right)^{p-1}\left(\|\nabla v\|_{L^p(\Omega;\mathbb{R}^m)}^p + \|v\|_{L^p(\Omega)}^p\right)^{1/p} \\ &= \|u\|_{W^{1,p}(\Omega)}^{p-1}\|v\|_{W^{1,p}(\Omega)}.\end{aligned}$$

[22] Hint: First, consider rather the modified (but equivalent) equation $(u - G) - \lambda \mathrm{div}\, a(\nabla u) + \lambda c(u) = g - G$ with the boundary condition $\nu \cdot a(\nabla u) + b(u) = b(b^{-1}(h))$. Then test it by $v = (u - G)^+$ and realize that $\nabla v = \chi_{\{u>G\}}\nabla u$, cf. (1.50). This yields

$$\begin{aligned}&\int_\Omega \left((u-G)^+\right)^2 + \lambda c(u)(u-G)^+ \mathrm{d}x + \int_{\{x\in\Omega;\ u(x)>G\}} \lambda a(\nabla u)\cdot\nabla u \mathrm{d}x \\ &\quad + \int_\Gamma \lambda\Big(b(u) - b(b^{-1}(h))\Big)(u-G)^+ \mathrm{d}S = \int_\Omega (g-G)(u-G)^+ \mathrm{d}x \le 0.\end{aligned}$$

Note that $(b(u) - b(b^{-1}(h)))(u-G)^+ \ge 0$ since $b(x,\cdot)$ is monotone. Also note that $p^* \ge 2$ is to be used.

[23] Hint: Take an m-accretive mapping A_0, some other accretive mapping A and (u,f) such that

Exercise 3.38 (*Accretivity of Laplacean* in $W^{1,q}$). Show the m-accretivity of $A = -\Delta$ with $\mathrm{dom}(A) := \{u \in W_0^{1,q}(\Omega);\ \Delta u \in W_0^{1,q}(\Omega)\}$ in $X := W_0^{1,q}(\Omega)$ if Ω is a C^2-domain and if $q^* \geq 2$ and $q \leq 2^*$.[24]

Exercise 3.39 (Renorming of $L^p(\Omega;\mathbb{R}^m)$ and $W_0^{1,p}(\Omega)$). Consider an equivalent norm on $L^p(\Omega;\mathbb{R}^m)$ given by $\|v\|_{L^p(\Omega;\mathbb{R}^m)} := (\int_\Omega \sum_{i=1}^m |v_i|^p \mathrm{d}x)^{1/p}$ and derive that $\|v\|_{L^p(\Omega;\mathbb{R}^m)^*} = (\int_\Omega \sum_{i=1}^m |v_i|^{p'} \mathrm{d}x)^{1/p'}$ and that the duality mapping J is given by[25]

$$\big[J(u)\big]_i(x) = u_i(x)\big|u_i(x)\big|^{p-2}/\big\|u\big\|^{p-2}_{L^p(\Omega;\mathbb{R}^m)}, \qquad i = 1,\dots,m. \tag{3.51}$$

Moreover, consider $W_0^{1,p}(\Omega)$ normed by $\|v\|_{W_0^{1,p}(\Omega)} := (\int_\Omega \sum_{i=1}^n |\nabla v_i|^p \mathrm{d}x)^{1/p}$ and derive that J now involves the "anisotropic" p-Laplacean, cf. Example 4.31, namely

$$J(u) = -\mathrm{div}\bigg(\Big|\frac{\partial u}{\partial x_1}\Big|^{p-2}\frac{\partial u}{\partial x_1},\dots,\Big|\frac{\partial u}{\partial x_n}\Big|^{p-2}\frac{\partial u}{\partial x_n}\bigg)\Big/\big\|u\big\|^{p-2}_{W_0^{1,p}(\Omega)}. \tag{3.52}$$

Exercise 3.40. Derive the very weak formulation (3.45)–(3.46) by applying Green's formula twice to (3.43).

Exercise 3.41. Considering $f \in L^1(\Omega)$, show that the infimum of the functional $u \mapsto \int_\Omega \frac{1}{2}|\nabla u|^2 - fu\,\mathrm{d}x$ on $W_0^{1,2}(\Omega)$ is $-\infty$ if and only if $f \notin L^{2^{*\prime}}(\Omega)$.[26]

Exercise 3.42 (*Selectivity for the distributional solution*). Show that, if the data are smooth enough, then any $u \in C^2(\Omega)$ satisfying (3.45)–(3.46) solves (3.43).[27]

$(u, f) \in \mathrm{graph}(A) \supset \mathrm{graph}(A_0)$, and take $v \in \mathrm{dom}(A_0)$ such that $v + A_0(v) = u + f$, and then from (3.8) for $\lambda = 1$ deduce that $\|u - v\| \leq \|u + A(u) - v - A(v)\| = \|u + f - v - A_0(v)\| = 0$, hencefore $u = v$ and also $(u, f) \in \mathrm{graph}(A_0)$.

[24]Hint: Taking into account (3.18), show that $\langle J(u-v), A(u)-A(v)\rangle = \int_\Omega \mathrm{div}(|\nabla(u-v)|^{q-2}\nabla(u-v))\Delta(u-v)\,\mathrm{d}x = (q-1)|\nabla(u-v)|^{q-2}|\Delta(u-v)|^2\mathrm{d}x \geq 0$. Further, show that the weak solution to $u - \Delta u - g \in W_0^{1,q}(\Omega)$ is in $W_0^{1,q}(\Omega)$ because, by Proposition 2.104 used for $g \in L^2(\Omega)$ if $q^* \geq 2$ and modified for zero-Dirichlet boundary condition, it holds $u \in W^{2,2}(\Omega) \cap W_0^{1,2}(\Omega) \subset W_0^{1,q}(\Omega)$ if $2^* \geq q$.

[25]Hint: Derive the corresponding Hölder inequality as in (1.19) but now from the Young inequality of the form $\int_\Omega \sum_{i=1}^m u_i v_i \mathrm{d}x \leq \int_\Omega \sum_{i=1}^m (\frac{1}{p}|u_i|^p + \frac{1}{p'}|v_i|^{p'})\,\mathrm{d}x$. From this, derive the norm of the dual space. Eventually, modify the calculations in Proposition 3.13.

[26]Hint: Assuming $\inf_{u\in W_0^{1,2}(\Omega)} \int_\Omega \frac{1}{2}|\nabla u|^2 - fu\,\mathrm{d}x \geq -C > -\infty$, realize that

$$\|f\|_{W_0^{1,2}(\Omega)^*} = \sup_{\|u\|_{W_0^{1,2}(\Omega)}\leq 1} \int_\Omega fu\,\mathrm{d}x \leq \sup_{\|u\|_{W_0^{1,2}(\Omega)}\leq 1} \int_\Omega \frac{1}{2}|\nabla u|^2\,\mathrm{d}x + C = \frac{1}{2} + C < +\infty$$

and therefore necessarily $f \in L^{2^{*\prime}}(\Omega)$.

[27]Hint: Assume in particular the measures μ and η to have respectively densities $g \in C(\bar\Omega)$ and

3.4 Bibliographical remarks

The concept of accretivity for nonlinear mappings was invented essentially by Kato [225, 226]; independently Browder [74] invented it in a stronger variant requiring (3.6) to hold for any $f \in J(u-v)$. Although in the theory of partial differential equations the accretivity concept is not the dominant one in comparison with monotonicity, there are a lot of monographs fully or at least partly devoted to accretive mappings, mainly Barbu [37], Browder [76], Cazenave and Haraux [90, Chap.2-3], Cioranescu [95, Chap.VI], Deimling [118, Chap.13], Hu and Papageorgiou [209, Part I, Sect.III.7], Ito and Kappel [212, Chap.1], Miyadera [287, Chap.2], Pavel [326], Showalter [383, Chap.4], Vainberg [414, Chap.VII], Yosida [425, Sect.XIV.6-7], and Zeidler [427, Chap.31 and 57]. A generalization for *set-valued accretive* mappings[28] exists, too; cf. [118, 209].

The duality mapping was introduced by Beurling and Livingston [51], cf. also [22, 64], and the above listed monographs.

The transposition method exposed in Section 3.2.5 was invented by Stampacchia [393], and thoroughly developed especially by Lions and Magenes [262] for linear problems. Semilinear equations with measures in the right-hand side are investigated by Amann and Quittner [16], Attouch, Bouchitté, and Mambrouk [24], or Brézis [67]. A counterexample of nonuniqueness is due to Serrin [381].

Quasilinear equations with measures in the right-hand side were attacked by Boccardo and Gallouët [56, 57], showing that there is a unique weak solution $u \in W^{1,q}(\Omega)$ with $q < n(p-1)/(n-1)$ with p referring to the growth of the principal part. In particular, for the problem (3.43) it gives $u \in W^{1,q}(\Omega)$ with any $q < n/(n-1)$ which is just embedded into $L^p(\Omega)$ with $p < n/(n-2)$ as taken in (3.44). A nonexistence result for c of a growth bigger than (3.44) in case $n \geq 3$ is due to Brézis and Benilan [67] while a counterexample of nonstability is in [24]. Other definitions of solutions have been scrutinized by Boccardo, Gallouët and Orsina [58], Dal Maso, Murat, Orsina, and Prignet [117], and Rakotoson [345]. For this topic, see also Dolzmann, Hungerbühler, and Müller [125] or the monograph by Malý and Ziemer [271, Sect.4.4].

$h \in C(\Gamma)$, apply the Green formula twice to the residuum in (3.45), and use (3.46) to obtain:

$$\begin{aligned} 0 &= \textstyle\int_\Omega (c(u)-g)v - u\,\mathrm{div}(\mathbb{A}^\top \nabla v)\,\mathrm{d}x - \int_\Gamma hv\,\mathrm{d}S \\ &= \textstyle\int_\Omega \big(c(u)-g-\mathrm{div}(\mathbb{A}\nabla u)\big)v\,\mathrm{d}x + \int_\Gamma (\mathbb{A}\nabla u{\cdot}\nu - h)v - u(\mathbb{A}^\top\nabla v{\cdot}\nu)\,\mathrm{d}S \\ &= \textstyle\int_\Omega \big(c(u)-g-\mathrm{div}(\mathbb{A}\nabla u)\big)v\,\mathrm{d}x + \int_\Gamma (\mathbb{A}\nabla u{\cdot}\nu + b_1 u - h)v\,\mathrm{d}S. \end{aligned}$$

Then use $v \in C_0(\Omega)$ to get $c(u) - \mathrm{div}(\mathbb{A}\nabla u)) = g$ on Ω, and eventually a general v to get the boundary condition $\nu^\top \mathbb{A}\nabla u + b_1 u = h$ on Γ.

[28] A set-valued mapping $A : \mathrm{dom}(A) \rightrightarrows V$, $\mathrm{dom}(A) \subset V$, is called accretive if $\forall u, v \in \mathrm{dom}(A)$ $\forall u_1 \in A(u)$, $v_1 \in A(v)$ $\exists j^* \in J(u-v)$: $\langle j^*, u_1 - v_1\rangle \geq 0$. Moreover, an accretive mapping $A : \mathrm{dom}(A) \rightrightarrows V$, $\mathrm{dom}(A) \subset V$, is called m-accretive if $\mathbf{I} + A$ is surjective, i.e. $\forall f \in V$ $\exists v \in V$: $v + A(v) \ni f$. It holds that A is m-accretive if and only if $\mathbf{I} + \lambda A$ is surjective for some (or for all) $\lambda > 0$.

Chapter 4

Potential problems: smooth case

Again, we consider V a reflexive and separable Banach space. Here we shall deal with the case that $A : V \to V^*$ has the form

$$A = \Phi' \tag{4.1}$$

for some functional (called a *potential*) $\Phi : V \to \mathbb{R}$, having the Gâteaux differential[1] denoted by $\Phi' : V \to V^*$. The methods based on the hypothesis (4.1) are called *variational methods.*[2]

4.1 Abstract theory

Definition 4.1. Let $\Phi : V \to \mathbb{R}$. Then:

(i) Φ is called *coercive* if $\lim_{\|u\|\to\infty} \Phi(u)/\|u\| = +\infty$.

(ii) $\Phi : V \to \mathbb{R}$ is called *weakly coercive* if $\lim_{\|u\|\to\infty} \Phi(u) = +\infty$.

(iii) $u \in V$ is a *critical point* for Φ if $\Phi'(u) = 0$.

Obviously, solutions to the equation $A(u) = f$ are just the critical points of the functional $u \mapsto \Phi(u) - \langle f, u\rangle$. The coercive potential problems can be treated by a so-called *direct method* based on the Bolzano-Weierstrass Theorem 1.8.

Theorem 4.2 (Direct method). *Let $\Phi : V \to \mathbb{R}$ be Gâteaux differentiable and weakly lower semicontinuous, and $A = \Phi'$. Then:*

---

[1]Let us recall that, by definition (1.10), Φ has Gâteaux differential at u if the directional derivative $\mathrm{D}\Phi(u, h) = \lim_{\varepsilon\searrow 0}(\Phi(u + \varepsilon h) - \Phi(u))/\varepsilon$ does exist for any $h \in V$ and $\mathrm{D}\Phi(u, \cdot)$ is a linear and continuous functional, denoted just by $\Phi'(u) \in V^*$.

[2]Sometimes, the notion "variational methods" is used in a wider sense for the setting using an operator $A : V \to V^*$, in contrast to non-variational methods as in Chapter 3 where, e.g., even the direct method may completely fail in general, cf. Exercise 3.41.

(i) *If Φ is weakly coercive, the equation $A(u) = 0$ has a solution.*

(ii) *If Φ is coercive then, for any $f \in V^*$, the equation $A(u) = f$ has a solution.*

(iii) *If Φ is strictly convex, then $A(u) = f$ has at most one solution.*

Proof. Take a minimizing sequence $\{u_k\}_{k\in\mathbb{N}}$ for Φ, i.e.

$$\lim_{k\to\infty} \Phi(u_k) = \inf_{v\in V} \Phi(v); \tag{4.2}$$

such a sequence does exist by the definition of the infimum[3]. As Φ is weakly coercive, $\{u_k\}_{k\in\mathbb{N}}$ is bounded. As V is assumed reflexive and separable (hence also V^* is separable, cf. Proposition 1.3), by the Banach Theorem 1.7 it has a weakly convergent subsequence, say $u_k \rightharpoonup u$. As Φ is weakly lower semicontinuous[4], $\Phi(u) \le \liminf_{k\to\infty} \Phi(u_k) = \lim_{k\to\infty} \Phi(u_k) = \min_{v\in V} \Phi(v)$. Suppose $\Phi'(u) \neq 0$, then for some $h \in V$ we would have $\langle \Phi'(u), h\rangle = \mathrm{D}\Phi(u,h) < 0$ so that, for a sufficiently small $\varepsilon > 0$, we would have

$$\Phi(u + \varepsilon h) = \Phi(u) + \varepsilon\big\langle \Phi'(u), h\big\rangle + o(\varepsilon) < \Phi(u), \tag{4.3}$$

a contradiction. Thus $A(u) = \Phi'(u) = 0$.

If Φ is coercive, then $u \mapsto \Phi(u) - \langle f, u\rangle$ is weakly coercive for any $f \in V^*$. Yet, this functional obviously has the gradient $A - f$.

If Φ is also convex, then any solution to $A(u) = f$ must minimize $\Phi - f$. Having two solutions u_1 and u_2 and supposing $u_1 \neq u_2$, then, by strict convexity, we get

$$\big[\Phi{-}f\big]\Big(\frac{1}{2}u_1 + \frac{1}{2}u_2\Big) < \frac{1}{2}\big[\Phi{-}f\big](u_1) + \frac{1}{2}\big[\Phi{-}f\big](u_2) = \min_{v\in V}\big[\Phi{-}f\big](v), \tag{4.4}$$

a contradiction showing $u_1 = u_2$. $\square$

Corollary 4.3. *Let $\Phi := \Phi_1 + \Phi_2$ be coercive with Φ_1 convex, and Gâteaux differentiable, and with Φ_2 weakly continuous, and Gâteaux differentiable. Then, for any $f \in V^*$, the equation $A(u) = f$ has a solution.*

Proof. Φ_1 convex and smooth implies that Φ_1 is weakly lower semicontinuous: indeed, by convexity always $\Phi_1(u) + \langle \Phi_1'(u), v - u\rangle \le \Phi_1(v)$, cf. (4.12) below, so that

$$\Phi_1(u) \le \liminf_{v\rightharpoonup u} \big(\Phi_1(v) + \langle \Phi_1'(u), u - v\rangle\big) = \liminf_{v\rightharpoonup u} \Phi_1(v) \tag{4.5}$$

because $\lim_{v\rightharpoonup u}\langle \Phi_1'(u), u - v\rangle = 0$. Thus $\Phi_1 + \Phi_2$ is weakly lower semicontinuous. Then one can use the previous Theorem 4.2(ii). $\square$

[3] Note also that $\inf_{v\in V} \Phi(v) > -\infty$ otherwise, by weak coercivity and weak lower semicontinuity of Φ, there would exist v such that $\Phi(v) = -\infty$, which contradicts $\Phi : V \to \mathbb{R}$.

[4] Recall our convention that by (semi)continuity, see (1.6), we mean what is sometimes called "sequential" (semi)continuity while the general concept of (semi)continuity works with generalized sequences (nets). For our purposes, the sequential concept is relevant. Additionally, as Φ is coercive and V separable, both modes of lower semicontinuity of Φ coincide with each other because the weak topology on bounded sets is metrizable.

Theorem 4.4 (RELATIONS BETWEEN A AND Φ). *Let $\Phi : V \to \mathbb{R}$ be Gâteaux differentiable, and $A = \Phi'$. Then:*

(i) *If A is coercive and monotone, then also Φ is coercive.*

(ii) *If A is pseudomonotone, then Φ is weakly lower semicontinuous.*

(iii) *If A is (strictly) monotone, then Φ is (strictly) convex, weakly lower semicontinuous and locally Lipschitz continuous.*

(iv) *Conversely, if Φ is convex (resp. strictly convex), then A is monotone (resp. strictly monotone).*

Proof. The point (i): First, let us realize that[5]

$$\Phi(u) = \Phi(0) + \int_0^1 \langle A(tu), u\rangle \,\mathrm{d}t \tag{4.6}$$

because, denoting $\varphi(t) = \Phi(tu)$, one has

$$\begin{aligned}\Phi(u) - \Phi(0) = \varphi(1) - \varphi(0) &= \int_0^1 \varphi'(t)\,\mathrm{d}t = \int_0^1 \lim_{\varepsilon\to 0} \frac{\Phi(tu{+}\varepsilon u) - \Phi(tu)}{\varepsilon}\,\mathrm{d}t \\ &= \int_0^1 \mathrm{D}\Phi(tu,u)\,\mathrm{d}t = \int_0^1 \langle \Phi'(tu), u\rangle\,\mathrm{d}t = \int_0^1 \langle A(tu), u\rangle\,\mathrm{d}t.\end{aligned} \tag{4.7}$$

Then, supposing for simplicity $\Phi(0) = 0$, one gets[6]

$$\begin{aligned}\Phi(u) &= \int_0^1 \langle A(tu), u\rangle\,\mathrm{d}t = \int_0^1 \frac{\langle A(tu) - A(0), tu - 0\rangle}{t}\,\mathrm{d}t + \langle A(0), u\rangle \\ &\geq \int_{1/2}^1 \Big(\frac{\langle A(tu), tu\rangle}{t} - \langle A(0), u\rangle\Big)\,\mathrm{d}t + \langle A(0), u\rangle \\ &= \int_{1/2}^1 \frac{\langle A(tu), tu\rangle}{t}\mathrm{d}t + \frac{\langle A(0), u\rangle}{2} \geq \int_{1/2}^1 \frac{\zeta(\|tu\|)\|tu\|}{t}\,\mathrm{d}t + \frac{\langle A(0), u\rangle}{2} \\ &\geq \int_{1/2}^1 \frac{\zeta(\|\frac{1}{2}u\|)\|tu\|}{t}\,\mathrm{d}t + \frac{\langle A(0), u\rangle}{2} \geq \frac{1}{2}\Big(\zeta(\|\tfrac{1}{2}u\|) - \|A(0)\|_*\Big)\|u\|,\end{aligned}$$

where $\zeta(\cdot)$ is a nondecreasing function with $\lim_{s\to+\infty} \zeta(s) = +\infty$ from Definition 2.5. Thus we proved a super-linear growth of Φ.

The point (ii): Suppose the contrary, i.e. Φ is not weakly lower semicontinuous at some point, say at 0; i.e. there are some $\delta > 0$ and a sequence $\{u_k\}_{k\in\mathbb{N}} \subset V$, $u_k \rightharpoonup 0$, such that:

$$\forall k \in \mathbb{N}: \quad \delta \leq \Phi(0) - \Phi(u_k). \tag{4.8}$$

[5]Note that, as $\varphi : t \mapsto \Phi(tu)$ is convex and finite, hence $\varphi' : t \mapsto \langle A(tu), u\rangle$, being nondecreasing, is a Borel function, hence measurable. The integral is finite as Φ is finite, cf. (4.7).

[6]Note that $\int_0^1 t^{-1}\langle A(tu)-A(0), tu-0\rangle\,\mathrm{d}t \geq \int_{1/2}^1 t^{-1}\langle A(tu)-A(0), tu-0\rangle\,\mathrm{d}t$ as A is monotone.

For $v, u \in V$, put $\varphi(t) = \Phi(u+tv)$. By the mean value theorem, there is $t \in (0,1)$ such that $\varphi(1) - \varphi(0) = \varphi'(t)$, i.e.

$$\Phi(v+u) - \Phi(u) = \big\langle \Phi'(u+tv), v \big\rangle. \tag{4.9}$$

Take $\varepsilon > 0$. By (4.8) and using (4.9) with $v := \varepsilon u_k$ and $u := 0$, we have

$$\delta \le \Phi(0) - \Phi(\varepsilon u_k) + \Phi(\varepsilon u_k) - \Phi(u_k) = \Phi(\varepsilon u_k) - \Phi(u_k) - \varepsilon\big\langle \Phi'(\varepsilon t_{k,\varepsilon} u_k), u_k \big\rangle,$$

where $t_{k,\varepsilon} \in (0,1)$ depends on ε and on k. As $\{u_k\}_{k\in\mathbb{N}}$, being weakly convergent, is bounded[7] and Φ', being pseudomonotone, is bounded on bounded subsets, the last term is $\mathscr{O}(\varepsilon)$. Consider $\varepsilon > 0$ fixed and so small that this last term is less than $\delta/2$ for all $k \in \mathbb{N}$, hence, for a suitable $t_k \in (0,1)$, one has

$$\frac{\delta}{2} \le \Phi(\varepsilon u_k) - \Phi(u_k) = \big\langle \Phi'(w_k), (\varepsilon{-}1) u_k \big\rangle = \frac{(\varepsilon{-}1)\big\langle \Phi'(w_k), w_k \big\rangle}{1 - t_k(1{-}\varepsilon)} \tag{4.10}$$

where we abbreviated $w_k := u_k - t_k(1{-}\varepsilon)u_k$ and where (4.9) was used for $v := (\varepsilon - 1)u_k$ and $u := u_k$. From (4.10), we have

$$\limsup_{k\to\infty} \big\langle \Phi'(w_k), w_k \big\rangle = \limsup_{k\to\infty} \frac{1 - t_k(1{-}\varepsilon)}{1 - \varepsilon} \big(\Phi(u_k) - \Phi(\varepsilon u_k)\big) \le \frac{1}{1{-}\varepsilon}\Big(\frac{-\delta}{2}\Big) < 0.$$

As $u_k \rightharpoonup 0$, we have also $w_k \rightharpoonup 0$. By using the pseudomonotonicity (2.3b) for $A = \Phi'$, $v := 0$, and the sequence $\{w_k\}_{k\in\mathbb{N}}$, we have $\liminf_{k\to\infty} \langle \Phi'(w_k), w_k \rangle \ge \langle \Phi'(0), 0 \rangle = 0$, which is the sought contradiction.

The point (iii): Monotonicity of A implies its local boundedness, see Lemma 2.15, and thus Φ, having a locally bounded derivative, is locally Lipschitz continuous. Denote $\varphi_w(t) := \Phi(u{+}tw)$. As A is assumed monotone, $\varphi_w'(t) = \langle A(u{+}tw), w \rangle$ is nondecreasing because obviously, for $t_2 > t_1$,

$$\begin{aligned} \varphi_w'(t_2) - \varphi_w'(t_1) &= \big\langle A(u{+}t_2 w) - A(u{+}t_1 w), w \big\rangle \\ &= \frac{\big\langle A(u{+}t_2 w) - A(u{+}t_1 w), (u{+}t_2 w) - (u{+}t_1 w) \big\rangle}{t_2 - t_1} \ge 0. \end{aligned} \tag{4.11}$$

Then we have

$$\begin{aligned} \Phi(v) - \Phi(u) &= \varphi_{v-u}(1) - \varphi_{v-u}(0) = \int_0^1 \varphi_{v-u}'(t)\,\mathrm{d}t \\ &\ge \int_0^1 \varphi_{v-u}'(0)\,\mathrm{d}t = \varphi_{v-u}'(0) = \big\langle A(u), v - u \big\rangle. \end{aligned} \tag{4.12}$$

Put $z := \frac{1}{2}u + \frac{1}{2}v$. By (4.12) we have

$$\Phi(u) \ge \Phi(z) + \big\langle A(z), u - z \big\rangle \quad \text{and} \quad \Phi(v) \ge \Phi(z) + \big\langle A(z), v - z \big\rangle. \tag{4.13}$$

[7] Here we use the Banach-Steinhaus principle, see Theorem 1.1.

Adding these inequalities and recalling the lower semicontinuity of Φ (cf. Exercise 4.18), one gets the convexity of Φ:

$$\Phi(u) + \Phi(v) \geq 2\Phi(z) + \big\langle A(z), u{-}z\big\rangle + \big\langle A(z), v{-}z\big\rangle = 2\Phi(z). \tag{4.14}$$

Strict monotonicity of A implies that φ_w is increasing for $w \neq 0$, and then (4.12) and (4.13) hold with strict inequalities provided $v \neq u$. Eventually, from (4.12), the weak lower semicontinuity of Φ has already been proved in (4.5).

The point (iv): Φ convex implies $\varphi_w : t \mapsto \Phi(u + tw)$ convex, and therefore $\varphi_w'(t) = \langle A(u + tw), w\rangle$ is nondecreasing. Thus, for $w = v - u$, we get

$$\big\langle A(v) - A(u), v{-}u\big\rangle = \varphi_{v-u}'(1) - \varphi_{v-u}'(0) \geq 0, \tag{4.15}$$

so that A is monotone. If Φ is strictly convex, so is φ_w if $w \neq 0$, and thus φ_w' is increasing, hence A strictly monotone. □

Remark 4.5. The coercivity of Φ does not imply coercivity of A.[8]

Corollary 4.6. *Let $A = \Phi'$, and one of the following two sets of conditions holds:*

(i) *A is pseudomonotone and Φ is coercive, or*

(ii) *A is monotone and coercive.*

Then the equation $A(u) = f$ has a solution for any $f \in V^*$.

Proof. As to (i), Theorem 4.4(ii) implies the weak lower semicontinuity of the coercive potential Φ of A. Then use Theorem 4.2(ii).

As to (ii), Theorem 4.4(i) and (iii) implies the coercivity and the weak lower semicontinuity of the potential Φ of A. Then again use Theorem 4.2(ii). □

Remark 4.7. Comparing Corollary 4.6 with Theorem 2.6, we can see that now we required in addition the potentiality, but got a more constructive proof avoiding the Brouwer fixed-point Theorem 1.10. Similarly, comparison with Browder-Minty Theorem 2.18 yields that potentiality makes no need for any explicit assumption of radial continuity[9] of A.

Remark 4.8 (Potentiality criteria). If A is hemicontinuous and Gâteaux differentiable, it has a potential (given then by the formula (4.6)) if and only if $\langle [A'(u)](v), w\rangle = \langle [A'(u)](w), v\rangle$ for any $u, v, w \in V$. In the general case, the following integral criterion is sufficient and necessary for potentiality of A:

$$\int_0^1 \langle A(tu), u\rangle \,\mathrm{d}t - \int_0^1 \langle A(tv), v\rangle \,\mathrm{d}t = \int_0^1 \langle A(v + t(u{-}v)), u{-}v\rangle \,\mathrm{d}t. \tag{4.16}$$

[8]Indeed, any Φ coercive can be changed in a neighbourhood of $\{n\,u;\ u \neq 0,\ n \in \mathbb{N}\}$ to be locally constant; then $A(nu) = 0$ so that A will not be coercive while Φ remains coercive.

[9]In fact, this is just an "optical" illusion as every monotone and potential mapping is even demicontinuous, cf. e.g. Gajewski at al. [168, Ch.III,Lemma 4.12].

Remark 4.9 (Iterative methods). The existence of a potential suggests iterative methods for minimization of Φ to solve the equation $A(u) = f$. E.g., if V is uniformly convex and V^* strictly convex, the abstract *steepest-descent*-like method looks as

$$u_{k+1} := u_k + \varepsilon_k J^{-1}\big(f - \Phi'(u_k)\big) \tag{4.17}$$

with $\varepsilon_k > 0$ sufficiently small; e.g. $\varepsilon_k := \min\big(1, 2/(\varepsilon + \ell\|u_k\| + \ell\|Au_k - f\|_*)\big)$ for ℓ the Lipschitz constant of A and $\varepsilon > 0$ guarantees global strong convergence in case A is strictly monotone and d-monotone, Lipschitz continuous, and coercive.[10] Note that, if $V \equiv V^*$ is a Hilbert space, $J^{-1}(f - \Phi'(u_k)) = f - \Phi'(u_k)$ is just the steepest descent of the landscape given by the graph of $\Phi - f$, which gave the name to this method; cf. also (4.17) with (2.43) or with Example 2.93.

Remark 4.10 (*Ritz' method* [350]). Assuming $\{V_k\}_{k\in\mathbb{N}}$ is a nondecreasing sequence of finite-dimensional subspaces of V whose union is dense, see (2.7), we can consider a sequence of problems

$$\text{Find } u_k \in V_k : \quad \Phi(u_k) = \min_{v\in V_k} \Phi(v)\ . \tag{4.18}$$

Note that u_k is simultaneously a Galerkin approximation to the equation $A(u) = 0$ with $A = \Phi'$, see (2.8). The Ritz method can be combined with (4.17) to get a computer implementable strategy, although much more efficient algorithms than (4.17) are usually implemented.

Remark 4.11 (Quadratic case). A very special case is that Φ is quadratic:

$$\Phi(u) = \Big\langle \frac{1}{2}Au - f, u\Big\rangle \tag{4.19}$$

with $A \in \mathscr{L}(V, V^*)$, $A^* = A$. Then A is weakly continuous, so the existence of a solution to $Au = f$ follows simply by Section 2.5 if A (or, equivalently, Φ) is coercive, which here reduces to positive definiteness of A, i.e. $\langle Av, v\rangle \geq \varepsilon\|v\|^2$ for some $\varepsilon > 0$ and all $v \in V$, cf. also the Lax-Milgram Theorem 2.19 where however $A^* = A$ was redundant. Here, alternatively, $Au = f$ is just equivalent with $\Phi'(u) = 0$ and a direct method can be applied, too.

4.2 Application to boundary-value problems

The method of mappings possessing a potential has powerful applications in boundary-value problems. However, the requirement (4.1) brings a certain restriction on problems that can be treated in this way. Considering again the boundary-value problem (2.49) for the quasilinear 2nd-order equation (2.45), i.e.

$$\left.\begin{aligned} -\mathrm{div}\big(a(x,u,\nabla u)\big) + c(x,u,\nabla u) &= g && \text{in } \Omega,\\ u|_\Gamma &= u_{\mathrm{D}} && \text{on } \Gamma_{\mathrm{D}},\\ \nu\cdot a(x,u,\nabla u) + b(x,u) &= h && \text{on } \Gamma_{\mathrm{N}}, \end{aligned}\right\} \tag{4.20}$$

[10]See Gajewski at al. [168, Thm.III.4.2] using the (S)-property which is here implied by d-monotonicity with uniform convexity of V, cf. Remark 2.21.

we will assume $a_i(x,\cdot,\cdot)$, $i=1,\dots,n$, and $c(x,\cdot,\cdot)$ smooth in the sense $W^{1,1}_{\mathrm{loc}}(\mathbb{R}\times\mathbb{R}^n)$, and impose the *symmetry conditions*

$$\frac{\partial a_i(x,r,s)}{\partial s_j}=\frac{\partial a_j(x,r,s)}{\partial s_i}\,,\qquad \frac{\partial a_i(x,r,s)}{\partial r}=\frac{\partial c(x,r,s)}{\partial s_i}\tag{4.21}$$

for all $1\le i\le n$, $1\le j\le n$, and for a.a. $(x,r,s)\in\Omega\times\mathbb{R}\times\mathbb{R}^n$. In other words, (4.21) says that the Jacobian matrix of the mapping

$$(r,s)\mapsto\big(c(x,r,s),a(x,r,s)\big):\mathbb{R}^{1+n}\to\mathbb{R}^{1+n}\tag{4.22}$$

is symmetric for a.a. $x\in\Omega$. Let us emphasize that (4.21) is not necessary for A to have a potential[11]. Yet, we will prove that, if (4.21) holds, then the mapping A defined by (2.59) has a potential $\Phi:W^{1,p}(\Omega)\to\mathbb{R}$ in the form

$$\Phi(u)\;=\;\int_\Omega\varphi(x,u,\nabla u)\,\mathrm{d}x+\int_{\Gamma_\mathrm{N}}\psi(x,u)\,\mathrm{d}S\,,\qquad\text{where}\tag{4.23a}$$

$$\varphi(x,r,s)=\int_0^1 s\cdot a(x,tr,ts)+r\,c(x,tr,ts)\,\mathrm{d}t\,,\tag{4.23b}$$

$$\psi(x,r)\;=\;\int_0^1 r\,b(x,tr)\,\mathrm{d}t\,;\tag{4.23c}$$

for the derivation of (4.23b,c) from the formula (4.6), cf. Exercise 4.22. In this situation, (4.20) is called the *Euler-Lagrange equation* for (4.23).

Lemma 4.12 (CONTINUITY OF Φ). *Let (2.55) with $\epsilon=0$ hold. Then Φ is continuous.*

Proof. The particular terms of Φ are continuous as a consequence of the continuity of the mapping $u\mapsto\nabla u:W^{1,p}(\Omega)\to L^p(\Omega;\mathbb{R}^n)$, of the embedding $W^{1,p}(\Omega)$ into $L^{p^*}(\Omega)$ and of the trace operator $u\mapsto u|_\Gamma:W^{1,p}(\Omega)\to L^{p^\#}(\Gamma_\mathrm{N})$ provided the continuity of the Nemytskiĭ operators $\mathscr{N}_\varphi:L^{p^*}(\Omega)\times L^p(\Omega;\mathbb{R}^n)\to L^1(\Omega)$ and $\mathscr{N}_\psi:L^{p^\#}(\Gamma_\mathrm{N})\to L^1(\Gamma_\mathrm{N})$ would be ensured. As to $\mathscr{N}_\varphi$, the needed growth condition on φ looks as

$$\exists\widehat\gamma\in L^1(\Omega)\;\;\exists\widehat C\in\mathbb{R}:\qquad|\varphi(x,r,s)|\le\widehat\gamma(x)+\widehat C|r|^{p^*}+\widehat C|s|^p.\tag{4.24}$$

In view of (4.23b), the condition (4.24) is indeed ensured by (2.55a,c) even weak-

[11] E.g. if $n=1$ and $a=a(x,s)$ and $c=c(x,r)$, then the basic Carathéodory hypothesis is obviously sufficient; $\varphi(x,\cdot,\cdot)$ is just the sum of the primitive functions of $a(x,\cdot)$ and $c(x,\cdot)$. In general, (4.21) holding only in the sense of distributions suffices, see Nečas [305, Theorem 3.2.12].

ened by putting $\epsilon = 0$ because of the estimate

$$
\begin{aligned}
|\varphi(x,r,s)| &\le \int_0^1 |s\cdot a(x,tr,ts)|\mathrm{d}t + \int_0^1 |rc(x,tr,ts)|\mathrm{d}t \\
&\le \int_0^1 \Big(|s|\big(\gamma_a(x) + C|tr|^{p^*/p'} + C|ts|^{p-1}\big) \\
&\quad + |r|\big(\gamma_c(x) + C|tr|^{p^*-1} + C|ts|^{p/p^{*\prime}}\big)\Big)\mathrm{d}t \\
&\le 2\frac{|s|^p}{p} + \frac{\gamma_a(x)^{p'}}{p'} + \frac{C^{p'}|r|^{p^*}}{p'(p^*+1)} + \frac{C|s|^p}{p} \\
&\quad + 2\frac{|r|^{p^*}}{p^*} + \frac{\gamma_c(x)^{p^{*\prime}}}{p^{*\prime}} + \frac{C|r|^{p^*}}{p^*} + \frac{C^{p^{*\prime}}|s|^p}{p^{*\prime}(p+1)},
\end{aligned}
\tag{4.25}
$$

which obviously requires $\gamma_a \in L^{p'}(\Omega)$ and $\gamma_c \in L^{p^{*\prime}}(\Omega)$ as indeed used in (2.55a) and (2.55c), respectively. Then (4.24) obviously follows.

The growth conditions for ψ, i.e. $|\psi(x,r)| \le \widehat{\gamma}(x) + \widehat{C}|r|^{p^\#}$ with some $\widehat{\gamma} \in L^1(\Gamma)$, can be treated analogously, resulting in (2.55b) with $\epsilon = 0$. □

Lemma 4.13 (Differentiability of Φ). *Let (2.55) for $\epsilon = 0$ and (4.21) hold, let $a(x,\cdot,\cdot) \in W^{1,1}_{\mathrm{loc}}(\mathbb{R}\times\mathbb{R}^n;\mathbb{R}^n)$ and $c(x,\cdot,\cdot) \in W^{1,1}_{\mathrm{loc}}(\mathbb{R}\times\mathbb{R}^n)$ for a.a. $x \in \Omega$. Then Φ is Gâteaux differentiable and $\Phi' = A$ with A given by (2.59).*

Proof. The directional derivative $\mathrm{D}\Phi(u,v)$ of Φ at u in the direction v is

$$
\begin{aligned}
\mathrm{D}\Phi(u,v) &:= \lim_{\varepsilon\to 0}\frac{\Phi(u+\varepsilon v) - \Phi(u)}{\varepsilon} = \frac{\mathrm{d}}{\mathrm{d}\varepsilon}\Phi(u+\varepsilon v)\Big|_{\varepsilon=0} \\
&= \frac{\mathrm{d}}{\mathrm{d}\varepsilon}\bigg(\int_\Omega \varphi(x, u+\varepsilon v, \nabla u + \varepsilon\nabla v)\,\mathrm{d}x + \int_{\Gamma_{\mathrm{N}}} \psi(x, u+\varepsilon v)\,\mathrm{d}S\bigg)\bigg|_{\varepsilon=0} \\
&= \int_\Omega \sum_{i=1}^n \frac{\partial\varphi(x,u,\nabla u)}{\partial s_i}\frac{\partial v}{\partial x_i} + \frac{\partial\varphi(x,u,\nabla u)}{\partial r}v\,\mathrm{d}x + \int_{\Gamma_{\mathrm{N}}} \frac{\partial\psi(x,u)}{\partial r}v\,\mathrm{d}S,
\end{aligned}
\tag{4.26}
$$

where we have changed the order of integration and differentiation by Theorem 1.29. This requires existence of a common (with respect to ε) integrable majorant of the collections $\{\varphi'_s(u+\varepsilon v, \nabla u+\varepsilon\nabla v)\cdot\nabla v\}_{0<\varepsilon\le\varepsilon_0}$ and $\{\varphi'_r(u+\varepsilon v, \nabla u+\varepsilon\nabla v)v\}_{0<\varepsilon\le\varepsilon_0}$ and $\{\psi'_r(u+\varepsilon v)v\}_{0<\varepsilon\le\varepsilon_0}$ for some $\varepsilon_0 > 0$, where we abbreviated $\partial\varphi/\partial s_i =: \varphi'_{s_i}$ etc. Assume, for a moment, that

$$\exists\gamma \in L^{p'}(\Omega)\ \ \exists C\in\mathbb{R}^+ : \quad |\varphi'_s(x,r,s)| \le \gamma(x) + C|r|^{p^*/p'} + C|s|^{p-1}, \tag{4.27a}$$

$$\exists\gamma\in L^{p^{\#\prime}}(\Gamma_{\mathrm{N}})\ \exists C\in\mathbb{R}^+ : \quad |\psi'_r(x,r)| \le \gamma(x) + C|r|^{p^\#-1}, \tag{4.27b}$$

$$\exists\gamma\in L^{p^{*\prime}}(\Omega)\ \ \exists C\in\mathbb{R}^+ : \quad |\varphi'_r(x,r,s)| \le \gamma(x) + C|r|^{p^*-1} + C|s|^{p/p^{*\prime}}. \tag{4.27c}$$

As for the first collection, for any $\varepsilon \in [0, \varepsilon_0]$ and a suitable C_p depending on p and C from by (4.27a), we have the estimate

$$\begin{aligned}|\varphi_s'(u+\varepsilon v, \nabla u+\varepsilon\nabla v)\cdot\nabla v| &\le \big(\gamma(x) + C|u+\varepsilon v|^{p^*/p'} + C|\nabla u+\varepsilon\nabla v|^{p-1}\big)|\nabla v|\\ &\le \big(\gamma(x) + C_p|u|^{p^*/p'} + \varepsilon_0^{p^*/p'}C_p|v|^{p^*/p'} + C_p|\nabla u|^{p-1} + C_p\varepsilon_0^{p-1}|\nabla v|^{p-1}\big)|\nabla v|\end{aligned}$$

which is the sought integrable majorant. The other two terms can be handled analogously, exploiting respectively (4.27b,c).

Moreover, $\mathrm{D}\Phi(u,\cdot) : W^{1,p}(\Omega) \to \mathbb{R}$ is obviously linear and, by (4.27), also continuous. Hence Φ has the Gâteaux differential.

The required form of the Gâteaux differential follows from the identities

$$\begin{aligned}\frac{\partial\varphi(x,r,s)}{\partial s_i} &= \int_0^1 a_i(x,tr,ts) + t\Big(\sum_{j=1}^n s_j\frac{\partial a_j}{\partial s_i}(x,tr,ts) + r\frac{\partial c}{\partial s_i}(x,tr,ts)\Big)\mathrm{d}t\\ &= \int_0^1 a_i(x,tr,ts) + t\Big(\sum_{j=1}^n s_j\frac{\partial a_i}{\partial s_j}(x,tr,ts) + r\frac{\partial a_i}{\partial r}(x,tr,ts)\Big)\mathrm{d}t\\ &= \int_0^1 \frac{\mathrm{d}}{\mathrm{d}t}\Big(t\,a_i(x,tr,ts)\Big)\mathrm{d}t = \Big[t\,a_i(x,tr,ts)\Big]_{t=0}^1 = a_i(x,r,s),\end{aligned}\tag{4.28}$$

where (4.23b) with (4.21) has been used; note that by Theorem 1.29 the change of the order of integration $\int_0^1 \mathrm{d}t$ and differentiation $\frac{\partial}{\partial s_i}$ in (4.28) requires a common integrable majorant of $\{|t \mapsto \frac{\partial}{\partial s_i}[s\cdot a(x,tr,ts)]|\}_{|s|\le M}$ and of $\{|t \mapsto \frac{\partial}{\partial s_i}[r\,c(x,tr,ts)]|\}_{|s|\le M}$ in $L^1(0,1)$ for any $M \in \mathbb{R}$, which holds because $a(x,r,\cdot) \in W^{1,1}_{\mathrm{loc}}(\mathbb{R}^n;\mathbb{R}^n)$ and $c(x,r,\cdot) \in W^{1,1}_{\mathrm{loc}}(\mathbb{R}^n)$ is assumed. Similarly also

$$\begin{aligned}\frac{\partial\varphi(x,r,s)}{\partial r} &= \int_0^1 c(x,tr,ts) + t\Big(r\frac{\partial c}{\partial r}(x,tr,ts) + \sum_{i=1}^n s_i\frac{\partial a_i}{\partial r}(x,tr,ts)\Big)\mathrm{d}t\\ &= \int_0^1 c(x,tr,ts) + t\Big(r\frac{\partial c}{\partial r}(x,tr,ts) + \sum_{i=1}^n s_i\frac{\partial c}{\partial s_i}(x,tr,ts)\Big)\mathrm{d}t\\ &= \int_0^1 \frac{\mathrm{d}}{\mathrm{d}t}\Big(t\,c(x,tr,ts)\Big)\mathrm{d}t = \Big[t\,c(x,tr,ts)\Big]_{t=0}^1 = c(x,r,s).\end{aligned}\tag{4.29}$$

The fact that $\partial\psi(x,r)/\partial r = b(x,r)$ can be derived by the easier way, realizing that (4.23c) defines, in fact, the primitive function of $b(x,\cdot)$, cf. (4.6). Note that (4.27a) then coincides with the former condition (2.55a), while (4.27b), (4.27c) is (2.55b,c) but weakened with $\epsilon = 0$, as indeed assumed. $\square$

In the following lemma, we will distinguish whether lower-order terms have critical growth (and then their monotonicity helps) or whether their growth is sub-critical (the cases (i) and (ii) in the following lemma).

Lemma 4.14 (Weak lower semicontinuity of Φ). *Let the assumptions of Lemma 4.13 and one of the following conditions be valid:*

(i) (2.55) *holds for* $\epsilon > 0$ *and the assumptions of Lemma* 2.32 *hold,*

(ii) (2.55) *holds for* $\epsilon > 0$ *and* $a(x,r,\cdot) : \mathbb{R}^n \to \mathbb{R}^n$ *is monotone,*

(iii) (2.55) *holds with* ϵ=0 *and the mappings* (4.22) *and* $b(x,\cdot)$:$\mathbb{R}\to\mathbb{R}$ *are monotone.*

Then Φ *is weakly lower semicontinuous.*

Proof. As to the case (i), by Lemmas 2.31–2.32, the gradient A of Φ is pseudomonotone; here we used also Lemma 4.13. By Theorem 4.4(ii), Φ is weakly lower semicontinuous.

The case (ii): by (4.28), monotonicity of $a(x,r,\cdot)$ is just monotonicity of $\varphi'_s(x,r,\cdot)$ from which convexity of $\varphi(x,r,\cdot)$ follows as in the proof of Theorem 4.4(iii). Likewise in (4.25), φ satisfies the growth condition (4.24) but now with $p^*-\epsilon$ instead of p^*. Hence, considering $u_k \rightharpoonup u$ in $W^{1,p}(\Omega)$, by the compact embedding $W^{1,p}(\Omega) \Subset L^{p^*-\epsilon}(\Omega)$, $\varphi(u_k,\nabla u) \to \varphi(u,\nabla u)$ in $L^1(\Omega)$. Similarly, by (4.28) and (2.55a) with $\epsilon > 0$, we have $\varphi'_s(u_k,\nabla u) \to \varphi'_s(u,\nabla u)$ in $L^{p'}(\Omega;\mathbb{R}^n)$. Altogether, by the convexity of $\varphi(x,r,\cdot)$,

$$\liminf_{k\to\infty}\int_\Omega \varphi(x,u_k,\nabla u_k)\,\mathrm{d}x \ge \lim_{k\to\infty}\int_\Omega \varphi(x,u_k,\nabla u)\,\mathrm{d}x \\ + \lim_{k\to\infty}\int_\Omega \varphi'_s(x,u_k,\nabla u)\cdot(\nabla u_k - \nabla u)\,\mathrm{d}x = \int_\Omega \varphi(x,u,\nabla u)\,\mathrm{d}x. \tag{4.30}$$

Moreover, by compactness of the trace operator $u \mapsto u|_\Gamma : W^{1,p}(\Omega) \to L^{p^\#-\epsilon}(\Gamma)$ and by the continuity of the Nemytskiĭ mapping $\mathscr{N}_\psi : L^{p^\#-\epsilon}(\Gamma_\mathrm{N}) \to L^1(\Gamma_\mathrm{N})$, we get the weak continuity of the boundary term in (4.23); the growth of ψ, i.e. $|\psi(x,r)| \le \widehat{\gamma}(x) + \widehat{C}|r|^{p^\#-\epsilon}$, can be estimated as in (4.25).

As to the case (iii), monotonicity of $[c,a](x,\cdot,\cdot) : \mathbb{R}^{1+n} \to \mathbb{R}^{1+n}$ implies convexity of $\varphi(x,\cdot,\cdot) : \mathbb{R}^{1+n} \to \mathbb{R}$, which can be seen similarly as in the proof of Theorem 4.4(iii). By monotonicity of $b(x,\cdot)$, the overall functional Φ is convex. By Lemma 4.13, Φ is smooth so, by (4.5), also weakly lower semicontinuous. □

Lemma 4.15 (Coercivity of Φ). *Let us assume* $\mathrm{meas}_{n-1}(\Gamma_\mathrm{D}) > 0$ *and* [12]

$$a(x,r,s)\cdot s + c(x,r,s)r \ge \varepsilon_1|s|^p + \varepsilon_2|r|^q - k_0(x)|s| - k_1(x)|r|, \tag{4.31a}$$
$$b(x,r)r \ge -k_2(x)|r| \tag{4.31b}$$

with some $\varepsilon_0,\varepsilon_1 > 0$, $p \ge q > 1$, *and* $k_0 \in L^{p'}(\Omega)$, $k_1 \in L^{p^{*\prime}}(\Omega)$, *and* $k_2 \in L^{p^{\#\prime}}(\Gamma)$. *Then* Φ *is coercive on* $W^{1,p}(\Omega)$. *Besides,* Φ *is coercive on* $V = \{v \in W^{1,p}(\Omega);\ v|_{\Gamma_\mathrm{D}} = 0\}$ *even if* $q = 0$.

[12] Cf. (4.31) with (2.92). Note that if one assumes, e.g. $b(x,r)r \ge -k_2(x)$, one would have $\int_0^1 k_2(x)/t\,\mathrm{d}t$ which is not finite, however.

Proof. In view of (4.23b), one has

$$\varphi(x,r,s) = \int_0^1 s\cdot a(x,tr,ts) + r\,c(x,tr,ts)\,\mathrm{d}t = \int_0^1 \frac{ts\cdot a(x,tr,ts)+tr\,c(x,tr,ts)}{t}\mathrm{d}t$$
$$\geq \int_0^1 \frac{\varepsilon_1|ts|^p + \varepsilon_2|tr|^q - k_0|ts| - k_1|tr|}{t}\mathrm{d}t = \frac{\varepsilon_1}{p}|s|^p + \frac{\varepsilon_2}{q}|r|^q - k_0|s| - k_1|r|.$$

Similarly, (4.31b) with (4.23c) implies $\psi(x,r) \geq -k_2|r|$. Then

$$\Phi(u) \geq \int_\Omega \Big(\frac{\varepsilon_1}{p}|\nabla u|^p + \frac{\varepsilon_2}{q}|u|^q - k_0|\nabla u| - k_1|u|\Big)\,\mathrm{d}x - \int_\Gamma k_2|u|\mathrm{d}S$$
$$\geq \varepsilon\|u\|^q_{W^{1,p}(\Omega)} - C - \|k_0\|_{L^{p'}(\Omega)}\|\nabla u\|_{L^p(\Omega;\mathbb{R}^n)}$$
$$-\|k_1\|_{L^{p^*\prime}(\Omega)}\|u\|_{L^{p^*}(\Omega)} - \|k_2\|_{L^{p^\#\prime}(\Gamma)}\|u\|_{L^{p^\#}(\Gamma)}$$
$$\geq \varepsilon\|u\|^q_{W^{1,p}(\Omega)} - C - \Big(\|k_0\|_{L^{p'}(\Omega)} + N_1\|k_1\|_{L^{p^*\prime}(\Omega)} + N_2\|k_2\|_{L^{p^\#\prime}(\Gamma)}\Big)\|u\|_{W^{1,p}(\Omega)}$$

with ε and C depending on $p, q, \varepsilon_1, \varepsilon_2$ and C_{P} from the Poincaré inequality (1.55), where N_1 and N_2 stand here respectively for the norms of the embedding $W^{1,p}(\Omega) \subset L^{p^*}(\Omega)$ and of the trace operator $u \mapsto u|_\Gamma : W^{1,p}(\Omega) \to L^{p^\#}(\Gamma)$. As $q > 1$, the functional Φ is coercive in the sense that $\Phi(u)/\|u\|_{W^{1,p}(\Omega)} \to +\infty$ for $\|u\|_{W^{1,p}(\Omega)} \to +\infty$.

For $q = 0$, we get coercivity on V by using Poincaré's inequality (1.57). □

Proposition 4.16 (Direct method for boundary-value problem (4.20)). *Let* (4.21), *the assumptions of Lemmas* 4.13–4.15 *hold, and f be defined by* (2.60), *i.e.* $\langle f, v\rangle := \int_\Omega gv\,\mathrm{d}x + \int_{\Gamma_{\mathrm{N}}} hv\,\mathrm{d}S$. *Then $\Phi - f$ has a minimizer on* $\{v \in W^{1,p}(\Omega);\ v|_{\Gamma_{\mathrm{D}}} = u_{\mathrm{D}}\}$, *and every such minimizer solves the boundary-value problem* (4.20) *in the weak sense.*

Proof. Let us first transform the problem on the linear space $V := \{v \in W^{1,p}(\Omega);\ v|_{\Gamma_{\mathrm{D}}} = 0\}$, cf. (2.52), as we did in Proposition 2.27: define $\Phi_0 : v \mapsto \Phi(v+w)$ with $w \in W^{1,p}(\Omega)$ such that $w|_{\Gamma_{\mathrm{D}}} = u_{\mathrm{D}}$ and $f \in V^*$ again by (2.60). Denoting $A_0 := \Phi_0'$, we have $A_0 : V \to V^*$ and $A_0(v) = A(v+w)$ with $A := \Phi'$. The reflexivity[13] of $W^{1,p}(\Omega)$ ensures also reflexivity of its closed[14] subspace V. The weak lower semicontinuity and coercivity of Φ, proved respectively in Lemma 4.14 and 4.15, is inherited by Φ_0, and therefore the existence of a minimizer u_0 of Φ_0 on V follows by the compactness argument. Then, in view of (4.26), $\mathrm{D}\Phi(u_0, v) = 0$ for any $v \in V$ just means that $u_0 \in V$ solves $A_0(u_0) = f$. Then $u := u_0 + w$ solves $A(u) = f$, i.e. it is the sought weak solution to the boundary-value problem (4.20) cf. Proposition 2.27.

Observing that $V + w = \{v \in W^{1,p}(\Omega);\ v|_{\Gamma_{\mathrm{D}}} = u_{\mathrm{D}}\}$, one can see that u minimizes $\Phi - f$ on $V + w$. □

[13] Recall that $1 < p < +\infty$ is supposed.

[14] Closedness of V follows from the continuity of the trace operator $u \mapsto u|_{\Gamma_{\mathrm{D}}}$.

Remark 4.17. In contrast to the non-potential case, Lemma 4.14(ii) allows us to treat nonlinearities of the type $c(\nabla u)$ without requiring strict monotonicity (2.68a) of $a(x,r,\cdot)$. Of course, the price for it is the severe restriction (4.21).

4.3 Examples and exercises

Exercise 4.18. Show that lower semicontinuous function $f : X \to \mathbb{R} \cup \{+\infty\}$ satisfying $\frac{1}{2}f(u_1) + \frac{1}{2}f(u_2) \geq f(\frac{1}{2}u_1 + \frac{1}{2}u_2)$ is convex.[15]

Example 4.19 (*Duality mapping*[16]). If V^* is strictly convex, the duality mapping $J : V \to V^*$ has a potential $\Phi(u) = \frac{1}{2}\|u\|^2$. Indeed, we have

$$\begin{aligned}\big\langle J(v), v-u\big\rangle &\geq \|v\|^2 - \|u\|\,\|v\| \geq \|v\|^2 - \frac{1}{2}\big(\|u\|^2 + \|v\|^2\big)\\ &= \frac{1}{2}\|v\|^2 - \frac{1}{2}\|u\|^2 \geq \|u\|\,\|v\| - \|u\|^2 \geq \big\langle J(u), v-u\big\rangle. \end{aligned} \tag{4.32}$$

Then put $v = u + th$. We get $\langle J(u+th), th\rangle \geq \frac{1}{2}\|u+th\|^2 - \frac{1}{2}\|u\|^2 \geq \langle J(u), th\rangle$. Divide it by $t \neq 0$, then let $t \to 0$. By the radial continuity[17] of J, we come to

$$\big\langle J(u), h\big\rangle = \lim_{t\to 0}\frac{1}{t}\Big(\frac{1}{2}\|u+th\|^2 - \frac{1}{2}\|u\|^2\Big) =: \mathrm{D}\Phi(u,h). \tag{4.33}$$

Exercise 4.20. Consider (4.21) and the situation in Exercise 2.89 with (2.153), i.e. $\ell_b^- < \min(1, \varepsilon_c)/N^2$. Show that Φ is strictly convex.

Exercise 4.21 (*Abstract Ritz approximation*). Consider the Ritz approximation (4.18) and show that $u_k \rightharpoonup u$ (for a subsequence) where u solves $A(u) = 0$, $A = \Phi'$ and, in addition, minimizes Φ over V. Besides, show that $\Phi(u_k) \to \Phi(u)$.[18] Moreover, assuming $\Phi = \Phi_1 + \Phi_0$ with Φ_0 weakly continuous and Φ_1 such that $\Phi_1(u_k) \to \Phi_1(u)$ and $u_k \rightharpoonup u$ imply $u_k \to u$, show that $u_k \to u$.[19]

---

[15] Hint: By iterating, show that $\lambda f(u_1)+(1-\lambda)f(u_2) \geq f(\lambda u_1+(1-\lambda)u_2)$ not only for $\lambda = 1/2$ but also for $\lambda = 1/4$ and $3/4$, and then for any dyadic number in $[0,1]$, i.e. $\lambda = k2^{-l}$, $l \in \mathbb{N}$, $k = 0, 1, \ldots, 2^l$. Such numbers are dense in $[0,1]$, and the general case $\lambda \in [0,1]$ then uses the lower semicontinuity of f.

[16] The observation that J has a potential is due to Asplund [22].

[17] Recall that we proved even the (norm,weak*)-continuity of J if V^* is strictly convex, see Lemma 3.2(ii).

[18] Hint: For any $v \in V$ take $v_k \in V_k$ such that $v_k \to v$. Then $\Phi(u_k) \leq \Phi(v_k)$ and, by weak lower semicontinuity and strong continuity of Φ, it holds that

$$\Phi(u) \leq \liminf_{k\to\infty}\Phi(u_k) \leq \liminf_{k\to\infty}\Phi(v_k) = \lim_{k\to\infty}\Phi(v_k) = \Phi(v).$$

For a special case $v := u$, it follows that $\Phi(u_k) \to \Phi(u)$.

[19] Hint: From $\Phi(u_k) \to \Phi(u)$ (already proved) and $\Phi_0(u_k) \to \Phi_0(u)$ (which follows from weak continuity of Φ_0), deduce $\Phi_1(u_k) \to \Phi_1(u)$.

Exercise 4.22. Derive (4.23) from (4.6), assuming existence of a potential and using Fubini's theorem 1.19.[20]

Example 4.23. (*p-Laplacean.*) The operator $A(u){:=}-\mathrm{div}\big(|\nabla u|^{p-2}\nabla u\big)$ on $W_0^{1,p}(\Omega)$ has the potential

$$\Phi(u) = \frac{1}{p}\int_\Omega |\nabla u(x)|^p\,\mathrm{d}x. \tag{4.34}$$

It just suffices to evaluate (4.23b) with $\varphi = \varphi(s)$:

$$\varphi(s){=}\int_0^1 s{\cdot}a(x,ts)\,\mathrm{d}t = \int_0^1 s{\cdot}|ts|^{p-2}ts\,\mathrm{d}t = \int_0^1 |s|^p t^{p-1}\mathrm{d}t = |s|^p\left[\frac{t^p}{p}\right]_{t=0}^1 = \frac{|s|^p}{p}.$$

Exercise 4.24. Verify (4.21) for $a_i(x,s) := |s|^{p-2}s_i$.[21] Show that $a\in W^{1,1}_{\rm loc}(\mathbb{R}^n)$.[22]

Example 4.25 (More general potentials[23]). Consider a coefficient $\sigma:\mathbb{R}^+\to\mathbb{R}^+$ depending on the magnitude of ∇u and the quasilinear mapping

$$u\mapsto -\mathrm{div}\big(\sigma(|\nabla u|^2)\nabla u\big). \tag{4.35}$$

In application, the concrete form of the function $\sigma(\cdot)>0$ may reflect some phenomenology resulting from experiments. Obviously, it fits with our concept for $a_i(x,r,s) := \sigma(|s|^2)s_i$ and $c\equiv 0$. The symmetry condition (4.21) is satisfied and

$$\varphi(x,r,s)\equiv\varphi(s) = \frac{1}{2}\int_0^{|s|^2}\sigma(\xi)\,\mathrm{d}\xi. \tag{4.36}$$

The monotonicity of the mapping (4.35) is related to positive definiteness of the second derivative $\varphi''(s)$, i.e. $\varphi''(s;\tilde s,\tilde s) = 2\sigma'(|s|^2)(s\cdot\tilde s)^2 + \sigma(|s|^2)|\tilde s|^2\geq 0$ for any $s,\tilde s\in\mathbb{R}^n$. This is trivially true if $\sigma'(|s|^2)\geq 0$. When estimating $(s\cdot\tilde s)^2\geq -|s|^2|\tilde s|^2$, one can see that this condition is certainly satisfied if

$$\forall\xi\geq 0:\qquad \sigma(\xi)\geq 2\xi\max\big(-\sigma'(\xi),0\big). \tag{4.37}$$

Hence σ may increase arbitrarily but must have a limited decay.

---

[20]Hint: $\int_0^1\langle A(tu),u\rangle\mathrm{d}t = \int_0^1\big(\int_\Omega a(tu,t\nabla u){\cdot}\nabla u{+}c(tu,t\nabla u)u\mathrm{d}x{+}\int_{\Gamma_{\rm N}}b(tu)u\mathrm{d}S\big)\,\mathrm{d}t{=}\int_\Omega\Big(\int_0^1 a(tu,t\nabla u){\cdot}\nabla u\mathrm{d}t + \int_0^1 c(tu,t\nabla u)u\,\mathrm{d}t\Big)\mathrm{d}x + \int_{\Gamma_{\rm N}}\int_0^1 b(tu)u\,\mathrm{d}t\mathrm{d}S = \int_\Omega\varphi(u,\nabla u)\,\mathrm{d}x + \int_{\Gamma_{\rm N}}\psi(u)\,\mathrm{d}S$.

[21]Hint: The symmetry of the matrix $a_s'(x,s)$ follows by the direct calculations:

$$\frac{\partial a_i(x,s)}{\partial s_j} = \frac{\partial|s|^{p-2}s_i}{\partial s_j} = s_i(p-2)|s|^{p-4}s_j + |s|^{p-2}\delta_{ij}.$$

[22]Hint: Indeed, $s_i(p-2)|s|^{p-4}s_j = \mathscr{O}(|s|^{p-2})$ for $s\to 0$. Hence this term is integrable also around the origin if $p>1$, as assumed.

[23]Cf. also Málek et al. [268, p.15] or Zeidler [427, Vol.II/B, Lemma 25.26].

Exercise 4.26 (*Regularizations of p-Laplacean*). Having in mind (4.36) with the coefficient σ in the analytical form

$$\sigma(\xi) := \varepsilon_1 + \big(\varepsilon_2 + \xi\big)^{(p-2)/2} \quad \text{or} \tag{4.38a}$$

$$\sigma(\xi) := \varepsilon_1 + \big(\varepsilon_2 + \sqrt{\xi}\big)^{p-2}, \quad \varepsilon_1, \varepsilon_2 \geq 0, \tag{4.38b}$$

show that $\varphi''(s; \tilde{s}, \tilde{s}) \geq 0$ so that (4.35) creates a monotone potential mapping.[24] One obviously gets the p-Laplacean when putting $\varepsilon_1 = \varepsilon_2 = 0$ in (4.38) while $\varepsilon_2 > 0$ makes its regularization around 0 as shown on Figure 9. The effect of $\varepsilon_1 > 0$ is just a vertical shift of σ and has already been considered in Exercise 2.89.

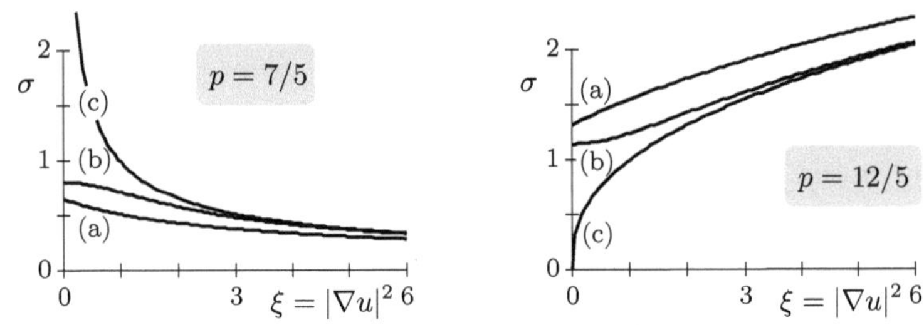

Figure 9. *Various dependence of the coefficient* σ *as a function of* $|\nabla u|^2$*;*
(a) = *the case* (4.38a) *with* $\varepsilon_1 = 0$ *and* $\varepsilon_2 = 2$,
(b) = *the case* (4.38b) *with* $\varepsilon_1 = 0$ *and* $\varepsilon_2 = 2$,
(c) = *the case* (4.38) *with* $\varepsilon_1 = \varepsilon_2 = 0$, *i.e. the p-Laplacean.*

Exercise 4.27 (Convergence of the *finite-element method*). Consider the boundary-value problem (2.147). Show that it has the potential

$$\Phi(u) = \int_\Omega \left(\frac{|\nabla u(x)|^p}{p} + \int_0^{u(x)} c(x, r)\,\mathrm{d}r \right) \mathrm{d}x + \int_\Gamma \left(\int_0^{u(x)} b(x, r)\,\mathrm{d}r \right) \mathrm{d}S \tag{4.39}$$

and note that no smoothness of $b(x,\cdot)$ and $c(x,\cdot)$ is required for (4.39). Assume Ω polyhedral, take a finite-dimensional V_k as in Example 2.67, and consider further an approximation by the *Ritz method*: minimize Φ on V_k to get some $u_k \in V_k$ satisfying the Galerkin identity (2.8) with $A = \Phi'$. Show that $u_k \rightharpoonup u$ where u minimizes Φ over $V = W^{1,p}(\Omega)$.[25] Assume a subcritical growth of $b(x,\cdot)$ and $c(x,\cdot)$ and deduce the strong convergence (in terms of subsequences)[26]

$$u_k \to u \quad \text{in } W^{1,p}(\Omega). \tag{4.40}$$

[24] Hint: Realize that, for $p \geq 2$, the coefficient σ is nondecreasing and positive (hence (4.37) holds trivially) while, for $p \leq 2$, (4.37) can be verified by calculations.

[25] Hint: Combine Exercise 4.21 with density of $\bigcup_{k\in\mathbb{N}} V_k$ in $W^{1,p}(\Omega)$ as in Example 2.67.

[26] Hint: Use Exercise 4.21 with $\Phi_1(u) = \frac{1}{p}\int_\Omega |\nabla u(x)|^p \mathrm{d}x$ and then just use Theorem 1.2 and uniform convexity of $L^p(\Omega; \mathbb{R}^n)$.

Exercise 4.28 (Nonmonotone terms with critical growth). Consider the equation $-\Delta u + c(u) = g$ with $c(r) = r^5 - r^2$ in $\Omega \subset \mathbb{R}^3$ with Dirichlet boundary conditions, $n = 3$, and show existence of a weak solution in $W^{1,2}(\Omega)$.[27]

Remark 4.29 (*Strong convergence of Ritz' method*). In fact, only a strict convexity of the nonlinearity $s \mapsto a(x, s)$ is sufficient for (4.40).[28] This is a nontrivial effect that, in this concrete potential case, the d-monotonicity needed in the abstract non-potential case, cf. Remark 2.21, can be considerably weakened.

Example 4.30 (*Advection* $\vec{v} \cdot \nabla u$ does not have any potential). Following Exercise 2.91, we consider $A : W^{1,2}(\Omega) \to W^{1,2}(\Omega)^*$ defined by $\langle A(u), v\rangle = \int_\Omega (\vec{v} \cdot \nabla u)\, v\mathrm{d}x$ with a given vector field $\vec{v}$ with, say, $\operatorname{div} \vec{v} = 0$ and $\vec{v}|_\Gamma = 0$. Using Green's formula, we can evaluate

$$\int_0^1 \langle A(tu), u\rangle \mathrm{d}t = \int_0^1 \int_\Omega tu\vec{v}\cdot\nabla u \,\mathrm{d}x\mathrm{d}t = \int_0^1 t \int_\Omega \vec{v}\cdot\nabla \frac{u^2}{2} \mathrm{d}x\mathrm{d}t = -\int_\Omega (\operatorname{div} \vec{v}) \frac{u^2}{4} \mathrm{d}x = 0.$$

By (4.6), a potential Φ of A would have to be constant so that $\Phi' = 0$, but obviously $A \neq 0$. This shows that A cannot have any potential. Realize that, of course, the condition (4.21) indeed fails.

Exercise 4.31 (*Anisotropic p-Laplacean*). Consider $\Phi(u) := \frac{1}{p} \int_\Omega \sum_{i=1}^n |\frac{\partial}{\partial x_i} u|^p \mathrm{d}x$ on $V := W_0^{1,p}(\Omega)$. Show that

$$\Phi'(u) = -\sum_{i=1}^n \frac{\partial}{\partial x_i} \left(\left|\frac{\partial u}{\partial x_i}\right|^{p-2} \frac{\partial u}{\partial x_i} \right) = (p-1) \sum_{i=1}^n \left|\frac{\partial u}{\partial x_i}\right|^{p-2} \frac{\partial^2 u}{\partial x_i^2}$$

and that Φ' is monotone and, if $p \geq 2$, uniformly monotone.[29]

Exercise 4.32 (*Higher-order Euler-Lagrange equation*). Consider the 4th-order equation as in Exercise 2.98, i.e.

$$\operatorname{div}\operatorname{div}\big(a(x, u, \nabla u, \nabla^2 u)\big) - \operatorname{div}\big(b(x, u, \nabla u, \nabla^2 u)\big) + c(x, u, \nabla u, \nabla^2 u) = g \quad (4.41)$$

here naturally with $(a, b, c) : \Omega\times\mathbb{R}\times\mathbb{R}^n\times\mathbb{R}^{n\times n} \to \mathbb{R}^{n\times n}\times\mathbb{R}^n\times\mathbb{R}$, and show that

[27] Hint: Realize that $2^* = 6$ for $n = 3$ and combine convex continuous functional $\Phi_1(u) := \frac{1}{6} \int_\Omega u^6 \mathrm{d}x$ with nonconvex but weakly continuous functional $\Phi_2(u) := -\frac{1}{3} \int_\Omega u^3 \mathrm{d}x$ on $W^{1,2}(\Omega)$, and realize coercivity of $\Phi(u) := \frac{1}{2} \int_\Omega |\nabla u|^2 \mathrm{d}x + \Phi_1(u) + \Phi_2(u)$.

[28] The proof, however, is rather nontrivial and uses so-called Young measures generated by minimizing sequences (here $\{u_k\}_{k\in\mathbb{N}}$) which must be composed from Dirac measures if $a(x, \cdot)$ is strictly convex; cf. Pedregal [331, Theorem 3.16]. See also Visintin [417].

[29] Hint: Modify (2.139). For (uniform) monotonicity, modify (2.141) or (2.142).

the *symmetry condition* like (4.21) now looks as

$$\frac{\partial a_{ij}(x,r,s,S)}{\partial S_{kl}} = \frac{\partial a_{kl}(x,r,s,S)}{\partial S_{ij}}, \quad \frac{\partial a_{ij}(x,r,s,S)}{\partial R_k} = \frac{\partial b_k(x,r,s,S)}{\partial S_{ij}}, \tag{4.42a}$$

$$\frac{\partial a_{ij}(x,r,s,S)}{\partial r} = \frac{\partial c(x,r,s,S)}{\partial S_{ij}}, \quad \frac{\partial b_i(x,r,s,S)}{\partial R_j} = \frac{\partial b_j(x,r,s,S)}{\partial R_i}, \tag{4.42b}$$

$$\frac{\partial b_i(x,r,s,S)}{\partial r} = \frac{\partial c(x,r,s,S)}{\partial R_i}, \tag{4.42c}$$

for $i,j,k,l{=}1,\dots,n$, i.e. symmetry of the Jacobian of the mapping

$$(c(x,\cdot,\cdot,\cdot),b(x,\cdot,\cdot,\cdot),a(x,\cdot,\cdot,\cdot)):\mathbb{R}{\times}\mathbb{R}^n{\times}\mathbb{R}^{n\times n}\to\mathbb{R}{\times}\mathbb{R}^n{\times}\mathbb{R}^{n\times n},$$

and then the potential is $\int_\Omega \varphi(x,u,\nabla u,\nabla^2 u)\,\mathrm{d}x$ with φ given as in (4.23b) now by

$$\varphi(x,r,s) = \int_0^1 S{:}a(x,tr,ts,tS) + s{\cdot}b(x,tr,ts,tS) + r\,c(x,tr,ts,tS)\,\mathrm{d}t\,. \tag{4.43}$$

Exercise 4.33 (*p-biharmonic operator*). Consider a_{ij} in the previous Exercise 4.32 given by (2.113) and $b_i = c = 0$, verify (4.42), and evaluate (4.43) to show that the p-biharmonic operator $\Delta(|\Delta|^{p-2}\Delta)$ on $V := W_0^{2,p}(\Omega)$ has the potential $\Phi(u) := \frac{1}{p}\int_\Omega |\Delta u|^p \mathrm{d}x$. For $p=2$, consider also $\Phi(u) = \frac{1}{2}\int_\Omega |\nabla^2 u|^2 \mathrm{d}x$.[30]

Exercise 4.34 (*Singular perturbations*). Consider the problem (2.167), use the estimates (2.168) and the potentiality of the operator $\varepsilon\,\mathrm{div}^2\nabla^2 - \Delta_p$, and show the weak convergence by passing to the limit in the underlying minimization problem.[31] Modify it by considering a quasilinear regularizing term as in Example 2.46.

4.4 Bibliographical remarks

Calculus of variations and related variational problems have been cultivated intensively since the 17th century by Fermat, Newton, Leibniz, Bernoulli, Euler,

---

[30]Hint: Realize that $\partial a_{ij}/\partial S_{kl} = 0$ for $i \neq j$ or $k \neq l$, and that (4.43) gives $\varphi(S) = |\sum_{k=1}^n S_{kk}|^p/p$. Further realize, for $p=2$, that $\int_\Omega |\nabla^2 u|^2 \mathrm{d}x = \int_\Omega |\Delta|^2 \mathrm{d}x$ under the Dirichlet boundary conditions, cf. Example 2.46.

[31]Hint: Using that u_ε minimizes the functional $u \mapsto \int_\Omega \frac{1}{p}|\nabla u|^p + \frac{\varepsilon}{2}|\nabla^2 u|^2 - gu\,\mathrm{d}x$ and that it converges weakly to some u in $W^{1,p}(\Omega)$, show

$$\begin{aligned}\int_\Omega \frac{1}{p}|\nabla u|^p \mathrm{d}x &\le \liminf_{\varepsilon\to 0}\int_\Omega \frac{1}{p}|\nabla u_\varepsilon|^p \mathrm{d}x \le \liminf_{\varepsilon\to 0}\int_\Omega \frac{1}{p}|\nabla u_\varepsilon|^p + \frac{\varepsilon}{2}|\nabla^2 u_\varepsilon|^2 \mathrm{d}x \\ &\le \lim_{\varepsilon\to 0}\int_\Omega \frac{1}{p}|\nabla v|^p + \frac{\varepsilon}{2}|\nabla^2 v|^2 - g(v{-}u_\varepsilon)\,\mathrm{d}x = \int_\Omega \frac{1}{p}|\nabla v|^p - g(v{-}u)\,\mathrm{d}x\end{aligned}$$

for any $v \in W_0^{2,2}(\Omega) \cap W^{1,p}(\Omega)$. By continuity, $\int_\Omega \frac{1}{p}|\nabla u|^p - gu\,\mathrm{d}x \le \int_\Omega \frac{1}{p}|\nabla v|^p - gv\,\mathrm{d}x$ holds for any $v \in W_0^{1,p}(\Omega)$.

Lagrange, Legendre, or Jacobi, often related to direct applications in physics and always bringing inspiration to development of mathematics.

The exposition here is narrowly focused on coercive problems leading to elliptic boundary-value problems. As to the abstract theory presented in Sect. 4.1, for further reading we refer to Blanchard, Brüning [53, Chap.2,3], Dacorogna [112, Chap.3], Gajewski, Gröger, Zacharias [168, Sect.III.4], Vainberg [414, Chap.II-IV], Zeidler [427, Parts II/B & III]. Reading about the problems having the potential of the type $\int_\Omega \varphi(u, \nabla u)\,\mathrm{d}x$ may include in particular Dacorogna [112, Chap.3], Evans [138, Chap.8], Gilbarg and Trudinger [Sect.11.5][178], Jost and Li-Jost [218], Ladyzhenskaya and Uraltseva [250, Chap.5].

Vectorial problems leading to systems of equations requiring special techniques, cf. also Sect. 6.1, are addressed e.g. by Dacorogna [112, Chap.IV], Evans [138, Chap.8], Giaquinta, Modica and J. Souček [177, Part II, Sect.1.4], Giusti [180, Chap.5], Morrey [291], Müller [297], and Pedregal [331, Chap.3].

There are many other variational techniques relying on critical points different from the global minimizers used here and more sophisticated principles, sometimes able to cope also with side conditions. Let us mention the celebrated mountain-pass technique by Ambrosetti and Rabinowitz [17] or Lyusternik and Schnirelman theory [263]. The monographs devoted to such advanced techniques are, e.g., Blanchard and Brüning [53], Chabrowski [91], Fučík, Nečas, Souček [158], Giaquinta and Hildebrandt [176], Giaquinta, Modica and J. Souček [177], Giusti [180], Kuzin and Pohozaev [247], Struwe [400], and Zeidler [427, Part III].

Chapter 5

Nonsmooth problems; variational inequalities

Many problems in physics and in other applications cannot be formulated as equations but have some more complicated structure, usually of a so-called complementarity problem. From the abstract viewpoint, the equations are replaced by inclusions involving set-valued mappings. We confine ourselves to a rather simple case (but still having wide applications) which involves set-valued mappings whose "set-valued part" can be described as a subdifferential of a convex but nonsmooth potential. Recall that we consider, if not said otherwise, V reflexive.

5.1 Abstract inclusions with a potential

A set-valued mapping $A : V \rightrightarrows V^*$ is called *monotone* if, for all $f_1 \in A(u_1)$ and $f_2 \in A(u_2)$, it holds that $\langle f_1 - f_2, u_1 - u_2 \rangle \geq 0$. We admit $A(u) = \emptyset$ and denote the definition domain of A by $\mathrm{dom}(A) := \{u \in V;\ A(u) \neq \emptyset\}$. Naturally, $A : V \rightrightarrows V^*$ is called *maximal monotone* if the graph of A is maximal (with respect to the ordering by inclusion) in the class of monotone graphs (i.e. graphs of monotone set-valued mappings) in $V \times V^*$. By the Kuratowski-Zorn lemma, any monotone set-valued mapping admits a maximal monotone extension, cf. Figure 10a,b. Besides, we call $A : V \rightrightarrows V^*$ *coercive* if

$$\lim_{\|u\|\to\infty} \inf_{f \in A(u)} \frac{\langle f, u \rangle}{\|u\|} = +\infty. \tag{5.1}$$

Here we shall consider some functional (again called a potential) $\Phi : V \to \bar{\mathbb{R}} := \mathbb{R} \cup \{+\infty, -\infty\}$ such that the (set-valued) mapping A represents a certain

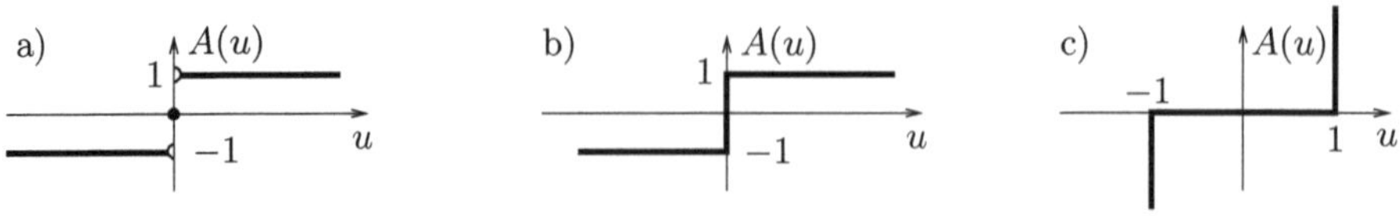

Figure 10. a) *a monotone but not maximal monotone mapping* $A:\mathbb{R}\to\mathbb{R}$,
b) *a maximal monotone extension* $A:\mathbb{R}\rightrightarrows\mathbb{R}$ *of the mapping from* a),
c) *another maximal monotone* $A:\mathbb{R}\rightrightarrows\mathbb{R}$, *inverse to the mapping from* b); *note that it is a normal-cone mapping to the interval* $[-1,1]$, *cf.* (5.3).

generalization of the gradient of Φ. We call $\Phi : V \to \bar{\mathbb{R}}$ *proper* if it is not identically equal to $+\infty$ and does not take the value $-\infty$. Except Remark 5.8, we confine ourselves to the case that Φ is convex, and then $A : V \rightrightarrows V^*$ will be the *subdifferential* of Φ, i.e. $A = \partial\Phi$, defined by

$$\partial\Phi(u) := \Big\{ f\in V^*;\ \forall v\in V:\ \Phi(v) + \langle f, u-v\rangle \geq \Phi(u)\Big\}, \tag{5.2}$$

cf. Figure 11. It is indeed a generalization of the gradient because, if A is also the Gâteaux differential, then $\partial\Phi(u) = \{\Phi'(u)\}$.[1] If Φ is finite and continuous at u, then $\partial\Phi(u) \neq \emptyset$,[2] otherwise emptiness of $\partial\Phi(u)$ is possible not only on $V\setminus\mathrm{dom}(\Phi)$ but also on $\mathrm{dom}(\Phi)$ as well as situations when $\mathrm{dom}(\partial\Phi)$ is not closed, cf. Figure 11.

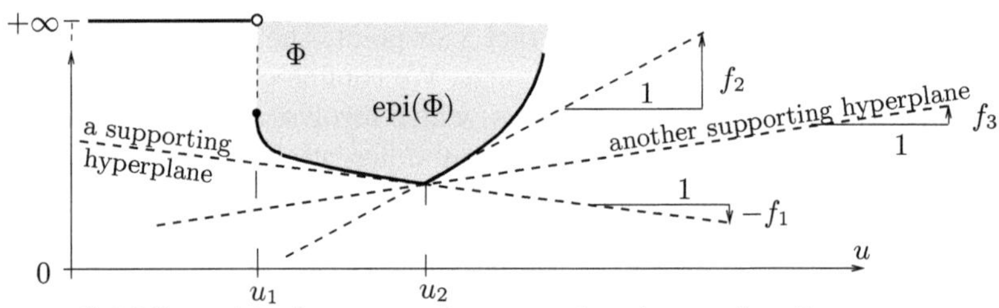

Figure 11. *Subdifferential of a convex lower semicontinuous function; an example for* $\partial\Phi(u_1) = \emptyset$, $\partial\Phi(u_2) = [f_1, f_2] \ni f_3$, *and* $\mathrm{dom}(\Phi)=[u_1,+\infty) \neq \mathrm{dom}(\partial\Phi)=(u_1,+\infty)$ *because* $\lim_{u\searrow u_1}\Phi'(u) = -\infty$.

Example 5.1 (*Normal-cone* mapping)**.** If K is a closed convex subset of V and $\Phi(u) = \delta_K(u)$ is the so-called *indicator function*, i.e. $\delta_K(u) = 0$ if $u \in K$ and $\delta_K(u) = +\infty$ if $u \notin K$, then, by the definition (5.2), we can easily see that

$$\partial\delta_K(u) = N_K(u) := \begin{cases} \{f\in V^*;\ \forall v\in K:\ \langle f, v-u\rangle \leq 0\} & \text{for } u\in K,\\ \emptyset & \text{for } u\notin K,\end{cases} \tag{5.3}$$

where $N_K(u)$ is the normal cone to K at u; cf. Figure 1 on p. 6.

Example 5.2 (*Potential of the duality mapping*)**.** For $\Phi(u) = \frac{1}{2}\|u\|^2$, it holds $\partial\Phi(u) = J(u)$, the duality mapping.[3] For V^* strictly convex, cf. Example 4.19.

[1] Cf. Exercise 5.34 below.

[2] This can be proved by the Hahn-Banach Theorem 1.5.

[3] The inclusion $\partial\Phi(u) \supset J(u)$ follows from $\frac{1}{2}\|v\|^2 - \frac{1}{2}\|u\|^2 \geq \|v\|\,\|u\| - \|u\|^2 \geq \langle f, v-u\rangle$ for $f \in J(u)$, cf. also (4.32). Conversely, $f \in J(u)$ implies $\frac{1}{2}\|u+th\|^2 - \frac{1}{2}\|u\|^2 \geq t\langle f, h\rangle$. Then, likewise (4.33), $\mathrm{D}\Phi(u,h) \geq \langle f, h\rangle$ for any $h \in X$, hence inevitably $f \in \partial\Phi(u)$.

Theorem 5.3 (CONVEX CASE[4]). *Let $A : V \rightrightarrows V^*$ have a proper convex potential $\Phi : V \to \bar{\mathbb{R}}$, i.e. $A = \partial\Phi$. Then:*

(i) *A is closed-valued, convex-valued, and monotone.*

(ii) *If Φ is lower semicontinuous, then A is maximal monotone.*

(iii) *If Φ is also coercive, then A is surjective in the sense that the inclusion*

$$A(u) \ni f \tag{5.4}$$

has a solution for any $f \in V^*$.

Proof. Closedness and convexity of the set $\partial\Phi(u)$ is obvious. To show monotonicity of the mapping $\partial\Phi$, we use the definition (5.2) so that, for any $f_i \in \partial\Phi(u_i)$ with $i = 1, 2$, one has

$$\Phi(u_2) \ge \Phi(u_1) + \langle f_1, u_2 - u_1\rangle \quad \text{and} \quad \Phi(u_1) \ge \Phi(u_2) + \langle f_2, u_1 - u_2\rangle. \tag{5.5}$$

By a summation, one gets $\langle f_1 - f_2, u_1 - u_2\rangle \ge 0$.

As to (ii), take $(u_0, f_0) \in V \times V^*$ and assume that $\langle f_0 - f, u_0 - u\rangle \ge 0$ for any $(f, u) \in \mathrm{Graph}(A)$. As V is assumed reflexive, we can consider it, after a possible renorming due to Asplund's theorem, as strictly convex together with its dual. Then, we consider (f, u) such that $J(u) + f = J(u_0) + f_0$, $f \in A(u)$, $J : V \to V^*$ the duality mapping, i.e. $[J + A](u) \ni J(u_0) + f_0$; such u does exist due to the point (iii) below applied to the convex coercive functional $v \mapsto \frac{1}{2}\|v\|^2 + \Phi(v)$, cf. also Example 4.19. Then $0 \le \langle f_0 - f, u_0 - u\rangle = \langle J(u) - J(u_0), u_0 - u\rangle$ and, by the strict monotonicity of J, cf. Lemma 3.2(iv), we get $u_0 = u$ so that $(f_0, u_0) \in \mathrm{Graph}(A)$.

The point (iii) can be proved by the *direct method*: Φ convex and lower semicontinuous implies that Φ is weakly lower semicontinuous; cf. Exercise 5.30. Then, by coercivity of Φ and reflexivity of V, the functional $\Phi - f$, being also coercive, possesses a minimizer u, see Theorem 4.2. Then $\partial\Phi(u) \ni f$ because otherwise $\partial\Phi(u) \not\ni f$ would imply, by the definition (5.2), that

$$\exists v \in V : \quad \Phi(v) + \langle f, u - v\rangle < \Phi(u) \tag{5.6}$$

so that $[\Phi - f](v) = \Phi(v) - \langle f, v\rangle < \Phi(u) - \langle f, u\rangle = [\Phi - f](u)$, a contradiction. □

Theorem 5.4 (SPECIAL NONCONVEX CASE). *Let $\Phi = \Phi_1 + \Phi_2 : V \to \bar{\mathbb{R}}$ be coercive, Φ_1 be a proper convex lower semicontinuous functional and Φ_2 be a weakly lower semicontinuous and Gâteaux differentiable functional, and let $A_1 = \partial\Phi_1$ and $A_2 = \Phi_2'$. Then, for any $f \in V^*$, there is $u \in V$ solving the inclusion*

$$A_1(u) + A_2(u) \ni f\ . \tag{5.7}$$

Remark 5.5 (Alternative formulations: inequalities). The inclusion (5.7), written as $\partial\Phi_1(u) \ni f - A_2(u)$, represents, in view of (5.2), a problem involving the variational inequality

$$\text{Find } u \in V : \ \forall v \in V : \quad \Phi_1(v) + \big\langle A_2(u), v - u\big\rangle \ge \Phi_1(u) + \langle f, v - u\rangle. \tag{5.8}$$

[4] In fact, (i)-(ii) holds even for non-reflexive spaces; see Rockafellar [352].

Proof of Theorem 5.4. Coercivity and weak lower semicontinuity of Φ with reflexivity implies the existence of a minimizer u of $\Phi - f$. In particular, $\Phi(u) < +\infty$ and hence also $\Phi_1(u) < +\infty$.

Suppose that (5.7) does not hold, i.e. $\partial\Phi_1(u) \not\ni f - \Phi_2'(u)$. By negation of (5.8), this just means that

$$\exists v \in V: \ \Phi_1(v) + \langle f - \Phi_2'(u), u - v\rangle < \Phi_1(u). \tag{5.9}$$

For $0 < \varepsilon \leq 1$, put $v_\varepsilon = u + \varepsilon(v-u)$. As Φ_1 is convex, it has the directional derivative $\mathrm{D}\Phi_1(u, v-u)$ and

$$\begin{aligned}\mathrm{D}\Phi_1(u, v-u) &:= \lim_{\varepsilon\searrow 0}\frac{\Phi_1(v_\varepsilon)-\Phi_1(u)}{\varepsilon} = \inf_{\varepsilon>0}\frac{\Phi_1(v_\varepsilon)-\Phi_1(u)}{\varepsilon}\\ &\leq \Phi_1(v_1) - \Phi_1(u) = \Phi_1(v) - \Phi_1(u) < +\infty.\end{aligned} \tag{5.10}$$

Note that $\mathrm{D}\Phi_1(u, v-u)$ is finite because it is bounded from below by $-\mathrm{D}\Phi_1(u, u-v) > -\infty$ by similar argument as (5.10). In particular,

$$\Phi_1(v_\varepsilon) = \Phi_1(u) + \varepsilon\mathrm{D}\Phi_1(u, v-u) + o_1(\varepsilon) \tag{5.11}$$

with some o_1 such that $\lim_{\varepsilon\searrow 0} o_1(\varepsilon)/\varepsilon$ is 0. Moreover, as Φ_2' is smooth, hence $\langle\Phi_2'(u), v-u\rangle = \mathrm{D}\Phi_2(u, v-u)$, by the definition of Gâteaux differential, it holds that

$$\Phi_2(v_\varepsilon) = \Phi_2(u) + \varepsilon\langle\Phi_2'(u), v-u\rangle + o_2(\varepsilon) \tag{5.12}$$

with some o_2 such that $\lim_{\varepsilon\to 0} o_2(\varepsilon)/\varepsilon = 0$. Thus, adding (5.11) and (5.12) and using (5.10), we get

$$\begin{aligned}\Phi_1(v_\varepsilon) + \Phi_2(v_\varepsilon) - \langle f, v_\varepsilon\rangle &= \Phi_1(u) + \Phi_2(u) - \langle f, u\rangle\\ &\quad+\varepsilon\Big(\mathrm{D}\Phi_1(u, v-u) + \mathrm{D}\Phi_2(u, v-u) - \langle f, v-u\rangle\Big) + o_1(\varepsilon) + o_2(\varepsilon)\\ &\leq \Phi_1(u) + \Phi_2(u) - \langle f, u\rangle\\ &\quad+\varepsilon\Big(\Phi_1(v) - \Phi_1(u) + \langle\Phi_2'(u), v-u\rangle - \langle f, v-u\rangle\Big) + o_1(\varepsilon) + o_2(\varepsilon).\end{aligned} \tag{5.13}$$

By (5.9), the multiplier of ε is negative, and therefore this term dominates $o_1(\varepsilon) + o_2(\varepsilon)$ if $\varepsilon>0$ is sufficiently small. Hence, for a small $\varepsilon>0$, the functional $\Phi_1 + \Phi_2 - f$ takes at v_ε a lower value than at u, a contradiction. $\square$

Remark 5.6 (Special cases)**.** If $\Phi_1 := \Phi_0 + \delta_K$ with both $\Phi_0 : V \to \mathbb{R}$ and $K \subset V$ convex, then, for $A = A_2$, (5.8) turns into the variational inequality:

$$\text{Find } u \in K: \quad \forall v \in K: \qquad \big\langle A(u), v-u\big\rangle + \Phi_0(v) - \Phi_0(u) \geq \langle f, v-u\rangle. \tag{5.14}$$

Often, $\Phi_0 = 0$ and then (5.14) can equally be written in the frequently used form

$$f - A(u) \in N_K(u), \tag{5.15}$$

which is a special case of (5.7).

Corollary 5.7. *Let $A = A_1 + A_2 : V \rightrightarrows V^*$ have the set-valued part $A_1 : V \rightrightarrows V^*$ monotone, coercive, and possessing a proper weakly lower semicontinuous potential Φ_1, and the single-valued part $A_2 : V \to V^*$ be pseudomonotone and possessing a (smooth) potential Φ_2 with an affine minorant. Then, for any $f \in V^*$, the inclusion* (5.4) *has a solution.*

Proof. Denote Φ_1 and Φ_2 the respective potentials. Then A_1 coercive and monotone implies Φ_1 coercive; cf. Exercise 5.32. Moreover, as Φ_2 has an affine minorant, $\Phi_1+\Phi_2$ is also coercive. Furthermore, Φ_2 is weakly lower semicontinuous, see Theorem 4.4(ii). Then we can use Theorem 5.4. □

Remark 5.8 (Hemivariational inequalities). In case of a general nonconvex locally Lipschitz Φ, the so-called *Clarke (generalized) gradient* is defined by:

$$\partial_C\Phi(u) := \Big\{ f \in V^*;\ \forall v \in V:\ \ \mathrm{D}^\circ\Phi(u,v) \geq \langle f, v\rangle \Big\} \tag{5.16}$$

where $\mathrm{D}^\circ\Phi(u,v)$ denotes the generalized directional derivative defined by

$$\mathrm{D}^\circ\Phi(u,v) := \limsup_{\substack{\tilde u \to u \\ \varepsilon \searrow 0}} \frac{\Phi(\tilde u + \varepsilon v) - \Phi(\tilde u)}{\varepsilon}\ ; \tag{5.17}$$

see Clarke [96] for more details. Inclusions involving Clarke's gradients are called *hemivariational inequalities.* In the special case of Theorem 5.4, we have $\partial_C\Phi = \partial\Phi_1 + \Phi_2'$ provided Φ_1 is locally Lipschitz continuous.

5.2 Application to elliptic variational inequalities

We will illustrate the previous theory on the 2nd-order elliptic variational inequality forming a so-called *unilateral problem* with an obstacle (determined by a function w) distributed over Ω and with Newton-type boundary conditions:

$$\left.\begin{array}{l} \left.\begin{array}{rcl} -\mathrm{div}\, a(x,u,\nabla u) + c(x,u,\nabla u) & \geq & g\,, \\ u & \geq & w, \\ \big(\mathrm{div}\, a(x,u,\nabla u) - c(x,u,\nabla u) + g\big)(u-w) & = & 0 \end{array}\right\} \ \text{in } \Omega, \\ \left.\begin{array}{rcl} \nu\cdot a(u,\nabla u) + b(x,u) & \geq & h, \\ u & \geq & w, \\ \big(\nu\cdot a(x,u,\nabla u) + b(x,u,\nabla u) - h\big)(u-w) & = & 0 \end{array}\right\} \ \text{on } \Gamma. \end{array}\right\} \tag{5.18}$$

The equalities in (5.18) express *transversality* of residua from the corresponding inequalities, while the triple composed from these two inequalities and one transversality relation is called a *complementarity problem.*

An interpretation illustrated in Figure 12 in a two-dimensional case is that u is a vertical deflection of an elastic membrane[5] elastically fixed on the contour Γ and stretched above a nonpenetrable obstacle given by the graph of w.

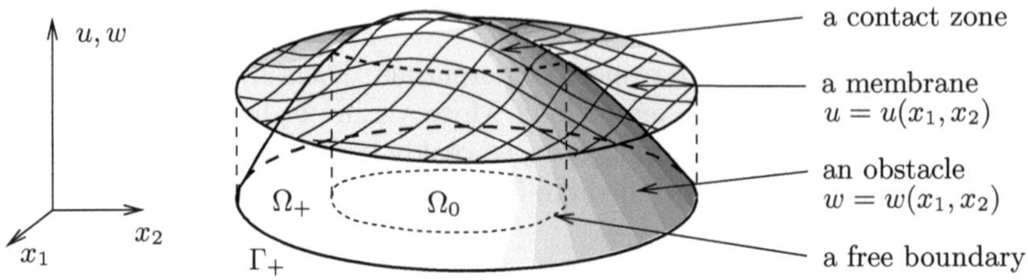

Figure 12. A schematic situation of unilateral problems on $\Omega \subset \mathbb{R}^2$: a deflected elastic membrane being in a partial contact with a rigid obstacle.

The abstract inequality (5.14) with $\Phi_0 = 0$, A given by (2.59) and f by (2.60) leads to the weak formulation, resulting in a variational inequality:

$$\begin{aligned} \text{Find } u\in K: \ \forall v\in K: \ & \int_\Omega a(u,\nabla u)\cdot\nabla(v-u)+c(u,\nabla u)(v-u)\,\mathrm{d}x \\ & + \int_\Gamma b(u)(v-u)\,\mathrm{d}S \geq \int_\Omega g(v-u)\,\mathrm{d}x + \int_\Gamma h(v-u)\,\mathrm{d}S \end{aligned} \tag{5.19}$$

where

$$K := \big\{v\in W^{1,p}(\Omega);\ \ v\geq w \text{ in } \Omega\big\}\,. \tag{5.20}$$

Assuming the symmetry condition (4.21), in view of Lemma 4.13, we can consider the abstract inequality (5.14) alternatively with $A = 0$, f from (2.60), and Φ_0 defined as in (4.23a) and φ and ψ are defined by (4.23b,c), which results in minimization over $W^{1,p}(\Omega)$ of the potential

$$\begin{aligned} \Phi - f : u \ \mapsto\ & \Phi_0(u) - \langle f, u\rangle + \delta_K(u) \\ := & \int_\Omega \big(\varphi(u,\nabla u) - gu\big)\,\mathrm{d}x + \int_\Gamma \big(\psi(u) - hu\big)\,\mathrm{d}S \ + \ \delta_K(u). \end{aligned} \tag{5.21}$$

An important question is whether the weak formulation (5.19) is *consistent* and *selective*; in other words, whether (5.18) contains enough information and is not overdetermined. The positive answer is:

Proposition 5.9 (Weak vs. classical formulations). *Let $w \in C(\bar\Omega)$ and $u \in C^2(\bar\Omega)$. Then the inequality (5.19) is satisfied if and only if (5.18) holds.*

Proof. Let us denote

$$\Omega_+ := \big\{x\in\Omega;\ u(x) > w(x)\big\}, \qquad \Omega_0 := \big\{x\in\Omega;\ u(x) = w(x)\big\},$$
$$\Gamma_+ := \big\{x\in\Gamma;\ u(x) > w(x)\big\}, \qquad \Gamma_0 := \big\{x\in\Gamma;\ u(x) = w(x)\big\},$$

[5]To be more precise, this interpretation refers to $a(x,r,s) = \alpha s$ with $\alpha > 0$ the elasticity coefficient, $c = 0$, and g a tangential outer force per unit area.

cf. Figure 12. For u solving (5.18) and for any $v \in W^{1,p}(\Omega)$ such that $v \geq w$, by Green's formula, we can write

$$\begin{aligned}
&\int_\Omega a(u,\nabla u)\cdot\nabla(v-u)+c(u,\nabla u)(v-u)\,\mathrm{d}x+\int_\Gamma b(u)(v-u)\,\mathrm{d}S\\
&\quad=\int_\Omega \big(\operatorname{div}a(u,\nabla u)-c(u,\nabla u)\big)(u-v)\,\mathrm{d}x+\int_\Gamma \big(\nu\cdot a(u,\nabla u)+b(u)\big)(v-u)\,\mathrm{d}S\\
&\quad=\int_{\Omega_+} \big(\operatorname{div}a(u,\nabla u)-c(u,\nabla u)\big)(u-v)\,\mathrm{d}x\\
&\qquad+\int_{\Omega_0} \big(\operatorname{div}a(u,\nabla u)-c(u,\nabla u)\big)(u-v)\,\mathrm{d}x\\
&\qquad+\int_{\Gamma_+} \big(\nu\cdot a(u,\nabla u)+b(u)\big)(v-u)\,\mathrm{d}S+\int_{\Gamma_0} \big(\nu\cdot a(u,\nabla u)+b(u)\big)(v-u)\,\mathrm{d}S\\
&\quad=:\int_{\Omega_+} I_1(x)\,\mathrm{d}x+\int_{\Omega_0} I_2(x)\,\mathrm{d}x+\int_{\Gamma_+} I_3(x)\,\mathrm{d}S+\int_{\Gamma_0} I_4(x)\,\mathrm{d}S.
\end{aligned}$$

Now, by (5.18), we have $I_1 = g(v-u)$ in Ω_+, $I_2 \geq g(v-u)$ because $\operatorname{div} a(u,\nabla u)-c(u,\nabla u) \leq -g$ and $u-v = w-v \leq 0$ in Ω_0, $I_3 = h(v-u)$ on Γ_+, and finally $I_4 \geq h(v-u)$ because $\nu\cdot a(u,\nabla u)+b(u) \geq h$ and $v-u = v-w \geq 0$ on Γ_0. Hence, altogether

$$\int_{\Omega_+} I_1\,\mathrm{d}x+\int_{\Omega_0} I_2\,\mathrm{d}x+\int_{\Gamma_+} I_3\,\mathrm{d}S+\int_{\Gamma_0} I_4\,\mathrm{d}S \geq \int_\Omega g(v-u)\,\mathrm{d}x+\int_\Gamma h(v-u)\,\mathrm{d}S$$

so that (5.19) has been obtained.

Conversely, if the solution u to (5.19) is regular enough, we can take z smooth such that $\operatorname{supp}(z) \subset \Omega_+$. Then, for a sufficiently small $|\varepsilon|$, $v := u+\varepsilon z \in K$ so that, by putting it into (5.19), one gets

$$\int_\Omega a(u,\nabla u)\cdot\nabla z+c(u,\nabla u)z\,\mathrm{d}x \geq \int_\Omega gz\,\mathrm{d}x. \tag{5.22}$$

Considering also $-z$ instead of z, we get an equality in (5.22). Then, by using the Green formula, we get $\operatorname{div} a(u,\nabla u)-c(u,\nabla u)+g = 0$ a.e. in Ω_+. The inequality $u \geq w$ is directly involved in (5.19). Since, for $z \geq 0$ with $\operatorname{supp}(z) \subset \Omega_0$, always $v = u+z \in K$, we get by putting such v into (5.19) the inequality $\int_\Omega gz\,\mathrm{d}x \leq \int_\Omega a(u,\nabla u)\cdot\nabla z+c(u,\nabla u)z\,\mathrm{d}x = \int_\Omega(-\operatorname{div} a(u,\nabla u)+c(u,\nabla u))z\,\mathrm{d}x$, which gives $\operatorname{div} a(u,\nabla u)-c(u,\nabla u)+g \leq 0$ a.e. in Ω_0. Altogether, the complementarity relations in Ω constituting (5.18) have been verified.

The complementarity relations on Γ can be verified analogously by taking test functions having nonvanishing traces on Γ. If $u(x) > w(x)$ for some $x \in \Gamma$, then, taking a sufficiently small neighbourhood N of x, we have $u > w$ on $\Omega\cap N$ (i.e. $N\cap\Omega_0 = \emptyset$), and then $v := u+\varepsilon z \in K$ for a sufficiently small $|\varepsilon|$ provided

$\mathrm{supp}(z) \subset \Omega \cap N$. Putting it into (5.19), one gets

$$\int_{\Omega\cap N} a(u,\nabla u)\cdot\nabla z + c(u,\nabla u)z\,\mathrm{d}x + \int_{\Gamma\cap N} b(u)z\,\mathrm{d}S \geq \int_{\Omega\cap N} gz\,\mathrm{d}x + \int_{\Gamma\cap N} hz\,\mathrm{d}S.$$

By using the Green formula, we get

$$\int_{\Omega\cap N} \big(-\operatorname{div} a(u,\nabla u) + c(u,\nabla u) - g\big)z\,\mathrm{d}x + \int_{\Gamma\cap N} \big(\nu\cdot a(u,\nabla u) + b(u) - h\big)z\,\mathrm{d}S \geq 0. \tag{5.23}$$

Considering also $-z$ instead of z, we get equality in (5.23). As we already know that $\operatorname{div} a(u,\nabla u) - c(u,\nabla u) + g = 0$ in $\Omega\cap N \subset \Omega_+$, we get $\nu\cdot a(u,\nabla u) + b(u) - h = 0$ on $\Gamma \cap N$. In general, we can take $z \geq 0$ arbitrary but such that it will be small in Ω yet still with prescribed values on Γ. This will push the first integral in (5.23) to zero while the second one then yields $\nu\cdot a(u,\nabla u) + b(u) - h \geq 0$ on Γ. □

A theoretically and to some extent also numerically[6] efficient method of *regularization* (or approximation) for problems like (5.18) is the so-called *penalty method.* In the potential case, its L^α-variant leads to approximation of the functional from (5.21) by the functional

$$\Phi_\varepsilon(u) = \int_\Omega \Big(\varphi(u,\nabla u) + \frac{\big((w-u)^+\big)^\alpha}{\alpha\varepsilon}\Big)\mathrm{d}x + \int_\Gamma \psi(u)\,\mathrm{d}S \tag{5.24}$$

where $v^+ := \max(0,v)$. The idea is then to minimize $\Phi_\varepsilon - f$ over the whole $W^{1,p}(\Omega)$, which corresponds to the boundary-value problem

$$\left.\begin{aligned} -\operatorname{div} a(u,\nabla u) + c(u,\nabla u) - \frac{1}{\varepsilon}\big((w-u)^+\big)^{\alpha-1} &= g \quad \text{in } \Omega,\\ \nu\cdot a(u,\nabla u) + b(x,u) &= h \quad \text{on } \Gamma. \end{aligned}\right\} \tag{5.25}$$

Now, we need to modify the coercivity $\langle A(v), v-w\rangle \geq \delta\|v\|^p_{W^{1,p}(\Omega)} + C$ with $\delta > 0$ and some C (depending possibly on w). E.g., we can modify (4.31) to

$$a(x,r,s)\cdot(s-\nabla w(x)) + c(x,r,s)(r-w(x)) \geq \varepsilon_1|s|^p + \varepsilon_2|r|^q - k_0(x) - k_1(x)|s| - k_2(x)|r|, \tag{5.26a}$$

$$b(x,r)(r-w(x)) \geq -k_3(x) - k_4(x)|r| \tag{5.26b}$$

with some $\varepsilon_0, \varepsilon_1 > 0$, $p \geq q > 1$, and $k_0 \in L^1(\Omega)$, $k_1 \in L^{p'}(\Omega)$, $k_2 \in L^{p^{*\prime}}(\Omega)$, $k_3 \in L^1(\Gamma)$, and $k_4 \in L^{p^{\#\prime}}(\Gamma)$ depending on a fixed w.

[6] If also a discretization as in Exercise 4.27 is applied, one can implement the resulting minimization problem on computers although its numerical solution is not always easy if $\varepsilon > 0$ in (5.24) has to be chosen small.

Proposition 5.10 (Convergence of the penalty method). *Assume $1<\alpha\le p^*$, and a, b and c satisfy the qualifications in Lemmas 4.13–4.14, in particular the symmetry (4.21) and coercivity in the sense (5.26) hold. Let also $w\in W^{1,p}(\Omega)$. Then:*

(i) *The boundary-value problem (5.25) has always a weak solution $u_\varepsilon\in W^{1,p}(\Omega)$.*

(ii) *The sequence $\{u_\varepsilon\}_{\varepsilon>0}$ is bounded in $W^{1,p}(\Omega)$. If Φ_0 is convex, then the values of Φ_0 converge, i.e. $\lim_{\varepsilon\to0}\Phi_0(u_\varepsilon):=\lim_{\varepsilon\to0}\int_\Omega\varphi(u_\varepsilon,\nabla u_\varepsilon)\,\mathrm{d}x+\int_\Gamma\psi(u_\varepsilon)\,\mathrm{d}S\to\int_\Omega\varphi(u,\nabla u)\,\mathrm{d}x+\int_\Gamma\psi(u)\,\mathrm{d}S=:\Phi_0(u)$, and $\{u_\varepsilon\}_{\varepsilon>0}$ converges for $\varepsilon\to0$ (in terms of subsequences) weakly to a solution u of (5.19).*

(iii) *If, in addition, the mapping A induced by (a,c) is d-monotone with respect to the seminorm $v\mapsto\|\nabla v\|_{L^p(\Omega;\mathbb{R}^n)}$, then $u_\varepsilon\to u$ (a subsequence) in $W^{1,p}(\Omega)$ strongly.*

Proof. For $\varepsilon>0$ fixed, existence of a weak solution $u_\varepsilon\in W^{1,p}(\Omega)$ to (5.25) follows by the direct method by using Proposition 4.16.

To prove a-priori estimates, put $v:=u_\varepsilon-w$ into the weak formulation of (5.25). Using also (5.26) and the estimates as in the proof of Lemma 4.15, for a suitable δ and C, one gets

$$\begin{aligned}\delta\|u_\varepsilon\|^q_{W^{1,p}(\Omega)}&+\frac{1}{\varepsilon}\|(w-u_\varepsilon)^+\|^\alpha_{L^\alpha(\Omega)}-C\\&\le\int_\Omega a(u_\varepsilon,\nabla u_\varepsilon)\cdot\nabla(u_\varepsilon-w)+c(u_\varepsilon,\nabla u_\varepsilon)(u_\varepsilon-w)+\frac{|(w-u_\varepsilon)^+|^\alpha}{\varepsilon}\,\mathrm{d}x\\&+\int_\Gamma b(u_\varepsilon)(u_\varepsilon-w)\,\mathrm{d}S=\int_\Omega g(u_\varepsilon-w)\,\mathrm{d}x+\int_\Gamma h(u_\varepsilon-w)\,\mathrm{d}S.\end{aligned}\tag{5.27a}$$

The integrals on the right-hand side can be estimated as

$$\begin{aligned}\int_\Omega g(u_\varepsilon-w)\,\mathrm{d}x&+\int_\Gamma h(u_\varepsilon-w)\,\mathrm{d}S=\langle f,u_\varepsilon-w\rangle\\&\le\|f\|_{W^{1,p}(\Omega)^*}\big(\|u_\varepsilon\|_{W^{1,p}(\Omega)}+\|w\|_{W^{1,p}(\Omega)}\big)\\&<\frac{\delta}{2}\|u_\varepsilon\|^q_{W^{1,p}(\Omega)}+C_\delta\|f\|^{q'}_{W^{1,p}(\Omega)^*}+\|f\|_{W^{1,p}(\Omega)^*}\|w\|_{W^{1,p}(\Omega)}\end{aligned}\tag{5.27b}$$

with f determined by (and estimated through) the data (g,h), cf. (2.60) and (2.62), and with $C_\delta=q'\,\sqrt[q-1]{q\delta/2}$, cf. (1.22). In this way, we show $\{u_\varepsilon\}_{\varepsilon>0}$ bounded in $W^{1,p}(\Omega)$ and, up to a subsequence, $u_\varepsilon\rightharpoonup u$ in $W^{1,p}(\Omega)$.

We have also $\frac{1}{\varepsilon}\|(w-u_\varepsilon)^+\|^\alpha_{L^\alpha(\Omega)}\le C$ so that $\|(w-u_\varepsilon)^+\|_{L^\alpha(\Omega)}=\mathscr{O}(\sqrt[\alpha]{\varepsilon})$. Using the weak continuity of $v\mapsto\|v^+\|_{L^\alpha(\Omega)}$ if $\alpha<p^*$ (or the weak lower semicontinuity if $\alpha=p^*$, cf. Exercise 5.30 below), one gets

$$\|(w-u)^+\|_{L^\alpha(\Omega)}\le\liminf_{\varepsilon\to0}\|(w-u_\varepsilon)^+\|_{L^\alpha(\Omega)}\le\limsup_{\varepsilon\to0}\|(w-u_\varepsilon)^+\|_{L^\alpha(\Omega)}=0.$$

Thus $u\in K$ has been proved.

The convergence of $\Phi_0(u_\varepsilon)$ to $\Phi_0(u)$ can be seen from the estimate

$$\big[\Phi_0-f\big](u) \le \liminf_{\varepsilon\to0}\big[\Phi_0-f\big](u_\varepsilon) \le \limsup_{\varepsilon\to0}\big[\Phi_0-f\big](u_\varepsilon) \le \big[\Phi_0-f\big](u) \tag{5.28}$$

because Φ_0 is weakly lower semicontinuous (see Lemma 4.14) and always $[\Phi_0-f](u_\varepsilon) \le [\Phi_\varepsilon-f](u_\varepsilon) \le [\Phi_\varepsilon-f](u) = [\Phi_0-f](u)$ because u_ε minimizes $\Phi_\varepsilon-f$ with $\Phi_\varepsilon := \Phi_0 + \frac{1}{\alpha\varepsilon}\|(w-\cdot)^+\|^\alpha_{L^\alpha(\Omega)}$ and because $u \in K$; here we used the assumption that Φ_0 is convex so that each solution to the Euler-Lagrange equation minimizes its potential Φ_0-f. Then (5.28) yields $\lim_{\varepsilon\to0}[\Phi_0-f](u_\varepsilon) = [\Phi_0-f](u)$, from which $\lim_{\varepsilon\to0}\Phi_0(u_\varepsilon) = \Phi_0(u)$ follows.

The fact that u solves (5.19), i.e. minimizes (5.21), follows directly from the proved facts that $u \in K$ and $[\Phi_0-f](u) \le \min(\Phi_0 - f + \delta_K)$, proved in (5.28) if one realizes also

$$[\Phi_0-f](u_\varepsilon) \le [\Phi_\varepsilon-f](u_\varepsilon) \le \min[\Phi_\varepsilon-f] \le \min(\Phi_0 - f + \delta_K) \tag{5.29}$$

because u_ε minimizes $\Phi_\varepsilon-f$ and because always $\Phi_\varepsilon \le \Phi_0 + \delta_K$.

Now, we are going to prove the strong convergence. By multiplying the equation in (5.25) by $(u - u_\varepsilon)$, applying Green's formula and using the boundary conditions in (5.25), one gets

$$\int_\Omega a(u_\varepsilon,\nabla u_\varepsilon)\cdot\nabla(u-u_\varepsilon) + c(u_\varepsilon,\nabla u_\varepsilon)(u-u_\varepsilon) - \frac{1}{\varepsilon}\big|(w-u_\varepsilon)^+\big|^{\alpha-1}(u-u_\varepsilon)\,\mathrm{d}x \\ + \int_\Gamma b(u_\varepsilon)(u-u_\varepsilon)\,\mathrm{d}S = \int_\Omega g(u-u_\varepsilon)\,\mathrm{d}x + \int_\Gamma h(u-u_\varepsilon)\,\mathrm{d}S. \tag{5.30}$$

Since $u \ge w$, the term $\frac{1}{\varepsilon}|(w-u_\varepsilon)^+|^{\alpha-1}(u-u_\varepsilon) \ge 0$ a.e. on Ω and, by omitting it, one gets

$$\int_\Omega a(u_\varepsilon,\nabla u_\varepsilon)\cdot\nabla(u-u_\varepsilon) + c(u_\varepsilon,\nabla u_\varepsilon)(u-u_\varepsilon)\,\mathrm{d}x \\ + \int_\Gamma b(u_\varepsilon)(u-u_\varepsilon)\,\mathrm{d}S \ge \int_\Omega g(u-u_\varepsilon)\,\mathrm{d}x + \int_\Gamma h(u-u_\varepsilon)\,\mathrm{d}S. \tag{5.31}$$

Then, if (a, c) induces a d-monotone mapping as assumed, by (5.31) we get

$$\Big(d\big(\|\nabla u_\varepsilon\|_{L^p(\Omega;\mathbb{R}^n)}\big) - d\big(\|\nabla u\|_{L^p(\Omega;\mathbb{R}^n)}\big)\Big)\big(\|\nabla u_\varepsilon\|_{L^p(\Omega;\mathbb{R}^n)} - \|\nabla u\|_{L^p(\Omega;\mathbb{R}^n)}\big) \\ \le \int_\Omega (a(u_\varepsilon,\nabla u_\varepsilon)-a(u,\nabla u))\cdot\nabla(u_\varepsilon-u) + (c(u_\varepsilon,\nabla u_\varepsilon)-c(u,\nabla u))(u_\varepsilon-u)\,\mathrm{d}x \\ \le \int_\Omega \big(g - c(u,\nabla u)\big)(u_\varepsilon-u) - a(u,\nabla u)\cdot\nabla(u_\varepsilon-u)\,\mathrm{d}x \\ + \int_\Gamma \big(h - b(u_\varepsilon)\big)(u_\varepsilon-u)\,\mathrm{d}S \to 0, \tag{5.32}$$

because subsequently $u_\varepsilon - u \rightharpoonup 0$ in $L^{p^*}(\Omega)$, $h - b(u_\varepsilon) \to h - b(u)$ in $L^{p^{\#\prime}}(\Gamma)$ and $u_\varepsilon - u \rightharpoonup 0$ in $L^{p^\#}(\Gamma)$, and $\nabla u_\varepsilon - \nabla u \rightharpoonup 0$ in $L^p(\Omega;\mathbb{R}^n)$. From (5.32), one deduces

$\|\nabla u_\varepsilon\|_{L^p(\Omega;\mathbb{R}^n)} \to \|\nabla u\|_{L^p(\Omega;\mathbb{R}^n)}$. From $\nabla u_\varepsilon \rightharpoonup \nabla u$ in $L^p(\Omega;\mathbb{R}^n)$ proved above, and from the uniform convexity of $L^p(\Omega;\mathbb{R}^n)$, by Theorem 1.2 we get $\nabla u_\varepsilon \to \nabla u$. □

In case Φ_0 is not convex, weak solutions u_ε to the boundary-value problem (5.25) do not necessarily minimize $\Phi_\varepsilon - f$. Anyhow, the proof of the strong convergence in Proposition 5.10(iii) holds, while the proof of mere weak convergence is to be made through the *Minty trick*. For example, if (c, a) is monotone but b not (hence Φ_0 indeed need not be convex), one can prove:

Proposition 5.11 (Convergence of the penalty method II). *Let* $1<\alpha\le p^*$, *and* a, b *and* c *satisfy the qualifications in Lemmas* 4.13–4.14 *and let* (c, a) *induce a monotone mapping* (4.22) *for a.a.* $x \in \Omega$. *Then* u_ε *from Proposition* 5.10(i) *converges for* $\varepsilon \to 0$ *(in terms of subsequences) weakly to a solution* u *of* (5.19).

Proof. We now want to use the Minty trick. To this goal, we multiply (5.25) by $(v - u_\varepsilon)$, apply Green's formula, and use the boundary conditions in (5.25) to get (5.30) with v in place of u. Considering $v \ge w$, one can see that $\frac{1}{\varepsilon}|(w - u_\varepsilon)^+|^{\alpha-1}(v - u_\varepsilon)$ is non-negative a.e. on Ω. Thus we arrive at (5.31) with v in place of u. Then, using monotonicity of $[c, a](x, \cdot, \cdot)$, we obtain

$$\begin{aligned}
0 &\le \int_\Omega (a(u_\varepsilon, \nabla u_\varepsilon) - a(v, \nabla v))\cdot\nabla(u_\varepsilon - v) + (c(u_\varepsilon, \nabla u_\varepsilon) - c(v, \nabla v))(u_\varepsilon - v)\,\mathrm{d}x \\
&\le \int_\Omega \big(g - c(v, \nabla v)\big)(u_\varepsilon - v) - a(v, \nabla v)\cdot\nabla(u_\varepsilon - v)\,\mathrm{d}x + \int_\Gamma \big(h - b(u_\varepsilon)\big)(u_\varepsilon - v)\,\mathrm{d}S \\
&\to \int_\Omega \big(g - c(v, \nabla v)\big)(u - v) - a(v, \nabla v)\cdot\nabla(u - v)\,\mathrm{d}x + \int_\Gamma \big(h - b(u)\big)(u - v)\,\mathrm{d}S;
\end{aligned} \tag{5.33}$$

here we used compactness of the trace operator $W^{1,p}(\Omega) \to L^{p^\# - \epsilon}(\Gamma)$ for $b(u_\varepsilon) \to b(u)$ in $L^{p^{\#'}}(\Gamma)$. Then we modify Minty's trick: instead of $v := u + \eta z$ used for equations with $\eta > 0$, cf. the proof of Lemma 2.13, we now put $v := \eta z + (1-\eta)u$ for $\eta \in (0, 1]$ and $z \ge w$; note that such v lives in K. As $v - u = \eta z + (1-\eta)u - u = \eta(z - u)$, this gives

$$\begin{aligned}
0 \le \int_\Omega &\big(g - c\big(\eta z + (1-\eta)u\,,\, \eta\nabla z + (1-\eta)\nabla u\big)\big)(\eta u - \eta v) \\
&- a\big(\eta z + (1-\eta)u\,,\, \eta\nabla z + (1-\eta)\nabla u\big)\cdot\nabla(\eta z - \eta u)\,\mathrm{d}x + \int_\Gamma \big(h - b(u)\big)(\eta z - \eta u)\,\mathrm{d}S.
\end{aligned}$$

Then we divide it by $\eta > 0$, and pass to the limit with $\eta \to 0$. It gives just the desired inequality (5.19). □

Remark 5.12. Strenghtening the growth condition (2.55) and the qualification of

w e.g. by considering $w \in W^{1,\infty}(\Omega)$ and

$$|a(x,r,s)| \le \gamma(x) + C|r|^{q-\epsilon} + C|s|^{p-1} \qquad \text{for some } \gamma \in L^{p'}(\Omega)\,, \tag{5.34a}$$

$$|\,b(x,r)\,| \ \le\ \gamma(x) + C|r|^{q-\epsilon} \qquad \text{for some } \gamma \in L^{p^{\#\prime}}(\Gamma), \tag{5.34b}$$

$$|c(x,r,s)| \le \gamma(x) + C|r|^{q-\epsilon} + C|s|^{p-\epsilon} \qquad \text{for some } \gamma \in L^{p^{*\prime}}(\Omega) \tag{5.34c}$$

and for some $\epsilon > 0$, one can replace the optimal but rather cumbersome coercivity condition (5.26) depending on w by the former condition (4.31). Then, instead of (5.27), we can estimate for some $\delta > 0$:

$$\begin{aligned}
&\delta\|u_\varepsilon\|^q_{W^{1,p}(\Omega)} + \frac{1}{\varepsilon}\|(w-u_\varepsilon)^+\|^\alpha_{L^\alpha(\Omega)} - C \\
&\le \int_\Omega \varepsilon_1|\nabla u_\varepsilon|^p + \varepsilon_2|u_\varepsilon|^q - k_0|\nabla u_\varepsilon| - k_1|u_\varepsilon| + \frac{|(w-u_\varepsilon)^+|^\alpha}{\varepsilon}\,\mathrm{d}x - \int_\Gamma k_2|u_\varepsilon|\,\mathrm{d}S \\
&\le \int_\Omega a(u_\varepsilon,\nabla u_\varepsilon)\cdot\nabla u_\varepsilon + c(u_\varepsilon,\nabla u_\varepsilon)u_\varepsilon + \frac{|(w-u_\varepsilon)^+|^\alpha}{\varepsilon}\,\mathrm{d}x + \int_\Gamma b(u_\varepsilon)u_\varepsilon\,\mathrm{d}S \\
&= \int_\Omega a(u_\varepsilon,\nabla u_\varepsilon)\cdot\nabla w + c(u_\varepsilon,\nabla u_\varepsilon)w + g(u_\varepsilon - w)\,\mathrm{d}x + \int_\Gamma b(u_\varepsilon)w + h(u_\varepsilon - w)\,\mathrm{d}S \\
&\le C_{\delta_1} + \delta_1\int_\Omega |\nabla u_\varepsilon|^p + |u_\varepsilon|^q\,\mathrm{d}x + \|f\|_{W^{1,p}(\Omega)^*}\big(\|u_\varepsilon\|_{W^{1,p}(\Omega)} + \|w\|_{W^{1,p}(\Omega)}\big)
\end{aligned}$$

for any small δ_1 and C_{δ_1} depending on δ_1 and γ's and C from (5.34) with ε_1, ε_2, k_0, k_1, and k_2 from (4.31). Then the a-priori estimates $\|u_\varepsilon\|_{W^{1,p}(\Omega)} \le C$ and $\|(w-u_\varepsilon)^+\|_{L^\alpha(\Omega)} \le C/\sqrt[\alpha]{\varepsilon}$ easily follow.

Remark 5.13 (*Free boundary problems*). The boundary $\bar\Omega_+ \cap \bar\Omega_0$, which is not known a-priori and is thus a part of the solution to (5.18), is called a *free boundary*, cf. Figure 12. We thus speak about free-boundary problems, abbreviated often as FBPs.

Remark 5.14 (Dual approaches). Having in mind the constraint $u \ge w$ as in (5.18), one can write δ_K used in (5.21) as

$$\delta_K(u) = \sup_{0\le\lambda\in L^{p^*\prime}(\Omega)} \int_\Omega (w-u)\lambda\,\mathrm{d}x. \tag{5.35}$$

Then, defining the so-called *Lagrangean* $L(u,\lambda) := \Phi(u) - \langle f,u\rangle + \int_\Omega (w-u)\lambda\,\mathrm{d}x$ and realizing that $L(\cdot,\lambda)$ is convex while $L(u,\cdot)$ is concave, we have

$$\begin{aligned}
\min_{u\in K}[\Phi - f] &= \min_{u\in W^{1,p}(\Omega)}\Big(\Phi(u) - \langle f,u\rangle + \sup_{0\le\lambda\in L^{p^*\prime}(\Omega)}\int_\Omega (w-u)\lambda\,\mathrm{d}x\Big) \\
&= \min_{u\in W^{1,p}(\Omega)}\ \sup_{0\le\lambda\in L^{p^*\prime}(\Omega)}\Big(\Phi(u) - \langle f,u\rangle + \int_\Omega (w-u)\lambda\,\mathrm{d}x\Big) \\
&\ge \sup_{0\le\lambda\in L^{p^*\prime}(\Omega)}\ \min_{u\in W^{1,p}(\Omega)} L(u,\lambda);
\end{aligned}$$

the last inequality holds because always $\min_{v\in W^{1,p}(\Omega)} L(v,\lambda) \le L(u,\lambda) \le \sup_{0\le\xi} L(u,\xi) = \Phi(u)$ for any u and λ.[7] Thus the problem now consists in seeking a saddle point of the Lagrangean L. The problem of finding a supremum over $\{\lambda \ge 0\}$ of the concave function

$$\Psi(\lambda) := \min_{u\in W^{1,p}(\Omega)} L(u,\lambda) \tag{5.36}$$

is referred to as the *dual problem* and can sometimes be easier to solve or/and gives useful additional information; e.g. the constraint in the dual problems are simpler and, having an approximate maximizer $\lambda^* \ge 0$ of Ψ and an approximate minimizer $u^* \ge w$ of Φ, we have a two-sided estimate $\Psi(\lambda^*) \le \min_{u\in K}[\Phi - f] \le \Phi(u^*)$. Cf. Exercise 5.51 for a concrete case of Ψ.

5.3 Some abstract non-potential inclusions

In this section we will again come back to the abstract level and deal with the inclusion of the type

$$\partial\Phi(u) + A(u) \ni f \tag{5.37}$$

with Φ convex and with A pseudomonotone but not necessarily having a potential. We will thus generalize Corollary 5.7 for the case that the smooth part has no potential. Simultaneously, we will illustrate a general-purpose regularization technique for the nonsmooth part of (5.37).[8]

Theorem 5.15. *Let $\Phi : V \to \mathbb{R} \cup \{+\infty\}$ be convex, lower semicontinuous, proper, and possess, for any $\varepsilon > 0$, a convex, Gâteaux differentiable regularization $\Phi_\varepsilon : V \to \mathbb{R}$ such that $\Phi'_\varepsilon : V \to V^*$ is bounded and radially continuous and $\Phi_\varepsilon \to \Phi$ in the sense*

$$\forall v \in V: \qquad \limsup_{\varepsilon\to 0} \Phi_\varepsilon(v) \le \Phi(v), \tag{5.38a}$$

$$v_\varepsilon \rightharpoonup v \quad \Longrightarrow \quad \liminf_{\varepsilon\to 0} \Phi_\varepsilon(v_\varepsilon) \ge \Phi(v), \tag{5.38b}$$

and $A : V \to V^*$ be pseudomonotone (non-potential, in general) and let, for some $\zeta : \mathbb{R}^+ \to \mathbb{R}^+$ such that $\lim_{s\to+\infty} \zeta(s) = +\infty$, the following uniform coercivity hold:

$$\exists v \in \mathrm{dom}\Phi \ \ \forall \varepsilon > 0 \ \ \forall u \in V: \qquad \frac{\Phi_\varepsilon(u) + \langle A(u), u - v\rangle}{\|u\|} \ge \zeta(\|u\|). \tag{5.39}$$

Then, for any $f \in V^*$, there is at least one $u \in V$ solving the inclusion (5.37).

[7] Let us remark that the opposite inequality would require a constraint qualification, here $p > n$ so that K from (5.20) would have a nonempty interior.

[8] For a usage of the regularization to potential problems see also Proposition 5.10. In (5.25), one uses $\|\cdot\|^\alpha_{L^\alpha}$-penalty term instead of $\|\cdot\|^2_{W^{1,p}}$ implied by usage of the formula (5.49), however.

Proof. By the coercivity (5.39) and the previous results, see Theorem 2.6 with Lemmas 2.9 and 2.11(i)[9], the regularized problem possesses a solution $u_\varepsilon \in V$, i.e.

$$\Phi'_\varepsilon(u_\varepsilon) + A(u_\varepsilon) = f. \tag{5.40}$$

Moreover, we show that the coercivity (5.39) is uniform with respect to ε, and hence u_ε will be a-priori bounded. Indeed, as Φ_ε is convex, in view of (5.8), the equation (5.40) means equivalently

$$\Phi_\varepsilon(v) + \langle A(u_\varepsilon) - f, v - u_\varepsilon\rangle \geq \Phi_\varepsilon(u_\varepsilon). \tag{5.41}$$

Moreover, for $v \in \mathrm{dom}(\Phi)$ and $\varepsilon > 0$ small enough, $\Phi_\varepsilon(v) \leq \Phi(v) + 1$ by (5.38a). Using subsequently (5.39), (5.41), and $\Phi_\varepsilon(v) \leq \Phi(v) + 1$, we get the estimate

$$\begin{aligned}\zeta(\|u_\varepsilon\|)\|u_\varepsilon\| &\leq \Phi_\varepsilon(u_\varepsilon) + \langle A(u_\varepsilon), u_\varepsilon - v\rangle \\ &\leq \Phi_\varepsilon(v) + \langle f, u_\varepsilon - v\rangle \leq \Phi(v) + 1 + \|f\|_*\big(\|u_\varepsilon\| + \|v\|\big).\end{aligned} \tag{5.42}$$

Hence, the sequence $\{u_\varepsilon\}_{\varepsilon>0}$ is bounded and, after taking possibly a subsequence, we can assume $u_\varepsilon \rightharpoonup u$.

Now, for $v \in V$ arbitrary, we will pass to the limit in (5.41). The right-hand side of (5.41) can be estimated by (5.38b) while (5.38a) can be used for the left-hand side to get:

$$\begin{aligned}&\Phi(v) - \langle f, v-u\rangle + \limsup_{\varepsilon\to0}\langle A(u_\varepsilon), v-u_\varepsilon\rangle \\ &\qquad \geq \limsup_{\varepsilon\to0}\Big(\Phi_\varepsilon(v) + \langle A(u_\varepsilon)-f, v-u_\varepsilon\rangle\Big) \geq \liminf_{\varepsilon\to0}\Phi_\varepsilon(u_\varepsilon) \geq \Phi(u).\end{aligned} \tag{5.43}$$

Passing to the limit with $\varepsilon \to 0$, we get $\Phi(v) + \langle A(u), v - u\rangle \geq \langle f, v - u\rangle + \Phi(u)$, which is just (5.37), provided we still prove

$$\limsup_{\varepsilon\to0}\langle A(u_\varepsilon), v - u_\varepsilon\rangle \leq \langle A(u), v - u\rangle. \tag{5.44}$$

To do this, we use the pseudomonotonicity of A: we are then to verify

$$\liminf_{\varepsilon\to0}\langle A(u_\varepsilon), u - u_\varepsilon\rangle \geq 0. \tag{5.45}$$

Using (5.41) for $v := u$, we have

$$\langle A(u_\varepsilon), u - u_\varepsilon\rangle \geq \langle f, u - u_\varepsilon\rangle + \Phi_\varepsilon(u_\varepsilon) - \Phi_\varepsilon(u) \tag{5.46}$$

so that, by using again (5.38),

$$\begin{aligned}\liminf_{\varepsilon\to0}\langle A(u_\varepsilon), u - u_\varepsilon\rangle &\geq \liminf_{\varepsilon\to0}\Big(\langle f, u - u_\varepsilon\rangle + \Phi_\varepsilon(u_\varepsilon) - \Phi_\varepsilon(u)\Big) \\ &\geq \lim_{\varepsilon\to0}\langle f, u - u_\varepsilon\rangle + \liminf_{\varepsilon\to0}\Phi_\varepsilon(u_\varepsilon) \\ &\quad - \limsup_{\varepsilon\to0}\Phi_\varepsilon(u) \geq 0 + \Phi(u) - \Phi(u) = 0,\end{aligned} \tag{5.47}$$

which proves (5.45). □

[9]The coercivity of $\Phi'_\varepsilon + A$ (even uniform in $\varepsilon > 0$) follows via the test of (5.40) by $u_\varepsilon - v$ from (5.42) below.

Remark 5.16 (*Mosco's convergence*). One can weaken (5.38a) to[10]

$$\forall v\in V \ \ \exists v_\varepsilon \to v \quad \Longrightarrow \quad \limsup_{\varepsilon\to 0} \Phi_\varepsilon(v_\varepsilon) \le \Phi(v) \tag{5.48}$$

and then (5.38b) with (5.48) is called Mosco's convergence [293] of Φ_ε to Φ; cf. Exercise 5.35 below. This is advantageous in particular if the regularization is combined with the Galerkin method.

A concrete regularization Φ_ε of Φ can be obtained by the formula

$$\Phi_\varepsilon(u) := \inf_{v\in V} \frac{\|u-v\|^2}{2\varepsilon} + \Phi(v); \tag{5.49}$$

here Φ_ε is called the *Yosida approximation* of the functional Φ. Note that, for $\Phi = \delta_K$, we have obviously $\Phi_\varepsilon(u) = \frac{1}{2\varepsilon}\mathrm{dist}(u,K)^2$. Realize the coincidence with the penalty method (5.24) for $q = 2$ and for $\|\cdot\|$ being the L^2-norm.

Lemma 5.17 (Yosida approximation). *Let $\Phi : V \to \bar{\mathbb{R}}$ be convex, proper, lower semicontinuous. Then:*

(i) *Each Φ_ε is convex and lower semicontinuous and the family $\{\Phi_\varepsilon\}_{\varepsilon>0}$ approximates Φ in the sense* (5.38).

(ii) *If V and V^* are strictly convex and reflexive, then each Φ_ε is Gâteaux differentiable and the differential $\Phi'_\varepsilon : V \to V^*$ is demicontinuous and bounded.*

Proof. Denote $\Psi_\varepsilon(u,v) = \frac{1}{2\varepsilon}\|u-v\|^2 + \Phi(v)$.

(i) The lower semicontinuity: take $u_k \to u$ and consider a minimizer v_k for $\Psi_\varepsilon(u_k,\cdot)$, i.e.[11]

$$\big\|u_k - v_k\big\|^2 = 2\varepsilon\big(\Phi_\varepsilon(u_k) - \Phi(v_k)\big). \tag{5.50}$$

As $\{\Phi_\varepsilon(u_k)\}_{k\in\mathbb{N}}$ is bounded from above[12] and Φ, being proper, has an affine minorant, (5.50) implies that $\{v_k\}_{k\in\mathbb{N}}$ is bounded. Considering $v_k \rightharpoonup v$ (a subsequence), by estimating the limit inferior in (5.50) we obtain

$$\begin{aligned}\frac{\|u-\widetilde{v}\|^2}{2\varepsilon} + \Phi(\widetilde{v}) &= \lim_{k\to\infty} \frac{\|u_k-\widetilde{v}\|^2}{2\varepsilon} + \Phi(\widetilde{v}) \ge \liminf_{k\to\infty} \Phi_\varepsilon(u_k)\\ &= \liminf_{k\to\infty} \frac{\|u_k-v_k\|^2}{2\varepsilon} + \Phi(v_k) \ge \frac{\|u-v\|^2}{2\varepsilon} + \Phi(v)\end{aligned} \tag{5.51}$$

[10] By weakening (5.38a) to $\exists v_\varepsilon \rightharpoonup v$: $\limsup_{\varepsilon\to 0}\Phi_\varepsilon(v_\varepsilon) \le \Phi(v)$, we would get the so-called Γ-convergence. This would, however, not be sufficient for passing to the limit in (5.47). Instead of the recovery sequence for $\{\Phi_\varepsilon\}_{\varepsilon>0}$ itself, we should rather use recovery sequences for $\{\Phi_\varepsilon\}_{\varepsilon>0}$ together with $\{A_\varepsilon\}_{\varepsilon>0}$ in the spirit of [281], namely here we might impose the condition: for all $u_\varepsilon \rightharpoonup u$ and $v \in V$ there is a "mutual recovery sequence" $\{v_\varepsilon\}_{\varepsilon>0}$ such that $\limsup_{\varepsilon\to 0}\Phi_\varepsilon(v_\varepsilon) + \langle A(u_\varepsilon){-}f, v_\varepsilon{-}u_\varepsilon\rangle - \Phi_\varepsilon(u_\varepsilon) \le \Phi(v) + \langle A(u){-}f, v{-}u\rangle - \Phi(u)$. The limit passage in (5.41) is then obvious.

[11] Existence of v_k follows by coercivity of $\Psi_\varepsilon(u_k,\cdot)$ (because Φ, being proper, has an affine minorant) and by its weak lower semicontinuity, cf. Exercise 5.30 and the proof of Theorem 4.2.

[12] Note that $\Phi_\varepsilon(u_k) \le \|u_k - w\|^2/\varepsilon + \Phi(w) \to \|u-w\|^2/\varepsilon + \Phi(w) < +\infty$ for $w \in \mathrm{dom}(\Phi)$ fixed.

for any $\widetilde{v} \in V$. Hence v minimizes $\Psi_\varepsilon(u,\cdot)$ so that $\|u-v\|^2/(2\varepsilon)+\Phi(v)=\Phi_\varepsilon(u)$, which showes the lower semicontinuity of Φ_ε.

We now prove that Φ_ε is convex: taking $u_1, u_2 \in V$ and v_1 a minimizer for $\Psi_\varepsilon(u_1,\cdot)$ and v_2 a minimizer for $\Psi_\varepsilon(u_2,\cdot)$, we have

$$\begin{aligned}\Phi_\varepsilon\Big(\frac{u_1+u_2}{2}\Big) &= \inf_{v\in V}\Psi_\varepsilon\Big(\frac{u_1+u_2}{2},v\Big) \le \Psi_\varepsilon\Big(\frac{u_1+u_2}{2},\frac{v_1+v_2}{2}\Big)\\ &\le \frac{1}{2}\Psi_\varepsilon(u_1,v_1)+\frac{1}{2}\Psi_\varepsilon(u_2,v_2)=\frac{1}{2}\Phi_\varepsilon(u_1)+\frac{1}{2}\Phi_\varepsilon(u_2).\end{aligned} \tag{5.52}$$

By the obvious fact $\Phi_\varepsilon \le \Phi$, (5.38a) immediately follows. To prove (5.38b), let us realize that $\Phi_\varepsilon \ge \Phi_\delta$ provided $0<\varepsilon\le\delta$. Then, for $v_\varepsilon \rightharpoonup v$ and for any $\delta>0$, the convexity and lower semicontinuity of Φ_δ implies

$$\liminf_{\varepsilon\to 0}\Phi_\varepsilon(v_\varepsilon) \ge \liminf_{\varepsilon\to 0}\ \Phi_\delta(v_\varepsilon)\ge \Phi_\delta(v). \tag{5.53}$$

Now, (5.38b) follows if one shows $\lim_{\delta\to 0}\Phi_\delta(v)=\Phi(v)$. First, let $v\in\mathrm{dom}(\Phi)$. Let v_δ be a minimizer for $\Psi_\delta(v,\cdot)$, i.e.

$$\|v-v_\delta\|^2 = 2\delta\big(\Phi_\delta(v)-\Phi(v_\delta)\big). \tag{5.54}$$

As $\{\Phi_\delta(v)\}_{\delta>0}$ is bounded from above by $\Phi(v)<+\infty$ and Φ has an affine minorant, (5.54) implies $\{v_\delta\}_{\delta>0}$ bounded. Then one can claim that $\{\Phi(v_\delta)\}_{\delta>0}$ is bounded from below, and then (5.54) gives $v_\delta\to v$. By (5.54), always

$$\Phi(v)\ge\Phi_\delta(v)\ge\Phi(v_\delta). \tag{5.55}$$

By the lower semicontinuity of Φ,

$$\Phi(v)\ge\limsup_{\delta\to 0}\Phi_\delta(v)\ge\liminf_{\delta\to 0}\Phi_\delta(v)\ge\liminf_{\delta\to 0}\Phi(v_\delta)\ge\Phi(v), \tag{5.56}$$

showing that $\lim_{\delta\to 0}\Phi_\delta(v) = \Phi(v)$. Second, consider $v\notin\mathrm{dom}(\Phi)$. Assume $\lim_{\delta\to 0}\Phi(v_\delta)\neq\Phi(v)$, i.e. $\Phi_\delta(v)\le C$ for some $C<+\infty$. Yet, then (5.54) again gives $v_\delta\to v$ and (5.56) implies $\Phi(v)\le C$, a contradiction.

(ii) Let u_ε be a solution to the minimization problem in (5.49). From the optimality conditions, we get

$$\frac{1}{\varepsilon}J(u_\varepsilon-u)+\partial\Phi(u_\varepsilon)\ni 0, \tag{5.57}$$

cf. Examples 4.19 and 5.2. This gives $u_\varepsilon = (\mathbf{I}{+}\varepsilon J^{-1}\partial\Phi)^{-1}(u)$ and also $\Phi(u_\varepsilon)-\Phi(w)\le\frac{1}{\varepsilon}\langle J(u{-}u_\varepsilon),u_\varepsilon{-}w\rangle$ for any w. In particular, considering some v, we will use it for $w:=v_\varepsilon:=(\mathbf{I}{+}\varepsilon J^{-1}\partial\Phi)^{-1}(v)$ to estimate (while abbreviating $\bar{u}_\varepsilon=u{-}u_\varepsilon$

and $\bar{v}_\varepsilon = v - v_\varepsilon$):

$$\begin{aligned}\Phi_\varepsilon(u) - \Phi_\varepsilon(v) &= \Phi(u_\varepsilon) - \Phi(v_\varepsilon) + \frac{\|\bar{u}_\varepsilon\|^2}{2\varepsilon} - \frac{\|\bar{v}_\varepsilon\|^2}{2\varepsilon} \\ &\le \Big\langle \frac{J(\bar{u}_\varepsilon)}{\varepsilon}, u_\varepsilon - v_\varepsilon \Big\rangle + \frac{\|\bar{u}_\varepsilon\|^2}{2\varepsilon} - \frac{\|\bar{v}_\varepsilon\|^2}{2\varepsilon} \\ &= \Big\langle \frac{J(\bar{u}_\varepsilon)}{\varepsilon}, u - v \Big\rangle - \Big\langle \frac{J(\bar{u}_\varepsilon)}{\varepsilon}, \bar{u}_\varepsilon - \bar{v}_\varepsilon \Big\rangle + \frac{\|\bar{u}_\varepsilon\|^2}{2\varepsilon} - \frac{\|\bar{v}_\varepsilon\|^2}{2\varepsilon} \\ &\le \Big\langle \frac{J(\bar{u}_\varepsilon)}{\varepsilon}, u - v \Big\rangle - \frac{\|\bar{u}_\varepsilon\|^2}{2\varepsilon} - \frac{\|\bar{v}_\varepsilon\|^2}{2\varepsilon} + \frac{\|\bar{u}_\varepsilon\|\,\|\bar{v}_\varepsilon\|}{\varepsilon} \le \Big\langle \frac{J(\bar{u}_\varepsilon)}{\varepsilon}, u - v \Big\rangle. \qquad (5.58)\end{aligned}$$

In particular, $\frac{1}{\varepsilon}J(u - u_\varepsilon) \in \partial\Phi_\varepsilon(u)$. By the same arguments, $\Phi_\varepsilon(v) - \Phi_\varepsilon(u) \le \langle \frac{1}{\varepsilon}J(u - u_\varepsilon), v - u\rangle$. Denoting

$$A_\varepsilon(u) := \frac{1}{\varepsilon}J(u - u_\varepsilon), \qquad (5.59)$$

we obtain

$$\big\langle A_\varepsilon(v) - A_\varepsilon(u), v - u \big\rangle \ge \Phi_\varepsilon(v) - \Phi_\varepsilon(u) - \big\langle A_\varepsilon(u), v - u \big\rangle \ge 0, \qquad (5.60)$$

the last inequality being due to just (5.58). By putting $v = u + tw$ and dividing it by t and assuming that A_ε is demicontinuous, one would get A_ε:

$$\lim_{t\to 0} \frac{\Phi_\varepsilon(u + tw) - \Phi_\varepsilon(u)}{t} = \big\langle A_\varepsilon(u), w \big\rangle \qquad (5.61)$$

which would show that Φ_ε is Gâteaux differentiable.

It thus remains to prove the demicontinuity and also the boundedness of A_ε. Taking some $u_0 \in \mathrm{dom}(\Phi)$ and $f \in \partial\Phi(u_0)$, testing (5.59) by $u_\varepsilon - u_0$, and using (5.57), i.e. $A_\varepsilon(u) \in \partial\Phi(u_\varepsilon)$, and the monotonicity of $\partial\Phi$, we get

$$\begin{aligned}\big\langle J(u_\varepsilon - u), u_\varepsilon - u_0 \big\rangle &= \varepsilon\big\langle A_\varepsilon(u), u_0 - u_\varepsilon \big\rangle \\ &\le \varepsilon\Big(\big\langle A_\varepsilon(u), u_0 - u_\varepsilon \big\rangle + \big\langle f - A_\varepsilon(u), u_0 - u_\varepsilon \big\rangle\Big) = \varepsilon\big\langle f, u_0 - u_\varepsilon \big\rangle. \qquad (5.62)\end{aligned}$$

Hence

$$\begin{aligned}\|u_\varepsilon - u\|^2 &= \big\langle J(u_\varepsilon - u), u_\varepsilon - u_0 \big\rangle + \big\langle J(u_\varepsilon - u), u_0 - u \big\rangle \\ &\le \varepsilon\|f\|_*\,\|u_0 - u_\varepsilon\| + \|u_\varepsilon - u\|\,\|u - u_0\|. \qquad (5.63)\end{aligned}$$

This implies that $u \mapsto u_\varepsilon$ is bounded (i.e. maps bounded sets into bounded sets) and, in view of (5.59), also A_ε is bounded.

Now consider $u_k \to u$ in V and the corresponding $u_{k\varepsilon} := (u_k)_\varepsilon$. Again by (5.59), we have $J(u_{k\varepsilon} - u_k) + \varepsilon A_\varepsilon(u_k) = 0$ and also $J(u_{l\varepsilon} - u_l) + \varepsilon A_\varepsilon(u_l) = 0$. Subtracting it and testing by $u_{k\varepsilon} - u_{l\varepsilon}$, we obtain

$$\begin{aligned}L_{kl}^{(1)} + L_{kl}^{(2)} &:= \big\langle J(u_{k\varepsilon} - u_k) - J(u_{l\varepsilon} - u_l), (u_{k\varepsilon} - u_k) - (u_{l\varepsilon} - u_l) \big\rangle \\ &\quad + \varepsilon\big\langle A_\varepsilon(u_k) - A_\varepsilon(u_l), u_{k\varepsilon} - u_{l\varepsilon} \big\rangle \\ &= \big\langle J(u_{k\varepsilon} - u_k) - J(u_{l\varepsilon} - u_l), u_l - u_k \big\rangle =: R_{kl}\,. \qquad (5.64)\end{aligned}$$

We have $L_{kl}^{(1)} \geq 0$ because J is monotone and also $L_{kl}^{(2)} \geq 0$ because $A_\varepsilon(u_k) \in \partial\Phi(u_{k\varepsilon})$ and $\partial\Phi$ is monotone. Moreover, $\lim_{k,l\to\infty} R_{kl} = 0$ because $\lim_{k,l\to\infty}(u_k - u_l) = 0$ while both $J(u_{k\varepsilon}-u_k)$ and $J(u_{l\varepsilon}-u_l)$ are bounded because the mapping $u \mapsto u_\varepsilon$ has already been shown bounded. Thus $\lim_{k,l\to\infty} L_{kl}^{(1)} = 0$ and $\lim_{k,l\to\infty} L_{kl}^{(2)} = 0$.

Considering (if needed) a subsequence, indexed for simplicity again by k, such that $u_{k\varepsilon} \rightharpoonup \tilde{u}$ in V, $A_\varepsilon(u_k) \rightharpoonup f$ and $J(u_{k\varepsilon}-u_k) \rightharpoonup j^*$ in V^*, and $\langle A_\varepsilon(u_k), u_{k\varepsilon}\rangle \to \xi$ in $\mathbb{R}$ for $k \to \infty$. From $\lim_{k,l\to\infty} L_{kl}^{(2)} = 0$ we get

$$\begin{aligned} 0 &= \lim_{l\to\infty}\Big(\lim_{k\to\infty}\big\langle A_\varepsilon(u_k)-A_\varepsilon(u_l), u_{k\varepsilon}-u_{l\varepsilon}\big\rangle\Big) \\ &= \lim_{l\to\infty}\Big(\xi - \langle f, u_{l\varepsilon}\rangle - \big\langle A_\varepsilon(u_l), \tilde{u}-u_{l\varepsilon}\big\rangle\Big) = 2\xi - 2\langle f, \tilde{u}\rangle. \end{aligned} \tag{5.65}$$

Hence $\langle A_\varepsilon(u_k), u_{k\varepsilon}\rangle \to \langle f, \tilde{u}\rangle$. Since Φ is monotone and $A_\varepsilon(u_k) \in \Phi(u_{k\varepsilon})$, we have $0 \leq \langle A_\varepsilon(u_k) - y, u_{k\varepsilon} - z\rangle \to \langle f - y, \tilde{u} - z\rangle$. As it holds for any (y, z) such that $y \in \partial\Phi(z)$ and as $\partial\Phi$ is maximal monotone, cf. Theorem 5.3(ii), we have $f \in \partial\Phi(\tilde{u})$.

Furthermore, from $A_\varepsilon(u_k) = J(u_k - u_{k\varepsilon})/\varepsilon$ and from the definition (3.1) of J, we obtain

$$\langle \varepsilon A_\varepsilon(u_k), u_{k\varepsilon} - u_k\rangle = \|u_{k\varepsilon} - u_k\|^2 = \varepsilon^2\|A_\varepsilon(u_k)\|_*. \tag{5.66}$$

In the limit, $\|\tilde{u} - u\|^2 \leq \liminf_{k\to\infty} \|u_{k\varepsilon} - u_k\|^2 = \lim_{k\to\infty}\langle \varepsilon A_\varepsilon(u_k), u_{k\varepsilon} - u_k\rangle = \langle \varepsilon f, \tilde{u}-u\rangle \leq \varepsilon\|f\|_*\|\tilde{u}-u\|$. Hence, in particular, $\|\tilde{u}-u\| \leq \varepsilon\|f\|_*$. Conversely, again by using (5.66), $\varepsilon\|f\|_*^2 \leq \varepsilon \liminf_{k\to\infty} \|A_\varepsilon(u_k)\|_*^2 = \lim_{k\to\infty}\langle A_\varepsilon(u_k), u_k - u_{k\varepsilon}\rangle = \langle f, u - \tilde{u}\rangle \leq \|f\|_*\|u - \tilde{u}\|$, hence $\varepsilon\|f\|_* \leq \|u - \tilde{u}\|$. Hence altogether we proved $\varepsilon\|f\|_* = \|u - \tilde{u}\|$ and thus also

$$\varepsilon^2\|f\|_*^2 = \|u - \tilde{u}\|^2 \leq \langle \varepsilon f, \tilde{u} - u\rangle \leq \varepsilon\|f\|_*\|u - \tilde{u}\| = \varepsilon^2\|f\|_*^2 \tag{5.67}$$

hence $\langle \varepsilon f, \tilde{u} - u\rangle = \|f\|_*^2$. Altogether, we proved $J(\tilde{u} - u) + \varepsilon f = 0$.

Therefore, $\tilde{u} = u_\varepsilon$ and $f = A_\varepsilon(u)$. Since this limit is identified uniquely, we have $A_\varepsilon(u_k) \rightharpoonup f = A_\varepsilon(u)$ for the whole sequence. □

Remark 5.18 (*Yosida approximation* of monotone mappings)**.** The differential Φ'_ε can be understood as the so-called Yosida approximation of $\partial\Phi$. In general, a monotone mapping $A_\varepsilon : V \to V^*$ defined by[13]

$$A_\varepsilon(u) = \frac{J\big(u - (\mathbf{I}{+}\varepsilon J^{-1}A)^{-1}(u)\big)}{\varepsilon}, \tag{5.68}$$

is called the Yosida approximation of the monotone (generally non-potential set-valued) mapping $A : V \rightrightarrows V^*$; for $V = \mathbb{R}^n$ see also (2.164b). Let us note that in the

[13] Note that J is single-valued as V^* is supposed strictly convex; cf. Lemma 3.2(ii).

proof of Lemma 5.17, we actually proved that, if V is strictly convex together with V^* and if A is maximal monotone, then $A_\varepsilon : V \to V^*$ is monotone, bounded, and demicontinuous. Moreover, it can be proved that w-$\lim_{\varepsilon\to 0} A_\varepsilon(u)$ is the element of $A(u)$ having the minimal norm, and that, if V is a Hilbert space, A_ε is even Lipschitz continuous.

Corollary 5.19. *Let V be reflexive, $\Phi : V \to \mathbb{R} \cup \{+\infty\}$ convex, proper, and lower semicontinuous, $A : V \to V^*$ be pseudomonotone and*

(i) *$\Phi(v) \geq \|v\|^\alpha$ for some $\alpha > 1$, and $\frac{\langle A(u),u-v\rangle}{\|u\|}$ be bounded from below for some $v \in \mathrm{Dom}\Phi$, or*
(ii) *Φ be bounded from below and A be coercive in the sense:*

$$\exists v \in \mathrm{Dom}(\Phi): \qquad \lim_{\|u\|\to\infty} \frac{\langle A(u), u-v\rangle}{\|u\|} = +\infty. \tag{5.69}$$

Then for any $f \in V^*$ there is at least one $u \in V$ solving (5.37).

Proof. By Asplund's theorem, V can be renormed (if needed) so that both V and V^* are strictly convex. Then, by Lemma 5.17, Φ possesses the smooth regularizing family $\{\Phi_\varepsilon\}_{\varepsilon>0}$ with the property (5.38) and with a bounded and demicontinuous Φ'_ε. Let us verify (5.39):

The case (i): for any $0 < \varepsilon \leq \varepsilon_0$ with ε_0 fixed, one gets[14]

$$\begin{aligned}\Phi_\varepsilon(u) \geq \Phi_{\varepsilon_0}(u) &= \inf_{v\in V} \frac{\|u-v\|^2}{2\varepsilon_0} + \Phi(v) \geq \inf_{v\in V} \frac{(\|u\|-\|v\|)^2}{2\varepsilon_0} + \|v\|^\alpha \\ &\geq \big[|\cdot|^\alpha\big]_{\varepsilon_0}(\|u\|) \geq r_{\varepsilon_0}^\alpha, \qquad \text{where} \quad r_{\varepsilon_0} + \alpha\varepsilon_0 r_{\varepsilon_0}^{\alpha-1} = \|u\|, \end{aligned} \tag{5.70}$$

where $[|\cdot|^\alpha]_{\varepsilon_0}$ is the Yosida approximation of the scalar function $|\cdot|^\alpha$; the last estimate follows likewise the second estimate in (5.55) while the equation for r_{ε_0} is an analog of (5.57). This relation allows for an estimate $r_{\varepsilon_0} \geq \varepsilon_0\|u\|^{\min(1,1/(\alpha-1))} - 1/\varepsilon_0$ for some $\varepsilon_0 > 0$, hence (5.70) yields a uniform (and superlinear) growth at least as $\|u\|^{\min(\alpha,\alpha')}$. Then, adding it with the assumed estimate $\langle A(u), u-v\rangle/\|u\| \geq -C$ yields (5.39).

The case (ii): Without loss of generality, we can assume $\Phi \geq 0$. Then $\Phi_\varepsilon \geq 0$, too. Then, adding $\Phi_\varepsilon(u)$ to the numerator in (5.69) gives (5.39).

Then the assertion follows by Theorem 5.15. $\square$

Theorem 5.20 (MONOTONE CASE: UNIQUENESS AND STABILITY). *Let the assumption of Corollary* 5.19 *be valid with A monotone and radially continuous. Then:*

(i) *If A is strictly monotone or Φ is strictly convex, then the solution u to* (5.37) *is unique and the mapping $f \mapsto u$ is demicontinuous.*
(ii) *If, in addition, A is d-monotone and V uniformly convex, then $f \mapsto u$ is continuous.*
(iii) *If A is uniformly monotone, then $f \mapsto u$ is uniformly continuous.*

[14] We used $\|u-v\| \geq \max(\|u\|-\|v\|, \|v\|-\|u\|)$ so that also $\|u-v\|^2 \geq (\|u\|-\|v\|)^2$.

Proof. Take $u_1, u_2 \in V$ two solutions to (5.37), i.e., for $i = 1, 2$,

$$\Phi(v) + \langle A(u_i), v - u_i\rangle \geq \Phi(u_i) + \langle f, v - u_i\rangle. \tag{5.71}$$

For $i = 1$ take $v = u_2$:

$$\Phi(u_2) + \langle A(u_1), u_2 - u_1\rangle \geq \Phi(u_1) + \langle f, u_2 - u_1\rangle\ . \tag{5.72}$$

Analogously, for $i = 2$ take $v = u_1$. Adding the obtained inequalities, we get

$$\langle A(u_1) - A(u_2), u_2 - u_1\rangle \geq 0, \tag{5.73}$$

from which we get $u_1 = u_2$ if A is strictly monotone.

If A is merely monotone, we can use strict convexity of Φ as follows: we consider again (5.71) but now for $v = \frac{1}{2}u_1 + \frac{1}{2}u_2$, and use monotonicity of A to get

$$\begin{aligned}\Phi(u_i) - \Phi(v) + \langle f, v{-}u_i\rangle &\leq \langle A(u_i), v{-}u_i\rangle \\ &= \langle A(v), v{-}u_i\rangle + \langle A(u_i){-}A(v), v{-}u_i\rangle \leq \langle A(v), v{-}u_i\rangle\end{aligned}$$

for $i = 1, 2$. Summing it together yields

$$\Phi(u_1) + \Phi(u_2) - 2\Phi(v) \leq \langle A(v) - f, 2v - u_1 - u_2\rangle = 0 \tag{5.74}$$

because obviously $2v{-}u_1{-}u_2 = 0$. Yet, (5.74) implies $u_1{=}u_2$ if Φ is strictly convex.

For the demicontinuity, we use again the *Minty trick*: take $f_i \to f$ and u_i the solution corresponding to f_i. By the a-priori estimate as in Corollary 5.19, $\{u_i\}_{i\in\mathbb{N}}$ is bounded. Then $u_i \rightharpoonup u$ (for a moment, possibly as a subsequence). Then, by the monotonicity of A, by (5.71) written for f_i instead of f, and by the weak lower semicontinuity of Φ, we get

$$\begin{aligned}0 &\leq \limsup_{i\to\infty}\langle A(v){-}A(u_i), v{-}u_i\rangle \\ &\leq \limsup_{i\to\infty}\Big(\langle A(v), v{-}u_i\rangle - \Phi(u_i) + \Phi(v) - \langle f_i, v{-}u_i\rangle\Big) \\ &\leq \langle A(v), v{-}u\rangle - \Phi(u) + \Phi(v) - \langle f, v{-}u\rangle.\end{aligned} \tag{5.75}$$

In particular, $u \in \mathrm{dom}(\Phi)$. Then, for $w \in \mathrm{dom}(\Phi)$, the convex combination $v_\varepsilon := \varepsilon w + (1-\varepsilon)u$ belongs to $\mathrm{dom}(\Phi)$. Similarly as in the proof of Proposition 5.11, using (5.75) with $v := v_\varepsilon$ and realizing that, by the convexity of Φ, we have $\Phi(v_\varepsilon) - \Phi(u) \leq \varepsilon(\Phi(w) - \Phi(u))$, we obtain

$$\begin{aligned}0 &\leq \big\langle A(v_\varepsilon), v_\varepsilon{-}u\big\rangle + \Phi(v_\varepsilon) - \Phi(u) - \big\langle f, v_\varepsilon{-}u\big\rangle \\ &\leq \big\langle A(v_\varepsilon), \varepsilon(w{-}u)\big\rangle + \varepsilon\big(\Phi(w) - \Phi(u)\big) - \big\langle f, \varepsilon(w{-}u)\big\rangle.\end{aligned} \tag{5.76}$$

Dividing it by $\varepsilon > 0$, we come to

$$\langle A(v_\varepsilon), w - u\rangle + \Phi(w) - \Phi(u) \geq \langle f, w - u\rangle. \tag{5.77}$$

Then, for $\varepsilon \searrow 0$ by using the radial continuity of A, we get that u solves $\langle A(u), w-u\rangle + \Phi(w) - \Phi(u) \geq \langle f, w-u\rangle$, i.e. u solves (5.37). As we proved such u to be unique, even the whole sequence $\{u_i\}_{i\in\mathbb{N}}$ converges weakly to u.

The norm (resp. uniform) continuity in the d-monotone (resp. uniform-monotone) case is a simple modification of (5.71)–(5.72) for $f = f_{1,2}$ so that (5.73) turns into $\langle A(u_1) - A(u_2), u_2 - u_1\rangle \leq \langle f_1 - f_2, u_2 - u_1\rangle$ and then one can proceed as in (2.33) (resp. in (2.34)). □

Remark 5.21. A special case: $\Phi = \delta_K$, $K \subset V$ convex, closed. Then (5.37) turns into (5.15), i.e. into the problem

$$\text{Find } u \in K: \quad \forall v \in K: \qquad \langle A(u), v-u\rangle \geq \langle f, v-u\rangle. \tag{5.78}$$

Remark 5.22 (Another penalty functional)**.** Considering the constraint of the type $u \in K$, one may be tempted to consider another norm than $L^\alpha(\Omega)$ used in (5.25). Inspired by (5.49), one can consider the functional $u \mapsto \frac{1}{2\varepsilon}\big(\inf_{v\in K}\int_\Omega |u-v|^p + |\nabla(u-v)|^p\big)^{2/p}$, which however leads to a nonlocal term in the approximating equation related, in fact, to the formula (5.68) for $A = N_K$. A certain caution is advisable: e.g. penalization of $K = \{v \geq 0 \text{ on } \Omega,\ v|_\Gamma = 0\}$ by $\frac{1}{2\varepsilon}\int_\Omega |\nabla(u^-)|^2 dx$ is not suitable because this functional is not convex.

Remark 5.23 (Abstract *Galerkin approximation* of variational inequalities[15])**.** We can adapt the finite-dimensional approximation from Section 2.1. Considering (5.78), instead of $u_k \in V_k$ solving $I_k^*(A(u_k) - f) = 0$, we will now start with $u_k \in K_k \subset K$ solving $u_k - P_k(u_k + J_k^{-1}I_k^*(f - A(u_k)))$ where $I_k : V_k \to V$ is the inclusion, $J_k : V_k \to V_k^*$ is the duality mapping, $P_k : V_k \to K_k$ is the projector with respect to the Euclidean inner product in V_k (which is thus considered as possibly renormed) and $K_k \subset V_k$ is a convex closed approximation of K whose union is dense in K, cf. also Exercise 5.42 below. In other words, $u_k \in K_k$ satisfies

$$\forall v \in K_k: \qquad \big\langle A(u_k), v - u_k\big\rangle \geq \langle f, v-u_k\rangle. \tag{5.79}$$

The existence of u_k again follows by the Brouwer fixed-point Theorem 1.10. Thus one can show that, if A is pseudomonotone and coercive on K in the sense

$$\exists v \in K: \qquad \lim_{\|u\|\to\infty} \frac{\big\langle A(u), u-v\big\rangle}{\|u\|} = +\infty, \tag{5.80}$$

cf. (5.69), then, for any $f \in V^*$, (5.78) has a solution. Actually, this is an alternative proof of Corollary 5.19 in the special case $\Phi = \delta_K$.

Remark 5.24 (Epigraphical approach)**.** In fact, (5.78) is a universal form for (5.8) if one makes the so-called *Mosco transformation* [292]: replace V by $V \times \mathbb{R}$, put $K := \text{epi}(\Phi_1) \subset V \times \mathbb{R}$, define the pseudomonotone mapping $A : V \times \mathbb{R} \to V^* \times \mathbb{R}$

[15] See Brézis [65] or Lions [261, Sect. II.8.2].

by $A(u,a):=(A_2(u),1)$, and the right-hand side $(f,0)$. Indeed, if $(u,a)\in K$ solves the problem (5.78) for such data, i.e. if $\Phi_1(u)\le a$ and, for all $(v,b)\in V\times\mathbb{R}$,

$$\Phi_1(v)\le b \;\Rightarrow\; \big\langle(A_2(u),1)\,,\,(v,b)-(u,a)\big\rangle\ge\big\langle(f,0)\,,\,(v,b)-(u,a)\big\rangle, \tag{5.81}$$

then $a=\Phi_1(u)$ and u solves (5.8).[16] The previous Remark 5.23 allows us to give an alternative proof of Corollary 5.19 under the following coercivity condition:

$$\exists v\in\mathrm{dom}(\Phi_1): \qquad \lim_{\|u\|\to\infty}\frac{\Phi_1(u)+\langle A_2(u),u-v\rangle}{\|u\|}=+\infty, \tag{5.82}$$

which covers (and generalizes) both cases (i) and (ii) in Corollary 5.19.

5.4 Excursion to quasivariational inequalities

There is a sensible generalization of (5.8) by allowing the convex functional Φ_1 to depend on the solution u itself, i.e. $\Phi_1=\boldsymbol{\Phi}(u,\cdot)$ for some $\boldsymbol{\Phi}:V\times V\to\mathbb{R}$. For $A=A_2$ monotone (not necessarily potential) we come to a so-called *quasivariational inequality*

$$\forall v\in V: \quad \boldsymbol{\Phi}(u,v)+\langle A(u),v-u\rangle\ge\boldsymbol{\Phi}(u,u)+\langle f,v-u\rangle. \tag{5.83}$$

To prove the existence of a solution to (5.83), various fixed-point theorems are usually used. Here, we use the Kakutani's Theorem 1.11. We denote by $M(w)$ the set of the solutions u to the following auxiliary variational inequality:

$$\forall v\in V: \quad \boldsymbol{\Phi}(w,v)+\langle A(u),v-u\rangle\ge\boldsymbol{\Phi}(w,u)+\langle f,v-u\rangle. \tag{5.84}$$

Lemma 5.25. *Let A be monotone, bounded, radially continuous, $w\mapsto\boldsymbol{\Phi}(w,\cdot)$ be weakly continuous in Mosco's sense, i.e. for all $v,w\in V$,*

$$\forall w_k\rightharpoonup w\;\;\exists v_k\to v: \quad \limsup_{\substack{w_k\rightharpoonup w\\ v_k\to v}}\boldsymbol{\Phi}(w_k,v_k)\le\boldsymbol{\Phi}(w,v), \tag{5.85a}$$

$$\forall w_k\rightharpoonup w\;\;\forall v_k\rightharpoonup v: \quad \liminf_{\substack{w_k\rightharpoonup w\\ v_k\rightharpoonup v}}\boldsymbol{\Phi}(w_k,v_k)\ge\boldsymbol{\Phi}(w,v), \tag{5.85b}$$

and, for any $w\in V$, $\boldsymbol{\Phi}(w,\cdot)\ge0$ be convex, $\mathrm{dom}(\boldsymbol{\Phi}(w,\cdot))\ni0$, *and A be coercive. Then $M(w):=\{u\in V;\ u$ solves (5.84)$\}$ is nonempty, closed and convex, and $M:V\rightrightarrows V$ is weakly upper semicontinuous, i.e.*[17]

$$\left.\begin{array}{r} w_k\rightharpoonup w,\\ u_k\rightharpoonup u,\\ u_k\in M(w_k)\end{array}\right\}\quad\Rightarrow\quad u\in M(w). \tag{5.86}$$

[16] Indeed, choosing $(v,b):=(u,\Phi_1(u))$ in (5.81), we get $\Phi_1(u)\ge a$, hence $\Phi_1(u)=a$. By this and by putting $(v,b):=(v,\Phi_1(v))$ into (5.81), we get just (5.8) with v arbitrary.

[17] Let us recall the generally applied "sequential" concept, i.e. (5.86) defines "sequential" weak upper semicontinuity. This is because the generally assumed separability and reflexivity of V (hence of V^* too) makes the weak topology metrizable if restricted to bounded sets and then the "sequential" concept can be applied equally as the usual general-topology concept.

Proof. By (5.85b), in particular, $\mathbf{\Phi}(w,\cdot)$ is weakly lower semicontinuous and then, by pseudomonotonicity of A and the coercivity, (5.84) has a solution; cf. Corollary 5.19. Hence $M(w)\neq\emptyset$.

Take u_1 and u_2 two solutions to (5.84), i.e. after a trivial re-arrangement:

$$\mathbf{\Phi}(w,v)-\mathbf{\Phi}(w,u_1)+\langle f,u_1-v\rangle\geq\langle A(u_1),u_1-v\rangle, \tag{5.87a}$$
$$\mathbf{\Phi}(w,v)-\mathbf{\Phi}(w,u_2)+\langle f,u_2-v\rangle\geq\langle A(u_2),u_2-v\rangle. \tag{5.87b}$$

Then we add it together, divide it by 2, and subtract the trivial identity $\langle A(v),u-v\rangle=\frac{1}{2}\langle A(v),u_1-v\rangle+\frac{1}{2}\langle A(v),u_2-v\rangle$ where $u=\frac{1}{2}u_1+\frac{1}{2}u_2$. Using subsequently the convexity of $\mathbf{\Phi}(w,\cdot)$, (5.87), and the monotonicity of A, we get

$$\begin{aligned}\mathbf{\Phi}(w,v)-\mathbf{\Phi}(w,u)+\big\langle f-A(v),u-v\big\rangle &\geq\mathbf{\Phi}(w,v)-\frac{1}{2}\mathbf{\Phi}(w,u_1)-\frac{1}{2}\mathbf{\Phi}(w,u_2)\\ &\quad+\Big\langle f,\frac{1}{2}u_1+\frac{1}{2}u_2-v\Big\rangle-\frac{1}{2}\big\langle A(v),u_1-v\big\rangle-\frac{1}{2}\big\langle A(v),u_2-v\big\rangle\\ &\geq\frac{1}{2}\big\langle A(u_1)-A(v),u_1-v\big\rangle+\frac{1}{2}\big\langle A(u_2)-A(v),u_2-v\big\rangle\geq 0.\end{aligned}$$

This is essentially the desired inequality if one replaces $A(v)$ by $A(u)$, which can be however made by Minty's trick by putting $v=\varepsilon z+(1-\varepsilon)u$ with $0<\varepsilon\leq 1$ and proceed as in (5.76)–(5.77). This shows that $u\in M(w)$. As u_1 and u_2 were arbitrary, by Proposition 1.6, $M(w)$ is shown convex if closed. This closedness follows from (5.86). To show it, take $w_k\rightharpoonup w$ and $u_k\rightharpoonup u$ such that $u_k\in M(w_k)$. In view of (5.84) for w_k instead of w, this means for any $v_k\in V$:

$$\mathbf{\Phi}(w_k,v_k)+\langle A(u_k),v_k-u_k\rangle\geq\mathbf{\Phi}(w_k,u_k)+\langle f,v_k-u_k\rangle. \tag{5.88}$$

Now we consider $v\in V$ arbitrary and a suitable sequence $\{v_k\}_{k\in\mathbb{N}}$ converging to v so that (5.85a) holds. By Lemma 2.9, A is pseudomonotone, and thus we can pass to the limit in (5.88) entirely similarly as in the proof of Theorem 5.15 with Remark 5.16, which gives (5.84).[18] Hence $u\in M(w)$, as claimed in (5.86). In particular, for $w_k\equiv w$ we get that $M(w)$ is closed. $\square$

Theorem 5.26. *Let the assumptions of Lemma* 5.25 *be fulfilled with* $\mathbf{\Phi}(w,0)\leq C(1+\|w\|)$ *with* $C<+\infty$. *Then* (5.83) *has a solution.*

Proof. By (5.84) with $v=0$ and by the assumed coercivity of A, for any $u\in M(w)$, $w\in V$, we have the a-priori estimate:

$$\begin{aligned}\zeta(\|u\|)\|u\| &\leq \langle A(u),u\rangle\leq\mathbf{\Phi}(w,u)+\langle A(u),u\rangle\\ &\leq \mathbf{\Phi}(w,0)+\langle f,u\rangle\leq C\big(1+\|w\|\big)+\|f\|_*\|u\|\end{aligned}$$

[18] To make this limit passage more direct without using the pseudomonotone argument, one needs additionally continuity of A, cf. Exercise 5.38 below.

with some $\zeta:\mathbb{R}^+\to\mathbb{R}^+$ such that $\lim_{r\to\infty}\zeta(r)=+\infty$. Divided by $\|u\|$, this gives

$$\zeta(\|u\|)\le C\frac{1+\|w\|}{\|u\|}+\|f\|_* \tag{5.89}$$

from which we can see that $M(B)\subset B$ for a sufficiently large ball $B\subset V$.

By Lemma 5.25, we can thus use the Kakutani fixed-point Theorem 1.11 for the ball B endowed with a weak topology to show the existence of $u\in V$ such that $M(u)\ni u$. Such u obviously solves (5.83). □

Example 5.27. For the typical case $\mathbf{\Phi}(w,u)=\delta_{K(w)}(u)$, (5.83) turns into the quasivariational inequality:

$$\text{Find } u\in K(u):\quad \forall v\in K(u):\quad \langle A(u),v-u\rangle\ge\langle f,v-u\rangle. \tag{5.90}$$

Then (5.85a) means that the set-valued mapping $K:V\rightrightarrows V$ is so-called (weak,norm)-lower semicontinuous in the Kuratowski sense[19] while (5.85b) is just (weak,weak)-upper semicontinuity[20].

Example 5.28. Let us consider $V=W_0^{1,2}(\Omega)$, $A=-\Delta$, and

$$\mathbf{\Phi}(w,v):=\int_\Omega\varphi(x,w(x),v(x))\,\mathrm{d}x \tag{5.91}$$

with $\varphi:\Omega\times\mathbb{R}\times\mathbb{R}\to\mathbb{R}$ a Carathéodory mapping satisfying the growth condition

$$\exists a\in L^1(\Omega),\ b\in\mathbb{R}^+:\qquad |\varphi(x,r_1,r_2)|\le a(x)+b\big(|r_1|^{2^*-\epsilon}+|r_2|^{2^*-\epsilon}\big). \tag{5.92}$$

Then $\mathbf{\Phi}:W^{1,2}(\Omega)\times W^{1,2}(\Omega)\to\mathbb{R}$ is (weak$\times$weak)-continuous; use the compact embedding $W^{1,2}(\Omega)\Subset L^{2^*-\epsilon}(\Omega)$ and then the continuity of the Nemytskiĭ mapping $\mathscr{N}_\varphi:L^{2^*-\epsilon}(\Omega)\times L^{2^*-\epsilon}(\Omega)\to L^1(\Omega)$.

Supposing additionally that $\varphi(x,r_1,\cdot)\ge0$ is convex and $\varphi(x,r_1,0)\le\gamma(x)+C|r_1|$ with some $\gamma\in L^1(\Omega)$ and $C\in\mathbb{R}$, we can prove the existence of a solution $u\in W_0^{1,2}(\Omega)$ to the quasivariational inequality

$$\int_\Omega\varphi(x,u,v)+\nabla u\cdot\nabla(v-u)\,\mathrm{d}x\ge\int_\Omega\varphi(x,u,u)+g(v-u)\,\mathrm{d}x \tag{5.93}$$

for any $v\in W_0^{1,2}(\Omega)$ which corresponds, in the classical formulation, to the problem:

$$\left.\begin{aligned}-\Delta u+\partial_{r_2}\varphi(u,u)&\ni g &&\text{in }\Omega,\\ u&=0&&\text{on }\Gamma.\end{aligned}\right\} \tag{5.94}$$

[19]This is, by definition: $\forall u_k\rightharpoonup u\in V\ \forall v\in K(u)\ \exists v_k\in K(u_k)$: $v_k\to v$.

[20]This is, by definition, just (5.86) with K in place of M.

5.5 Exercises

Exercise 5.29. Specify the potential Φ of A from Figure 10b and c.[21]

Exercise 5.30. By using Proposition 1.6, show that any convex lower semicontinuous functional $\Phi : V \to \mathbb{R} \cup \{+\infty\}$ is weakly lower semicontinuous.[22]

Exercise 5.31. Show $\partial(\Phi_1+\Phi_2)(u) \subset \partial\Phi_1(u)+\partial\Phi_2(u)$ for $\Phi_1, \Phi_2 : V\to\mathbb{R}$ convex.[23]

Exercise 5.32. Show that Φ convex and $\partial\Phi : V \rightrightarrows V^*$ coercive imply Φ coercive.[24]

Exercise 5.33. Assuming Φ convex and lower semicontinuous, prove that the graph of the multivalued mapping $\partial\Phi : V \rightrightarrows V^*$ is (weak$\times$norm)- and (norm$\times$weak)-closed.[25]

Exercise 5.34. Show that, if $\Phi : V \to \mathbb{R}$ is Gâteaux differentiable and convex, then $\partial\Phi(u) = \{\Phi'(u)\}$.[26]

Exercise 5.35. Modify the proof of Theorem 5.15 if $\Phi_\varepsilon \to \Phi$ only in Mosco's sense, i.e. (5.38b)–(5.48).[27]

Exercise 5.36. Modify Theorem 5.20(i) for the case $A = A_1+A_2$ with A_1 monotone and radially continuous and A_2 totally continuous, to obtain upper semicontinuity of the set-valued mapping $f \mapsto \{u \in V;\ u \text{ solves } (5.37)\}$ as $(V^*,\text{norm})\rightrightarrows (V,\text{weak})$. If $A_2 = 0$, show that this set-valued mapping has convex values.[28]

Exercise 5.37. Verify the convergence $\Phi_\varepsilon \to \Phi$ in the sense (5.38) for $\Phi = \delta_K$ and $\Phi_\varepsilon(u) = \frac{1}{2\varepsilon}\mathrm{dist}(u,K)^2$ directly, without using Lemma 5.17.

Exercise 5.38. Strengthening the assumptions in Lemma 5.25 by requiring A not only demicontinuous (as Lemma 2.16 says) but even continuous, perform the limit

[21] Hint: The absolute value $|\cdot|$ and the indicator function $\delta_{[-1,1]}(\cdot)$, up to a constant, of course.

[22] Hint: Assume the contrary, i.e. $l := \lim_{k\to\infty}\Phi(u_k) < \Phi(u)$ for some $u_k \rightharpoonup u$, and realize that the level set $L = \{v \in V;\ \Phi(v) \le l\}$ is convex and closed because Φ is convex and lower semicontinuous. By Proposition 1.6, L is weakly closed, so that $L \ni \text{w-}\lim_{k\to\infty} u_k = u$, i.e. $\Phi(u) \le l$, a contradiction.

[23] Hint: It follows directly from the definition (5.2).

[24] Hint: Modify the proof of Theorem 4.4(i).

[25] Hint: Assume either $u_k \rightharpoonup u$ and $f_k \to f$ or $u_k \to u$ and $f_k \rightharpoonup f$, and make a limit passage in the inequality in (5.2).

[26] Hint: By (5.10), $\langle\Phi'(u), v-u\rangle = \mathrm{D}\Phi(u, v-u) \le \Phi(v) - \Phi(u)$, hence $\Phi'(u) \in \partial\Phi(u)$. Conversely, consider $f \in \partial\Phi(u)$, i.e. $\Phi(v) - \Phi(u) \ge \langle f, v-u\rangle$ for all v, and in particular for $v := u + \varepsilon w$, hence $(\Phi(u+\varepsilon w) - \Phi(u))/\varepsilon \ge \langle f, w\rangle$. For $\varepsilon \searrow 0$, deduce $\mathrm{D}\Phi(u,w) \ge \langle f, w\rangle$. Since $\mathrm{D}\Phi(u,w) = \langle\Phi'(u), w\rangle$, hence $f{=}\Phi'(u)$.

[27] Hint: Put v_ε instead of v into (5.41) to be used for (5.43), and then use $\limsup_{\varepsilon\to0}\langle A(u_\varepsilon), v_\varepsilon - u_\varepsilon\rangle = \lim_{\varepsilon\to0}\langle A(u_\varepsilon), v_\varepsilon - v\rangle + \limsup_{\varepsilon\to0}\langle A(u_\varepsilon), v - u_\varepsilon\rangle$ where the first right-hand-side term is zero if $v_\varepsilon \to v$, as assumed in (5.48).

[28] Hint: For two solutions u_1 and u_2, write $2\langle f-A(v), u-v\rangle = \langle f-A(v), u_1-v\rangle+\langle f-A(v), u_2-v\rangle \ge \langle A(u_1)-A(v), u_1-v\rangle+\Phi(u_1)-\Phi(v)+\langle A(u_2)-A(v), u_2-v\rangle+\Phi(u_2)-\Phi(v) \ge 2\Phi(u)-2\Phi(v)$, cf. also (2.28) for $\Phi = 0$, and the use the Minty trick as in the proof of Theorem 5.20(i).

passage in (5.88) by Minty's trick.[29]

Exercise 5.39 (Two-sided obstacles). Consider $w_1, w_2 \in W^{1,p}(\Omega)$, $w_1 < w_2$ in Ω, $K = \{v \in W^{1,p}(\Omega);\ w_1 \le v \le w_2\}$, and the variational inequality (5.19). Formulate the complementarity problem like (5.18) for this case and modify the proof of Proposition 5.9 accordingly.

Exercise 5.40 (Obstacle on Γ). For some $w \in W^{1-1/p,p}(\Gamma)$, consider the so-called *Signorini-type problem*, i.e. (in the classical formulation) a problem involving the complementarity *Signorini-type boundary conditions* on Γ only:

$$\left.\begin{array}{rcll} -\mathrm{div}\big(|\nabla u|^{p-2}\nabla u\big) + |u|^{q-2}u &=& g & \text{in } \Omega, \\ \left.\begin{array}{rcl} |\nabla u|^{p-2}\frac{\partial u}{\partial \nu} + b(x,u) &\ge& h, \\ u &\ge& w, \\ \big(|\nabla u|^{p-2}\frac{\partial u}{\partial \nu} + b(x,u) - h\big)(u-w) &=& 0 \end{array}\right\} & & & \text{on } \Gamma \end{array}\right\} \tag{5.95}$$

with b qualified as in (5.26b) and $1 < q \le p^*$. Assemble the weak formulation which involves the convex set

$$K := \big\{u \in W^{1,p}(\Omega);\ u(x) \ge w(x) \text{ for a.a. } x \in \Gamma\big\} \tag{5.96}$$

and show the relation with the classical formulation.[30] Modify (5.25): consider $q \le p^\#$ and the penalty term in the form $(\varepsilon\alpha)^{-1}\int_\Gamma |(w-u)^+|^\alpha\,\mathrm{d}S$. Show the a-priori estimates $\|u_\varepsilon\|_{W^{1,p}(\Omega)} = \mathscr{O}(1)$ and $\|(w-u_\varepsilon)^+\|_{L^\alpha(\Gamma)} = \mathscr{O}(\sqrt[\alpha]{\varepsilon})$ and the convergence for $\varepsilon \to 0$.[31] Perform the analysis for $q > p^*$ by modifing the space V.[32] Further, modify the equation by adding $\vec{c}(u)\cdot\nabla u$ or $c(\nabla u)$ as in Exercise 5.46.

Exercise 5.41 (Dirichlet boundary condition). Instead of (5.96), consider

$$K := \big\{u \in W^{1,p}(\Omega);\ u \ge w \text{ on } \Omega,\ u|_\Gamma = u_{\mathrm{D}} \text{ on } \Gamma\big\}, \tag{5.97}$$

which is nonempty if $w|_\Gamma \le u_{\mathrm{D}}$. Modify Section 5.2.

Exercise 5.42 (*Ritz' method*). Consider Ω polygonal, Φ from (4.23), $K = \big\{v \in W^{1,p}(\Omega);\ v \ge w \text{ in } \Omega\big\}$ as in (5.20), V_k a finite-dimensional subspace of $W^{1,p}(\Omega)$ constructed by the piece-wise affine finite elements, cf. Example 2.67. As in (2.61),

[29] Hint: Write $0 \le \langle A(v_k) - A(u_k), v_k - u_k\rangle \le \langle A(v_k), v_k - u_k\rangle + \Phi(w_k, v_k) - \Phi(w_k, u_k)$, use (5.85) together with $A(v_k) \to A(v)$ so that $\langle A(v_k), v_k - u_k\rangle \to \langle A(v), v - u\rangle$ and finish by the Minty trick as in the proof of Theorem 5.20(i).

[30] Hint: Just simplify the proof of Proposition 5.9.

[31] Hint: Use the test by $v = u_\varepsilon - w$ with a suitable extension $w \in W^{1,p}(\Omega)$ of the originally given $w \in W^{1-1/p,p}(\Gamma)$. Realize how (5.26a) can be verified, namely $|s|^{p-2}s\cdot(s-\nabla w(x)) \ge |s|^p - |s|^{p-1}|\nabla w(x)| \ge \frac{1}{p}|s|^p - \frac{1}{p}|\nabla w(x)|^p$ and $|r|^{q-2}r\cdot(r-w(x)) \ge |r|^q - |r|^{q-1}|w(x)| \ge \frac{1}{q}|r|^q - \frac{1}{q}|w(x)|^q$. For convergence, modify the proof of Proposition 5.10.

[32] Hint: Like (2.128), consider $V = W^{1,p}(\Omega) \cap L^q(\Omega)$. Assuming $w \in V$, make estimation like in Remark 5.12; in particular, estimate $\int_\Omega |u_\varepsilon|^{q-2}u_\varepsilon w\,\mathrm{d}x \le \frac{1}{q}\|u_\varepsilon\|^q_{L^q(\Omega)} + \frac{1}{q}\|w\|^q_{L^q(\Omega)}$.

define $\Phi_0(u) = \Phi(u+w)$ and prove existence of a minimizer $u_k \in V_k$ of Φ_0 subject to $u_k \geq 0$ a.e. in Ω. Further, prove[33]

$$\mathrm{cl}\Big(\bigcup_{k\in\mathbb{N}} \{v\in V_k;\ v\geq 0\}\Big) = \{v\in W^{1,p}(\Omega);\ v\geq 0\}, \tag{5.98}$$

derive a-priori estimates, and show convergence by a direct method, i.e. without Minty's trick, as in (5.28)–(5.29), of u_k to u_0, a minimizer of Φ_0 on $\{v \in W^{1,p}(\Omega);\ v \geq 0\}$. Show that $u = u_0 + w$ solves the original variational problem, more precisely it minimizes of Φ on K.

Exercise 5.43 (*Galerkin method*). Consider $A(u) := -\mathrm{div}(\mathbb{A}\nabla u)$ with $\mathbb{A} \in \mathbb{R}^{n\times n}$ positive definite (but, in general, nonsymmetric hence the problem is non-potential) on a polygonal domain Ω, and the unilateral problem

$$\left.\begin{array}{r}\left.\begin{array}{rl}-\mathrm{div}(\mathbb{A}\nabla u) &\geq g\\ u &\geq w\\ \big(\mathrm{div}(\mathbb{A}\nabla u)+g\big)(u-w) &= 0\end{array}\right\}\quad \text{on } \Omega,\\ \begin{array}{rl} u &= 0\end{array}\qquad\qquad \text{on } \Gamma.\end{array}\right\} \tag{5.99}$$

Use the transformation (2.61) to get a problem like (5.99) but with $g+\mathrm{div}(\mathbb{A}\nabla w)$ and 0 in place of g and w, respectively. Make the approximation by a finite-dimensional subspace V_k of $W^{1,p}(\Omega)$, use (5.98), derive a-priori estimates and show convergence either by Minty's trick or by a direct limit passage.[34]

Exercise 5.44 (*Regularization of the elliptic variational inequality I*). Consider (5.18) without symmetry (4.21), thus without any potential, and the regularization (5.25). Assume $a(x,r,\cdot)$ monotone and $b = b(r)$ and $c = c(r)$ having a sub-critical growth, i.e. (2.56b,c) holds. Show a-priori estimates $\|u_\varepsilon\|_{W^{1,p}(\Omega)} = \mathscr{O}(1)$ and $\|(w-u_\varepsilon)^+\|_{L^2(\Omega)} = \mathscr{O}(\sqrt{\varepsilon})$.[35] Further, show the convergence $u_\varepsilon \rightharpoonup u$ in $W^{1,p}(\Omega)$ with u being a weak solution to (5.18) by using the *Minty trick*.[36]

---

[33] Hint: Realize density of smooth non-negative functions in $\{v\in W^{1,p}(\Omega);\ v\geq 0\}$, which can be proved by applying a convolution with a mollifier (for $n = 1$ see also (7.11) in Sect. 7.1 and Figure 16). Note that the general constraint $v \geq w$ would not be preserved by mollifying v, which is why the shift $\Phi_0(u) = \Phi(u+w)$ was made.

[34] Hint: Use $\limsup_{k\to\infty} \int_\Omega (\nabla v - \nabla u_k)^\top \mathbb{A}\nabla u_k\,\mathrm{d}x \leq \int_\Omega (\nabla v - \nabla u_0)^\top \mathbb{A}\nabla u_0\,\mathrm{d}x$ if $u_k \to u_0$.

[35] Hint: This is essentially as in the proof of Proposition 5.10.

[36] Hint: modify the proof of Proposition 5.11. By $a(x,r,\cdot)$, instead of (5.33), arrive to

$$\begin{aligned}0 &\leq \int_\Omega \big(a(u_\varepsilon,\nabla u_\varepsilon) - a(u_\varepsilon,\nabla v)\big)\cdot\nabla(u_\varepsilon - v)\,\mathrm{d}x\\ &\leq \int_\Omega \big(g - c(u_\varepsilon)\big)(u_\varepsilon - v) - a(u_\varepsilon,\nabla v)\cdot\nabla(u_\varepsilon - v)\,\mathrm{d}x + \int_\Gamma \big(h - b(u_\varepsilon)\big)(u_\varepsilon - v)\,\mathrm{d}S\\ &\to \int_\Omega \big(g - c(u)\big)(u - v) - a(u,\nabla v)\cdot\nabla(u - v)\,\mathrm{d}x + \int_\Gamma \big(h - b(u)\big)(u - v)\,\mathrm{d}S,\end{aligned}$$

where the limit passage in lower-order terms used $c(u_\varepsilon) \to c(u)$ in $L^{p^{*\prime}}(\Omega)$ and $a(u_\varepsilon,\nabla v) \to a(u,\nabla v)$ in $L^{p'}(\Omega;\mathbb{R}^n)$ by compactness of the embedding $W^{1,p}(\Omega) \subset L^{p^*-\epsilon}(\Omega)$ and continuity of the Nemytskiĭ mappings $\mathscr{N}_c$ and $\mathscr{N}_{a(\cdot,\cdot,\nabla v)}$, and also $b(u_\varepsilon) \to b(u)$ in $L^{p^{\#\prime}}\Gamma)$ by compactness of the trace operator $W^{1,p}(\Omega) \to L^{p^\#-\epsilon}(\Gamma)$ and continuity of the Nemytskiĭ mapping $\mathscr{N}_b$.

Exercise 5.45. Modify the approach from Exercise 5.44 by starting directly with the monotonicity of

$$(r,s)\mapsto\Big(-\frac{1}{\varepsilon}((w(x){-}r)^+)^{\alpha-1},a(x,\tilde r,s)\Big):\mathbb{R}{\times}\mathbb{R}^n\to\mathbb{R}{\times}\mathbb{R}^n \tag{5.100}$$

for any $\tilde r$ fixed instead of the monotonicity of $a(x,r,\cdot):\mathbb{R}^n\to\mathbb{R}^n$.

Exercise 5.46 (*Regularization of the elliptic variational inequality II*). Consider

$$\left.\begin{array}{rcll} -\Delta u_\varepsilon+c(\nabla u_\varepsilon)+\dfrac{1}{\varepsilon}u_\varepsilon^+ &=& g & \text{in }\Omega,\\ u_\varepsilon &=& 0 & \text{on }\Gamma, \end{array}\right\} \tag{5.101}$$

where $u^+=\max(0,u)$, with c continuous of sub-linear growth, i.e. $|c(s)|\le C(1+|s|^{1-\epsilon})$ as in Exercise 2.86 for $p=2$. Show a-priori estimates $\|u_\varepsilon\|_{W^{1,2}(\Omega)}=\mathscr{O}(1)$ and $\|u_\varepsilon^+\|_{L^2(\Omega)}=\mathscr{O}(\sqrt{\varepsilon})$.[37] Further show convergence to the weak solution to the complementarity problem:[38]

$$\left.\begin{array}{rcll} \left.\begin{array}{rcl} -\Delta u+c(\nabla u) &\le& g\,,\\ u &\le& 0,\\ \big(\Delta u-c(\nabla u)+g\big)u &=& 0, \end{array}\right\} & & \text{in }\Omega,\\ u=0 & & \text{on }\Gamma. \end{array}\right\} \tag{5.102}$$

Exercise 5.47 (Bingham-fluid-like model). Consider the potential:[39]

$$\Phi(u)=\int_\Omega \varepsilon|\nabla u|^2+|\nabla u|+\delta|u-\bar u|^2+fu\,\mathrm{d}x \tag{5.103}$$

with $\varepsilon>0$ a regularization parameter, $\bar u\in L^2(\Omega)$ given. Show the existence of a unique minimizer $u\in W_0^{1,2}(\Omega)$. Formulate the corresponding variational inequality.

[37] Hint: Test (5.101) by u_ε.

[38] Hint: By the a-priori estimates we can select a subsequence $u_\varepsilon\rightharpoonup u$ in $W^{1,2}(\Omega)$, $u\le 0$ a.e. in Ω. For any $v\in W^{1,2}(\Omega)$, $v\le 0$, by using (5.101),

$$\begin{aligned}\int_\Omega|\nabla u_\varepsilon-\nabla v|^2\mathrm{d}x &\le \int_\Omega|\nabla u_\varepsilon-\nabla v|^2+\frac{1}{\varepsilon}(u_\varepsilon^+-v^+)(u_\varepsilon-v)\,\mathrm{d}x\\ &=\int_\Omega(g-c(\nabla u_\varepsilon))(u_\varepsilon-v)-\nabla v{\cdot}\nabla(u_\varepsilon-v)-\frac{1}{\varepsilon}v^+(u_\varepsilon-v)\,\mathrm{d}x.\end{aligned}$$

Note that the last term vanishes as $v\le 0$. In particular, take $v:=u$ to see that $u_\varepsilon\to u$ in $W_0^{1,2}(\Omega)$. Thus pass to the limit in the nonlinear Nemytskiĭ mapping $\mathscr{N}_c$, i.e. $c(\nabla u_\varepsilon)\to c(\nabla u)$. Then, by (5.101) tested by $v-u_\varepsilon$, make a limit passage in

$$\int_\Omega\nabla u_\varepsilon\cdot\nabla(v-u_\varepsilon)+(c(\nabla u_\varepsilon)-g)(v-u_\varepsilon)\mathrm{d}x=-\frac{1}{\varepsilon}\int_\Omega u_\varepsilon^+(v-u_\varepsilon)\,\mathrm{d}x\ge 0,$$

provided $v\le 0$, which gives the weak formulation of (5.102).

[39] For $\delta=0$, this is a scalar version of a so-called Bingham-fluid model, while in case $n=2$ and $f=0$ this problem has another interpretation in image enhancement/reconstruction.

Exercise 5.48 (Plasticity-like model[40]). Consider the problem: find $u \in W_0^{1,\infty}(\Omega)$ such that $|\nabla u| \leq 1$ a.e. in Ω and

$$\forall v \in W_0^{1,\infty}(\Omega), \ \ |\nabla v| \leq 1 \text{ (a.e.)}: \quad \int_\Omega a(\nabla u)\cdot\nabla(v-u)\mathrm{d}x \geq \int_\Omega g(v-u)\,\mathrm{d}x. \quad (5.104)$$

Moreover, assume $a(s)\cdot s \geq |s|^p$ and $|a(s)| \leq C(1+|s|^{p-1})$, and $a(\cdot)$ monotone, and use a penalization by the functional $u \mapsto \int_\Omega \big((|\nabla u|-1)^+\big)^2 \mathrm{d}x/(2\varepsilon)$. Show that it leads to the approximate problem (in the classical formulation):[41]

$$\left.\begin{aligned} -\mathrm{div}\Big(a(\nabla u_\varepsilon) + \frac{\big(|\nabla u_\varepsilon|-1\big)^+}{\varepsilon|\nabla u_\varepsilon|}\nabla u_\varepsilon\Big) &= g && \text{in } \Omega, \\ u_\varepsilon &= 0 && \text{on } \Gamma. \end{aligned}\right\} \quad (5.105)$$

Further, assume $g \in L^{\max(2^*,p^*)'}(\Omega)$, and show existence of a weak solution $u_\varepsilon \in W_0^{1,\max(2,p)}(\Omega)$ to (5.105), a-priori estimates by testing (5.105) by u_ε[42] and convergence $u_\varepsilon \rightharpoonup u$ in $W_0^{1,\max(2,p)}(\Omega)$ where $u \in W_0^{1,\infty}(\Omega)$ satisfies (5.104)[43]. Eventually, modify the whole procedure for $a \equiv 0$.

Exercise 5.49 (Quasivariational inequality). Verify (5.85) for the case $\mathbf{\Phi}(w,u) = \delta_{K(w)}(u)$ provided $K(w) := \{v \in W_0^{1,p}(\Omega);\ |\nabla v(x)| \leq \mathfrak{m}(w(x)) \text{ for a.a. } x \in \Omega\}$ with $p > n$ and $\mathfrak{m}: \mathbb{R} \to \mathbb{R}^+$ continuous, $\mathfrak{m}(\cdot) \geq \varepsilon > 0$.[44] Assuming $a(s)\cdot s \geq |s|^p$ and

---

[40]This complementarity problem is related to a stress field in an elastic/plastic (or, rather, inelastic) bar undergoing a torsion via a Haar-Karman principle; $n = 2$ and $\Omega \subset \mathbb{R}^2$ is then the cross-section. See e.g. Elliott and Ockendon [135, Sect.IV.6], Friedman [155], or Glowinski et al. [182, p.6 & Chap.3]. Alternatively, the variant $a \equiv 0$ is related to (a steady-state of) a sand flow; in the evolution variant see Aronsson, Evans, Wu [19]. For the classical formulation of (5.104) see (5.108) below.

[41]Hint: The directional derivative at u_ε in the direction v is $\frac{1}{\varepsilon}\int_\Omega(|\nabla u_\varepsilon|-1)^+ \frac{\nabla u_\varepsilon}{|\nabla u_\varepsilon|}\cdot\nabla v\,\mathrm{d}x$.

[42]Hint: Test (5.105) by u_ε and realize the estimate $s\cdot s \geq (|s|-1)^+|s|$ for $s \in \mathbb{R}^n$, so that

$$\int_\Omega \frac{(|\nabla u_\varepsilon|-1)^+}{\varepsilon|\nabla u_\varepsilon|}\nabla u_\varepsilon\cdot\nabla u_\varepsilon \mathrm{d}x \geq \frac{1}{\varepsilon}\int_\Omega \Big(\big(|\nabla u_\varepsilon|-1\big)^+\Big)^2 \mathrm{d}x.$$

[43]Hint: Test (5.105) by $v - u_\varepsilon$ with $|\nabla v| \leq 1$ a.e. in Ω, realize that

$$\frac{(|\nabla u_\varepsilon|-1)^+}{\varepsilon|\nabla u_\varepsilon|}\nabla u_\varepsilon\cdot(\nabla v - \nabla u_\varepsilon) \leq 0 \qquad \text{a.e. on } \Omega$$

and continue the proof as in (5.32) by showing the strong convergence in $W^{1,p}(\Omega)$. To show that $|\nabla u| \leq 1$ a.e. in Ω, estimate the limit inferior in the estimate

$$\|(|\nabla u_\varepsilon|-1)^+\|_{L^2(\Omega)} = \mathscr{O}(\sqrt{\varepsilon}).$$

[44]Hint: For (5.85a), one needs to show: $\forall w_k \rightharpoonup w$ in $W_0^{1,p}(\Omega)$ $\forall u \in W_0^{1,p}(\Omega)$, $|\nabla u| \leq \mathfrak{m}(w)$ $\exists u_k$: $u_k \to u$ in $W_0^{1,p}(\Omega)$ and $|\nabla u_k| \leq \mathfrak{m}(w_k)$ for all $k \in \mathbb{N}$. By the compact embedding $W_0^{1,p}(\Omega) \Subset L^\infty(\Omega)$, realize that $\mathfrak{m}(w_k) \to \mathfrak{m}(w)$ in $L^\infty(\Omega)$, take $u_k := \lambda_k u$ with $\lambda_k := \min_\Omega \big(\mathfrak{m}(w_k)/\mathfrak{m}(w)\big) \to 1$. For (5.85b), one needs to pass to the limit in $|\nabla u_k| \leq \mathfrak{m}(w_k)$ if

$|a(s)| \le C(1+|s|^{p-1})$, show the existence of a weak solution in $W^{1,p}(\Omega)$. Note that, for $\mathfrak{m}=1$, one arrives at Exercise 5.48.

Exercise 5.50. Modify Example 5.28 for $\mathbf{\Phi}(w,v) := \int_\Omega \varphi(x,w(x),\nabla v(x))\,\mathrm{d}x$ with $\varphi = \varphi(x,r,s)$. Thus solve the inclusion $-\Delta u - \mathrm{div}(\partial_s\varphi(u,\nabla u)) \ni g$ with the boundary condition $\frac{\partial}{\partial\nu}u + \nu\cdot\partial_s\varphi(u,\nabla u)) \ni 0$, which modifies (5.94).

Exercise 5.51 (Dual problem)**.** Consider Exercise 5.41 with $u_{\mathrm{D}} = 0$ and the p-Laplacean, i.e. the complementarity problem

$$\left.\begin{array}{ll}\left.\begin{array}{ll}-\mathrm{div}\big(|\nabla u|^{p-2}\nabla u\big) \ge g, & u \ge w\,,\\ \big(\mathrm{div}\big(|\nabla u|^{p-2}\nabla u\big)+g\big)\big(u{-}w\big)=0 & \end{array}\right\} & \text{in }\Omega,\\ \qquad\qquad\qquad\qquad\qquad u=0 & \text{on }\Gamma,\end{array}\right\} \tag{5.106}$$

and, using (8.261) on p.299, show that the dual problem uses the convex functional Ψ from (5.36) in the form[45]

$$\Psi(\lambda) = \int_\Omega w\lambda - \frac{1}{p'}\big|\nabla\Delta_p^{-1}(g{-}\lambda)\big|^p \mathrm{d}x, \qquad \lambda\in W_0^{1,p}(\Omega)^* \cong W^{-1,p'}(\Omega). \tag{5.107}$$

Exercise 5.52 (Plasticity-like model II)**.** Consider the potential

$$\Phi(u) := \int_\Omega \varphi(u,\nabla u) - gu + \delta_{K(\cdot)}(\nabla u)\,\mathrm{d}x \quad \text{ with } \; K(x) = \big\{s{\in}\mathbb{R}^n;\ |s| \le m(x)\big\}$$

on $W_0^{1,p}(\Omega)$. Realizing that $\delta_{K(x)}(s) = \sup_{\lambda\in\mathbb{R}^n} s{\cdot}\lambda - m(x)|\lambda|$ and defining, like in Remark 5.14, the *Lagrangean* by $L(u,\lambda) := \int_\Omega \varphi(u,\nabla u) - gu + \lambda{\cdot}\nabla u - m|\lambda|\,\mathrm{d}x$, identify the underlying variational inequality as the classical formulation of the conditions for (u,λ) to be a critical point of L,[46] i.e. the conditions $L'_u(u,\lambda)=0$ and $\partial_\lambda L(u,\lambda) \ni 0$, and show, in particular, that the classical formulation of (5.104) looks like[47]

$$\left.\begin{array}{ll}\left.\begin{array}{l}-\mathrm{div}\big(a(\nabla u)+\lambda\big) = g,\\ \lambda \in N_{B_1}(\nabla u),\\ |\nabla u| \le 1\end{array}\right\} & \text{in }\Omega,\\ u = 0 & \text{on }\Gamma,\end{array}\right\} \tag{5.108}$$

with B_1 denoting here the unit ball in $\mathbb{R}^n$.

$(u_k,w_k) \rightharpoonup (u,w)$ in $W_0^{1,p}(\Omega)^2$. Again, by $W_0^{1,p}(\Omega) \Subset L^\infty(\Omega)$, $\mathfrak{m}(w_k) \to \mathfrak{m}(w)$ in $L^\infty(\Omega)$. Moreover, for every $M \subset \Omega$ measurable, by weak lower semicontinuity of convex continuous functions, $\int_M \mathfrak{m}(w_k)\,\mathrm{d}x = \lim_{k\to\infty}\int_M \mathfrak{m}(w_k)\,\mathrm{d}x \ge \liminf_{k\to\infty}\int_M |\nabla u_k|\,\mathrm{d}x \ge \int_M |\nabla u|\,\mathrm{d}x$, from which $\mathfrak{m}(w) \ge |\nabla u|$ a.e. in Ω.

[45] Hint: Read (8.261) as $\min_{u\in W_0^{1,p}(\Omega)} \int_\Omega \frac{1}{p}|\nabla u|^p - \xi u\,\mathrm{d}x = -\sup_{u\in W_0^{1,p}(\Omega)} \int_\Omega \xi u - \frac{1}{p}|\nabla u|^p \mathrm{d}x = -\frac{1}{p'}\|\nabla\Delta_p^{-1}\xi\|^p_{L^p(\Omega;\mathbb{R}^n)}$ and then substitute $\xi := g-\lambda$.

[46] Hint: $L'_u(u,\lambda)=0$ means $\mathrm{div}\big(a(u,\nabla u){+}\lambda\big){-}c(u,\nabla u){+}g = 0$ with a and c from (4.23) while $\partial_\lambda L(u,\lambda) \ni 0$ has a local character $m|s|{-}\nabla u(x){\cdot}(s{-}\lambda(x)) \ge m|\lambda(x)|$ for all $s{\in}\mathbb{R}^n$ and a.a. $x{\in}\Omega$, which equivalently means $\lambda \in B_m(\nabla u)$ a.e. with B_m being the ball in $\mathbb{R}^n$ of the radius m.

[47] Hint: use just a special data $m=1$ and $\varphi = \varphi(\nabla u)$ so that $a = \varphi'$.

5.6 Some applications to free-boundary problems

Variational inequalities are often directly fitted with various unilateral problems naturally arising in sciences, as the unilateral contact problem on Figure 12. Sometimes, (quasi)variational inequalities arise from concrete free-boundary problems only after sophisticated transformations, which is illustrated in this section in concrete cases.

5.6.1 Porous media flow: a potential variational inequality

We consider the simplest model of a porous, permeable, isotropic, and homogeneous medium undergoing a flow (a seepage) of an incompressible fluid in a wet, fully saturated domain while the rest is completely dry. Another simplification concerns a geometry consisting in a cylindrical vertically oriented domain; to be more specific, let us consider two reservoirs adjacent to this domain which can be then considered as a dam. In addition, we consider nonpermeability of the flat horizontal support and of the sides which are not adjacent to any reservoir, and no source on the free boundary (i.e. no contribution by rain water).[48] We use the notation (cf. Fig. 13 on p. 165):

v velocity of the flow,
π a piesometric head; we consider[49] $\pi = x_3 + p$, where p is a pressure,
$\varphi : \mathbb{R}^2 \to \mathbb{R}$ a function whose graph is the free boundary $x_3 = \varphi(x_1, x_2)$,
$h_{\rm U}$ the altitude of the upper reservoir,
$h_{\rm L}$ the altitude of the lower reservoir,
k the permeability coefficient.

The seepage flow is then governed by Darcy's law together, of course, with the continuity equation, i.e. respectively[50]

$$v = -k\nabla\pi\,, \tag{5.109a}$$

$$\text{div}\ v = 0\,. \tag{5.109b}$$

This gives $k\Delta\pi = 0$ so that

$$\Delta p = \Delta(\pi - x_3) = \Delta\pi = 0. \tag{5.110}$$

On the free boundary, whose position is not known a-priori, there are two conditions

$$p = 0 \qquad \text{and} \qquad v \cdot \nu = 0, \tag{5.111}$$

[48] See the monographs by Baiocchi and Capelo [30, Chapter 8], Chipot [99, Chapter 4], Crank [111, Chapter 2], Duvaut and Lions [130, Appendix 2], Elliott and Ockendon [135, Sect. IV.4], Friedman [155], Rodrigues [354, Sect. 2.3], where more general situations can be found, too.

[49] More generally, one should consider $\pi = x_3 + p/(\varrho g)$ with ϱ the mass density and g gravity acceleration. Here we put $\varrho g = 1$ for simplicity.

[50] Note that (5.109) arises from the so-called Darcy-Brinkman system $\varrho(v{\cdot}\nabla)v - \mu\Delta v + v/k + \nabla\pi = 0$ and div $v = 0$ when viscosity μ and inertia by mass density ϱ are neglected. This system modifies the Navier-Stokes equation, see (6.49) below, for the flow in porous media where k depends on porosity. Cf. also (8.208) on p. 281 for the evolution variant.

which would seemingly create an overdetermination if it were not the fact that the position of the free boundary itself is not determined in advance. In (5.111), ν is the unit normal to the free boundary oriented from the dry region to the wet one, which, in terms of φ, means

$$\nu = \frac{\left(\frac{\partial\varphi}{\partial x_1}, \frac{\partial\varphi}{\partial x_2}, -1\right)}{\sqrt{\left|\frac{\partial\varphi}{\partial x_1}\right|^2 + \left|\frac{\partial\varphi}{\partial x_2}\right|^2 + 1}}. \tag{5.112}$$

Comparing (5.109a) with the second condition in (5.111) yields $\frac{\partial}{\partial\nu}p = -\frac{\partial}{\partial\nu}x_3 = -\nu_3$, so that (5.112) then results in

$$\frac{\partial p}{\partial x_1}\frac{\partial\varphi}{\partial x_1} + \frac{\partial p}{\partial x_2}\frac{\partial\varphi}{\partial x_2} - \frac{\partial p}{\partial x_3} = 1. \tag{5.113}$$

The other boundary conditions are outlined in the left-hand part of Figure 13. We apply the so-called *Baiocchi transformation*:

$$u(x) \equiv u(x_1, x_2, x_3) := \begin{cases} \displaystyle\int_{x_3}^{\varphi(x_1,x_2)} p(x_1, x_2, \xi)\,\mathrm{d}\xi & \text{for } x_3 \le \varphi(x_1, x_2), \\ 0 & \text{for } x_3 > \varphi(x_1, x_3). \end{cases} \tag{5.114}$$

Obviously, $\frac{\partial}{\partial x_3}u = -p$. In view of (5.110), we get:

$$\Delta u = g(x) \qquad \text{on } \Omega_+ := \{x \in \Omega;\ u(x) > 0\}; \tag{5.115}$$

where we implicitly assume $p > 0$ so that Ω_+ represents the wet region. To determine g, let us apply $\frac{\partial}{\partial x_1}$ and $\frac{\partial}{\partial x_2}$ to (5.114), which gives

$$\begin{aligned} \frac{\partial u}{\partial x_i} &= \int_{x_3}^{\varphi(x_1,x_2)} \frac{\partial p(x_1, x_2, \xi)}{\partial x_i}\mathrm{d}\xi \\ &\quad + \frac{\partial\varphi}{\partial x_i}p(x_1, x_2, \varphi(x_1, x_2)) = \int_{x_3}^{\varphi(x_1,x_2)} \frac{\partial p(x_1, x_2, \xi)}{\partial x_i}\mathrm{d}\xi \end{aligned} \tag{5.116}$$

for $i = 1, 2$ because $p(x_1, x_2, \varphi(x_1, x_2)) = 0$. Applying again $\frac{\partial}{\partial x_i}$ and using both (5.110) and (5.113), we obtain

$$\begin{aligned} \frac{\partial^2 u}{\partial x_1^2} + \frac{\partial^2 u}{\partial x_2^2} &= \int_{x_3}^{\varphi(x_1,x_2)} \frac{\partial^2 p(x_1, x_2, \xi)}{\partial x_1^2} + \frac{\partial^2 p(x_1, x_2, \xi)}{\partial x_2^2}\,\mathrm{d}\xi \\ &\quad + \frac{\partial\varphi}{\partial x_1}\frac{\partial p}{\partial x_1}(x_1, x_2, \varphi(x_1, x_2)) + \frac{\partial\varphi}{\partial x_2}\frac{\partial p}{\partial x_1}(x_1, x_2, \varphi(x_1, x_2)) \\ &= \int_{x_3}^{\varphi(x_1,x_2)} -\frac{\partial^2 p(x_1, x_2, \xi)}{\partial x_3^2}\,\mathrm{d}\xi + 1 + \frac{\partial p}{\partial x_3}(x_1, x_2, \varphi(x_1, x_2)) \\ &= \left[-\frac{\partial p}{\partial x_3}(x_1, x_2, \xi)\right]_{\xi=x_3}^{\varphi(x_1,x_2)} + 1 + \frac{\partial p}{\partial x_3}(x_1, x_2, \varphi(x_1, x_2)) \\ &= 1 + \frac{\partial p}{\partial x_3} = 1 - \frac{\partial^2 u}{\partial x_3^2}. \end{aligned} \tag{5.117}$$

Comparing it with (5.115), we get $g = 1$.

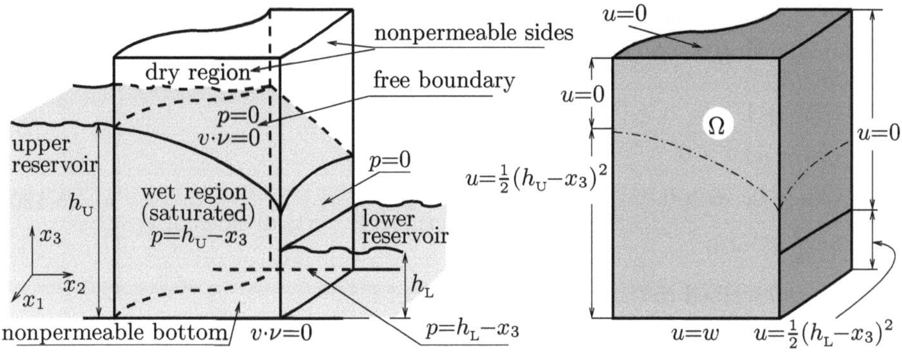

Figure 13. Geometric configuration of the dam problem and boundary conditions; original (left) and transformed (right).

The boundary conditions on the vertical sides are either the Dirichlet or the Neumann ones:[51]

$$u = \begin{cases} 0 & \text{on the upper side,} \\ \frac{1}{2}\big((h_U - x_3)^+\big)^2 & \text{on the side adjacent to the upper reservoir,} \\ \frac{1}{2}\big((h_L - x_3)^+\big)^2 & \text{on the side adjacent to the lower reservoir,} \\ w & \text{on the bottom, nonpermeable side,} \end{cases} \tag{5.118a}$$

$$\frac{\partial u}{\partial \nu} = 0 \qquad \text{on the vertical nonpermeable sides.} \tag{5.118b}$$

The Dirichlet boundary condition at the bottom part uses continuity of u and $\frac{\partial^2}{\partial x_1^2}u + \frac{\partial^2}{\partial x_2^2}u = 0$,[52] which implies that the function $w = w(x_1, x_2)$ occurring in (5.118) can be determined as the unique solution to the following 2-dimensional boundary-value problem:

$$\left.\begin{aligned} \frac{\partial^2 w}{\partial x_1^2} + \frac{\partial^2 w}{\partial x_2^2} &= 0 && \text{on the bottom side,} \\ w &= \tfrac{1}{2}h_U^2 && \text{on the bottom edge adjacent to the upper reservoir,} \\ w &= \tfrac{1}{2}h_L^2 && \text{on the bottom edge adjacent to the lower reservoir,} \\ \frac{\partial w}{\partial \nu} &= 0 && \text{on the bottom edges adjacent to the nonpermeable sides.} \end{aligned}\right\} \tag{5.119}$$

These boundary conditions are outlined in the right-hand part of Figure 13.

[51] On the upper-reservoir side, $\frac{\partial}{\partial x_3}u = -p = h_U - x_3$ implies $u = \frac{1}{2}(h_U - x_3)^2$, and similar condition but with h_L instead of h_U takes place on the lower-reservoir side.

[52] This follows from (5.117) by using $\frac{\partial^2}{\partial x_3^2}u = -\frac{\partial}{\partial x_3}p = -\frac{\partial}{\partial x_3}\pi + 1 = 1$ because $\frac{\partial}{\partial x_3}\pi = v{\cdot}\nu = 0$ on the bottom side.

As $p \geq 0$ should hold from physical reasons, u should be nonincreasing along the x_3-direction, hence $u \geq 0$. In the dry region one has $u = 0$, hence $1 - \Delta u = 1 \geq 0$, while in the wet region we derived $1 - \Delta u = 0$ in (5.117). Altogether, we get the following complementarity problem[53]

$$\left.\begin{array}{ll} \left.\begin{array}{l} -\Delta u + 1 \geq 0, \ u \geq 0, \\ (\Delta u - 1)\, u = 0, \end{array}\right\} & \text{in } \Omega, \\ \left.\begin{array}{l} \frac{\partial}{\partial \nu} u \geq 0, \ \ u \geq 0, \\ u\,(\frac{\partial}{\partial \nu} u) \ = 0 \end{array}\right\} & \text{on nonpermeable vertical sides of } \Gamma, \\ u|_\Gamma \text{ prescribed in (5.118a)} & \text{on the rest of } \Gamma. \end{array}\right\} \tag{5.120}$$

The corresponding weak formulation admits a unique solution $u \in K := \{v \in W^{1,2}(\Omega);\ v|_\Gamma \text{ satisfying } (5.118a)\}$, which can be proved straightforwardly by the direct method as in Theorem 5.3(iii) with $\Phi(u) := \int_\Omega \frac{1}{2}|\nabla u|^2 - u \, dx$ for $u \geq 0$ a.e. in Ω, otherwise $\Phi(u) = +\infty$. Therefore, the original problem has a unique (very weak) solution $p = -\frac{\partial}{\partial x_3} u \in L^2(\Omega)$.

5.6.2 Continuous casting: a non-potential variational inequality

A great amount of steel is nowadays casted continuously: hot liquid steel is continuously filled from the top into a mold cooled by water (cf. Figure 14(left)), partly solidifies but keeping still a hot liquid kernel, and continuously extracted by rollers and further cooled down to a complete solidification and then cut to a final product. We shall present only a very simple steady-state model of this advanced technology.[54] The following notation will be used, cf. also Figure 14 below:

θ_0 temperature of the liquid phase (melting temperature),
θ_1 final temperature of the cooled outlet,[55]
$\theta_2(x)$ temperature of the environment,

[53] Note that the Neumann condition (5.118b) is replaced by the complementarity condition on the nonpermeable vertical sides of Γ, but these are equivalent with each other if u is regular enough because, in the dry region, $u = 0$ implies $\nabla u = 0$ hence $\nu \cdot \nabla u = \frac{\partial}{\partial \nu} u = 0$ on the dry boundary while on the wet boundary $u > 0$ and $u\,(\frac{\partial}{\partial \nu} u) = 0$ imply $\frac{\partial}{\partial \nu} u = 0$.

[54] Our simplifications involve, in particular, calm liquid phase on the melting temperature (i.e. we neglect convection in the liquid part like in Section 6.2), linear heat equation (i.e. we neglect Stefan-Boltzmann radiation on the boundary like (2.125) and temperature dependence of c and κ), solidification at a single temperature (i.e. no over-cooling effects, no mutual influence of the melting temperature and chemical composition of a steel which is, in fact, a mixture of iron and other elements such as carbon, etc.), known temperature θ_2 at the mold side (temperature distribution in the mold is not solved), etc. Besides a huge amount of papers, the reader is referred to a monograph by Rodrigues [354, Sect. 2.5].

[55] This will represent a Dirichlet boundary condition on the bottom end (cf. Figure 14(left)) which, however, is rather artificial and simplifies the heat convection in the continuation of the casted workpiece. Yet, this does not essentially influence the process in the upper part if v_3 is large enough and the bottom end is far enough from the mold.

$b \geq 0$ the heat-convection coefficient,
$\vec{v} = (0, 0, v_3)$ extraction velocity,
$\kappa > 0$ the heat-conductivity coefficient,
$c > 0$ the heat-capacity coefficient,
$\ell \geq 0$ the latent heat,
$x_3 = \varphi(x_1, x_2)$ a free boundary between the liquid and the solid phases.

Naturally, we assume $\theta_1 < \theta_0$, $\theta_2(x) \leq \theta_0$, and v_3, κ, c and ℓ positive. The equation for the temperature θ in the steady-state extraction regime is:

$$c\vec{v} \cdot \nabla\theta = \kappa\Delta\theta \qquad \text{if } \theta < \theta_0. \tag{5.121}$$

The so-called *Stefan condition* on the free boundary expresses that the normal heat flux $-\kappa\nabla\theta \cdot \nu = \kappa\frac{\partial}{\partial\nu}\theta$ is spent as the heat needed for the phase change, here the solidification, $\ell\vec{v} \cdot \nu$:

$$-\kappa\frac{\partial\theta}{\partial\nu} = -\ell\vec{v} \cdot \nu, \tag{5.122}$$

where ν is the unit normal oriented from the liquid phase to the solid one. As also $\theta = \theta_0$ on the free boundary, we have seemingly too many conditions on it but, as in Section 5.6.1, again the position of the free boundary itself is unknown and is to be determined just in this way that both (5.122) and $\theta = \theta_0$ are fulfilled. As the heat equation is considered only in one phase (here solid) while the temperature of the other is assumed constant, this problem is called a *one-phase Stefan problem.*

The other boundary conditions are outlined in the left-hand part of Figure 14, in particular the conditions on the vertical boundary reflect the cooling by convection:

$$-\kappa\frac{\partial\theta}{\partial\nu} = b(x)\big(\theta - \theta_2(x)\big). \tag{5.123}$$

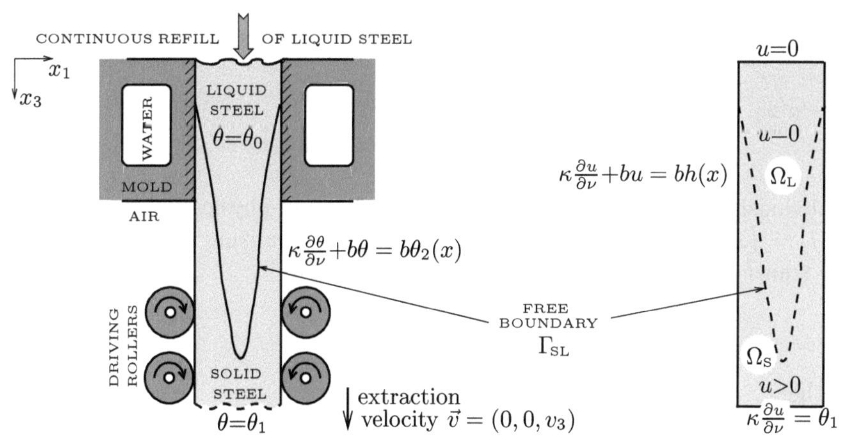

Figure 14. Geometric configuration (as a cross-section) of the continuous casting problem and boundary conditions; original (left) and transformed (right).

In terms of the auxiliary function $\varphi = \varphi(x_1, x_2)$ describing the free boundary in the sense that $\Gamma_{\rm SL} = \{x = (x_1, x_2, x_3) \in \Omega;\ x_3 = \varphi(x_1, x_2)\}$, the condition (5.122) on the free boundary reads as

$$\frac{\partial\theta}{\partial x_3} - \frac{\partial\theta}{\partial x_1}\frac{\partial\varphi}{\partial x_1} - \frac{\partial\theta}{\partial x_2}\frac{\partial\varphi}{\partial x_2} = \frac{\ell v_3}{\kappa}. \tag{5.124}$$

Formally, we have

$$\kappa\Delta\theta - c\vec{v}\cdot\nabla\theta = \ell v_3 \frac{\partial}{\partial x_3}\chi_{\Omega_{\rm S}} \tag{5.125}$$

in the sense of distributions. Indeed, for $\Omega_{\rm L} := \Omega \setminus \Omega_{\rm S}$ and $\Gamma_{\rm SL} := \bar{\Omega}_{\rm S} \cap \bar{\Omega}_{\rm L}$ (=the free boundary), and for any $v \in \mathscr{D}(\Omega)$, by using Green's formula twice and that $\nabla\theta = 0$ on $\Omega_{\rm L}$, it holds that

$$\langle\Delta\theta, v\rangle = -\int_\Omega \nabla\theta\cdot\nabla v\,{\rm d}x = -\int_{\Omega_{\rm S}} \nabla\theta\cdot\nabla v\,{\rm d}x = \int_{\Omega_{\rm S}} \Delta\theta v\,{\rm d}x - \int_{\Gamma_{\rm SL}} \frac{\partial\theta}{\partial\nu} v\,{\rm d}S \tag{5.126}$$

and, again by using Green's formula twice,

$$\Big\langle\frac{\partial}{\partial x_3}\chi_{\Omega_{\rm S}}, v\Big\rangle = -\int_\Omega \chi_{\Omega_{\rm S}}\frac{\partial v}{\partial x_3}\,{\rm d}x = -\int_{\Omega_{\rm S}} \frac{\partial v}{\partial x_3}\,{\rm d}x = \int_{\Gamma_{\rm SL}} v\,\nu_3\,{\rm d}S \tag{5.127}$$

so that, by using successively (5.126), (5.121), (5.122), and (5.127), one obtains

$$\begin{aligned}\big\langle\kappa\Delta\theta - c\vec{v}\cdot\nabla\theta, v\big\rangle &= \int_{\Omega_{\rm S}} (\kappa\Delta\theta - c\vec{v}\cdot\nabla\theta)v\,{\rm d}x - \int_{\Gamma_{\rm SL}} \frac{\partial\theta}{\partial\nu} v\,{\rm d}S \\ &= -\int_{\Gamma_{\rm SL}} \ell\vec{v}\nu v\,{\rm d}S = -\ell v_3 \int_{\Gamma_{\rm SL}} v\,\nu_3{\rm d}S = \ell v_3\Big\langle\frac{\partial}{\partial x_3}\chi_{\Omega_{\rm S}}, v\Big\rangle.\end{aligned} \tag{5.128}$$

Then we use the *Baiocchi transformation*:

$$u(x) \equiv u(x_1, x_2, x_3) := \begin{cases} \displaystyle\int_{x_3}^{\varphi(x_1,x_2)} \theta(x_1, x_2, \xi)\,{\rm d}\xi & \text{for } x_3 \le \varphi(x_1, x_2), \\ 0 & \text{for } x_3 > \varphi(x_1, x_2). \end{cases} \tag{5.129}$$

Then obviously $\theta = -\frac{\partial}{\partial x_3}u$ and, assuming that $\theta > 0$ in physically relevant situations, $\Omega_{\rm S} := \{x\in\Omega;\ u(x) > 0\} = \{x\in\Omega;\ \theta(x) < \theta_0\}$. Realizing that $\vec{v} = (0, 0, v_3)$, (5.125) transforms by integration in the x_3-direction to

$$\kappa\Delta u - c\vec{v}\cdot\nabla u = \ell v_3\chi_{\Omega_{\rm S}}. \tag{5.130}$$

Altogether:

$$\kappa\Delta u - c\vec{v}\cdot\nabla u = \begin{cases} \ell v_3 \\ 0 < \ell v_3 \end{cases} \quad\text{and}\quad u\begin{cases} > 0 & \text{on } \Omega_{\rm S}, \\ = 0 & \text{on } \Omega_{\rm L}. \end{cases} \tag{5.131}$$

Since the Baiocchi transformation commutes with "$\frac{\partial}{\partial\nu}$" on the vertical lines, the boundary condition (5.123) transforms to

$$-\frac{\partial u}{\partial \nu} = bu - h(x), \qquad h(x_1,x_2,x_3) = \int_{x_3}^{0} b\theta_2(x_1,x_2,\xi)\,\mathrm{d}\xi, \tag{5.132}$$

provided b is independent of x which we have to assume from now on. The other boundary conditions are outlined on Figure 14(right). This means we get the complementarity problem

$$\left.\begin{array}{l} \left.\begin{array}{l} -\kappa\Delta u + c\vec{v}\cdot\nabla u \ge -\ell v_3, \quad u\ge 0,\\ \big(\kappa\Delta u - c\vec{v}\cdot\nabla u + \ell v_3\big)\,u = 0, \end{array}\right\} \text{ in } \Omega,\\ \left.\begin{array}{l} \dfrac{\partial u}{\partial\nu} + bu \ge h, \quad u\ge 0,\\ \Big(\dfrac{\partial u}{\partial\nu} + bu - h\Big)\,u = 0 \end{array}\right\} \text{ on } \Gamma. \end{array}\right\} \tag{5.133}$$

As in Proposition 5.9, we arrive at the variational inequality formulation:

$$\int_\Omega \kappa\nabla u\cdot\nabla(v{-}u) + \big(c\vec{v}\cdot\nabla u + \ell v_3\big)(v{-}u)\,\mathrm{d}x + \int_\Gamma (bu-h)(v{-}u)\,\mathrm{d}S \ge 0 \tag{5.134}$$

for $v\in K := \{v\in W^{1,2}(\Omega);\ v\ge 0,\ v(x_1,x_2,0)=0\}$. This variational inequality involves a pseudomonotone (non-potential) operator and has a solution $u\in K$ by Corollary 5.19; thus we get $w = \frac{\partial}{\partial x_3}u \in L^2(\Omega)$ a *very weak solution*. Moreover, this operator is even uniformly monotone because, by Green's formula,

$$\begin{aligned}\int_\Omega \vec{v}\cdot\nabla(u_1{-}u_2)(u_1{-}u_2)\,\mathrm{d}x &= \frac{1}{2}\int_\Omega \vec{v}\cdot\nabla(u_1{-}u_2)^2\mathrm{d}x\\ &= -\frac{1}{2}\int_\Omega \mathrm{div}(\vec{v})(u_1{-}u_2)^2\mathrm{d}x + \frac{1}{2}\int_\Gamma (\vec{v}\cdot\nu)(u_1{-}u_2)^2\mathrm{d}S \ge 0; \end{aligned} \tag{5.135}$$

note that the last volume integral vanishes since $\mathrm{div}(\vec{v})=0$ while the last boundary integral is non-negative since $(u_1{-}u_2)^2 = 0$ on top, $(\vec{v}{\cdot}\nu)=0$ on vertical sides, and both $(\vec{v}\cdot\nu)\ge 0$ and $(u_1-u_2)^2\ge 0$ on the bottom. Then we can use Theorem 5.20 which gives even uniqueness of this solution and continuous dependence on ℓ, v_3, and h. Example 4.30 showed that this problem is indeed non-potential.

5.7 Bibliographical remarks

Subdifferentials of convex functions has been scrutinized in many monographs from so-called convex analysis, among them Hu and Papageorgiou [209, Sect.3.4], Rockafellar and Wetts [353], or Zeidler [427, Chap.47].

Variational inequalities are addressed in many monographs: Baiocchi, Capelo [30], Chipot [99], Elliott, Ockendon [135], Friedman [155], Glowinski, Lions,

Trémolières [182], Goeleven, Motreanu [183], Kinderlehrer, Stampacchia [232], Lions [261, Chap.2,Sect.8 and Chap.3,Sect.5], Malý, Ziemer [271, Chap.5-6], Pascali, Sburlan [325], Rodrigues [354], Růžička [376, Sect. 3.3.4], Troianiello [409], and Zeidler [427, Chap.54]. A fundamental paper is by Brézis [65, Chap.I]. Applications to mechanics, in particular to contact problems, is in Duvaut, Lions [130], Eck, Jarušek, Krbec [132], Hlaváček, Haslinger, Nečas, Lovíšek [204], Nečas, Hlaváček [308], or Kikuchi and Oden [231]. Variational inequalities in the context of their optimal control are in Barbu [38, Chap.3] and Outrata, Kočvara, and Zowe [321].

Quasivariational inequalities have been thoroughly exposed in the monograph by Baiocchi and Capelo [30], cf. also Aubin [27, Sect. 9.11]. Important application is the ground-water propagation through a dam of a general, nonrectangular shape, see Baiocchi and Capelo [30, Chap.8], Chipot [99, Chap.8], or Crank [111, Sect.2.3.7]. A related subject (not mentioned here) is the so-called *implicit variational inequalities*: find u such that, for all $w \in V$, it holds that $A(u,w) - A(u,u) + F(E(u),w) - F(E(u),u) \geq 0$. Typically, it involves problems like mechanical contacts with friction that have a dual formulation as quasivariational inequalities. Transformation between it and quasivariational inequality is in Mosco [294].

For hemivariational inequalities, introduced essentially by Panagiotopoulos [322] with a certain motivation in continuum mechanics, see also the monographs by Goeleven and Motreanu [183], Haslinger, Mietinen, and Panagiotopoulos [199] and Naniewicz and Panagiotopoulos [298].

A generalization for the monotone set-valued part being non-potential does exist, too, being based on the concept of the maximal monotone set-valued mappings. An analog of Browder-Minty's theorem says that any maximal monotone and coercive $A : V \rightrightarrows V^*$ is surjective, i.e. the inclusion $A(u) \ni f$ has at least one solution for any $f \in V^*$; cf. Hu and Papageorgiou [209, Sect.3.1-2] or Zeidler [427, Chap.32]. Set-valued generalization does exist also for *pseudomonotone* mappings[56], being invented by Browder [76]. Set-valued generalization of *mappings of type (M)*[57] is due to Kenmochi [227]. For the surjectivity of pseudomonotone set-valued mappings we refer to Browder and Hess [77]; a thorough exposition is in the handbook by Hu and Papageorgiou [209, Part I, Chap.III].

[56] A set-valued mapping $A : V \rightrightarrows V^*$ is called pseudomonotone if
1) $\forall u \in V$: $A(u)$ is nonempty, bounded, closed, and convex,
2) $\forall U \subset V$ finite-dimensional subspace: $A|_U$ is (norm,weak*)-upper semicontinuous,
3) if $u_k \rightharpoonup u$, $f_k \in A(u_k)$, $\limsup_{k\to\infty} \langle f_k, u_k - u\rangle \leq 0$, then $\forall v \in V\ \exists f \in A(u)$: $\liminf_{k\to\infty} \langle f_k, u_k - v\rangle \geq \langle f, u - v\rangle$.

[57] A set-valued mapping $A : V \rightrightarrows V^*$ is called of type (M) if $A|_U$ is weakly* upper semicontinuous for all $U \subset V$ finite-dimensional, $A(u)$ is nonempty, bounded, closed, and convex, and if $f_k \in A(u_k)$, and $(u_k, f_k) \stackrel{*}{\rightharpoonup} (u, f)$ in $V \times V^*$ and $\limsup_{k\to\infty} \langle f_k, u_k\rangle \leq \langle f, u\rangle$, then $f \in A(u)$.

Chapter 6

Systems of equations: particular examples

No general theory for systems of nonlinear equations exists. Systems usually require a combination of specific, sometimes very sophisticated tricks, possibly with a fixed-point technique finely fitted to a particular structure. Although certain general approaches can be adopted,[1] a pragmatic observation is that systems are much more difficult than single equations and sometimes only partial results (typically for small data) can be obtained with current knowledge. Even worse, many natural systems arising from physical problems still remain unsolved with respect to even the existence of a solution; in particular cases, however, this may be related with an oscillatory-like or explosion-like character of related evolutionary systems which thus lack any steady states that would solve these stationary systems.

We confine ourselves to only a few illustrative examples having a straightforward physical interpretation and using the previously exposed theory in a nontrivial but still rather uncomplicated manner.

6.1 Minimization-type variational method: polyconvex functionals

For the "Lagrangean" $\varphi:\Omega\times\mathbb{R}^m\times\mathbb{R}^{m\times n}\to\mathbb{R}$ we consider the system of nonlinear equations ($j=1,\dots,m$):

$$\left.\begin{aligned} -\sum_{i=1}^{n}\frac{\partial}{\partial x_i}\frac{\partial\varphi}{\partial S_{ij}}(x,u,\nabla u)+\frac{\partial\varphi}{\partial R_j}(x,u,\nabla u) &= g_j &&\text{on }\Omega,\\ \sum_{i=1}^{n}\nu_i\frac{\partial\varphi}{\partial S_{ij}}(x,u,\nabla u)+b_j(x,u) &= h_j &&\text{on }\Gamma, \end{aligned}\right\} \tag{6.1}$$

[1] Cf. Ladyzhenskaya and Uraltseva [250, Chap.8].

where, instead of the notation $(r,s) \in \mathbb{R} \times \mathbb{R}^n$, we used here $(R,S) \in \mathbb{R}^m \times \mathbb{R}^{m\times n}$, and then $\varphi = \varphi(x,R,S)$, like we already did in Sect. 2.4.4; thus $S \in \mathbb{R}^{m\times n}$ denotes here a matrix, hopefully without confusion with S occuring in the surface measure $\mathrm{d}S$. The weak formulation of (6.1) is obtained by multiplying the equation in (6.1) by v_j, integrating over Ω, summing it for $j = 1,\dots,m$, and using Green's formula:

$$\int_\Omega \Big(\frac{\partial\varphi}{\partial S}(x,u,\nabla u) : \nabla v + \sum_{j=1}^m \frac{\partial\varphi}{\partial R_j}(x,u,\nabla u)v_j\Big)\,\mathrm{d}x + \sum_{j=1}^m \int_\Gamma b_j(x,u)v_j\,\mathrm{d}S = \sum_{j=1}^m \Big(\int_\Omega g_j v_j\,\mathrm{d}x + \int_\Gamma h_j v_j\,\mathrm{d}S\Big) \tag{6.2}$$

for all $v \in C^1(\Omega;\mathbb{R}^m)$, where $S : \widetilde{S} := \sum_{i=1}^n \sum_{j=1}^m S_{ij}\widetilde{S}_{ij}$. Assuming still

$$\frac{\partial b_i}{\partial R_j} = \frac{\partial b_j}{\partial R_i}, \qquad i,j = 1,\dots,m, \tag{6.3}$$

the left-hand-side of the boundary-value problem (6.1) has a potential

$$\Phi(u) = \int_\Omega \varphi(x,u,\nabla u)\,\mathrm{d}x + \int_\Gamma \psi(x,u)\,\mathrm{d}S, \tag{6.4}$$

where $\psi(x,R)$ is defined by the formula (cf. (4.23c)):

$$\psi(x,R) = \int_0^1 R\cdot b(x,tR)\,\mathrm{d}t\ , \qquad R\in\mathbb{R}^m, \quad b:\Gamma\times\mathbb{R}^m \to \mathbb{R}^m. \tag{6.5}$$

Although it is, in general, not possible to pass to a limit through a nonlinearity by a weak convergence, cf. Remark 2.39, it is sometimes possible in special nonlinearities (here the determinant) if special sequences (here generated by gradients[2]) are considered:

Lemma 6.1. *Let $u_k \rightharpoonup u$ in $W^{1,p}(\Omega;\mathbb{R}^m)$, $p > n$, $n = m$. Then*

$$\det \nabla u_k \rightharpoonup \det \nabla u \quad \text{in } L^{p/n}(\Omega). \tag{6.6}$$

Proof. It is a well-known fact from matrix algebra that, for $S \in \mathbb{R}^{m\times n}$ with $m = n$, it holds that

$$(\det S)\,\mathbb{I} = (\operatorname{cof} S)^\top S, \tag{6.7}$$

where the "cofactor" $[\operatorname{cof} S]_{ij}$ is the determinant of the matrix arising from S by omitting the ith row and j^{th} column but multiplied by $(-1)^{i+j}$. Putting $S := \nabla u$ and summing it for $i,j = 1,\dots,n$, this allows us to show

$$n \det \nabla u = \sum_{i,j=1}^n \frac{\partial u^i}{\partial x_j}(\operatorname{cof}\nabla u)^i_j = \sum_{i,j=1}^n \frac{\partial}{\partial x_j}\big(u^i(\operatorname{cof}\nabla u)^i_j\big) - u^i\frac{\partial}{\partial x_j}(\operatorname{cof}\nabla u)^i_j. \tag{6.8}$$

[2]Such sequences are rotation free due to the well-known fact that $\mathrm{rot}(\nabla u) \equiv 0$. This constraint causes sometimes surprising effects, e.g. concerning higher integrability, cf. Müller [296].

The last term vanishes because of Piola's identity $\sum_{j=1}^{n}\frac{\partial}{\partial x_j}(\mathrm{cof}\nabla u)^i_j = 0$ for all $i = 1,\dots,n$.[3]

Then, by using subsequently (6.8), twice Green's formula, and again (6.8), one gets

$$\lim_{k\to\infty}\int_\Omega(\det\nabla u_k)v\,\mathrm{d}x = \frac{1}{n}\lim_{k\to\infty}\int_\Omega\sum_{i,j=1}^{n}\frac{\partial}{\partial x_j}\Big(u_k^i\big(\mathrm{cof}\nabla u_k\big)^i_j\Big)v$$
$$= -\frac{1}{n}\lim_{k\to\infty}\sum_{i,j=1}^{n}\int_\Omega u_k^i\big(\mathrm{cof}\nabla u_k\big)^i_j\frac{\partial v}{\partial x_j}\,\mathrm{d}x$$
$$= -\frac{1}{n}\sum_{i,j=1}^{n}\int_\Omega u^i(\mathrm{cof}\nabla u)^i_j\frac{\partial v}{\partial x_j}\,\mathrm{d}x = \int_\Omega(\det\nabla u)v\,\mathrm{d}x \tag{6.9}$$

for any $v\in\mathscr{D}(\Omega)$ because $u_k\to u$ in $L^\infty(\Omega;\mathbb{R}^n)\Supset W^{1,p}(\Omega;\mathbb{R}^n)$ and $\mathrm{cof}\nabla u_k\rightharpoonup\mathrm{cof}\nabla u$ in $L^{p/(n-1)}(\Omega;\mathbb{R}^{n\times n})$ which is obvious for $n=2$ while it follows by induction if $n\geq 3$. □

Lemma 6.2 (Weak lower semicontinuity). *Let $m=n$, let φ be coercive in the sense $\varphi(x,R,S)\geq\varepsilon|S|^p$ for $p>n$ and $\varphi(x,R,\cdot)$ be polyconvex in the sense*

$$\varphi(x,R,S)=\mathfrak{f}(x,R,S,\det S) \tag{6.10}$$

with some $\mathfrak{f}:\Omega\times\mathbb{R}^m\times\mathbb{R}^{m\times n}\times\mathbb{R}\to\mathbb{R}$ such that $\mathfrak{f}(x,R,\cdot,\cdot)$ is convex and smooth[4]*, and satisfy the following growth conditions:*

$$\exists\gamma\in L^1(\Omega)\ :\qquad \big|\varphi(x,R,S)\big|\leq C\big(\gamma(x)+\beta(|R|)+|S|^p\big), \tag{6.11a}$$
$$\exists\gamma\in L^{p'}(\Omega)\ :\qquad \Big|\frac{\partial\mathfrak{f}}{\partial S}\Big|\leq C\big(\gamma(x)+\beta(|R|)+|S|^{p-1}\big), \tag{6.11b}$$
$$\exists\gamma\in L^{(p/n)'}(\Omega):\qquad \Big|\frac{\partial\mathfrak{f}}{\partial\det S}\Big|\leq C\big(\gamma(x)+\beta(|R|)+|S|^{p-n}\big), \tag{6.11c}$$

for some $C\in\mathbb{R}^+$ and $\beta:\mathbb{R}\to\mathbb{R}$ continuous (with arbitrary growth), and moreover $b(x,\cdot):\mathbb{R}^n\to\mathbb{R}^n$ is monotone for a.a. $x\in\Gamma$ and satisfies the growth condition

$$\exists\gamma\in L^1(\Gamma);\qquad |b(x,R)|\leq\gamma(x)+\beta(|R|). \tag{6.12}$$

Then Φ is weakly lower semicontinuous.[5]

---

[3] See Ciarlet [93, proof of Thm.1.7-1] or Evans [138, Sect.8.1.4, Lemma 1] for technical details.

[4] Differentiability of $\mathfrak{f}(x,R,\cdot,\cdot)$ is just a technical assumption which can be avoided when selecting (in a measurable way) subgradients of $\mathfrak{f}(x,R,\cdot,\cdot)$ in place of partial derivatives used here.

[5] Recall again our convention that by semicontinuity, see (1.6), we mean the "sequential" semicontinuity. Here, however, it is even equivalent with general-topological semicontinuity which uses generalized sequences (nets) since Φ is coercive and V separable.

Proof. Take a sequence $u_k \rightharpoonup u$ in $W^{1,p}(\Omega)$. By Banach-Steinhaus Theorem 1.1, $\{u_k\}_{k\in\mathbb{N}}$ is bounded. Without loss of generality we can suppose that $\Phi(u_k) \to \liminf_{k\to\infty} \Phi(u_k)$. We are to show that $\Phi(u) \le \lim_{k\to\infty} \Phi(u_k)$.

By the compact embedding $W^{1,p}(\Omega) \Subset L^\infty(\Omega;\mathbb{R}^m)$ (recall that $n < p$ is assumed) we have $u_k \to u$ uniformly on Ω. Let us put

$$\Omega_\varepsilon := \Big\{x\in\Omega;\ \ |\nabla u(x)| \le \frac{1}{\varepsilon}\Big\}. \tag{6.13}$$

Then $\lim_{\varepsilon\to 0}\mathrm{meas}_n(\Omega\setminus\Omega_\varepsilon) = 0$. Using subsequently non-negativity of φ, polyconvexity of $\varphi(x,R,\cdot)$ hence convexity of $\mathfrak{f}(x,R,\cdot,\cdot)$, one obtains

$$\begin{aligned}
&\liminf_{k\to\infty}\int_\Omega \varphi(x,u_k,\nabla u_k)\,\mathrm{d}x \ge \liminf_{k\to\infty}\int_{\Omega_\varepsilon}\varphi(x,u_k,\nabla u_k)\,\mathrm{d}x\\
&\quad = \liminf_{k\to\infty}\int_{\Omega_\varepsilon}\mathfrak{f}(x,u_k,\nabla u_k,\mathrm{det}\nabla u_k)\,\mathrm{d}x \ge \lim_{k\to\infty}\int_{\Omega_\varepsilon}\mathfrak{f}(x,u_k,\nabla u,\mathrm{det}\nabla u)\,\mathrm{d}x\\
&\qquad + \lim_{k\to\infty}\int_{\Omega_\varepsilon}\frac{\partial\mathfrak{f}}{\partial S}(x,u_k,\nabla u,\mathrm{det}\nabla u):(\nabla u_k-\nabla u)\,\mathrm{d}x\\
&\qquad + \lim_{k\to\infty}\int_{\Omega_\varepsilon}\frac{\partial\mathfrak{f}}{\partial\mathrm{det}S}(x,u_k,\nabla u,\mathrm{det}\nabla u)(\mathrm{det}\nabla u_k-\mathrm{det}\nabla u)\,\mathrm{d}x\\
&\quad = \int_{\Omega_\varepsilon}\mathfrak{f}(x,u,\nabla u,\mathrm{det}\nabla u)\,\mathrm{d}x = \int_{\Omega_\varepsilon}\varphi(x,u,\nabla u)\,\mathrm{d}x;
\end{aligned}$$

we used the convergences $\mathfrak{f}(x,u_k,\nabla u,\mathrm{det}\nabla u) \to \mathfrak{f}(x,u,\nabla u,\mathrm{det}\nabla u)$ in $L^1(\Omega_\varepsilon)$, $\frac{\partial}{\partial S}\mathfrak{f}(x,u_k,\nabla u,\mathrm{det}\nabla u) \to \frac{\partial}{\partial S}\mathfrak{f}(x,u,\nabla u,\mathrm{det}\nabla u)$ in $L^{p'}(\Omega_\varepsilon)$ and $\frac{\partial}{\partial\mathrm{det}S}\mathfrak{f}(x,u_k,\nabla u,\mathrm{det}\nabla u) \to \frac{\partial}{\partial\mathrm{det}S}\mathfrak{f}(x,u,\nabla u,\mathrm{det}\nabla u)$ in $L^{(p/n)'}(\Omega_\varepsilon)$ by continuity of the respective Nemytskiĭ mappings, and eventually $\nabla u_k - \nabla u \rightharpoonup 0$ and, by Lemma 6.1, $\mathrm{det}\nabla u_k - \mathrm{det}\nabla u \rightharpoonup 0$. Finally we pass to the limit with $\varepsilon \to 0$ by using Lebesgue's Theorem 1.14 to show that the last integral approaches $\int_\Omega \varphi(x,u,\nabla u)\,\mathrm{d}x$.

As to the boundary integral, (6.12) makes $u \mapsto \int_\Gamma \psi(x,u)\,\mathrm{d}S$ with ψ from (6.5) continuous and even smooth, and monotonicity of $b(x,\cdot)$ makes $\psi(x,\cdot)$ convex, hence the weak lower-semicontinuity follows as in (4.5).

Altogether, the weak lower-semicontinuity of Φ was thus proved. □

Proposition 6.3 (Existence: the direct method). *Let the assumptions of Lemma 6.2 be valid and, moreover, b be coercive in the sense $b(x,R)\cdot R \ge \varepsilon|R|^q - k$ with $\varepsilon > 0$ and $q > 1$. Then (6.1) has a weak solution.*

Proof. Analogous to Proposition 4.16 with f defined by $\langle f,v\rangle := \int_\Omega g\cdot v\,\mathrm{d}x + \int_\Gamma h\cdot v\,\mathrm{d}S$, but simplified due to absence of Dirichlet boundary conditions here. □

Remark 6.4 (*Polyconvexity*). The formula (6.10) gives a good generality only if $m = n = 2$. In general, one should assume

$$\varphi(x,R,S) = \mathfrak{f}\Big(x,R,(\mathrm{adj}_i S)_{i=1}^{\min(n,m)}\Big) \tag{6.14}$$

with some $\mathfrak{f} : \Omega \times \mathbb{R}^m \times \prod_{i=1}^{\min(n,m)} \mathbb{R}^{k(i,n,m)} \to \mathbb{R}$, where $k(i,n,m)$ is the number of all minors of the i-th order, such that $\mathfrak{f}(x,R,\cdot)$ is convex, where $\mathrm{adj}_i S$ denotes the determinants of all ($i{\times}i$)-submatrices. Then, following Ball [31], $\varphi(x,R,\cdot)$ is called *polyconvex*. In particular, $\mathrm{adj}_1 S = S$ and $\mathrm{adj}_{\min(n,m)-1} S = \mathrm{cof} S$ and, if $m = n$, $\mathrm{adj}_{\min(n,m)} S = \det S$. Then Lemma 6.1 is to be generalized for $\mathrm{adj}_i \nabla u_k \rightharpoonup \mathrm{adj}_i \nabla u$ in $L^{p/i}(\Omega; \mathbb{R}^{k(i,n,m)})$ provided $p > i \le \min(m,n)$, and Lemma 6.2 as well as Proposition 6.3 is to be modified for (6.14) in place of (6.10).

Remark 6.5 (*Quasiconvexity*). Polyconvexity of $\varphi(x,R,\cdot)$ is only sufficient for the weak lower semicontinuity of Φ but not necessary if $\min(n,m) \ge 2$. The precise condition (i.e. sufficient and necessary) is the so-called $W^{1,p}$-quasiconvexity, defined in a rather non-explicit way by

$$\varphi(x,R,S) = \inf_{v \in W_0^{1,p}(O;\mathbb{R}^m)} \frac{1}{|O|} \int_O \varphi\big(x,R,S{+}\nabla v(\xi)\big)\,\mathrm{d}\xi \tag{6.15}$$

where $O \subset \mathbb{R}^n$ is a (in fact, arbitrary) Lipschitz domain. This condition, whose inevitable nonlocality has been proved by Kristensen [242], cannot be verified efficiently except for very special cases, as e.g. polyconvexity which is a (strictly) stronger condition. Henceforth, another mode, a so-called *rank-one convexity*, was introduced by Morrey [291] by requiring $t \mapsto \varphi(x,R,S + ta \otimes b) : \mathbb{R} \to \mathbb{R}$ to be convex for any $a \in \mathbb{R}^n$, $b \in \mathbb{R}^m$, $[a \otimes b]_{ij} := a_i b_j$. Since Morrey [290] invented quasiconvexity, the question of coincidence with rank-one convexity was open for many decades and eventually answered negatively by Šverák [401] at least if $m \ge 3$ and $n \ge 2$. For smooth $\varphi(x,R,\cdot)$, the rank-1 convexity is equivalent with the so-called Legendre-Hadamard condition $\varphi''_S(x,R,S)(\widetilde{S},\widetilde{S}) \ge 0$ for all $S,\widetilde{S} \in \mathbb{R}^{m\times n}$ with $\widetilde{S} = a \otimes b$, $a \in \mathbb{R}^m$, $b \in \mathbb{R}^n$. Obviously, polyconvexity (and thus all mentioned notions) is weaker than usual convexity, and for $\min(n,m) = 1$ all mentioned modes coincide with usual convexity of $\varphi(x,R,\cdot)$.

Remark 6.6 (*Symmetry conditions*[6]). Considering the general system of m quasilinear equations

$$-\mathrm{div}\big(a^j(x,u,\nabla u)\big) + c^j(x,u,\nabla u) = g^j, \qquad j = 1,\dots,m, \tag{6.16}$$

the symmetry condition (4.21) which, here together with (6.3), ensures existence of a potential in the form (6.4) bears now the form

$$\frac{\partial a_i^l(x,R,S)}{\partial S_{jk}} = \frac{\partial a_k^j(x,R,S)}{\partial S_{li}}, \tag{6.17a}$$

$$\frac{\partial a_i^l(x,R,S)}{\partial R_j} = \frac{\partial c^j(x,R,S)}{\partial S_{li}}, \tag{6.17b}$$

$$\frac{\partial c^j(x,R,S)}{\partial R_l} = \frac{\partial c^l(x,R,S)}{\partial R_j} \tag{6.17c}$$

[6] See e.g. Nečas [305, Sect.3.2].

for all $i, k = 1, \dots, n$ and $j, l = 1, \dots, m$ and for a.a. $(x, R, S) \in \Omega \times \mathbb{R}^m \times \mathbb{R}^{m\times n}$. Then, as in (4.23b), φ occurring in (6.4) is given (up to a constant) by the formula

$$\varphi(x,R,S) = \int_0^1 \sum_{j=1}^{m} \Big(\sum_{i=1}^{n} a_i^j(x,tR,tS)S_{ji} + c^j(x,tR,tS)R_j \Big)\,\mathrm{d}t \tag{6.18}$$

and (6.16) coincides with the equation in (6.1) because $\partial\varphi/\partial S_{ij} = a_i^j$ and $\partial\varphi/\partial R_j = c^j$. Like (4.21)–(4.22), now (6.17) expresses just symmetry of the Jacobian of the mapping $(R,S) \mapsto (c(x,R,S), a(x,R,S)) : \mathbb{R}^m \times \mathbb{R}^{m\times n} \to \mathbb{R}^m \times \mathbb{R}^{m\times n}$. In this case, (6.16) is the *Euler-Lagrange equation* for the potential having the "density" (6.18).

Example 6.7 (*Elasticity*: large strains). Systems (6.1) with $\varphi(x,R,S) = \phi(x, \mathbb{I}+S)$ occur in steady-state elasticity where $n = m$, $u : \Omega \to \mathbb{R}^n$ means displacement of a body occupying in an undeformed state the reference domain Ω while $y(x) := x + u(x)$ defines the deformation at $x \in \Omega$. The deformed body then occupies the domain $y(\Omega) \subset \mathbb{R}^n$ and $\phi(x,F)$ expresses the specific stored energy at $x \in \Omega$ and at the deformation gradient $F = \mathbb{I} + S$. The direct method used in Proposition 6.3 expresses minimization of overall stored energy and energy contained in an elastic support on the boundary (through ψ) which is a variational principle that sometimes (but not always) governs steady states of loaded elastic bodies. The so-called frame-indifference principle requires $\phi(x,\cdot)$ in fact to depend only on the so-called (right) Cauchy-Green stretch tensor

$$C = F^\top F = (S+\mathbb{I})^\top(S+\mathbb{I}) = \mathbb{I} + S^\top + S + S^\top S. \tag{6.19}$$

The often considered potential

$$\phi(x,F) := \frac{1}{2}E^\top \mathbb{C} E, \qquad E := \frac{1}{2}(C - \mathbb{I}), \quad C = F^\top F, \tag{6.20}$$

where E is called the Green-Lagrange *strain tensor*, describes the so-called Saint Venant-Kirchhoff's material with $\mathbb{C} = [\mathbb{C}_{ijkl}]$ the positive-definite elastic-moduli tensor. This 4th-order tensor $\mathbb{C}$ has a lot of symmetries leading to only few independent entries.[7] Unfortunately, the choice (6.20) leads to $\varphi(x,R,\cdot)$ which is even not rank-one convex, however. An example of a polyconvex energy $\varphi(x,R,\cdot)$ is Mooney-Rivlin's material described by[8]

$$\phi(x,F) := c_1\mathrm{tr}(E) + c_2\mathrm{tr}\big(\mathrm{cof}(C) - \mathbb{I}\big) + \phi_0\big(\det(F)\big), \tag{6.21}$$

with C, E again from (6.20), $c_1, c_2 > 0$ and ϕ_0 a convex function; $\mathrm{tr}(\cdot)$ in (6.21) denotes the trace of a matrix. This is a special case of a so-called Ogden's material.[9]

[7] Number of independent entries of $\mathbb{C}$ in case of anisotropic crystals: 3 (cubic), 6 (tetragonal), 9 (orthorombic), 13 (monoclinic), or 21 (triclinic). Polycrystalic materials can be considered isotropic and leads to 2 independent entries only, cf. (6.23).

[8] Note that $\det(F) = \det(F^\top) = \sqrt{\det(F^\top F)} = \sqrt{\det(C)}$ actually depends only on C hence (6.21) leads indeed to a frame-indifferent potential.

[9] Ogden's material allows for more general nonlinearities, cf. e.g. Zeidler [427, Sect.61.8 and 62.14]. In this way, the coercivity in Lemma 6.2 can be satisfied.

Example 6.8 (Elasticity: *small strains*). If the displacement u is small, one can neglect the higher-order term $S^\top S$ in (6.19) so that the Green-Lagrange strain tensor E from (6.20) turns into a so-called small-strain tensor $e(u) := \frac{1}{2}\nabla u + \frac{1}{2}(\nabla u)^\top$, i.e.

$$e_{ij}(u) = \frac{1}{2}\frac{\partial u_i}{\partial x_j} + \frac{1}{2}\frac{\partial u_j}{\partial x_i}. \tag{6.22}$$

In fact, only the gradient of u is to be small rather than u itself. Then the St.Venant-Kirchhoff's potential ϕ from (6.20) with $e(u)$ substituted for E turns φ into a quadratic form of the displacement gradient ∇u. For isotropic material, it looks as

$$\varphi(x, R, S) = \varphi(S) = \mu\Big|\frac{1}{2}S + \frac{1}{2}S^\top\Big|^2 + \frac{\lambda}{2}\big(\mathrm{tr}(S)\big)^2, \tag{6.23}$$

i.e. $\varphi(\nabla u) = \mu|e(u)|^2 + \frac{1}{2}\lambda(\mathrm{div}\, u)^2$, where $\mu > 0$ and $\lambda \geq 0$ stand here for the so-called Lamé constants describing the elastic response on shear and compression, respectively. In particular, φ is then convex and (6.1) reduces to a so-called *Lamé system* of linear elasticity whose weak formulation (6.2) then results in[10]

$$\int_\Omega \sigma(e(u)){:}e(v)\,\mathrm{d}x + \int_\Gamma b(u)\cdot v\,\mathrm{d}S = \int_\Omega g\cdot v\,\mathrm{d}x + \int_\Gamma h\cdot v\,\mathrm{d}S \tag{6.24}$$

where $\sigma(e(u))$ denotes the stress tensor

$$\big[\sigma(e(u))\big]_{ij} = 2\mu e_{ij}(u) + \lambda(\mathrm{div}\, u)\delta_{ij} \tag{6.25}$$

with δ_{ij} the Kronecker symbol.

Remark 6.9 (Bibliographical notes). In fact, polyconvexity provides existence proof on $W^{1,p}(\Omega;\mathbb{R}^3)$ even for $p \geq 3$ and faster growth of φ admitting $\varphi \to +\infty$ if $\det(\mathbb{I} + \nabla u) \searrow 0$, see Ball [31]. For advanced study of deep and difficult topics around quasiconvexity the reader is referred to monographs by Dacorogna [112, Chap.IV], Evans [138, Chap.8], Giaquinta, Modica and J. Souček [177, Part II, Sect.1.4], Giusti [180, Chap.5], Morrey [291], Müller [297], and Pedregal [331, Chap.3]. Existence of a minimizer of (6.4) on $W^{1,p}(\Omega;\mathbb{R}^m)$ is due to Acerbi and Fusco [2]. For mathematical aspects of nonlinear elasticity see monographs by Ciarlet [93], Pedregal [332], or Zeidler [427, Vol.4]. Although elasticity theory has received attention throughout centuries, there are still many open fundamental problems in nonlinear elasticity especially when $\varphi(x, R, \cdot)$ has faster growth than polynomial[11], see [32]. E.g., a question about injectivity of $y : \Omega \to \mathbb{R}^n$, in particular avoiding self-contact, has been pointed out by Ciarlet and Nečas [94]. Linear elasticity is exposed e.g. in Duvaut and Lions [130] or Nečas and Hlaváček [308], and related unilateral problems in Hlaváček et al. [204].

[10]Note that $\sigma(e(u)){:}\nabla v = \sigma(e(u)){:}e(v) + \frac{1}{2}\sigma(e(u)) : (\nabla v - (\nabla v)^\top)$ and the last term vanishes because $\sigma(e(u))$ is symmetric and thus orthogonal to antisymmetric matrices.

[11]It is quite natural to assume especially $\varphi(x, R, S) \to +\infty$ when $\det(\mathbb{I} + S) \searrow 0$.

6.2 Buoyancy-driven viscous flow

It is an every-day experience that a warmer fluid in the gravity field tends to run up while a cooler fluid falls down, in special situations known as *Bénard's problem*[12]. These processes obviously involve mutually coupled velocity and temperature fields. *Oberbeck-Boussinesq's model* for (a steady-state of) this process involving incompressible viscous *non-Newtonean fluid*[13] occupying a fixed domain Ω is governed by the following system[14]:

$$(u\cdot\nabla)u - \operatorname{div}\sigma\big(e(u)\big) + \nabla\pi = g(1-\alpha\theta), \tag{6.26a}$$

$$\operatorname{div} u = 0\,, \tag{6.26b}$$

$$u\cdot\nabla\theta - \kappa\Delta\theta = 0\,, \tag{6.26c}$$

with $e(u) := \frac{1}{2}(\nabla u)^\top + \frac{1}{2}\nabla u$ as in (6.22) and where we denoted
$u : \Omega \to \mathbb{R}^n$ a velocity field,
$\pi : \Omega \to \mathbb{R}$ a pressure field,
$\theta : \Omega \to \mathbb{R}$ a temperature field,
$\kappa > 0$ the heat-conductivity coefficient,
α a coefficient of mass density variation with respect to temperature,
$\sigma = \sigma(e) =$ the viscous stress tensor,
$g =$ an external (e.g. gravity) force.
We have to specify boundary conditions. Let us consider, e.g., no-slip for u and Newton's condition for θ with some $b_1 > 0$:

$$u = 0\,, \qquad \kappa\frac{\partial\theta}{\partial\nu} + b_1\theta = h \qquad \text{on } \Gamma. \tag{6.27}$$

Let us assume that, for some $0 < c_1 \le c_3$, $0 \ge c_2$, and $q := 2^*p^*/(2^*p^* - p^* - 2^*)$:

$$\forall e \in \mathbb{R}^{n\times n}_{\rm sym}: \qquad \sigma(e){:}e \ge c_1|e|^p, \qquad |\sigma(e)| \le c_2|e|^{p-1} + c_3, \tag{6.28a}$$

$$\forall e_1, e_2 \in \mathbb{R}^{n\times n}_{\rm sym},\ e_1 \ne e_2: \qquad \big(\sigma(e_1) - \sigma(e_2)\big):(e_1 - e_2) > 0, \tag{6.28b}$$

$$g \in L^q(\Omega;\mathbb{R}^n), \qquad h \in L^{2^{\#'}}(\Gamma), \tag{6.28c}$$

where $\mathbb{R}^{n\times n}_{\rm sym}$ denotes the set of $n\times n$ symmetric matrices. An example for (6.28a)–(6.28b) is $\sigma(e) = |e|^{p-2}e$.

We will employ a fixed-point technique, which illustrates quite a typical approach to systems of equations and works here easily also for $p \le 2$ in contrast to techniques based on joint coercivity of this system, cf. Exercise 6.19 below. Let us denote

$$W^{1,p}_{0,\rm div}(\Omega;\mathbb{R}^n) := \{v \in W^{1,p}_0(\Omega;\mathbb{R}^n);\ \operatorname{div} v = 0\} \tag{6.29}$$

[12] Cf. Straughan [392, Chap.3].

[13] The adjective "non-Newtonean" refers to a non-constant viscosity; cf. Remark 6.15.

[14] This model is derived from a full compressible system on assumptions of small u, nearly constant θ, and negligible dissipative and adiabatic heat. Non-Newtonean fluids in this context have been used in Málek at al. [270]. See [392] for an extensive reference list. For a more general model see e.g. [223, 363].

and consider a mapping M from $W^{1,p}_{0,\mathrm{div}}(\Omega;\mathbb{R}^n)$ to itself, defined by

$$M := M_2\circ\big(M_1\times\mathrm{id}\big) : v\mapsto u, \qquad M_1 : v\mapsto\theta, \quad M_2 : (v,\theta)\mapsto u, \tag{6.30}$$

with u and θ being the weak solutions to

$$(v\cdot\nabla)u - \mathrm{div}\,\sigma\big(e(u)\big) + \nabla\pi = g(1-\alpha\theta), \quad \mathrm{div}\,u = 0, \qquad u|_\Gamma = 0\ , \tag{6.31a}$$

$$v\cdot\nabla\theta - \kappa\Delta\theta = 0\ , \qquad \kappa\frac{\partial\theta}{\partial\nu} + b_1\theta|_\Gamma = h\ . \tag{6.31b}$$

Note that the system (6.31) is decoupled: first, one can solve (6.31b) to get θ and then, knowing both v and θ, one can solve (6.31a). If σ is linear, then the problem (6.31a) arising via the "frozen" velocity v in the convective term is also linear and is then called the *Oseen equation.*

Lemma 6.10 (A-PRIORI ESTIMATES). *Let $p>1$, $p\ge\max(n/2, 3n/(n+2))$. There is R dependent on c_1, c_2, c_3, g and h but not on $v\in W^{1,p}_{0,\mathrm{div}}(\Omega;\mathbb{R}^n)$ such that*

$$\|\theta\|_{W^{1,2}(\Omega)}\le R \qquad \textit{and} \qquad \|u\|_{W^{1,p}(\Omega;\mathbb{R}^n)}\le R. \tag{6.32}$$

Proof. To estimate the temperature, we test (6.31b) by θ and, for $p\ge n/2$,[15] use Green's Theorem 1.31 and the identity

$$\int_\Omega (v\cdot\nabla\theta)\,\theta\,\mathrm{d}x = \frac12\int_\Omega v\cdot\nabla\theta^2\,\mathrm{d}x = -\frac12\int_\Omega(\mathrm{div}\,v)\,\theta^2\,\mathrm{d}x = 0 \tag{6.33}$$

so that, by the Poincaré inequality (1.56) considered with $p=2=q$,

$$\begin{aligned} C_{\mathrm{P}}^{-1}\min(\kappa,b_1)\|\theta\|^2_{W^{1,2}(\Omega)} &\le \int_\Omega(v\cdot\nabla\theta)\,\theta\,\mathrm{d}x + \kappa\int_\Omega|\nabla\theta|^2\,\mathrm{d}x + b_1\int_\Gamma\theta^2\,\mathrm{d}S \\ = \int_\Gamma h\theta\,\mathrm{d}S &\le \|h\|_{L^{2^{\#\prime}}(\Gamma)}\|\theta\|_{L^{2^\#}(\Gamma)} \le \frac{N^2}{4\varepsilon}\|h\|^2_{L^{2^{\#\prime}}(\Gamma)} + \varepsilon\|\theta\|^2_{W^{1,2}(\Omega)}\,, \end{aligned} \tag{6.34}$$

where C_{P} comes from (1.56) and N is the norm of the trace operator $W^{1,2}(\Omega)\to L^{2^\#}(\Gamma)$. For $\varepsilon>0$ small enough, it gives $\|\theta\|_{W^{1,2}(\Omega)}\le R$ with R independent of v.

To estimate the velocity, we test (6.31a) by u and use, for $p\ge 3n/(n+2)$, Green's Theorem 1.31 and the identities $\int_\Omega\nabla\pi\cdot u\,\mathrm{d}x = -\int_\Omega\pi\mathrm{div}\,u\,\mathrm{d}x = 0$ and[16]

$$\begin{aligned} \int_\Omega\big((v\cdot\nabla)u\big)\cdot u\,\mathrm{d}x &= \int_\Omega\sum_{k=1}^n\sum_{j=1}^n v_k\frac{\partial u_j}{\partial x_k}u_j\,\mathrm{d}x \\ = -\int_\Omega\Big(\sum_{k=1}^n\sum_{j=1}^n u_j\frac{\partial v_k}{\partial x_k}u_j + u_jv_k\frac{\partial u_j}{\partial x_k}\Big)\,\mathrm{d}x &= -\int_\Omega\big((v\cdot\nabla)u\big)\cdot u\,\mathrm{d}x \end{aligned} \tag{6.35}$$

[15]To ensure integrability of $(v\cdot\nabla\theta)\,\theta$ in (6.33) when $\theta\in L^{2^*}(\Omega)$ and $\nabla\theta\in L^2(\Omega;\mathbb{R}^n)$, one needs $v\in L^n(\Omega;\mathbb{R}^n)$, i.e. one needs $p^*\ge n$, which is just equivalent to $p\ge n/2$.

[16]Note that $\sum_{k=1}^n\partial v_k/\partial x_k = \mathrm{div}\,v = 0$.

so that

$$\int_\Omega ((v\cdot\nabla)u)\cdot u\,\mathrm{d}x = 0 \tag{6.36}$$

and, by Korn's inequality (1.59) and by (6.28a),

$$\begin{aligned} c_1 C_{\mathrm{K}}^{-p}\|u\|^p_{W_0^{1,p}(\Omega;\mathbb{R}^n)} &\le c_1\|e(u)\|^p_{L^p(\Omega;\mathbb{R}^{n\times n})} \le \int_\Omega \sigma(e(u))\!:\!e(u)\,\mathrm{d}x \\ &= \int_\Omega ((v\!\cdot\!\nabla)u)\cdot u + \sigma(e(u))\!:\!e(u) + \nabla\pi\!\cdot\! u\,\mathrm{d}x = \int_\Omega g(1-\alpha\theta)\cdot u\,\mathrm{d}x \\ &\le \|g\|_{L^q(\Omega;\mathbb{R}^n)}\big(\mathrm{meas}_n(\Omega)^{1/2^*} + \alpha\|\theta\|_{L^{2^*}(\Omega)}\big)\|u\|_{L^{p^*}(\Omega;\mathbb{R}^n)}. \end{aligned} \tag{6.37}$$

From this, the second estimate in (6.32) follows by Young's inequality and the already obtained estimate of θ. □

Lemma 6.11 (Uniqueness and continuity[17]). *Let $p > 3n/(n+2)$. Given v, the solution (u,θ) to* (6.31) *is unique. Besides, $M : v \mapsto u$ is weakly continuous.*

Proof. Uniqueness of temperature θ follows from the a-priori estimate (6.34) because (6.31b) is linear in terms of θ. The weak continuity of M_1 is obvious when one realizes that $v^k\!\cdot\!\nabla\theta_k \rightharpoonup v\!\cdot\!\nabla\theta$ weakly in $L^1(\Omega)$ because $v^k \rightharpoonup v$ weakly in $W^{1,p}(\Omega;\mathbb{R}^n)$ (hence strongly in $L^{p^*-\epsilon}(\Omega;\mathbb{R}^n)$) and $\nabla\theta_k \rightharpoonup \nabla\theta$ weakly in $L^2(\Omega;\mathbb{R}^{n\times n})$.

For the uniqueness of the velocity, we take u^1, u^2 two weak solutions of (6.31a), and test the difference of the weak formulation of (6.31a) by $u^{12} := u^1 - u^2$. Using (6.36) for u^{12} instead of u, it gives:

$$\int_\Omega \big(\sigma(e(u^1)) - \sigma(e(u^2))\big)\!:\!e(u^{12})\,\mathrm{d}x = -\int_\Omega ((v\cdot\nabla)u^{12})\cdot u^{12}\,\mathrm{d}x = 0$$

so that, by strict monotonicity (6.28b), it holds that $e(u^{12}) = 0$ a.e. in Ω and then, by Korn's inequality (1.59), $u^{12} = 0$.

To show the weak continuity of $M_2 : (v,\theta) \mapsto u$, one can use monotonicity of σ and consider a weakly converging sequence $(v^k,\theta_k) \rightharpoonup (v,\theta)$ and corresponding solutions u^k:

$$\begin{aligned} 0 &\le \int_\Omega \big(\sigma(e(u^k)) - \sigma(e(w))\big)\!:\!e(u^k - w)\,\mathrm{d}x \\ &= \int_\Omega \big(g(1-\alpha\theta_k) - (v^k\cdot\nabla)u^k\big)(u^k - w) - \sigma(e(w))\!:\!e(u^k - w)\,\mathrm{d}x \\ &\to \int_\Omega \big(g(1-\alpha\theta) - (v\cdot\nabla)u\big)(u - w) - \sigma(e(w))\!:\!e(u - w)\,\mathrm{d}x \end{aligned} \tag{6.38}$$

and then use Minty's trick. Note that (6.38) used the compact embedding $W_0^{1,p}(\Omega;\mathbb{R}^n) \Subset L^{(p^*-\epsilon)}(\Omega;\mathbb{R}^n)$ which allowed for the limit passage in the term $\int_\Omega (v^k\cdot\nabla u^k)u^k\mathrm{d}x$ if $p^{-1} + 2(p^*-\epsilon)^{-1} \le 1$ which requires $p > 3n/(n+2)$. □

[17] A more involved technique allows for improving existence for non-Newtonean fluids (6.26a,b) itself even for $p > 2n/(n+2)$, cf. [122, 151].

Proposition 6.12 (EXISTENCE). *Let* (6.28) *hold. Then the system* (6.26) *has at least one weak solution.*

Proof. It follows from Schauder's fixed-point Theorem 1.9 (cf. Exercise 2.55) for M on the ball in $B := \{v \in W^{1,p}_{0,\mathrm{div}}(\Omega;\mathbb{R}^n);\ \|v\|_{W^{1,p}(\Omega;\mathbb{R}^n)} \le R\}$ with a sufficiently large radius R from (6.32) depending only on the data g, h, σ, and α, endowed by the weak topology which makes it compact. □

An alternative to the no-slip (i.e. $u = 0$) boundary conditions (6.27) is a partial-slip condition (with $0 \le \gamma_1 \le \gamma$ phenomenological coefficients) which, in our thermally coupled system, reads as:

$$u_\mathrm{n} = 0, \qquad \vec{\sigma}_\mathrm{t} + \gamma u_\mathrm{t} = 0, \tag{6.39a}$$

$$\kappa\frac{\partial\theta}{\partial\nu} + b_1\theta = h + \gamma_1\big|u_\mathrm{t}\big|^2, \tag{6.39b}$$

where $u_\mathrm{t} := u - u_\mathrm{n}$ is the tangential velocity and $u_\mathrm{n} := (u\cdot\nu)\,\nu$ is the normal velocity, and similarly for the so-called traction force $\vec{\sigma}$ defined as $[\vec{\sigma}]_i = \sum_{j=1}^n \sigma_{ij}(e(u))\nu_j$. The two conditions in (6.39a) form the so-called *Navier boundary condition.*[18] For $\gamma = 0$, it expresses a no-stick (or ideally slippery) boundary while for $\gamma \to +\infty$ it approximates the no-slip boundary. Instead of $W^{1,p}_{0,\mathrm{div}}(\Omega;\mathbb{R}^n)$ defined in (6.29), the weak formulation now uses the linear space $\{v \in W^{1,p}(\Omega;\mathbb{R}^n);\ \mathrm{div}\, v = 0 \text{ in } \Omega,\ \nu\cdot v = 0 \text{ on } \Gamma\}$ and leads to a weak formulation of (6.26a,b) as the integral identity

$$\int_\Omega ((u\cdot\nabla)\,u)\cdot z + \sigma(e(u)):e(z)\,\mathrm{d}x + \int_\Gamma \gamma u_\mathrm{t}\cdot z_\mathrm{t}\,\mathrm{d}S = \int_\Omega g(1-\alpha\theta)\,\mathrm{d}x \tag{6.40}$$

In contrast to the no-slip or the no-stick cases, for $0 < \gamma < +\infty$ the condition (6.39a) dissipates energy which may partly (namely with the ratio $\gamma_1/\gamma \in [0,1]$) contribute to a heat production on the boundary. This just gives rise to the last term in (6.39b). If $\gamma_1 < \gamma$, the resting portion $1-\gamma_1/\gamma$ of the mechanical dissipated energy is indeed lost from the system in this model.

For $\gamma_1 > 0$, we have an additional coupling, which might be not entirely easy to treat and which may even cause doubt about existence of solutions if the fluid is not dissipative enough (i.e. here if $p \le 3$, which is related to the quadratic growth of the boundary term $\gamma_1|u_\mathrm{t}|^2$) and simultaneously if the external supply of energy is not small. Let us confine ourselves to the gravity-type force g such that $\mathrm{curl}\, g = 0$, where the *rotation* is defined as $\mathrm{curl}\,\cdot := \nabla\times\cdot$ where $\times : \mathbb{R}^3\times\mathbb{R}^3 \to \mathbb{R}^3$ is the vector product on $\mathbb{R}^3$.

Proposition 6.13 (EXISTENCE WITH NAVIER'S BOUNDARY CONDITIONS). *Let* (6.28) *hold with* $p \ge 4n/(n+4)$ *and let one of the following two conditions holds*

[18] The Navier's conditions (6.39a) are "mathematically" very natural in comparison with mere Dirichlet condition $u = 0$, as pointed out by Frehse and Málek [150].

(i) $p>3$ *(i.e. the fluid is "dissipative enough"), or*

(ii) $1<p<3$ *and the heat flux h be "small enough" in $L^{2^{\#'}}(\Gamma)$.*

Then the system (6.26) with the boundary conditions (6.39) has weak solutions.

Proof. Again, use the Schauder fixed-point theorem based, instead of (6.31) on two decoupled problems:

$$\left.\begin{array}{lll}(v{\cdot}\nabla)u-\operatorname{div}\sigma\big(e(u)\big)+\nabla\pi=g(1{-}\alpha\theta), & \operatorname{div}u=0 & \text{on }\Omega,\\ u_{\rm n}=0 \qquad\text{and}\qquad \big[\sigma(e(u))\nu\big]_{\rm t}{+}\gamma u_{\rm t}=0 & & \text{on }\Gamma,\end{array}\right\}\tag{6.41a}$$

$$\left.\begin{array}{ll}v{\cdot}\nabla\theta-\kappa\Delta\theta=0 & \text{on }\Omega,\\ \kappa\frac{\partial\theta}{\partial\nu}+b_1\theta|_\Gamma=h+\gamma_1\big|v_{\rm t}\big|^2 & \text{on }\Gamma.\end{array}\right\}\tag{6.41b}$$

Note that, for $v\in W^{1,p}(\Omega;\mathbb{R}^n)$, the additional boundary heat source $\gamma_1|v_{\rm t}|^2$ belongs to $L^{p^\#/2}(\Gamma)\subset L^{2^{\#'}}(\Gamma)$ provided $p^\#/2\geq 2^{\#'}$, which is fulfilled only if $p\geq 4n/(n{+}4)$, so that (6.41b) possesses a conventional weak solution $\theta\in W^{1,2}(\Omega)$ which is unique and depends continuously on v. Also, $\operatorname{curl}g=0$ means existence of a potential φ so that $g=\nabla\varphi$, and then $\int_\Omega g{\cdot}u\,{\rm d}x=\int_\Omega\nabla\varphi{\cdot}u\,{\rm d}x=\int_\Gamma\varphi u{\cdot}\nu\,{\rm d}S-\int_\Omega\varphi\operatorname{div}u\,{\rm d}x=0$. From (6.41a) tested by u, one gets

$$\|u\|_{W^{1,p}(\Omega;\mathbb{R}^n)}^{p-1}\leq C_1\|\theta\|_{W^{1,2}(\Omega)}\tag{6.42}$$

with C_1 depending on c_1 from (6.28a) and on g and α. The estimate (6.34) is now modified by using $\int_\Gamma(h{+}\gamma_1|v_{\rm t}|^2)\theta\,{\rm d}S\leq 2(\|h\|_{L^{2^{\#'}}(\Gamma)}{+}\gamma_1\|v\|_{L^{2^{\#'}2}(\Gamma)}^2)\|\theta\|_{L^{2^\#}(\Gamma)}$ so that it gives

$$\|\theta\|_{W^{1,2}(\Omega)}\leq C_0\big(\|h\|_{L^{2^{\#'}}(\Gamma)}^2+\|v\|_{W^{1,p}(\Omega;\mathbb{R}^n)}^2\big)\tag{6.43}$$

for some C_0 depending also on the norms of the trace operators $W^{1,p}(\Omega)\to L^{2^{\#'}2}(\Gamma)$ and $W^{1,2}(\Omega)\to L^{2^\#}(\Gamma)$. For $p>3$, we can combine (6.42) and (6.43) to obtain

$$\|u\|_{W^{1,p}(\Omega;\mathbb{R}^n)}^{p-1}\leq C_0C_1\big(\|h\|_{L^{2^{\#'}}(\Gamma)}^2+\|v\|_{W^{1,p}(\Omega;\mathbb{R}^n)}^2\big)\leq C_p+\frac{1}{2}\|v\|_{W^{1,p}(\Omega;\mathbb{R}^n)}^{p-1}\tag{6.44}$$

with some C_p depending on p, C_0C_1, and h. This suggests that we choose v from the ball of the radius $(2C_p)^{1/(p-1)}$ to obtain the property that M from (6.30) maps it into itself. Thus (i) has been proved.

As to (ii), we can exploit the first inequality in (6.44) to see that, for sufficiently small $\|h\|_{L^{2^{\#'}}(\Gamma)}$, we can again find r such that $\|v\|_{W^{1,p}(\Omega;\mathbb{R}^n)}\leq r$ will imply $\|u\|_{W^{1,p}(\Omega;\mathbb{R}^n)}\leq r$. □

Remark 6.14 (*Coupling through dissipative/adiabatic heat effects*). The Oberbeck-Boussinesq model (6.26) is only one of various options and there are various modifications appearing in literature.[19] Typically, some enhancement of the heat equation (6.26c) as

$$u\cdot\nabla\theta - \kappa\Delta\theta = \alpha_1\sigma(e(u)):e(u) - \alpha_2\theta u \tag{6.45}$$

is considered with some phenomenological coefficients α_1 and α_2; usually $0 \le \alpha_1 \le 1$ and $0 \le \alpha_2 \le \alpha$ is considered. This makes another nonlinear coupling which violates coercivity of the whole system, so that one can expect existence of a solution possibly for small data only. Indeed, assuming $p > n$, $n = 2$ or 3, a smooth domain Ω, and $b_1(x) \ge b_0 > 0$ with $b_1 \in W^{1,2^\#2/(2^\#-2)}(\Gamma)$ and $c_3 = 0$ in (6.28a), one can use the fixed point of the mapping $(v,\vartheta) \mapsto (u,\theta)$ determined by

$$\left.\begin{array}{ll}(v\cdot\nabla)u - \operatorname{div}\sigma\big(e(u)\big) + \nabla\pi = g(1-\alpha\vartheta), \quad \operatorname{div} u = 0 & \text{on } \Omega,\\ u = 0 & \text{on } \Gamma,\end{array}\right\} \tag{6.46a}$$

$$\left.\begin{array}{ll}v\cdot\nabla\theta - \kappa\Delta\theta = \alpha_1\sigma(e(v)):e(v) - \alpha_2\vartheta v & \text{on } \Omega,\\ \kappa\frac{\partial\theta}{\partial\nu} + b_1\theta|_\Gamma = h & \text{on } \Gamma.\end{array}\right\} \tag{6.46b}$$

We need to use L^1-theory for the heat equation from Chapter 3. Namely, we use the interpolation for (3.47) with $\mathbb{A} := \kappa\mathbb{I}$, and $g := \alpha_1\sigma(e(v)):e(v) - \alpha_2\vartheta v$ combined with a bootstrap argument of the advective term $v\cdot\nabla\theta$ with $v \in W^{1,p}_{0,\mathrm{div}}(\Omega;\mathbb{R}^n)$ to obtain the estimate[20]

$$\|\theta\|_{L^q(\Omega)} \le K\Big(1+\|v\|^r_{W^{1,p}(\Omega;\mathbb{R}^n)}\Big)\Big(\big\|\alpha_1\sigma(e(v)):e(v)-\alpha_2\vartheta v\big\|_{L^1(\Omega)} + \big\|h\big\|_{L^1(\Gamma)}\Big) \tag{6.47}$$

with $q < n/(n-2)$ and $r > n/2-1$ for a distributional solution θ to the boundary-value problem (6.46b). Combining (6.47) with the obvious estimate $\|u\|^{p-1}_{W^{1,p}(\Omega;\mathbb{R}^n)} \le C\|\theta\|_{L^q(\Omega)}$ and assuming (6.28a) with $c_3 = 0$, one obtains an estimate of the type

$$\begin{aligned}\|u\|^{p-1}_{W^{1,p}(\Omega;\mathbb{R}^n)} &\le \alpha_1 C\|v\|^p_{W^{1,p}(\Omega;\mathbb{R}^n)} + \alpha_1 C\|v\|^{p+r}_{W^{1,p}(\Omega;\mathbb{R}^n)}\\ &+ \alpha_2\|v\|_{W^{1,p}(\Omega;\mathbb{R}^n)}\Big(1+\|v\|^r_{W^{1,p}(\Omega;\mathbb{R}^n)}\Big)\|\vartheta\|_{L^1(\Gamma)} + \|v\|^r_{W^{1,p}(\Omega;\mathbb{R}^n)}\|h\|_{L^q(\Omega)}.\end{aligned} \tag{6.48}$$

[19] Cf. e.g. [223, 343] for a genesis of various possibilities. The starting point is always the complete compressible fluid system of $n{+}2$ conservation laws for mass, impulse, and energy. Then, the so-called incompressible limit represents a small perturbation around a stationary homogeneous state, i.e. around constant mass density, constant temperature, and zero velocity.

[20] The mapping B used in the proof of Lemma 3.29 allows for the estimates $\|B\|_{\mathscr{L}(W^{1,2}(\Omega)^*,W^{1,2}(\Omega))} \le C$ and $\|B\|_{\mathscr{L}(L^2(\Omega)^*,W^{2,2}(\Omega))} \le C(1+\|v\|_{L^\infty(\Omega;\mathbb{R}^n)})$ when realizing that $v \in L^\infty(\Omega;\mathbb{R}^n)$ and $v\cdot\nabla\theta \in L^2(\Omega)$. Thus, by interpolation, also $\|B\|_{\mathscr{L}(W^{\lambda,2}(\Omega)^*,W^{2-\lambda,2}(\Omega))} \le C(1+\|v\|_{L^\infty(\Omega;\mathbb{R}^n)})^{1-\lambda}$, which yields (6.47) by transposition used for $\lambda < 2-n/2$ as in the proof of Lemma 3.29, together with the embedding $W^{\lambda,2}(\Omega) \subset L^{2n/(n-2\lambda)}(\Omega)$.

Then, if the external heating h is small in $L^1(\Gamma)$-norm, one can find a sufficiently small ball in $W^{1,p}_{0,\mathrm{div}}(\Omega;\mathbb{R}^n)\times L^q(\Gamma)$ mapped to itself by the mapping $(v,\vartheta)\mapsto(u,\theta)$ with u being the weak solution to (6.46a) and θ being the distributional solution to (6.46b). For such sort of existence analysis see also [299, 363] even for $p\le n$.

Remark 6.15 (*Navier-Stokes equations*). For $\sigma(e)=2\mu e$, (6.26a,b) turns (when neglecting buoyancy, i.e. $\alpha=0$, and thus also temperature variation) into the system[21]

$$(u\cdot\nabla)u-\mu\Delta u+\nabla\pi=g,\qquad \operatorname{div}u=0, \tag{6.49}$$

which is called a steady-state Navier-Stokes system.[22] The coefficient μ is called a kinematic viscosity coefficient and fluids exhibiting such constant viscosity are called *Newtonean fluids.* The no-stick boundary conditions (6.39a) have now an alternative (though not fully equivalent) option: $u{\cdot}\nu=0$, $(\operatorname{curl}u){\cdot}\nu=0$, and $(\operatorname{curl}^2u){\cdot}\nu=0$ on Γ, see [42]. For other conditions we refer also to [105].

Exercise 6.16. Write a weak formulation for (6.31), use divergence-free test functions.[23] Similarly, derive the weak formulation (6.40) of the Navier boundary value problem (6.26a,b)–(6.39a).[24]

Exercise 6.17. Prove Proposition 6.12 when assuming only mere monotonicity of σ instead of the strict monotonicity (6.28b).[25]

Exercise 6.18 (Temperature-dependent viscosity[26]). Modify the proofs of Lemmas 6.10 and 6.11 if $\sigma=\sigma(\theta,e)$ in (6.26a) and (6.31a), assuming continuity of σ and the properties (6.28a,b) holding for $\sigma(\theta,\cdot)$ uniformly for θ.

---

[21] Indeed, taking into account $\operatorname{div}u=\sum_{i=1}^n\partial u_i/\partial x_i=0$ and $\sigma(e(u))=2\mu e(u)$, one gets

$$\operatorname{div}2\mu e(u)=2\mu\sum_{i=1}^n\frac{\partial}{\partial x_i}\Big(\frac12\frac{\partial u_i}{\partial x_j}+\frac12\frac{\partial u_j}{\partial x_i}\Big)=\mu\Big(\frac{\partial}{\partial x_j}\Big(\sum_{i=1}^n\frac{\partial u_i}{\partial x_i}\Big)+\sum_{i=1}^n\frac{\partial^2u_j}{\partial x_i\partial x_i}\Big)=\mu\Delta u.$$

[22] Cf. Constantin and Foias [106], Galdi [170], Sohr [391], or Temam [403] for a thorough treatment.

[23] Hint: Realize that, by Green's Theorem 1.31, $\int_\Omega\nabla\pi\cdot z\,\mathrm{d}x=\int_\Omega-\pi\operatorname{div}v\,\mathrm{d}x=0$ and that symmetric and antisymmetric matrices are mutually orthogonal so that σ symmetric implies

$$\int_\Omega-\operatorname{div}(\sigma(e(u))){\cdot}z\,\mathrm{d}x=\int_\Omega\sigma(e(u)){:}\nabla z\,\mathrm{d}x=\int_\Omega\sigma(e(u)){:}e(z)\,\mathrm{d}x.$$

cf. also (6.40). As to (6.31b), the advective term leads either to $\int_\Omega v{\cdot}\nabla\theta z\,\mathrm{d}x$ or to $\int_\Omega-\theta v{\cdot}\nabla z\,\mathrm{d}x$.

[24] Hint: After applying Green's formula, treat the resulting boundary term as

$$\int_\Gamma\sigma(e(u)){:}(z\otimes\nu)\,\mathrm{d}S=\int_\Gamma\vec\sigma{\cdot}z\,\mathrm{d}S=\int_\Gamma(\vec\sigma_{\rm n}+\vec\sigma_{\rm t}){\cdot}(z_{\rm n}+z_{\rm t})\,\mathrm{d}S=\int_\Gamma\vec\sigma_{\rm n}{\cdot}z_{\rm n}+\vec\sigma_{\rm t}{\cdot}z_{\rm t}\,\mathrm{d}S=-\int_\Gamma\gamma u_{\rm t}{\cdot}z_{\rm t}\,\mathrm{d}S$$

when realizing that normal and tangential vectors are mutually orthogonal, and when using $z_{\rm n}=0$ and the latter condition in (6.39a).

[25] Hint: Prove that the mapping M_2 defined by (6.30), which is now set-valued, has convex values (cf. Theorem 2.14(i)) and is upper semi-continuous, and then use Kakutani's fixed-point Theorem 1.11 instead of Schauder's.

[26] For a non-Newtonean model with viscosity dependent also on temperature we refer to e.g. [270]. The Newtonean case was treated in [300] even for the coupled system as in Remark 6.14.

Exercise 6.19. Considering the boundary-value problem (6.26)–(6.27) and $p>2$, verify the assumptions of Brézis theorem 2.6 for $V=W^{1,p}_{0,\mathrm{div}}(\Omega;\mathbb{R}^n)\times W^{1,2}(\Omega)$.[27] Prove existence of u and of θ solving (6.31) by the Galerkin method. Considering $\sigma(e)=|e|^{p-1}e$, show strong convergence of this Galerkin approximation.[28] Do the same for the problem (6.26)–(6.39) for $p>3$ by considering $V=\{(v,\theta)\in W^{1,p}(\Omega;\mathbb{R}^n)\times W^{1,2}(\Omega);\ \mathrm{div}\,v=0 \text{ in } \Omega,\ \nu{\cdot}v=0 \text{ on } \Gamma\}$.[29]

Exercise 6.20. Uniqueness of the obtained solution does not hold in general by natural reasons. Nevertheless, prove uniqueness of the weak solution to (6.26)–(6.27) for g and h small enough provided σ is strongly monotone, i.e. $(\sigma(e_1){-}\sigma(e_2)){:}(e_1{-}e_2)\ge c_3|e_1{-}e_2|^2$ for some $c_3>0$, and provided also $n\le 4$ and $q\ge n/2$ for (6.28c).[30]

Exercise 6.21 (*Nonlinear heat* transfer). Instead of (6.26c), consider the nonlinear heat transfer $\mathfrak{c}(\theta)u{\cdot}\nabla\theta-\mathrm{div}(\kappa(\theta)\nabla\theta)=0$ with $\mathfrak{c}$ and κ denoting the heat capacity and heat-transfer coefficient depending possibly on temperature, cf. also (2.136). Then apply the *enthalpy transformation* to replace it by $u{\cdot}\nabla w-\Delta\beta(w)=0$, cf. Remark 3.25, and, denoting still by γ the inverse of the primitive function to $\mathfrak{c}$, write the resulting system as

$$(u{\cdot}\nabla)u-\mathrm{div}\,\sigma\big(e(u)\big)+\nabla\pi=g(1-\alpha\gamma(w)),\tag{6.50a}$$

$$\mathrm{div}\,u=0\,,\tag{6.50b}$$

$$u{\cdot}\nabla w-\Delta\beta(w)=0\,.\tag{6.50c}$$

[27] Hint: Show the coercivity of the underlying operator on V by testing (6.31a) and (6.31b) respectively by u and θ, imitating and merging the estimates (6.34) and (6.38). Realize that, if the assumption $p>2$ were not hold, there would be difficulties with estimating the term $\int_\Omega g\alpha\theta u\,\mathrm{d}x$.

[28] Hint: Employ d-monotonicity of the highest-order parts of (6.31) together with uniform convexity of the underlying space V.

[29] Hint: Imitate and merge the estimates (6.42) and (6.43).

[30] Hint: Consider two solutions (u^1,θ_1) and (u^2,θ_2), test the difference of weak formulations of (6.26a,b) by $u^{12}:=u^1-u^2$ and of (6.26c) by $\theta_{12}:=\theta_1-\theta_2$, sum them up, realize that small data imply both $\|u^2\|_{W^{1,p}(\Omega;\mathbb{R}^n)}$ and $\|\theta_2\|_{W^{1,2}(\Omega)}$ small, and estimate

$$\begin{aligned}
&\int_\Omega \kappa|\theta_{12}|^2+c_3|e(u^{12})|^2\,\mathrm{d}x+\int_\Gamma b_1|\theta_{12}|^2\,\mathrm{d}S\\
&\quad\le\int_\Omega g\alpha\theta_{12}{\cdot}u^{12}-\big((u^1{\cdot}\nabla)u^1{-}(u^2{\cdot}\nabla)u^2\big)u^{12}-\big(u^1{\cdot}\nabla\theta_1{-}u^2{\cdot}\nabla\theta_2\big)\theta_{12}\,\mathrm{d}x\\
&\quad=\int_\Omega g\alpha\theta_{12}{\cdot}u^{12}-\big((u^1{\cdot}\nabla)u^{12}{+}(u^{12}{\cdot}\nabla)u^2\big)u^{12}-\big(u^1{\cdot}\nabla\theta_{12}{+}u^{12}{\cdot}\nabla\theta_2\big)\theta_{12}\,\mathrm{d}x\\
&\quad=\int_\Omega g\alpha\theta_{12}{\cdot}u^{12}-\big((u^{12}{\cdot}\nabla)u^2\big)u^{12}-\big(u^{12}{\cdot}\nabla\theta_2\big)\theta_{12}\,\mathrm{d}x\\
&\quad\le\alpha\|g\|_{L^q(\Omega;\mathbb{R}^n)}\|\theta_{12}\|_{L^{2^*}(\Omega)}\|u^{12}\|_{L^{2^*}(\Omega;\mathbb{R}^n)}+C_1\|u^{12}\|^2_{L^{2^*}(\Omega;\mathbb{R}^n)}\|\nabla u^2\|_{L^p(\Omega;\mathbb{R}^{n\times n})}\\
&\qquad\qquad+C_2\|\nabla\theta_2\|_{L^2(\Omega;\mathbb{R}^n)}\|u^{12}\|_{L^{2^*}(\Omega;\mathbb{R}^n)}\|\theta_{12}\|_{L^{2^*}(\Omega)}
\end{aligned}$$

with C_1,C_2 depending on p and n, and finally by Young, Poincaré and Korn inequalities conclude that $\theta_{12}=0$ and $u^{12}=0$. The condition $n\le4$ is needed for Hölder's inequality leading to C_2.

Considering again the no-slip boundary conditions (6.27), now in the form $u = 0$ and $\frac{\partial}{\partial\nu}\beta(w)+b_1\gamma(w) = h$, design suitably a mapping whose fixed point does exist and is a weak solution to the system (6.50).[31]

6.3 Reaction-diffusion system

Let us investigate the so-called steady-state *Lotka-Volterra system:*[32]

$$\left.\begin{aligned} &-d_1\Delta u = u(a_1 - b_1 u - c_1 v) + g_1 && \text{in } \Omega,\\ &-d_2\Delta v = v(a_2 - b_2 v - c_2 u) + g_2 && \text{in } \Omega,\\ &u|_\Gamma = 0, \quad v|_\Gamma = 0 && \text{on } \Gamma. \end{aligned}\right\} \tag{6.51}$$

This system has applications in ecology:

$u, v \ge 0$ are the unknown concentrations of two species,
a_1, a_2 are the birth (or, if a's are negative, death) rates,
$b_1, b_2 > 0$ are related to the carrying capacities of the environment,
c_1, c_2 are the interaction rates,
$d_1, d_2 > 0$ are diffusion coefficients related to migrations,
$g_1, g_2 \ge 0$ are the outer supply rates.

If both c_1 and c_2 are positive, (6.51) describes a *competition-in-ecology* type model while, if both c_1 and c_2 are negative, (6.51) refers to a *cooperation-in-ecology* type model. Eventually, if $c_1 > 0$ and $c_2 < 0$, we get a *predator/prey model*; u is then the prey species concentration while v is the predator concentration.

The peculiarity of this system is its non-coercivity similarly like in the case of the systems in the previous section 6.2 for $p < 2$. Thus, again we will employ the fixed-point technigue, specifically here by involving the mapping $(\bar u, \bar v) \mapsto (u, v)$ where $(u, v) \in W^{1,2}(\Omega)^2$ is the weak solution to the following two equations:

$$\left.\begin{aligned} &-d_1\Delta u = u(a_1 - b_1 u^+ - c_1 \bar v) + g_1 && \text{in } \Omega,\\ &-d_2\Delta v = v(a_2 - b_2 v^+ - c_2 \bar u) + g_2 && \text{in } \Omega,\\ &u|_\Gamma = 0, \quad v|_\Gamma = 0 && \text{on } \Gamma. \end{aligned}\right\} \tag{6.52}$$

Existence of a weak solution u and v to these de-coupled equations can be shown, e.g., by a direct method, cf. Proposition 4.16, under assumptions made below. Let us agree to consider $\|v\|_{W_0^{1,2}(\Omega)} := \|\nabla v\|_{L^2(\Omega;\mathbb{R}^n)}$.

[31] Hint: Consider the mapping $(v, \omega) \mapsto (u, w)$ with (u, w) solving the system

$$(v\cdot\nabla)u - \operatorname{div}\sigma\big(e(u)\big) + \nabla\pi = g(1 - \alpha\gamma(w)), \qquad \operatorname{div} u = 0, \qquad v\cdot\nabla w - \operatorname{div}\big(\beta'(\omega)\nabla w\big) = 0\,.$$

Realize that, for any (v, ω), this system is decoupled and partly linearized, and (u, w) is indeed determined uniquely. Then use Schauder's fixed point theorem.

[32] Original studies of oscillation in biological or ecological systems (not necessarily in the presence of diffusion) originated in Lotka [264] and Volterra [423] and later received intensive scrutiny, cf. e.g. Pao [324, Sect.12.4–6].

Lemma 6.22 (NON-NEGATIVITY OF u AND v). *Let $\bar u,\bar v \ge 0$ a.e., and $a_1 < d_1N^{-2}+c_1^-\operatorname{ess\,sup}_{x\in\Omega}\bar v(x)$ and $a_2 < d_2N^{-2}+c_2^-\operatorname{ess\,sup}_{x\in\Omega}\bar u(x)$ with N the norm of the embedding $W_0^{1,2}(\Omega)\subset L^2(\Omega)$. Then $u,v\ge 0$ a.e. in Ω.*

Proof. Let us first consider $c_1 \ge 0$ and test the first equation in (6.52) by u^- and notice that $b_1u^+u^-=0$, which gives

$$d_1\|u^-\|^2_{W_0^{1,2}(\Omega)} \le \int_\Omega d_1|\nabla u^-|^2 + c_1\bar v(u^-)^2 - g_1u^-\,\mathrm{d}x = a_1\|u^-\|^2_{L^2(\Omega)}. \tag{6.53}$$

If $N^2a_1 < d_1$, we can absorb the last term in the left-hand side, which then immediately gives $u^-=0$ a.e., so $u\ge 0$ a.e. in Ω. For $c_1<0$, we must estimate

$$\begin{aligned} d_1\|u^-\|^2_{W_0^{1,2}(\Omega)} &\le \int_\Omega d_1|\nabla u^-|^2 - g_1u^-\,\mathrm{d}x = \int_\Omega a_1(u^-)^2 \\ &\quad - c_1\bar v(u^-)^2\,\mathrm{d}x \le \big(a_1+|c_1|\,\|\bar v\|_{L^\infty(\Omega)}\big)\|u^-\|^2_{L^2(\Omega)}. \end{aligned} \tag{6.54}$$

Analogous considerations work to show $v\ge 0$. □

Lemma 6.23 (UPPER BOUNDS). *Let, in addition to the assumptions in Lemma 6.22, also $g_1,g_2\in L^\infty(\Omega)$, $b_1>-c_1^-$ and $b_2>-c_2^-$. Then there is a constant K sufficiently large such that $\bar u,\bar v\in[0,K]$ a.e. in Ω implies $u,v\le K$ a.e. in Ω.*

Proof. We test the first equation in (6.52) by $(u-K)^+$ and use (1.50), which gives

$$\begin{aligned} d_1\|(u-K)^+\|^2_{W_0^{1,2}(\Omega)} &\le \int_\Omega d_1|\nabla(u-K)^+|^2 \\ &\quad + \big(b_1u^2-a_1u-g_1+c_1\bar v u\big)(u-K)^+\,\mathrm{d}x = 0. \end{aligned} \tag{6.55}$$

We take K so large that $r\mapsto b_1r^2-a_1r-g_1(x)+c_1\bar v(x)r$ is non-negative on $[K,+\infty)$, namely $b_1K^2-a_1K-\operatorname{ess\,sup}_{x\in\Omega}g_1(x)+c_1^-K^2\ge 0$ and $2b_1K-a_1+2c_1^-K\ge 0$. Such K does exist whenever $b_1+c_1^->0$. Then (6.55) yields $u\le K$ a.e. in Ω. Analogous considerations are for v. □

Lemma 6.24 (CONTINUITY OF $(\bar u,\bar v)\mapsto(u,v)$). *Let*

$$a_1<d_1N^{-2}+c_1^-K,\quad a_2<d_2N^{-2}+c_2^-K,\quad b_1>-c_1^-,\quad b_2>-c_2^- \tag{6.56}$$

with K so large that Lemma 6.23 is in effect. Then the solution (u,v) to (6.52) is unique and the mapping $(\bar u,\bar v)\mapsto(u,v)$ is weakly continuous as $L^2(\Omega)^2\to W^{1,2}(\Omega)^2$ if both arguments satisfy $0\le\bar u\le K$ and $0\le\bar v\le K$.

Proof. Uniqueness of the solution to (6.52): consider two solutions $u_1,u_2\in W^{1,2}(\Omega)$ to the first equation in (6.52) and test the difference by $u_1-u_2=:u_{12}$. It gives

$$d_1\|u_{12}\|^2_{W_0^{1,2}(\Omega)} \le \int_\Omega d_1|\nabla u_{12}|^2 + b_1\big(u_1u_1^+ - u_2u_2^+\big)u_{12} + c_1\bar v u_{12}^2\mathrm{d}x = a_1\int_\Omega u_{12}^2\mathrm{d}x,$$

which gives $u_{12} = 0$ if $a_1 < d_1 N^2$ and $c_1 \geq 0$, or if $c_1 < 0$ and $a_1 - c_1 K < d_1 N^{-2}$.

Now, consider a sequence $\{\bar{v}_k\}_{k\in\mathbb{N}}$ converging to $\bar{v}$ weakly in $L^2(\Omega)$. The corresponding solutions u_k are bounded in $W^{1,2}(\Omega)$, hence (up to a subsequence) u_k converges to some u weakly in $W^{1,2}(\Omega)$. Using the compact embedding $W^{1,2}(\Omega) \subset L^2(\Omega)$ in the integral identity $\int_\Omega \nabla u_k \nabla z - u_k(a_1 - b_1 u_k^+ - c_1\bar{v}_k)z - g_1 z \,\mathrm{d}x = 0$ for any $z \in W^{1,2}(\Omega) \cap L^\infty(\Omega)$, we get that u is the weak solution to the first equation in (6.52). As this solution is unique, even the whole sequence $\{u_k\}_{k\in\mathbb{N}}$ converges to it. The weak continuity of $\bar{v} \mapsto u : L^2(\Omega) \to W^{1,2}(\Omega)$ has thus been shown. The mapping $\bar{u} \mapsto v$ can be treated analogously. □

Proposition 6.25 (Existence of a solution to (6.51)). *Let $g_1, g_2 \in L^\infty(\Omega)$, and the birth rates a_1 and a_2 be small enough and the carrying capacities b_1 and b_2 be large enough as specified in* (6.56) *with K sufficiently large as specified in the proof of Lemma* 6.23. *Then there is a solution (u, v) to* (6.51) *such that $0 \leq u(\cdot) \leq K$, $0 \leq v(\cdot) \leq K$ a.e. on Ω.*

Proof. We apply the Schauder fixed-point theorem (cf. Exercise 2.55 modified for the weak* topology) to the mapping $(\bar{u}, \bar{v}) \mapsto (u, v)$ defined by (6.52) on the compact convex set $\{(u, v) \in L^\infty(\Omega)^2;\ 0 \leq u(\cdot) \leq K,\ 0 \leq v(\cdot) \leq K$ a.e. on $\Omega\}$ equipped with the weak topology of $L^2(\Omega)^2$. Note that this set is mapped into itself if K is taken suitably as mentioned in Lemma 6.24. As the resulting fixed point (u, v) is non-negative, it solves the original system (6.51), too. □

Remark 6.26. Existence of steady states especially in non-cooperative ecological systems is not automatic hence it is not surprising that Proposition 6.25 works only under rather strong data qualification.

Remark 6.27. It should be emphasized that the mere existence of solutions to the steady-state Lotka-Volterra system (6.51) is only a basic ambition in this analysis. The research in this area focuses on more advanced questions such as multiplicity of the solutions, their stability both with respect to data perturbations and whether they attract trajectories of the evolution variant of this system, cf. (12.57), etc.

6.4 Thermistor

We will address the steady-state of electric and temperature fields in an isotropic homogeneous electrically conductive medium occupying the domain Ω whose conductivity (both electrical and thermal) depends on temperature which is, vice versa, influenced by the produced Joule's heat. Electrical devices using these effects to link temperature with electrical properties are called thermistors. Anyhow, a filament in each bulb, working with temperatures ranging many hundreds of degrees, is addressed by the following system, too:

$$-\mathrm{div}\big(\kappa(\theta)\nabla\theta\big) = \sigma(\theta)|\nabla\phi|^2 \quad \text{on } \Omega\,, \tag{6.57a}$$

$$-\mathrm{div}\big(\sigma(\theta)\nabla\phi\big) = 0 \quad \text{on } \Omega\,, \tag{6.57b}$$

with the following interpretation:

ϕ is the electrostatic potential,
θ is the temperature,
σ the electric conductivity (depending on θ),
κ the heat conductivity (depending again on θ).

Therefore, (6.57a) is the heat equation, $-\kappa(\theta)\nabla\theta$ denoting the heat flux governed by the *Fourier's law* while (6.57b) is the Kirchhoff's continuity equation for the electric current j being governed by Ohm's law $j = \sigma(\theta)\nabla\phi$. The (specific) power of the electric current is the scalar product of j with the intensity $\nabla\phi$ of the electric field, i.e. the so-called Joule heat $j \cdot \nabla\phi = \sigma(\theta)|\nabla\phi|^2$ being the source term in the right-hand side of (6.57a).

Of course, the system (6.57) is to be completed by boundary conditions: e.g. the Dirichlet one on $\Gamma_{\rm D}$ with $\mathrm{meas}_{n-1}(\Gamma_{\rm D}) > 0$ (=electrodes) and zero Neumann condition on $\Gamma_{\rm N} = \Gamma \setminus \Gamma_{\rm D}$ (=an isolated part), i.e.

$$\theta|_{\Gamma_{\rm D}} = \theta_{\rm D}, \quad \phi|_{\Gamma_{\rm D}} = \phi_{\rm D} \quad \text{on } \Gamma_{\rm D}\,, \qquad \frac{\partial\theta}{\partial\nu} = \frac{\partial\phi}{\partial\nu} = 0 \quad \text{on } \Gamma_{\rm N}. \tag{6.58}$$

Let us note that the right-hand side of (6.57) has higher homogeneity than its left-hand side, so that again we meet the phenomenon of loss of coercivity of the whole system. Inspite of this, we get existence of solutions under rather general data qualification. The basic trick[33] relies on the special feature that the Joule heat $\sigma(\theta)|\nabla\phi|^2$ with $\phi \in W^{1,2}(\Omega)$ is not only in $L^1(\Omega)$ but also in $W^{1,2}(\Omega)^*$ if (6.57b) holds, and consists in the transformation of (6.57) into the system[34]

$$\mathrm{div}\big(\kappa(\theta)\nabla\theta + \sigma(\theta)\phi\nabla\phi\big) = 0, \tag{6.59a}$$

$$\mathrm{div}\big(\sigma(\theta)\nabla\phi\big) = 0. \tag{6.59b}$$

Now, again the crucial point is to design a fixed-point scheme suitably. Here, an advantageous option is to decouple the system (6.59) as follows:

$$\mathrm{div}\big(\kappa(\vartheta)\nabla\theta + \sigma(\vartheta)\phi\nabla\phi\big) = 0, \tag{6.60a}$$

$$\mathrm{div}\big(\sigma(\vartheta)\nabla\phi\big) = 0, \tag{6.60b}$$

this means we consider the mapping

$$M := M_2 \circ \big(M_1 \times \mathrm{id}\big) : \vartheta \mapsto \theta, \qquad M_1 : \vartheta \mapsto \phi, \qquad M_2 : (\phi, \vartheta) \mapsto \theta, \tag{6.61}$$

where, for ϑ given, ϕ solves in a weak sense (6.60b) and then θ solves in a weak sense (6.60b), considering naturally the boundary conditions (6.58). The existence and uniqueness of such solutions has been proved in Chapter 2.

[33] Without this trick, one must use regularity to guarantee strong convergence of ϕ's in $W^{1,p}(\Omega)$ and then $|\nabla\phi|^2$ in a suitable $L^{p/2}(\Omega) \subset W^{1,p}(\Omega)^*$ or, possibly after Kirchhoff's transformation, the L^1-theory like in Remark 6.14 together with the strong convergence as in Exercise 6.33.

[34] Realize that, by the formula $\mathrm{div}(av) = a\,\mathrm{div}\,v + \nabla a \cdot v$ and by (6.57b), one indeed has

$$\mathrm{div}\big(\sigma(\theta)\phi\nabla\phi\big) = \mathrm{div}\big(\sigma(\theta)\nabla\phi\big)\phi + \sigma(\theta)\nabla\phi \cdot \nabla\phi = \sigma(\theta)\nabla\phi \cdot \nabla\phi = \text{ Joule's heat.}$$

Lemma 6.28 (A-PRIORI ESTIMATES). *Let us assume* $\sigma : \mathbb{R} \to \mathbb{R}$ *and* $\kappa : \mathbb{R} \to \mathbb{R}$ *continuous,* $0 < c_\sigma \le \sigma(\cdot) \le C_\sigma$, $0 < c_\kappa \le \kappa(\cdot) \le C_\kappa$, $\theta_{\rm D} = \theta_0|_\Gamma$, *and* $\phi_{\rm D} = \phi_0|_\Gamma$ *for some* $\theta_0, \phi_0 \in W^{1,2}(\Omega) \cap L^\infty(\Omega)$, *and* $\vartheta \in W^{1,2}(\Omega)$ *arbitrary, and let* ϕ *and* θ *solve* (6.60)–(6.58) *in the weak sense. Then, for some constants* C_1, C_2, *and* C_3 *independent of* ϑ,

$$\big\|\phi\big\|_{L^\infty(\Omega)} \le C_1 , \tag{6.62a}$$

$$\big\|\nabla\phi\big\|_{L^2(\Omega;\mathbb{R}^n)} \le C_2, \tag{6.62b}$$

$$\big\|\theta\big\|_{W^{1,2}(\Omega)} \le C_3 . \tag{6.62c}$$

Proof. The estimate (6.62b) follows by using the test function $v = \phi - \phi_0 \in W^{1,2}(\Omega)$ in a weak formulation of (6.60b); note that obviously $v|_{\Gamma_{\rm D}} = 0$.

The estimate (6.62a) follows by testing (6.60b) by $v = (\phi \mp \|\phi_0\|_{L^\infty(\Omega)})^\pm$ as in Exercise 2.76; note that again $v|_{\Gamma_{\rm D}} = 0$.

The estimate (6.62c) can be obtained by using the test function $v = \theta - \theta_0 \in W_0^{1,2}(\Omega)$ in a weak formulation of (6.60a); note that obviously $v|_{\Gamma_{\rm D}} = 0$. By Hölder's inequality, this leads to the estimate

$$\begin{aligned}
c_\kappa\|\nabla\theta\|^2_{L^2(\Omega;\mathbb{R}^n)} &\le \int_\Omega \kappa(\vartheta)\nabla\theta\cdot\nabla\theta\,{\rm d}x = \int_\Omega \kappa(\vartheta)\nabla\theta\cdot\nabla\theta_0 - \sigma(\vartheta)\phi\nabla\phi\cdot\nabla(\theta-\theta_0)\,{\rm d}x\\
&\le \big\|\kappa(\vartheta)\big\|_{L^\infty(\Omega)}\big\|\nabla\theta\big\|_{L^2(\Omega;\mathbb{R}^n)}\big\|\nabla\theta_0\big\|_{L^2(\Omega;\mathbb{R}^n)}\\
&\quad + \big\|\sigma(\vartheta)\big\|_{L^\infty(\Omega)}\big\|\phi\big\|_{L^\infty(\Omega)}\big\|\nabla\phi\big\|_{L^2(\Omega;\mathbb{R}^n)}\big\|\nabla\theta - \nabla\theta_0\big\|_{L^2(\Omega;\mathbb{R}^n)}\\
&\le \big(C_\kappa\big\|\nabla\theta_0\big\|_{L^2(\Omega;\mathbb{R}^n)} + C_\sigma C_1 C_2\big)\big\|\nabla\theta\big\|_{L^2(\Omega;\mathbb{R}^n)} + C_\sigma C_1 C_2\big\|\nabla\theta_0\big\|_{L^2(\Omega;\mathbb{R}^n)}.
\end{aligned}$$ □

Proposition 6.29. *The system* (6.57)–(6.58) *has a weak solution* $(\theta,\phi) \in W^{1,2}(\Omega)^2$.

Proof. We take a sufficiently large ball in $W^{1,2}(\Omega)$, namely

$$B = \big\{\theta \in W^{1,2}(\Omega);\ \|\theta\|_{W^{1,2}(\Omega)} \le C_3\big\}, \tag{6.63}$$

and apply Schauder's fixed-point Theorem 1.9 (cf. Exercise 2.55) for the mapping M defined by (6.61) on B endowed by the weak topology which makes it compact. For this, we have to prove the weak continuity of the mapping $M : \vartheta \mapsto \theta : W^{1,2}(\Omega) \to W^{1,2}(\Omega)$. Supposing $\vartheta_k \rightharpoonup \vartheta$ weakly in $W^{1,2}(\Omega)$ hence strongly in $L^{p^*-\epsilon}(\Omega)$, we get

$$\phi_k \rightharpoonup \phi \qquad \big(\text{weakly in } W^{1,2}(\Omega)\big) \tag{6.64}$$

hence strongly in $L^{p^*-\epsilon}(\Omega)$, from which we then get

$$\theta_k \rightharpoonup \theta \qquad \big(\text{weakly in } W^{1,2}(\Omega)\big). \tag{6.65}$$

For (6.64), we used $\sigma(\vartheta_k) \to \sigma(\vartheta)$ in any $L^q(\Omega)$, $q < +\infty$, and made the limit passage in the identity

$$0 = \int_\Omega \sigma(\vartheta_k)\nabla\phi_k \cdot \nabla v\,{\rm d}x \to \int_\Omega \sigma(\vartheta)\nabla\phi \cdot \nabla v\,{\rm d}x \tag{6.66}$$

for $v \in W^{1,\infty}(\Omega)$ which is a dense subset in $W^{1,2}(\Omega)$. Also, we used uniqueness of ϕ solving (6.60b) for ϑ fixed, which is obvious since (6.60b) is a linear equation. Furthermore, for (6.65) we used $\kappa(\vartheta_k) \to \kappa(\vartheta)$ in any $L^q(\Omega)$, $q < +\infty$ and in strong convergence $\phi_k \to \phi$, cf. (6.64), and made the limit passage in the identity

$$0 = \int_\Omega \kappa(\vartheta_k)\nabla\theta_k \cdot \nabla v + \sigma(\vartheta_k)\phi_k\nabla\phi_k \cdot \nabla v \,\mathrm{d}x$$
$$\to \int_\Omega \kappa(\vartheta)\nabla\theta \cdot \nabla v + \sigma(\vartheta)\phi\nabla\phi \cdot \nabla v \,\mathrm{d}x, \tag{6.67}$$

which holds for v smooth enough, say $W^{1,\infty}(\Omega)$. Again, we use also uniqueness of θ solving (6.60a) for ϕ and ϑ fixed.

Then, by the Schauder theorem, M has a fixed point $\theta \in B$, and then obviously the couple (θ, ϕ) with $\phi = M_1(\theta)$ solves (6.57). □

Exercise 6.30 (Other boundary conditions). Assume κ constant, and modify the boundary conditions (6.58) as

$$\kappa\frac{\partial\theta}{\partial\nu} + b\theta = b\theta_\mathrm{e} \qquad \text{and} \qquad \sigma(\theta)\frac{\partial\phi}{\partial\nu} = j_\mathrm{e} \qquad \text{on } \Gamma,$$

with $b \geq 0$, and the external temperature $\theta_\mathrm{e} \in L^\infty(\Gamma)$ and the prescribed electric current $j_\mathrm{c} \in L^{2^{\#'}}(\Gamma)$. Show solvability of such a modified system provided $\int_\Gamma j_\mathrm{e}\mathrm{d}S = 0$ and provided κ is constant.[35] Additionally, modify the linear heat transfer by considering the Stefan-Boltzmann conditions, cf. (2.125); this would apply, e.g., to a lamp filament working in high temperatures where the heat/light radiation mechanism intentionally dominates the usual heat convection.

[35]Hint: For M_1, realize that the zero-current condition $\int_\Gamma j_\mathrm{e}\mathrm{d}S = 0$ is necessary to ensure existence of ϕ and that only $\nabla\phi$ is determined uniquely while ϕ is determined only up to constants, and then this nonuniqueness is smeared out by M_2. Note that (6.60b) has a variational structure of minimizing the convex potential $\phi \mapsto \int_\Omega \frac{1}{2}\sigma(\vartheta)|\nabla\phi|^2\mathrm{d}x - \int_\Gamma j_\mathrm{e}\phi\,\mathrm{d}S$ on $\{v \in W^{1,2}(\Omega);\ \int_\Omega \phi\,\mathrm{d}x = 0\}$. The zero-current condition is necessary to ensure that this functional is finite. Show coercivity by estimating

$$\int_\Omega \frac{1}{2}\sigma(\vartheta)|\nabla\phi|^2\mathrm{d}x - \int_\Gamma j_\mathrm{e}\phi\,\mathrm{d}S \geq \frac{1}{2}\inf\sigma(\cdot)\|\nabla\phi\|^2_{L^2(\Omega)} - \|j_\mathrm{e}\|_{L^{2^{\#'}}(\Gamma)}\|\phi\|_{L^{2^\#}(\Gamma)}$$
$$\geq \frac{1}{4}\inf\sigma(\cdot)C_\mathrm{P}^{-2}\Big(\|\phi\|^2_{W^{1,2}(\Omega)} - \Big|\int_\Omega \phi\,\mathrm{d}x\Big|^2\Big) - N\|j_\mathrm{e}\|_{L^{2^{\#'}}(\Gamma)}\|\phi\|_{W^{1,2}(\Omega)},$$

where C_P is the constant from the Poincaré inequality (1.58) used for $p = 2$, and N is the norm of the trace operator $u \mapsto u|_\Gamma : W^{1,2}(\Omega) \to L^{2^\#}(\Gamma)$. Thus we can see that this functional is coercive on the mentioned subspace of $W^{1,2}(\Omega)$. Even it is strictly convex. As its differential is even strongly monotone, the continuity of $\vartheta \to \nabla\phi$ follows. As the L^∞-estimate of ϕ cannot now be expected, proceed directly by using L^1-theory for the linear equation $\mathrm{div}\big(\kappa(\vartheta)\nabla\theta\big) + \sigma(\vartheta)|\nabla\phi|^2 = 0$, use a-priori estimates of $\theta \in W^{\lambda,2}(\Omega) \Subset L^{n/(n-2)-\epsilon}(\Omega)$ with $\lambda < 2 - n/2$, cf. Proposition 3.31, and (not relying on uniqueness but only on convexity of the set of solutions for fixed ϑ) employ Kakutani's fixed point theorem 1.11.

Exercise 6.31. Prove (6.66) and (6.67) directly for $v \in W^{1,2}(\Omega)$.[36]

Exercise 6.32. Again, from natural reasons, uniqueness of the weak solution to the whole system (6.59) can be expected only for small data. Prove this uniqueness, assuming σ and κ Lipschitz continuous and both $\theta_{\rm D}$ and $\phi_{\rm D}$ are small enough.[37]

Exercise 6.33. Prove strong convergence in (6.64) and in (6.65).[38]

6.5 Semiconductors

Semiconductor devices, such as diodes, bipolar and unipolar transistors, thyristors, etc., and their systems in integrated circuits, have formed a technological base of fast industrial and post-industrial development of mankind in the 2nd half of the 20th century.[39] Mathematical modelling of particular semiconductor devices uses various models. The basic, so-called *drift-diffusion model* has been formulated by Roosbroeck [355] and, in the steady-state isothermal variant, is governed by the following system[40]

$$\operatorname{div}\big(\varepsilon\nabla\phi\big) = n - p + c_{\rm D} \qquad \text{in } \Omega, \tag{6.68a}$$

$$\operatorname{div}\big(\nabla n - n\nabla\phi\big) = r(n,p) \qquad \text{in } \Omega, \tag{6.68b}$$

$$\operatorname{div}\big(\nabla p + p\nabla\phi\big) = r(n,p) \qquad \text{in } \Omega, \tag{6.68c}$$

[36]Hint: Consider the Nemytskiĭ mapping $\mathscr{N}_a$ determined by the integrand $a{:}\Omega\times\mathbb{R} \to \mathbb{R}^n{:}(x,r) \mapsto \sigma(r)\nabla v(x)$ and, verifying (1.48), show its continuity as $\mathscr{N}_a{:}L^{p^*-\epsilon}(\Omega) \to L^2(\Omega;\mathbb{R}^n)$. For (6.67), consider $a{:}(x,r) \mapsto \kappa(r)\nabla v(x)$.

[37]Hint: Imitate the strategy of Exercise 6.20. In particular, the term $\operatorname{div}(\sigma(\theta)\phi\nabla\phi)$ results to

$$\int_\Omega \big(\sigma(\theta_1)\phi_1\nabla\phi_1 - \sigma(\theta_2)\phi_2\nabla\phi_2)\big)\cdot\nabla\theta_{12}\,\mathrm{d}x = \int_\Omega \Big(\sigma(\theta_1) - \sigma(\theta_2)\big)\phi_1\nabla\phi_1 + \sigma(\theta_2)\phi_{12}\nabla\phi_1 + \sigma(\theta_2)\phi_2\nabla\phi_{12}\Big)\cdot\nabla\theta_{12}\,\mathrm{d}x$$

and then use Hölder inequality to estimate it "on the right-hand side".

[38]Hint: Use uniform monotonicity and $c_\sigma\|\nabla\phi_k - \nabla\phi\|_{L^2(\Omega;\mathbb{R}^n)} \le \int_\Omega \sigma(\vartheta_k)|\nabla\phi_k - \nabla\phi|^2\mathrm{d}x = \int_\Omega \sigma(\vartheta_k)\nabla\phi_k\cdot\nabla(\phi_k - \phi) - \sigma(\vartheta_k)\nabla\phi\cdot\nabla(\phi_k - \phi)\,\mathrm{d}x = -\int_\Omega \sigma(\vartheta_k)\nabla\phi\cdot\nabla(\phi_k - \phi)\,\mathrm{d}x \to 0$. In the case (6.65), use the weak lower-semicontinuity of $(\vartheta_k,\theta_k) \mapsto \int_\Omega \kappa(\vartheta_k)|\nabla\theta_k|^2\mathrm{d}x$:

$$\lim_{k\to\infty} c_\kappa\|\nabla\theta_k - \nabla\theta\|_{L^2(\Omega;\mathbb{R}^n)} \le \limsup_{k\to\infty}\int_\Omega \kappa(\vartheta_k)|\nabla\theta_k - \nabla\theta|^2\mathrm{d}x = -\liminf_{k\to\infty}\int_\Omega \kappa(\vartheta_k)\nabla\theta_k\cdot\nabla(\theta_k - \theta)\,\mathrm{d}x - \lim_{k\to\infty}\int_\Omega \kappa(\vartheta_k)\nabla\theta\cdot\nabla(\theta_k - \theta)\,\mathrm{d}x \le 0.$$

[39]This was reflected by Nobel prizes awarded for discovery of the transistor effect to W.B. Shockley, J. Bardeen, and W.H. Brattain in 1956, for invention of integrated circuits to J.S. Kilby in 2000, and for semiconductor heterostructures to Z.I. Alferov and H. Kroemer also in 2000.

[40]For more details, the reader is referred to the monographs by Markowich [275], Markowich, Ringhofer, and Schmeiser [276], Mock [289], or Selberherr [380], or to papers, e.g., by Gajewski [162], Gröger [191], Jerome [216], or Mock [288]. The model (6.68) can be derived from particle-type models on the assumption that the average distance between two subsequent collisions tends to zero; cf. [276].

where we use the conventional notation[41]

n a concentration of the negative-charge carriers (i.e. of the electrons),
p a concentration of the positive-charge carriers (the so-called holes),
ϕ the electrostatic potential,
$c_{\rm D}=c_{\rm D}(x)$ a given profile of concentration of dopants (=donors−acceptors),
$\varepsilon > 0$ a given permitivity,
$r = r(n,p)$ generation and recombination rate, cf. Example 6.37 below.

The so-called *Poisson equation*[42] (6.68a) is the rest of Maxwell's equations when neglecting magnetic-field effects, which says that divergence of the electric induction $\varepsilon\nabla\phi$ has as the source the total electric charge $n - p + c_{\rm D}$. The equation (6.68b) is the continuity equation for the phenomenological electron current[43] $j_n = \nabla n - n\nabla\phi$ with the source $r = r(n,p)$. The equation (6.68c) has a similar meaning for the phenomenological hole current $j_p = -\nabla p - p\nabla\phi$.

Of course, (6.68) is to be completed by boundary conditions: let us consider, for simplicity, the Dirichlet one on $\Gamma_{\rm D}$ with $\mathrm{meas}_{n-1}(\Gamma_{\rm D}) > 0$ (which describes conventional electrodes) and zero Neumann on $\Gamma_{\rm N} = \Gamma\setminus\Gamma_{\rm D}$ (an isolated part), i.e.

$$\phi|_{\Gamma_{\rm D}} = \phi_{\rm D},\quad n|_{\Gamma_{\rm D}} = n_{\rm D},\quad p|_{\Gamma_{\rm D}} = p_{\rm D}\quad\text{on }\Gamma_{\rm D},\qquad \frac{\partial\phi}{\partial\nu} = \frac{\partial n}{\partial\nu} = \frac{\partial p}{\partial\nu} = 0\quad\text{on }\Gamma_{\rm N}. \tag{6.69}$$

Examples of geometry of typical semiconductor devices, a bi-polar and a uni-polar transistors, are in Figure 15.[44]

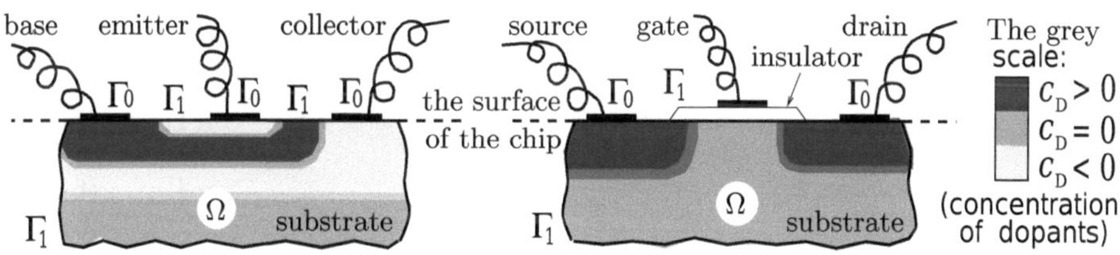

Figure 15. *Schematic geometry of a bi-polar transistor (left) and a uni-polar field-effect transistor (so-called FET) (right) which are basic elements of integrated circuits manufactured by an epitaxial technology. The grey scale refers to the level of dopants (hence the left figure refers to a so-called p-n-p transistor).*

[41]Hopefully, "n" and "p" used in this section causes no confusion with the dimension n of the domain $\Omega\subset\mathbb{R}^n$ used also here, or the integrability in $L^p(\Omega)$ spaces used in other parts.

[42]More precisely, the Poisson equation is $\Delta u = g$. For $g = 0$ it is called the Laplace equation.

[43]For simplicity, we consider diffusivity and mobility constant (and equal 1). Dependence especially on $\nabla\phi$ is, however, often important and may even create instability of steady-states on which operational regimes of special devices, so-called Gunn's diodes, made from binary semiconductors (e.g. GaAs) are based; such diodes have no steady state under some voltage and therefore must oscillate (typically on very high frequencies ranging GHz). For mathematical analysis of such system see Frehse and Naumann [152] or Markowich, Ringhofer, and Schmeiser [276, Sect.4.8].

[44]Transistors have always three electrodes. In the bi-polar transistor, Figure 15(left), $\Gamma_{\rm D}$ has therefore three disjoint components. In the unipolar transistor, Figure 15(right), $\Gamma_{\rm D}$ has only two components, the third electrode, called a gate, is realized through Newton-type boundary conditions $\varepsilon\frac{\partial\phi}{\partial\nu} = (\phi-\phi_{\rm G})$ instead of the Neumann one (6.69), with $\phi_{\rm G}$ denoting the electrostatic potential of the gate, cf. Exercise 6.38.

A substantial trick consists in a nonlinear transformation: we introduce a new variable set (ϕ, u, v) related to (ϕ, n, p) by

$$n = \mathrm{e}^{\phi} u\,, \qquad p = \mathrm{e}^{-\phi} v\,, \tag{6.70}$$

and abbreviate

$$s(\phi, u, v) := r(\mathrm{e}^{\phi} u, \mathrm{e}^{-\phi} v) \quad \text{and} \quad \sigma(\phi, u, v) = \frac{s(\phi, u, v)}{uv - 1}. \tag{6.71}$$

Let us remark that $-\ln(u)$ and $\ln(v)$ are called quasi-Fermi potentials of electrons and holes, respectively. Obviously, (6.70) transforms the currents j_n and j_p to

$$j_n = \nabla n - n\nabla\phi = \mathrm{e}^{\phi}\nabla u + \mathrm{e}^{\phi}\nabla\phi u - \mathrm{e}^{\phi}\nabla\phi\, u = \mathrm{e}^{\phi}\nabla u\,, \tag{6.72a}$$

$$j_p = -\nabla p - p\nabla\phi = -\mathrm{e}^{-\phi}\nabla v + \mathrm{e}^{-\phi}\nabla\phi v - \mathrm{e}^{-\phi}\nabla\phi\, v = -\mathrm{e}^{-\phi}\nabla v, \tag{6.72b}$$

and thus the system (6.68) transforms to

$$\mathrm{div}(\varepsilon\nabla\phi) = \mathrm{e}^{\phi} u - \mathrm{e}^{-\phi} v + c_{\mathrm{D}}, \tag{6.73a}$$

$$\mathrm{div}(\mathrm{e}^{\phi}\nabla u) = s(\phi, u, v), \tag{6.73b}$$

$$\mathrm{div}(\mathrm{e}^{-\phi}\nabla v) = s(\phi, u, v), \tag{6.73c}$$

while the boundary conditions (6.69) transform to

$$\phi|_{\Gamma_{\mathrm{D}}} = \phi_{\mathrm{D}}, \quad u|_{\Gamma_{\mathrm{D}}} = u_{\mathrm{D}} := \mathrm{e}^{-\phi_{\mathrm{D}}} n_{\mathrm{D}}, \quad v|_{\Gamma_{\mathrm{D}}} = v_{\mathrm{D}} := \mathrm{e}^{\phi_{\mathrm{D}}} p_{\mathrm{D}} \qquad \text{on } \Gamma_{\mathrm{D}}, \tag{6.74a}$$

$$\frac{\partial\phi}{\partial\nu} = \frac{\partial u}{\partial\nu} = \frac{\partial v}{\partial\nu} = 0 \qquad \text{on } \Gamma_{\mathrm{N}}. \tag{6.74b}$$

We will again use the fixed-point technique, designed by means of a mapping $M(\bar{u}, \bar{v}) = M_2(M_1(\bar{u}, \bar{v}), \bar{u}, \bar{v})$, where $M_1 : (u, v) \mapsto \phi =$ the weak solution to (6.73a) with (6.74), and $M_2 : (\phi, \bar{u}, \bar{v}) \mapsto (u, v) =$ the weak solutions to:

$$\mathrm{div}(\mathrm{e}^{\phi}\nabla u) = \sigma(\phi, \bar{u}, \bar{v})(u\bar{v} - 1), \tag{6.75a}$$

$$\mathrm{div}(\mathrm{e}^{-\phi}\nabla v) = \sigma(\phi, \bar{u}, \bar{v})(\bar{u}v - 1), \tag{6.75b}$$

with the boundary conditions (6.74). We assume $\phi_{\mathrm{D}}, n_{\mathrm{D}}, p_{\mathrm{D}} \in L^\infty(\Gamma_{\mathrm{D}})$, $n_{\mathrm{D}}(\cdot) \geq \delta$, $p_{\mathrm{D}}(\cdot) \geq \delta$ with some $\delta > 0$, so that one can take $K \geq 1$ such that u_{D} and v_{D} are in $[\mathrm{e}^{-K}, \mathrm{e}^{K}]$. Moreover, let $\phi_{\mathrm{D}} = \phi_0|_\Gamma$, $u_{\mathrm{D}} = u_0|_\Gamma$, $v_{\mathrm{D}} = v_0|_\Gamma$ for some $\phi_0, u_0, v_0 \in W^{1,2}(\Omega) \cap L^\infty(\Omega)$.

Lemma 6.34 (A-PRIORI ESTIMATES). *Let $\sigma : \mathbb{R} \times \mathbb{R}^+ \times \mathbb{R}^+ \to \mathbb{R}$ be a positive continuous function. For $u, v \in [\mathrm{e}^{-K}, \mathrm{e}^{K}]$,* (6.73a) *with the boundary condition from* (6.74) *has a unique weak solution $\phi \in W^{1,2}(\Omega) \cap L^\infty(\Omega)$ satisfying*

$$\phi(x) \in [\phi_{\min}, \phi_{\max}] \quad \textit{for a.a. } x \in \Omega, \tag{6.76}$$

with $\phi_{\min} \in \mathbb{R}$ so small and $\phi_{\max} \in \mathbb{R}$ so large that

$$\phi_{\min} \le \inf_{x\in\Gamma_{\rm D}} \phi_{\rm D}(x)\,, \qquad e^{\phi_{\min}+K} - e^{-\phi_{\min}-K} + \sup_{x\in\Omega} c_{\rm D}(x) \le 0, \tag{6.77a}$$

$$\phi_{\max} \ge \sup_{x\in\Gamma_{\rm D}} \phi_{\rm D}(x)\,, \qquad e^{\phi_{\max}-K} - e^{-\phi_{\max}+K} + \inf_{x\in\Omega} c_{\rm D}(x) \ge 0. \tag{6.77b}$$

Moreover, for $\phi \in L^\infty(\Omega)$ and $\bar u, \bar v \in [e^{-K}, e^{K}]$, (6.74)–(6.75) *have unique weak solutions u and v satisfying, for some C_K depending on K,*

$$\|u\|_{W^{1,2}(\Omega)} \le C_K\,, \qquad \|v\|_{W^{1,2}(\Omega)} \le C_K\,, \tag{6.78a}$$

$$u(x), v(x) \in [e^{-K}, e^{K}] \quad \textit{for a.a. } x\in\Omega. \tag{6.78b}$$

Proof. Use the direct method for the strictly convex and coercive potential $\phi \mapsto \int_\Omega \frac{1}{2}\varepsilon|\nabla\phi|^2 + ue^{\phi} + ve^{-\phi} - c_{\rm D}\phi\,{\rm d}x$ on the affine manifold $\{\phi \in W^{1,2}(\Omega);\ \phi|_{\Gamma_{\rm D}} = \phi_{\rm D}\}$; note that this functional can take the value $+\infty$. We thus get a unique weak[45] solution $\phi = M_1(u,v)$ to the equation (6.73a) with the boundary condition from (6.74). The $W^{1,2}$-estimate can be obtained by a test of (6.73a) by $\phi-\phi_0$: realizing that always $e^{-\phi}(\phi-\phi_0)^+ \le e^{\|\phi_0\|_{L^\infty(\Omega)}}$ and $-e^{\phi}(\phi-\phi_0)^- \le e^{\|\phi_0\|_{L^\infty(\Omega)}}$, we have by Green's Theorem 1.31

$$\begin{aligned}
\int_\Omega \varepsilon|\nabla\phi|^2\,{\rm d}x &\le \int_\Omega \varepsilon|\nabla\phi|^2 + e^{\phi}(\phi-\phi_0)^+u - e^{-\phi}(\phi-\phi_0)^-v\,{\rm d}x \\
&= \int_\Omega c_{\rm D}(\phi_0-\phi) - e^{\phi}(\phi-\phi_0)^-u + e^{-\phi}(\phi-\phi_0)^+v + \varepsilon\nabla\phi\cdot\nabla\phi_0\,{\rm d}x \\
&\le \|c_{\rm D}\|_{L^\infty(\Omega)}\big(\|\phi\|_{L^1(\Omega)} + \|\phi_0\|_{L^1(\Omega)}\big) \\
&\quad + {\rm meas}_n(\Omega)e^{K+\|\phi_0\|_{L^\infty(\Omega)}} + \varepsilon\|\nabla\phi\|_{L^2(\Omega;\mathbb{R}^n)}\|\nabla\phi_0\|_{L^2(\Omega;\mathbb{R}^n)},
\end{aligned}$$

from which an a-priori bound for ϕ in $W^{1,2}(\Omega)$ follows. The upper bound in (6.76) can be shown by a comparison likewise in Exercise 2.76, here we use the test function $z := (\phi-\phi_{\max})^+$. Note that the first condition in (6.77b) implies $z|_{\Gamma_{\rm D}} = 0$ hence it is indeed a legal test function for the weakly formulated boundary-value problem (6.73a)-(6.74). This test gives

$$\int_\Omega \varepsilon\nabla\phi\cdot\nabla(\phi-\phi_{\max})^+ + \big(e^{\phi}u - e^{-\phi}v + c_{\rm D}\big)(\phi-\phi_{\max})^+\,{\rm d}x = 0. \tag{6.79}$$

Now, we realize that the first term in (6.79) is always non-negative, cf. (1.50), and that, if $u \ge e^{-K}$ and $v \le e^{K}$, then necessarily $e^{\phi}u - e^{-\phi}v + c_{\rm D} > 0$ wherever $(\phi-\phi_{\max})^+ > 0$ with $\phi_{\max}$ satisfying the second inequality in (6.77b). We can therefore see that (6.79) yields $(\phi-\phi_{\max})^+ \le 0$ a.e. in Ω. The lower bound in (6.76) can be shown similarly by testing (6.73a) by $z := (\phi-\phi_{\min})^-$.

[45] This sort of solution is called a variational solution. If, however, we show a-posteriori boundedness in $L^\infty(\Omega)$, cf. (6.76), this solution is the weak solution. One can also imagine the monotone nonlinearity $r \mapsto u(x)e^r - v(x)e^{-r}$ in (6.73a) modified, for a moment, out of $[\phi_{\min}, \phi_{\max}]$ to have a subcritical polynomial growth.

The unique weak solution u to the linear boundary-value problem (6.75a)–(6.74) obviously does exist. The a-priori estimate can be obtained by testing (6.75a) by $u - u_0$:

$$\begin{aligned}
e^{\phi_{\min}} \int_\Omega |\nabla u|^2 dx &\le \int_\Omega e^\phi |\nabla u|^2 + \sigma(\phi, \bar{u}, \bar{v}) u^2 \bar{v} \, dx \\
&= \int_\Omega e^\phi \nabla u \cdot \nabla u_0 + \sigma(\phi, \bar{u}, \bar{v})(u u_0 \bar{v} + u - u_0) \, dx \\
&\le e^{\phi_{\max}} \|\nabla u\|_{L^2(\Omega;\mathbb{R}^n)} \|\nabla u_0\|_{L^2(\Omega;\mathbb{R}^n)} \\
&\quad + C_\sigma \Big(\|u\|_{L^2(\Omega)} \|u_0\|_{L^2(\Omega)} e^K + \|u\|_{L^1(\Omega)} + \|u_0\|_{L^1(\Omega)} \Big) \qquad (6.80)
\end{aligned}$$

where $C_\sigma := \sup_{[\phi_{\min},\phi_{\max}]\times[e^{-K},e^K]^2} \sigma(\cdot,\cdot,\cdot)$ so that u is bounded in $W^{1,2}(\Omega)$. The upper bound for u in (6.78b) can be shown again by a comparison, now by choosing $z := (u - e^K)^+$ as a test function for (6.75a). As $u|_{\Gamma_D} = u_D \le e^K$ due to the choice of K, $z|_{\Gamma_D} = 0$ hence it is indeed a legal test function for the weakly formulated boundary-value problem (6.75a)-(6.74). This test gives

$$\int_\Omega e^\phi \nabla u \cdot \nabla (u - e^K)^+ + \sigma(\phi, \bar{u}, \bar{v})(u\bar{v} - 1)(u - e^K)^+ \, dx = 0. \qquad (6.81)$$

As in (6.79), the first term in (6.81) is always non-negative and, if $\bar{v} \ge e^{-K}$, the second term is positive wherever $u - e^K > 0$, and we can therefore see that (6.81) yields $u \le e^K$ a.e. in Ω. The lower bound in (6.78b) can be proved similarly by testing (6.75a) by $z := (u - e^{-K})^-$.

Analogous considerations hold for v. □

Lemma 6.35 (Continuity). *Let $\sigma : \mathbb{R} \times \mathbb{R}^+ \times \mathbb{R}^+ \to \mathbb{R}^+$ be continuous.*

(i) *The mapping $M_1 : (u, v) \mapsto \phi : L^2(\Omega)^2 \to W^{1,2}(\Omega)$ is weakly continuous if restricted on $\{(u, v);\ (6.78b)$ holds$\}$.*

(ii) *The mapping $M_2 : (\phi, \bar{u}, \bar{v}) \mapsto (u, v) : L^2(\Omega)^3 \to W^{1,2}(\Omega)^2$ is demicontinuous if restricted on $\{(\phi, u, v);\ (6.76)$ and $(6.78b)$ hold$\}$.*

Proof. Assume $u_k \rightharpoonup u$ and $v_k \rightharpoonup v$ in $L^2(\Omega)$. Then, consider $\phi_k = M_1(u_k, v_k)$ and (possibly for a subsequence) $\phi_k \rightharpoonup \phi$ in $W^{1,2}(\Omega)$. Then $\phi_k \to \phi$ in $L^{2^*-\epsilon}(\Omega)$, and also $e^{\phi_k} u_k \rightharpoonup e^\phi u$ and $e^{\phi_k} v_k \rightharpoonup e^\phi v$ in $L^2(\Omega)$ provided ϕ_k is bounded in $L^\infty(\Omega)$, as it really is due to (6.76).[46] Then one can pass to the limit in the identity $\int_\Omega \varepsilon \nabla \phi_k \cdot \nabla z + e^{\phi_k} u_k z - e^{\phi_k} v_k z + c_D z \, dx = 0$, showing that $\phi = M_1(u, v)$ and, in fact, the whole sequence converges.

As to (ii), considering $(\phi_k, \bar{u}_k, \bar{v}_k) \to (\phi, \bar{u}, \bar{v})$ in $L^2(\Omega)^3$, by the a-priori estimate (6.78a) the corresponding sequence (u_k, v_k) converges (at least as a subsequence) weakly in $W^{1,2}(\Omega)^2$ to some (u, v). Passing to the limit in the integral

[46] Realize that we can imagine that the nonlinearities $(\xi, r) \mapsto e^\xi r$ and $(\xi, r) \mapsto e^{-\xi} r$ are modified for $\xi \notin [\phi_{\min}, \phi_{\max}]$ to have a linear growth.

identities[47]

$$\int_\Omega e^{\phi_k}\nabla u_k\cdot\nabla z+\sigma(\phi_k,\bar u_k,\bar v_k)(u_k\bar v_k-1)z\,dx\ =0,\tag{6.82a}$$

$$\int_\Omega e^{-\phi_k}\nabla v_k\cdot\nabla z+\sigma(\phi_k,\bar u_k,\bar v_k)(\bar u_k v_k-1)z\,dx=0\tag{6.82b}$$

for all $z\in W^{1,\infty}(\Omega)$, $z|_{\Gamma_D}=0$, we can see that (u,v) solves (6.75) with the boundary conditions (6.74). As $\sigma\ge0$ and also $\bar u\ge0$ and $\bar v\ge0$, this (u,v) must be unique and thus the whole sequence converges to it. □

Proposition 6.36 (Existence). *Under the above assumptions on σ, c_D, n_D, p_D, and ϕ_D, the system* (6.73)–(6.74) *has a weak solution.*

Proof. Use Schauder fixed-point Theorem 1.9 for the mapping $M=M_2\circ M_1$ on $S:=\{(u,v)\in L^\infty(\Omega)\times L^\infty(\Omega);\ (6.78b)\text{ holds}\}$ equipped with the norm topology of $L^2(\Omega)$. Realize that, by (6.78a) and Rellich-Kondrachov's Theorem 1.21, $M(S)$ is indeed relatively compact. □

Example 6.37 (Shockley-Read-Hall model). The generation/recombination rate is often modelled by

$$r=r(n,p)=\frac{np-c_{\rm int}^2}{\tau_n(n+c_{\rm int})+\tau_p(p+c_{\rm int})}\tag{6.83}$$

with $c_{\rm int}>0$ an intrinsic concentration and $\tau_n>0$ and $\tau_p>0$ the electron and the hole live-time, respectively. Assuming, without loss of generality if suitable physical units are chosen, that $c_{\rm int}=1$, the model (6.83) indeed gives σ as a positive continuous function as required in Lemma 6.34, namely

$$\sigma(\phi,u,v)=\frac{1}{\tau_n(e^{\phi}u+1)+\tau_p(e^{-\phi}v+1)}\,.\tag{6.84}$$

Exercise 6.38 (Newton boundary conditions for ϕ). Modify Lemma 6.34 for combining the boundary Dirichlet/Neumann boundary conditions (6.69) with the Newton one: $\varepsilon\frac{\partial\phi}{\partial\nu}=(\phi-\phi_G)$ on some part of Γ_N with $\phi_G\in L^\infty(\Gamma_N)$; this part of Γ_N corresponds to the so-called gate of an FET-transistor on Figure 15(right).

Exercise 6.39. Strengthen Lemma 6.35 by proving the total continuity of M_1 and the continuity of M_2.[48]

[47] As $\nabla z\in L^\infty(\Omega;\mathbb{R}^n)$, we can use $e^{\phi_k}\to e^{\phi}$ and $e^{-\phi_k}\to e^{-\phi}$ in $L^2(\Omega)$ if (6.76) holds. By (6.78b), we can assume $\sigma(\phi_k,\bar u_k,\bar v_k)\to\sigma(\phi,\bar u,\bar v)$ in any $L^r(\Omega)$, $r<+\infty$, which allows us to pass to the limit in the last terms in (6.82a,b). Eventually, the resulting identities can be extended for z from $W^{1,\infty}(\Omega)$ onto the whole $W^{1,2}(\Omega)$.

[48] Hint: Show $\phi_k\to\phi$ in $W^{1,2}(\Omega)$ due to the strong monotonicity of the Laplacean with the Dirichlet boundary condition on Γ_D by testing the difference of weak formulations determining

Remark 6.40 (Uniqueness). The weak solution to (6.68)–(6.69), whose existence was proved in Proposition 6.36, is unique only on special occasions. In general, there are even semiconductor devices such as thyristors whose operational regimes just exploit non-uniqueness of steady states.

---

respectively ϕ_k and ϕ by $\phi_k - \phi$, which gives

$$\int_\Omega \varepsilon|\nabla\phi_k - \nabla\phi|^2 \mathrm{d}x = \int_\Omega \big(\mathrm{e}^{\phi_k}u_k - \mathrm{e}^{\phi}u - \mathrm{e}^{-\phi_k}v_k + \mathrm{e}^{-\phi}v\big)(\phi_k - \phi)\,\mathrm{d}x \to 0.$$

As to u_k, use the uniform (with respect to k) strong monotonicity of $u \mapsto -\mathrm{div}(\mathrm{e}^{\phi_k}\nabla u)$ likewise in Exercise 2.75, and test (6.82a) by $z := u_k - u$:

$$\begin{aligned} \mathrm{e}^{\phi_{\min}}\big\|\nabla(u_k - u)\big\|^2_{L^2(\Omega;\mathbb{R}^n)} &\leq \int_\Omega \mathrm{e}^{\phi_k}\big|\nabla(u_k - u)\big|^2 \mathrm{d}x \\ &= \int_\Omega \sigma(\phi_k, \bar{u}_k, \bar{v}_k)(u_k\bar{v}_k - 1)(u_k - u) - \mathrm{e}^{\phi_k}\nabla u \cdot \nabla(u_k - u)\,\mathrm{d}x \to 0. \end{aligned}$$

Eventually, $v_k \to v$ is similar.

Part II

EVOLUTION PROBLEMS

Chapter 7

Special auxiliary tools

In evolution problems, one scalar variable, denoted by t and having a meaning of time, takes a special role, which is also reflected by mathematical analysis. In particular, here we first present a few useful assertions about spaces of abstract functions on a "time" interval $I := [0, T]$, introduced already in Section 1.5, but now possessing additionally derivatives with respect to time. Always, T will denote a fixed finite time horizon.[1]

7.1 Sobolev-Bochner space $W^{1,p,q}(I; V_1, V_2)$

For V_1 a Banach space and V_2 a locally convex space, $V_1 \subset V_2$, let us define

$$W^{1,p,q}(I; V_1, V_2) := \Big\{ u \in L^p(I; V_1);\ \frac{\mathrm{d}u}{\mathrm{d}t} \in L^q(I; V_2) \Big\} \tag{7.1}$$

with $\frac{\mathrm{d}}{\mathrm{d}t}u$ denoting the *distributional derivative* of u understood as the abstract linear operator $\frac{\mathrm{d}}{\mathrm{d}t}u \in \mathscr{L}(\mathscr{D}(I), (V_2, \text{weak}))$ defined by

$$\frac{\mathrm{d}u}{\mathrm{d}t}(\varphi) = -\int_0^T u \frac{\mathrm{d}\varphi}{\mathrm{d}t}\,\mathrm{d}t \tag{7.2}$$

for any $\varphi \in \mathscr{D}(I)$, where $\mathscr{D}(I)$ stands for infinitely differentiable functions with a compact support in $(0, T)$. Mostly, both V_1 and V_2 will be Banach spaces, and then $W^{1,p,q}(I; V_1, V_2)$ itself is a Banach space if equipped with the norm $\|u\|_{W^{1,p,q}(I;V_1,V_2)} := \|u\|_{L^p(I;V_1)} + \|\frac{\mathrm{d}}{\mathrm{d}t}u\|_{L^q(I;V_2)}$. Sometimes, $V_1 = V_2$ will occur and then we will briefly write

$$W^{1,p}(I; V) := W^{1,\infty,p}(I; V, V). \tag{7.3}$$

[1]For more detailed study, the reader is referred e.g. to monographs by Gajewski at al. [168, Sect.IV.1] or Zeidler [427, Chap.23].

Occasionally, we use also spaces having 2nd-order time derivatives valued in V_3:

$$W^{2,\infty,p,q}(I;V_1,V_2,V_3) := \Big\{u\in L^\infty(I;V_1);\\ \frac{\mathrm{d}u}{\mathrm{d}t}\in L^p(I;V_2) \ \text{ and } \ \frac{\mathrm{d}^2u}{\mathrm{d}t^2}\in L^q(I;V_3)\Big\}. \tag{7.4}$$

As to V_2 in (7.1), certain degrees of generality will be found useful for the Rothe and the Galerkin method below, namely replacement of $L^q(I;V_2)$ by $\mathscr{M}(I;V_2)$ or considering V_2 a metrizable locally convex space, respectively. As to the former generalization, we just replace L^p with $\mathscr{M}$ in (7.1) and equip it with the norm counting the total variation of $u(\cdot)$ in V_2, i.e. $\|u\|_{L^p(I;V_1)}+\|\frac{\mathrm{d}}{\mathrm{d}t}u\|_{\mathscr{M}(I;V_2)}$; cf. (7.40) below. As to the latter generalization, without loss of generality, we can assume the topology of V_2 generated by a countable collection of seminorms $\{|\cdot|_\ell\}_{\ell\in\mathbb{N}}$. Then (7.1) defines a locally convex space of functions $u\in L^p(I;V_1)$ such that

$$\Big|\frac{\mathrm{d}u}{\mathrm{d}t}\Big|_{q,\ell} := \bigg(\int_0^T\Big|\frac{\mathrm{d}u}{\mathrm{d}t}\Big|_\ell^q\mathrm{d}t\bigg)^{1/q} < +\infty \tag{7.5}$$

for any $\ell\in\mathbb{N}$; we then consider $W^{1,p,q}(I;V_1,V_2)$ equipped with the topology generated by the functionals $u\mapsto\|u\|_{L^p(I;V_1)}+|\frac{\mathrm{d}}{\mathrm{d}t}u|_{q,\ell}$, $\ell\in\mathbb{N}$.

Lemma 7.1. *Let $p,q\ge 1$ and let $V_1\subset V_2$ continuously. Then $W^{1,p,q}(I;V_1,V_2)\subset C(I;V_2)$ continuously.*

Proof. Let us confine ourselves on V_2 a Banach space, the generalization for a locally convex space being clear. Let $u\in W^{1,p,q}(I;V_1,V_2)$. Then $\frac{\mathrm{d}}{\mathrm{d}t}u$ is integrable, and we can put $v(t):=\int_0^t\frac{\mathrm{d}u}{\mathrm{d}\vartheta}\mathrm{d}\vartheta$. Then

$$\big\|v(t_2)-v(t_1)\big\|_{V_2} = \Big\|\int_{t_1}^{t_2}\frac{\mathrm{d}u}{\mathrm{d}t}\mathrm{d}t\Big\|_{V_2} \le \int_{t_1}^{t_2}\Big\|\frac{\mathrm{d}u}{\mathrm{d}t}\Big\|_{V_2}\mathrm{d}t. \tag{7.6}$$

This shows that $t\mapsto v(t):I\to V_2$ is continuous. Yet, $v=u+c$, $c\in V_2$ because $\frac{\mathrm{d}}{\mathrm{d}t}v=\frac{\mathrm{d}}{\mathrm{d}t}u$. Thus u is continuous, too. Moreover, we can estimate

$$\|v(t)\|_{V_2} \le \int_0^T\Big\|\frac{\mathrm{d}u}{\mathrm{d}t}\Big\|_{V_2}\mathrm{d}t = \Big\|\frac{\mathrm{d}u}{\mathrm{d}t}\Big\|_{L^1(I;V_2)} \le N_q\Big\|\frac{\mathrm{d}u}{\mathrm{d}t}\Big\|_{L^q(I;V_2)}, \tag{7.7}$$

and then

$$\begin{aligned}\|c\|_{V_2} &= T^{-1/p}\bigg(\int_0^T\|c\|_{V_2}^p\mathrm{d}t\bigg)^{1/p} = T^{-1/p}\big\|u-v\big\|_{L^p(I;V_2)}\\ &\le T^{-1/p}\big\|u\big\|_{L^p(I;V_2)}+T^{-1/p}\big\|v\big\|_{L^p(I;V_2)}\\ &\le T^{-1/p}N_{12}\big\|u\big\|_{L^p(I;V_1)}+T^{-1/p}T^{1/p}N_q\Big\|\frac{\mathrm{d}u}{\mathrm{d}t}\Big\|_{L^q(I;V_2)},\end{aligned} \tag{7.8}$$

where N_q and N_{12} are the norms of the embeddings $L^q(0,T) \subset L^1(0,T)$ and $V_1 \subset V_2$, respectively.[2] From this, we get

$$\begin{aligned}\|u\|_{C(I;V_2)} &= \sup_{t\in I}\Big\|\int_0^t \frac{\mathrm{d}u}{\mathrm{d}\vartheta}\mathrm{d}\vartheta - c\Big\|_{V_2} \le \int_0^T \Big\|\frac{\mathrm{d}u}{\mathrm{d}\vartheta}\Big\|_{V_2}\mathrm{d}\vartheta + \|c\|_{V_2} \\ &\le N_q\Big\|\frac{\mathrm{d}u}{\mathrm{d}t}\Big\|_{L^q(I;V_2)} + T^{-1/p}N_{12}\|u\|_{L^p(I;V_1)} + N_q\Big\|\frac{\mathrm{d}u}{\mathrm{d}t}\Big\|_{L^q(I;V_2)} \\ &\le \max\big(T^{-1/p}N_{12}, 2N_q\big)\|u\|_{W^{1,p,q}(I;V_1,V_2)}. \end{aligned} \tag{7.9}$$

□

Lemma 7.2. *Let $p,q \ge 1$ and let $V_1 \subset V_2$ continuously. Then $C^1(I;V_1)$ is contained densely in $W^{1,p,q}(I;V_1,V_2)$.*

Proof. Take $u \in W^{1,p,q}(I;V_1,V_2)$ and, for $\varepsilon > 0$, put

$$u_\varepsilon(t) := \int_0^T \varrho_\varepsilon\big(t + \xi_\varepsilon(t) - s\big)u(s)\,\mathrm{d}s, \qquad \xi_\varepsilon(t) := \varepsilon\frac{T-2t}{T}, \tag{7.10}$$

where $\varrho_\varepsilon : \mathbb{R} \to \mathbb{R}$ is a positive, C^∞-function supported on $[-\varepsilon,\varepsilon]$ and satisfying $\int_{\mathbb{R}} \varrho_\varepsilon(t)\,\mathrm{d}t = 1$. Such functions are called *mollifiers*. To be more specific, we can take

$$\varrho_\varepsilon(t) := \begin{cases} c\,\varepsilon^{-1}\mathrm{e}^{t^2/(t^2-\varepsilon^2)} & \text{for } |t| < \varepsilon, \\ 0 & \text{elsewhere,} \end{cases} \tag{7.11}$$

with c a suitable constant so that $\int_{\mathbb{R}} \varrho_1(t)\,\mathrm{d}t = 1$. Note that the function ξ_ε (converging to 0 for $\varepsilon \to 0$) is just to shift slightly the kernel in the convolution integral in (7.10) so that only values of u inside $[0,T]$ are taken into account, cf. Figure 16.

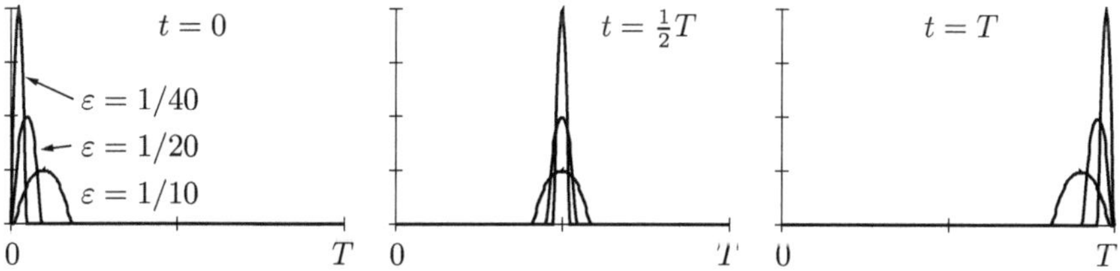

Figure 16. Example of the "mollifying" kernel $s \mapsto \varrho_\varepsilon\big(t + \xi_\varepsilon(t) - s\big)$ in the convolutory integral (7.10) for three values $t = 0$, $T/2$, and T, and for three values $\varepsilon = T/10$, $T/20$, and $T/40$.

Denoting ϱ'_ε the derivative of ϱ_ε, we can write the formula

$$\begin{aligned}\frac{\mathrm{d}u_\varepsilon}{\mathrm{d}t} &= \Big(1 - \frac{2\varepsilon}{T}\Big)\int_0^T \varrho'_\varepsilon\big(t + \xi_\varepsilon(t) - s\big)u(s)\,\mathrm{d}s \\ &= \frac{T-2\varepsilon}{T}\int_0^T \varrho_\varepsilon\big(t + \xi_\varepsilon(t) - s\big)\frac{\mathrm{d}u}{\mathrm{d}s}(s)\,\mathrm{d}s. \end{aligned} \tag{7.12}$$

[2] It holds that $N_q = \|1\|_{L^{q'}(I)} = T^{1-1/q}$, cf. also Exercise 2.68.

In particular, the first equality in (7.12) shows that $u_\varepsilon \in C^1(I;V_1)$. We can estimate

$$\int_0^T \big\|u_\varepsilon(t) - u(t)\big\|_{V_1}^p \mathrm{d}t = \int_0^T \bigg\| \int_{t-\varepsilon+\xi_\varepsilon(t)}^{t+\varepsilon+\xi_\varepsilon(t)} \varrho_\varepsilon\big(t+\xi_\varepsilon(t)-s\big)\big(u(s)-u(t)\big)\,\mathrm{d}s\bigg\|_{V_1}^p \mathrm{d}t$$
$$\le \int_0^T \bigg(\int_{\xi_\varepsilon(t)-\varepsilon}^{\xi_\varepsilon(t)+\varepsilon} \varrho_\varepsilon\big(h-\xi_\varepsilon(t)\big)\big\|u(t+h)-u(t)\big\|_{V_1}\mathrm{d}h\bigg)^p \mathrm{d}t$$
$$\le \bigg(\int_{-\varepsilon}^{\varepsilon} \varrho_\varepsilon(h)^{p'}\mathrm{d}h\bigg)^{p-1} \int_0^T\int_{\xi_\varepsilon(t)-\varepsilon}^{\xi_\varepsilon(t)+\varepsilon} \big\|u(t+h)-u(t)\big\|_{V_1}^p \mathrm{d}h\mathrm{d}t$$
$$\le \frac{2^{p-1}c^p}{\varepsilon}\int_0^T\int_{\xi_\varepsilon(t)-\varepsilon}^{\xi_\varepsilon(t)+\varepsilon} \big\|u(t+h)-u(t)\big\|_{V_1}^p \mathrm{d}h\mathrm{d}t \le 2^p c^p \sup_{|h|\le\varepsilon}\int_0^T \big\|u(t+h)-u(t)\big\|_{V_1}^p \mathrm{d}t.$$

Then we use $\lim_{\varepsilon\to 0}\sup_{|h|\le\varepsilon}\|u(\cdot+h)-u\|_{L^p(I;V_1)}^p$ which is easy to see for u piecewise constant while the general case follows by using additionally Proposition 1.36 uniformly for the collection $\{u(\cdot+h)\}_{|h|\le\varepsilon}$.[3]

Analogously, one can show that the last integral in (7.12) approaches $\frac{\mathrm{d}}{\mathrm{d}t}u$ in $L^q(I;V_2)$. Yet, (7.12) says that this integral equals just to $\frac{\mathrm{d}}{\mathrm{d}t}u_\varepsilon$ up to a factor $T/(T-2\varepsilon)$ converging to 1 when $\varepsilon\to 0$. Hence even $\frac{\mathrm{d}}{\mathrm{d}t}u_\varepsilon$ itself converges to $\frac{\mathrm{d}}{\mathrm{d}t}u$ in $L^q(I;V_2)$. □

7.2 Gelfand triple, embedding $W^{1,p,p'}(I;V,V^*)\subset C(I;H)$

A basic abstract setting for evolution problems relies on the following construction. Let H be a Hilbert space identified with its own dual, $H\equiv H^*$, and the embedding $V\subset H$ be continuous and dense. Note that then $H\subset V^*$ continuously; indeed, the adjoint mapping i^* (which is continuous) to the embedding $i: V\to H$ maps $H^*\equiv H$ into V^* and is injective, i.e.[4]

$$u_1\ne u_2 \quad\Longrightarrow\quad i^*u_1\ne i^*u_2 \quad\Longleftrightarrow\quad \exists v\in V:\ \langle u_1,v\rangle\ne\langle u_2,v\rangle. \tag{7.13}$$

Let us agree to identify i^*u with u if $u\in H$. Thus we may indeed consider $H\subset V^*$ and the duality pairing between V^* and V as a continuous extension of the inner product on H, denoted by $(\cdot,\cdot)$, i.e. for $u\in H$ and $v\in V$ we have[5]

$$\big(u,v\big) = \big\langle u,v\big\rangle_{H^*\times H} = \big\langle u,iv\big\rangle_{H^*\times H} = \big\langle i^*u,v\big\rangle_{V^*\times V} = \big\langle u,v\big\rangle_{V^*\times V}\,. \tag{7.14}$$

The indices in (7.14) indicate the spaces paired by the duality. The triple $V\subset H\subset V^*$ is called an evolution triple, or sometimes *Gelfand's triple*, and the Hilbert

[3] See e.g. Gajewski et al. [168, Chap.IV, Lemma 1.5].

[4] The equivalence in (7.13) just expresses that the functionals u_1 and u_2 on H must have different traces (=restrictions) on any dense subset of H, in particular on V.

[5] The equalities in (7.14) follow subsequently from the identification of H with H^*, the embedding $V\subset H$, the definition of the adjoint operator i^*, and the identification of i^*u with u.

space H a *pivot*. Moreover, the embedding $H \subset V^*$ is dense. Occasionally, we will need V or H separable, hence let us assume it generally without restriction of applicability to partial differential equations.

Lemma 7.3 (BY-PART INTEGRATION FORMULA). *Let $V \subset H \cong H^* \subset V^*$, and $p' = p/(p-1)$ be the conjugate exponent to p, cf. (1.20). Then $W^{1,p,p'}(I;V,V^*) \subset C(I;H)$ continuously and the following by-part integration formula holds for any $u, v \in W^{1,p,p'}(I;V,V^*)$ and any $0 \le t_1 \le t_2 \le T$:*

$$\big(u(t_2), v(t_2)\big) - \big(u(t_1), v(t_1)\big) = \int_{t_1}^{t_2} \Big\langle \frac{\mathrm{d}u}{\mathrm{d}t}, v(t)\Big\rangle_{V^*\times V} + \Big\langle u(t), \frac{\mathrm{d}v}{\mathrm{d}t}\Big\rangle_{V\times V^*} \mathrm{d}t. \tag{7.15}$$

Proof.[6] Note that (7.15) holds for $u, v \in C^1(I;V)$ by classical calculus, by using $\frac{\mathrm{d}}{\mathrm{d}t}(u,v) = (\frac{\mathrm{d}u}{\mathrm{d}t}, v) + (u, \frac{\mathrm{d}v}{\mathrm{d}t}) = \langle \frac{\mathrm{d}u}{\mathrm{d}t}, v\rangle_{V^*\times V} + \langle u, \frac{\mathrm{d}v}{\mathrm{d}t}\rangle_{V\times V^*}$ (here (7.14) have been employed) and integrating it over $[t_1, t_2]$.

Put $q = \min(2, p)$. For $u \in C^1(I;V)$, we can use (7.15) with $v := u$, $t_2 := t$ and t_1 such that $\|u(t_1)\|_H^q = \frac{1}{T}\int_0^T \|u(\vartheta)\|_H^q \mathrm{d}\vartheta$, i.e. the mean value. Thus we get

$$\begin{aligned}
\|u(t)\|_H^q &= \|u(t_1)\|_H^q + \big(\|u(t)\|_H^q - \|u(t_1)\|_H^q\big) \\
&\le \frac{1}{T}\int_0^T \|u(\vartheta)\|_H^q \mathrm{d}\vartheta + \Big|\|u(t)\|_H^2 - \|u(t_1)\|_H^2\Big|^{q/2} \\
&= \frac{1}{T}\int_0^T \|u(\vartheta)\|_H^q \mathrm{d}\vartheta + \Big|2\int_{t_1}^t \Big\langle \frac{\mathrm{d}u}{\mathrm{d}\vartheta}, u(\vartheta)\Big\rangle \mathrm{d}\vartheta\Big|^{q/2} \\
&\le \frac{1}{T}\|u\|_{L^q(I;H)}^q + 2^{q/2}\Big(\Big\|\frac{\mathrm{d}u}{\mathrm{d}t}\Big\|_{L^{p'}(I;V^*)} \|u\|_{L^p(I;V)}\Big)^{q/2} \\
&= \frac{1}{T}\|u\|_{L^q(I;H)}^q + \Big\|\frac{\mathrm{d}u}{\mathrm{d}t}\Big\|_{L^{p'}(I;V^*)}^q + \|u\|_{L^p(I;V)}^q,
\end{aligned} \tag{7.16}$$

where we used, besides Hölder's inequality, also the inequalities $a^q - b^q \le |a^2 - b^2|^{q/2}$, which holds for $a, b \ge 0$ and $q \in [1, 2]$,[7] and $(a+b)^{q/2} \le a^{q/2} + b^{q/2}$. Then we use still the estimate

$$\|u\|_{L^q(I;H)} \le \begin{cases} N_1\|u\|_{L^p(I;V)} & \text{if } p < 2, \\ N_1N_2\|u\|_{L^p(I;V)} & \text{if } p \ge 2, \end{cases} \tag{7.17}$$

where N_1 and N_2 are the norms of the embedding $V \subset H$ and $L^p(I) \subset L^2(I)$, respectively. As the estimate (7.16) is uniform with respect to t, the continuity

[6] This proof generalizes that one by Renardy and Rogers [349, p.380] for $p \ne 2$. For p general, see e.g. Gajewski [168, Sect. IV.1.5] or Zeidler [427, Proposition 23.23] where a bit different technique was used.

[7] This can be proved simply by analyzing the function $(1-\xi^q)^2/|1-\xi^2|^q$ with $\xi = a/b > 0$. This function is either constant=1 for $q = 2$ or, if $q < 2$, decreasing on [0,1] and increasing on $[1, +\infty)$ and always below 1 (except for $\xi = 0$ where it equals 1).

of the embedding $W^{1,p,q}(I;V,V^*)\subset C(I;H)$ has been proved if one confines to functions from $C^1(I;V)$.

Yet, the desired embedding as well as the formula (7.15) can be obtained by the density argument for all functions from $W^{1,p,q}(I;V,V^*)$; cf. Lemma 7.2. The fact that $u:I\to H$ is continuous follows from (7.15): if used by v constant and letting $t_2\to t_1$, we get $(u(t_2),v)\to(u(t_1),v)$, hence $u(\cdot)$ is weakly continuous, and by $v=u$ we get $\|u(t_2)\|_H\to\|u(t_1)\|_H$, hence by Theorem 1.2 $u(t_2)\to u(t_1)$ in the norm topology of H. □

The following approximation property will occasionally be used.

Lemma 7.4. *Let $1\le p<+\infty$. For any $u\in L^p(I;V)\cap L^\infty(I;H)$ and any $u_0\in H$, there is a sequence $\{u_\varepsilon\}_{\varepsilon>0}\subset W^{1,\infty,\infty}(I;V,H)$ such that*

$$u=\lim_{\varepsilon\to0}u_\varepsilon \qquad \textit{in}\ \ L^p(I;V), \tag{7.18a}$$

$$\limsup_{\varepsilon\to0}\int_0^T\Big(\frac{\mathrm{d}u_\varepsilon}{\mathrm{d}t},u_\varepsilon-u\Big)\,\mathrm{d}t\le0, \tag{7.18b}$$

$$u_0=\lim_{\varepsilon\to0}u_\varepsilon(0) \qquad \textit{in}\ \ H, \tag{7.18c}$$

$$\|u_\varepsilon\|_{L^\infty(I;H)}\le\|u\|_{L^\infty(I;H)},\quad u(t)=\operatorname*{w-lim}_{\varepsilon\to0}u_\varepsilon(t)\ \ \textit{in } H \textit{ for a.a. } t\in I. \tag{7.18d}$$

Proof. [8] As $V\subset H$ densely, we can take $\{u_{0\varepsilon}\}_{\varepsilon>0}\subset V$ such that $\lim_{\varepsilon\to0}u_{0\varepsilon}=u_0$ in H. Then we make the prolongation of u by $u_{0\varepsilon}$ for $t<0$, let us denote it by $\bar u_\varepsilon$, and define

$$u_\varepsilon(t):=\frac{1}{\varepsilon}\int_0^{+\infty}\mathrm{e}^{-s/\varepsilon}\bar u_\varepsilon(t-s)\,\mathrm{d}s. \tag{7.19}$$

In other words, u_ε is a convolution of $\bar u_\varepsilon$ with the kernel $\varrho_\varepsilon(t):=\chi_{[0,+\infty)}\varepsilon^{-1}\mathrm{e}^{-t/\varepsilon}$. A simple calculation gives $u_\varepsilon(0)=\varepsilon^{-1}u_{0\varepsilon}\int_0^{+\infty}\mathrm{e}^{-s/\varepsilon}\,\mathrm{d}s=u_{0\varepsilon}$ hencefore (7.18c) is proved. Also, $\{u_\varepsilon(t)\}_{\varepsilon>0}$ is bounded in H and thus converges weakly, for $t\in I$ fixed, as a subsequence to some $\tilde u(t)$. Simultaneously, for $u^*\in H$, the whole sequence $\{\langle u^*,u_\varepsilon(t)\rangle\}_{\varepsilon>0}$ converges to $\{\langle u^*,u(t)\rangle\}_{\varepsilon>0}$ at each left Lebesgue point of $\langle u^*,u(\cdot)\rangle$. Using separability of H, we get that $\tilde u(t)=u(t)$ for a.a. $t\in I$, i.e. (7.18d). Also,

$$\begin{aligned}\big\|u_\varepsilon\big\|_{L^\infty(I;V)}&=\max_{t\in I}\Big\|\frac{1}{\varepsilon}\int_0^t\mathrm{e}^{-s/\varepsilon}u(t-s)\,\mathrm{d}s+\frac{1}{\varepsilon}\int_t^{+\infty}\mathrm{e}^{-s/\varepsilon}u_{0\varepsilon}(t-s)\,\mathrm{d}s\Big\|_V\\&\le\big\|\rho_\varepsilon\big\|_{L^\infty(\mathbb{R})}\big\|u\big\|_{L^1(I;V)}+\mathrm{e}^{-t/\varepsilon}\big\|u_{0\varepsilon}\big\|_V\le\frac{T^{1-1/p}}{\varepsilon}\big\|u\big\|_{L^p(I;V)}+\big\|u_{0\varepsilon}\big\|_V.\end{aligned}$$

[8] Cf. Showalter [383, Sect.III.7]. For $p\ge2$ see also Lions [261, Ch.II, Sect.9.2].

Moreover, on I, it holds

$$\begin{aligned}\frac{\mathrm{d}u_\varepsilon}{\mathrm{d}t} &= \frac{\mathrm{d}}{\mathrm{d}t}\frac{1}{\varepsilon}\int_0^{+\infty} \mathrm{e}^{-s/\varepsilon}\bar{u}_\varepsilon(t-s)\,\mathrm{d}s = \frac{\mathrm{d}}{\mathrm{d}t}\frac{1}{\varepsilon}\int_{-\infty}^{t} \mathrm{e}^{(\xi-t)/\varepsilon}\bar{u}_\varepsilon(\xi)\,\mathrm{d}\xi \\ &= \frac{u(t)}{\varepsilon} - \frac{1}{\varepsilon^2}\int_{-\infty}^{t} \mathrm{e}^{(\xi-t)/\varepsilon}\bar{u}_\varepsilon(\xi)\,\mathrm{d}\xi = \frac{u(t)}{\varepsilon} - \frac{1}{\varepsilon^2}\int_0^{+\infty} \mathrm{e}^{-s/\varepsilon}\bar{u}_\varepsilon(t-s)\,\mathrm{d}s = \frac{u-u_\varepsilon}{\varepsilon}\end{aligned} \tag{7.20}$$

where the substitution $t - s = \xi$ has been used twice. In particular, $\frac{\mathrm{d}}{\mathrm{d}t}u_\varepsilon \in L^\infty(I;H)$, and one can test (7.20) by $u_\varepsilon - u$, which gives

$$\int_0^T \Big(\frac{\mathrm{d}u_\varepsilon}{\mathrm{d}t}, u_\varepsilon - u\Big)\,\mathrm{d}t = -\frac{1}{\varepsilon}\int_0^T \|u_\varepsilon - u\|_H^2\,\mathrm{d}t \le 0 \tag{7.21}$$

and therefore (7.18b) is certainly valid. Eventually, (7.18a) can be obtained by the calculations in Lemma 7.2 modified for the kernel ϱ_ε specified here. □

Remark 7.5. The formula (7.15) for $u = v$ gives

$$\frac{1}{2}\|u(t_2)\|_H^2 - \frac{1}{2}\|u(t_1)\|_H^2 = \int_{t_1}^{t_2} \Big\langle \frac{\mathrm{d}u}{\mathrm{d}t}, u(t)\Big\rangle_{V^*\times V}\,\mathrm{d}t\,. \tag{7.22}$$

From this formula, one can also see that the function $t \mapsto \frac{1}{2}\|u(t)\|_H^2$ is absolutely continuous. Hence, its derivative exists a.e. on I and the *chain rule* holds:

$$\frac{1}{2}\frac{\mathrm{d}}{\mathrm{d}t}\|u(t)\|_H^2 = \Big\langle \frac{\mathrm{d}u}{\mathrm{d}t}, u(t)\Big\rangle_{V^*\times V} \qquad \text{for a.a. } t\in I. \tag{7.23}$$

7.3 Aubin-Lions lemma

We saw already in Part I that limit passage in lower-order terms needs compactness arguments. This will be the application of the results presented in this section. Let us first prove one auxiliary inequality which is sometimes referred to as *Ehrling's lemma*[9] if V_3 is a Banach space. Here, however, we admit V_3 a locally convex space, which will simplify application to the Galerkin method in Section 8.4.

Lemma 7.6 (Ehrling [134], generalized)**.** *Let V_1, V_2 be Banach spaces, and V_3 be a metrizable Hausdorff locally convex space, $V_1 \Subset V_2$ (compact embedding), and $V_2 \subset V_3$ (continuous embedding). Then, for any $p \ge 1$,*

$$\forall \varepsilon > 0\ \ \exists a > 0\ \ \exists K\in\mathbb{N}\ \ \forall v\in V_1: \qquad \|v\|_{V_2}^p \le \varepsilon\|v\|_{V_1}^p + a\sum_{\ell=1}^K |v|_\ell^p. \tag{7.24}$$

[9]The Ehrling lemma says: if V_1, V_2, V_3 are Banach spaces, a linear operator $A : V_1 \to V_2$ is compact and a linear operator $B : V_2 \to V_3$ is injective. Then $\forall\varepsilon > 0\ \exists C < +\infty\ \forall u \in V_1$: $\|Au\|_{V_2} \le \varepsilon\|u\|_{V_1} + C\|BAu\|_{V_3}$; cf. e.g. Alt [10, p.335]. In the original paper, Ehrling [134] formulated this sort of assertion in less generality.

Proof. Suppose the contrary. Thus we get $\varepsilon > 0$ such that for all $a > 0$ and $K \in \mathbb{N}$ there is $v_{a,K} \in V_1$: $\|v_{a,K}\|_{V_2}^p > \varepsilon\|v_{a,K}\|_{V_1}^p + a\sum_{\ell=1}^K |v_{a,K}|_\ell^p$. Putting $w_{a,K} = v_{a,K}/\|v_{a,K}\|_{V_1}$, we get

$$\|w_{a,K}\|_{V_2}^p \geq \varepsilon + a\sum_{\ell=1}^K |w_{a,K}|_\ell^p \tag{7.25}$$

and also $\|w_{a,K}\|_{V_2} \leq N_{12}$, the norm of the embedding $V_1 \subset V_2$. From (7.25) we get $(\sum_{\ell=1}^K |w_{a,K}|_\ell^p)^{1/p} \leq a^{-1/p}N_{12}$ and therefore also $|w_{a,K}|_\ell \leq a^{-1/p}N_{12}$ for any a and any $K \geq \ell$, and thus $\lim_{a,K\to+\infty} |w_{a,K}|_\ell = 0$ for any $\ell \in \mathbb{N}$. As $\{w_{a,K}\}_{a>0,K\in\mathbb{N}}$ is bounded in V_1 and the embedding $V_1 \subset V_2$ is compact, we have (up to a subsequence) $w_{a,K} \to w$ in V_2 if $a, K \to +\infty$. Hence also $|w_{a,K} - w|_\ell \to 0$ for any $\ell \in \mathbb{N}$ because $V_2 \subset V_3$ continuous. Clearly,

$$|w|_\ell \leq |w_{a,K}|_\ell + |w_{a,K} - w|_\ell \to 0$$

so that $|w|_\ell = 0$. Hence $w = 0$ because V_3 is assumed a Hausdorff space, so that $w_{a,K} \to 0$ in V_2, which contradicts (7.25). Thus (7.24) is proved. □

Lemma 7.7 (Aubin and Lions, generalized[10]). *Let V_1, V_2 be Banach spaces, and V_3 be a metrizable Hausdorff locally convex space, V_1 be separable and reflexive, $V_1 \Subset V_2$ (a compact embedding), $V_2 \subset V_3$ (a continuous embedding), $1 < p < +\infty$, $1 \leq q \leq +\infty$. Then*

$$W^{1,p,q}(I; V_1, V_3) \Subset L^p(I; V_2) \qquad (a\ compact\ embedding). \tag{7.26}$$

Proof. We will consider V_3 equipped with a collection of seminorm $\{|\cdot|_\ell\}_{\ell\in\mathbb{N}}$.

We are to prove that bounded sets in $W^{1,p,q}(I; V_1, V_3)$ are relatively compact in $L^p(I; V_2)$. Take $\{u_k\}$ a bounded sequence in $W^{1,p,q}(I; V_1, V_3)$.[11] In particular, as V_1 is reflexive and separable and $p \in (1, +\infty)$, the Bochner space $L^p(I; V_1)$ is reflexive, cf. Proposition 1.38, and thus we have (up to a subsequence)

$$u_k \rightharpoonup u \quad \text{in } L^p(I; V_1). \tag{7.27}$$

As always $L^q(I; V_3) \subset L^1(I; V_3)$, we have

$$\left\{\frac{\mathrm{d}u_k}{\mathrm{d}t}\right\}_{k\in\mathbb{N}} \quad \text{bounded in } L^1(I; V_3). \tag{7.28}$$

We may consider $u = 0$ in (7.27) without loss of generality. Putting $v := u_k(t)$ into (7.24) and integrating it over I, we get

$$\|u_k\|_{L^p(I;V_2)}^p \leq \varepsilon\|u_k\|_{L^p(I;V_1)}^p + a\sum_{\ell=1}^K |u_k|_{p,\ell}^p \tag{7.29}$$

[10] For the original version with V_3 a Banach space see Aubin [25] and Lions [261]. For a generalization, see also Dubinskiĭ [127] and Simon [386]. For a nonmetrizable V_3, see [359].

[11] As we address compactness in a Banach space $L^p(I; V_2)$, we can work in terms of sequences only, which agrees with our definition of compactness of sets as a "sequential" compactness.

where $|u|_{p,\ell} := (\int_0^T |u(t)|_\ell^p \mathrm{d}t)^{1/p}$, cf. (7.5). The first right-hand-side term can be made arbitrarily small by taking $\varepsilon > 0$ small independently of k because $\sup_{k\in\mathbb{N}} \|u_k\|_{L^p(I;V_1)} < +\infty$. Hence, take $\varepsilon > 0$ fixed, which then fixes also a and K. Then, for ℓ arbitrary but fixed, we are to push to zero the term $|u_k|_{p,\ell}^p = \int_0^{T/2} |u_k(t)|_\ell^p \mathrm{d}t + \int_{T/2}^T |u_k(t)|_\ell^p \mathrm{d}t$ and we may investigate only, say, the first integral. Take $\delta > 0$, we can assume $\delta \le T/2$. For $t \in I/2 := [0, T/2]$ we can decompose

$$u_k = \tilde{u}_k + z_k\,, \qquad \text{with} \quad \tilde{u}_k(t) := \frac{1}{\delta}\int_0^\delta u_k(t+\vartheta)\mathrm{d}\vartheta\,, \tag{7.30}$$

i.e. $\tilde{u}_k$, being absolutely continuous, represents the "mollified u_k". Thus, using by-part integration, we have the formula for the remaining z_k:

$$\begin{aligned} z_k(t) &= \int_0^\delta \Big(\frac{\vartheta}{\delta} - 1\Big)\frac{\mathrm{d}}{\mathrm{d}\vartheta}u_k(t+\vartheta)\mathrm{d}\vartheta \\ &= \Big[\Big(\frac{\vartheta}{\delta} - 1\Big)u_k(t+\vartheta)\Big]_{\vartheta=0}^{\delta} - \int_0^\delta \frac{1}{\delta}u_k(t+\vartheta)\mathrm{d}\vartheta = u_k(t) - \tilde{u}_k(t)\,. \end{aligned} \tag{7.31}$$

Then

$$\int_0^{T/2} |u_k(t)|_\ell^p \mathrm{d}t \le 2^{p-1}\int_0^{T/2} |\tilde{u}_k(t)|_\ell^p \mathrm{d}t + 2^{p-1}\int_0^{T/2} |z_k(t)|_\ell^p \mathrm{d}t =: I_{1,\ell} + I_{2,\ell}. \tag{7.32}$$

We can estimate

$$I_{2,\ell} \le \int_0^{T/2}\bigg(\int_0^\delta \Big(1 - \frac{\vartheta}{\delta}\Big)\Big|\frac{\mathrm{d}}{\mathrm{d}t}u_k(t+\vartheta)\Big|_\ell \mathrm{d}\vartheta\bigg)^p \mathrm{d}t = \bigg\|\Big|\frac{\mathrm{d}u_k}{\mathrm{d}t}\Big|_\ell \star \psi_\delta\bigg\|_{L^p(I/2)}^p =: I_{3,\ell}^p, \tag{7.33}$$

where "$\star$" denotes the convolution and $\psi_\delta(t) := (t/\delta + 1)\chi_{[-\delta,0]}(t)$. The following estimate can be proved[12]:

$$\|f \star g\|_{L^p(\mathbb{R})} \le \|f\|_{L^1(\mathbb{R})}\|g\|_{L^p(\mathbb{R})}. \tag{7.34}$$

For $f = |\frac{\mathrm{d}}{\mathrm{d}t}u_k|_\ell$ and $g = \psi_\delta$, we get

$$I_{3,\ell} \le \bigg\|\Big|\frac{\mathrm{d}u_k}{\mathrm{d}t}\Big|_\ell\bigg\|_{L^1(I/2+\delta)} \|\psi_\delta\|_{L^p(\mathbb{R})}. \tag{7.35}$$

By (7.28) and by $\|\psi_\delta\|_{L^p(\mathbb{R})} \le \sqrt[p]{\delta}$, we have $I_{3,\ell} = \mathscr{O}(\sqrt[p]{\delta})$ hence $I_{2,\ell} = \mathscr{O}(\delta)$. In particular, $I_{2,\ell}$ can be made arbitrarily small if $\delta > 0$ is small enough.

[12] One can use the trivial estimates $\|f \star g\|_{L^1(\mathbb{R})} \le \|f\|_{L^1(\mathbb{R})}\|g\|_{L^1(\mathbb{R})}$ and $\|f \star g\|_{L^\infty(\mathbb{R})} \le \|f\|_{L^1(\mathbb{R})}\|g\|_{L^\infty(\mathbb{R})}$ and, as $g \mapsto f \star g$ is a linear operator, obtain (7.34) by interpolation by the classical Riesz-Thorin convexity theorem.

Let us now take $\delta > 0$ fixed. By (7.27) with $u = 0$ and by the definition (7.30) of $\tilde{u}$, we have $\tilde{u}_k(t) \rightharpoonup 0$ in V_1 for every t, hence also $\tilde{u}_k(t) \to 0$ in V_2 because of the compactness of the embedding $V_1 \subset V_2$. Then also $|\tilde{u}_k(t)|_\ell \to 0$ because $V_2 \subset V_3$ is continuous. As $\{u_k\}_{k\in\mathbb{N}}$ is bounded in $L^p(I;V_1)$, it is bounded in $L^1(I;V_2)$ too. Thus

$$\big|\tilde{u}_k(t)\big|_\ell \le C_{1,\ell}\|\tilde{u}_k(t)\|_{V_1} \le \frac{C_{1,\ell}}{\delta}\int_0^\delta \|u_k(t+\vartheta)\|_{V_1}\mathrm{d}\vartheta \le \frac{C_{1,\ell}T^{1-1/p}}{\delta}\|u_k\|_{L^p(I;V_1)},$$

where $C_{1,\ell} = \sup_v |v|_\ell/\|v\|_{V_1}$ is finite because the embedding $V_1 \subset V_3$ is assumed continuous. Thus $\tilde{u}_k(t)$ is bounded in V_3 independently of k and t. By Lebesgue dominated convergence Theorem 1.14, $I_{1,\ell} := \int_0^{T/2} |\tilde{u}_k(t)|_\ell^p \mathrm{d}t \to 0$ for $k \to \infty$.

In view of (7.29), the assertion is proved. □

The following modification of Aubin-Lions' Lemma 7.7 is useful in the situation that we have another Banach space; let us denote it by H, and an L^∞-estimate valued in H at our disposal. This enables us to improve integrability in the target space from p to p/λ:

Lemma 7.8 (Interpolation). *Let V_1, V_2, V_3 be as in Lemma 7.7, and H and V_4 other Banach spaces such that $V_1 \Subset V_2 \subset V_4 \subset H$ and (V_2, V_4, H) forms an interpolation triple in the sense*

$$\big\|v\big\|_{V_4} \le C\big\|v\big\|_H^{1-\lambda}\big\|v\big\|_{V_2}^\lambda \tag{7.36}$$

for some $\lambda \in (0,1)$; then

$$W^{1,p,q}(I;V_1,V_3) \cap L^\infty(I;H) \Subset L^{p/\lambda}(I;V_4). \tag{7.37}$$

Proof. By (7.36), we have the estimate

$$\begin{aligned}\int_0^T \big\|v(t)\big\|_{V_4}^{p/\lambda}\mathrm{d}t &\le C^{p/\lambda}\int_0^T \big\|v(t)\big\|_{V_2}^p\big\|v(t)\big\|_H^{(1-\lambda)p/\lambda}\mathrm{d}t \\ &\le C^{p/\lambda}\big\|v\big\|_{L^\infty(I;H)}^{(1-\lambda)p/\lambda}\big\|v\big\|_{L^p(I;V_2)}^p\end{aligned} \tag{7.38}$$

with C the constant from (7.36). Then by (7.38) we get

$$\big\|u_k - u\big\|_{L^{p/\lambda}(I;V_4)} \le C\big\|u_k - u\big\|_{L^\infty(I;H)}^{1-\lambda}\big\|u_k - u\big\|_{L^p(I;V_2)}^\lambda \to 0 \tag{7.39}$$

because $\|u_k - u\|_{L^\infty(I;H)}^{1-\lambda}$ is bounded by assumption while $\|u_k - u\|_{L^p(I;V_2)}^\lambda$ converges to 0 by Lemma 7.7. □

Still one modification of Aubin-Lions' Lemma 7.7 will be found useful.

Corollary 7.9 (GENERALIZATION FOR $\frac{du}{dt}$ A MEASURE[13]). *Assuming $V_1 \Subset V_2 \subset V_3$ (the compact and the continuous embeddings between Banach spaces, respectively), V_1 reflexive, the Banach space V_3 having a predual space V_3', i.e. $V_3 = (V_3')^*$, and $1 < p < +\infty$, it holds that*

$$W^{1,p,\mathscr{M}}(I;V_1,V_3) := \Big\{u \in L^p(I;V_1);\ \frac{\mathrm{d}u}{\mathrm{d}t} \in \mathscr{M}(I;V_3)\Big\} \Subset L^p(I;V_2). \tag{7.40}$$

Proof. By Hahn-Banach's Theorem 1.5, we can extend $\frac{\mathrm{d}}{\mathrm{d}t}u_k$ from $\mathscr{M}(I;V_3) = C(I;V_3')^*$ to $L^\infty(I;V_3')^*$ while preserving its norm, so that we can test $\frac{\mathrm{d}}{\mathrm{d}t}u_k$ by the discontinuous function ψ_δ as done in (7.31), and eventually get the same estimate as (7.35) but with $\|\frac{\mathrm{d}}{\mathrm{d}t}u_k\|_{L^\infty(I/2+\delta;V_3')^*}$ instead of $\|\,\|\frac{\mathrm{d}}{\mathrm{d}t}u_k\|_{V_3}\|_{L^1(I/2+\delta)}$. □

The case $p = +\infty$ has been excluded in Lemma 7.7 as well as in Corollary 7.9. There is, however, a simple modification for weakly continuous functions.

Lemma 7.10 (ARZELÀ-ASCOLI-TYPE MODIFICATION). *Let V_1, V_2 and V_3 be as in Lemma 7.7 with V_3 a separable reflexive Banach space and let $1 < q \le +\infty$. Then*

$$C(I;(V_1,\text{weak})) \cap W^{1,q}(I;V_3) \Subset C(I;V_2). \tag{7.41}$$

Proof. Consider a bounded sequence $\{u_k\}_{k\in\mathbb{N}}$ in $C(I;(V_1,\text{weak}))\cap W^{1,q}(I;V_3)$. It suffices to make the proof for $q < +\infty$. As $W^{1,q}(I;V_3)$ is reflexive, we can take a subsequence (denoted again as $\{u_k\}_{k\in\mathbb{N}}$ for simplicity) such that $u_k \rightharpoonup u$ in $W^{1,q}(I;V_3)$. By Lemma 7.1, $W^{1,q}(I;V_3) \subset C(I;V_3)$, hence $u_k(t) \rightharpoonup u(t)$ in V_3 for any $t \in I$. Then also $u_k(t) \rightharpoonup u(t)$ in V_1, and by the compact embedding $V_1 \Subset V_2$ also $u_k(t) \to u(t)$ strongly in V_2 for any $t \in I$. The sequence $\{u_k : I \to V_3\}_{k\in\mathbb{N}}$ is equicontinuous because

$$\begin{aligned}\big\|u_k(t_1)-u_k(t_2)\big\|_{V_3} &\le \Big\|\int_{t_1}^{t_2}\frac{\mathrm{d}u_k}{\mathrm{d}t}\,\mathrm{d}t\Big\|_{V_3} \le \int_{t_1}^{t_2}\Big\|\frac{\mathrm{d}u_k}{\mathrm{d}t}\Big\|_{V_3}\mathrm{d}t\\ &\le \Big\|\frac{\mathrm{d}u_k}{\mathrm{d}t}\Big\|_{L^q(I;V_3)}\|1\|_{L^{q'}([t_1,t_2])} = |t_1-t_2|^{1-1/q}\Big\|\frac{\mathrm{d}u_k}{\mathrm{d}t}\Big\|_{L^q(I;V_3)}\end{aligned}$$

for any $0 \le t_1 < t_2 \le T$. Assume that a selected sequence $\{u_k\}_{k\in\mathbb{N}}$ does not converge to u in $C(I;V_2)$. Thus $\|u_k-u\|_{C(I;V_2)} \ge \varepsilon > 0$ for some ε and for all k (from the already selected subsequence), and we would get $\|u_k(t_k)-u(t_k)\|_{V_2} \ge \varepsilon$ for some t_k. By compactness of $I := [0,T]$, we can further select a subsequence and some $t \in I$ so that $t_k \to t$. Then we have $u(t_k) \to u(t)$ in V_2. By the above proved equicontinuity, we have also $u_k(t_k) \rightharpoonup u(t)$ in V_3. By the boundedness of $\{u_k(t_k)\}_{k\in\mathbb{N}}$ in $V_1 \Subset V_2$, we have also $u_k(t_k) \to u(t)$ in V_2. Then $\|u_k(t_k)-u(t_k)\|_{V_2} \to \|u(t)-u(t)\|_{V_2} = 0$, a contradiction. □

[13] For $p = +\infty$ in $L^p(I;V_1)$, this assertion has been stated in [310].

Chapter 8

Evolution by pseudomonotone or weakly continuous mappings

As already advertised in the previous Chapter 7, evolution problems involve one variable, a time t, having a certain specific character and thus a specific treatment is useful, although some methods (applicable under special circumstances, see Sections 8.9 and 8.10) can wipe this specific character off. Conventional methods we will scrutinize in this chapter deal with this one-dimensional variable t by two ways:

(i) discretize t, and then thus created auxiliary approximate problems are based on our knowledge from Part I,

(ii) keep t continuous but approximate the rest by a Galerkin method similarly as we did in Section 2.1, and then the approximate problems are based on ordinary differential equations and Section 1.6.

8.1 Abstract initial-value problems

In this chapter, we will focus on the setting that the evolution is governed by the abstract initial-value problem (the so-called abstract *Cauchy problem*):[1]

$$\frac{\mathrm{d}u}{\mathrm{d}t} + A(t, u(t)) = f(t) \quad \text{for a.a. } t \in I, \qquad u(0) = u_0 \,. \tag{8.1}$$

The latter equality in (8.1) is called an *initial condition.*. We will address especially the case that $A : I \times V \to V^*$, $I := [0, T]$ a fixed bounded time interval, and $V \subset H \cong H^* \subset V^*$ is a Gelfand triple, V a separable reflexive Banach space

[1]In fact, making A time dependent allows us to consider $f = 0$ without loss of generality. Anyhow, it is often convenient to distinguish f. Also, the adjective "Cauchy" in concrete partial differential equations often refers to initial-value problems on unbounded domains without boundary conditions.

embedded continuously and densely (and, for treatment of non-monotone lower-order terms, also compactly) into a Hilbert space $H \cong H^*$.

Definition 8.1 (*Strong solution*). We call $u \in W^{1,p,p'}(I; V, V^*)$ with $p' := p/(p-1)$ a strong solution to (8.1) if the first equality in (8.1) holds in V^* while the second one in H.

The main feature of the concept of Definition 8.1 is that $\frac{\mathrm{d}}{\mathrm{d}t}u$ lives in the dual space to the space where u lives, i.e. here $L^p(I; V)$. In particular, the initial condition $u(0) = u_0 \in H$ has a good sense simply due to Lemma 7.3. Sometimes, this information about $\frac{\mathrm{d}}{\mathrm{d}t}u$ is not available, however. Then, in some cases, it suffices to take $L^p(I; V)$ in Definition 8.1 smaller, e.g. $L^p(I; V) \cap L^q(I; H)$ as in Remark 8.12 or $W^{1,p,p'}(I; V, V^*) \cap L^\infty(I; H)$, but sometimes even this is not possible. Such situations are indeed difficult but some results can still be achieved with the following definition, working with a dense subspace $Z \subset V$ and considering $A : I \times V \to Z^*$.

Definition 8.2 (*Weak solution*). We call $u \in L^p(I; V) \cap C(I; (H, \text{weak}))$ a weak solution to (8.1) if the integral identity

$$\int_0^T \langle A(t, u(t)) - f(t), v(t) \rangle_{Z^* \times Z} - \left\langle \frac{\mathrm{d}v}{\mathrm{d}t}, u(t) \right\rangle_{V^* \times V} \mathrm{d}t + \big(v(T), u(T)\big) = \big(v(0), u_0\big) \tag{8.2}$$

holds for all $v \in W^{1,\infty,\infty}(I; Z, V^*)$; the parenthesis $(\cdot, \cdot)$ is the inner product in H.

Sometimes, one can consider a modification of Definition 8.2 by requiring $v(T) = 0$ and then $u \in L^p(I; V)$ only. A justification of Definition 8.2 is its selectivity (Lemma 8.4 with a uniqueness in qualified cases in Theorem 8.36 below) and the following assertion of consistency:

Lemma 8.3 (Consistency of the weak solution). *Any strong solution u to (8.1) with $f \in L^{p'}(I; V^*)$ is also a weak solution (considering an arbitrary dense $Z \subset V$).*

Proof. Note that $t \mapsto A(t, u(t)) = f(t) - \frac{\mathrm{d}}{\mathrm{d}t}u \in L^{p'}(I; V^*)$. Considering $v \in W^{1,\infty,\infty}(I; Z, V^*) \subset W^{1,p,p'}(I; V, V^*)$ and testing (8.1) by $v = v(t)$, we obtain (8.2) after integration over I by using the by-part formula (7.15) and the initial condition $u(0) = u_0$. □

Lemma 8.4 (Selectivity of the weak solution). *Let $f(t) \in V^*$ for a.a. $t \in I$. Then any weak solution u which also belongs to $W^{1,p,p'}(I; V, V^*)$ and for which $A(t, u(t)) \in V^*$ for a.a. $t \in I$ is also the strong solution due to Definition* 8.1.

Proof. The qualification of u allows us to use the by-part formula (7.15) to the term $\langle \frac{\mathrm{d}}{\mathrm{d}t}v, u \rangle$ in (8.2), which results in

$$\int_0^T \left\langle \frac{\mathrm{d}u}{\mathrm{d}t} + A(t, u(t)) - f(t), v(t) \right\rangle_{V^* \times V} \mathrm{d}x = \big(u_0 - u(0), v(0)\big). \tag{8.3}$$

Using $v(0) = 0$, the right-hand-side term vanishes, and realizing that the left-hand-side integral is the duality pairing in $L^{p'}(I;V^*) \times L^p(I;V)$ we obtain $\frac{\mathrm{d}}{\mathrm{d}t}u + A(t,u(t)) = f(t)$ for a.a. $t \in I$, cf. Exercise 8.49 and realize that $\{g \in W^{1,p'}(I);\ g(0) = 0\}$ is dense in $L^p(I)$. Putting this into (8.3), we obtain $(u_0 - u(0), v(0)) = 0$. Taking now v general, we get still $u(0) = u_0$. □

It should be emphasized that various adjectives such as "weak" or "strong" may address different issues, depending on the context. This is a general state of the art in theory of partial differential equations, especially evolutionary, which is reflected also here. Therefore, referring to a specific definition is always advisable. For readers convenience, the used terminology is summarized in Table 2.

abstract level	level of concrete differential equations or variational inequalities
strong solution	weak solution
weak solution	very weak solution

Table 2. Terminology about solutions to evolution problems.

8.2 Rothe method

Let us begin with the problem (8.1) in the autonomous case, i.e. $A : V \to V^*$ is independent of time t:

$$\frac{\mathrm{d}u}{\mathrm{d}t} + A\big(u(t)\big) = f(t)\ , \qquad u(0) = u_0\ . \tag{8.4}$$

In this section we present the so-called *Rothe method* [356] consisting in discretization in time. Let $\tau > 0$ be a time step; for simplicity, we take an equidistant partition of I and suppose T/τ is an integer. Moreover, we will work with a sequence of such time steps $\{\tau_l\}_{l\in\mathbb{N}}$ such that $\lim_{l\to\infty}\tau_l = 0$ and, again for simplicity, assume that $\tau_l = 2^{-l}T$ so that the partitions are "nested" in the sense that the subsequent partition always refines the previous one. Let us further agree to omit the index l and write simply $\tau \to 0$ instead of $\tau_l \to 0$ for $l \to \infty$. Moreover, we must approximate values of f at particular points $t = k\tau$, $0 \le k \le T/\tau$. One possible way is to apply a mollifier as (7.10) to f instead of u; let us denote by $f_\varepsilon \in C(I;V^*)$ the resulting smoothened right-hand side. Then, choosing a suitable $\varepsilon = \varepsilon(\tau)$, cf. Lemma 8.7 below, we put $f_\tau^k = f_{\varepsilon(\tau)}(k\tau)$. Then we define $u_\tau^k \in V$, $k = 1,\dots,T/\tau$, by the following recursive formula:

$$\frac{u_\tau^k - u_\tau^{k-1}}{\tau} + A(u_\tau^k) = f_\tau^k\ , \qquad f_\tau^k := f_{\varepsilon(\tau)}(k\tau)\ , \qquad u_\tau^0 = u_{0\tau}\ , \tag{8.5}$$

called sometimes an *implicit Euler formula.*[2] Sometimes, the recursion (8.5) is started simply from the original initial condition for (8.4), i.e. $u_{0\tau} = u_0$, but in

[2]The adjective "implicit" emphasizes that the unknown u_τ^k occurs in A and cannot be explicitly expressed, and to distinguish it from an explicit Euler formula $u_\tau^k - u_\tau^{k-1} + \tau A(u_\tau^{k-1}) = \tau f_\tau^k$

general $u_{0\tau}$ may be only a suitable approximation of u_0, cf. (8.38) below. Furthermore, we define the piecewise affine interpolant $u_\tau \in C(I;V)$ by

$$u_\tau(t) = \Big(\frac{t}{\tau} - (k-1)\Big)u_\tau^k + \Big(k - \frac{t}{\tau}\Big)u_\tau^{k-1} \qquad \text{for } (k-1)\tau < t \le k\tau \tag{8.6}$$

and the piecewise constant interpolant $\bar{u}_\tau \in L^\infty(I;V)$ by

$$\bar{u}_\tau(t) = u_\tau^k \qquad \text{for } (k-1)\tau < t \le k\tau, \quad k = 0,\dots,T/\tau. \tag{8.7}$$

Let us note that u_τ has a derivative $\frac{\mathrm{d}}{\mathrm{d}t}u_\tau \in L^\infty(I;V)$ which is piecewise constant; cf. Figure 17, whereas $\bar{u}_\tau$ has only a distributional derivative composed from Dirac masses, cf. (8.34). Analogously, $\bar{f}_\tau$ abbreviates the piece-wise constant function defined by

$$\bar{f}_\tau(t) := f_\tau^k := f_{\varepsilon(\tau)}(k\tau) \qquad \text{for} \quad t \in \big[(k-1)\tau, k\tau\big], \quad k = 1,\dots,T/\tau. \tag{8.8}$$

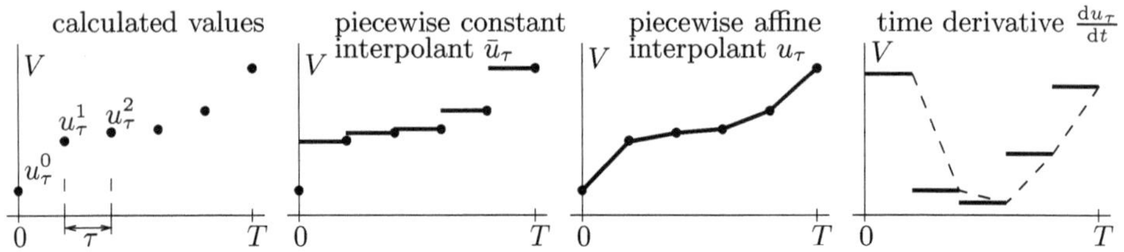

Figure 17. Illustration of Rothe's interpolants $\bar{u}_\tau$ and u_τ constructed from a sequence $\{u_\tau^k\}_{k=0}^{T/\tau}$, and the time derivative $\frac{\mathrm{d}}{\mathrm{d}t}u_\tau$; the dashed line on the last picture shows the interpolated derivative $[\frac{\mathrm{d}}{\mathrm{d}t}u_\tau]^{\mathrm{i}}$ which will be used in Chapter 11.

Let us consider a seminorm on V, denoted by $|\cdot|_V$, such that the following "abstract Poincaré-type" inequality holds:

$$\exists C_{\mathrm{P}} \in \mathbb{R}^+ \;\; \forall v \in V: \qquad \|v\|_V \le C_{\mathrm{P}}\big(|v|_V + \|v\|_H\big). \tag{8.9}$$

A trivial case of (8.9) is $|\cdot|_V := \|\cdot\|_V$ with $C_{\mathrm{P}} = 1$. Referring to such seminorm, we will call A *semi-coercive*[3]

$$\big\langle A(u), u\big\rangle \ge c_0|u|_V^p - c_1|u|_V - c_2\|u\|_H^2\,; \tag{8.10}$$

this condition essentially determines the power $p < +\infty$ in the functional setting of the problem. In some special cases, typically for $|\cdot|_V = \|\cdot\|_V$ and $c_1 = 0$, the mappings A satisfying (8.10) are also called *weakly coercive*, cf. e.g. [314], and, in the case $p = 2$, it also generalizes the so-called *Gårding inequality* (designed originally for specific linear differential operators [172]). In this Chapter, we will assume $p > 1$.

which is, however, not suitable if V is infinite-dimensional. For semi-implicit formulae see Remark 8.14 below, while a multilevel formula is in Remark 8.20.

[3]Note that, even if $|\cdot|_V = \|\cdot\|_V$ would be considered, $\langle A(u), u\rangle$ may tend to $-\infty$ for $\|u\|_V \to \infty$ provided $p < 2$. Thus (8.10) is indeed much weaker than the "full" coercivity (2.5). For some considerations, (8.10) can be even weakened by considering $c_2\|u\|_H^2 \ln(\|u\|_H^2)$ instead of $c_2\|u\|_H^2$ and then using a generalized Gronwall inequality, though e.g. Lemma 8.5 would not hold.

Lemma 8.5 (Existence of Rothe's sequence). *Let $A : V \to V^*$ be pseudomonotone and semi-coercive, $f \in L^1(I; V^*)$, and $u_{0\tau} \in V^*$. Then, for a sufficiently small $\tau > 0$ (namely $\tau < 1/c_2$), the Rothe solution $u_\tau \in C(I; V^*)$ does exist.*

Proof. Let us notice that the identity mapping $\mathbf{I} : V \to V$ is monotone as a mapping $V \to V^*$ because the embedding $V \subset H \subset V^*$ implies

$$\langle \mathbf{I}u - \mathbf{I}v, u - v\rangle_{V^* \times V} = \langle u - v, u - v\rangle_{H^* \times H} = (u - v, u - v) = \|u - v\|_H^2 \geq 0.$$

Thus $\frac{1}{\tau}\mathbf{I} + A : V \to V^*$ is pseudomonotone because $\mathbf{I}$ is monotone, bounded, radially continuous (hence pseudomonotone by Lemma 2.9) and the sum of two pseudomonotone mappings is again pseudomonotone, see Lemma 2.11(i). Also, $\frac{1}{\tau}\mathbf{I} + A : V \to V^*$ is coercive for τ small enough. Indeed,

$$\langle u + \tau A(u), u\rangle = (u, u) + \tau\langle A(u), u\rangle \geq \tau c_0 |u|_V^p - \tau c_1 |u|_V + (1 - \tau c_2)\|u\|_H^2$$

which can, for $\tau c_2 \leq 1$, be estimated from below by $\varepsilon(|u|_V + \|u\|_H)^{\min(2,p)} - 1/\varepsilon \geq \varepsilon(\|u\|_V/C_{\rm P})^{\min(2,p)} - 1/\varepsilon$ for $\varepsilon > 0$ small enough, so that the coercivity (2.5) is fulfilled. Thus, by Theorem 2.6, the equation $[\frac{1}{\tau}\mathbf{I} + A](u) = \frac{1}{\tau}u_\tau^{k-1} + f_\tau^k$ has some solution $u =: u_\tau^k \in V$. $\square$

The mapping $A : V \to V^*$ induces a superposition mapping (i.e. a special case of the abstract Nemytskiĭ mapping) $\mathscr{A}$ by

$$[\mathscr{A}(u)](t) := A(u(t)) \; . \tag{8.11}$$

In the following, we will need to have some information about the time derivative which can be ensured by assuming that

$$\mathscr{A} : L^p(I; V) \cap L^\infty(I; H) \to L^{q'}(I; Z^*) \;\; \text{bounded} \tag{8.12}$$

with some $q \geq 1$ and $Z \subset V$ densely. A simple example for a condition that guarantees (8.12) is the growth condition on A:

$$\exists \mathfrak{C} : \mathbb{R} \to \mathbb{R} \text{ increasing } \forall v \in V : \qquad \|A(v)\|_{Z^*} \leq \mathfrak{C}(\|v\|_H)\big(1 + \|v\|_V^{p/q'}\big). \tag{8.13}$$

In concrete cases, (8.12) may involve rather fine estimates, cf. Example 8.63(2) below.

Lemma 8.6 (Basic a-priori estimates). *Let (8.9) hold, A be pseudomonotone and semi-coercive, let*

$$f = f_1 + f_2 \qquad \textit{with } f_1 \in L^{p'}(I; V^*), \; f_2 \in L^{q'}(I; H), \tag{8.14}$$

let the mollified approximation $f_{\varepsilon(\tau)} = (f_1 + f_2)_{\varepsilon(\tau)} := f_{1\varepsilon(\tau)} + f_{2\varepsilon(\tau)}$ used in (8.5) and the corresponding interpolant $\bar{f}_\tau = \bar{f}_{1\tau} + \bar{f}_{2\tau}$ constructed by (8.5) satisfy

$$\big\|f_{1\varepsilon(\tau)}\big\|_{C(I;V^*)} \leq \frac{K_1}{\sqrt{\tau}} \quad \textit{and} \quad \big\|f_{2\varepsilon(\tau)}\big\|_{C(I;H)} \leq \frac{K_2}{\sqrt{\tau}}, \tag{8.15a}$$

$$\big\|\bar{f}_{1\tau}\big\|_{L^{p'}(I;V^*)} \leq K_1 \quad \textit{and} \quad \big\|\bar{f}_{1\tau}\big\|_{L^{q'}(I;H)} \leq K_2 \tag{8.15b}$$

for some (but arbitrarily chosen) K_1, K_2 *and let* $\{u_{0\tau}\}_{\tau>0}$ *be bounded in* H*. Then, for any* $0 < \tau \le \tau_0$ *with* τ_0 *small enough so that*

$$2\tau_0 c_2 + \sqrt{\tau_0} C_{\rm P} K_1 + \sqrt{\tau_0} K_2 < 1 \tag{8.16}$$

with $C_{\rm P}$ *from* (8.9)*, the following a-priori estimates*

$$\|u_\tau\|_{C(I;H)} \le C_1 , \tag{8.17a}$$
$$\|\bar{u}_\tau\|_{L^\infty(I;H)\,\cap\,L^p(I;V)} \le C_1 , \tag{8.17b}$$
$$\big\|u_\tau|_{[\tau_0,T]}\big\|_{L^p([\tau_0,T];V)} \le C_1 , \tag{8.17c}$$
$$\Big\|\frac{{\rm d}u_\tau}{{\rm d}t}\Big\|_{L^2(I;H)} \le \frac{C_2}{\sqrt{\tau}} , \tag{8.17d}$$

hold with some C_1 *and* C_2 *depending on* p, $C_{\rm P}$, $\|f_1\|_{L^{p'}(I;V^*)}$, $\|f_2\|_{L^1(I;H)}$, *and* $\sup_{\tau>0}\|u_{0\tau}\|_H$ *only. Moreover, if* $u_{0\tau} \in V$*, then*

$$\|u_\tau\|_{L^p(I;V)} \le \sqrt[p]{C_1^p + \tau\|u_{0\tau}\|_V^p} . \tag{8.18}$$

Eventually, if also (8.12) *and* (8.14) *hold with* $q \ge p$*, then*

$$\Big\|\frac{{\rm d}u_\tau}{{\rm d}t}\Big\|_{L^{q'}(I;Z^*)} \le C_3, \tag{8.19a}$$
$$\Big\|\frac{{\rm d}\bar{u}_\tau}{{\rm d}t}\Big\|_{\mathscr{M}(I;Z^*)} \le T^{1/q} C_3. \tag{8.19b}$$

Before starting a rigorous proof, let us sketch the heuristics in an easily observable way neglecting (otherwise necessary) technicalities related to the approximate problem. First, "multiply" the equation $\frac{\rm d}{{\rm d}t}u + A(u) = f$ by the solution u itself, and then use $\langle \frac{\rm d}{{\rm d}t}u, u\rangle = \frac{1}{2}\frac{\rm d}{{\rm d}t}\|u\|_H^2$, cf. also (7.23), the semi-coercivity (8.10) and Young's inequality. This gives

$$\frac{\rm d}{{\rm d}t}\Big(\frac{1}{2}\|u\|_H^2 + c_0\int_0^t |u(\theta)|_V^p\,{\rm d}\theta\Big) = \frac{1}{2}\frac{\rm d}{{\rm d}t}\|u\|_H^2 + c_0|u|_V^p \le \Big\langle \frac{{\rm d}u}{{\rm d}t} + A(u), u\Big\rangle$$
$$+ c_1|u|_V + c_2\|u\|_H^2 = c_1|u|_V + c_2\|u\|_H^2 + \langle f, u\rangle =: R_1 + R_2 + R_3, \tag{8.20}$$

where $|u|_V$, $\|u\|_H$, f, etc. abbreviate $|u(t)|_V$, $\|u(t)\|_H$, $f(t)$, etc., respectively. By (8.9), the last term can be estimated as

$$\begin{aligned} R_3 &:= \langle f, u\rangle = \langle f_1 + f_2, u\rangle \le \|f_1\|_{V^*}\|u\|_V + \|f_2\|_H\|u\|_H \\ &\le C_{\rm P}\|f_1\|_{V^*}\big(|u|_V + \|u\|_H\big) + \|f_2\|_H\Big(\frac{1}{2} + \frac{1}{2}\|u\|_H^2\Big) \\ &\le C_{\rm P}^{p'} C_\varepsilon\|f_1\|_{V^*}^{p'} + C_{\rm P}^p\varepsilon|u|_V^p + \big(C_{\rm P}\|f_1\|_{V^*} + \|f_2\|_H\big)\Big(\frac{1}{2} + \frac{1}{2}\|u\|_H^2\Big) \end{aligned} \tag{8.21}$$

with C_ε from (1.22). The term $C_{\rm P}\varepsilon|u|_V^p$ can then be absorbed in the left-hand side if $\varepsilon < \frac{1}{2}c_0/C_{\rm P}$ is chosen, while the other terms have integrable coefficients as functions of t just by the assumption (8.14). Similarly, $R_1 \le c_1(C_\varepsilon + \varepsilon|u|_V^p)$ with C_ε again from (1.22), and then $c_1\varepsilon|u|_V^p$ can be absorbed in the left-hand side if also $\varepsilon < c_0/(2c_1)$. Altogether, we obtain

$$\frac{\mathrm{d}}{\mathrm{d}t}\Big(\frac{1}{2}\|u\|_H^2 + \Big(c_0 - \varepsilon C_{\rm P} - \varepsilon c_1\Big)\int_0^t |u(\theta)|_V^p\,\mathrm{d}\theta\Big) \le C_\varepsilon(c_1 + C_{\rm P}\|f_1\|_{V^*}^{p'}) \\ +c_2\|u\|_H^2 + \big(C_{\rm P}\|f_1\|_{V^*} + \|f_2\|_H\big)\Big(\frac{1}{2} + \frac{1}{2}\|u\|_H^2\Big). \tag{8.22}$$

Then we can use directly Gronwall's inequality (1.66). In such a way, we obtain $\|u(t)\|_H^2 + \int_0^t |u(\theta)|_V^p\,\mathrm{d}\theta$ bounded uniformly with respect to $t \in I$, which yields already (8.17a). Then, for $t = T$, we get also the bound for $\int_0^T |u(t)|_V^p\,\mathrm{d}t$, so that still (8.17b) follows by using also (8.10) and the already obtained estimate (8.17a). The "dual" estimate (8.19a) is then essentially determined by (8.17b) through the condition (8.12). Assuming we know $\frac{\mathrm{d}}{\mathrm{d}t}u = f - A(u)$, which we indeed will know for the discrete problem, and using Hölder's inequality, we have

$$\begin{aligned}\Big\|\frac{\mathrm{d}u}{\mathrm{d}t}\Big\|_{L^{q'}(I;Z^*)} &= \sup_{\|v\|_{L^q(I;Z)}\le 1} \int_0^T \Big\langle\frac{\mathrm{d}u}{\mathrm{d}t}, v\Big\rangle\,\mathrm{d}t \\ &= \sup_{\|v\|_{L^q(I;Z)}\le 1} \int_0^T \langle f - A(u), v\rangle\,\mathrm{d}t \\ &\le \sup_{\|v\|_{L^q(I;Z)}\le 1} \Big(\|f\|_{L^{q'}(I;Z^*)} + \|A(u)\|_{L^{q'}(I;Z^*)}\Big)\|v\|_{L^q(I;Z)}\end{aligned} \tag{8.23}$$

and then (8.12) with the already proved estimate of u in $L^p(I;V)\cap L^\infty(I;H)$ is used.

On the other hand, the estimates (8.17c,d) and (8.18) explicitly involve τ and τ_0 and cannot be seen by such heuristical considerations.

Proof of Lemma 8.6. Let us now make the proof of Lemma 8.6 with full rigor. Multiply (8.5) by u_τ^k. This yields

$$\Big\langle\frac{u_\tau^k - u_\tau^{k-1}}{\tau}, u_\tau^k\Big\rangle + \big\langle A(u_\tau^k), u_\tau^k\big\rangle = \big\langle f_\tau^k, u_\tau^k\big\rangle\,. \tag{8.24}$$

Then sum (8.24) for $k = 1,\dots,l$, multiply by τ, and use the identity $(u_\tau^k - u_\tau^{k-1}, u_\tau^k) = \|u_\tau^k\|_H^2 - (u_\tau^{k-1}, u_\tau^k) = \frac{1}{2}\|u_\tau^k\|_H^2 - \frac{1}{2}\|u_\tau^{k-1}\|_H^2 + \frac{1}{2}\|u_\tau^k - u_\tau^{k-1}\|_H^2$ which follows from (1.4)[4] and which implies the estimate

[4]Indeed, using $u := u_\tau^{k-1}$ and $v := u_\tau^k$ in (1.4) yields $(u_\tau^{k-1}, u_\tau^k) = \frac{1}{4}\|u_\tau^k + u_\tau^{k-1}\|_H^2 - \frac{1}{4}\|u_\tau^k - u_\tau^{k-1}\|_H^2 = \frac{1}{4}\|u_\tau^k\|_H^2 + \frac{1}{2}(u_\tau^{k-1}, u_\tau^k) + \frac{1}{4}\|u_\tau^{k-1}\|_H^2 - \frac{1}{4}\|u_\tau^k - u_\tau^{k-1}\|_H^2$ which further yields the identity in question.

$$\sum_{k=1}^{l}\langle u_\tau^k - u_\tau^{k-1}, u_\tau^k\rangle = \sum_{k=1}^{l}\big(u_\tau^k - u_\tau^{k-1}, u_\tau^k\big) = \sum_{k=1}^{l}\frac{1}{2}\|u_\tau^k\|_H^2 - \frac{1}{2}\|u_\tau^{k-1}\|_H^2 \\ + \frac{1}{2}\|u_\tau^k - u_\tau^{k-1}\|_H^2 \ \geq\ \frac{1}{2}\|u_\tau^l\|_H^2 - \frac{1}{2}\|u_\tau^0\|_H^2. \tag{8.25}$$

This gives

$$\frac{1}{2}\big\|u_\tau^l\big\|_H^2 - \frac{1}{2}\big\|u_\tau^0\big\|_H^2 + \tau\sum_{k=1}^{l}\big\langle A(u_\tau^k), u_\tau^k\big\rangle \leq \tau\sum_{k=1}^{l}\langle f_\tau^k, u_\tau^k\rangle. \tag{8.26}$$

Following the scheme (8.20)–(8.22), by (8.10) and by Hölder's inequality, we can further estimate

$$\frac{1}{2}\|u_\tau^l\|_H^2 + c_0\tau\sum_{k=1}^{l}|u_\tau^k|_V^p \leq \frac{1}{2}\|u_{0\tau}\|_H^2 + \tau\sum_{k=1}^{l}\Big(\langle f_\tau^k, u_\tau^k\rangle + c_1|u_\tau^k|_V + c_2\|u_\tau^k\|_H^2\Big) \\ \leq \frac{1}{2}\|u_{0\tau}\|_H^2 + \tau\sum_{k=1}^{l}\Big(\varepsilon(c_1 + C_{\rm P})|u_\tau^k|_V^p + C_\varepsilon(C_1 + C_{\rm P}\big\|f_{1\tau}^k\big\|_{V^*}^{p'}) \\ + \frac{1}{2}C_{\rm P}\big\|f_{1\tau}^k\big\|_{V^*} + \frac{1}{2}\big\|f_{2\tau}^k\big\|_H + \Big(c_2 + \frac{1}{2}C_{\rm P}\big\|f_{1\tau}^k\big\|_{V^*} + \frac{1}{2}\big\|f_{2\tau}^k\big\|_H\Big)\big\|u_\tau^k\big\|_H^2\Big) \tag{8.27}$$

with $\varepsilon > 0$ small and C_ε correspondingly large, cf. (1.22). If $\varepsilon < c_0/(c_1 + C_{\rm P})$, we can absorb the term with $\varepsilon(c_1 + C_{\rm P})|u_\tau^k|_V^p$ in the respective term in the left-hand side. Then discrete Gronwall's inequality (1.70) can be used; note that (1.70) here requires

$$\tau < \frac{1}{2c_2 + \max_{k=1,\dots,T/\tau}\Big(C_{\rm P}\big\|f_{1\tau}^k\big\|_{V^*} + \big\|f_{2\tau}^k\big\|_H\Big)} \tag{8.28}$$

which is indeed satisfied if $\tau \leq \tau_0$ with τ_0 small as specified in (8.16). Also note that

$$\tau\sum_{k=1}^{l}\big\|f_{1\tau}^k\big\|_{V^*}^{p'} \leq \tau\sum_{k=1}^{T/\tau}\big\|f_{1\tau}^k\big\|_{V^*}^{p'} = \big\|\bar f_{1\tau}\big\|_{L^{p'}(I;V^*)}^{p'} \leq K_1^{p'}, \tag{8.29a}$$

$$\tau\sum_{k=1}^{l}\big\|f_{2\tau}^k\big\|_H \leq \tau\sum_{k=1}^{T/\tau}\big\|f_{2\tau}^k\big\|_H = \big\|\bar f_{2\tau}\big\|_{L^1(I;H)} \leq T^{1/q}K_2 \tag{8.29b}$$

for K_1 and K_2 from (8.15b). By this way, we get already the estimate of $\|\bar u_\tau\|_{L^\infty(I;H)\cap L^p(I;V)}$ and of $\|u_\tau\|_{L^\infty(I;H)}$ if $\tau \leq \tau_0$. Certainly[5]

$$\|u_\tau\|_{L^p([\tau,T];V)} \leq \|\bar u_\tau\|_{L^p(I;V)}, \tag{8.30}$$

[5] It can easily be proved for $p = 1$ and for $p = +\infty$. For $1 < p < +\infty$, we get it by interpolation.

which already gives (8.17c) for $\tau \le \tau_0$.

Moreover, by a finer usage of (8.25) exploiting also the "forgotten" term $\|u_\tau^k - u_\tau^{k-1}\|_H^2$, we get still the boundedness of $\sum_{k=1}^{l} \|u_\tau^k - u_\tau^{k-1}\|_H^2$, from which (8.17d) follows.

As to (8.18), we can formally extend $\bar{u}_\tau$ for $t \le 0$ by $u_{0\tau}$, and then, like (8.30), we have

$$\|u_\tau\|_{L^p(I;V)} \le \|\bar{u}_\tau\|_{L^p([-\tau,T];V)} = \Big(\int_{-\tau}^{0} \|u_{0\tau}\|_V^p\,\mathrm{d}t + \int_0^T \|\bar{u}_\tau\|_V^p \mathrm{d}t\Big)^{1/p}$$
$$= \Big(\tau\|u_{0\tau}\|_V^p + \|\bar{u}_\tau\|_{L^p(I;V)}^p\Big)^{1/p}. \tag{8.31}$$

Then (8.17b) gives (8.18).

As to (8.19a), in view of (8.5), as in (8.23), we have

$$\begin{aligned}\int_0^T \Big\langle \frac{\mathrm{d}u_\tau}{\mathrm{d}t}, v\Big\rangle \,\mathrm{d}t &= \int_0^T \langle \bar{f}_\tau - A(\bar{u}_\tau), v\rangle\,\mathrm{d}t \\ &\le \Big(\|\bar{f}_\tau\|_{L^{q'}(I;Z^*)} + \|A(\bar{u}_\tau)\|_{L^{q'}(I;Z^*)}\Big)\|v\|_{L^q(I;Z)} \\ &\le \bigg(\|f\|_{L^{q'}(I;Z^*)} + \Big(\int_0^T \|A(\bar{u}_\tau)\|_{Z^*}^{q'}\,\mathrm{d}t\Big)^{1/q'}\bigg)\|v\|_{L^q(I;Z)} \end{aligned}\tag{8.32}$$

where $\bar{f}_\tau$ is from (8.8). As $L^{q'}(I;Z^*)$ is isometrically isomorphic with $L^q(I;Z)^*$, cf. Proposition 1.38, it holds that

$$\begin{aligned}\Big\|\frac{\mathrm{d}u_\tau}{\mathrm{d}t}\Big\|_{L^{q'}(I;Z^*)} &= \sup_{\|v\|_{L^q(I;Z)}\le 1} \int_0^T \Big\langle \frac{\mathrm{d}u_\tau}{\mathrm{d}t}, v\Big\rangle\,\mathrm{d}t \qquad (8.33)\\ &\le N\|f\|_{L^{p'}(I;V^*)} + \sup_{\|v\|_{L^p(I;V)\cap L^\infty(I;H)}\le C_1} \|\mathscr{A}(v)\|_{L^{q'}(I;Z^*)}\end{aligned}$$

where N denotes the norm of the embedding $L^{p'}(I;V^*) \subset L^{q'}(I;Z^*)$; here the assumption $q \ge p$ is used. The a-priori bound (8.17b) then gives (8.19a). Then (8.19b) follows easily:

$$\begin{aligned}\Big\|\frac{\mathrm{d}\bar{u}_\tau}{\mathrm{d}t}\Big\|_{\mathscr{M}(I;Z^*)} &= \Big\|\sum_{k=1}^{T/\tau}(u_\tau^k - u_\tau^{k-1})\delta_{(k-1)\tau}\Big\|_{\mathscr{M}(I;Z^*)} = \tau\sum_{k=1}^{T/\tau}\Big\|\frac{u_\tau^k - u_\tau^{k-1}}{\tau}\Big\|_{Z^*} \\ &= \Big\|\frac{\mathrm{d}u_\tau}{\mathrm{d}t}\Big\|_{L^1(I;Z^*)} \le T^{1/q}\Big\|\frac{\mathrm{d}u_\tau}{\mathrm{d}t}\Big\|_{L^{q'}(I;Z^*)} \le T^{1/q}C_3\end{aligned}\tag{8.34}$$

where $\delta_{(k-1)\tau}$ denotes the Dirac distribution at time $t = (k-1)\tau$. $\square$

For the convergence, the following approximation property will be found useful:

Lemma 8.7 (CONVERGENCE OF $\bar{f}_\tau$). *If $f \in L^q(I;X)$ for $1 \le q < +\infty$ and X a Banach space, then $\bar{f}_\tau$ defined by (8.8) with f_ε being a mollified f satisfying $\|f_\varepsilon\|_{C^1(I;X)} \le L(\varepsilon)$ for some $L : \mathbb{R}^+ \to \mathbb{R}^+$ (possibly unbounded) and with $\varepsilon = \varepsilon(\tau)$ so that both $\lim_{\tau\to 0} \varepsilon(\tau) = 0$ and $\sup_{\tau>0} \sqrt{\tau} L(\varepsilon(\tau)) < +\infty$ converges to f in $L^q(I;X)$ when $\tau \to 0$. Also $\|\bar{f}_\tau\|_{C(I;X)} \le K\tau^{-1/2}$, cf. also (8.16), holds for some K, too.*

Proof. As $f_\varepsilon : I \to X$ is Lipschitz continuous with the Lipschitz constant $L(\varepsilon)$, it holds that $\|f_\varepsilon - \bar{f}_\tau\|_{C(I;X)} \le L(\varepsilon)\tau$. Choosing $\varepsilon = \varepsilon(\tau)$ so that $\tau L(\varepsilon(\tau)) \to 0$, we have

$$\begin{aligned}\|\bar{f}_\tau - f\|_{L^q(I;X)} &\le \|\bar{f}_\tau - f_{\varepsilon(\tau)}\|_{L^q(I;X)} + \|f_{\varepsilon(\tau)} - f\|_{L^q(I;X)} \\ &\le T^{1/q}\|\bar{f}_\tau - f_{\varepsilon(\tau)}\|_{L^\infty(I;X)} + \|f_{\varepsilon(\tau)} - f\|_{L^q(I;X)} \\ &\le T^{1/q}\tau L(\varepsilon(\tau)) + \|f_{\varepsilon(\tau)} - f\|_{L^q(I;X)} \to 0\end{aligned}$$

because $f_\varepsilon \to f$ in $L^q(I;X)$, cf. the proof of Lemma 7.2. Eventually, $\|\bar{f}_\tau\|_{C(I;X)} \le L(\varepsilon(\tau)) \le K\tau^{-1/2}$ is obvious with $K := \sup_{\tau>0} \sqrt{\tau} L(\varepsilon(\tau))$. □

When one uses Lemma 8.7 for $f_1 \in L^{p'}(I;V^*)$ and also for $f_2 \in L^{q'}(I;H)$ in place of $f \in L^q(I;X)$, it will justify (8.15a) for $\varepsilon = \varepsilon(\tau)$ decreasing to 0 sufficiently slowly for $\tau \to 0$ while also (8.15b) holds because of the proved convergence. An additional useful auxiliary assertion addresses *pseudomonotonicity* of $\mathscr{A}$.

Lemma 8.8 (PAPAGEORGIOU [323], HERE MODIFIED[6]). *Let $A : V \to V^*$ be pseudomonotone in the sense of (2.3), satisfying (8.10) and (8.13) with $q = p$ and $Z = V$. Then $\mathscr{A}$ is pseudomonotone on $\mathscr{W} := W^{1,p,\mathscr{M}}(I;V,V^*) \cap L^\infty(I;H)$ in the sense analogous to (2.3), i.e.,*

$$\mathscr{A} \text{ is bounded, and} \tag{8.35a}$$

$$\left.\begin{array}{r} u_k \overset{*}{\rightharpoonup} u \;\text{ in } \mathscr{W} \\ \limsup\limits_{k\to\infty} \langle \mathscr{A}(u_k), u_k - u\rangle \le 0 \end{array}\right\} \Rightarrow \begin{array}{l} \forall v \in \mathscr{W} : \\ \langle \mathscr{A}(u), u - v\rangle \le \liminf\limits_{k\to\infty} \langle \mathscr{A}(u_k), u_k - v\rangle. \end{array} \tag{8.35b}$$

Proof. The boundedness (8.35a) of $\mathscr{A} : \mathscr{W} \to \mathscr{W}^*$ follows just by (8.4).

To prove (8.35b), let us now take $u_k \overset{*}{\rightharpoonup} u$ in $\mathscr{W}$. By *Helly's selection principle* for mappings valued in a separable reflexive V^* (see e.g. [39, Chap.1, Thm.3.5]), there is a subsequence (denoted by the same indexes k for simplicity) and $\tilde{u} : I \to V^*$ with a bounded variation such that $u_k(t) \rightharpoonup \tilde{u}(t)$ in V^* for all $t \in I$. Yet, $\tilde{u} = u$ a.e. on I; indeed, for any $v \in L^\infty(I;V^*)$ we have $\langle u_k, v\rangle \to \langle u, v\rangle$ and, by Lebesgue Theorem 1.14, simultaneously $\langle u_k, v\rangle \to \langle \tilde{u}, v\rangle$. Thus, using the boundedness of $\{u_k(t)\}_{k\in\mathbb{N}}$ in H for a.a. $t \in I$, we have even $u_k(t) \rightharpoonup u(t)$ in H for a.a. $t \in I$.

[6] In [323], a non-autonomous case like in Lemma 8.29 below has been addressed but in the case $\mathscr{W} := W^{1,p,p'}(I;V,V^*)$. See also [209, Part II, Chap.I, Thm.2.35].

Denote $\xi_k(t) := \langle A(u_k(t)), u_k(t) - u(t)\rangle$ and, in accord with (8.35b), assume $\limsup_{k\to\infty}\int_0^T \xi_k(t)\,\mathrm{d}t \le 0$. By (8.10) and (8.13) with $q = p$ and $Z = V$, we have

$$\xi_k(t) \ge c_0\big|u_k(t)\big|_V^p - \zeta_k(t), \quad \text{with}$$
$$\zeta_k(t) := c_1\big|u_k(t)\big|_V + c_2\big\|u_k(t)\big\|_H^2 + \mathfrak{C}\big(\|u_k(t)\|_H\big)\big(1+\|u_k(t)\|_V^{p-1}\big)\big\|u(t)\big\|_V. \tag{8.36}$$

Now, for a moment, assume $\liminf_{k\to\infty}\xi_k(t) < 0$ with t fixed. Then, for a subsequence such that $\lim_{k\to\infty}\xi_k(t) < 0$, the estimate (8.36) implies that $\{u_k(t)\}_{k\in\mathbb{N}}$ is bounded in V, hence (again for a subsequence depending possibly on t) $u_k(t) \rightharpoonup u(t)$ in V because also $u_k(t) \rightharpoonup u(t)$ in H. By pseudomonotonicity (2.3) of A used with $u_k = u_k(t)$ and $u = v = u(t)$, we obtain $\liminf_{k\to\infty}\xi_k(t) \ge 0$. This holds for a.a. $t \in I$.

As $\xi_k \ge -\zeta_k$ and $\{\zeta_k\}_{k\in\mathbb{N}}$ is uniformly integrable[7], by the generalized Fatou Theorem 1.18 it holds

$$0 \le \int_0^T \liminf_{k\to\infty}\xi_k(t)\,\mathrm{d}t \le \liminf_{k\to\infty}\int_0^T \xi_k(t)\,\mathrm{d}t \le \limsup_{k\to\infty}\int_0^T \xi_k(t)\,\mathrm{d}t \le 0, \tag{8.37}$$

and therefore $\lim_{k\to\infty}\int_0^T \xi_k(t)\,\mathrm{d}t = 0$.

Since $\liminf_{k\to\infty}\xi_k(t) \ge 0$, we have $\xi_k^-(t) \to 0$ a.e. and thus, by Vitali's Theorem 1.17, we also have $\lim_{k\to\infty}\int_0^T \xi_k^-(t)\,\mathrm{d}t = 0$ because $0 \ge \xi_k^- \ge -\zeta_k$ and because $\{\zeta_k\}_{k\in\mathbb{N}}$ is uniformly integrable. Altogether, $\lim_{k\to\infty}\int_0^T |\xi_k(t)|\mathrm{d}t = \lim_{k\to\infty}\int_0^T \xi_k(t) - 2\xi_k^-(t)\,\mathrm{d}t = 0$. Hence, possibly in terms of a subsequence, $\lim_{k\to\infty}\xi_k(t) = 0$ for a.a. $t \in I$. Taking $v \in \mathscr{W}$, by the pseudomonotonicity of A, we have $\liminf_{k\to\infty}\langle A(u_k(t)), u_k(t) - v(t)\rangle \ge \langle A(u(t)), u(t) - v(t)\rangle$ for a.a. $t \in I$, and eventually again by the generalized Fatou Theorem 1.18 it holds

$$\liminf_{k\to\infty}\big\langle \mathscr{A}(u_k), u_k - v\big\rangle = \liminf_{k\to\infty}\int_0^T \big\langle A(u_k(t)), u_k(t) - v(t)\big\rangle\,\mathrm{d}t$$
$$\ge \int_0^T \liminf_{k\to\infty}\big\langle A(u_k(t)), u_k(t) - v(t)\big\rangle\mathrm{d}t \ge \int_0^T \big\langle A(u(t)), u(t) - v(t)\big\rangle\mathrm{d}t = \big\langle \mathscr{A}(u), u - v\big\rangle$$

because $\langle A(u_k(\cdot)), u_k(\cdot) - v(\cdot))\rangle$ has a uniformly integrable minorant, namely $-\zeta_k$ as in (8.36) but with v in place of u. □

As we required $\frac{\mathrm{d}}{\mathrm{d}t}u \in L^{p'}(I;V^*)$ in Definition 8.1, it is reasonable to have both $\mathscr{A}(u)$ and f in $L^{p'}(I;V^*)$, i.e. $q = p$ in (8.12) and (8.14). As always $H \subset V^*$, we can consider $f_2 = 0$ in (8.14) without loss of generality as far as values of f

[7]The uniform integrability or, through Dunford-Pettis' Theorem 1.16, rather equi-absolute-continuity of the collection $\{\|u_k(\cdot)\|_V^{p-1}\|u(\cdot)\|_V\}_{k\in\mathbb{N}}$ can easily be seen by Hölder inequality; indeed, $\int_J \|u_k(t)\|_V^{p-1}\|u(t)\|_V\mathrm{d}t \le (\int_J \|u_k(t)\|_V^p\mathrm{d}t)^{1/p'} (\int_J \|u(t)\|_V^p\mathrm{d}t)^{1/p} \le (\int_0^T \|u_k(t)\|_V^p\mathrm{d}t)^{1/p'} (\int_J \|u(t)\|_V^p\mathrm{d}t)^{1/p}$ can be made small uniformly if $\mathrm{meas}_1(J)$ is small because of absolute-continuity of $\|u(\cdot)\|_V^p \in L^1(I)$ and of boundedness of the collection $\{\|u_k(\cdot)\|_V^p\}_{k\in\mathbb{N}}$ in $L^1(I)$.

concerns. For a more general integrability of f in time considering $f_2 \neq 0$ with $q > p$ in (8.14), cf. Remark 8.12 below.

Theorem 8.9 (Existence of a strong solution). *Let $A : V \to V^*$ be pseudomonotone and semi-coercive, satisfy the growth condition (8.12) with $q = p$ and $Z = V$, $f \in L^{p'}(I; V^*)$ and $u_0 \in H$. Then the Cauchy problem (8.4) possesses a strong solution $u \in W^{1,p,p'}(I; V, V^*)$ which can, in addition, be attained in the weak topology of $W^{1,p,p'}(I; V, V^*)$ by a subsequence of Rothe functions $\{u_\tau\}_{\tau>0}$ constructed by considering $\lim_{\tau\to0} \varepsilon(\tau) = 0$ for $f_{\varepsilon(\tau)}$ in (8.5) satisfying (8.15a) (now with $K_2 = 0$) and $\{u_{0\tau}\}_{\tau>0} \subset V$ such that*

$$\|u_{0\tau}\|_V = \mathscr{O}(\tau^{-1/p}) \qquad \text{and} \qquad u_{0\tau} \to u_0 \ \text{ in } H; \tag{8.38}$$

note that such $\{u_{0\tau}\}_{\tau>0}$ always exists because V is assumed dense in H.

Proof. Combining (8.18) with (8.38), we have u_τ bounded in $L^p(I; V)$, and counting also (8.17b) and (8.19a), we can take a subsequence and some $u \in L^\infty(I; H) \cap L^p(I; V)$, $\tilde{u} \in L^\infty(I; H) \cap L^p(I; V)$, and $\dot{u} \in L^{p'}(I; V^*)$ such that

$$u_\tau \overset{*}{\rightharpoonup} u \qquad \text{in } L^\infty(I; H) \cap L^p(I; V)\,, \tag{8.39a}$$

$$\bar{u}_\tau \overset{*}{\rightharpoonup} \tilde{u} \qquad \text{in } L^\infty(I; H) \cap L^p(I; V)\,, \tag{8.39b}$$

$$\frac{\mathrm{d}u_\tau}{\mathrm{d}t} \rightharpoonup \dot{u} \qquad \text{in } L^{p'}(I; V^*). \tag{8.39c}$$

We want to show that

$$u = \tilde{u} \qquad \& \qquad \frac{\mathrm{d}u}{\mathrm{d}t} = \dot{u}\,. \tag{8.40}$$

Let us show that $u_\tau - \bar{u}_\tau \rightharpoonup 0$ in $L^p(I; H)$. Take $\chi_{[\tau_0 k_1, \tau_0 k_2]} v$ for some $\tau_0 > 0$, $k_1 < k_2$ and $v \in H$; linear combinations of all such functions are dense in $L^{p'}(I; H)$ due to Proposition 1.36. Then, for $\tau \leq \tau_0$,

$$\begin{aligned}
\big\langle u_\tau - \bar{u}_\tau, \chi_{[\tau_0 k_1, \tau_0 k_2]} v\big\rangle &= \int_{\tau_0 k_1}^{\tau_0 k_2} \big\langle u_\tau(t) - \bar{u}_\tau(t), v\big\rangle \, \mathrm{d}t \\
&= \sum_{k=k_1\tau_0/\tau+1}^{k_2\tau_0/\tau} \int_{(k-1)\tau}^{k\tau} \Big\langle (u_\tau^k - u_\tau^{k-1}) \frac{t - k\tau}{\tau}, v\Big\rangle \, \mathrm{d}t = \frac{\tau}{2} \sum_{k=k_1\tau_0/\tau+1}^{k_2\tau_0/\tau} \big\langle u_\tau^k - u_\tau^{k-1}, v\big\rangle \\
&= \frac{\tau}{2} \big\langle u_\tau^{k_2\tau_0/\tau} - u_\tau^{k_1\tau_0/\tau}, v\big\rangle = \frac{\tau}{2} \big\langle u_\tau(\tau_0 k_2) - u_\tau(\tau_0 k_1), v\big\rangle = \mathscr{O}(\tau),
\end{aligned}$$

where we eventually used that $u_\tau(\tau_0 k_2) - u_\tau(\tau_0 k_1)$ is bounded in H by (8.17b). Thus $u_\tau - \bar{u}_\tau \rightharpoonup 0$ in $L^p(I; H)$, and thus also in $L^p(I; V)$ because of (8.39). Moreover, by using subsequently (8.39c), (7.15), and (8.39a), we get

$$\langle \dot{u}, \varphi\rangle \leftarrow \Big\langle \frac{\mathrm{d}u_\tau}{\mathrm{d}t}, \varphi\Big\rangle = -\Big\langle u_\tau, \frac{\mathrm{d}\varphi}{\mathrm{d}t}\Big\rangle \to -\Big\langle u, \frac{\mathrm{d}\varphi}{\mathrm{d}t}\Big\rangle \tag{8.41}$$

for any $\varphi \in \mathscr{D}(I;V)$, which, in particular, implies $\dot{u} = \frac{\mathrm{d}}{\mathrm{d}t}u$ in the sense of distributions[8]. Thus (8.40) must hold.

The initial condition

$$u(0) = u_0 \tag{8.42}$$

is satisfied because $u_\tau \rightharpoonup u$ in $W^{1,p,p'}(I;V,V^*)$ and by the continuity (hence also weak continuity) of $u \mapsto u(0) : W^{1,p,p'}(I;V,V^*) \to H$ (see Lemma 7.3) we have $u_\tau(0) \rightharpoonup u(0)$ in H so that

$$u_0 \leftarrow u_{0\tau} = u_\tau(0) \rightharpoonup u(0)\ , \tag{8.43}$$

which immediately implies (8.42).

From (8.5) one can see that $\frac{\mathrm{d}}{\mathrm{d}t}u_\tau + A(\bar{u}_\tau(t)) = \bar{f}_\tau$ with $\bar{f}_\tau$ from (8.8). Thus, for any $v \in L^p(I;V)$, one has

$$\int_0^T \Big\langle \frac{\mathrm{d}u_\tau}{\mathrm{d}t}, v(t) \Big\rangle + \big\langle A(\bar{u}_\tau(t)), v(t) \big\rangle\, \mathrm{d}t = \int_0^T \big\langle \bar{f}_\tau(t), v(t) \big\rangle\, \mathrm{d}t\ . \tag{8.44}$$

In terms of $\langle \cdot, \cdot \rangle$ as the duality between $L^{p'}(I;V^*)$ and $L^p(I;V)$, one can rewrite (8.44) into $\langle \frac{\mathrm{d}}{\mathrm{d}t}u_\tau, v\rangle + \langle \mathscr{A}(\bar{u}_\tau), v\rangle = \langle \bar{f}_\tau, v\rangle$. Putting $v - \bar{u}_\tau$ instead of v, one obtains

$$\big\langle \mathscr{A}(\bar{u}_\tau), v - \bar{u}_\tau \big\rangle = \big\langle \bar{f}_\tau, v - \bar{u}_\tau \big\rangle - \Big\langle \frac{\mathrm{d}u_\tau}{\mathrm{d}t}, v - \bar{u}_\tau \Big\rangle =: I_\tau^{(1)} - I_\tau^{(2)}. \tag{8.45}$$

As $\bar{f}_\tau \to f$ in $L^{p'}(I;V^*)$ due to Lemma 8.7 and $\bar{u}_\tau \to u$ weakly in $L^p(I;V)$ due to (8.39b) with (8.40), obviously

$$\lim_{\tau \to 0} I_\tau^{(1)} = \langle f, v - u \rangle. \tag{8.46}$$

By (8.39a,c) with (8.40), the weak continuity of the mapping $u \mapsto u(T)$: $W^{1,p,p'}(I;V,V^*) \to H$ (see Lemma 7.3), and by the weak lower semicontinuity of $\|\cdot\|_H^2$, one gets (using also (8.25))

$$\begin{aligned}
\limsup_{\tau \to 0} I_\tau^{(2)} &\leq \limsup_{\tau \to 0} \Big(\Big\langle \frac{\mathrm{d}u_\tau}{\mathrm{d}t}, v \Big\rangle - \Big\langle \frac{\mathrm{d}u_\tau}{\mathrm{d}t}, u_\tau \Big\rangle \Big) \\
&\leq \limsup_{\tau \to 0} \Big(\Big\langle \frac{\mathrm{d}u_\tau}{\mathrm{d}t}, v \Big\rangle - \frac{1}{2}\|u_\tau(T)\|_H^2 + \frac{1}{2}\|u_{0\tau}\|_H^2 \Big) \\
&= \lim_{\tau \to 0} \Big\langle \frac{\mathrm{d}u_\tau}{\mathrm{d}t}, v \Big\rangle - \frac{1}{2} \liminf_{\tau \to 0} \|u_\tau(T)\|_H^2 + \frac{1}{2} \lim_{\tau \to 0} \|u_{0\tau}\|_H^2 \\
&\leq \Big\langle \frac{\mathrm{d}u}{\mathrm{d}t}, v \Big\rangle - \|u(T)\|_H^2 + \frac{1}{2}\|u_0\|_H^2 = \Big\langle \frac{\mathrm{d}u}{\mathrm{d}t}, v - u \Big\rangle
\end{aligned} \tag{8.47}$$

[8] Let us recall that $\frac{\mathrm{d}}{\mathrm{d}t}u \in \mathscr{L}(\mathscr{D}(I), (V^*, \text{weak}))$ is defined by the Bochner integral $[\frac{\mathrm{d}}{\mathrm{d}t}u](\phi) = -\int_0^T u(\frac{\mathrm{d}}{\mathrm{d}t}\phi)\,\mathrm{d}t$ for any $\phi \in \mathscr{D}(I)$. The equality $\dot{u} = \frac{\mathrm{d}}{\mathrm{d}t}u$ can be got from (8.41) by putting $\varphi(t) = \phi(t)v$ for $\phi \in \mathscr{D}(I)$ and $v \in V$ arbitrary.

where the last identity employs (8.42) and (7.22). Altogether, (8.46) and (8.47) lead to

$$\liminf_{\tau\to 0}\big\langle\mathscr{A}(\bar u_\tau),v-\bar u_\tau\big\rangle\ge\Big\langle f-\frac{\mathrm{d}u}{\mathrm{d}t},v-u\Big\rangle. \tag{8.48}$$

In particular, for $v{:=}\,u$ we have got $\limsup_{\tau\to0}\langle\mathscr{A}(\bar u_\tau),\bar u_\tau-u\rangle\le 0$. By Lemma 8.8, i.e. the pseudomonotonicity of $\mathscr{A}$, we can conclude that, for any $v\in L^p(I;V)$,

$$\liminf_{\tau\to 0}\big\langle\mathscr{A}(\bar u_\tau),\bar u_\tau-v\big\rangle\ge\big\langle\mathscr{A}(u),u-v\big\rangle. \tag{8.49}$$

Joining (8.48) and (8.49), one gets $\langle\mathscr{A}(u),u-v\rangle\le\langle f,u-v\rangle-\langle\frac{\mathrm{d}}{\mathrm{d}t}u,u-v\rangle$. As it holds for v arbitrary, we can conclude that

$$\big\langle\mathscr{A}(u),v\big\rangle=\langle f,v\rangle-\Big\langle\frac{\mathrm{d}u}{\mathrm{d}t},v\Big\rangle.$$

As v is arbitrary, $\mathscr{A}(u)=f-\frac{\mathrm{d}}{\mathrm{d}t}u$ must hold for a.a. $t\in I$, cf. Exercise 8.49. □

Remark 8.10 (Error in $u_\tau-\bar u_\tau$). If (8.13) holds, then $u=\tilde u$ in (8.40) can alternatively be proved by a simple direct calculation:

$$\begin{aligned}\|u_\tau-\bar u_\tau\|^{q'}_{L^{q'}(I;Z^*)}&=\sum_{k=1}^{T/\tau}\int_{(k-1)\tau}^{k\tau}\Big\|(u_\tau^k-u_\tau^{k-1})\frac{t-k\tau}{\tau}\Big\|^{q'}_{Z^*}\mathrm{d}t\\&=\frac{\tau}{q'+1}\sum_{k=1}^{T/\tau}\|u_\tau^k-u_\tau^{k-1}\|^{q'}_{Z^*}=\frac{\tau^{q'+1}}{q'+1}\sum_{k=1}^{T/\tau}\Big\|\frac{u_\tau^k-u_\tau^{k-1}}{\tau}\Big\|^{q'}_{Z^*}\\&=\frac{\tau^{q'}}{q'+1}\Big\|\frac{\mathrm{d}u_\tau}{\mathrm{d}t}\Big\|^{q'}_{L^{q'}(I;Z^*)}=\mathscr{O}(\tau^{q'})\end{aligned} \tag{8.50}$$

where the bound (8.19a) has been used. Therefore, $\|u_\tau-\bar u_\tau\|_{L^{q'}(I;Z^*)}=\mathscr{O}(\tau)$ and thus also $\|u_\tau-\bar u_\tau\|_{L^1(I;Z^*)}=\mathscr{O}(\tau)$. As certainly $L^p(I;V)\subset L^1(I;Z^*)$, we can estimate the limit $\|u-\tilde u\|_{L^1(I;Z^*)}=0$ and thus the first equality in (8.40) is proved once again. Using (8.17d), the calculation (8.50) yields the error $u_\tau-\bar u_\tau$ estimated in a stronger norm[9]:

$$\|u_\tau-\bar u_\tau\|_{L^2(I;H)}=\frac{\tau}{\sqrt3}\Big\|\frac{\mathrm{d}u_\tau}{\mathrm{d}t}\Big\|_{L^2(I;H)}=\mathscr{O}(\sqrt{\tau}). \tag{8.51}$$

Remark 8.11 (Strong convergence $\bar u_\tau\to u$ in $L^p(I;V)$). Let us subtract $\frac{\mathrm{d}}{\mathrm{d}t}u_\tau+A(\bar u_\tau(t))=\bar f_\tau$ from $\frac{\mathrm{d}}{\mathrm{d}t}u+A(u(t))=f$ and test it by $\bar u_\tau-u$. By using the inequality $\langle\frac{\mathrm{d}}{\mathrm{d}t}u_\tau(t),\bar u_\tau(t)-u_\tau(t)\rangle=\|\frac{\mathrm{d}}{\mathrm{d}t}u_\tau\|_H^2(k\tau-t)\ge0$ for any $t\in((k-1)\tau,k\tau)$, we obtain

$$\begin{aligned}\Big\langle\frac{\mathrm{d}u_\tau}{\mathrm{d}t}-\frac{\mathrm{d}u}{\mathrm{d}t},\bar u_\tau-u\Big\rangle&=\Big\langle\frac{\mathrm{d}u_\tau}{\mathrm{d}t},\bar u_\tau-u_\tau\Big\rangle+\Big\langle\frac{\mathrm{d}u_\tau}{\mathrm{d}t}-\frac{\mathrm{d}u}{\mathrm{d}t},u_\tau-u\Big\rangle\\&\quad+\Big\langle\frac{\mathrm{d}u}{\mathrm{d}t},u_\tau-\bar u_\tau\Big\rangle\ge\frac12\frac{\mathrm{d}}{\mathrm{d}t}\|u_\tau-u\|_H^2+\Big\langle\frac{\mathrm{d}u}{\mathrm{d}t},u_\tau-\bar u_\tau\Big\rangle\end{aligned}$$

[9]See e.g. Feistauer [147, Theorem 8.7.25].

for a.a. $t \in I$; here the dualities are between V^* and V. After integration over I, this gives

$$\frac{1}{2}\big\|u_\tau(T)-u(T)\big\|_H^2+\langle\mathscr{A}(\bar{u}_\tau)-\mathscr{A}(u),\bar{u}_\tau-u\rangle \\ \leq\frac{1}{2}\|u_{0\tau}-u_0\|_H^2+\Big\langle\bar{f}_\tau-f+\frac{\mathrm{d}u}{\mathrm{d}t},\bar{u}_\tau-u_\tau\Big\rangle\to 0 \tag{8.52}$$

by using respectively $u_{0\tau}\to u_0$ in H, $\bar{f}_\tau\to f$ in $L^{p'}(I;V^*)$ due to Lemma 8.7, and $u_\tau-\bar{u}_\tau\rightharpoonup u-\tilde{u}=0$ in $L^p(I;V)$ due to (8.39a,b)–(8.40). This gives $\bar{u}_\tau\to u$ in $L^p(I;V)$ if $A=A_1+A_2$ and V is uniformly convex, and $\mathscr{A}_1$ is assumed d-monotone in the sense

$$\big\langle\mathscr{A}_1(u)-\mathscr{A}_1(v),u-v\big\rangle_{L^{p'}(I;V^*)\times L^p(I;V)} \\ \geq\Big(d\big(\|u\|_{L^p(I;V)}\big)-d\big(\|v\|_{L^p(I;V)}\big)\Big)\Big(\|u\|_{L^p(I;V)}-\|v\|_{L^p(I;V)}\Big) \tag{8.53}$$

for some $d:\mathbb{R}\to\mathbb{R}$ increasing, and $\mathscr{A}_2$ is totally continuous. Then we can use uniform convexity of $L^p(I;V)$, cf. Proposition 1.37, and Theorem 1.2; details are left as an exercise, cf. also (8.118).

Remark 8.12 (The case $f\in L^{p'}(I;V^*)+L^{q'}(I;H)$, $p<q<+\infty$[10]). In this case, instead of $u\in W^{1,p,p'}(I;V,V^*)$, Definition 8.1 should require $u\in L^p(I;V)\cap L^q(I;H)$ with $\frac{\mathrm{d}}{\mathrm{d}t}u\in L^{p'}(I;V^*)+L^{q'}(I;H)$. The Banach space of such u's is again embedded into $C(I;H)$ and also the by-part formula (7.15) extends to hold. Then the proof of Theorem 8.4 bears slight modifications; e.g. the estimate $\|\frac{\mathrm{d}}{\mathrm{d}t}u_\tau\|_{L^{p'}(I;V^*)}\leq C_3$ resulting from (8.19a) is now $\|\frac{\mathrm{d}}{\mathrm{d}t}u_\tau\|_{L^{p'}(I;V^*)+L^{q'}(I;H)}\leq C_3$, the convergence (8.39c) takes $L^{p'}(I;V^*)+L^{q'}(I;H)$, and the dualities $\langle\frac{\mathrm{d}}{\mathrm{d}t}u,u\rangle$ and $\langle f,u\rangle$ refer to (1.9).

Theorem 8.13 (Weak solution). *Let $1<p\leq q\leq+\infty$, $p<+\infty$, $A:V\to V^*$ be pseudomonotone*[11] *and semicoercive such that $\mathscr{A}$ is weakly* continuous from $W^{1,p,\mathscr{M}}(I;V,Z^*)\cap L^\infty(I;H)$ to $L^\infty(I;Z)^*$ and satisfy (8.12) with some $Z\subset V$ densely, let $u_0\in H$ and $u_{0\tau}$ satisfy (8.38), and let f satisfy (8.14), and (8.15a) hold, too. Then there is a weak solution u due to the Definition 8.2 and, moreover, $\frac{\mathrm{d}}{\mathrm{d}t}u\in L^{q'}(I;Z^*)$.*

Proof. We test $\frac{\mathrm{d}}{\mathrm{d}t}u_\tau+\mathscr{A}(\bar{u}_\tau)=\bar{f}_\tau$, which arises from (8.5) if the notation (8.6), (8.7), and (8.8) applies, by v. Using the by-part formula (7.15), we can write

$$0=\Big\langle\frac{\mathrm{d}u_\tau}{\mathrm{d}t}+\mathscr{A}(\bar{u}_\tau)-\bar{f}_\tau,v\Big\rangle \\ =\big\langle\mathscr{A}(\bar{u}_\tau)-\bar{f}_\tau,v\big\rangle-\Big\langle\frac{\mathrm{d}v}{\mathrm{d}t},u_\tau\Big\rangle+(v(T),u_\tau(T))-(v(0),u_{0\tau}) \tag{8.54}$$

[10] For this approach, we refer to Gajewski et al. [168, Chap.VI with Sect.IV.1.5].

[11] In Theorem 8.31 we will still put off this assumption.

for any $v \in W^{1,p,p'}(I;V,V^*)$; as $A: V \to V^*$ is assumed bounded, hence $\mathscr{A}(\bar{u}_\tau) \in L^\infty(I;V^*)$, and as also $\bar{f}_\tau \in L^\infty(I;V^*)$, the dualities in (8.54) can be understood as between $L^p(I;V)$ and its dual.

By using (8.17b) and (8.19b), we have the a-priori boundedness of $\{\bar{u}_\tau\}_{0<\tau\le\tau_0}$ in $\mathscr{W} = W^{1,p,\mathscr{M}}(I;V,Z^*)\cap L^\infty(I;H)$. Thus, after choosing a subsequence, we have $\bar{u}_\tau \stackrel{*}{\rightharpoonup} \tilde{u}$ in $\mathscr{W}$.

In view of (8.17a) and (8.18), we can select such a subsequence that also $u_\tau \stackrel{*}{\rightharpoonup} u$ in $L^p(I;V) \cap L^\infty(I;H)$. As in the proof of Theorem 8.9 we can see that $u = \tilde{u}$ and also that $\frac{\mathrm{d}}{\mathrm{d}t}u_\tau \stackrel{*}{\rightharpoonup} \frac{\mathrm{d}}{\mathrm{d}t}u$ in $\mathscr{M}(I;Z^*)$ (or in $L^{q'}(I;Z^*)$ if $q < +\infty$). As $\frac{\mathrm{d}}{\mathrm{d}t}v \in L^{p'}(I;V^*)$ is fixed, we have $\langle\frac{\mathrm{d}}{\mathrm{d}t}v, u_\tau\rangle \to \langle\frac{\mathrm{d}}{\mathrm{d}t}v, u\rangle$.

Now, we consider only $v \in W^{1,\infty,\infty}(I;Z,V^*)$. Then $\langle\bar{f}_\tau, v\rangle \to \langle f, v\rangle$ with the last duality between $L^{q'}(I;Z^*)$ and $L^q(I;Z)$. Also $\int_0^T \langle A(\bar{u}_\tau(t)), v(t)\rangle_{Z^*\times Z}\,\mathrm{d}t \to \int_0^T \langle A(u(t)), v(t)\rangle_{Z^*\times Z}\,\mathrm{d}t$ due to the assumed weak* continuity of $\mathscr{A}$. Hence (8.2) is proved at least if $v(T) = 0 = v(0)$. This says, in particular, that $\mathscr{A}(u) - f = -\frac{\mathrm{d}}{\mathrm{d}t}u$ in the sense of distributions on I. However, by (8.12), $u \in L^\infty(I;H) \cap L^p(I;V)$ implies $\mathscr{A}(u) \in L^{q'}(I;Z^*)$. By (8.14) with $q \ge p \ge 1$, also $f \in L^{q'}(I;Z^*)$. Hencefore, $\frac{\mathrm{d}}{\mathrm{d}t}u \in L^{q'}(I;Z^*)$ even if $q = +\infty$.

As $\{u_\tau(T)\}_{\tau>0}$ is bounded in H, hence it converges (possibly as further selected subsequence) to some u_T weakly in H. On the other hand, $u_\tau(T) = u_{0\tau} + \int_0^T \frac{\mathrm{d}}{\mathrm{d}t}u_\tau \mathrm{d}t$ converges to $u_0 + \int_0^T \frac{\mathrm{d}}{\mathrm{d}t}u\,\mathrm{d}t = u(T)$ in Z^*. Hence $u_T = u(T)$, the further selection was redundant, and the term $(u_\tau(T), v(T))$ converges to $(u(T), v(T))$. The convergence of $(v(0), u_{0\tau})$ to $(v(0), u_0)$ is obvious.

The weak continuity of the mapping $t \mapsto u(t) : I \to H$ required in Definition 8.2 follows from its boundedness and from having information about $\frac{\mathrm{d}}{\mathrm{d}t}u \in L^1(I;Z^*)$, hence it is absolutely continuous as $I \to Z^*$. □

Remark 8.14 (*Semi-implicit formulae*). Especially for further numerical applications it is often advantageous to consider a certain "linearization" $B(w,\cdot) : V \to V^*$ of A at a current point w, likewise (but not necessarily just as) in (2.73), and then to modify the fully implicit formula (8.5) as

$$\frac{u_\tau^k - u_\tau^{k-1}}{\tau} + B(u_\tau^{k-1}, u_\tau^k) = f_\tau^k\,, \qquad k \ge 1. \tag{8.55}$$

In any case, the compatibility $A(u) = B(u,u)$ is required and linearity of $B(w,\cdot)$ is an optional property from which some benefits may follow. The a-priori estimates and convergence analysis are to be made case by case, cf. Exercises 8.74 and 8.92. Besides a linearization, semi-implicit formulae can serve to decouple systems of equations, cf. e.g. (12.58) or Remark 8.25 and Exercise 12.28.

Remark 8.15 (Alternative estimation). Instead of the heuristics (8.20)–(8.21), one can use

$$\|u\|_H\frac{\mathrm{d}}{\mathrm{d}t}\|u\|_H + c_0|u|_V^p \le c_1|u|_V + c_2\|u\|_H^2 + \langle f, u\rangle$$
$$\le c_1|u|_V + c_2\|u\|_H^2 + C_0\|f_1\|_{V^*}^{p'} + \frac{c_0}{2}|u|_V^p + \big(C_{\mathrm{P}}\|f_1\|_{V^*} + \|f_2\|_H\big)\|u\|_H \quad (8.56)$$

with some C_0 depending on c_0 and p and on the Poincaré constant C_{P}. From this, one again obtains boundedness of u in $L^\infty(I;H) \cap L^p(I;V)$ by Gronwall's inequality when qualifying $f = f_1+f_2$ as in (8.14) and when using $\|u\|_H\frac{\mathrm{d}}{\mathrm{d}t}\|u\|_H = \frac{1}{2}\frac{\mathrm{d}}{\mathrm{d}t}\|u\|_H^2$. Yet (8.56) suggests a modification of (8.25) and (8.27) by using respectively the estimates $\langle u_\tau^k - u_\tau^{k-1}, u_\tau^k\rangle \ge \|u_\tau^k\|_H(\|u_\tau^k\|_H - \|u_\tau^{k-1}\|_H)$ and $\langle f_\tau^k, u_\tau^k\rangle = \langle f_{1\tau}^k + f_{2\tau}^k, u_\tau^k\rangle \le C_0\|f_{1\tau}^k\|_{V^*}^{p'} + \frac{1}{2}c_0|u_\tau^k|_V^p + (C_{\mathrm{P}}\|f_{1\tau}^k\|_{V^*} + \|f_{2\tau}^k\|_H)\|u_\tau^k\|_H$, which would turn (8.26) into the estimate

$$\|u_\tau^k\|_H\frac{\|u_\tau^k\|_H - \|u_\tau^{k-1}\|_H}{\tau} + \frac{c_0}{2}|u_\tau^k|_V^p \le c_1|u_\tau^k|_V + c_2\|u_\tau^k\|_H^2$$
$$+ C_0\|f_{1\tau}^k\|_{V^*}^{p'} + \big(C_{\mathrm{P}}\|f_{1\tau}^k\|_{V^*} + \|f_{2\tau}^k\|_H\big)\|u_\tau^k\|_H. \quad (8.57)$$

Then, after estimating still $c_1|u_\tau^k|_V \le \frac{1}{4}c_0|u_\tau^k|_V^p + C_p'$ with some constant C_p' depending on p and on c_0, one can apply the Nochetto-Savaré-Verdi variant of the discrete Gronwall inequality (1.71)–(1.72) with $y_k := \|u_\tau^k\|_H$, $z_k := \frac{1}{4}c_0|u_\tau^k|_V^p$, and $c = c_2$. One thus obtains the estimates (8.17a-c) under the restriction $\tau_0 < 1/c_2$ which is weaker than (8.16), and also the condition (8.15a) is redundant for this strategy. Having $\|u_\tau^k\|_H$ estimated, one can obtain also the estimate (8.17d) again by (8.24). Alternatively to (8.8), $\bar f_\tau$ can be defined as

$$\bar f_\tau(t) := f_\tau^k := \frac{1}{\tau}\int_{(k-1)\tau}^{k\tau} f(\vartheta)\,\mathrm{d}\vartheta \qquad \text{for} \quad t \in \big[(k-1)\tau, k\tau\big]. \quad (8.58)$$

Such $\bar f_\tau$ is called the zero-order *Clément quasi-interpolant* of f.[12] The convergence $\bar f_\tau \to f$ and thus also the condition (8.15b) can be proved.[13] As for (8.29a), by the Hölder inequality, we now have more explicitly

[12] "Zero-order" refers to the order of polynomials used to construct $\bar f_\tau$. For the first-order quasi-interpolation see Remark 8.19 below. The quasi-interpolation procedure was proposed in [98].

[13] For a general $f \in L^q(I;X)$, one can proceed as follows: Take $\eta > 0$. Using the convolution with a mollifier as in (7.10) with $\varepsilon > 0$ small enough, we get $f_\varepsilon \in C(I;X)$ such that $\|f_\varepsilon - f\|_{L^q(I;X)} \le \eta/3$. As I is compact, $f_\varepsilon : I \to X$ is uniformly continuous and thus there is $\tau_0 > 0$ sufficiently small such that $\|f_\varepsilon(t_1) - f_\varepsilon(t_2)\|_X \le T^{-1/q}\eta/3$ whenever $|t_1 - t_2| \le \tau_0$. Then also $\|f_\varepsilon(t) - \overline{[f_\varepsilon]}_\tau(t)\|_X \le T^{-1/q}\eta/3$ for any $0 < \tau \le \tau_0$ and $t \in I$, hencefore also $\|f_\varepsilon - \overline{[f_\varepsilon]}_\tau\|_{L^q(I;X)} \le T^{1/q}\|f_\varepsilon - \overline{[f_\varepsilon]}_\tau\|_{C(I;X)} \le \eta/3$. Eventually, as in (8.59), we have also $\|\overline{[f_\varepsilon]}_\tau - \bar f_\tau\|_{L^q(I;X)} \le \|f_\varepsilon - f\|_{L^q(I;X)} \le \eta/3$. Altogether, $\|\bar f_\tau - f\|_{L^q(I;X)} \le \|\bar f_\tau - \overline{[f_\varepsilon]}_\tau\|_{L^q(I;X)} + \|\overline{[f_\varepsilon]}_\tau - f_\varepsilon\|_{L^q(I;X)} + \|f_\varepsilon - f\|_{L^q(I;X)} \le \frac{1}{3}\eta + \frac{1}{3}\eta + \frac{1}{3}\eta = \eta$ for any $0 < \tau \le \tau_0$.

$$\tau\sum_{k=1}^{l}\|f_\tau^k\|_{V^*}^{p'} \le \tau\sum_{k=1}^{T/\tau}\|f_\tau^k\|_{V^*}^{p'} = \tau\sum_{k=1}^{T/\tau}\Big\|\frac{1}{\tau}\int_{(k-1)\tau}^{k\tau} f(t)\,\mathrm{d}t\Big\|_{V^*}^{p'}$$
$$\le \frac{1}{\sqrt[p-1]{\tau}}\sum_{k=1}^{T/\tau}\Big(\int_{(k-1)\tau}^{k\tau}\|f(t)\|_{V^*}\mathrm{d}t\Big)^{p'}$$
$$\le \sum_{k=1}^{T/\tau}\int_{(k-1)\tau}^{k\tau}\|f(t)\|_{V^*}^{p'}\mathrm{d}t = \|f\|_{L^{p'}(I;V^*)}^{p'} \tag{8.59}$$

for any $l = 1,\dots,k$. Similar considerations hold for (8.29b), too.

8.3 Further estimates

The strategy (8.20)–(8.22) of testing the equation (8.4) by u can be modified to get better results under modified (mostly stronger) data qualification; yet, note that the condition (8.13) on the growth of A need not be assumed in this section. This additional quality of the solution is referred to as its certain *regularity*. Now, besides (8.8), we can also use the approximation of f due to Clément's quasi-interpolation (8.58).

Theorem 8.16 (Regularity). *Let $A : V \to V^*$ be pseudomonotone and semicoercive* (8.10) *with some $p > 1$ (its value is, in fact, not important now), and*

$$u_0 \in V, \tag{8.60a}$$
$$f \in L^2(I;H), \tag{8.60b}$$
$$A = A_1 + A_2 \text{ with } A_1 = \Phi',\quad \Phi : V \to \mathbb{R}\ \text{ convex}, \tag{8.60c}$$
$$\Phi(u) \ge c_0|u|_V^q - c_1\|u\|_H^2\,, \qquad \|A_2(u)\|_H \le C\big(1 + |u|_V^{q/2} + \|u\|_H\big) \tag{8.60d}$$

for some $c_0, q > 0$.[14] *Then:*

(i) *The Rothe sequence $\{u_\tau\}_{\tau_0\ge\tau>0}$ constructed by* (8.5) *with f_τ^k from* (8.58), *with $u_{0\tau} = u_0$ and with $\tau_0 < \frac{1}{2}c_0C^{-2}/\max(1, 4c_1T)$ is bounded in $W^{1,\infty,2}(I;V,H)$.*

(ii) *Moreover, it has a weakly* convergent subsequence in this space, and if also $V \Subset H$ (a compact embedding), and*

$$A_1 : V \to V^* \quad \text{is bounded and radially continuous,} \tag{8.61a}$$
$$A_2 : V \to V^* \quad \text{is totally continuous,} \tag{8.61b}$$

then every $u \in W^{1,\infty,2}(I;V,H)$ obtained as the weak limit of a subsequence $\{u_\tau\}_{\tau>0}$ solves the abstract Cauchy problem* (8.4).

(iii) *Also, $u \in C(I;(V,\text{weak}))$.*

[14] Often, but not necessarily, $q = p$ with p referring to (8.10). In fact, q in only an intermediate exponent occuring in (8.60d), cf. also Exercise 8.62 below.

Let us note that, for any Banach space V_2 such that $V \Subset V_2 \subset H$, the mentioned weak* convergence in $W^{1,\infty,2}(I;V,H)$ together with the fact that $W^{1,\infty,2}(I;V,H) \subset C(I;(V,\text{weak}))$ implies the strong convergence $u_\tau \to u$ in $C(I;V_2)$; this can be seen from Lemma 7.10 for $V_1 = V$ and $V_3 = H$.

Let us first make heuristics of the proof of (i) for a non-discretized problem: test the equation $\frac{\mathrm{d}}{\mathrm{d}t}u + A(u) = f$ by $\frac{\mathrm{d}}{\mathrm{d}t}u$, use $\langle A(u), \frac{\mathrm{d}}{\mathrm{d}t}u\rangle = \frac{\mathrm{d}}{\mathrm{d}t}\Phi(u) + \langle A_2(u), \frac{\mathrm{d}}{\mathrm{d}t}u\rangle$, which formally[15] gives

$$\Big\|\frac{\mathrm{d}u}{\mathrm{d}t}\Big\|_H^2 + \frac{\mathrm{d}}{\mathrm{d}t}\Phi(u) = \Big\langle f - A_2(u), \frac{\mathrm{d}u}{\mathrm{d}t}\Big\rangle \le \big\|A_2(u)\big\|_H^2 + \big\|f(t)\big\|_H^2 + \frac{1}{2}\Big\|\frac{\mathrm{d}u}{\mathrm{d}t}\Big\|_H^2. \tag{8.62}$$

Then we absorb the last term in the left-hand side and denote $U(t) := \int_0^t \|\frac{\mathrm{d}}{\mathrm{d}\vartheta}u\|_H^2 \mathrm{d}\vartheta$ so that $\|\frac{\mathrm{d}}{\mathrm{d}t}u\|_H^2 = \frac{\mathrm{d}}{\mathrm{d}t}U$ and, by Hölder's inequality,

$$\big\|u(t)\big\|_H^2 = \Big\|u_0 + \int_0^t \frac{\mathrm{d}u}{\mathrm{d}\vartheta}\mathrm{d}\vartheta\Big\|_H^2 \le 2\Big(\int_0^t \Big\|\frac{\mathrm{d}u}{\mathrm{d}\vartheta}\Big\|_H \mathrm{d}\vartheta\Big)^2 + 2\big\|u_0\big\|_H^2 \le 2tU(t) + 2\big\|u_0\big\|_H^2. \tag{8.63}$$

Thus

$$\begin{aligned}\frac{\mathrm{d}}{\mathrm{d}t}\Big(\frac{1}{2}U(t) + \Phi(u)\Big) &\le \big\|A_2(u)\big\|_H^2 + \big\|f(t)\big\|_H^2 \\ &\le 3C^2\big(1 + |u|_V^q + \|u\|_H^2\big) + \big\|f(t)\big\|_H^2 \\ &\le 3C^2\Big(1 + \frac{c_0+c_1}{c_0}\|u\|_H^2 + \frac{1}{c_0}\Phi(u)\Big) + \big\|f(t)\big\|_H^2 \\ &\le 3C^2\Big(1 + 2\frac{c_0+c_1}{c_0}\big(TU(t) + \|u_0\|_H^2\big) + \frac{1}{c_0}\Phi(u)\Big) + \big\|f(t)\big\|_H^2. \end{aligned} \tag{8.64}$$

Then we use the Gronwall inequality. Note that it needs $\Phi(u_0) < +\infty$, i.e. $u_0 \in V$. Eventually, we get $\Phi(u(t)) + U(t)$ bounded independently of $t \in I$, which implies $u \in L^\infty(I;V)$ and $\|\frac{\mathrm{d}}{\mathrm{d}t}u\|_{L^2(I;H)} = \sqrt{U(T)}$ bounded.

Proof of Theorem 8.16. Multiply (8.5) by $u_\tau^k - u_\tau^{k-1}$, and use

$$\big\langle A(u_\tau^k), u_\tau^k - u_\tau^{k-1}\big\rangle = \big\langle \Phi'(u_\tau^k), u_\tau^k - u_\tau^{k-1}\big\rangle + \big\langle A_2(u_\tau^k), u_\tau^k - u_\tau^{k-1}\big\rangle. \tag{8.65}$$

We can estimate

$$\langle \Phi'(u_\tau^k), u_\tau^k - u_\tau^{k-1}\rangle \ge \Phi(u_\tau^k) - \Phi(u_\tau^{k-1}) \tag{8.66}$$

because Φ is convex; in fact, up to the factor $1/\tau$, (8.66) is a *discrete* analog of the *chain rule* $\langle \Phi'(u), \frac{\mathrm{d}}{\mathrm{d}t}u\rangle = \frac{\mathrm{d}}{\mathrm{d}t}\Phi(u)$. Thus, dividing (8.65) still by τ and using

[15] Note that, if a-priori no other information about $\frac{\mathrm{d}}{\mathrm{d}t}u$ than $\frac{\mathrm{d}}{\mathrm{d}t}u \in L^{p'}(I;V^*)$ is known, this test cannot be rigorously made.

Young's inequality, we get

$$
\begin{aligned}
\Big\|\frac{u_\tau^k-u_\tau^{k-1}}{\tau}\Big\|_H^2+\frac{\Phi(u_\tau^k)-\Phi(u_\tau^{k-1})}{\tau} &= \Big\langle f_\tau^k - A_2(u_\tau^k), \frac{u_\tau^k-u_\tau^{k-1}}{\tau}\Big\rangle \\
&\le \|f_\tau^k\|_H^2+\|A_2(u_\tau^k)\|_H^2+\frac{1}{2}\Big\|\frac{u_\tau^k-u_\tau^{k-1}}{\tau}\Big\|_H^2 \\
&\le \|f_\tau^k\|_H^2+3C^2\big(1+|u_\tau^k|_V^q+\|u_\tau^k\|_H^2\big)+\frac{1}{2}\Big\|\frac{u_\tau^k-u_\tau^{k-1}}{\tau}\Big\|_H^2 \\
&\le \|f_\tau^k\|_H^2+3C^2\Big(1+\frac{c_0+c_1}{c_0}\|u_\tau^k\|_H^2+\frac{\Phi(u_\tau^k)}{c_0}\Big)+\frac{1}{2}\Big\|\frac{u_\tau^k-u_\tau^{k-1}}{\tau}\Big\|_H^2. \qquad (8.67)
\end{aligned}
$$

We first absorb the last term in the left-hand side, and then, denoting $U_\tau^k := \tau^{-1}\sum_{l=1}^k \|u_\tau^l - u_\tau^{l-1}\|_H^2$, we further estimate

$$
\frac{U_\tau^k-U_\tau^{k-1}}{2\tau}+\frac{\Phi(u_\tau^k)-\Phi(u_\tau^{k-1})}{\tau} \le \|f_\tau^k\|_H^2+3C^2\Big(1+2\frac{c_0+c_1}{c_0}\big(TU_\tau^k+\|u_0\|_H^2\big)+\frac{\Phi(u_\tau^k)}{c_0}\Big)
$$

and use the discrete Gronwall inequality (1.68) provided τ is small enough, namely $\tau < \frac{1}{2}c_0C^{-2}/\max(1,4c_1T)$, together with the estimate $\tau\sum_{k=1}^l \|f_\tau^l\|_H^2 = \int_0^{l\tau}\|\bar f_\tau\|_H^2 \mathrm{d}t \le \int_0^T \|f\|_H^2\mathrm{d}t$, which is bounded independently of τ. This bounds $\Phi(u_\tau)$ in $L^\infty(I)$, hence also u_τ in $L^\infty(I;V)$. Also U_τ^k is bounded, so in particular we get a bound for $U_\tau^{T/\tau} = \|\frac{\mathrm{d}}{\mathrm{d}t}u_\tau\|_{L^2(I;H)}^2$, as claimed in (i).

As to (ii), by Theorem 4.4(iv), A_1 is monotone, hence for any $v \in L^\infty(I;V)$ it holds that

$$
0 \le \big\langle \mathscr{A}_1(\bar u_\tau) - \mathscr{A}_1(v), \bar u_\tau - v\big\rangle = \Big\langle \bar f_\tau - \frac{\mathrm{d}u_\tau}{\mathrm{d}t} - \mathscr{A}_2(\bar u_\tau) - \mathscr{A}_1(v), \bar u_\tau - v\Big\rangle. \qquad (8.68)
$$

Let now $\{u_\tau\}_{\tau>0}$ refer to a selected subsequence converging weakly* in $W^{1,\infty,2}(I;V,H)$ and u be its limit. Then $\frac{\mathrm{d}}{\mathrm{d}t}u_\tau \rightharpoonup \frac{\mathrm{d}}{\mathrm{d}t}u$ weakly in $L^2(I;H)$. Moreover, by the Aubin-Lions lemma, $W^{1,\infty,2}(I;V,H) \Subset L^2(I;H)$ and therefore $u_\tau \to u$ in $L^2(I;H)$. As $\|u_\tau - \bar u_\tau\|_{L^2(I;H)} = 3^{-1/2}\tau\|\frac{\mathrm{d}}{\mathrm{d}t}u_\tau\|_{L^2(I;H)} = \mathscr{O}(\tau)$, cf. (8.51), also $\bar u_\tau \to u$ in $L^2(I;H)$. Then $\lim_{\tau\to0}\langle\frac{\mathrm{d}}{\mathrm{d}t}u_\tau, \bar u_\tau\rangle = \langle\frac{\mathrm{d}}{\mathrm{d}t}u, u\rangle$.[16] In particular,[17] $\bar u_\tau(t) \to u(t)$ in H for a.a. $t \in I$. As $\{\bar u_\tau(t)\}_{0<\tau\le\tau_0}$ is bounded also in V, we also know that $\bar u_\tau(t) \rightharpoonup u(t)$ weakly in V for a.a. $t \in I$. By (8.61b), $\langle A_2(\bar u_\tau(t)), \bar u_\tau(t)\rangle \to \langle A_2(u(t)), u(t)\rangle$ for a.a. $t \in I$. By using (8.60d) and boundedness of $\{\bar u_\tau\}_{0<\tau\le\tau_0}$ in $L^\infty(I;V)$, we can see that $A_2(\bar u_\tau(t))$ is bounded in H independently of τ and t. Hence $\langle A_2(\bar u_\tau(t)), \bar u_\tau(t)\rangle$ is bounded independently of both t and τ, hence by Lebesgue's Theorem 1.14, $\int_0^T\langle A_2(\bar u_\tau), \bar u_\tau\rangle\,\mathrm{d}t \to \int_0^T\langle A_2(u), u\rangle\,\mathrm{d}t$.

[16] Alternatively, we could use the inequality (8.47) if Lemma 7.3 and the by-part integration formula (7.22) employ the space $W^{1,2,2}(I;H,H)$ instead of $W^{1,p,p'}(I;V,V^*)$.

[17] For a moment, we can select a subsequence to guarantee this; see Theorem 1.7. When the limit of $\langle\mathscr{A}_2(\bar u_\tau), \bar u_\tau\rangle$ is uniquely identified, we can avoid this further selection, however.

Therefore, (8.68) implies

$$0 \le \lim_{\tau\to 0} \Big\langle \bar{f}_\tau - \frac{\mathrm{d}u_\tau}{\mathrm{d}t} - \mathscr{A}_2(\bar{u}_\tau) - \mathscr{A}_1(v), \bar{u}_\tau - v \Big\rangle = \Big\langle f - \frac{\mathrm{d}u}{\mathrm{d}t} - \mathscr{A}_2(u) - \mathscr{A}_1(v), u - v \Big\rangle.$$

Now, we proceed by *Minty's trick* by putting $v := u + \varepsilon z$ for $z \in L^\infty(I;V)$ arbitrary and $\varepsilon > 0$, which gives $\langle f - \frac{\mathrm{d}}{\mathrm{d}t}u - \mathscr{A}_2(u) - \mathscr{A}_1(u+\varepsilon z), \varepsilon z\rangle \le 0$. Then we divide it by $\varepsilon > 0$, and pass $\varepsilon \to 0$ by using (8.61a).[18] As z is arbitrary, we conclude $f - \frac{\mathrm{d}}{\mathrm{d}t}u - \mathscr{A}_2(u) - \mathscr{A}_1(u) = 0$ a.e. on I.

As to (iii), in particular we have obtained $u \in W^{1,2}(I;H) \subset C(I;H)$ due to Lemma 7.1. Thus, $u(\vartheta) \to u(t)$ in H for $\vartheta \to t$. Since $u \in L^\infty(I;V)$, $\{u(\vartheta)\}_{\vartheta\in I}$ is bounded in V and hence (possibly up to a subsequence) $u(\vartheta) \rightharpoonup v$ in V. Yet, since $V \subset H$, $v = u(t)$. As this limit is thus determined uniquely, the whole sequence (or net) must converge to $u(t)$ weakly. Hence, $u \in C(I;(V,\text{weak}))$. □

Remark 8.17 (Asymptotics for a special case $A = \Phi'$ and f constant). In this special case, $t \mapsto [\Phi - f](u(t))$ is non increasing because it fulfills $\frac{\mathrm{d}}{\mathrm{d}t}[\Phi - f](u(t)) = -\|\frac{\mathrm{d}}{\mathrm{d}t}u\|_H^2 \le 0$. Moreover, it can be shown that $t \mapsto [\Phi - f](u(t))$ is convex and $u(t)$ tends weakly to the minimum of $\Phi - f$ for $t \to \infty$.[19]

Theorem 8.18 (Regularity II). *Let $A : V \to V^*$ be pseudomonotone and semi-coercive* (8.10) *with some $p > 1$ (whose value is again not important now), and*

$$f \in W^{1,2}(I;V^*), \tag{8.69a}$$
$$u_0 \in V \quad \textit{such that} \quad A(u_0) - f(0) \in H, \tag{8.69b}$$
$$\big\langle A(u_1) - A(u_2), u_1 - u_2 \big\rangle \ge c_0|u_1 - u_2|_V^2 - c_2\|u_1 - u_2\|_H^2 \tag{8.69c}$$

with some $c_0 > 0$. Then:

(i) *The Rothe sequence $\{u_\tau\}_{\tau_0\ge\tau>0}$ constructed by the formula* (8.5) *with f_τ^k from* (8.58) *and with $u_{0\tau} = u_0$ is bounded in $W^{1,\infty}(I;H) \cap W^{1,2}(I;V)$ provided $\tau_0 < 1/(2c_2)$.*

(ii) *Moreover, it has a weakly* convergent subsequence in this space, and if also $V \Subset H$ (a compact embedding), $A = A_1 + A_2$ satisfying* (8.61) *with A_1 monotone, every $u \in W^{1,\infty}(I;H) \cap W^{1,2}(I;V)$ obtained as the weak* limit of a subsequence $\{u_\tau\}_{\tau>0}$ solves the abstract Cauchy problem* (8.4).

Heuristics of the proof of (i): apply $\frac{\mathrm{d}}{\mathrm{d}t}$ to the equation $\frac{\mathrm{d}}{\mathrm{d}t}u + A(u) = f$ and then test it by $\frac{\mathrm{d}}{\mathrm{d}t}u$, use $\langle \frac{\mathrm{d}}{\mathrm{d}t}A(u), \frac{\mathrm{d}}{\mathrm{d}t}u\rangle \ge c_0|\frac{\mathrm{d}}{\mathrm{d}t}u|_V^2 - c_2\|\frac{\mathrm{d}}{\mathrm{d}t}u\|_H^2$ (which is a continuous

---

[18] More in detail, we proceed as in (8.165) but the common integrable majorant here is now even in $L^\infty(I)$ because we have $u, z \in L^\infty(I;V)$ and A_1 maps bounded sets in V to bounded sets in V^* as assumed in (8.61a).

[19] More about such cases can be found, e.g., in Aubin and Cellina [28, Section 3.4] or Brezis [66, Section III.3]. See also Proposition 11.9 and Remark 8.22 below.

analog of (8.73) below). Using also (8.9), this gives

$$\begin{aligned}\frac{1}{2}\frac{\mathrm{d}}{\mathrm{d}t}\Big\|\frac{\mathrm{d}u}{\mathrm{d}t}\Big\|_H^2+c_0\Big|\frac{\mathrm{d}u}{\mathrm{d}t}\Big|_V^2&\le\Big\langle\frac{\mathrm{d}}{\mathrm{d}t}\Big(\frac{\mathrm{d}u}{\mathrm{d}t}+A(u)\Big),\frac{\mathrm{d}u}{\mathrm{d}t}\Big\rangle+c_2\Big\|\frac{\mathrm{d}u}{\mathrm{d}t}\Big\|_H^2\\&=\Big\langle\frac{\mathrm{d}f}{\mathrm{d}t},\frac{\mathrm{d}u}{\mathrm{d}t}\Big\rangle+c_2\Big\|\frac{\mathrm{d}u}{\mathrm{d}t}\Big\|_H^2\le\frac{1}{2\varepsilon}\Big\|\frac{\mathrm{d}f}{\mathrm{d}t}\Big\|_{V^*}^2+\frac{\varepsilon}{2}\Big\|\frac{\mathrm{d}u}{\mathrm{d}t}\Big\|_V^2+c_2\Big\|\frac{\mathrm{d}u}{\mathrm{d}t}\Big\|_H^2\\&\le\frac{1}{2\varepsilon}\Big\|\frac{\mathrm{d}f}{\mathrm{d}t}\Big\|_{V^*}^2+\varepsilon C_{\mathrm{P}}^2\Big|\frac{\mathrm{d}u}{\mathrm{d}t}\Big|_V^2+\varepsilon C_{\mathrm{P}}^2\Big\|\frac{\mathrm{d}u}{\mathrm{d}t}\Big\|_H^2+c_2\Big\|\frac{\mathrm{d}u}{\mathrm{d}t}\Big\|_H^2.\end{aligned}\tag{8.70}$$

Choosing $\varepsilon<c_0/C_{\mathrm{P}}^2$, (8.70) reads as

$$\frac{\mathrm{d}}{\mathrm{d}t}\Big(\frac{1}{2}\Big\|\frac{\mathrm{d}u}{\mathrm{d}t}\Big\|_H^2+\big(c_0-\varepsilon C_{\mathrm{P}}^2\big)\int_0^t\Big|\frac{\mathrm{d}u}{\mathrm{d}\theta}\Big|_V^2\mathrm{d}\theta\Big)\le\big(c_2+\varepsilon C_{\mathrm{P}}^2\big)\Big\|\frac{\mathrm{d}u}{\mathrm{d}t}\Big\|_H^2+\frac{1}{2\varepsilon}\Big\|\frac{\mathrm{d}f}{\mathrm{d}t}\Big\|_{V^*}^2\tag{8.71}$$

and then, by the Gronwall inequality, the first term gives the estimate of $\frac{\mathrm{d}}{\mathrm{d}t}u$ in $L^\infty(I;H)$ while the second one for $t=T$ gives $\frac{\mathrm{d}}{\mathrm{d}t}u$ in $L^2(I;V)$. Note that to apply Gronwall's inequality, we must have guaranteed $\frac{\mathrm{d}}{\mathrm{d}t}u(0)\in H$, i.e. $A(u_0)-f(0)\in H$.[20]

Proof of Theorem 8.18. Take (8.5) for k and for $k-1$, i.e.

$$\frac{u_\tau^k-u_\tau^{k-1}}{\tau}+A(u_\tau^k)=f_\tau^k\,,\qquad\frac{u_\tau^{k-1}-u_\tau^{k-2}}{\tau}+A(u_\tau^{k-1})=f_\tau^{k-1}\,,\tag{8.72}$$

subtract them, then test it by $u_\tau^k-u_\tau^{k-1}$, and divide it by τ^2. Thus we get

$$\begin{aligned}\frac{1}{2\tau}\Big\|\frac{u_\tau^k-u_\tau^{k-1}}{\tau}\Big\|_H^2&-\frac{1}{2\tau}\Big\|\frac{u_\tau^{k-1}-u_\tau^{k-2}}{\tau}\Big\|_H^2+c_0\Big|\frac{u_\tau^k-u_\tau^{k-1}}{\tau}\Big|_V^2-c_2\Big\|\frac{u_\tau^k-u_\tau^{k-1}}{\tau}\Big\|_H^2\\&\le\Big\langle\frac{u_\tau^k-2u_\tau^{k-1}+u_\tau^{k-2}}{\tau^2},\frac{u_\tau^k-u_\tau^{k-1}}{\tau}\Big\rangle+\Big\langle\frac{A(u_\tau^k)-A(u_\tau^{k-1})}{\tau},\frac{u_\tau^k-u_\tau^{k-1}}{\tau}\Big\rangle\\&=\Big\langle\frac{f_\tau^k-f_\tau^{k-1}}{\tau},\frac{u_\tau^k-u_\tau^{k-1}}{\tau}\Big\rangle\le\frac{1}{2\varepsilon}\Big\|\frac{f_\tau^k-f_\tau^{k-1}}{\tau}\Big\|_{V^*}^2+\frac{\varepsilon}{2}\Big\|\frac{u_\tau^k-u_\tau^{k-1}}{\tau}\Big\|_V^2\\&\le\frac{1}{2\varepsilon}\Big\|\frac{f_\tau^k-f_\tau^{k-1}}{\tau}\Big\|_{V^*}^2+\varepsilon C_{\mathrm{P}}^2\Big|\frac{u_\tau^k-u_\tau^{k-1}}{\tau}\Big|_V^2+\varepsilon C_{\mathrm{P}}^2\Big\|\frac{u_\tau^k-u_\tau^{k-1}}{\tau}\Big\|_H^2;\end{aligned}\tag{8.73}$$

the first inequality is due to (8.25) for $(u_\tau^k-u_\tau^{k-1})/\tau$ in place of u_τ^k and (8.69c) while the last one is due to the Young inequality. By extension of f for $t<0$ by putting $f(t)=f(0)$, we still have $f\in W^{1,2}([-\tau,T];V^*)$. Then $f_\tau^0:=f(0)$, and we get $(u_\tau^0-u_\tau^{-1})/\tau=f(0)-A(u_0)\in H$ by the assumption.[21] Absorbing the term with $\varepsilon<c_0/C_{\mathrm{P}}^2$ and then summing (8.73) for $k=1,\dots,l$, we obtain

$$\begin{aligned}\frac{1}{2}\Big\|\frac{u_\tau^l-u_\tau^{l-1}}{\tau}\Big\|_H^2&+\big(c_0-\varepsilon C_{\mathrm{P}}^2\big)\tau\sum_{k=1}^l\Big|\frac{u_\tau^k-u_\tau^{k-1}}{\tau}\Big|_V^2\\&\le\frac{1}{2}\big\|f(0)-A(u_0)\big\|_H^2+\big(c_2+\varepsilon C_{\mathrm{P}}^2\big)\tau\sum_{k=1}^l\Big(\Big\|\frac{u_\tau^k-u_\tau^{k-1}}{\tau}\Big\|_H^2+\frac{1}{2\varepsilon}d_\tau^k\Big)\end{aligned}\tag{8.74}$$

[20] It does not mean that $f(0)\in H$, however. In fact, $f(0)$ has a good sense only in V^*.

[21] In other words, this is the definition of u_τ^k for $k=-1$ needed here.

where we abbreviated $d_\tau^k := \|(f_\tau^k - f_\tau^{k-1})/\tau\|_{V^*}^2$. Then, provided $\tau \le \tau_0 < 1/(2c_2)$, $\varepsilon > 0$ can be chosen so small that the discrete Gronwall inequality (1.68) applies, which gives an a-priori bound for $\frac{\mathrm{d}}{\mathrm{d}t}u_\tau$ in $L^\infty(I;H)$ and in $L^2(I;V)$ provided $\tau\sum_{k=1}^l d^k$ can be bounded independently of τ and $l \le T/\tau$. This can be seen from the estimate

$$\begin{aligned}\tau\sum_{k=1}^{T/\tau} d_\tau^k &\le \tau\sum_{k=1}^{T/\tau}\frac{1}{\tau^2}\int_{(k-1)\tau}^{k\tau}\int_0^\tau\Big\|\frac{\mathrm{d}}{\mathrm{d}t}f(t-\vartheta)\Big\|_{V^*}^2\mathrm{d}\vartheta\mathrm{d}t\\ &\le \frac{1}{\tau}\sum_{k=1}^{T/\tau}\int_0^\tau\int_{(k-1)\tau-\vartheta}^{k\tau-\vartheta}\Big\|\frac{\mathrm{d}}{\mathrm{d}t}f(\xi)\Big\|_{V^*}^2\mathrm{d}\xi\mathrm{d}\vartheta\\ &\le \frac{1}{\tau}\sum_{k=1}^{T/\tau}\int_0^\tau\int_{(k-2)\tau}^{k\tau}\Big\|\frac{\mathrm{d}}{\mathrm{d}t}f(\xi)\Big\|_{V^*}^2\mathrm{d}\xi\mathrm{d}\vartheta \le 2\Big\|\frac{\mathrm{d}f}{\mathrm{d}t}\Big\|_{L^2(I;V^*)}^2\end{aligned}\tag{8.75}$$

where, for the first inequality in (8.75), we used, after the substitution $t-\vartheta=\xi$, also

$$\begin{aligned}d_\tau^k := \Big\|\frac{f_\tau^k - f_\tau^{k-1}}{\tau}\Big\|_{V^*}^2 &= \frac{1}{\tau^2}\Big\|\frac{1}{\tau}\int_{(k-1)\tau}^{k\tau} f(t)-f(t-\tau)\,\mathrm{d}t\Big\|_{V^*}^2\\ &= \frac{1}{\tau^4}\Big\|\int_{(k-1)\tau}^{k\tau}\int_0^\tau\frac{\mathrm{d}}{\mathrm{d}t}f(t-\vartheta)\,\mathrm{d}\vartheta\mathrm{d}t\Big\|_{V^*}^2\\ &\le \frac{1}{\tau^4}\Big(\int_{(k-1)\tau}^{k\tau}\int_0^\tau\Big\|\frac{\mathrm{d}}{\mathrm{d}t}f(t-\vartheta)\Big\|_{V^*}\mathrm{d}\vartheta\mathrm{d}t\Big)^2\\ &\le \frac{1}{\tau^2}\int_{(k-1)\tau}^{k\tau}\int_0^\tau\Big\|\frac{\mathrm{d}}{\mathrm{d}t}f(t-\vartheta)\Big\|_{V^*}^2\mathrm{d}\vartheta\mathrm{d}t\end{aligned}\tag{8.76}$$

where the last inequality uses Hölder's inequality.

Eventually, the convergence claimed in the point (ii) has been proved in Theorem 8.16. □

Remark 8.19 (*1st-order Clément's quasi-interpolation* [98]). Defining the 1st-order quasi-interpolant $f_\tau \in W^{1,\infty}(I;V^*)$ as the piecewise affine interpolation of the sequence $\{f_\tau^k\}_{k=0}^{T/\tau}$, one can interpret (8.75) as the estimate $\|\frac{\mathrm{d}}{\mathrm{d}t}f_\tau\|_{L^2(I;V^*)}^2 \le 2\|\frac{\mathrm{d}}{\mathrm{d}t}f\|_{L^2(I;V^*)}^2$.

Remark 8.20 (*Multilevel formulae*). The two-level formula (8.5) is not the only option to be used for theoretical investigation and for further numerical applications. An example of an alternative option is the 3-level *Gear's formula* [174]:

$$\frac{3u_\tau^k - 4u_\tau^{k-1} + u_\tau^{k-2}}{2\tau} + A(u_\tau^k) = f_\tau^k\,, \qquad k \ge 2, \tag{8.77}$$

while for $k = 1$ one is to use (8.5). This formula approximates the time derivative with a higher order, may yield a better error estimate than (8.5) if a solution is enough regular, and may simultaneously have good stability properties, as shown for a linear case in [357]. A mere convergence can be shown quite simply: use the test by $\delta_\tau^k := (u_\tau^k - u_\tau^{k-1})/\tau$ as in the proof of Theorem 8.16(i) and the estimate

$$\Big\langle\frac{3u_\tau^k-4u_\tau^{k-1}+u_\tau^{k-2}}{2\tau},\frac{u_\tau^k-u_\tau^{k-1}}{\tau}\Big\rangle=\frac{3}{2}\big\|\delta_\tau^k\big\|_H^2-\frac{1}{2}\langle\delta_\tau^k,\delta_\tau^{k-1}\rangle\geq\big\|\delta_\tau^k\big\|_H^2-\frac{1}{8}\big\|\delta_\tau^{k-1}\big\|_H^2,$$

with the agreement that $\delta_\tau^k := 0$ for $k = 0$. Summation then gives $\frac{7}{8}\sum_{k=1}^l \big\|\delta_\tau^k\big\|_H^2 + \frac{1}{8}\big\|\delta_\tau^l\big\|_H^2 - \frac{1}{8}\big\|\delta_\tau^1\big\|_H^2$, which is to be used to modify (8.67). This gives the a-priori estimate of u_τ in $W^{1,2}(I;H)\cap L^\infty(I;V)$ as in Theorem 8.16. The convergence can then be proved when realizing that (8.77) can be written in the form

$$\frac{3}{2}\frac{\mathrm{d}u_\tau}{\mathrm{d}t}-\frac{1}{2}\frac{\mathrm{d}u_\tau^{\mathrm{R}}}{\mathrm{d}t}+\mathscr{A}(\bar{u}_\tau)=\bar{f}_\tau \tag{8.78}$$

with the "retarded" Rothe function u_τ^{R} defined by $u_\tau^{\mathrm{R}}(t) := u_\tau(t-\tau)$ for $t \in [\tau, T]$ while $u_\tau^{\mathrm{R}}(t) := u_\tau(t)$ for $t \in [0, \tau]$. Modification of the proof of Theorem 8.16(ii) is left as an Exercise 8.58. Higher-level formulae do exist, too, and exhibit stability (and thus the a-priori estimates and convergence) but up to the level 7, i.e. at most u_τ^{k-6} is involved; we refer to Thomée [405, Chap.10].

Remark 8.21 (*Non-autonomous case*). The Rothe method can be generalized to the time-dependent, so-called non-autonomous case (8.1); more precisely, we will consider $A : I \times V \to V^*$ as a Carathéodory mapping such that the corresponding Nemytskiĭ mapping, denoted by $\mathscr{A} := \mathscr{N}_A$ like (8.11), i.e.

$$\big[\mathscr{A}(v)\big](t) := A(t, v(t)), \tag{8.79}$$

satisfies (8.12). For example, $\mathscr{A} : L^p(I;V)\cap L^\infty(I;H) \to L^{p'}(I;V^*)$ is bounded if, instead of (8.13), the following growth condition holds:

$$\exists\gamma\in L^{p'}(I),\ \mathfrak{C}{:}\mathbb{R}{\to}\mathbb{R}\text{ increasing}:\quad \big\|A(t,v)\big\|_{V^*}\leq\mathfrak{C}\big(\|v\|_H\big)\big(\gamma(t)+\|v\|_V^{p-1}\big). \tag{8.80}$$

Then the Rothe sequence can be defined by the recursive formula

$$\frac{u_\tau^k-u_\tau^{k-1}}{\tau}+A_\tau^k(u_\tau^k)=f_\tau^k,\qquad u_\tau^0=u_0\,,\qquad\text{with}$$
$$A_\tau^k(u):=\frac{1}{\tau}\int_{(k-1)\tau}^{k\tau}A(t,u)\,\mathrm{d}t\,,\quad f_\tau^k:=\frac{1}{\tau}\int_{(k-1)\tau}^{k\tau}f(t)\,\mathrm{d}t. \tag{8.81}$$

Then, e.g., $\tau\sum_{k=1}^l\langle A_\tau^k(u_\tau^k),u_\tau^k\rangle = \int_0^{l\tau}\langle A(t,\bar{u}_\tau(t)),\bar{u}_\tau(t)\rangle\mathrm{d}t$. The modification of Lemmas 8.5 and 8.6 and Theorem 8.9 would require auxiliary smoothing like in (8.5), also Lemma 8.8 holds with its proof just straightforwardly modified, while the modification of Theorems 8.16 and 8.18 requires additional smoothness of $A(\cdot, u)$.

Remark 8.22 (Infinite time horizon). By a subsequent continuation, one can pass $T \to +\infty$ and obtain respective results on $I := [0, +\infty)$. E.g. Theorem 8.9 gives $u \in L^p_{\rm loc}(I; V) \cap W^{1,p'}_{\rm loc}(I; V^*)$ if $f \in L^{p'}_{\rm loc}(I; V^*)$, Theorem 8.16 gives $u \in L^\infty_{\rm loc}(I; V) \cap W^{1,2}_{\rm loc}(I; H)$ if $f \in L^2_{\rm loc}(I; H)$, and Theorem 8.18 gives $u \in W^{1,\infty}_{\rm loc}(I; H) \cap W^{1,2}_{\rm loc}(I; V)$ if $f \in W^{1,2}_{\rm loc}(I; V^*)$.

Remark 8.23 (Alternative estimation). One can also assume, alternatively to (8.60b), that $f \in W^{1,1}(I; V^*)$ in Theorem 8.16. The estimation strategy (8.62) is to be integrated over $[0, t]$ and modified by a by-part integration and by the Poincaré-type inequality (8.9) as

$$\begin{aligned}
&\int_0^t \Big\|\frac{\mathrm{d}u}{\mathrm{d}t}\Big\|_H^2 \mathrm{d}t + \Phi(u(t)) = \Phi(u_0) + \int_0^t \Big\langle f - A_2(u), \frac{\mathrm{d}u}{\mathrm{d}t}\Big\rangle \mathrm{d}t \\
&= \big\langle f(t), u(t)\big\rangle - \int_0^t \Big\langle A_2(u), \frac{\mathrm{d}u}{\mathrm{d}t}\Big\rangle + \Big\langle \frac{\mathrm{d}f}{\mathrm{d}t}, u\Big\rangle \,\mathrm{d}t \\
&\le \big\|f(t)\big\|_{V^*}\big\|u(t)\big\|_V + \int_0^t \big\|A_2(u)\big\|_H^2\Big\|\frac{\mathrm{d}u}{\mathrm{d}t}\Big\|_H^2 + \Big\|\frac{\mathrm{d}f}{\mathrm{d}t}\Big\|_{V^*}\big\|u\big\|_V \mathrm{d}t + \Phi(u_0) - \big\langle f(0), u_0\big\rangle \\
&\le C_{\rm P}\big\|f(t)\big\|_{V^*}\Big(\big|u(t)\big|_V + \big\|u(t)\big\|_H\Big) + \int_0^t \frac{1}{2}\big\|A_2(u)\big\|_H^2 + \frac{1}{2}\Big\|\frac{\mathrm{d}u}{\mathrm{d}t}\Big\|_H^2 \\
&\qquad + C_{\rm P}\Big\|\frac{\mathrm{d}f}{\mathrm{d}t}\Big\|_{V^*}\Big(\big|u(t)\big|_V + \big\|u(t)\big\|_H\Big)\,\mathrm{d}t + \Phi(u_0) - \big\langle f(0), u_0\big\rangle \\
&\le C_q\big\|f(t)\big\|_{V^*}\Big(1 + \big\|u(t)\big\|_H^2\Big) + \varepsilon\big|u(t)\big|_V^q + C_{q,\varepsilon}\big\|f(t)\big\|_{V^*}^{q'} + \int_0^t \frac{1}{2}\big\|A_2(u)\big\|_H^2 \\
&\qquad + \frac{1}{2}\Big\|\frac{\mathrm{d}u}{\mathrm{d}t}\Big\|_H^2 + C_q\Big\|\frac{\mathrm{d}f}{\mathrm{d}t}\Big\|_{V^*}\Big(1 + \big|u\big|_V^q + \big\|u\big\|_H^2\Big)\,\mathrm{d}t + \Phi(u_0) - \big\langle f(0), u_0\big\rangle
\end{aligned} \tag{8.82}$$

with some C_q and $C_{q,\varepsilon}$ depending on $C_{\rm P}$ and q and also on $\varepsilon > 0$. Then, choosing $\varepsilon > 0$ small enough and using also (8.60d) and (8.63), the strategy (8.64) modifies as

$$\begin{aligned}
\frac{1}{2}U(t) + \frac{c_0 - \varepsilon}{c_0}\Phi(u(t)) &\le C_q\big\|f(t)\big\|_{V^*}\Big(1 + \big\|u(t)\big\|_H^2\Big) \\
&\quad + \frac{\varepsilon c_1}{c_0}\big\|u(t)\big\|_H^2 + C_{q,\varepsilon}\big\|f(t)\big\|_{V^*}^{q'} + \int_0^t \frac{1}{2}\big\|A_2(u)\big\|_H^2 + \frac{1}{2}\Big\|\frac{\mathrm{d}u}{\mathrm{d}t}\Big\|_H^2 \\
&\quad + C_q'\Big\|\frac{\mathrm{d}f}{\mathrm{d}t}\Big\|_{V^*}\Big(1 + \frac{1}{c_0}\Phi(u) + \frac{c_0 + c_1}{c_0}\big\|u\big\|_H^2\Big)\,\mathrm{d}t + \Phi(u_0) - \big\langle f(0), u_0\big\rangle
\end{aligned} \tag{8.83}$$

and, after estimating $\|u\|_H^2 \le 2TU + 2\|u_0\|_H^2$, cf. (8.63), is to be finished by Gronwall's inequality. Also the qualification (8.69a) can be modified by requiring $f \in W^{1,1}(I; H)$ and the strategy (8.70) can use the estimate

$$\Big\langle \frac{\mathrm{d}f}{\mathrm{d}t}, \frac{\mathrm{d}u}{\mathrm{d}t}\Big\rangle \le \Big\|\frac{\mathrm{d}f}{\mathrm{d}t}\Big\|_H\Big(\frac{1}{4} + \Big\|\frac{\mathrm{d}u}{\mathrm{d}t}\Big\|_H^2\Big) \tag{8.84}$$

and then treated by Gronwall's inequality. Of course, these modifications can be combined with the original strategies so that $f \in L^2(I;H) + W^{1,1}(I;V^*)$ and $f \in W^{1,2}(I;V^*) + W^{1,1}(I;H)$ can be considered for Theorems 8.16 and 8.18, respectively. Implementing these strategies into the proofs of Theorems 8.16 and 8.18 needs either usage of the finer version of the discrete Gronwall inequality like in Remark 8.15 or, instead of usage (8.58), the scheme (8.5) with a controlled smoothening of f like in (8.15). This technical difficulty related with time discretisation does not occur when using the Galerkin method, as presented in the following Section 8.4.

Remark 8.24 (Semiconvex potential Φ)**.** The convexity of Φ assumed in (8.60) can be relaxed by assuming Φ only *semi-convex* (with respect to the norm of H) in the sense

$$\exists K \in \mathbb{R}^+ : \qquad v \mapsto \Phi(v) + K\|v\|_H^2 : V \to \mathbb{R}^+ \ \text{ is convex.} \tag{8.85}$$

Then, for τ sufficiently small, namely $0 < \tau \le (2K)^{-2}$, the estimation (8.65)–(8.67) can be modified by using

$$\begin{aligned}
&\Big\langle \frac{u_\tau^k - u_\tau^{k-1}}{\tau} + \Phi'(u_\tau^k), u_\tau^k - u_\tau^{k-1}\Big\rangle \\
&\quad = \Big\langle \frac{u_\tau^k}{\sqrt{\tau}} + \Phi'(u_\tau^k), u_\tau^k - u_\tau^{k-1}\Big\rangle - \Big\langle \frac{u_\tau^{k-1}}{\sqrt{\tau}}, u_\tau^k - u_\tau^{k-1}\Big\rangle + \frac{1-\sqrt{\tau}}{\tau}\big\|u_\tau^k - u_\tau^{k-1}\big\|_H^2 \\
&\quad \ge \frac{1}{2\sqrt{\tau}}\big\|u_\tau^k\big\|_H^2 + \Phi(u_\tau^k) - \frac{1}{2\sqrt{\tau}}\big\|u_\tau^{k-1}\big\|_H^2 - \Phi(u_\tau^{k-1}) \\
&\qquad\qquad - \Big\langle \frac{u_\tau^{k-1}}{\sqrt{\tau}}, u_\tau^k - u_\tau^{k-1}\Big\rangle + \frac{1-\sqrt{\tau}}{\tau}\big\|u_\tau^k - u_\tau^{k-1}\big\|_H^2 \\
&\quad = \Phi(u_\tau^k) - \Phi(u_\tau^{k-1}) + \tau\Big(1 - \frac{\sqrt{\tau}}{2}\Big)\Big\|\frac{u_\tau^k - u_\tau^{k-1}}{\tau}\Big\|_H^2.
\end{aligned} \tag{8.86}$$

This refines Rothe's method and, in this respect, brings it closer to the Galerkin method which, when executing the estimation strategy (8.62), does not need any convexity of Φ at all; cf. also Exercise 8.56 below.

Remark 8.25 (*Decoupling by semi-implicit Rothe method*)**.** The test by the time difference used in Theorem 8.16 allows for interesting effects in systems. For notational simplicity, let us demonstrate it for a system of three equations governed by $\Phi(u, v, z)$, i.e.,

$$\frac{\mathrm{d}u}{\mathrm{d}t} + \Phi'_u(u,v,z) = f_1, \qquad \frac{\mathrm{d}v}{\mathrm{d}t} + \Phi'_v(u,v,z) = f_2, \qquad \frac{\mathrm{d}z}{\mathrm{d}t} + \Phi'_z(u,v,z) = f_3. \tag{8.87}$$

Instead of a fully implicit discretisation, we can consider the semi-implicit scheme:

$$\frac{u_\tau^k - u_\tau^{k-1}}{\tau} + \Phi'_u(u_\tau^k, v_\tau^{k-1}, z_\tau^{k-1}) = f_{1,\tau}^k, \tag{8.88a}$$

$$\frac{v_\tau^k - v_\tau^{k-1}}{\tau} + \Phi'_v(u_\tau^k, v_\tau^k, z_\tau^{k-1}) = f_{2,\tau}^k, \tag{8.88b}$$

$$\frac{z_\tau^k - z_\tau^{k-1}}{\tau} + \Phi'_z(u_\tau^k, v_\tau^k, z_\tau^k) = f_{3,\tau}^k. \tag{8.88c}$$

Testing particular equations by time differences, as in (8.65), leads to the estimates

$$\begin{aligned}
&\Big\|\frac{u_\tau^k - u_\tau^{k-1}}{\tau}\Big\|^2 + \frac{\Phi(u_\tau^k, v_\tau^{k-1}, z_\tau^{k-1}) - \Phi(u_\tau^{k-1}, v_\tau^{k-1}, z_\tau^{k-1})}{\tau} \leq \Big\langle f_{1,\tau}^k, \frac{u_\tau^k - u_\tau^{k-1}}{\tau}\Big\rangle,\\
&\Big\|\frac{v_\tau^k - v_\tau^{k-1}}{\tau}\Big\|^2 + \frac{\Phi(u_\tau^k, v_\tau^k, z_\tau^{k-1}) - \Phi(u_\tau^k, v_\tau^{k-1}, z_\tau^{k-1})}{\tau} \leq \Big\langle f_{2,\tau}^k, \frac{v_\tau^k - v_\tau^{k-1}}{\tau}\Big\rangle,\\
&\Big\|\frac{z_\tau^k - z_\tau^{k-1}}{\tau}\Big\|^2 + \frac{\Phi(u_\tau^k, v_\tau^k, z_\tau^k) - \Phi(u_\tau^k, v_\tau^k, z_\tau^{k-1})}{\tau} \leq \Big\langle f_{3,\tau}^k, \frac{z_\tau^k - z_\tau^{k-1}}{\tau}\Big\rangle
\end{aligned} \tag{8.89}$$

provided $\Phi(\cdot, v, z)$, $\Phi(u, \cdot, z)$, and $\Phi(u, v, \cdot)$ are convex. Summation then yields a notable "telescopical cancellation effect": four terms, namely $\pm\Phi(u_\tau^k, v_\tau^k, z_\tau^{k-1})$ and $\pm\Phi(u_\tau^k, v_\tau^{k-1}, z_\tau^{k-1})$, mutually cancel and one obtains:

$$\begin{aligned}
&\Big\|\frac{u_\tau^k - u_\tau^{k-1}}{\tau}\Big\|^2 + \Big\|\frac{v_\tau^k - v_\tau^{k-1}}{\tau}\Big\|^2 + \Big\|\frac{z_\tau^k - z_\tau^{k-1}}{\tau}\Big\|^2\\
&\quad + \frac{\Phi(u_\tau^k, v_\tau^k, z_\tau^k) - \Phi(u_\tau^{k-1}, v_\tau^{k-1}, z_\tau^{k-1})}{\tau}\\
&\qquad \leq \Big\langle f_{1,\tau}^k, \frac{u_\tau^k - u_\tau^{k-1}}{\tau}\Big\rangle + \Big\langle f_{2,\tau}^k, \frac{v_\tau^k - v_\tau^{k-1}}{\tau}\Big\rangle + \Big\langle f_{3,\tau}^k, \frac{z_\tau^k - z_\tau^{k-1}}{\tau}\Big\rangle,
\end{aligned} \tag{8.90}$$

and then one can proceed as in (8.67). Let us note that, in contrast to (8.87), the time-discrete system (8.88) is decoupled and that, in contrast to a fully implicit discretisation and Proposition 11.6, only a separate convexity (instead of the joint convexity) of Φ is needed now. Even more, in the spirit of Remark 8.24, only separate semi-convexity of Φ is sufficient if the strategy (8.86) is applied to each inequality in (8.89). Of course, this approach applies for an arbitrary number of equations in (8.87); for systems of two specific differential equations cf. also [220, 252]. For f constant, denoting $w = (u, v, z)$, $A_1(w) = (\Phi'_u(w) - f_1, 0, 0)$, $A_2(w) = (0, \Phi'_v(w) - f_2, 0)$, and $A_3(w) = (0, 0, \Phi'_z(w) - f_3)$, the system (8.87) takes the form $\frac{\mathrm{d}w}{\mathrm{d}t} + A_1(w) + A_2(w) + A_3(w) = 0$ and the scheme (8.88) is revealed as a *fractional-step method* or also a so-called Lie-Trotter (or *sequential*) *splitting* combined *with the implicit Euler formula*, cf. [142, 272, 411]:

$$\frac{w_\tau^{k-1+j/3} - w_\tau^{k-4/3+j/3}}{\tau} + A_j(w_\tau^{k-1+j/3}) = 0, \qquad j = 1, 2, 3, \tag{8.91}$$

where, for j=1, we put $w_\tau^{k-1} = (u_\tau^{k-1}, v_\tau^{k-1}, z_\tau^{k-1})$. Then one finds that $w_\tau^{k-2/3} = (u_\tau^k, v_\tau^{k-1}, z_\tau^{k-1})$, $w_\tau^{k-1/3} = (u_\tau^k, v_\tau^k, z_\tau^{k-1})$, and eventually $w_\tau^k = (u_\tau^k, v_\tau^k, z_\tau^k)$.

8.4 Galerkin method

An alternative method to analyze evolution problems, consisting in discretization of V, is referred to as a *Faedo-Galerkin method* [141], or mostly briefly as *Galerkin method* likewise in case of steady-state problems where, however, it led directly to finite-dimensional problems. Similarly as in the proof of Theorem 2.6, we consider a sequence of finite-dimensional subspaces $V_k \subset V$ satisfying (2.7), i.e. $V_k \subset V_{k+1}$ and $\bigcup_{k\in\mathbb{N}} V_k$ dense in V. As also V is dense in H, for $u_0 \in H$ we can consider a sequence $\{u_{0k}\}_{k\in\mathbb{N}}$ converging to u_0 in H and such that $u_{0k} \in V_k$. Now, we can very naturally consider A also time-dependent, using the convention (8.79), even in a more general setting $A : I \times V \to Z^*$ for some $Z \subset V$ densely provided $\bigcup_{k\in\mathbb{N}} V_k \subset Z$, although mostly, in particular for the purpose of strong solutions, the case $Z = V$ is general enough. Then the Galerkin sequence $\{u_k\}_{k\in\mathbb{N}}$ of approximate solutions $u_k \in W^{1,p'}(I; V_k) = W^{1,\infty,p'}(I; V_k, V_k)$, cf. (7.3), to the Cauchy problem (8.1) is defined by

$$\forall v\in V_k\ \forall(\text{a.a.})\,t\in I: \quad \Big(\frac{\mathrm{d}u_k}{\mathrm{d}t}, v\Big) + \big\langle A(t, u_k(t)), v\big\rangle_{Z^*\times Z} = \big\langle f(t), v\big\rangle_{V^*\times V}, \tag{8.92a}$$

$$u_k(0) = u_{0k}, \tag{8.92b}$$

where $(\cdot,\cdot)$ is the inner product in H as before. In fact, we have $V_k \subset Z \subset V \subset H \subset V^* \subset Z^*$, hence $\frac{\mathrm{d}}{\mathrm{d}t}u_k$ is valued also in Z^* and it will be useful to introduce the seminorms $|\cdot|_k$, $k \in \mathbb{N}$, on Z^* defined by

$$|\xi|_k = \sup_{v\in V_k,\ \|v\|_Z\le 1} \big\langle \xi, v\big\rangle_{Z^*\times Z}. \tag{8.93}$$

In accord with (7.5), $|\cdot|_{q',k}$ denotes the seminorm on $L^{q'}(I; Z^*)$ defined by

$$\big|\xi\big|_{q',k} := \Big(\int_0^T \big|\xi(t)\big|_k^{q'}\,\mathrm{d}t\Big)^{1/q'} = \sup_{\substack{\|v\|_{L^q(I;Z)}\le 1\\ v(t)\in V_k \text{ for a.a. } t\in I}} \int_0^T \big\langle \xi(t), v(t)\big\rangle_{Z^*\times Z}\,\mathrm{d}t. \tag{8.94}$$

Lemma 8.26 (GALERKIN APPROXIMATIONS, A-PRIORI ESTIMATES). *Let $f \in L^{p'}(I; V^*) + L^{q'}(I; H)$, $u_0 \in H$, and $A : I\times V \to Z^*$ be a Carathéodory mapping such that $\mathscr{A}$ satisfies (8.80) with $p \le q \le +\infty$ and its restriction $A : I \times Z \to Z^*$ is semi-coercive in the sense*

$$\exists c_0 > 0,\quad c_1\in L^{p'}(I),\quad c_2\in L^1(I)\ \ \forall v \in Z:$$
$$\big\langle A(t, v), v\big\rangle_{Z^*\times Z} \ge c_0|v|_V^p - c_1(t)|v|_V - c_2(t)\|v\|_H^2 \tag{8.95}$$

with $|\cdot|_V$ referring again to (8.9). If $V_k \subset Z$ and if $\{u_{0k}\}_{k\in\mathbb{N}}$ is bounded in H, then there is a solution u_k to (8.92) satisfying

$$\big\|u_k\big\|_{L^\infty(I;H)\,\cap\, L^p(I;V)} \le C_1\,, \tag{8.96a}$$

$$\Big|\frac{\mathrm{d}u_k}{\mathrm{d}t}\Big|_{q',l} \le C_2 \quad \forall k \ge l, \tag{8.96b}$$

for any l; note that C_2 does not depend on l. If, in addition, there is a selfadjoint projector $P_k : H \to H$ such that $P_k(V) = V_k$ and $\|P_k|Z\|_{\mathscr{L}(Z,Z)}$ is bounded independently of k, then also

$$\left\|\frac{\mathrm{d}u_k}{\mathrm{d}t}\right\|_{L^{q'}(I;Z^*)} \le C_3 . \tag{8.97}$$

Proof. Taking a base $\{v_{ki}\}_{i=1,\dots,n_k}$, $n_k := \dim(V_k)$, in V_k and assuming $u_k(t) = \sum_{i=1}^{n_k} c_{ki}(t)v_{ki}$, (8.92) represents an initial-value problem for a system of n_k ordinary differential equations for the coefficients $(c_i)_{i=1,\dots,n_k}$. Due to Theorem 1.44, it has a solution on some sufficiently short time interval $[0,t_k)$.[22] As the test functions for (8.92a) are the spaces V_k where also the approximate solution u_k is sought, we are authorized to put $v = u_k(t)$ in (8.92). Then, as in (8.20)–(8.21), we get the estimate:

$$\begin{aligned}\frac{1}{2}\frac{\mathrm{d}}{\mathrm{d}t}\big\|u_k(t)\big\|_H^2 + c_0\big|u_k(t)\big|_V^p &\le c_1(t)\big|u_k(t)\big|_V + c_2(t)\big\|u_k(t)\big\|_H^2 + \big\langle f(t), u_k(t)\big\rangle \\ &\le C_\varepsilon c_1(t)^{p'} + \varepsilon\big|u_k(t)\big|_V^p + c_2(t)\big\|u_k(t)\big\|_H^2 + C_{\rm P} C_\varepsilon\big\|f_1\big\|_{V^*}^{p'} \\ &\quad + C_{\rm P}\varepsilon\big|u_k\big|_V^p + \big(C_{\rm P}\big\|f_1\big\|_{V^*} + \big\|f_2\big\|_H\big)\Big(\frac{1}{2} + \frac{1}{2}\big\|u_k\big\|_H^2\Big)\end{aligned} \tag{8.98}$$

with C_ε from (1.22) and with $f = f_1 + f_2$, $f_1 \in L^{p'}(I;V^*)$, $f_2 \in L^{q'}(I;H)$. In particular, by Gronwall's inequality (1.66) as used in (8.22), we have an $L^\infty(0,t_k)$-estimate so that $u_k(t)$ must live in a ball of V_k which is compact, and hence we can prolong the solution on the whole interval I because, assuming the contrary, we would get a limit time inside I not allowing for any further local solution, a contradiction[23]. Besides, this a-priori estimate yields that u_k is bounded in $L^\infty(I;H) \cap L^p(I;V)$ independently of k, as claimed in (8.96a).

If $k \ge l$, using (8.92), the estimate (8.96b) follows similarly like (8.23):

$$\begin{aligned}\left|\frac{du_k}{dt}\right|_{q',l} &= \sup_{\substack{\|v\|_{L^q(I;Z)}\le 1\\ v(\cdot)\in V_l \text{ a.e.}}} \int_0^T \big\langle f(t) - A(t,u_k(t)), v(t)\big\rangle_{Z^*\times Z}\,\mathrm{d}t \\ &\le \sup_{\|v\|_{L^q(I;Z)}\le 1} \int_0^T \big\langle f(t) - A(t,u_k(t)), v(t)\big\rangle_{Z^*\times Z}\,\mathrm{d}t = \big\|f - \mathscr{A}(u_k)\big\|_{L^{q'}(I;Z^*)}\end{aligned}$$

which is bounded due to (8.12) and the already proved estimate (8.96a), cf. also (8.33). Moreover, realizing $P_k u_k = u_k$ and $P_k^* = P_k$ and using again (8.92), and

[22] Note that, since A is a Carathéodory mapping, $I \times \mathbb{R}^{n_k} \to \mathbb{R}^{n_k} : (t, c_1, \dots, c_{n_k}) \mapsto (\langle A(t, \sum_{i=1}^{n_k} c_{ki}v_{ki}), v_{kj}\rangle)_{j=1,\dots,n_k}$ is a Carathéodory mapping, as needed for Theorem 1.44.

[23] In special cases, e.g. (8.80) for $p \le 2$ and $\mathfrak{C}(\cdot)$ bounded, the right-hand side of the underlying system of ordinary differential equations has at most a linear growth, so the global existence follows directly by Theorem 1.45.

also (8.12) and (8.96a), we can modify the estimate (8.33) as

$$\Big\langle\frac{\mathrm{d}u_k}{\mathrm{d}t},v\Big\rangle=\Big\langle P_k\frac{\mathrm{d}u_k}{\mathrm{d}t},v\Big\rangle=\Big\langle\frac{\mathrm{d}u_k}{\mathrm{d}t},P_kv\Big\rangle=\int_0^T\big\langle f(t)-A(t,u_k(t)),P_kv(t)\big\rangle\,\mathrm{d}t$$
$$\le\Big(\|\mathscr{A}(u_k)\|_{L^{q'}(I;Z^*)}+\|f\|_{L^{q'}(I;Z^*)}\Big)\|P_kv\|_{L^q(I;Z)}$$
$$\le\Big(N_0\mathfrak{C}(C_1)\big(\|\gamma\|_{L^{q'}(I)}+C_1^{p-1}\big)+\|f\|_{L^{q'}(I;Z^*)}\Big)\big\|P_k\big\|_{\mathscr{L}(Z,Z)}\|v\|_{L^q(I;Z)}\tag{8.99}$$

where γ and $\mathfrak{C}$ are from (8.80) and N_0 is the norm of the embedding $Z\subset V$ and thus also of the embedding $V^*\subset Z^*$. From this, (8.97) with $C_3:=(N_0\mathfrak{C}(C_1)$ $(\|a\|_{L^{q'}(I)}+C_1^{p-1})+\|f\|_{L^{q'}(I;V^*)})\sup_{k\in\mathbb{N}}\|P_k\|_{\mathscr{L}(Z,Z)}$ follows similarly as (8.19a). $\square$

Remark 8.27 (The projector P_k). Taking a base $\{v_{ki}\}_{i=1,\dots,\dim(V_k)}$ of V_k orthogonal with respect to the inner product $(\cdot,\cdot)$ in H, by putting

$$P_ku:=\sum_{i=1}^{\dim(V_k)}(u,v_{ki})v_{ki}\tag{8.100}$$

we obtain a selfadjoint projector $P_k:H\to H$, and $P_kH=P_kV=V_k$. It remains, however, to be proved in particular cases that $\|P_k\|_{\mathscr{L}(Z,Z)}$ is bounded independently of k for a suitable Z, cf. also Remark 8.44 below.

Lemma 8.28. *Let the collection $\{V_k\}_{k\in\mathbb{N}}$ satisfy (2.7). Then $\bigcup_{k\in\mathbb{N}}L^\infty(I;V_k)$ is dense in $L^p(I;V)$ for any $1\le p\le+\infty$.*

Proof. As $L^\infty(I;V)$ is dense in $L^p(I;V)$, it suffices to prove it for $p=+\infty$. Take $v\in L^\infty(I;V)$. As v is Bochner measurable, there is a sequence $\{v_k\}_{k\in\mathbb{N}}$ of simple functions such that $v_k(t)\to v(t)$ for a.a. $t\in I$. Besides, the construction of v_k can be performed so that $v_k(I)\subset v(I)$, hence $\|v_k\|_{L^\infty(I;V)}\le\|v\|_{L^\infty(I;V)}$, and $v_k\to v$ in $L^\infty(I;V)$. Now, taking v_k fixed and realizing (2.7), each of the (finite number of) values of v_k can be approximated by a value in V_l if l is sufficiently large, obtaining some $v_{kl}\in L^\infty(I;V_l)$ such that $\lim_{l\to\infty}v_{kl}=v_k$ in $L^\infty(I;V)$. Thus $\lim_{k\to\infty}\lim_{l\to\infty}v_{kl}=v$ and, by a suitable diagonalization, we get a sequence of v_{kl} attaining v. $\square$

Lemma 8.29 (Papageorgiou [323], here generalized). *Let the Carathéodory mapping $A:I\times V\to V^*$ satisfy (8.80) and (8.95) with $Z=V$ and $q:=p$, and let $A(t,\cdot)$ be pseudomonotone for a.a. $t\in I$. Then $\mathscr{A}$ is pseudomonotone on $\mathscr{W}:=W^{1,p,p'}(I;V,V^*_{\rm lcs})\cap L^\infty(I;H)$.*[24]

Proof. Just an obvious modification of the proof of Lemma 8.8. $\square$

[24] Recall that $V^*_{\rm lcs}$ denotes the dual space V^* considered as the locally convex space equipped with the collection of seminorms $\{|\cdot|_k\}_{k\in\mathbb{N}}$ which induces the seminorms on $L^{p'}(I;V^*)$ by the formula (8.94) with $q:=p$. The pseudomonotonicity is again understood in the sense of (8.35).

Theorem 8.30 (CONVERGENCE TO STRONG SOLUTIONS). *Let the assumptions of Lemmas 8.28–8.29 be fulfilled, let $u_{0k} \to u_0$ in H with $u_{0k} \in V_k$. Then $u_k \rightharpoonup u$ in $L^p(I;V)$ (possibly in terms of subsequences) and u is a strong solution to the Cauchy problem* (8.1).

Proof. By (8.96a) and the reflexivity of $L^p(I;V)$, we can take a subsequence and some $u \in L^p(I;V)$ such that

$$u_k \overset{*}{\rightharpoonup} u \qquad \text{in } \ L^p(I;V) \cap L^\infty(I;H). \tag{8.101}$$

Moreover, referring to the embedding $I_k : V_k \to V$ from the proof of Theorem 2.6, we have $I_l^* \frac{\mathrm{d}}{\mathrm{d}t} u_k \rightharpoonup \xi_l$ in any $L^{p'}(I;V_l^*)$ and ξ_{l+1} can be assumed as an extension of ξ_l from $L^p(I;V_l)$ to $L^p(I;V_{l+1})$.[25] By (8.96b), $\|\xi_l\|_{L^{p'}(I;V_l^*)} \le C_2$ independently of $l \in \mathbb{N}$. Hence, by density of $\bigcup_{l\in\mathbb{N}} L^p(I;V_l)$ in $L^p(I;V)$ (cf. Lemma 8.28) and by a (uniquely defined) continuous extension, we get eventually a functional $\dot u \in L^p(I;V)^* \cong L^{p'}(I;V^*)$ whose norm can again be upper-bounded by C_2. Moreover, $\dot u = \frac{\mathrm{d}}{\mathrm{d}t} u$ because $\dot u|_{L^p(I;V_l)} = \xi_l = \frac{\mathrm{d}}{\mathrm{d}t} u|_{L^p(I;V_l)}$ for any l, cf. also (8.41).

Note that the initial condition $u(0) = u_0$ is satisfied because $u_k(0) = u_{0k}$ and because of $u_{0k} \to u_0$ in H and of the weak continuity of the mapping $u \mapsto u(0) : W^{1,p,p'}(I;V,V^*_{\rm lcs}) \to V^*_{\rm lcs}$ by Lemma 7.1. Hence, $u_k(0) \rightharpoonup u(0)$ in $V^*_{\rm lcs}$. Simultaneously, $u_k(0) = u_{0k} \to u_0 = u(0)$ in H; cf. also (8.43).

For $v \in W^{1,p,p'}(I;V,V^*)$ let us take a sequence $v_k \in L^p(I;V_k)$ such that $v_k \to v$ in $L^p(I;V)$; such a sequence does exist due to Lemma 8.28.

From (8.92) one can see that, for any $z \in L^p(I;V_k)$, one has

$$\int_0^T \Big(\frac{\mathrm{d}u_k}{\mathrm{d}t}, z\Big) + \big\langle A(t,u_k(t)), z(t)\big\rangle_{V^*\times V} \,\mathrm{d}t = \int_0^T \big\langle f(t), z(t)\big\rangle_{V^*\times V} \,\mathrm{d}t. \tag{8.102}$$

In terms of $(\cdot,\cdot)$ as the inner product in the Hilbert space $L^2(I;H)$ and $\langle\cdot,\cdot\rangle$ as the duality on $L^{p'}(I;V^*) \times L^p(I;V)$, one can rewrite (8.102) into $(\frac{\mathrm{d}}{\mathrm{d}t}u_k, z) + \langle \mathscr{A}(u_k), z\rangle = \langle f, z\rangle$. Putting $z := v_k - u_k$, one gets

$$\langle \mathscr{A}(u_k), v_k - u_k\rangle = \langle f, v_k - u_k\rangle - \Big(\frac{\mathrm{d}u_k}{\mathrm{d}t}, v_k - u_k\Big) =: I_k^{(1)} - I_k^{(2)}. \tag{8.103}$$

As $u_k \rightharpoonup u$ in $L^p(I;V)$, obviously $\lim_{k\to\infty} I_k^{(1)} = \langle f, v - u\rangle$.

By (8.96a), $u_k(T)$ is bounded in H, so we can assume $u_k(T) \rightharpoonup \zeta$ in H. By

[25] This is a bit technical argument: having selected a subsequence such that $I_1^* \frac{\mathrm{d}}{\mathrm{d}t} u_k \rightharpoonup \xi_1$ in $L^{p'}(I;V_1^*)$, we can select further a subsequence such that $I_2^* \frac{\mathrm{d}}{\mathrm{d}t} u_k \rightharpoonup \xi_2$ in any $L^{p'}(I;V_2^*)$. This does not violate the convergence we have already for $l = 1$. Then we can continue for $l = 3, 4, \ldots$, and eventually to make a diagonalization like in the proof of Banach Theorem 1.7, cf. Exercise 2.51.

(8.96b), we have

$$I_l^*\zeta = I_l^* \lim_{k\to\infty} u_k(T) = \lim_{k\to\infty} I_l^* u_k(T) = \lim_{k\to\infty} \int_0^T I_l^* \frac{\mathrm{d}u_k}{\mathrm{d}t}\,\mathrm{d}t + I_l^* u_{0k}$$
$$= \int_0^T \xi_l \,\mathrm{d}t + I_l^* u_0 = I_l^*\Big(\int_0^T \frac{\mathrm{d}u}{\mathrm{d}t}\,\mathrm{d}t + u(0)\Big) = I_l^* u(T).$$

As it holds for any $l \in \mathbb{N}$, by the density of $\bigcup_{k\in\mathbb{N}} V_k$ in H, we get $\zeta = u(T)$.

Now we use (7.22) for $u_k \in W^{1,p'}(I;V_k)$.[26] Then, by the weak lower semicontinuity of $\|\cdot\|_H^2$ and by using also $\|u_{0k}\|_H \to \|u_0\|_H$ and $u_0 = u(0)$, we can estimate

$$\begin{aligned}\limsup_{k\to\infty} I_k^{(2)} &= \lim_{k\to\infty} \Big(\frac{\mathrm{d}u_k}{\mathrm{d}t}, v_k\Big) - \frac{1}{2}\liminf_{k\to\infty} \|u_k(T)\|_H^2 + \frac{1}{2}\lim_{k\to\infty} \|u_{0k}\|_H^2 \\ &\le \Big\langle \frac{\mathrm{d}u}{\mathrm{d}t}, v\Big\rangle - \frac{1}{2}\|u(T)\|_H^2 + \frac{1}{2}\|u(0)\|_H^2 = \Big\langle \frac{\mathrm{d}u}{\mathrm{d}t}, v-u\Big\rangle, \end{aligned} \tag{8.104}$$

cf. also (8.47). Altogether,

$$\liminf_{k\to\infty} \big\langle \mathscr{A}(u_k), v_k - u_k\big\rangle \ge \Big\langle f - \frac{\mathrm{d}u}{\mathrm{d}t}, v-u\Big\rangle. \tag{8.105}$$

Using still the boundedness of $\{u_k\}_{k\in\mathbb{N}}$ in $\mathscr{W}$ and the growth assumption (8.80), we can see that $\{\mathscr{A}(u_k)\}_{k\in\mathbb{N}}$ is bounded in $L^{p'}(I;V^*) = L^p(I;V)^*$. As $v_k \to v$ in $L^p(I;V)$, we have

$$\begin{aligned}\limsup_{k\to\infty} \big\langle \mathscr{A}(u_k), u_k - v\big\rangle &= \limsup_{k\to\infty} \big\langle \mathscr{A}(u_k), u_k - v_k\big\rangle \\ &\quad + \lim_{k\to\infty} \big\langle \mathscr{A}(u_k), v_k - v\big\rangle \le \Big\langle \frac{\mathrm{d}u}{\mathrm{d}t} - f, v-u\Big\rangle. \end{aligned} \tag{8.106}$$

In particular, for $v := u$ we have got $\limsup_{k\to\infty}\langle \mathscr{A}(u_k), u_k - u\rangle \le 0$. By Lemma 8.29, i.e. the pseudomonotonicity of $\mathscr{A}$, we can conclude that

$$\liminf_{k\to\infty} \big\langle \mathscr{A}(u_k), u_k - v\big\rangle \ge \big\langle \mathscr{A}(u), u-v\big\rangle \tag{8.107}$$

for any $v \in L^p(I;V)$. Joining (8.106) and (8.107), one gets $\langle \mathscr{A}(u), u-v\rangle \le \langle f, u-v\rangle - \langle \frac{\mathrm{d}}{\mathrm{d}t}u, u-v\rangle$. As it holds for v arbitrary, we can conclude that

$$\langle \mathscr{A}(u), v\rangle = \langle f, v\rangle - \Big\langle \frac{\mathrm{d}u}{\mathrm{d}t}, v\Big\rangle. \tag{8.108}$$

As v is arbitrary, $\mathscr{A}(u) = f - \frac{\mathrm{d}}{\mathrm{d}t}u$ holds a.e. on I, cf. Exercise 8.49. □

[26] Note that $V_k \subset H$ need not be dense for it.

In the above proof, one could obtain $\dot{u} \in L^p(I;V)^*$ by an alternative, although less constructive, argumentation: As $\frac{du_k}{dt} \in L^{p'}(I;V_k) \subset L^{p'}(I;V^*) \cong L^p(I;V)^*$, one can consider an extension $\dot{u}_k \in L^p(I;V)^*$ of $\frac{du_k}{dt}$ according to the Hahn-Banach Theorem 1.5 with $\|\dot{u}_k\|_{L^p(I;V)^*} = \|\frac{du_k}{dt}\|_{L^{p'}(I;V_k)} \le C_2$ with C_2 from (8.96b). Then, up to a subsequence, $\dot{u}_k \rightharpoonup \dot{u}$ in $L^{p'}(I;V^*)$. Although in general $\dot{u}_k \neq \frac{du_k}{dt}$, in the limit one has $\dot{u} = \frac{du}{dt}$. This can be seen from an analog of (8.41):

$$\big\langle \dot{u}, v\big\rangle_{L^p(I;V)^* \times L^p(I;V)} \leftarrow \big\langle \dot{u}_k, v\big\rangle_{L^p(I;V)^* \times L^p(I;V)} = \int_0^T \Big(\frac{du_k}{dt}, v\Big)\,dt$$
$$= -\int_0^T \Big(u_k, \frac{dv}{dt}\Big)\,dt \to -\int_0^T \Big\langle u, \frac{dv}{dt}\Big\rangle\,dt = -\Big\langle u, \frac{dv}{dt}\Big\rangle,$$

which holds for any $v \in C_0^1(I;V_l)$ for a sufficiently large k, namely $k \ge l$. Eventually, one uses density of $\bigcup_{l\in\mathbb{N}} C_0^1(I;V_l)$ in $C_0^1(I;V)$, cf. Lemma 8.28, to see that $\dot{u}$ is indeed the distributional derivative of u.

If the projectors P_k from Lemma 8.26 are at our disposal, the situation is even simpler because, due to the estimate (8.97), one can choose directly $\dot{u}_k = \frac{du_k}{dt}$.

Theorem 8.31 (Weak solution). *Let the assumptions of Lemma* 8.26 *which guarantee* (8.96) *be satisfied, in particular,* $f \in L^{p'}(I;V^*) + L^{q'}(I;H)$, *and let* A *satisfy*

$$\exists \gamma \in L^{q'}(I),\ \mathfrak{C}{:}\mathbb{R}\to\mathbb{R} \text{ increasing: } \ \|A(t,u)\|_{Z^*} \le \mathfrak{C}(\|u\|_H)\big(\gamma(t) + \|u\|_V^{p/q'}\big), \tag{8.109}$$

with $1 < q \le +\infty$ *and with some Banach space* Z *embedded into* V *densely, and induce* $\mathscr{A}$ *weakly* continuous from* $W^{1,p,\mathscr{M}}(I;V,Z^*) \cap L^\infty(I;H)$ *to* $L^\infty(I;Z)^*$, *and let* $u_{0k} \to u_0 \in H$. *Then there is a weak solution* u *due to the Definition* 8.2 *and, moreover,* $\frac{d}{dt}u \in L^{q'}(I;Z^*)$.

Proof. By Lemma 8.26, we have the a-priori estimate (8.96a) at our disposal, hence we choose a subsequence $u_k \stackrel{*}{\rightharpoonup} u$ in $L^\infty(I;H) \cap L^p(I;V)$. Besides, as in the proof of Theorem 8.30, $\frac{d}{dt}u$ has a sense in $L^{q'}(I;Z^*)$ if $q < +\infty$ or in $\mathscr{M}(I;Z^*)$ if $q = +\infty$, and $\frac{d}{dt}u_k$ converges to $\frac{d}{dt}u|_{L^q(I;V_l)}$ in each $L^{q'}(I;V_l^*)$ if $q < +\infty$ or in $\mathscr{M}(I;V_l^*)$ if $q = +\infty$.

Now, paraphrasing the proof of Theorem 8.13, we consider for $l \le k$ fixed, $v_l \in W^{1,\infty,\infty}(I;V_l,Z^*)$, put $v = v_l(t)$ into (8.92a), integrate it over $[0,T]$, and use the by-part integration (7.15),[27] one obtains

$$\int_0^T \langle A(t,u_k) - f, v_l\rangle - \Big\langle \frac{dv_l}{dt}, u_k\Big\rangle\,dt + \big(u_k(T), v_l(T)\big) = \big(u_{0k}, v_l(0)\big)\,; \tag{8.110}$$

note that (8.96a) and (8.109) guarantees $\mathscr{A}(u_k) \in L^{q'}(I;Z^*)$. As $\{u_k(T)\}_{k\in\mathbb{N}}$ is bounded in H, hence it converges (possibly as further selected subsequence) to

[27] We use (7.15) with V_k instead of V, realizing that $\frac{d}{dt}u_k \in L^{q'}(I;V_k^*)$ and $\frac{d}{dt}v_l \in L^\infty(I;V_k^*)$.

some u_T weakly in H. On the other hand, $u_k(T) = u_{0k} + \int_0^T \frac{\mathrm{d}}{\mathrm{d}t}u_k\,\mathrm{d}t$ converges to $u_0+\int_0^T \frac{\mathrm{d}}{\mathrm{d}t}u\,\mathrm{d}t = u(T)$ in Z^*. Hence $u_T = u(T)$, the further selection was redundant, and $\lim_{k\to\infty}(u_k(T), v_l(T)) = (u(T), v_l(T))$. The convergence of $\lim_{k\to\infty}(u_{0k}, v_l(0))$ to $(u_0, v_l(0))$ is obvious. Using the weak* continuity of $\mathscr{A}$, we can pass to the limit in (8.110) with $k \to \infty$, obtaining $\int_0^T \langle A(t,u)-f, v_l\rangle - \langle\frac{\mathrm{d}}{\mathrm{d}t}v_l, u\rangle\,\mathrm{d}t+(u(T), v_l(T)) = (u_0, v_l(0))$. Taking arbitrary $v \in W^{1,\infty,\infty}(I; Z, V^*)$, by Lemma 7.2 we can consider $\widetilde{w} \in C^1(I; Z)$ such that $\widetilde{w} \to v$ in $L^q(I; Z)$ (here we rely on $q < +\infty$) and also $\frac{\mathrm{d}}{\mathrm{d}t}\widetilde{w} \to \frac{\mathrm{d}}{\mathrm{d}t}v$ in $L^{p'}(I; V^*)$. Then, e.g. by a piecewise affine interpolation and subsequent approximation from Z to V_l, we can further approximate $\widetilde{w}$ by $v_l \in W^{1,\infty}(I; V_l)$ in $W^{1,p'}(I; Z)$. By a suitable diagonalization, passing eventually with $l\to\infty$, one gets $\langle\mathscr{A}(u)-f, v_l\rangle_{L^1(I;Z^*)\times L^\infty(I;Z)} \to \langle\mathscr{A}(u)-f, v\rangle_{L^1(I;Z^*)\times L^\infty(I;Z)}$ and also $\langle\frac{\mathrm{d}}{\mathrm{d}t}v_l, u\rangle_{L^{p'}(I;V^*)\times L^p(I;V)} \to \langle\frac{\mathrm{d}}{\mathrm{d}t}v, u\rangle_{L^{p'}(I;V^*)\times L^p(I;V)}$, so that (8.2) follows.

Moreover, it says that $\mathscr{A}(u) - f = -\frac{\mathrm{d}}{\mathrm{d}t}u$ in the sense of distributions on I. However, by (8.109) ensuring here (8.12), $u \in L^\infty(I; H)\cap L^p(I; V)$ implies $\mathscr{A}(u) \in L^{q'}(I; Z^*)$. By the assumptions of Lemma 8.26 also $f \in L^{q'}(I; Z^*)$. Hencefore, $\frac{\mathrm{d}}{\mathrm{d}t}u \in L^{q'}(I; Z^*)$.

The weak continuity of the mapping $t \mapsto u(t) : I \to H$ required in Definition 8.2 follows as in the proof of Theorem 8.13. □

Remark 8.32 (Monotone case: convergence via Minty's trick[28]). If $A(t,\cdot) : V \to V^*$ is monotone and radially continuous, under the additional growth condition (8.80), we can use Lemma 2.9 to show that $\mathscr{A}$ is pseudomonotone (cf. Example 8.52) which is then employed in the proof of convergence as in Theorem 8.9 or 8.30. Alternatively, we can use Lemma 8.8 or 8.29. In this monotone case, however, these chains of arguments can be made shorter and more explicit: By the a-priori estimates, we can select a subsequence such that

$$u_k \overset{*}{\rightharpoonup} u \quad \text{in } W^{1,p,p'}(I; V, V_{\rm lcs}^*) \cap L^\infty(I; H). \tag{8.111}$$

We use also $u_k(T) \rightharpoonup u(T)$ in H and $v_k \in L^p(I; V_k)$, $v \in L^p(I; V)$, $v_k \to v$ in $L^p(I; V)$ as in the proof of Theorem 8.30. By (8.102), we have

$$\begin{aligned} 0 \le I_k &:= \big\langle\mathscr{A}(u_k) - \mathscr{A}(v), u_k - v\big\rangle \\ &= \big\langle\mathscr{A}(u_k), u_k - v_k\big\rangle + \big\langle\mathscr{A}(u_k), v_k - v\big\rangle - \big\langle\mathscr{A}(v), u_k - v\big\rangle \\ &= \Big\langle f - \frac{\mathrm{d}u_k}{\mathrm{d}t}, u_k - v_k\Big\rangle + \big\langle\mathscr{A}(u_k), v_k - v\big\rangle - \big\langle\mathscr{A}(v), u_k - v\big\rangle \\ &= \frac{1}{2}\big\|u_{0k}\big\|_H^2 - \frac{1}{2}\big\|u_k(T)\big\|_H^2 + \big\langle f, u_k - v_k\big\rangle \\ &\quad + \Big\langle\frac{\mathrm{d}u_k}{\mathrm{d}t}, v_k\Big\rangle + \big\langle\mathscr{A}(u_k), v_k - v\big\rangle - \big\langle\mathscr{A}(v), u_k - v\big\rangle. \end{aligned} \tag{8.112}$$

[28] Cf. also the proof of Theorem 8.16(ii).

Using $\liminf_{k\to\infty}\|u_k(T)\|_H^2 \geq \|u(T)\|_H^2$ as in (8.104) and using also $|\langle\mathscr{A}(u_k), v_k - v\rangle| \leq \sup_{l\in\mathbb{N}}\|\mathscr{A}(u_l)\|_{L^{p'}(I;V^*)}\|v_k - v\|_{L^p(I;V)} \to 0$, we obtain

$$0 \leq \limsup_{k\to\infty} I_k \leq \frac{1}{2}\big\|u_0\big\|_H^2 - \frac{1}{2}\big\|u(T)\big\|_H^2 + \langle f, u-v\rangle + \Big\langle\frac{\mathrm{d}u}{\mathrm{d}t}, v\Big\rangle$$
$$- \big\langle\mathscr{A}(v), u-v\big\rangle = \Big\langle f - \frac{\mathrm{d}u}{\mathrm{d}t}, u-v\Big\rangle - \big\langle\mathscr{A}(v), u-v\big\rangle. \qquad (8.113)$$

Then we use the Minty-trick Lemma 2.13; put $v = u+\varepsilon w$ into (8.113), divide it by $\varepsilon > 0$, pass ε to 0 while using the radial continuity of $\mathscr{A}$; the last argument exploits the radial continuity of A and the Lebesgue dominated-convergence Theorem 1.14, cf. (8.165) below.

In case $\mathscr{A}$ is even d-monotone and V is uniformly convex, by using (8.113) for $v := u$ and by uniform convexity of $L^p(I;V)$, cf. Proposition 1.37, we get even the convergence $u_k \to u$ in $L^p(I;V)$; cf. also Remark 8.11.

Remark 8.33 (Various concepts of *pseudomonotonicity*). There is certain freedom in the choice of $\mathscr{W}$. In general, the smaller the space $\mathscr{W}$ (or the finer its topology), the bigger the collection of a-priori estimates exploited, and thus the weaker the conditions imposed on $\mathscr{A}$ by requiring its pseudomonotonicity as $\mathscr{W} \to \mathscr{W}^*$ by (8.35). The choice of $\mathscr{W}$ from Lemma 8.8 was essentially similar as in Lemma 8.29, only fitted to the particular method. We could also consider $\mathscr{W} := L^p(I;V)\cap L^\infty(I;H)$ but this would enable us to treat only monotone operators, cf. Example 8.52 below or Exercise 8.64 still for another $\mathscr{W}$ of this type. In the Galerkin method, smaller $\mathscr{W}$ (or finer topology on it) needs more difficult proof of density of V_l-valued functions in $\mathscr{W}$, which can, however, be overcome by an additional condition requiring boundedness of $\mathscr{A}$ as a mapping into a smaller space than $\mathscr{W}^*$. This we indeed made in Theorem 8.30 where (8.80) implies boundedness of $\mathscr{A} : W^{1,p,p'}(I;V,V^*)\to L^{p'}(I;V^*) \subset W^{1,p,p'}(I;V,V^*)^*$ and then it suffices to have an approximation in $L^p(I;V)$, cf. Lemma 8.28. Weakening the growth assumption so that $\mathscr{A}$ is bounded as a mapping $W^{1,p,p'}(I;V,V^*)\to(L^p(I;V)\cap L^\infty(I;H))^*$, as will be used in the setting of Proposition 8.39 below, would need a better approximation, namely in $L^p(I;V)\cap L^\infty(I;H)$, cf. Exercise 8.55.

8.5 Uniqueness and continuous dependence on data

Weakening of concepts of solutions is always a dangerous process in the sense that, if done in a too "insensitive" way, one can loose selectivity of the definition of such solution: then a solution is not unique even in well qualified cases.[29] Therefore, the question about uniqueness of the solution has its own theoretical importance. In addition, the analysis of uniqueness of a solution is usually closely related to another interesting question, namely its continuous dependence on the data, i.e. a well-posedness of the problem.

[29] See [370] for examples of such situations.

Theorem 8.34 (Uniqueness of the strong solution). *Let A satisfy, besides assumptions guaranteeing existence of a strong solution to (8.1), also*

$$\exists c \in L^1(I)\ \forall u, v \in V\ \forall(\text{a.a.})t \in I: \ \big\langle A(t,u) - A(t,v), u - v\big\rangle \geq -c(t)\|u-v\|_H^2. \quad (8.114)$$

Then the Cauchy problem (8.4) possesses a unique strong solution u.

Proof. Take two strong solutions $u_1, u_2 \in W^{1,p,p'}(I; V, V^*)$. Then take (8.4) for u_1 and u_2 such that $u_1(0) = u_0 = u_2(0)$, subtract it, and put $v := u_1 - u_2$, and integrate over $(0, t)$. By (7.22), one gets

$$\begin{aligned} 0 &= \int_0^t \Big\langle \frac{\mathrm{d}(u_1 - u_2)}{\mathrm{d}t}, u_1 - u_2 \Big\rangle \,\mathrm{d}\vartheta + \int_0^t \big\langle \mathscr{A}(u_1) - \mathscr{A}(u_2), u_1 - u_2 \big\rangle \,\mathrm{d}\vartheta \\ &\geq \frac{1}{2}\big\|u_1(t) - u_2(t)\big\|_H^2 - \frac{1}{2}\big\|u_1(0) - u_2(0)\big\|_H^2 - \int_0^t c(\vartheta)\|u_1(\vartheta) - u_2(\vartheta)\|_H^2 \mathrm{d}\vartheta. \quad (8.115) \end{aligned}$$

Using the fact that $u_1(0) - u_2(0) = 0$ and the Gronwall inequality (1.66) with $y(t) := \frac{1}{2}\|u_1(t) - u_2(t)\|_H^2$, $C := 0$, $b := 0$, and $a(t) := c(t)$, we obtain $y(t) \leq 0$, and therefore $u_1(t) - u_2(t) = 0$ for any $t \in I$. □

Theorem 8.35 (Continuous dependence on f and u_0). *Let A satisfy assumptions guaranteeing existence of a strong solution to (8.1) and (8.114). Then:*

(i) *The mapping $(f, u_0) \mapsto u : L^{p'}(I; V^*) \times H \to W^{1,p,p'}(I; V, V^*) \cap L^\infty(I; H)$, where u denotes the unique solution to the investigated problem, is (norm,weak*)-continuous, i.e. it is demicontinuous.*

(ii) *The mapping $(f, u_0) \mapsto u : L^2(I; H) \times H \to C(I; H)$ is Lipschitz continuous.*

(iii) *The mapping $(f, u_0) \mapsto u : L^1(I; H) \times H \to C(I; H)$ is uniformly continuous and locally Lipschitz continuous.*

(iv) *Moreover, V is uniformly convex, the splitting $A = A_1 + A_2$ holds with $\mathscr{A}_1$ satisfying the d-monotonicity (8.53) and $\mathscr{A}_2$ being totally continuous as a mapping $\mathscr{W} \to \mathscr{W}^*$ with $\mathscr{W} := W^{1,p,p'}(I; V, V^*) \cap L^\infty(I; H)$. Then $(f, u_0) \mapsto u : L^{p'}(I; V^*) \times H \to L^p(I; V)$ is continuous.*

Proof. As to (i), the a-priori estimates and uniqueness imply immediately the weak* convergence in $W^{1,p,p'}(I; V, V^*) \cap L^\infty(I; H)$ by paraphrasing the proof of Theorem 8.9.

As to (ii), let us take two solutions $u_1, u_2 \in W^{1,p,p'}(I; V, V^*)$ corresponding to two right-hand sides $f_1, f_2 \in L^2(I; H)$ and two initial conditions $u_{01}, u_{02} \in H$, abbreviate $u_{12} := u_1 - u_2$, $f_{12} := f_1 - f_2$, and $u_{012} := u_{01} - u_{02}$, and then take again (8.4) for u_1, u_2, subtract it, put $v := u_{12}$, and integrate over $[0, t]$. Likewise

(8.115), by (7.22) and Hölder's inequality, one gets

$$\begin{aligned}\frac{1}{2}\big\|u_{12}(t)\big\|_H^2-\frac{1}{2}\big\|u_{012}\big\|_H^2-\int_0^t c_2(\vartheta)\big\|u_{12}(\vartheta)\big\|_H^2\,\mathrm{d}\vartheta\\ \le\int_0^t\Big\langle\frac{\mathrm{d}u_{12}}{\mathrm{d}\vartheta},u_{12}(\vartheta)\Big\rangle\,\mathrm{d}\vartheta+\int_0^t\big\langle\mathscr{A}(u_1)-\mathscr{A}(u_2),u_{12}\big\rangle\,\mathrm{d}\vartheta\\ =\int_0^t\big\langle f_{12},u_{12}\big\rangle\,\mathrm{d}\vartheta\le\int_0^t\frac{1}{2}\big\|f_{12}\big\|_H^2+\frac{1}{2}\big\|u_{12}\big\|_H^2\,\mathrm{d}\vartheta.\end{aligned}\tag{8.116}$$

Using the Gronwall inequality (1.66) with $y(t):=\|u_{12}(t)\|_H^2$, $C=\|u_{012}\|_H^2$, $a(t):=1+2c_2(t)$, $b(t):=\|f_{12}(t)\|_H^2$, one gets

$$\begin{aligned}\big\|u_{12}(t)\big\|_H^2\le\Big(\big\|u_{012}\big\|_H^2+\int_0^t\big\|f_{12}(\vartheta)\big\|_H^2\mathrm{e}^{-\int_0^\vartheta 1+2c_2(\theta)\mathrm{d}\theta}\mathrm{d}\vartheta\Big)\\ \times\,\mathrm{e}^{\int_0^\vartheta 1+2c_2(\theta)\mathrm{d}\theta}\le\Big(\big\|u_{012}\big\|_H^2+\big\|f_{12}\big\|_{L^2(I;H)}^2\Big)\mathrm{e}^{1+2\|c_2\|_{L^1(I)}}.\end{aligned}\tag{8.117}$$

As to (iii), it suffices to modify (8.116) as $\int_0^t\langle f_{12},u_{12}\rangle\,\mathrm{d}\vartheta\le\int_0^t\|f_{12}\|_H(\frac{1}{2}+\frac{1}{2}\|u_{12}\|_H^2)$ which allows for usage of the Gronwall inequality (1.66) with $a(t):=\|f_{12}(t)\|_H+2c_2(t)$ and $b(t):=\|f_{12}(t)\|_H$ to modify (8.117) to get $\|u_{12}(t)\|_H^2\le(\|u_{012}\|_H^2+\|f_{12}\|_{L^1(I;H)})\mathrm{e}^{\|f_{12}\|_{L^1(I;H)}+2\|c_2\|_{L^1(I)}}$.

To prove (iv), one can just modify (8.52) so that

$$\begin{aligned}\frac{1}{2}\big\|u_{12}(T)\big\|_H^2+\big\langle\mathscr{A}_1(u_1)-\mathscr{A}_1(u_2),u_{12}\big\rangle\le\frac{1}{2}\big\|u_{012}\big\|_H^2\\ +\big\langle f_{12},u_{12}\big\rangle+\big\langle\mathscr{A}_2(u_2)-\mathscr{A}_2(u_1),u_{12}\big\rangle=:I_1+I_2+I_3.\end{aligned}\tag{8.118}$$

Considering $u_{02}\to u_{01}$ in H and $f_2\to f_1$ in $L^{p'}(I;V^*)$, by Step (i), we know $u_2\to u_1$ in $\mathscr{W}$ weakly*. Then obviously $I_1\to 0$ and $I_2\to 0$. Total continuity of $\mathscr{A}_2$ eventually gives also $I_3\to 0$ and d-monotonicity of $\mathscr{A}_1$ gives $u_2\to u_1$ in $L^p(I;V)$. □

Now we come to uniqueness of the weak solution, which is an important assertion justifying Definition 8.2 whose *selectivity* is otherwise not entirely obvious. The serious difficulty consists in lack of regularity of the weak solution which does not allow for using it as a test function. Hence, we must use a suitable smoothing procedure and the proof is much more technical than in the case of the strong solution. Here we have at our disposal the procedure (7.18) which, unfortunately, still forces us to impose growth qualification on A corresponding to the strong solution, so the only extension is in the right-hand side f which is not required to live in $L^{p'}(I;V^*)$ for Definition 8.2.

Theorem 8.36 (Uniqueness of the weak solution). *Let $A(t,\cdot)$ satisfy* (8.114) *and* (8.12) *with $q=p$ and $Z=V$ be considered. Then the weak solution according to Definition* 8.2 *is unique.*

Proof. Take $u_1, u_2 \in L^p(I;V) \cap L^\infty(I;H)$ two weak solutions, i.e. both u_1 and u_2 satisfy (8.2). Let us sum (8.2) for u_1 and u_2, smoothen $u_{12} := u_1 - u_2$ by a regularization procedure with the properties (7.18) with considering $u_0 = 0$ there, let us denote the result as u_{12}^ε, and then use the test function v as u_{12}^ε "continuously cut" at some $\vartheta \in (0,T]$, namely

$$v(t) := \begin{cases} u_{12}^\varepsilon(t) & \text{if } t \le \vartheta, \\ \dfrac{\vartheta+\varepsilon-t}{\varepsilon} u_{12}^\varepsilon(\vartheta) & \text{if } \vartheta < t < \vartheta+\varepsilon, \\ 0 & \text{if } t \ge \vartheta+\varepsilon. \end{cases} \tag{8.119}$$

This gives

$$\int_0^\vartheta \big\langle A(t,u_1(t)) - A(t,u_2(t)), u_{12}^\varepsilon(t)\big\rangle - \Big\langle u_{12}(t), \frac{\mathrm{d}u_{12}^\varepsilon}{\mathrm{d}t}\Big\rangle\,\mathrm{d}t$$
$$+ \int_\vartheta^{\vartheta+\varepsilon} \frac{\vartheta+\varepsilon-t}{\varepsilon} \big\langle A(t,u_1(t)) - A(t,u_2(t)), u_{12}^\varepsilon(\vartheta)\big\rangle + \Big\langle u_{12}(t), \frac{u_{12}^\varepsilon(\vartheta)}{\varepsilon}\Big\rangle\,\mathrm{d}t = 0. \tag{8.120}$$

By (8.12) with $q = p$ and $Z = V$, we have $\mathscr{A}(u_i) \in L^{p'}(I;V^*)$ for $i = 1,2$. By (7.18a) and (8.114), $\lim_{\varepsilon\to 0} \int_0^\vartheta \langle A(t,u_1(t)) - A(t,u_2(t)), u_{12}^\varepsilon(t)\rangle\,\mathrm{d}t = \int_0^\vartheta \langle A(t,u_1(t)) - A(t,u_2(t)), u_{12}(t)\rangle\,\mathrm{d}t \ge -\int_0^\vartheta c(t)\|u_{12}(t)\|_H^2\mathrm{d}t$. By this argument also $\lim_{\varepsilon\to 0} \int_\vartheta^{\vartheta+\varepsilon} \frac{\vartheta+\varepsilon-t}{\varepsilon} \langle A(t,u_1) - A(t,u_2), u_{12}^\varepsilon(\vartheta)\rangle\mathrm{d}t = 0$. We further consider $\vartheta \in (0,T]$ as a right Lebesgue point for $u_{12} : I \to V$ to guarantee $\lim_{\varepsilon\to 0} \varepsilon^{-1} \int_\vartheta^{\vartheta+\varepsilon} u_{12}(t)\,\mathrm{d}t = u_{12}(\vartheta)$, and simultaneously a left Lebesgue point for $\langle u^*, u_{12}(\cdot)\rangle : I \to \mathbb{R}$ for any $u^* \in H$ to guarantee (7.18d) at $t = \vartheta$; here we use a general assumption that H and V are separable hence the set of such ϑ's is dense in I, cf. Theorem 1.35. Then, by using (7.18b-d),

$$\begin{aligned}
&\liminf_{\varepsilon\to 0} \Big(-\int_0^\vartheta \Big\langle u_{12}, \frac{\mathrm{d}u_{12}^\varepsilon}{\mathrm{d}t}\Big\rangle\,\mathrm{d}t + \int_\vartheta^{\vartheta+\varepsilon} \Big\langle u_{12}(t), \frac{u_{12}^\varepsilon(\vartheta)}{\varepsilon}\Big\rangle \mathrm{d}t \Big) \\
&\quad \ge \liminf_{\varepsilon\to 0} \Big(-\int_0^\vartheta \Big\langle u_{12}^\varepsilon, \frac{\mathrm{d}u_{12}^\varepsilon}{\mathrm{d}t}\Big\rangle\,\mathrm{d}t + \Big\langle \frac{1}{\varepsilon}\int_\vartheta^{\vartheta+\varepsilon} u_{12}\,\mathrm{d}t, u_{12}^\varepsilon(\vartheta)\Big\rangle \Big) \\
&\quad - \limsup_{\varepsilon\to 0} \int_0^\vartheta \Big\langle u_{12}^\varepsilon - u_{12}, \frac{\mathrm{d}u_{12}^\varepsilon}{\mathrm{d}t}\Big\rangle\,\mathrm{d}t \\
&\quad \ge \liminf_{\varepsilon\to 0} \Big(\frac{1}{2}\|u_{12}^\varepsilon(0)\|_H^2 + \frac{1}{2}\|u_{12}^\varepsilon(\vartheta)\|_H^2 \Big) \\
&\quad + \lim_{\varepsilon\to 0} \Big(\Big\langle \frac{1}{\varepsilon}\int_\vartheta^{\vartheta+\varepsilon} u_{12}\,\mathrm{d}t, u_{12}^\varepsilon(\vartheta)\Big\rangle - \|u_{12}^\varepsilon(\vartheta)\|_H^2 \Big) \ge \frac{1}{2}\|u_{12}(\vartheta)\|_H^2.
\end{aligned} \tag{8.121}$$

Now we are ready to lower-bound the limit inferior of (8.120), which gives $\frac{1}{2}\|u_{12}(\vartheta)\|_H^2 - \int_0^\vartheta c(t)\|u_{12}(t)\|_H^2\mathrm{d}t \le 0$ for a.a. ϑ, from which $u_{12} = 0$ follows by the Gronwall inequality (1.66). $\square$

8.6 Application to quasilinear parabolic equations

For Ω a bounded, Lipschitz, time-independent domain in $\mathbb{R}^n$ with the boundary Γ, we will use the notation $Q := I \times \Omega$ and $\Sigma := I \times \Gamma$ and consider the initial-boundary-value problem (with Newton-type boundary conditions) for the quasilinear parabolic 2nd-order equation:

$$\left.\begin{aligned} \frac{\partial u}{\partial t} - \sum_{i=1}^{n} \frac{\partial}{\partial x_i} a_i(t,x,u,\nabla u) + c(t,x,u,\nabla u) &= g(t,x) \quad \text{for } (t,x)\in Q, \\ \nu(x)\cdot a(t,x,u,\nabla u) + b(t,x,u) &= h(t,x) \quad \text{for } (t,x)\in\Sigma, \\ u(0,x) &= u_0(x) \quad \text{for } x\in\Omega, \end{aligned}\right\} \tag{8.122}$$

where again $\nabla u := (\frac{\partial}{\partial x_1}u, \ldots, \frac{\partial}{\partial x_n}u)$ and $\nu = (\nu_1, \ldots, \nu_n)$ denotes the unit outward normal to Γ. In accord with Convention 2.23, we occasionally omit the arguments (t,x) in (8.122), writing shortly, e.g., $a_i(t,x,u,\nabla u)$ instead of $a_i(t,x,u(t,x),\nabla u(t,x))$. Also, recall the notation $a = (a_1, \ldots, a_n)$. The conventional setting will mostly be based on

$$V := W^{1,p}(\Omega), \qquad H := L^2(\Omega). \tag{8.123}$$

The desired reflexivity of V and the compact embedding $V \Subset H$ then need

$$p > \max\left(1, \frac{2n}{n+2}\right), \tag{8.124}$$

cf. (1.34). Note that it brings no restriction on $p > 1$ provided $n = 1$ or 2, but, e.g., for $n = 3$ it requires $p > 6/5$; cf. Remark 8.42 for the opposite case. This fits with the abstract formulation (8.1) if $A : I \times W^{1,p}(\Omega) \to W^{1,p}(\Omega)^*$ and $f(t) \in W^{1,p}(\Omega)^*$ are defined, for any $v \in W^{1,p}(\Omega)$, by

$$\begin{aligned} \langle A(t,u), v\rangle := \int_\Omega & a(t,x,u(x),\nabla u(x)) \cdot \nabla v(x) \\ & + c(t,x,u(x),\nabla u(x))v(x)\,\mathrm{d}x + \int_\Gamma b(t,x,u(x))v(x)\,\mathrm{d}S, \end{aligned} \tag{8.125a}$$

$$\langle f(t), v\rangle := \int_\Omega g(t,x)v(x)\,\mathrm{d}x + \int_\Gamma h(t,x)v(x)\,\mathrm{d}S. \tag{8.125b}$$

The strong formulation of the initial-value problem (8.1) now leads to

$$\begin{aligned} \left\langle \frac{\partial u}{\partial t}, v\right\rangle_{W^{1,p}(\Omega)^*\times W^{1,p}(\Omega)} &+ \int_\Omega a(t,x,u,\nabla u)\cdot\nabla v(x) + c(t,x,u,\nabla u)v(x)\,\mathrm{d}x \\ &+ \int_\Gamma b(t,x,u)v(x)\,\mathrm{d}S = \int_\Omega g(t,\cdot)v\,\mathrm{d}x + \int_\Gamma h(t,\cdot)v\,\mathrm{d}S \end{aligned} \tag{8.126}$$

for a.a. $t \in I$, and $u(0,\cdot) = u_0$. Obviously, (8.126) can be obtained from (8.122) the following four steps:

1) multiplication of the first line in (8.122) by $v \in W^{1,p}(\Omega)$,
2) integration over Ω,
3) Green's theorem in space,
4) usage of the boundary conditions in (8.122).

As such, (8.126) is called a *weak formulation* of (8.122) and a *weak solution* u is then required to belong to $W^{1,p,p'}(I; W^{1,p}(\Omega), W^{1,p}(\Omega)^*)$.

Alternatively, a very weak formulation (corresponding to what is on the abstract level called the weak formulation, see (8.2) and Table 2 on p. 215) can be obtained by the following four steps:

1) multiplication of the first line in (8.122) by $v(t)$,
2) integration over Q,
3) Green's theorem in space and by-part integration in time,
4) usage of the boundary and the initial conditions from (8.122).

Thus we have

$$\int_\Omega u(T,x)v(T,x)\,\mathrm{d}x + \int_Q a(t,x,u,\nabla u)\cdot\nabla v + c(t,x,u,\nabla u)v - \frac{\partial v}{\partial t}u\,\mathrm{d}x\mathrm{d}t$$
$$+ \int_\Sigma b(t,x,u)v\,\mathrm{d}S\mathrm{d}t = \int_Q gv\,\mathrm{d}x\mathrm{d}t + \int_\Sigma hv\,\mathrm{d}S\mathrm{d}t + \int_\Omega u_0 v(0,\cdot)\,\mathrm{d}x. \quad (8.127)$$

The *very weak solution* $u \in L^p(I; W^{1,p}(\Omega))$ is then to satisfy (8.127) for all $v \in W^{1,\infty,\infty}(I; W^{1,\infty}(\Omega), L^{p^{*\prime}}(\Omega))$; here we require even $\frac{\partial}{\partial t}v \in L^\infty(I; L^{p^{*\prime}}(\Omega))$ in order to express the duality $\langle\frac{\partial}{\partial t}v, u\rangle$ in terms of a conventional Lebesgue integral but by a density argument it extends for test functions used in Definition 8.2 too.

In this section, we focus on the weak formulation (8.126) while the very weak formulation (8.127) will be addressed in Section 8.7. We are to design the growth conditions on a, b, and c to guarantee the integrals in (8.126) to have a good sense and to be in $L^1(I)$ as a function of t. Let us realize that, by (1.33) and (1.63),

$$L^p(I; W^{1,p}(\Omega)) \cap L^\infty(I; L^2(\Omega)) \subset L^{p^{\circledast}}(Q) \quad (8.128)$$

for a suitable $p^{\circledast} > p$. To determine this exponent optimally, i.e. as big as possible, we use Gagliardo-Nirenberg's Theorem 1.24 which allows here for the interpolation $W^{1,p}(\Omega) \cap L^2(\Omega) \subset L^q(\Omega)$, namely

$$\big\|v\big\|_{L^q(\Omega)} \le C_{\mathrm{GN}}\big\|v\big\|^{\lambda}_{W^{1,p}(\Omega)}\big\|v\big\|^{1-\lambda}_{L^2(\Omega)} \quad \text{if} \quad \frac{1}{q} \ge \lambda\frac{n-p}{np} + \frac{1-\lambda}{2} =: \frac{1}{q_1(\lambda)}. \quad (8.129)$$

Note that the function $\lambda \mapsto q_1(\lambda)$ is non-decreasing if $W^{1,p}(\Omega) \subset L^2(\Omega)$, i.e. if (8.124) holds. Then we can further estimate

$$\big\|v\big\|^q_{L^q(Q)} = \int_0^T \big\|v\big\|^q_{L^q(\Omega)}\,\mathrm{d}t \le C^q_{\mathrm{GN}} \int_0^T \|v\|^{q\lambda}_{W^{1,p}(\Omega)}\|v\|^{q(1-\lambda)}_{L^2(\Omega)}\,\mathrm{d}t$$
$$\le C^q_{\mathrm{GN}}\|v\|^{q(1-\lambda)}_{L^\infty(I;L^2(\Omega))}\int_0^T \|v\|^{q\lambda}_{W^{1,p}(\Omega)}\,\mathrm{d}t. \quad (8.130)$$

The last term is bounded if $q\lambda \le p$, i.e. we need $\frac{1}{q} \ge \frac{\lambda}{p} =: \frac{1}{q_2(\lambda)}$. Obviously, the function $\lambda \mapsto q_2(\lambda)$ is decreasing. The aim is to choose q as big as possible, i.e. $\min(q_1(\lambda), q_2(\lambda))$ as big as possible, which suggest an optimal choice for λ such that $q_1(\lambda) = q_2(\lambda)$. A simple algebra reveals $\lambda = \frac{n}{n+2}$ and thus $q = p\frac{n+2}{n}$; note that indeed $\lambda \in (0,1)$, as required in Theorem 1.24. Such q plays the optimal role of the "anisotropic-interpolation" exponent $p^{\circledast}$. Thus we put

$$p^{\circledast} := \frac{np + 2p}{n}; \tag{8.131}$$

cf. also [120, Sect.I.3].

To design optimally the growth conditions of the boundary terms, one needs an analog of (8.128) for the trace operator

$$u \mapsto u|_{\Sigma} : L^p(I; W^{1,p}(\Omega)) \cap L^\infty(I; L^2(\Omega)) \to L^{p^{\oplus}}(\Sigma) \tag{8.132}$$

for a suitable $p^{\oplus} > p$. Optimal choice of this exponent is, however, quite technical. We need a version of the *Gagliardo-Nirenberg inequality* generalizing (8.129) for the fractional Sobolev-Slobodeckiĭ spaces. Namely, we exploit:

$$\big\|v\big\|_{W^{\beta,\pi}(\Omega)} \le C'_{\rm GN} \big\|v\big\|^{\lambda}_{W^{1,p}(\Omega)} \big\|v\big\|^{1-\lambda}_{L^2(\Omega)} \quad \text{if } \frac{1}{\pi} - \frac{\beta}{n} \ge \frac{1}{q_1(\lambda)} \quad \text{for } \pi = \max(2,p) \tag{8.133}$$

with $q_1(\lambda)$ again from (8.129). This special choice of π makes it relatively easy to show (8.133) by interpolating the Hilbert-type Sobolev-Slobodeckiĭ spaces[30] or by interpolating the Sobolev/Lebesgue spaces of the same exponent p.[31] Then one

[30] If $p \le 2$, we prove (8.133) for $\pi = 2$ by using embedding $W^{1,p}(\Omega) \subset W^{\gamma,2}(\Omega)$ with $\gamma = (np+2p-2n)/(2p)$, and then by interpolating the Hilbert spaces $W^{\gamma,2}(\Omega)$ and $L^2(\Omega)$, cf. (1.44), so that we obtain

$$\big\|v\big\|_{W^{\beta,p}(\Omega)} \le K_1 \|v\|^{\lambda}_{W^{\gamma,2}(\Omega)} \|v\|^{1-\lambda}_{L^2(\Omega)} \le K_2 \|v\|^{\beta/\gamma}_{W^{1,p}(\Omega)} \|v\|^{1-\beta/\gamma}_{L^2(\Omega)}$$

for $0 \le \beta \le \gamma$, which is just (8.133) with $\lambda = \beta/\gamma$. Cf. also [78, Lemma B.3] or [79, Lemma 2.1].

[31] If $p > 2$, we prove (8.133) for $\pi = p$ by interpolating $W^{1,p}(\Omega)$ and $L^p(\Omega)$, cf. (1.44) for $k = \beta$, $k_1 = 1$, $k_2 = 0$, to obtain $\|v\|_{W^{\beta,p}(\Omega)} \le K_1 \|v\|^{\beta}_{W^{1,p}(\Omega)} \|v\|^{1-\beta}_{L^p(\Omega)}$, and then we interpolate $L^p(\Omega)$ in between $W^{1,p}(\Omega)$ and $L^2(\Omega)$ by using the Gagliardo-Nirenberg inequality (1.39), i.e. $\|v\|_{L^p(\Omega)} \le K_2 \|v\|^{\mu}_{W^{1,p}(\Omega)} \|v\|^{1-\mu}_{L^2(\Omega)}$ with $\mu = q_1^{-1}(p)$ with q_1 from (8.129), i.e. $\mu = (2n-np)/(2n-np-2n)$. Thus

$$\|v\|_{W^{\beta,p}(\Omega)} \le K_1 K_2^{1-\beta} \|v\|^{\beta+(1-\beta)\mu}_{W^{1,p}(\Omega)} \|v\|^{(1-\beta)(1-\mu)}_{L^2(\Omega)}$$

which is just (8.133) for $\lambda = \beta + (1-\beta)\mu$; indeed, after some algebra, one can verify $1/p - \beta/n = 1/q_1(\lambda)$ for the above specified λ and μ.

can use the existence of the *trace operator* on Sobolev-Slobodeckiĭ spaces[32]

$$u \mapsto u|_\Gamma : W^{\beta,\pi}(\Omega) \to L^q(\Gamma) \qquad \text{for } q = \frac{(n-1)\pi}{n-\pi\beta}. \tag{8.134}$$

From (8.133), we have $q_1(\lambda) \geq \frac{n\pi}{n-\beta\pi}$, so that $q \leq \frac{n-1}{n} q_1(\lambda)$. Note that both β and π, determining an auxiliary space $W^{\beta,\pi}(\Omega)$, have been eliminated from this estimate of q. Altogether, we obtained an estimate $\|v\|_{L^q(\Gamma)} \leq C\|v\|_{L^2(\Omega)}^{1-\lambda}\|v\|_{W^{1,p}(\Omega)}^{\lambda}$. Like (8.130), we have

$$\begin{aligned}\|v\|_{L^q(\Sigma)}^q = \int_0^T \|v\|_{L^q(\Gamma)}^q \,\mathrm{d}t &\leq C^q \int_0^T \|v\|_{W^{1,p}(\Omega)}^{q\lambda}\|v\|_{L^2(\Omega)}^{q(1-\lambda)}\,\mathrm{d}t \\ &\leq C^q \|v\|_{L^\infty(I;L^2(\Omega))}^{q(1-\lambda)} \int_0^T \|v\|_{W^{1,p}(\Omega)}^{q\lambda}\,\mathrm{d}t. \end{aligned} \tag{8.135}$$

Like in the case of (8.130), we need $q \leq q_2(\lambda) := p/\lambda$ and our aim is to choose $\min(\frac{n-1}{n}q_1(\lambda), q_2(\lambda))$ as big as possible, which suggests an optimal choice for λ such that $\frac{n-1}{n}q_1(\lambda) = q_2(\lambda)$. From this, we obtain $\lambda = \frac{np}{np+2p-2}$; note that indeed $\lambda \in (0,1)$. From this, we determine $q = p/\lambda$ which plays the optimal role of the "anisotropic-interpolation" exponent $p^{\oplus}$ in (8.132). Thus we obtained

$$p^{\oplus} := \frac{np+2p-2}{n} \qquad \text{provided } p > \frac{2n+2}{n+2}; \tag{8.136}$$

cf. also [78, 83] for $p \leq 2$ and [79] for a general case.[33] To compare all introduced exponents, we have

$$\left.\begin{aligned} p < p^{\circledast} < p^*, \\ p < p^{\oplus} < p^{\#}, \end{aligned}\right\} \qquad \text{provided (8.124) holds,} \tag{8.137a}$$

$$p < p^{\oplus} < p^{\circledast}, \tag{8.137b}$$

$$p < p^{\#} \leq p^*; \tag{8.137c}$$

note that the last inequality is even strict if $p < n$ and that (8.124) is needed for the latter inequalities in (8.137a). If $p > 3n/(n+2)$, we have also $p^{\circledast} < p^{\#}$ so that then

[32] There is a trace and subsequent embedding operator $u \mapsto u|_\Gamma : W^{\beta,\pi}(\Omega) \to W^{\beta-1/\pi,\pi}(\Gamma) \subset L^q(\Gamma)$ for a certain $q \geq 2$ sufficiently small whenever $\beta - 1/\pi > 0$, i.e. whenever $\beta\pi > 1$; cf. [302] for $\pi = 2$ or [79] for a general π based on [408]. In analog with the exponent in Theorem 1.22, now considered for $\beta-1/\pi$, π, and $n-1$ respectively in place of k, p, and n, one can easily calculate q as specified in (8.134).

[33] The restriction on p in (8.136) comes from $q \geq 2$ needed in (8.134) for $\pi = 2$; note that it only yields $p^{\oplus} > 2$ and $\beta > 1/2$. This restriction is rather related with the particular ansatz used for π in (8.133), yet the extension of $p^{\oplus}$ for lower p's satisfying (8.124) seems to be unjustified in literature. Anyhow, some anisotropical trace space can be identified even for small p satisfying only (8.124) in [78, Lemma B.3].

$p < p^{\oplus} < p^{\circledast} < p^{\#} \le p^*$. As an example in the "physically" relevant 3-dimensional situation, for linear-like growth $p = 2$, the exponents worth remembering are:

$$n = 3 \qquad \Longrightarrow \qquad 2 < 2^{\oplus} = \frac{8}{3} < 2^{\circledast} = \frac{10}{3} < 2^{\#} = 4 < 2^* = 6. \tag{8.138}$$

The natural requirement we will assume through the following text is that

$$\left.\begin{array}{l} a : Q \times (\mathbb{R}\times\mathbb{R}^n) \to \mathbb{R}^n, \\ c : Q \times (\mathbb{R}\times\mathbb{R}^n) \to \mathbb{R}, \\ b : \Sigma \times \mathbb{R} \to \mathbb{R} \end{array}\right\} \text{ are Carathéodory mappings.} \tag{8.139}$$

The growth of a, c, and b fitted to (8.126) is to be designed so that the corresponding Nemytskiĭ mappings $\mathscr{N}_a$, $\mathscr{N}_c$, and $\mathscr{N}_b$ work as $L^{p^{\circledast}}(Q) \times L^p(Q;\mathbb{R}^n) \to L^{p'}(Q;\mathbb{R}^n)$, $L^{p^{\circledast}}(Q) \times L^p(Q;\mathbb{R}^n) \to L^{p^{\circledast\prime}}(Q)$, and $L^{p^{\oplus}}(\Sigma) \to L^{p^{\oplus\prime}}(\Sigma)$, respectively. This means

$$\exists \gamma \in L^{p'}(Q),\quad C \in \mathbb{R}: \quad |a(t,x,r,s)| \le \gamma(t,x) + C|r|^{p^{\circledast}/p'} + C|s|^{p-1}, \tag{8.140a}$$
$$\exists \gamma \in L^{p^{\circledast\prime}}(Q),\ C \in \mathbb{R}: \quad |c(t,x,r,s)| \le \gamma(t,x) + C|r|^{p^{\circledast}-1} + C|s|^{p/p^{\circledast\prime}}, \tag{8.140b}$$
$$\exists \gamma \in L^{p^{\oplus\prime}}(\Sigma),\ C \in \mathbb{R}: \quad |b(t,x,r)| \le \gamma(t,x) + C|r|^{p^{\oplus}-1}. \tag{8.140c}$$

Lemma 8.37 (Carathéodory property of A). *Let* (8.139) *and* (8.140) *be valid. Then $A : I \times W^{1,p}(\Omega) \to W^{1,p}(\Omega)^*$ defined by* (8.125a) *is a Carathéodory mapping.*

Proof. Note that, as $p^{\circledast} < p^*$ and $p^{\oplus} < p^{\#}$, (8.140) implies, in particular, that $a(t,\cdot) : \Omega \times (\mathbb{R} \times \mathbb{R}^n) \to \mathbb{R}^n$, $b(t,\cdot) : \Gamma \times \mathbb{R} \to \mathbb{R}$ and $c(t,\cdot) : \Omega \times (\mathbb{R} \times \mathbb{R}^n) \to \mathbb{R}$ satisfy the growth conditions (2.55) with $\epsilon = 0$ for a.a. $t \in I$. Taking t such that (2.55) applies with $\epsilon = 0$ and considering $u_k \to u$ in $W^{1,p}(\Omega)$, we can estimate

$$\begin{aligned} \big\|A(t,u_k) - A(t,u)\big\|_{W^{1,p}(\Omega)^*} &= \sup_{\|v\|_{W^{1,p}(\Omega)} \le 1} \int_\Omega \big(a(t,x,u_k(x),\nabla u_k(x)) \\ &\quad - a(t,x,u(x),\nabla u(x))\big) \cdot \nabla v(x) + \big(c(t,x,u_k(x),\nabla u_k(x)) \\ &\quad - c(t,x,u(x),\nabla u(x))\big) v(x)\,\mathrm{d}x + \int_\Gamma \Big(b\big(t,x,u_k(x)\big) - b\big(t,x,u(x)\big)\Big) v(x)\,\mathrm{d}S \\ &\le \big\|\mathscr{N}_{a(t,\cdot)}(u_k,\nabla u_k) - \mathscr{N}_{a(t,\cdot)}(u,\nabla u)\big\|_{L^{p'}(\Omega;\mathbb{R}^n)} \\ &\quad + N_1 \big\|\mathscr{N}_{c(t,\cdot)}(u_k,\nabla u_k) - \mathscr{N}_{c(t,\cdot)}(u,\nabla u)\big\|_{L^{p^{*\prime}}(\Omega)} \\ &\quad + N_2 \big\|\mathscr{N}_{b(t,\cdot)}(u_k) - \mathscr{N}_{b(t,\cdot)}(u)\big\|_{L^{p^{\#\prime}}(\Gamma)} \end{aligned}$$

where N_1 and N_2 stand respectively for the norms of the embedding $W^{1,p}(\Omega) \subset L^{p^*}(\Omega)$ and of the trace operator $u \mapsto u|_\Gamma : W^{1,p}(\Omega) \to L^{p^{\#}}(\Gamma)$. By continuity of the Nemytskiĭ mappings $\mathscr{N}_{a(t,\cdot)}$, $\mathscr{N}_{b(t,\cdot)}$, and $\mathscr{N}_{c(t,\cdot)}$, the continuity of $A(t,\cdot)$:

$W^{1,p}(\Omega) \to W^{1,p}(\Omega)^*$ follows. More specifically, here we used the embeddings and the continuity

$$L^{p^*}(\Omega)\times L^p(\Omega;\mathbb{R}^n) \subset L^{p^\circledast}(\Omega)\times L^p(\Omega;\mathbb{R}^n) \overset{\mathscr{N}_{c(t,\cdot)}}{\longrightarrow} L^{p^{\circledast\prime}}(\Omega) \subset L^{p^{*\prime}}(\Omega)$$

and an analogous chain for $\mathscr{N}_{b(t,\cdot)}$:

$$W^{1,p}(\Omega) \overset{\text{trace}}{\longrightarrow} L^{p^\#}(\Gamma) \subset L^{p^\oplus}(\Gamma) \overset{\mathscr{N}_{b(t,\cdot)}}{\longrightarrow} L^{p^{\oplus\prime}}(\Gamma) \subset L^{p^{\#\prime}}(\Gamma).$$

Also, $t \mapsto \langle A(t,u),v\rangle$ is measurable. As $W^{1,p}(\Omega)$ is separable, by Pettis' Theorem 1.34, $A(t,\cdot)$ is also Bochner measurable. Hence $A : I \times W^{1,p}(\Omega) \to W^{1,p}(\Omega)^*$ is a Carathéodory mapping, as claimed. □

For the usage of the Galerkin method, we consider a (nonspecified) sequence of finite-dimensional subspaces V_k of $W^{1,p}(\Omega)$ and the respective seminorms on $W^{1,p}(\Omega)^*$ creating a locally convex topology, referred to by the notation $[W^{1,p}(\Omega)]^*_{\rm lcs}$. We first confine ourselves to a lesser growth of the lower order terms, leading to the growth condition (8.80) and allowing for a more direct usage of the abstract theory from Sections 8.2 or 8.4.

Proposition 8.38 (PSEUDOMONOTONICITY OF $\mathscr{A}$). *Let the assumption* (8.139) *hold and $a : Q \times \mathbb{R} \times \mathbb{R}^n \to \mathbb{R}^n$ satisfy the Leray-Lions condition*

$$(a(t,x,r,s) - a(t,x,r,\tilde{s}))\cdot(s-\tilde{s}) \geq 0, \tag{8.141a}$$
$$(a(t,x,r,s) - a(t,x,r,\tilde{s}))\cdot(s-\tilde{s}) = 0 \quad\Longrightarrow\quad s=\tilde{s}, \tag{8.141b}$$

and a strengthened growth condition (8.140) *hold with some $\epsilon{>}0$ and $C{<}{+}\infty$:*

$$\exists\gamma\in L^{p'}(Q): \qquad |a(t,x,r,s)| \leq \gamma(t,x) + C|r|^{(p^\circledast-\epsilon)/p'} + C|s|^{p-1}, \tag{8.142a}$$
$$\exists\gamma\in L^{p'}(I;L^{p^{*\prime}}(\Omega)): \quad |c(t,x,r,s)| \leq \gamma(t,x){+}C|r|^{p^\circledast/p'}{+}C|s|^{p-1}, \tag{8.142b}$$
$$\exists\gamma\in L^{p'}(I;L^{p^{\#\prime}}(\Gamma)): \quad |b(t,x,r)| \leq \gamma(t,x) + C|r|^{p^\oplus/p'}. \tag{8.142c}$$

Eventually, let the coercivity

$$a(t,x,r,s)\cdot s + c(t,x,r,s)r \geq c_0|s|^p - c_1(t,x)|s| - c_2(t)r^2, \tag{8.143a}$$
$$b(t,x,r)r \geq 0 \tag{8.143b}$$

hold with some $c_0 > 0$, $c_1 \in L^{p'}(Q)$, and $c_2 \in L^1(I)$. Then $\mathscr{A} : \mathscr{W} {\to} \mathscr{W}^*$, with $\mathscr{W}$ from Lemma 8.8 *or* 8.29, *is pseudomonotone in the sense* (8.35).

Proof. The condition (8.124) implies $p^\circledast \leq p^*$ and $p^\oplus \leq p^\#$, and therefore (8.142) guarantees that $A(t,\cdot) : W^{1,p}(\Omega) \to W^{1,p}(\Omega)^*$ for a.a. $t \in I$.

Then we use Lemma 2.32 to show that $A(t,\cdot)$ is pseudomonotone; note that the coercivity (2.68b) is implied by (8.143) and (8.142b) similarly as in Remark 2.37. Considering the choice (8.123) together with the seminorm

$|v|_V := \|\nabla v\|_{L^p(\Omega;\mathbb{R}^n)}$, (8.143) implies the semi-coercivity assumption (8.95) with $Z = V = W^{1,p}(\Omega)$. Indeed, for any $v \in W^{1,p}(\Omega)$, we have

$$\begin{aligned}\langle A(t,v),v\rangle &= \int_\Omega a(v,\nabla v)\cdot\nabla v + c(v,\nabla v)v\,\mathrm{d}x + \int_\Gamma b(v)v\,\mathrm{d}S \\ &\geq c_0\big\|\nabla v\big\|_{L^p(\Omega;\mathbb{R}^n)}^p - \big\|c_1(t,\cdot)\big\|_{L^{p'}(\Omega)}\big\|\nabla v\big\|_{L^p(\Omega;\mathbb{R}^n)} - c_2(t)\big\|v\big\|_{L^2(\Omega)}^2 \end{aligned} \tag{8.144}$$

which verifies (8.95). Then, the inequality (8.9) just turns to be (1.55) with $q=2$.

We still have to verify the growth condition (8.80). As to (8.142a), we can here, for simplicity, consider even $\epsilon = 0$, i.e. (8.140a), and use an *interpolation* as follows:

$$\begin{aligned}\sup_{\|v\|_{W^{1,p}(\Omega)}\leq 1}\int_\Omega a(t,u,\nabla u)\cdot\nabla v\,\mathrm{d}x &\leq \sup_{\|v\|_{W^{1,p}(\Omega)}\leq 1}\Big(\big\|\gamma(t,\cdot)\big\|_{L^{p'}(\Omega)} \\ &\qquad + C\big\|\,|u|^{p^\circledast/p'}\big\|_{L^{p'}(\Omega)} + C\big\|\,|\nabla u|^{p-1}\big\|_{L^{p'}(\Omega)}\Big)\big\|\nabla v\big\|_{L^p(\Omega;\mathbb{R}^n)} \\ &\leq \big\|\gamma(t,\cdot)\big\|_{L^{p'}(\Omega)} + C\big\|u\big\|_{L^{p^\circledast}(\Omega)}^{p^\circledast/p'} + C\big\|\nabla u\big\|_{L^p(\Omega;\mathbb{R}^n)}^{p-1} \\ &\leq \big\|\gamma(t,\cdot)\big\|_{L^{p'}(\Omega)} + C\big\|u\big\|_{L^2(\Omega)}^{(1-\lambda)p^\circledast/p'}\big\|u\big\|_{W^{1,p}(\Omega)}^{\lambda p^\circledast/p'} + C\big\|\nabla u\big\|_{L^p(\Omega;\mathbb{R}^n)}^{p-1} \end{aligned} \tag{8.145}$$

provided $\lambda = n/(n{+}2)$ as used already in the derivation of (8.131). After an algebraic manipulation we come exactly to $\lambda p^\circledast/p' = p-1$, hence the right-hand side of (8.145) turns to

$$\begin{aligned}&\big\|\gamma(t,\cdot)\big\|_{L^{p'}(\Omega)} + C\big\|u\big\|_{L^2(\Omega)}^{(1-\lambda)p^\circledast/p'}\big\|u\big\|_{W^{1,p}(\Omega)}^{p-1} + C\big\|\nabla u\big\|_{L^p(\Omega;\mathbb{R}^n)}^{p-1} \\ &\qquad\leq \max\big(1, C\big\|u\big\|_{L^2(\Omega)}^{(1-\lambda)p^\circledast/p'}, C\big)\big(\big\|\gamma(t,\cdot)\big\|_{L^{p'}(\Omega)} + \big\|\nabla u\big\|_{L^p(\Omega;\mathbb{R}^n)}^{p-1}\big), \end{aligned} \tag{8.146}$$

which is already of the form (8.80). As to (8.142b), we estimate:

$$\begin{aligned}\sup_{\|v\|_{W^{1,p}(\Omega)}\leq 1}\int_\Omega c(t,u,\nabla u)v\,\mathrm{d}x &\leq \sup_{\|v\|_{W^{1,p}(\Omega)}\leq 1}\big(\big\|\gamma(t,\cdot)\big\|_{L^{p^{*\prime}}(\Omega)} \\ &\qquad + C\big\|\,|u|^{p^\circledast/p'}\big\|_{L^{p^{*\prime}}(\Omega)} + C\big\|\,|\nabla u|^{p-1}\big\|_{L^{p^{*\prime}}(\Omega)}\big)\big\|v\big\|_{L^{p^*}(\Omega)} \\ &\leq N\big(\big\|\gamma(t,\cdot)\big\|_{L^{p^{*\prime}}(\Omega)} + C\big\|u\big\|_{L^{p^\circledast p^{*\prime}/p'}(\Omega)}^{p^\circledast/p'} + C\big\|\nabla u\big\|_{L^{p^{*\prime}(p-1)}(\Omega;\mathbb{R}^n)}^{p-1}\big) \\ &\leq N\big(\big\|\gamma(t,\cdot)\big\|_{L^{p^{*\prime}}(\Omega)} + CN_1\big\|u\big\|_{W^{1,p}(\Omega)}^{\lambda p^\circledast/p'}\big\|u\big\|_{L^2(\Omega)}^{(1-\lambda)p^\circledast/p'} + CN_2^{p-1}\big\|\nabla u\big\|_{L^p(\Omega;\mathbb{R}^n)}^{p-1}\big) \end{aligned} \tag{8.147}$$

with N the norm of the embedding $W^{1,p}(\Omega) \subset L^{p^*}(\Omega)$, N_1 containing the constant from the Gagliardo-Nirenberg inequality and the norm of the embedding $L^{p^\circledast}(\Omega) \subset L^{p^\circledast p^{*\prime}/p'}(\Omega)$, and with N_2 the norm of the embedding $L^p(\Omega) \subset L^{p^{*\prime}(p-1)}(\Omega)$. Using again $\lambda = n/(n{+}2)$, we arrive to $\lambda p^\circledast/p' = p-1$ and thus (8.147) again complies with the growth condition (8.80).

Analogously, the boundary term can be estimated as

$$\begin{aligned}\sup_{\|v\|_{W^{1,p}(\Omega)}\le 1}\int_\Gamma b(u)v\,\mathrm{d}S &\le \sup_{\|v\|_{W^{1,p}(\Omega)}\le 1}\big\|\gamma(t,\cdot)+C|u|^{p^{\oplus}/p'}\big\|_{L^{p^{\#\prime}}(\Gamma)}\big\|v\big\|_{L^{p^{\#}}(\Gamma)}\\ &\le N\Big(\big\|\gamma(t,\cdot)\big\|_{L^{p^{\#\prime}}(\Gamma)}+C\big\|u\big\|^{p^{\oplus}/p'}_{L^{p^{\oplus}p^{\#\prime}/p'}(\Gamma)}\Big)\\ &\le N\Big(\big\|\gamma(t,\cdot)\big\|_{L^{p^{\#\prime}}(\Gamma)}+CN_1\big\|u\big\|^{\lambda p^{\oplus}/p'}_{W^{1,p}(\Omega)}\big\|u\big\|^{(1-\lambda)p^{\oplus}/p'}_{L^2(\Omega)}\Big)\end{aligned} \tag{8.148}$$

with N being the norm of the trace operator $W^{1,p}(\Omega)\to L^{p^{\#}}(\Gamma)$, and N_1 containing the constant from the inequality (8.135) and the norm of the embedding $L^{p^{\circledast}}(\Gamma)\subset L^{p^{\oplus}p^{\#\prime}/p'}(\Gamma)$. Using $\lambda = np/(np{+}2p{-}2)$ as we did for derivation of (8.136), we arrive to $\lambda p^{\oplus}/p' = p-1$ and thus (8.148) again complies with the growth condition (8.80).

The pseudomonotonicity of $\mathscr{A}$ now follows by Lemma 8.8 or 8.29. $\square$

For the optimal treatment of the lower-order terms, one should realize that the growth condition (8.80) is fitted to boundedness of $\mathscr{A}:\mathscr{W}\to L^{p'}(I;W^{1,p}(\Omega)^*)$ and is only sufficient for the boundedness of $\mathscr{A}:\mathscr{W}\to\mathscr{W}^*$. This weaker boundedness and also the related pseudomonotonicity can be ensured by a weaker condition than (8.140b,c), closer to natural growth conditions (8.140):

Proposition 8.39 (Pseudomonotonicity of $\mathscr{A}$: a general case). *Let the assumptions* (8.139), (8.141), (8.142a) *and* (8.143) *hold and, with some* $\epsilon>0$ *and* $C<+\infty$:

$$\exists\gamma\in L^{p^{\circledast\prime}+\epsilon}(Q):\quad |c(t,x,r,s)|\le\gamma(t,x){+}C|r|^{p^{\circledast}-\epsilon-1}{+}C|s|^{(p-\epsilon)/p^{\circledast\prime}}, \tag{8.149a}$$

$$\exists\gamma\in L^{p^{\oplus\prime}}(\Sigma):\quad |b(t,x,r)|\le\gamma(t,x){+}C|r|^{p^{\oplus}-\epsilon-1}. \tag{8.149b}$$

Then $\mathscr{A}:\mathscr{W}{\to}\mathscr{W}^*$, *with* $\mathscr{W}$ *from Lemma* 8.8 *or* 8.29, *is pseudomonotone.*

Proof. The boundedness of $\mathscr{A}:\mathscr{W}\to(L^p(I;W^{1,p}(\Omega))\cap L^\infty(I;L^2(\Omega)))^*\subset\mathscr{W}^*$ is to be proved by an easy combination of the growth conditions (8.142a) and (8.149) (even used with $\epsilon=0$) and the above interpolation. E.g. for the contribution of c, we can modify (8.147) to estimate:

$$\begin{aligned}\sup_{\substack{\|v\|_{L^p(I;W^{1,p}(\Omega))}\le 1\\ \|v\|_{L^\infty(I;L^2(\Omega))}\le 1}}\int_Q c(u,\nabla u)v\,\mathrm{d}x\mathrm{d}t &\le \sup_{\substack{\|v\|_{L^p(I;W^{1,p}(\Omega))}\le 1\\ \|v\|_{L^\infty(I;L^2(\Omega))}\le 1}}\Big(\|\gamma\|_{L^{p^{\circledast\prime}}(Q)}\\ &\qquad + C\big\|\,|u|^{p^{\circledast}-1}\big\|_{L^{p^{\circledast\prime}}(Q)}+C\big\|\,|\nabla u|^{p/p^{\circledast\prime}}\big\|_{L^{p^{\circledast\prime}}(Q)}\Big)\big\|v\big\|_{L^{p^{\circledast}}(Q)}\\ \le \sup_{\substack{\|v\|_{L^p(I;W^{1,p}(\Omega))}\le 1\\ \|v\|_{L^\infty(I;L^2(\Omega))}\le 1}} &C_{\rm GN}\Big(\|\gamma\|_{L^{p^{\circledast\prime}}(\Omega)}+C\big\|u\big\|^{p^{\circledast}-1}_{L^{p^{\circledast}}(\Omega)}\\ &\qquad +C\big\|\nabla u\big\|^{p/p^{\circledast\prime}}_{L^p(Q;\mathbb{R}^n)}\Big)\big\|v\big\|^{\lambda}_{L^p(I;W^{1,p}(\Omega))}\big\|v\big\|^{1-\lambda}_{L^\infty(I;L^2(\Omega))}\end{aligned}$$

$$\le C_{\rm GN}\big\|\gamma\big\|_{L^{p^*\prime}(\Omega)}+C_{\rm GN}^2C\big\|u\big\|_{L^p(I;W^{1,p}(\Omega))}^{\lambda(p^{\circledast}-1)}\big\|u\big\|_{L^\infty(I;L^2(\Omega))}^{(1-\lambda)(p^{\circledast}-1)}+C_{\rm GN}C\big\|\nabla u\big\|_{L^p(Q;\mathbb{R}^n)}^{p/p^{\circledast\prime}} \tag{8.150}$$

with $\lambda = n/(n+2)$ as used for the derivation of (8.131) and $C_{\rm GN}$ from (8.130). This shows the term bounded on bounded subsets of $\mathscr{W}$. Analogous calculations work for the contribution of the boundary term b.

Then the condition (2.3b) can be proved directly for $\mathscr{A}$ by paraphrasing Lemma 2.32, without using Lemma 8.8 or 8.29 and replacing the coercivity (8.143) by the coercivity of $a(t,x,r,\cdot)$ like (2.68b). Instead of the compactness of $u\mapsto(u,u|_\Gamma):W^{1,p}(\Omega)\to L^{p^*-\epsilon}(\Omega)\times L^{p^\#-\epsilon}(\Gamma)$ used in Lemma 2.32, we must use the "interpolated" Aubin-Lions lemma 7.8 (possibly with the modification by employing Corollary 7.9) with $V_2:=W^{1-\epsilon_1,p}(\Omega)$, $H:=L^2(\Omega)$, $V_4:=L^q(\Omega)$ for $q^{-1}=\frac{1}{2}(1-\lambda)+\lambda/((p-\epsilon_1)^{-1}-n^{-1})$, cf. (1.23). Here we use the compact embedding $W^{1,p}(\Omega)\Subset W^{1-\epsilon_1,p}(\Omega)$ for any $\epsilon_1>0$, see (1.42) for the definition of the Sobolev-Slobodetskiĭ space $W^{1-\epsilon_1,p}(\Omega)$. Thus we obtain $\mathscr{W}\Subset L^{p/\lambda}(I;L^q(\Omega))$. The optimal choice of $\lambda\in(0,1)$ is $\lambda=n/(n+2)-\epsilon_2$, which gives that $p/\lambda=q=p^{\circledast}-\epsilon$. This yields an "$\epsilon$-modification" of (8.130) and thus the concrete form of (7.39) as

$$\big\|u_k-u\big\|_{L^{p/\lambda}(I;L^{p^{\circledast}-\epsilon}(\Omega))}\le C_{\rm GN}\big\|u_k-u\big\|_{L^\infty(I;L^2(\Omega))}^{1-\lambda}\big\|u_k-u\big\|_{L^p(I;W^{1-\epsilon_1,p}(\Omega))}^{\lambda}\to 0 \tag{8.151}$$

for any weakly* converging sequence $u_k\overset{*}{\rightharpoonup}u$, which altogether yields

$$u_k\to u \qquad \text{in } L^{p^{\circledast}-\epsilon}(Q) \tag{8.152}$$

with $p^{\circledast}$ from (8.131) and $\epsilon>0$ provided $\epsilon_1>0$ is sufficiently small (with respect to $\epsilon>0$). Furthermore, we can modify analogously (8.135) and claim that

$$u_k|_\Sigma\to u|_\Sigma \qquad \text{in } L^{p^{\oplus}-\epsilon}(\Sigma). \tag{8.153}$$

Then, by the continuity of the Nemytskiĭ mappings $\mathscr{N}_{a(\cdot,\nabla v)}$ and $\mathscr{N}_b$, we get $a(u_k,\nabla v)\to a(u,\nabla v)$ in $L^{p'}(Q;\mathbb{R}^n)$, cf. (8.142a), and $b(u_k)\to b(u)$ in $L^{p^{\oplus\prime}}(\Sigma)$. Eventually, by arguments (2.85)–(2.90) on p.52 applied on Q instead of Ω, one obtains also $c(u_k,\nabla u_k)\to c(u,\nabla u)$ in $L^{p^{\circledast\prime}}(Q)$.[34] $\square$

Proposition 8.40 (Existence of a weak solution). *Let the assumptions of Proposition* 8.39 *be valid and let* $g\in L^{p'}(I;L^{p^{*\prime}}(\Omega))$, $h\in L^{p'}(I;L^{p^{\#\prime}}(\Gamma))$, *and* $u_0\in L^2(\Omega)$. *Then the initial-boundary-value problem* (8.122) *has a weak solution.*

Proof. It just follows from the abstract Theorem 8.9 or 8.30 possibly (i.e. if the growth of b and c is indeed higher than (8.142)) based directly on Proposition 8.39 instead of Lemma 8.8 or 8.29. $\square$

[34] To be more precise, $\epsilon>0$ in (8.153) to be chosen small enough depending on $\epsilon>0$ in (8.149b).

Remark 8.41 (Modifications). The above Propositions 8.38–8.40 bear various modifications. E.g., if $a(t,x,r,\cdot)$ is merely monotone (not strictly), then, as in Lemma 2.32, $c(t,x,r,\cdot)$ has to be affine but growth restriction (8.142b) can be slightly relaxed. Also the coercivity assumption (8.143) can be modified. E.g., $b(t,x,r)r \geq -c_3(x) - C|r|^{p-\epsilon}$ with $C,\epsilon > 0$ and $c_3 \in L^1(\Gamma)$ leads just to a simple modification in derivation of the above a-priori estimates. Moreover, we can consider $g \in L^1(I;L^2(\Omega))$, and thus also $g \in L^{p^{\circledast}{}'}(Q)$ because, since $L^p(I;W^{1,p}(\Omega)) \cap L^\infty(I;L^2(\Omega))$ is densely embeded into $L^{p^{\circledast}}(Q)$, also $L^{p^{\circledast}{}'}(Q) \subset L^{p'}(I;W^{1,p}(\Omega)^*) + L^1(I;L^2(\Omega)))$.[35] Similarly, also $h \in L^{p^{\oplus}{}'}(\Sigma)$ can be considered.

Remark 8.42 (The case $1 < p \leq 2n/(n+2)$). If (8.124) does not hold, the choice $V := W^{1,p}(\Omega) \cap L^2(\Omega)$ and $H := L^2(\Omega)$ guarantees trivially $V \subset H$. For example, the Laplacean $-\Delta_p$ remains semicoercive in the sense (8.10) if $|v|_V := \|\nabla v\|_{L^p(\Omega;\mathbb{R}^n)}$ is chosen. Now $V \not\Subset H$ but $V \Subset L^{2-\epsilon}(\Omega)$ for any $\epsilon > 0$, which can again be used for lower-order terms through Aubin-Lions' lemma.

Remark 8.43 (Full discretization). One can merge Rothe's and Galerkin's method, obtaining thus a full discretization in time and space which can be implemented at least conceptually[36] on computers. Let $\tau > 0$ be a time step and $l \in \mathbb{N}$ a spatial-discretization parameter.[37] Define $u_{l\tau}^k \in V_l \subset W^{1,p}(\Omega)$, $k = 1,\ldots,T/\tau$, by the following recursive formula:

$$\int_\Omega \frac{u_{l\tau}^k - u_{l\tau}^{k-1}}{\tau} v + a_\tau^k(x, u_{l\tau}^k, \nabla u_{l\tau}^k)\cdot\nabla v \\ + \big(c_\tau^k(x,u_{l\tau}^k,\nabla u_{l\tau}^k) - g_\tau^k\big)v\,\mathrm{d}x + \int_\Gamma \big(b_\tau^k(x,u_{l\tau}^k) - \tau_\tau^k\big)v\,\mathrm{d}S = 0 \quad (8.154)$$

for any $v \in V_l$, with the initial condition $u_{l\tau}^0 = u_{0l}$ where $u_{0l} \in V_l$ is defined[38] by $\int_\Omega (u_{0l} - u_0)v\mathrm{d}x = 0$ for any $v \in V_l$, and where the Clément zero-order quasi-interpolation of the coefficients is defined by

$$a_\tau^k(x,r,s) := \frac{1}{\tau}\int_{(k-1)\tau}^{k\tau} a(t,x,r,s)\,\mathrm{d}t, \quad b_\tau^k(x,r) := \frac{1}{\tau}\int_{(k-1)\tau}^{k\tau} b(t,x,r)\,\mathrm{d}t, \quad (8.155)$$

and analogously for c_τ^k. In the previous notation (8.81), we would define $A_\tau^k : W^{1,p}(\Omega) \to W^{1,p}(\Omega)^*$ by

$$\langle A_\tau^k(u), v\rangle := \int_\Omega a_\tau^k(x,u,\nabla u)\cdot\nabla v + c_\tau^k(x,u,\nabla u)v\,\mathrm{d}x + \int_\Gamma b_\tau^k(x,u)v\,\mathrm{d}S. \quad (8.156)$$

[35] The dual space to $L^p(I;W^{1,p}(\Omega)) \cap L^\infty(I;L^2(\Omega))$ contains $L^\infty(I;L^2(\Omega))^*$ but, in fact, $L^{p^{\circledast}}(Q)$ lives in a smaller space involving $L^1(I;L^2(\Omega))$ instead of $L^\infty(I;L^2(\Omega))^*$.

[36] At this point, various numerical-integration formulae usually have to be employed in (8.154) and (8.155). Also, we assume that the resulting system of algebraic equations can be solved numerically.

[37] With only a small loss of generality, V_l as a finite-element space with the mesh size $1/l$, cf. Example 2.67.

[38] In other words, u_{0l} is the $L^2(\Omega)$-orthogonal projection of u_0.

Remark 8.44 (Projectors P_k). The projectors $P_k(u) := \sum_{i=1}^k \left(\int_\Omega uv_i \mathrm{d}x\right) v_i$ (cf. (8.100)) that can alternatively be used in the abstract Galerkin method can now employ $v_i \in W_0^{r,2}(\Omega) \subset W^{1,p}(\Omega)$ (which requires $r \geq 1+n(p-2)/(2p)$) solving the eigenvalue problem

$$\Delta^r v_i = \lambda_i v_i \,. \tag{8.157}$$

Moreover, we can assume that v_i makes an orthonormal basis in $L^2(\Omega)$ and $v_i/\sqrt{\lambda_i}$ an orthonormal basis in $W_0^{r,2}(\Omega)$. Then the projector P_k is selfadjoint, and

$$\|P_k\|_{\mathscr{L}(L^2(\Omega),L^2(\Omega))} \leq 1 \qquad \& \qquad \|P_k\|_{\mathscr{L}(W_0^{r,2}(\Omega),W_0^{r,2}(\Omega))} \leq 1\,. \tag{8.158}$$

The second estimate then can be used to get the a-priori estimate[39]

$$\left\|\frac{\partial u_k}{\partial t}\right\|_{L^{p'}(I;W^{-r,2}(\Omega))} \leq C. \tag{8.159}$$

Remark 8.45 (Pseudomonotone memory: *integro-differential equations*). For a Carathéodory mapping $\mathfrak{f} : [Q \times Q] \times \mathbb{R} \times \mathbb{R}^n \to \mathbb{R}$ one can consider the nonlinear Uryson integral operator $(u, y) \mapsto \big((t, x) \mapsto \int_Q \mathfrak{f}(x, t, \xi, \vartheta, u(\xi, \vartheta), y(\xi, \vartheta))\mathrm{d}\xi\mathrm{d}\vartheta\big)$ which is, under certain not much restrictive conditions[40], totally continuous as a mapping $L^p(Q;\mathbb{R}^{1+n}) \to L^{p'}(Q)$ and, as such, it is pseudomonotone, cf. Corollary 2.12. Thus one can treat e.g. the integro-differential equation

$$\frac{\partial u}{\partial t} - \mathrm{div}\big(|\nabla u|^{p-2}\nabla u\big) + \int_Q \mathfrak{f}\big(x, t, \xi, \vartheta, u(\xi, \vartheta), \nabla u(\xi, \vartheta)\big)\,\mathrm{d}\xi\mathrm{d}\vartheta = g. \tag{8.160}$$

8.7 Application to semilinear parabolic equations

In this section we focus on the very weak formulation (8.127) in the special case when $a(t, x, r, \cdot) : \mathbb{R}^n \to \mathbb{R}^n$ and $c(t, x, r, \cdot) : \mathbb{R}^n \to \mathbb{R}$ are affine, i.e.

$$a_i(t, x, r, s) := \sum_{j=1}^n a_{ij}(t, x, r)s_j + a_{i0}(t, x, r), \quad i = 1, \ldots, n, \tag{8.161a}$$

$$c(t, x, r, s) := \sum_{j=1}^n c_j(t, x, r)s_j + c_0(t, x, r), \tag{8.161b}$$

[39] Unfortunately, $W^{1,p}(\Omega)$ is not an interpolant between $L^2(\Omega)$ and $W^{r,2}(\Omega)$ so that the interpolation theory to get the estimate $\|P_k\|_{\mathscr{L}(W^{1,p}(\Omega),W^{1,p}(\Omega))} \leq 1$ cannot be used.

[40] Namely, the growth condition $|\mathfrak{f}(x, t, \xi, \vartheta, r, s)| \leq \gamma_0(x, t, \xi, \vartheta) + \gamma_1(x, t)(|r|^p + |s|^p)$ with $\gamma_0 \in L^{p'}(Q; L^1(Q))$ and $\gamma_1 \in L^{p'}(Q)$ and the equicontinuity condition:

$$\forall c > 0: \quad \lim_{|A|\to 0} \sup_{\substack{\|u\|_{L^p(Q)}\leq c\\ \|y\|_{L^p(Q;\mathbb{R}^n)}\leq c}} \int_Q \left|\int_A \mathfrak{f}(x, t, \xi, \vartheta, u(\xi, \vartheta), y(\xi, \vartheta))\,\mathrm{d}\xi\mathrm{d}\vartheta|^{p'}\right|\mathrm{d}x\mathrm{d}t = 0.$$

We refer to Krasnoselskiĭ et al. [238, Theorem 19.3]. The latter condition is fulfilled, e.g., if the growth condition is slightly strengthened, namely $|\mathfrak{f}(x, t, \xi, \vartheta, r, s)| \leq \gamma_0(x, t, \xi, \vartheta) + \gamma_1(x, t)(|r|^{p-\epsilon} + |s|^{p-\epsilon})$ with some $\epsilon > 0$ and γ_0 in the form $\sum_{\text{finite}} \gamma_{0l}(\xi, \vartheta)\tilde{\gamma}_{0l}(x, t)$ with $\gamma_{0l} \in L^1(Q)$ and $\tilde{\gamma}_{0l} \in L^{p'}(Q)$.

with $a_{ij}, c_j : Q\times\mathbb{R} \to \mathbb{R}$ Carathéodory mappings whose growth is to be designed to induce the Nemytskiĭ mappings $\mathscr{N}_{(a_{i1},\dots,a_{in})}, \mathscr{N}_{(c_1,\dots,c_n)} : L^{2^{\circledast}-\epsilon}(Q) \to L^2(Q;\mathbb{R}^n)$ and $\mathscr{N}_{a_{i0}}, \mathscr{N}_{c_0} : L^{2^{\circledast}-\epsilon}(Q) \to L^1(Q)$ with $\epsilon > 0$ and with $2^{\circledast} := 2 + 4/n$, which corresponds to (8.131) with $p = 2$. Besides, the boundary nonlinearity $b : \Sigma\times\mathbb{R} \to \mathbb{R}$ is now to induce the Nemytskiĭ mapping $\mathscr{N}_b : L^2(I; L^{2^{\#}-\epsilon}(\Gamma)) \to L^1(\Sigma)$. This means, for $i, j = 1, \dots, n$,

$$\exists \gamma_1 \in L^2(Q),\ C \in \mathbb{R}: \quad |a_{ij}(t,x,r)| \le \gamma_1(t,x) + C|r|^{(2^{\circledast}-\epsilon)/2},$$

$$|c_j(t,x,r)| \le \gamma_1(t,x) + C|r|^{(2^{\circledast}-\epsilon)/2}, \tag{8.162a}$$

$$\exists \gamma_2 \in L^1(Q),\ C \in \mathbb{R}: \quad |a_{i0}(t,x,r)| \le \gamma_2(t,x) + C|r|^{2^{\circledast}-\epsilon},$$

$$|c_0(t,x,r)| \le \gamma_2(t,x) + C|r|^{2^{\circledast}-\epsilon}, \tag{8.162b}$$

$$\exists \gamma_3 \in L^1(\Sigma),\ C \in \mathbb{R}: \quad |b(t,x,r)| \le \gamma_3(t,x) + C|r|^{2^{\oplus}-\epsilon}. \tag{8.162c}$$

The exponent $p = 2$ is natural because the growth $a(t,x,r,\cdot)$ is now linear. Note that these requirements just guarantee that all integrals in (8.127) have a good sense if $v \in W^{1,\infty,\infty}(I; W^{1,\infty}(\Omega), L^{2^{*\prime}}(\Omega))$.

Lemma 8.46 (Weak continuity of $\mathscr{A}$). *Let* (8.161)–(8.162) *hold. Then* $\mathscr{A}$ *is weakly* continuous as a mapping*

$$W^{1,2,1}(I; W^{1,2}(\Omega), [W^{1,2}(\Omega)]^*_{\rm lcs}) \cap L^\infty(I; L^2(\Omega)) \to L^\infty(I; W^{1,\infty}(\Omega))^*.$$

Proof. By the Aubin-Lions lemma, we have the compact embedding $W^{1,2,1}(I; W^{1,2}(\Omega), [W^{1,2}(\Omega)]^*_{\rm lcs}) \Subset L^2(I; W^{1-\epsilon_1,2}(\Omega))$ for any $\epsilon_1 > 0$. Taking $\epsilon_1 > 0$ suitably small, for some $0 < \lambda \le 1$ we have the interpolation estimate $\|u\|_{L^{2^{\circledast}-\epsilon}(Q)} \le C\|u\|^{\lambda}_{L^2(I;W^{1-\epsilon_1,2}(\Omega))}\|u\|^{1-\lambda}_{L^\infty(I;L^2(\Omega))}$ for any $u \in L^2(I; W^{1-\epsilon_1,2} \cap L^\infty(I; L^2(\Omega))$; cf. also (8.151). Hence, having a weakly* convergent sequence $\{u_k\}_{k\in\mathbb{N}}$ in $L^2(I; W^{1,2}(\Omega)) \cap L^\infty(I; L^2(\Omega))$ with $\{\frac{\rm d}{{\rm d}t}u_k\}_{k\in\mathbb{N}}$ bounded in $L^1(I; [W^{1,2}(\Omega)]^*_{\rm lcs})$, this sequence converges strongly in $L^{2^{\circledast}-\epsilon}(Q)$. Then, by the continuity of the Nemytskiĭ mappings $\mathscr{N}_{(a_{i1},\dots,a_{in})}, \mathscr{N}_{(c_1,\dots,c_n)} : L^{2^{\circledast}-\epsilon}(Q) \to L^2(Q;\mathbb{R}^n)$ and $\mathscr{N}_{a_{i0}}, \mathscr{N}_{c_0} : L^{2^{\circledast}-\epsilon}(Q) \to L^1(Q)$, it holds that

$$\int_Q \sum_{i=1}^n \Big(\sum_{j=1}^n a_{ij}(u_k)\frac{\partial u_k}{\partial x_j} + a_{i0}(u_k)\Big)\frac{\partial v}{\partial x_i} + \Big(\sum_{j=1}^n c_j(u_k)\frac{\partial u_k}{\partial x_j} + c_0(u_k)\Big) v\,{\rm d}x{\rm d}t$$

$$\to \int_Q \sum_{i=1}^n \Big(\sum_{j=1}^n a_{ij}(u)\frac{\partial u}{\partial x_j} + a_{i0}(u)\Big)\frac{\partial v}{\partial x_i} + \Big(\sum_{j=1}^n c_j(u)\frac{\partial u}{\partial x_j} + c_0(u)\Big) v\,{\rm d}x{\rm d}t$$

for $k \to \infty$ and for any $v \in L^\infty(I; W^{1,\infty}(\Omega))$. As in (8.153), we have now $u_k|_\Sigma \to u|_\Sigma$ in $L^{2^{\oplus}-\epsilon}(\Sigma)$ and, by (8.162c), we have convergence also in the boundary term $\int_\Sigma b(u_k)v\,{\rm d}S{\rm d}t \to \int_\Sigma b(u)v\,{\rm d}S{\rm d}t$. $\square$

Proposition 8.47 (Existence of very weak solutions). *Let* (8.161)–(8.162) *hold for some* $\gamma_1 \in L^{2+\epsilon}(Q)$, $\gamma_2 \in L^{1+\epsilon}(Q)$, *and* $\gamma_3 \in L^{1+\epsilon}(\Sigma)$. *Moreover, let* $g \in L^2(I; L^{2^{*\prime}}(\Omega)) + L^1(I; L^2(\Omega))$, $h \in L^2(I; L^{2^{\#\prime}}(\Gamma))$, *and, for some* $\varepsilon > 0$, $\gamma_1 \in L^2(I)$, $\gamma_2 \in L^1(Q)$, $\gamma_3 \in L^1(I)$, $\gamma_4 \in L^1(\Gamma)$, *and for a.a.* $(t,x)\in Q$ *(resp.* $(t,x)\in\Sigma$ *for* (8.163)*c) and all* $(r,s) \in \mathbb{R}^{1+n}$, *it holds that*

$$\sum_{i=1}^{n}\Big(\sum_{j=1}^{n} a_{ij}(t,x,r)s_j + a_{i0}(t,x,r)\Big)s_i \geq \varepsilon|s|^2 - \gamma_1(t)|s|, \tag{8.163a}$$

$$\Big(\sum_{j=1}^{n} c_j(t,x,r)s_j + c_0(t,x,r)\Big)r \geq -\gamma_2(t,x) - \gamma_3(t)|r|^2 - C|s|^{2-\varepsilon}, \tag{8.163b}$$

$$b(t,x,r)r \geq -\gamma_4(x) - C|r|^{2-\varepsilon}. \tag{8.163c}$$

Then the initial-boundary value problem (8.122) *has a very weak solution.*

Proof. We can use the abstract Theorem 8.31 now with $V := W^{1,2}(\Omega)$, $Z := W^{1,\infty}(\Omega)$, and V_k some finite-dimensional subspaces of $W^{1,\infty}(\Omega)$ satisfying (2.7).[41] The semi-coercivity (8.95) is implied by (8.163) by routine calculations.[42] Moreover, (8.162) implies the growth condition (8.109) with $p = 2$ and $q < +\infty$, which ensures boundedness of $\mathscr{A}$ from $L^2(I; W^{1,2}(\Omega)) \cap L^\infty(I; L^2(\Omega))$ to $L^{1+\epsilon}(I; W^{1,\infty}(\Omega)^*)$ with some $\epsilon > 0$ (possibly different from ϵ in (8.162)), as required in Theorem 8.31. Indeed, using (8.162a,b) for simplicity heuristically with $\epsilon = 0$, we obtain

$$\begin{aligned}
&\sup_{\|v\|_{W^{1,\infty}(\Omega)}\leq 1}\int_\Omega \sum_{i=1}^{n}\Big(\sum_{j=1}^{n} a_{ij}(u)\frac{\partial u}{\partial x_j} + a_{i0}(u)\Big)\frac{\partial v}{\partial x_i}\,\mathrm{d}x \\
&\quad\leq \Big\|\sum_{i=1}^{n}\Big(\sum_{j=1}^{n} a_{ij}(u)\frac{\partial u}{\partial x_j} + a_{i0}(u)\Big)\frac{\partial v}{\partial x_i}\Big\|_{L^1(\Omega)} \\
&\quad\leq \sum_{i,j=1}^{n} \|a_{ij}(u)\|_{L^2(\Omega)}\|\nabla u\|_{L^2(\Omega;\mathbb{R}^n)} + \|a_{i0}(u)\|_{L^1(\Omega)} \\
&\quad\leq \sum_{i,j=1}^{n} \frac{1}{2}\|\gamma_1(t,\cdot)\|_{L^2(\Omega)}^2 + \|\gamma_2(t,\cdot)\|_{L^1(\Omega)} + \frac{2C+C^2}{2}\|u\|_{L^{2^\circledast}(\Omega)}^{2^\circledast} + \|\nabla u\|_{L^2(\Omega;\mathbb{R}^n)}^2.
\end{aligned}$$

Now we estimate by interpolation[43] $\|u\|_{L^{2^\circledast}(\Omega)}^{2^\circledast} \leq \|u\|_{L^2(\Omega)}^{(1-\lambda)2^\circledast}\|u\|_{W^{1,2}(\Omega)}^{2}$ for $\lambda =$

[41] Recall that one can consider finite-element subspaces as in Example 2.67.

[42] We have $\langle A(t,v),v\rangle = \int_\Omega \sum_{i=1}^n(\sum_{j=1}^n a_{ij}(v)\frac{\partial v}{\partial x_j} + a_{i0}(v))\frac{\partial v}{\partial x_i} + \sum_{j=1}^n c_j(v)\frac{\partial v}{\partial x_j}v + c_0(v)v\,\mathrm{d}x + \int_\Gamma b(v)v\,\mathrm{d}S \geq \varepsilon\|\nabla v\|_{L^2(\Omega;\mathbb{R}^n)}^2 - \frac{1}{2}\mathrm{meas}_n(\Omega)\gamma_1(t)^2 - \frac{1}{2}\|\nabla v\|_{L^2(\Omega;\mathbb{R}^n)}^2 - \|\gamma_2(t,\cdot)\|_{L^1(\Omega)} - \gamma_3(t)\|v\|_{L^2(\Omega)}^2 - C\|\nabla v\|_{L^{2-\varepsilon}(\Omega;\mathbb{R}^n)}^{2-\varepsilon} - \|\gamma_4\|_{L^1(\Gamma)} - C\|\nabla v\|_{L^{2-\varepsilon}(\Gamma)}^{2-\varepsilon}$ and then we can obtain (8.95) by Young inequality.

[43] By (1.23), $\|u\|_{L^{2^\circledast}(\Omega)}^{2^\circledast} \leq \|u\|_{L^2(\Omega)}^{(1-\lambda)2^\circledast}\|u\|_{L^{2^*}(\Omega)}^{\lambda 2^\circledast}$ provided $\frac{1}{2}(1-\lambda) + \lambda\frac{n-2}{2n} = 2^\circledast$, which yields $\lambda = n/(n+2)$ and $\lambda 2^\circledast = 2$ as used (8.131) for p=2.

$n/(n+2)$, from which already the estimate of the type (8.109) follows. The contribution of $\sum_{i=1}^n c_i(u)\frac{\partial}{\partial x_i}u + c_0(u)$ follows from essentially the same calculations. Under the condition (8.162c), the contribution of the boundary term b is analogous, based on the interpolation of the trace operator $\|u\|^{2^\circledast}_{L^{2^\circledast}(\Gamma)} \le \|u\|^{(1-\lambda)2^\circledast}_{L^2(\Omega)} \|u\|^2_{W^{1,2}(\Omega)}$ with $\lambda = n/(n+1)$ for which $\lambda 2^\circledast = 2$ as used already for (8.136). □

Corollary 8.48 (Weak solutions). *Let, in addition to the assumptions of Proposition* 8.47, *also* $g \in L^2(I; L^{2^{*\prime}}(\Omega))$ *and the growth condition* (8.142) *with* $p = 2$ *hold. Then there is a weak solution to the initial-boundary-value problem* (8.4).

Proof. It suffices to merge Proposition 8.47 and Lemma 8.4. □

8.8 Examples and exercises

This section completes the previous theory by assorted, and often physically motivated, examples together with some exercises accompanied mostly by brief hints.

8.8.1 General tools

Exercise 8.49. Assuming V separable, $\xi \in L^{p'}(I; V^*)$, and $\int_0^T \langle \xi(t), v(t)\rangle_{V^*\times V} \mathrm{d}t = 0$ for any v of the form $v(t) = g(t)z_i$, $g \in L^p(I)$, $\{z_i\}_{i\in\mathbb{N}}$ dense in V, show that $\xi(t) = 0$ for a.a. $t \in I$.[44] Cf. Proposition 1.38.

Exercise 8.50. Modify Theorem 8.18 for $c_0 = 0$ in (8.69c) but, on the other hand, assuming $f \in W^{1,1}(I; H)$ in (8.69a). Only the estimate $u \in W^{1,\infty}(I; H)$ can thus be obtained.

Exercise 8.51 (Continuous dependence on the data). Consider a sequence $(f_k, u_{0k}) \to (f, u_0)$ in $L^{p'}(I; V^*) \times H$ and prove the convergence of the corresponding solutions as claimed in Theorem 8.35(i).

Example 8.52 (The case of $A(t,\cdot) : V \to V^*$ monotone). If $A(t,\cdot)$ is monotone, radially continuous and satisfies the growth condition (8.80), then $\mathscr{A}$ is pseudomonotone even as a mapping $L^p(I; V) \cap L^\infty(I; H) \to L^{p'}(I; V^*)$, i.e. no bound on the time derivative is needed. Indeed, $\mathscr{A}$ is obviously monotone and is bounded because

$$\begin{aligned}
\|\mathscr{A}(u)\|_{L^{p'}(I;V^*)} &= \Big(\int_0^T \|A(t,u(t))\|^{p'}_{V^*} \mathrm{d}t\Big)^{1/p'} \\
&\le \Big(\int_0^T \mathfrak{C}(\|u(t)\|_H)^{p/(p-1)} \Big(\gamma(t) + \|u(t)\|_V^{p-1}\Big)^{p/(p-1)} \mathrm{d}t\Big)^{(p-1)/p} \\
&\le 2^{1/p}\mathfrak{C}\big(\|u(t)\|_{L^\infty(I;H)}\big)\big(\|\gamma\|_{L^{p'}(I)} + \|u\|^{p-1}_{L^p(I;V)}\big) \qquad (8.164)
\end{aligned}$$

[44]Hint: Fixing z_i, realize that $\int_0^T \langle \xi(t), v(t)\rangle_{V^*\times V} \mathrm{d}t = \int_0^T g(t)\langle \xi(t), z_i\rangle_{V^*\times V} \mathrm{d}t = 0$ for all g implies $\langle \xi(t), z_i\rangle_{V^*\times V} = 0$ for a.a. $t \in I$. This holds true even if z_i ranges over the countable set $\{z_i\}_{i\in\mathbb{N}}$. As this set is dense in V, $\xi(t) = 0$ for a.a. $t \in I$.

where γ and $\mathfrak{C}$ is from (8.80). Moreover, $\mathscr{A}$ is radially continuous because, for any $u, v \in L^p(I;V) \cap L^\infty(I;H)$ and for a.a. $t \in I$, $\langle A(t,u(t)+\varepsilon v(t)), v(t)\rangle \to \langle A(t,u(t)), v(t)\rangle$ because $A(t,\cdot)$ is radially continuous, and thus

$$\begin{aligned}\big\langle \mathscr{A}(t,u+\varepsilon v), v\big\rangle &= \int_0^T \big\langle A(t,u(t)+\varepsilon v(t)), v(t)\big\rangle\,\mathrm{d}t \\ &\to \int_0^T \big\langle A(t,u(t)), v(t)\big\rangle\,\mathrm{d}t = \big\langle \mathscr{A}(t,u), v\big\rangle, \qquad (8.165)\end{aligned}$$

by the Lebesgue Theorem 1.14, where we used also the fact that the collection $\{t \mapsto \langle A(t,u(t)+\varepsilon v(t)), v(t)\rangle\}_{\varepsilon\in[0,\varepsilon_0]}$ has a common integrable majorant because, in view of (8.80),

$$\begin{aligned}&\big|\langle A(t,u(t)+\varepsilon v(t)), v(t)\rangle\big| \le \big\|A(u(t)+\varepsilon v(t))\big\|_{V^*}^{p'} + \big\|v(t)\big\|_V^p \\ &\le \mathfrak{C}\big(\|u{+}\varepsilon v\|_{L^\infty(I;H)}\big)\big(\gamma(t) + \big\|u(t){+}\varepsilon v(t)\big\|_V^{p-1}\big)^{p'} + \big\|v(t)\big\|_V^p \\ &\le 2^{p'-1}\mathfrak{C}\big(\|u\|_{L^\infty(I;H)}{+}\varepsilon_0\|v\|_{L^\infty(I;H)}\big)\big(\gamma(t)^{p'}{+}\|u(t)\|_V^p{+}\varepsilon_0^{p'}\|v(t)\|_V^p\big) + \big\|v(t)\big\|_V^p.\end{aligned}$$

Then $\mathscr{A}$ is pseudomonotone by Lemma 2.9.

Example 8.53 (Totally continuous terms). Let $V_1 \Supset V$ and $A : I \times V_1 \to V^*$ be a Carathéodory mapping satisfying (8.80) modified by replacing V with V_1, i.e.

$$\exists\,\gamma\in L^{p'}(I),\ \ \mathfrak{C}{:}\mathbb{R}{\to}\mathbb{R}\text{ increasing}: \ \ \big\|A(t,v)\big\|_{V^*}{\le}\mathfrak{C}\big(\|v\|_H\big)\big(\gamma(t){+}\|v\|_{V_1}^{p-1}\big). \qquad (8.166)$$

Then the abstract Nemytskiĭ mapping $\mathscr{A} : \mathscr{W} \to L^{p'}(I;V^*)$ with $\mathscr{W}$ from Lemma 8.8 or 8.29, is totally continuous. Indeed, having a sequence $u_k \overset{*}{\rightharpoonup} u$ in $\mathscr{W}$, by Aubin-Lions Lemma 7.7 or its Corollary 7.9, $u_k \to u$ in $L^p(I;V_1)$. Then, using $\big\|A(t,u_k)\big\|_{V^*}{\le}C(\gamma(t){+}\|u_k\|_{V_1}^{p-1})$ with $C := \mathfrak{C}(\sup_{k\in\mathbb{N}}\|u_k\|_{L^\infty(I;H)})$ and Theorem 1.43, we obtain $\mathscr{A}(u_k) \to \mathscr{A}(u)$ in $L^{p'}(I;V^*)$.

Exercise 8.54. Assume A as in Example 8.52 and prove the convergence of the Rothe method directly by Minty's trick in parallel to Remark 8.32.

Exercise 8.55. Assuming (2.7) and relying upon $\bigcup_{k\in\mathbb{N}} C^1(I;V_k)$ being dense in $W^{1,p,p'}(I;V,V^*)$,[45] prove density of $\bigcup_{k\in\mathbb{N}} L^\infty(I;V_k)$ in $L^p(I;V) \cap L^\infty(I;H)$.[46]

Exercise 8.56. Consider the Galerkin approximation u_k to the abstract Cauchy problem (8.1) with data qualification (8.60), and prove the boundedness of $\{u_k\}_{k>0}$

[45]This density follows by Lemma 7.2 and by the famous Weierstrass theorem giving a possibility of approaching each function $C^1(I;V)$ by polynomials in t with coefficients in V, and eventually by approximating these coefficients in V_k with k sufficiently large; see e.g. Gajewski at al. [168, Sect.VI.1, Lemma 1.5] for details.

[46]Hint: Use approximation by $C^1(I;V_k)$ with k sufficiently large in the topology of $W^{1,p,p'}(I;V,V^*)$, and then continuity both of the embedding $W^{1,p,p'}(I;V,V^*) \subset L^\infty(I;H)$ by Lemma 7.3 and of the embedding $W^{1,p,p'}(I;V,V^*) \subset L^p(I;V)$. Cf. also Lemma 8.28.

in $W^{1,2}(I;H)\cap L^\infty(I;V)$.[47] Note that, now, Φ in (8.60c,d) need not be assumed convex.

Exercise 8.57. Consider u_k as in Exercise 8.56 and the data qualification (8.69), and prove the boundedness of $\{u_k\}_{k>0}$ in $W^{1,\infty}(I;H)\cap W^{1,2}(I;V)$.[48]

Exercise 8.58. Show the convergence of u_τ from Gear's formula (8.78). Modify the proof of Theorem 8.16(ii).[49]

Exercise 8.59. Modify Remark 8.32 for totally continuous perturbation mentioned in Example 8.53.[50]

Exercise 8.60. Prove the interpolation formula (1.63) by using (1.23) and Hölder's inequality.[51]

Exercise 8.61. Prove that all integrals in (8.126) and (8.127) have a good sense.

Exercise 8.62. Modify the estimation scenario (8.64) by requiring the (possibly non-polynomial) growth condition $\|A_2(u)\|_H \le C(1+\|u\|_H+\Phi(u)^{1/2})$ instead of (8.60d), assuming (without loss of generality) that $\Phi\ge 0$.

8.8.2 Parabolic equation of type $\frac{\partial}{\partial t}u-\mathrm{div}(|\nabla u|^{p-2}\nabla u)+c(u)=g$

The following examples are to be considered as a detailed scrutiny of estimation technique on a heuristical level. Rigorously, it works if we assume that a solution u with appropriate qualitative properties has been already obtained. Adaptation to the Galerkin method is simple, and to the Rothe method is, in view of Sections 8.2–8.3, also quite routine.

Example 8.63 (Monotone parabolic problem: a-priori estimates)**.** For $p\in(1,+\infty)$ and $q_1,q_2\ge 1$ specified later, let us consider the initial-boundary-value problem:

$$\left.\begin{aligned} \frac{\partial u}{\partial t}-\mathrm{div}\big(|\nabla u|^{p-2}\nabla u\big)+|u|^{q_1-2}u &= g && \text{in } Q,\\ |\nabla u|^{p-2}\frac{\partial u}{\partial \nu}+|u|^{q_2-2}u &= h && \text{on } \Sigma,\\ u(0,\cdot) &= u_0 && \text{in } \Omega, \end{aligned}\right\} \tag{8.167}$$

where $g\in L^{p'}(I;L^{p^{*\prime}}(\Omega))$ and $h\in L^{p'}(I;L^{p^{\#\prime}}(\Gamma))$. We will prove the a-priori estimates on the heuristic level.

[47] Hint: Modify the proof of Theorem 8.16(i).

[48] Hint: Modify the proof of Theorem 8.18(i).

[49] Hint: Realize that $\frac{\mathrm{d}}{\mathrm{d}t}u_\tau^{\mathrm{R}}-\frac{\mathrm{d}}{\mathrm{d}t}u_\tau\rightharpoonup 0$ in $L^2(I;H)$ due to the by-part formula $\langle u_\tau^{\mathrm{R}}-u_\tau,\frac{\mathrm{d}}{\mathrm{d}t}\varphi\rangle\to 0$ for any $\varphi\in\mathscr{D}(I;H)$ because of $\|u_\tau^{\mathrm{R}}-u_\tau\|_{L^2(I;H)}=\mathscr{O}(\tau)$ which is to be proved by a modification of (8.50) and by using the boundedness of $\{\frac{\mathrm{d}}{\mathrm{d}t}u_\tau\}_{0<\tau\le\tau_0}$ in $L^2(I;H)$.

[50] Hint: Generalize the proof of the "steady-state" Proposition 2.17 for the evolutionary case.

[51] Hint: By (1.23), $\|v(\cdot)\|_{L^q(\Omega)}\le\|v(\cdot)\|_{L^{q_1}(\Omega)}^{\lambda}\|v(\cdot)\|_{L^{q_2}(\Omega)}^{1-\lambda}$, and then integrate it over I and use Hölder's inequality with the (mutually conjugate) exponents $p_1/(\lambda p)$ and $p_2/((1-\lambda)p)$.

(1) Following the strategy (8.21) for $f = f_1$, we use a test by $u(t,\cdot)$ itself:

$$\begin{aligned}\frac{1}{2}\frac{\mathrm{d}}{\mathrm{d}t}\|u\|^2_{L^2(\Omega)} &+ \|\nabla u\|^p_{L^p(\Omega;\mathbb{R}^n)} + \|u\|^{q_1}_{L^{q_1}(\Omega)} + \|u\|^{q_2}_{L^{q_2}(\Gamma)} \\ &= \int_\Omega gu\mathrm{d}x + \int_\Gamma hu\,\mathrm{d}S \le N\big(\|g\|_{L^{p^*\prime}(\Omega)} + \|h\|_{L^{p^{\#}\prime}(\Gamma)}\big)\|u\|_{W^{1,p}(\Omega)} \\ &\le NC_{\rm P}\big(\|g\|_{L^{p^*\prime}(\Omega)} + \|h\|_{L^{p^{\#}\prime}(\Gamma)}\big)\big(\|\nabla u\|_{L^p(\Omega;\mathbb{R}^n)} + \|u\|_{L^2(\Omega)}\big) \\ &\le C_\varepsilon N^{p'}C_{\rm P}^{p'}\big(\|g\|_{L^{p^*\prime}(\Omega)} + \|h\|_{L^{p^{\#}\prime}(\Gamma)}\big)^{p'} + \varepsilon\|\nabla u\|^p_{L^p(\Omega;\mathbb{R}^n)} \\ &\quad + \frac{NC_{\rm P}}{2}\big(\|g\|_{L^{p^*\prime}(\Omega)} + \|h\|_{L^{p^{\#}\prime}(\Gamma)}\big)\big(1 + \|u\|^2_{L^2(\Omega)}\big) \end{aligned} \tag{8.168}$$

where N is greater than the norm of the embedding/trace operator $u \mapsto (u, u|_\Gamma) : W^{1,p}(\Omega) \to L^{p^*}(\Omega) \times L^{p^{\#}}(\Gamma)$, and where we used the Poincaré inequality in the form $\|u(t,\cdot)\|_{W^{1,p}(\Omega)} \le C_{\rm P}(\|\nabla u(t,\cdot)\|_{L^p(\Omega;\mathbb{R}^n)} + \|u(t,\cdot)\|_{L^2(\Omega)})$, cf. (1.55). This, after choosing $\varepsilon < 1$, using the Gronwall inequality, and integration over $[0,T]$, gives the a-priori estimate for u in $L^p(I;W^{1,p}(\Omega)) \cap L^\infty(I;L^2(\Omega))$.

(2) The estimate for $\frac{\partial}{\partial t}u$ in $L^{p'}(I;W^{1,p}(\Omega)^*)$ requires assumptions on q_1 and q_2. In detail, imitating the scenario (8.23), we estimate:

$$\begin{aligned}\Big\langle\frac{\partial u}{\partial t}, v\Big\rangle &= \int_Q gv - |\nabla u|^{p-2}\nabla u\cdot\nabla v - |u|^{q_1-2}uv\,\mathrm{d}x\mathrm{d}t \\ &+ \int_\Sigma (h - |u|^{q_2-2}u)v\,\mathrm{d}S\mathrm{d}t \le \|\nabla u\|^{p-1}_{L^p(Q;\mathbb{R}^n)}\|\nabla v\|_{L^p(Q;\mathbb{R}^n)} \\ &+ \big\||u|^{q_1-1}\big\|_{L^{p'}(I;L^{p^*\prime}(\Omega))}\|v\|_{L^p(I;L^{p^*}(\Omega))} \\ &+ \big\||u|^{q_2-1}\big\|_{L^{p'}(I;L^{p^{\#}\prime}(\Gamma))}\|v\|_{L^p(I;L^{p^{\#}}(\Gamma))} \\ &+ \|g\|_{L^{p'}(I;L^{p^*\prime}(\Omega))}\|v\|_{L^p(I;L^{p^*}(\Omega))} + \|h\|_{L^{p'}(I;L^{p^{\#}\prime}(\Gamma))}\|v\|_{L^p(I;L^{p^{\#}}(\Gamma))}.\end{aligned} \tag{8.169}$$

This needs $q_1 \le p$ and $q_2 \le p$. Thus we get the estimate of $\frac{\partial}{\partial t}u$ in $L^{p'}(I;W^{1,p}(\Omega)^*)$. A weaker bound for q_1 can be obtained by *interpolation* to exploit also the information $u \in L^\infty(I;L^2(\Omega))$:

$$\begin{aligned}\big\||u|^{q_1-1}\big\|_{L^{p'}(I;L^{p^*\prime}(\Omega))} &= \bigg(\int_0^T\Big(\int_\Omega |u(t,x)|^{(q_1-1)p^{*\prime}}\mathrm{d}x\Big)^{p'/p^{*\prime}}\mathrm{d}t\bigg)^{1/p'} \\ &= \|u\|^{q_1-1}_{L^{p'(q_1-1)}(I;L^{p^{*\prime}(q_1-1)}(\Omega))} \le C\|u\|^{(q_1-1)\lambda}_{L^p(I;W^{1,p}(\Omega))}\|u\|^{(q_1-1)(1-\lambda)}_{L^\infty(I;L^2(\Omega))}\end{aligned} \tag{8.170}$$

provided q_1 and $\lambda \in [0,1]$ satisfy

$$\frac{1}{p^{*\prime}(q_1-1)} \ge \frac{\lambda(n-p)}{np} + \frac{1-\lambda}{2} \quad\text{and}\quad \frac{1}{p'(q_1-1)} \ge \frac{\lambda}{p}. \tag{8.171}$$

These inequalities are upper bounds for q_1. If $p < 2n/(n+2)$, i.e. $p^* < 2$, the optimal choice of λ is simply $\lambda = 0$, and then (8.171) implies $q_1 \le 1 + 2/p^{*\prime}$.

If $p \geq 2n/(n+2)$, the optimal choice of λ then balances both bounds for $\frac{1}{q_1-1}$ occuring in (8.171), i.e. $p^{*\prime}(\lambda(n-p)/np + (1-\lambda)/2) = p'\lambda/p$. After some algebra, for $p \neq n$, one can see that this means $\lambda = np(p-1)/(np^2-np+2p^2-2(p-n)^+)$, and then (8.171) yields $q_1 \leq (np^2+2p^2-2(p-n)^+)/np$, while for $p = n$ a strict inequality holds. For $p < n$, we obtain simply $q_1 \leq (np+2p)/n = p^{\circledast}$, cf. (8.131).

The interpolation in the boundary term (hence relaxing the bound $q_2 \leq p$) can be made analogously by using (8.171) with q_2 and $p^{\#\prime}$ in place of q_1 and $p^{*\prime}$, respectively.

A certain alternative approach is the estimate of $\frac{\partial}{\partial t}u$ in a bigger space, namely $(L^p(I; W^{1,p}(\Omega)) \cap L^\infty(I; L^2(\Omega)))^*$. Then one can modify (8.169) by estimating

$$
\begin{aligned}
&\int_Q -|u|^{q_1-2}uv \,\mathrm{d}x\mathrm{d}t + \int_\Sigma -|u|^{q_2-2}uv \,\mathrm{d}S\mathrm{d}t \\
&\quad\leq \big\| |u|^{q_1-1}\big\|_{L^{p^{\circledast\prime}}(Q)} \big\|v\big\|_{L^{p^{\circledast}}(Q)} + \big\| |u|^{q_2-1}\big\|_{L^{p^{\oplus\prime}}(\Sigma)} \big\|v\big\|_{L^{p^{\oplus}}(\Sigma)} \\
&\quad\leq \big\|u\big\|^{q_1-1}_{L^{(q_1-1)p^{\circledast\prime}}(Q)} \big\|v\big\|_{L^{p^{\circledast}}(Q)} + \big\|u\big\|^{q_2-1}_{L^{(q_2-1)p^{\oplus\prime}}(\Sigma)} \big\|v\big\|_{L^{p^{\oplus}}(\Sigma)} \\
&\quad\leq N_1\big\|u\big\|^{q_1-1}_{L^{(q_1-1)p^{\circledast\prime}}(Q)} \big\|u\big\|^{\lambda_1}_{L^p(I;W^{1,p}(\Omega))} \big\|u\big\|^{1-\lambda_1}_{L^\infty(I;L^2(\Omega))} \\
&\qquad + N_2\big\|u\big\|^{q_2-1}_{L^{(q_2-1)p^{\oplus\prime}}(\Sigma)} \big\|u\big\|^{\lambda_2}_{L^p(I;W^{1,p}(\Omega))} \big\|u\big\|^{1-\lambda_2}_{L^\infty(I;L^2(\Omega))}
\end{aligned}
\tag{8.172}
$$

with $\lambda_1 = n/(n+2)$ and $\lambda_2 = np/(np+2p-2)$ as used for (8.130) and (8.135). The bound of u in $L^{p^{\circledast}}(Q)$ imposes the requirement $p^{\circledast\prime}(q_1-1) \leq p^{\circledast}$, which further yields the restriction $q_1 \leq p^{\circledast}$, and similarly we get also $q_2 \leq p^{\oplus}$. If $p < n$, this weakens the previous restrictions on q_1 and q_2.

Another approach (at least for usage of the Aubin-Lions lemma) consists in weakening the dual norm to estimate $\frac{\partial}{\partial t}u$ in $L^{p'}(I; W_0^{1,p}(\Omega)^*) \cong L^{p'}(I; W^{-1,p'}(\Omega))$; note that $L^2(\Omega) \subset W^{-1,p'}(\Omega)$ because $W_0^{1,p}(\Omega) \subset L^2(\Omega)$ densely. For such an estimate one takes v in (8.169) from $L^p(I; W_0^{1,p}(\Omega))$ so that the term with q_2 completely vanishes, hence no restriction on q_2 is imposed for this estimate.

(3) To make a test by $v := \frac{\partial}{\partial t}u(t,\cdot)$, we assume $g \in L^2(Q)$ and $h = 0$. Then

$$
\begin{aligned}
\Big\|\frac{\partial u}{\partial t}\Big\|^2_{L^2(\Omega)} &+ \frac{1}{p}\frac{\mathrm{d}}{\mathrm{d}t}\big\|\nabla u\big\|^p_{L^p(\Omega;\mathbb{R}^n)} + \frac{1}{q_1}\frac{\mathrm{d}}{\mathrm{d}t}\big\|u\big\|^{q_1}_{L^{q_1}(\Omega)} + \frac{1}{q_2}\frac{\mathrm{d}}{\mathrm{d}t}\big\|u\big\|^{q_2}_{L^{q_2}(\Gamma)} \\
&= \int_\Omega g(t,\cdot)\frac{\partial u}{\partial t}\,\mathrm{d}x \leq \frac{1}{2}\big\|g\big\|^2_{L^2(\Omega)} + \frac{1}{2}\Big\|\frac{\partial u}{\partial t}\Big\|^2_{L^2(\Omega)}.
\end{aligned}
\tag{8.173}
$$

Assuming $u_0 \in W^{1,p}(\Omega) \cap L^2(\Omega)$, which means $u_0 \in W^{1,p}(\Omega)$ if p satisfies (8.124), by the Gronwall inequality, we thus get the estimate for u in $L^\infty(I; W^{1,p}(\Omega) \cap L^{q_1}(\Omega)) \cap W^{1,2}(I; L^2(\Omega))$ and for $u|_\Sigma$ in $L^\infty(I; L^{q_2}(\Gamma))$ for $q_1, q_2 \geq 1$ arbitrary.

Alternatively, one can assume $g \in W^{1,1}(I; L^{p^{*\prime}}(\Omega))$ and $h \in W^{1,1}(I; L^{p^{\#\prime}}(\Gamma))$

and use the strategy from Remark 8.23. Then we can estimate

$$\begin{aligned}
\int_0^t\!\!\int_\Omega g\frac{\partial u}{\partial t}\,\mathrm{d}x\mathrm{d}t &= \int_\Omega g(t,x)u(t,x)-g(0,x)u_0(x)\,\mathrm{d}x - \int_0^t\!\!\int_\Omega u\frac{\partial g}{\partial t}\,\mathrm{d}x\mathrm{d}t \\
&\le \big\|g(t)\big\|_{L^{p^*\prime}(\Omega)}\big\|u(t)\big\|_{L^{p^*}(\Omega)} + \int_0^t \big\|u\big\|_{L^{p^*}(\Omega)}\Big\|\frac{\partial g}{\partial t}\Big\|_{L^{p^*\prime}(\Omega)}\mathrm{d}t + \int_\Omega g(0)u_0\,\mathrm{d}x \\
&\le C\big\|g(t)\big\|_{L^{p^*\prime}(\Omega)}\Big(\big\|\nabla u\big\|_{L^p(\Omega;\mathbb{R}^n)} + \big\|u\big\|_{L^2(\Omega)}\Big) \\
&\quad + \int_0^t C\Big(\big\|\nabla u\big\|_{L^p(\Omega;\mathbb{R}^n)} + \big\|u\big\|_{L^{q_1}(\Omega)}\Big)\Big\|\frac{\partial h}{\partial t}\Big\|_{L^{p^\#\prime}(\Gamma)}\mathrm{d}t + \int_\Omega g(0)u_0\,\mathrm{d}x \\
&\le C\big\|g(t)\big\|_{L^{p^*\prime}(\Omega)}\big\|\nabla u\big\|_{L^p(\Omega;\mathbb{R}^n)} + C^2\big\|g(t)\big\|^2_{L^{p^*\prime}(\Omega)} + \big\|u\big\|^2_{L^2(\Omega)} \\
&\quad + \int_0^t C'\Big(1+\big\|\nabla u\big\|^p_{L^p(\Omega;\mathbb{R}^n)} + \big\|u\big\|^{q_1}_{L^{q_1}(\Omega)}\Big)\Big\|\frac{\partial h}{\partial t}\Big\|_{L^{p^\#\prime}(\Gamma)}\mathrm{d}t + \int_\Omega g(0)u_0\,\mathrm{d}x
\end{aligned}$$

and then proceed by Gronwall's inequality, and similarly we can use

$$\begin{aligned}
\int_0^t\!\!\int_\Gamma h\frac{\partial u}{\partial t}\,\mathrm{d}S\mathrm{d}t &= \int_\Gamma h(t,x)u(t,x)-h(0,x)u_0(x)\,\mathrm{d}S - \int_0^t\!\!\int_\Gamma u\frac{\partial h}{\partial t}\,\mathrm{d}S\mathrm{d}t \\
&\le \big\|h(t)\big\|_{L^{p^\#\prime}(\Gamma)}\big\|u(t)\big\|_{L^{p^\#}(\Gamma)} + \int_0^t \big\|u\big\|_{L^{p^\#}(\Gamma)}\Big\|\frac{\partial h}{\partial t}\Big\|_{L^{p^\#\prime}(\Gamma)}\mathrm{d}t + \int_\Gamma h(0)u_0\,\mathrm{d}S \\
&\le C\big\|h(t)\big\|_{L^{p^\#\prime}(\Gamma)}\Big(\big\|\nabla u\big\|_{L^p(\Omega;\mathbb{R}^n)} + \big\|u\big\|_{L^2(\Omega)}\Big) \\
&\quad + \int_0^t C\Big(\big\|\nabla u\big\|_{L^p(\Omega;\mathbb{R}^n)} + \big\|u\big\|_{L^{q_1}(\Omega)}\Big)\Big\|\frac{\partial h}{\partial t}\Big\|_{L^{p^\#\prime}(\Gamma)}\mathrm{d}t + \int_\Gamma h(0)u_0\,\mathrm{d}S \\
&\le C\big\|h(t)\big\|_{L^{p^\#\prime}(\Gamma)}\big\|\nabla u\big\|_{L^p(\Omega;\mathbb{R}^n)} + C^2\big\|h(t)\big\|^2_{L^{p^\#\prime}(\Gamma)} + \big\|u\big\|^2_{L^2(\Omega)} \\
&\quad + \int_0^t C'\Big(1+\big\|\nabla u\big\|^p_{L^p(\Omega;\mathbb{R}^n)} + \big\|u\big\|^{q_1}_{L^{q_1}(\Omega)}\Big)\Big\|\frac{\partial h}{\partial t}\Big\|_{L^{p^\#\prime}(\Gamma)}\mathrm{d}t + \int_\Gamma h(0)u_0\,\mathrm{d}S.
\end{aligned}$$

(4) Further, we apply $\frac{\partial}{\partial t}$ to the weak formulation of the equation with the boundary conditions in (8.167), then use the test function $v = \frac{\partial}{\partial t}u$, and estimate

$$\begin{aligned}
\frac{\partial}{\partial t}\big(|\nabla u|^{p-2}\nabla u\big)\cdot\nabla\frac{\partial u}{\partial t} &= |\nabla u|^{p-2}\frac{\partial \nabla u}{\partial t}\cdot\frac{\partial \nabla u}{\partial t} \\
&\quad + \Big((p-2)|\nabla u|^{p-4}\nabla u\cdot\frac{\partial \nabla u}{\partial t}\Big)\Big(\nabla u\cdot\frac{\partial \nabla u}{\partial t}\Big) \\
&= |\nabla u|^{p-2}\Big|\frac{\partial \nabla u}{\partial t}\Big|^2 + \frac{p-2}{4}|\nabla u|^{p-4}\Big(\frac{\partial |\nabla u|^2}{\partial t}\Big)^2 \\
&\ge \frac{4}{p^2}\Big(\frac{\partial |\nabla u|^{p/2}}{\partial t}\Big)^2 + \frac{p-2}{4}\Big(\big(|\nabla u|^2\big)^{(p-4)/4}\frac{\partial |\nabla u|^2}{\partial t}\Big)^2 \\
&= \Big(\frac{4}{p^2} + \frac{4p-8}{p^2}\Big)\Big(\frac{\partial |\nabla u|^{p/2}}{\partial t}\Big)^2 \ge 0 \qquad (8.174)
\end{aligned}$$

if $p\geq 1$. Similar calculations work for the lower-order terms when "forgetting" ∇'s for $q_1\geq 1$ and $q_2\geq 1$.[52] Altogether, one gets

$$\begin{aligned}\frac{1}{2}\frac{\mathrm{d}}{\mathrm{d}t}\Big\|\frac{\partial u}{\partial t}\Big\|^2_{L^2(\Omega)}&+\frac{4p-4}{p^2}\Big\|\frac{\partial|\nabla u|^{p/2}}{\partial t}\Big\|^2_{L^2(\Omega)}\\&+\frac{4q_1-4}{q_1^2}\Big\|\frac{\partial|u|^{q_1/2}}{\partial t}\Big\|^2_{L^2(\Omega)}+\frac{4q_2-4}{q_2^2}\Big\|\frac{\partial|u|^{q_2/2}}{\partial t}\Big\|^2_{L^2(\Gamma)}\\&\leq\int_\Omega\frac{\partial g}{\partial t}\frac{\partial u}{\partial t}\mathrm{d}x+\int_\Gamma\frac{\partial h}{\partial t}\frac{\partial u}{\partial t}\mathrm{d}S=:I_1(t)+I_2(t).\end{aligned}\tag{8.175}$$

The integral I_1 can be estimated as $\|\frac{\partial}{\partial t}g\|_{L^2(\Omega)}(\frac{1}{4}+\|\frac{\partial}{\partial t}u\|^2_{L^2(\Omega)})$, cf. (8.84). It needs $\frac{\partial g}{\partial t}\in L^1(I;L^2(\Omega))$. Alternatively, we can estimate I_1 integrated over $[0,t]$ as

$$\begin{aligned}\int_0^t I_1(t)\,\mathrm{d}t=\int_0^t\!\!\int_\Omega\frac{\partial g}{\partial t}\frac{\partial u}{\partial t}\,\mathrm{d}x\mathrm{d}t=\int_\Omega\frac{\partial g}{\partial t}(t,x)u(t,x)\,\mathrm{d}x\\-\int_0^t\!\!\int_\Omega\frac{\partial^2 g}{\partial\vartheta^2}(\vartheta,x)\,u(\vartheta,x)\,\mathrm{d}x\mathrm{d}\vartheta-\int_\Omega\frac{\partial g}{\partial t}(0,x)u_0(x)\,\mathrm{d}x\end{aligned}\tag{8.176}$$

which is bounded if $g\in W^{2,1}(I;L^{p^{*\prime}}(\Omega))$, when the estimate of u in $L^\infty(I;W^{1,p}(\Omega))$ obtained already at Step (3) is employed. Similarly, the integral $\int_0^t I_2(t)\,\mathrm{d}t$ is to be treated by

$$\begin{aligned}\int_0^t I_2(t)\,\mathrm{d}t=\int_0^t\!\!\int_\Gamma\frac{\partial h}{\partial t}\frac{\partial u}{\partial t}\,\mathrm{d}S\mathrm{d}t=\int_\Gamma\frac{\partial h}{\partial t}(t,x)u(t,x)\,\mathrm{d}S\\-\int_0^t\!\!\int_\Gamma\frac{\partial^2 h}{\partial\vartheta^2}(\vartheta,x)\,u(\vartheta,x)\,\mathrm{d}S\mathrm{d}\vartheta-\int_\Gamma\frac{\partial h}{\partial t}(0,x)u_0(x)\,\mathrm{d}S\end{aligned}\tag{8.177}$$

which is bounded if $h\in W^{2,1}(I;L^{p^{\#\prime}}(\Gamma))$. Then, usage of the Gronwall inequality requires $g\in W^{1,2}(I;L^2(\Omega))$, and $u_0\in W^{2,q}(\Omega)\cap L^{2(q_1-1)}(\Omega)$ with $q\geq 2p^*/(p^*-2p+4)$,[53] and it gives the estimate u in $W^{1,\infty}(I;L^2(\Omega))$ and of $|\nabla u|^{p/2}$ in $W^{1,2}(I;L^2(\Omega))\subset L^\infty(I;L^2(\Omega))$, which yields $u\in L^\infty(I;W^{1,p}(\Omega))$.

If $p\geq 2$, the term $\frac{4p-4}{p^2}(\frac{\partial|\nabla u|^{p/2}}{\partial t})^2=(p-1)|\nabla u|^{p-2}|\frac{\partial}{\partial t}\nabla u|^2$ in (8.174) gives, through (1.46), an estimate of ∇u in the fractional space $L^p(\Omega;W^{2/p-\epsilon,p}(I;\mathbb{R}^n))$ $\cong W^{2/p-\epsilon,p}(I;L^p(\Omega;\mathbb{R}^n))$.

[52] Note that (8.174) then allows for a modification $\frac{\partial}{\partial t}(|u|^{q-2}u)\frac{\partial}{\partial t}u=(q-1)|u|^{q-2}(\frac{\partial}{\partial t}u)^2\geq 0$ for both $q=q_1\geq 1$ and $q=q_2\geq 1$.

[53] This condition implies $\Delta_p u_0\in L^2(\Omega)$, cf. also (8.69b), because of the obvious estimate $\int_\Omega|\Delta_p v|^2\mathrm{d}x\leq(p-1)^2\int_\Omega|\nabla v|^{2p-4}|\nabla^2 v|^2\mathrm{d}x\leq(p-1)^2\|\nabla v\|^{2p-4}_{L^{(2p-4)q'}(\Omega;\mathbb{R}^n)}\|\nabla^2 v\|^2_{L^{2q}(\Omega;\mathbb{R}^{n\times n})}\leq N\|v\|^{2(p-1)}_{W^{2,q}(\Omega)}$, cf. (2.139).

For $p = 2$, the term I_1 can be estimated more finely as

$$\begin{aligned} I_1(t) &\le \left\|\frac{\partial g}{\partial t}\right\|_{L^{2^*\prime}(\Omega)} \left\|\frac{\partial u}{\partial t}\right\|_{L^{2^*}(\Omega)} \le \frac{1}{4\varepsilon}\left\|\frac{\partial g}{\partial t}\right\|^2_{L^{2^*\prime}(\Omega)} + \varepsilon\left\|\frac{\partial u}{\partial t}\right\|^2_{L^{2^*}(\Omega)} \\ &\le \frac{1}{4\varepsilon}\left\|\frac{\partial g}{\partial t}\right\|^2_{L^{2^*\prime}(\Omega)} + N_1^2\varepsilon\left\|\frac{\partial u}{\partial t}\right\|^2_{L^2(\Omega)} + N_1^2\varepsilon\left\|\nabla\frac{\partial u}{\partial t}\right\|^2_{L^2(\Omega;\mathbb{R}^n)} \end{aligned} \tag{8.178}$$

where N_1 is the norm of the embedding $W^{1,2}(\Omega) \subset L^{2^*}(\Omega)$. Similarly, I_2 bears the estimate

$$\begin{aligned} I_2(t) &\le \left\|\frac{\partial h}{\partial t}\right\|_{L^{2^\#\prime}(\Gamma)} \left\|\frac{\partial u}{\partial t}\right\|_{L^{2^\#}(\Gamma)} \le \frac{1}{4\varepsilon}\left\|\frac{\partial h}{\partial t}\right\|^2_{L^{2^\#\prime}(\Gamma)} + \varepsilon\left\|\frac{\partial u}{\partial t}\right\|^2_{L^{2^\#}(\Gamma)} \\ &\le \frac{1}{4\varepsilon}\left\|\frac{\partial h}{\partial t}\right\|^2_{L^{2^\#\prime}(\Gamma)} + N_2^2\varepsilon\left\|\frac{\partial u}{\partial t}\right\|^2_{L^2(\Omega)} + N_2^2\varepsilon\left\|\nabla\frac{\partial u}{\partial t}\right\|^2_{L^2(\Omega;\mathbb{R}^n)} \end{aligned} \tag{8.179}$$

where N_2 is the norm of the trace operator $W^{1,2}(\Omega) \to L^{2^\#}(\Gamma)$. Then we take $\varepsilon > 0$ small, namely $(N_1^2 + N_2^2)\varepsilon < 1$, so that the last terms in (8.178)–(8.179) can be absorbed in the corresponding term arising on the left-hand side of (8.175); note that (8.174) equals $|\frac{\partial}{\partial t}\nabla u|^2$ if $p = 2$. Then we use Gronwall's inequality to handle the last-but-one terms in (8.178)–(8.179). Cf. also (8.70)–(8.71). Like in (8.69a), it requires $g \in W^{1,2}(I; L^{2^{*\prime}}(\Omega))$ and $h \in W^{1,2}(I; L^{2^{\#\prime}}(\Gamma))$ only. Then $u \in W^{1,2}(I; W^{1,2}(\Omega))$.

Assuming also u_0 regular enough, namely $u_0 \in W^{2,2}(\Omega) \cap L^{2(q_1-1)}(\Omega)$, and $g(0) \in L^2(\Omega)$, we have $\frac{\partial}{\partial t}u(0) \in L^2(\Omega)$, and we can apply Gronwall's inequality to (8.175). We thus get the estimate for u in $W^{1,2}(I; W^{1,2}(\Omega)) \cap W^{1,\infty}(I; L^2(\Omega))$.

(5) If both $\frac{\partial}{\partial t}u$ and f are functions (not only distributions), $\mathrm{div}(|\nabla u|^{p-2}\nabla u)$ is more regular than $W^{1,p}(\Omega)^*$, so that by elliptic regularity theory we obtain a spatial regularity. E.g., if $p = 2$, we can use the interior $W^{2,2}$- or $W^{3,2}$-regularity as established in Proposition 2.103; this needs $1 < q_1 \le (2n-2)/(n-2)$ (and also $q_1 \ge 2$ in the latter case). In combination with Steps (3) and (4), one thus obtains the last three lines in Table 3. If also Ω would be qualified, we could use Proposition 2.104 to get regularity up to the boundary.

Exercise 8.64 (Large q_1 or q_2: an alternative setting)**.** For $q_1 > p^*$ or $q_2 > p^\#$, we can take the space $\mathscr{W} := \{v \in L^p(I; W^{1,p}(\Omega)) \cap L^{q_1}(Q);\ v|_\Sigma \in L^{q_2}(\Sigma)\}$ endowed naturally by the norm $\|v\|_{\mathscr{W}} := \|v\|_{L^p(I;W^{1,p}(\Omega))} + \|v\|_{L^{q_1}(Q)} + \|v|_\Sigma\|_{L^{q_2}(\Sigma)}$. Prove that $\mathscr{W}$ is a Banach space[54], and that the monotone mapping $\mathscr{A}$ related to (8.167) maps $\mathscr{W}$ into $\mathscr{W}^*$ and that $\frac{\partial}{\partial t}u \in \mathscr{W}^*$ if u is a weak solution to (8.167).[55]

[54]Hint: Consider a Cauchy sequence $\{u_k\}_{k\in\mathbb{N}}$ in $\mathscr{W}$. Realize that, in particular, it converges to u in $L^p(I; W^{1,p}(\Omega))$ and in $L^{q_1}(Q)$ as these spaces are complete, and also $u_k|_\Sigma \to u|_\Sigma$ in $L^p(I; L^{p^\#}(\Gamma))$, and, as $\{u_k|_\Sigma\}_{k\in\mathbb{N}}$ is a Cauchy sequence in $L^{q_2}(\Sigma)$, also $u_k|_\Sigma \to u_\Sigma$ in the complete space $L^{q_2}(\Sigma)$, and thus $u_\Sigma = u|_\Sigma$. Cf. also Exercise 2.72.

[55]Hint: Show $\mathscr{A} : \mathscr{W} \to \mathscr{W}^*$ just by using Hölder's inequality. Further realize that $\|\frac{\partial}{\partial t}u\|_{\mathscr{W}^*} = \sup_{\|v\|_{\mathscr{W}}\le 1} \int_Q |\nabla u|^{p-1}\nabla u\cdot\nabla v + |u|^{q_1-1}uv - gv\,\mathrm{d}x + \int_\Sigma |u|^{q_2-1}uv - hv\,\mathrm{d}S$ and estimate it by the Hölder inequality.

qualification of				quality of u
g	h	u_0	p	
$L^{p'}(I;L^{p^{*\prime}}(\Omega))$	$L^{p'}(I;L^{p^{\#\prime}}(\Gamma))$	$L^2(\Omega)$	>1	$L^p(I;W^{1,p}(\Omega))$ $L^\infty(I;L^2(\Omega))$ $W^{1,p'}(I;W^{1,p}(\Omega)^*)$
$W^{1,1}(I;L^{p^{*\prime}}(\Omega))$ $+\,L^2(Q)$	$W^{1,1}(I;L^{p^{\#\prime}}(\Gamma))$	$W^{1,p}(\Omega)$ $\cap\, L^2(\Omega)$	>1	$L^\infty(I;W^{1,p}(\Omega))$ $W^{1,2}(I;L^2(\Omega))$
$W^{2,1}(I;L^{p^{*\prime}}(\Omega))$ $+W^{1,1}(I;L^2(\Omega))$	$W^{2,1}(I;L^{p^{\#\prime}}(\Gamma))$	$W^{2,q}(\Omega)$ with $q\ge\frac{2p^*}{p^*-2p+4}$	>1	$W^{1,\infty}(I;L^2(\Omega))$ $L^\infty(I;W^{1,p}(\Omega))$
			≥ 2	$W^{2/p-\epsilon,p}(I;W^{1,p}(\Omega))$
$L^2(Q)$	$W^{1,1}(I;L^{2^{\#\prime}}(\Gamma))$	$W^{1,2}(\Omega)$	$=2$	$L^2(I;W^{2,2}_{\rm loc}(\Omega))$
$W^{1,1}(I;L^2(\Omega))$	$W^{2,1}(I;L^{2^{\#\prime}}(\Gamma))$	$W^{2,2}(\Omega)$	$=2$	$L^\infty(I;W^{2,2}_{\rm loc}(\Omega))$
$W^{1,2}(I;L^{2^{*\prime}}(\Omega))$ $\cap L^2(I;W^{1,2}_{\rm loc}(\Omega))$	$W^{1,2}(I;L^{2^{\#\prime}}(\Gamma))$	$W^{2,2}(\Omega)$	$=2$	$L^2(I;W^{3,2}_{\rm loc}(\Omega))$

Table 3. Summary of Example 8.63; qualification of q_1 and q_2 not displayed.

Example 8.65 (Nonmonotone term: a-priori estimates). Consider the initial-boundary-value problem with a nonmonotone term $|u|^\mu$ instead of $|u|^{q_1-2}u$:

$$\left.\begin{aligned}\frac{\partial u}{\partial t}-{\rm div}\big(|\nabla u|^{p-2}\nabla u\big)+|u|^\mu &= g &&\text{in } Q,\\ |\nabla u|^{p-2}\frac{\partial u}{\partial\nu}+|u|^{q_2-2}u &= h &&\text{on } \Sigma,\\ u(0,\cdot) &= u_0 &&\text{on } \Omega,\end{aligned}\right\}\tag{8.180}$$

where again $g\in L^{p'}(I;L^{p^{*\prime}}(\Omega))$ and $h\in L^{p'}(I;L^{p^{\#\prime}}(\Gamma))$. We will show the a-priori estimates again on an heuristical level only, and specify μ.

(1) The test by $u(t,\cdot)$ itself now gives:

$$\begin{aligned}\frac12\frac{\rm d}{{\rm d}t}\|u\|^2_{L^2(\Omega)}+\|\nabla u\|^p_{L^p(\Omega;\mathbb{R}^n)} &\le \int_\Omega |u|^{\mu+1}+gu\,{\rm d}x+\int_\Gamma hu\,{\rm d}S\\ &\le \|u\|^{\mu+1}_{L^{\mu+1}(\Omega)}+N\big(\|g\|_{L^{p^{*\prime}}(\Omega)}+\|h\|_{L^{p^{\#\prime}}(\Gamma)}\big)\|u\|_{W^{1,p}(\Omega)}\end{aligned}\tag{8.181}$$

where N is as in (8.168). If $\mu\le1$, we can estimate

$$\|u\|^{\mu+1}_{L^{\mu+1}(\Omega)}\le\big({\rm meas}_n(\Omega)\big)^{(\mu-1)/2}\|u\|^{\mu+1}_{L^2(\Omega)}\le\big({\rm meas}_n(\Omega)\big)^{(\mu-1)/2}\big(\|u\|^2_{L^2(\Omega)}+c\big)$$

with some $c>0$,[56] and then use directly the Gronwall inequality. For superlinearly growing nonlinearities, i.e. $\mu>1$, $\|u\|^{\mu+1}_{L^{\mu+1}(\Omega)}$ can be absorbed in the left-hand side

[56] Cf. Exercise 2.58 for the norm of the embedding $L^{\mu+1}(\Omega)\subset L^2(\Omega)$.

by using Young's inequality through the estimate

$$\|u\|_{L^{\mu+1}(\Omega)}^{\mu+1} \leq N^{\mu+1}\|u\|_{W^{1,p}(\Omega)}^{\mu+1} \leq N^{\mu+1}\big(\varepsilon\|u\|_{W^{1,p}(\Omega)}^{p} + N_\varepsilon\big) \tag{8.182}$$

where N is the norm of $W^{1,p}(\Omega) \subset L^{\mu+1}(\Omega)$ and the last inequality uses

$$\mu < p-1; \tag{8.183}$$

note that this implies also $\mu+1 \leq p^*$ used for the first inequality. This condition makes the approach effective only if $p > 2$ (otherwise the previous approach via the Gronwall inequality can be used, too). The other terms can be estimated in the same way as in (8.168). Again, this gives the a-priori estimate for u in $L^p(I; W^{1,p}(\Omega)) \cap L^\infty(I; L^2(\Omega))$.

(2) The estimate for $\frac{\partial}{\partial t}u$ can be made just the same way as (8.169) or (8.172) now with $\mu+1$ in place of q_1.

(3) The test function $v := \frac{\partial}{\partial t}u(t,\cdot)$ needs again $g \in L^2(Q)$. For simplicity, we take $h = 0$; otherwise, cf. Example 8.63(3). Then

$$\Big\|\frac{\partial u}{\partial t}\Big\|_{L^2(\Omega)}^2 + \frac{1}{p}\frac{\mathrm{d}}{\mathrm{d}t}\|\nabla u\|_{L^p(\Omega;\mathbb{R}^n)}^p \leq \int_\Omega |u|^\mu\Big|\frac{\partial u}{\partial t}\Big|\,\mathrm{d}x + \int_\Omega g\frac{\partial u}{\partial t}\,\mathrm{d}x =: I_1(t) + I_2(t). \tag{8.184}$$

We can estimate $I_1(t) \leq \varepsilon\|\frac{\partial u}{\partial t}\|_{L^2(\Omega)}^2 + C_\varepsilon\|u(t)\|_{L^{2\mu}(\Omega)}^{2\mu}$. Assuming $\mu \leq 1$, we can estimate simply

$$\|u(t)\|_{L^{2\mu}(\Omega)}^{2\mu} \leq C\big(1 + \|u(t)\|_{L^2(\Omega)}^2\big) \leq C\bigg(1 + 2\|u_0\|_{L^2(\Omega)}^2 + 2T\int_0^t \Big\|\frac{\partial u}{\partial t}\Big\|_{L^2(\Omega)}^2 \mathrm{d}t\bigg),$$

cf. (8.63), and then use Gronwall inequality for (8.184). Thus we get the estimate for u in $L^\infty(I; W^{1,p}(\Omega)) \cap W^{1,2}(I; L^2(\Omega))$.

For $2 < p \leq 2n/(n-2)$, we can afford a super-linear growth $1 < \mu < p-1$. Relying on (8.181) with (8.183), we have the estimate $u \in L^\infty(I; L^2(\Omega))$ at our disposal. Then we can use the interpolation of $L^{2\mu}(\Omega)$ between $L^2(\Omega)$ and $W^{1,p}(\Omega)$, i.e.,

$$\|u\|_{L^{2\mu}(\Omega)} \leq C\|u\|_{L^2(\Omega)}^{1-\lambda}\|u\|_{W^{1,p}(\Omega)}^{\lambda} \tag{8.185}$$

provided

$$\frac{1}{2\mu} \geq \frac{1-\lambda}{2} + \frac{\lambda(n-p)}{np}; \tag{8.186}$$

cf. (1.39). Of course, we need $0 \leq \lambda \leq 1$. Then, we can estimate $\|u(t,\cdot)\|_{L^{2\mu}(\Omega)}^{2\mu} \leq C^{2\mu}\|u(t,\cdot)\|_{W^{1,p}(\Omega)}^{2\lambda\mu}\|u(t,\cdot)\|_{L^2(\Omega)}^{2(1-\lambda)\mu}$ and, assuming still

$$2\lambda\mu \leq p, \tag{8.187}$$

to treat it by the Gronwall inequality with help of the fact that $\|u(t,\cdot)\|_{L^2(\Omega)}$ is a-priori bounded uniformly for $t \in I$. Thus we get again the estimate for u in

$L^\infty(I;W^{1,p}(\Omega)) \cap W^{1,2}(I;L^2(\Omega))$. It is important that, for $p \le 2n/(n-2)$, the inequalities (8.186) and (8.187) are mutually consistent; we can just take $\lambda = p/(2\mu)$ which satisfies $\lambda \le 1$ if $\mu \ge p/2$, while for $1 < \mu < p/2$ we can first estimate $\|u\|_{L^{2\mu}(\Omega)}^{2\mu} \le C(1 + \|u\|_{L^{2\mu}(\Omega)}^p)$ and then follow as if $\mu = p/2$.

For $p > 2n/(n-2)$, we must slightly reduce the growth because the inequalities (8.186) and (8.187) represent a certain restriction on μ, namely $\mu \le np/(np - \lambda(np+2p-2n))$ and $\mu \le p/(2\lambda)$, respectively. The optimal choice of λ is to make these bounds equal to each other, which gives $\lambda = np/(np+2p)$ and thus

$$\mu \le p\frac{n+2}{2n}. \tag{8.188}$$

(4) Further, we apply $\frac{\partial}{\partial t}$ to the equation and then use the test function $v = \frac{\partial}{\partial t}u$. As in Example 8.63, we consider $p \ge 2$ and now $\mu = 1$; this is a model case for arbitrary Lipschitz nonlinearities that could be treated by a modification of this estimate. Then the term $|u|^\mu = |u|$ can be estimated by

$$\int_\Omega \Big(\frac{\partial}{\partial t}|u|\Big)\frac{\partial u}{\partial t}\,dx = \int_\Omega \mathrm{dir}(u)\Big|\frac{\partial u}{\partial t}\Big|^2\,dx \le \Big\|\frac{\partial u}{\partial t}\Big\|_{L^2(\Omega)}^2 \tag{8.189}$$

and then treated by the Gronwall inequality. The rest can be treated as in Example 8.63(4).

Exercise 8.66 (Limit passage)**.** Suppose u_k is the Galerkin solution for (8.180). Make the limit passage via Minty's trick by using only the basic a-priori estimates and monotonicity of the p-Laplacean.[57] Alternatively, use d-monotonicity of the p-Laplacean and prove convergence directly without the Minty trick.[58]

Exercise 8.67 (Weaker estimate for $\frac{\partial}{\partial t}u$)**.** Consider the problem (8.167) and derive the estimate of $\frac{\partial}{\partial t}u$ in $\mathscr{M}(I;W^{-k,2}(\Omega))$ for $k \in \mathbb{N}$ so large that $W_0^{k,2}(\Omega) \subset L^\infty(\Omega)$, i.e. $k > p/n$; this weaker estimate allows for bigger q_1 and still suffices for using Aubin-Lions' lemma as Corollary 7.9.[59]

Remark 8.68 (*Regularized p-Laplacean*)**.** For $p > 2$, one can consider the parabolic problem with a regularized p-Laplacean:

$$\frac{\partial u}{\partial t} - \mathrm{div}\Big(\big(\varepsilon + |\nabla u|^{p-2}\big)\nabla u\Big) = g, \quad u(0) = u_0, \quad u|_\Sigma = 0, \tag{8.190}$$

cf. (4.38). This allows us to use the estimate from Step (4) from Example 8.63 based on the uniform-like monotonicity (8.69c). Indeed, using also (8.174), we have

$$\frac{\partial}{\partial t}\Big(\big(\varepsilon + |\nabla u|^{p-2}\big)\nabla u\Big)\cdot\frac{\partial \nabla u}{\partial t} \ge \varepsilon\Big|\frac{\partial \nabla u}{\partial t}\Big|^2 + \frac{4p-4}{p^2}\Big(\frac{\partial}{\partial t}|\nabla u|^{p/2}\Big)^2 \ge \varepsilon\Big|\frac{\partial \nabla u}{\partial t}\Big|^2. \tag{8.191}$$

[57] Hint: Cf. Remark 8.32 modified by treating the non-monotone lower-order term $|u|^\mu$ by compactness as suggested in Exercise 8.59.

[58] Hint: Cf. Exercise 8.81 below.

[59] Hint: Modify Example 8.63(2) by considering $v \in C(I;W_0^{k,2}(\Omega))$ in (8.169).

Exercise 8.69. Consider again the regularized problem (8.190). Denoting u_ε its solution, prove the a-priori estimates $\|u_\varepsilon\|_{L^p(I;W^{1,p}(\Omega))} \le C$, $\|u_\varepsilon\|_{L^2(I;W^{1,2}(\Omega))} \le C/\sqrt{\varepsilon}$ and $\|\frac{\partial}{\partial t}u_\varepsilon\|_{L^{p'}(I;W^{1,p}(\Omega)^*)} \le C/\sqrt{\varepsilon}$, and then, passing $\varepsilon \to 0$, prove $u_\varepsilon \to u$ with u denoting the solution with $\varepsilon = 0$.[60]

Example 8.70 (*Dirichlet boundary conditions*)**.** Let us illustrate the Dirichlet condition for a simple parabolic equation with the p-Laplacean, i.e.

$$\frac{\partial u}{\partial t} - \mathrm{div}\big(|\nabla u|^{p-2}\nabla u\big) = 0, \qquad u(0,\cdot) = u_0, \qquad u|_\Sigma = u_{\rm D}|_\Sigma, \tag{8.192}$$

with some $u_{\rm D} : \bar{Q} \to \mathbb{R}$ prescribed. By multiplying the equation in (8.192) by $v \in W_0^{1,p}(\Omega)$ and using Green's formula, one gets the weak formulation:

$$\forall (\text{a.a.})\, t \in I\ \forall v \in W_0^{1,p}(\Omega): \quad \Big\langle \frac{\partial u}{\partial t}, v\Big\rangle + \int_\Omega |\nabla u(t,x)|^{p-2}\nabla u(t,x)\cdot\nabla v(x)\,\mathrm{d}x = 0$$

completed, of course, by $u(0,\cdot) = u_0$ and $u|_\Sigma = u_{\rm D}|_\Sigma$.

(1) We cannot test it by $v = u(t,\cdot)$ if $u_{\rm D}(t,\cdot)|_\Sigma \neq 0$. Instead, the basic a-priori estimate is obtained by a test by $v = [u - u_{\rm D}](t,\cdot)$:[61]

$$\begin{aligned} \frac{1}{2}\frac{\mathrm{d}}{\mathrm{d}t}\big\|u - u_{\rm D}\big\|^2_{L^2(\Omega)} + \big\|\nabla u\big\|^p_{L^p(\Omega;\mathbb{R}^n)} &= \int_\Omega |\nabla u|^{p-2}\nabla u\cdot\nabla u_{\rm D} - \frac{\partial u_{\rm D}}{\partial t}(u - u_{\rm D})\,\mathrm{d}x \\ &\le \varepsilon\big\|\nabla u\big\|^p_{L^p(\Omega;\mathbb{R}^n)} + C_\varepsilon\big\|\nabla u_{\rm D}\big\|^p_{L^p(\Omega;\mathbb{R}^n)} + \Big\|\frac{\partial u_{\rm D}}{\partial t}\Big\|_{L^2(\Omega)}\Big(1 + \big\|u - u_{\rm D}\big\|^2_{L^2(\Omega)}\Big). \end{aligned} \tag{8.193}$$

If $u_{\rm D} \in W^{1,p,1}(I; W^{1,p}(\Omega), L^2(\Omega))$ and $u_0 \in L^2(\Omega)$, by Gronwall's inequality we obtain $u - u_{\rm D}$ bounded in $L^\infty(I; L^2(\Omega))$. As $u_{\rm D} \in L^\infty(I; L^2(\Omega))$ due to Lemma 7.1, also u itself bounded in $L^\infty(I; L^2(\Omega))$. Integrating still (8.193) over $[0,T]$ we get u bounded in $L^p(I; W^{1,p}(\Omega))$. Then, from the equation (8.192) itself, one gets $\frac{\partial}{\partial t}u$ bounded in $L^{p'}(I; W_0^{1,p}(\Omega)^*)$.

---

[60] Hint: Use Minty's trick:

$$\begin{aligned} 0 &\le \int_Q \big(|\nabla u_\varepsilon|^{p-2}\nabla u_\varepsilon - |\nabla v|^{p-2}\nabla v + \varepsilon\nabla(u_\varepsilon - v)\big)\cdot\nabla(u_\varepsilon - v)\,\mathrm{d}x\mathrm{d}t \\ &= \int_Q \Big(g - \frac{\partial u_\varepsilon}{\partial t}\Big)(u_\varepsilon - v) - |\nabla v|^{p-2}\nabla v\cdot\nabla(u_\varepsilon - v) - \varepsilon\nabla v\cdot\nabla(u_\varepsilon - v)\,\mathrm{d}x\mathrm{d}t \end{aligned}$$

and realize that $\int_Q \varepsilon\nabla v\cdot\nabla(u_\varepsilon - v)\,\mathrm{d}x\mathrm{d}t = \mathscr{O}(\sqrt{\varepsilon})$. Then put $v = u + \delta w$, and pass $\delta > 0$ to zero.

[61] Equally, one can apply the shift (2.61), i.e. here $A_0(t,u) = A(u + u_{\rm D}(t))$. Then, writing $\tilde{u}(t) := u - u_{\rm D}(t)$, the original equation $\frac{\mathrm{d}}{\mathrm{d}t}u + A(u) = f$ is equivalent to $\frac{\mathrm{d}}{\mathrm{d}t}\tilde{u} + A_0(t,\tilde{u}) = f_0 := f - \frac{\mathrm{d}}{\mathrm{d}t}u_{\rm D}$ and its test by $v := \tilde{u}(t,\cdot) \in W_0^{1,p}(\Omega)$ is precisely (8.193) in the special case $f = 0$.

(2) Further estimates can be obtained by testing by $v = \frac{\partial}{\partial t}[u - u_{\rm D}](t,\cdot)$:

$$\begin{aligned}\Big\|\frac{\partial u}{\partial t}\Big\|^2_{L^2(\Omega)} + \frac{1}{p}\frac{\rm d}{{\rm d}t}\|\nabla u\|^p_{L^p(\Omega;\mathbb{R}^n)} &= \int_\Omega |\nabla u|^{p-2}\nabla u\cdot\nabla\frac{\partial u_{\rm D}}{\partial t} + \frac{\partial u}{\partial t}\frac{\partial u_{\rm D}}{\partial t}\\ &\le \big\|\,|\nabla u|^{p-1}\big\|_{L^{p'}(\Omega)}\Big\|\,\Big|\nabla\frac{\partial u_{\rm D}}{\partial t}\Big|\,\Big\|_{L^p(\Omega)} + \Big\|\frac{\partial u}{\partial t}\Big\|_{L^2(\Omega)}\Big\|\frac{\partial u_{\rm D}}{\partial t}\Big\|_{L^2(\Omega)}\\ &\le \Big(1+\|\nabla u\|^p_{L^p(\Omega;\mathbb{R}^n)}\Big)\Big\|\nabla\frac{\partial u_{\rm D}}{\partial t}\Big\|_{L^p(\Omega;\mathbb{R}^n)} + \frac{1}{4}\Big\|\frac{\partial u}{\partial t}\Big\|^2_{L^2(\Omega)} + \Big\|\frac{\partial u_{\rm D}}{\partial t}\Big\|^2_{L^2(\Omega)}. \end{aligned}\tag{8.194}$$

Here one needs $\frac{\partial}{\partial t}u_{\rm D} \in L^2(Q)\cap L^1(I;W^{1,p}(\Omega))$ and $u_0 \in W^{1,p}(\Omega)$ to get u bounded in $L^\infty(I;W^{1,p}(\Omega))\cap W^{1,2}(I;L^2(\Omega))$.[62]

(3) Still further estimates can be obtained by differentiating (8.192) in time and by testing it by $v = \frac{\partial}{\partial t}[u - u_{\rm D}](t,\cdot)$. In view of (8.174), the term thus arising on the right-hand side can be estimated as

$$\begin{aligned}\Big|\frac{\partial}{\partial t}(|\nabla u|^{p-2}\nabla u)\cdot\nabla\frac{\partial u_{\rm D}}{\partial t}\Big| &= \Big|\,|\nabla u|^{p-2}\frac{\partial\nabla u}{\partial t}\cdot\frac{\partial\nabla u_{\rm D}}{\partial t}\\ &\quad + \Big((p{-}2)|\nabla u|^{p-4}\nabla u\cdot\frac{\partial\nabla u}{\partial t}\Big)\Big(\nabla u\cdot\frac{\partial\nabla u}{\partial t}\Big)\Big|\\ &\le (p{-}1)|\nabla u|^{p-2}\Big|\frac{\partial\nabla u}{\partial t}\Big|\,\Big|\frac{\partial\nabla u_{\rm D}}{\partial t}\Big|\\ &\le \frac{p{-}1}{2}|\nabla u|^{p-2}\Big|\frac{\partial\nabla u}{\partial t}\Big|^2 + \frac{p{-}1}{2}|\nabla u|^{p-2}\Big|\frac{\partial\nabla u_{\rm D}}{\partial t}\Big|^2\\ &=: I_1 + I_2.\end{aligned}\tag{8.195}$$

The term I_1 can be absorbed in term arising from (8.174) on the left-hand side which just equals $\frac{1}{2}I_1$, while, assuming $u_{\rm D} \in W^{1,p}(I;W^{1,p}(\Omega))$, the term I_2 bears the estimate

$$\int_Q I_2\,{\rm d}x{\rm d}t \le \frac{p{-}1}{2}\|\nabla u\|^{p-2}_{L^p(Q;\mathbb{R}^n)}\Big\|\frac{\partial\nabla u_{\rm D}}{\partial t}\Big\|^2_{L^p(Q;\mathbb{R}^n)}.\tag{8.196}$$

Altogether, like (8.175), one gets

$$\frac{1}{2}\frac{\rm d}{{\rm d}t}\Big\|\frac{\partial u}{\partial t}\Big\|^2_{L^2(\Omega)} + \frac{2p{-}2}{p^2}\Big\|\frac{\partial|\nabla u|^{p/2}}{\partial t}\Big\|^2_{L^2(\Omega)} \le \frac{p{-}1}{2}\|\nabla u\|^{p-2}_{L^p(Q;\mathbb{R}^n)}\Big\|\frac{\partial\nabla u_{\rm D}}{\partial t}\Big\|^2_{L^p(Q;\mathbb{R}^n)}.$$

Integrating it on $[0,t]$ for a general $t \in I$, one obtains $\frac{\partial}{\partial t}u \in L^\infty(I;L^2(\Omega))$ and $\frac{\partial}{\partial t}|\nabla u|^{p/2} \in L^2(Q)$ if also $\frac{\partial}{\partial t}u(0) = {\rm div}(|\nabla u_0|^{p-2}\nabla u_0) \in L^2(\Omega)$; this last qualification is satisfied if $u_0 \in W^{2,q}(\Omega)$ with $q \ge 2p^*/(p^*{-}2p{+}4)$ as in Example 8.63(4).

[62] In view of Theorem 8.16(iii), one may think that $u(t){-}u_{\rm D}(t) \in W^{1,p}_0(\Omega)$ so that, in particular, $u_0|_\Gamma = u_{\rm D}$ on Γ. Yet, here we should rather consider the $W^{1,1}$-structure of the loading $u_{\rm D}$ (like in Remark 8.23) to be extended for a BV-loading that does not need to be continuous at $t = 0$ so that even $u_0|_\Gamma \ne u_{\rm D}$ on Γ gives the expected a-priori estimate.

Note that, in particular, u_D and u_0 qualify for using Step (1), hence certainly $\nabla u \in L^p(Q;\mathbb{R}^n)$, so that the right-hand side of (8.196) is indeed finite. As in Example 8.63(4), we also get $\nabla u \in W^{2/p-\epsilon,p}(I;L^p(\Omega;\mathbb{R}^n))$ if $p \geq 2$.

qualification of		quality of u
u_D	u_0	
$W^{1,p,1}(I;W^{1,p}(\Omega),L^2(\Omega))$	$L^2(\Omega)$	$W^{1,p,p'}(I;W^{1,p}(\Omega),W^{1,p}(\Omega)^*)$ $L^\infty(I;L^2(\Omega))$
$W^{1,2}(I;L^2(\Omega))\cap W^{1,1}(I;W^{1,p}(\Omega))$	$W^{1,p}(\Omega)$	$L^\infty(I;W^{1,p}(\Omega))$ $W^{1,2}(I;L^2(\Omega))$
$W^{1,p}(I;W^{1,p}(\Omega))$	$W^{2,q}(\Omega)$ $q\geq\frac{2p^*}{p^*-2p+4}$	$W^{1,\infty}(I;L^2(\Omega))$ $p\geq 2:\ W^{2/p-\epsilon,p}(I;W^{1,p}(\Omega))$

Table 4. Summary of Example 8.70.

8.8.3 Semilinear heat equation $\mathfrak{c}(u)\frac{\partial}{\partial t}u - \mathrm{div}(\kappa(u)\nabla u) = g$

In this subsection, we will scrutinize the heat transfer in nonlinear but homogeneous isotropic media, described by the *heat equation* for the unknown θ (instead of u) which has the interpretation as temperature:

$$\mathfrak{c}(\theta)\frac{\partial\theta}{\partial t} - \mathrm{div}(\kappa(\theta)\nabla\theta) = g \tag{8.197}$$

where

$\theta : Q \to \mathbb{R}$ is the unknown temperature,
$\kappa : \mathbb{R} \to \mathbb{R}^+$ is the heat conductivity,
$\mathfrak{c} : \mathbb{R} \to \mathbb{R}^+$ is the heat capacity,
g the volume heat sources,

cf. Example 2.71 for a steady-state variant.

Example 8.71 (Enthalpy transformation). Powerful tools for nonlinear differential equations are various transformations of independent variables. Here we can apply, besides the *Kirchhoff transformation*, also the *enthalpy transformation*:

$$\widehat{\mathfrak{c}}(r) := \int_0^r \mathfrak{c}(\varrho)\,\mathrm{d}\varrho \qquad \& \qquad \widehat{\kappa}(r) := \int_0^r \kappa(\varrho)\,\mathrm{d}\varrho\,. \tag{8.198}$$

Putting $u := \widehat{\mathfrak{c}}(\theta)$ (u is called enthalpy) and denoting $\beta(u) := [\widehat{\kappa}\circ\widehat{\mathfrak{c}}^{-1}](u)$, we have $\frac{\partial}{\partial t}u = [\widehat{\mathfrak{c}}]'(\theta)\frac{\partial}{\partial t}\theta = \mathfrak{c}(\theta)\frac{\partial}{\partial t}\theta$ and $\Delta(\beta(u)) = \mathrm{div}(\widehat{\kappa}'(\widehat{\mathfrak{c}}^{-1}(u))\nabla\widehat{\mathfrak{c}}^{-1}(u)) = \mathrm{div}(\widehat{\kappa}'(\theta)\nabla\theta) = \mathrm{div}(\kappa(\theta)\nabla\theta)$. This transforms the original equation (8.197) to $\frac{\partial}{\partial t}u - \Delta\beta(u) = g$. Considering the initial condition in terms of the enthalpy u_0 and the boundary condition as in Example 2.71, i.e. $\kappa(\theta)\frac{\partial}{\partial\nu}\theta = b_1(\theta_\mathrm{e}-\theta)+b_2(\theta_\mathrm{e}-|\theta|^3\theta)$ with θ_e an external temperature, we come to the following initial-boundary-value problem:

$$\left.\begin{aligned} \frac{\partial u}{\partial t} - \Delta\beta(u) &= g && \text{in } Q,\\ \frac{\partial\beta(u)}{\partial\nu} + \Big(b_1 + b_2|\widehat{\kappa}^{-1}(u)|^3\Big)\widehat{\kappa}^{-1}(u) &= h && \text{on } \Sigma,\\ u(0,\cdot) &= u_0 && \text{on } \Omega. \end{aligned}\right\} \tag{8.199}$$

Exercise 8.72 (Pseudomonotone approach). The nonlinear operator $-\Delta\beta(u) = -\mathrm{div}\big(\beta'(u)\nabla u\big)$ can be considered as pseudomonotone and treated by Propositions 8.38 and 8.40. Assuming $\beta\in C^1(\mathbb{R})$, verify (8.141), (8.142), and (8.143) in this special case.[63] Realize, in particular, the condition $0 < \inf\beta'(\mathbb{R}) \le \sup\beta'(\mathbb{R}) < +\infty$.

Exercise 8.73 (Weak-continuity approach). Realizing that the operator $-\mathrm{div}\big(\beta'(u)\nabla u\big)$ is semilinear in the sense (8.161), one can use Proposition 8.47. Assume, besides $\beta\in C^1(\mathbb{R})$, the growth restriction

$$0 < \varepsilon \le \beta'(r) \le C\big(1 + |r|^{(2^{\circledast}-\epsilon)/2}\big) \tag{8.200}$$

for some $\epsilon > 0$ and $2^{\circledast} = n + 4/n$, cf. (8.131). Verify (8.12) for $q = \infty$ and $Z = W^{1,\infty}(\Omega)$, and also (8.161), (8.162) and (8.163).[64]

Assuming $g \in L^2(I;L^{2^{*\prime}}(\Omega)) + L^1(I;L^2(\Omega))$, $h \in L^2(I;L^{2^{\#\prime}}(\Gamma))$ and $u_0 \in L^2(\Omega)$, prove the a-priori estimates of u in $L^2(I;W^{1,2}(\Omega))\cap L^\infty(I;L^2(\Omega))$, and of $\frac{\partial}{\partial t}u$ in $L^1(I;W^{1,\infty}(\Omega)^*)$, and the convergence of approximate solutions to a *very weak solution*.[65] Realize, in particular, that now the heat conductivity $\kappa(\cdot)$ need not be bounded, e.g. if $n = 3$, then $\kappa(r) = 1 + |r|^{q_1}$ with $q_1 < 5/3$ is admitted if $\mathfrak{c}(\cdot) \ge \varepsilon > 0$. Also note that $\frac{\partial}{\partial t}u$, living in $L^1(I;W^{1,\infty}(\Omega)^*)$ in general, is not in duality with u, hence the concept of the very weak solution is indeed essential

---

[63] Hint: Realize that here $a(t,x,r,s) = \beta'(r)s$, $b(t,x,r) = \big(b_1(t,x)+b_2(t,x)|\widehat{\kappa}^{-1}(r)|^3\big)\widehat{\kappa}^{-1}(r)$, and $c(t,x,r,s) = 0$. Then (8.141) needs $\beta' > 0$, but (8.142)–(8.143) needs $p = 2$ and β' bounded and away from zero.

[64] Hint: Obviously $a(t,x,r,s) = \beta'(r)s$ is of the form (8.161a) and then realize that the upper bound in (8.200) is just (8.162a) while (8.163) needs just the lower bound in (8.200). As to (8.12), use (8.200) and the interpolation between $L^2(\Omega)$ and $L^{2^*}(\Omega)$ with λ from (8.145) to estimate

$$\begin{aligned} &\sup_{\|v\|_{W^{1,\infty}(\Omega)}\le 1}\int_\Omega \beta'(u)\nabla u\cdot\nabla v\,\mathrm{d}x \le \|\beta'(u)\|_{L^2(\Omega)}\|\nabla u\|_{L^2(\Omega;\mathbb{R}^n)}\\ &\quad\le \big\|1+|u|^{2^{\circledast}/2}\big\|_{L^2(\Omega)}\|\nabla u\|_{L^2(\Omega;\mathbb{R}^n)} \le C\Big(\mathrm{meas}_n(\Omega)^{1/2} + \|u\|_{L^{2^{\circledast}}(\Omega)}^{2^{\circledast}/2}\Big)\|\nabla u\|_{L^2(\Omega;\mathbb{R}^n)}\\ &\quad\le C\Big(\mathrm{meas}_n(\Omega)^{1/2} + \|u\|_{L^2(\Omega)}^{\lambda 2^{\circledast}/2}\|u\|_{L^{2^*}(\Omega)}^{(1-\lambda)2^{p^*}/2}\Big)\|\nabla u\|_{L^2(\Omega;\mathbb{R}^n)}. \end{aligned}$$

[65] Hint: Considering e.g. Galerkin approximate solutions u_k, by Aubin-Lions Lemma 7.7, $u_k \to u$ holds in $L^{2^{\circledast}-\epsilon}(Q)$. By (8.200), we have $\beta'(u_k) \to \beta'(u)$ in $L^2(Q)$. Then make the limit passage $\int_\Omega \nabla\beta(u_k)\cdot\nabla v\,\mathrm{d}x = \int_\Omega \beta'(u_k)\nabla u_k\cdot\nabla v\,\mathrm{d}x \to \int_\Omega \beta'(u)\nabla u\cdot\nabla v\,\mathrm{d}x = \int_\Omega \nabla\beta(u)\cdot\nabla v\,\mathrm{d}x$.

if $\beta'(\cdot)$ is not bounded. Another occurrence of this effect is under an advection driven by a velocity field which is not regular enough, see Lemma 12.14.

Exercise 8.74 (*Semi-implicit time discretization*). Consider the linearization of the nonlinear "heat-transfer" operator related to (8.199), namely $\big[B(w,u)\big](v) := \int_\Omega \beta'(w)\nabla u\cdot\nabla v\,\mathrm{d}x + \int_\Gamma \big(b_1 u + b_2|w|^3 u\big)v\,\mathrm{d}S$, and then the semi-implicit formula (8.55), which leads to

$$\Big\langle\frac{\partial u_\tau}{\partial t}, v\Big\rangle + \int_\Omega \beta'(\bar u_\tau^{\mathrm{R}})\nabla\bar u_\tau{\cdot}\nabla v - \bar g_\tau v\,\mathrm{d}x = \int_\Gamma \big(b_1\bar u_\tau + b_2|\bar u_\tau^{\mathrm{R}}|^3\bar u_\tau\big)v\,\mathrm{d}S \tag{8.201}$$

for a.a. $t{\in}I$ and all $v{=}W^{1,2}(\Omega)$ with the 'retarded' Rothe function $\bar u_\tau^{\mathrm{R}}$ defined by

$$\bar u_\tau^{\mathrm{R}}(t,\cdot) := \begin{cases} \bar u_\tau(t-\tau,\cdot) & \text{for } t\in[\tau,T],\\ u_\tau(0,\cdot) & \text{for } t\in[0,\tau].\end{cases} \tag{8.202}$$

Assuming (8.200), make a basic a-priori estimate by a test by u_τ^k and prove the convergence for $\tau\to 0$.[66]

Example 8.75 (Heat equation with *advection*). The heat transfer in a medium moving with a prescribed velocity field $\vec v: Q\to\mathbb{R}^n$ is governed by the equation

$$\mathfrak{c}(\theta)\Big(\frac{\partial\theta}{\partial t} + \vec v\cdot\nabla\theta\Big) - \mathrm{div}\big(\kappa(\theta)\nabla\theta\big) = g. \tag{8.203}$$

In the enthalpy formulation from Example 8.71 it reads as $\frac{\partial}{\partial t}u{+}\vec v{\cdot}\nabla u{+}\Delta\beta(u) = g$. Assuming $\mathrm{div}\,\vec v\le 0$ and $(\vec v|_\Sigma)\cdot\nu\ge 0$ as in Exercise 2.91 and using (6.33), the mapping $A(t,u): W^{1,2}(\Omega)\to W^{1,\infty}(\Omega)^*$ defined by $\langle A(t,u),z\rangle := \int_\Omega \beta'(u)\nabla u\cdot\nabla z + (\vec v(t,\cdot)\cdot\nabla u)\,z\,\mathrm{d}x$ can be shown semi-coercive if β' satisfies (8.200):

$$\big\langle A(t,u),u\big\rangle = \int_\Omega \beta'(u)|\nabla u|^2 + \big(\vec v(t,\cdot)\cdot\nabla u\big)\,u\,\mathrm{d}x \ge \varepsilon\|\nabla u\|^2_{L^2(\Omega;\mathbb{R}^n)}. \tag{8.204}$$

In case $\mathrm{div}\,\vec v = 0$ and $(\vec v|_\Sigma){\cdot}\nu = 0$, the scalar variant of (6.35) yields $\int_\Omega(\vec v(t,\cdot)\cdot\nabla u)\,z\,\mathrm{d}x = -\int_\Omega(\vec v(t,\cdot)\cdot\nabla z)\,u\,\mathrm{d}x \le \|\vec v(t,\cdot)\|_{L^2(\Omega;\mathbb{R}^n)}\|\nabla z\|_{L^\infty(\Omega;\mathbb{R}^n)}\|u\|_{L^2(\Omega)}$ and, expanding Exercise 8.73, we can rely on the concept of a very weak solution $u\in W^{1,2,1}(I;W^{1,2}(\Omega),W^{1,\infty}(\Omega)^*)$ if $\vec v\in L^1(I;W^{1,2n/(n+2)}_{0,\mathrm{div}}(\Omega;\mathbb{R}^n))$. If $\beta'(\cdot)$ is bounded and $\vec v{\in}L^\infty(I;L^{2^*2/(2^*2-2^*-2)}(\Omega))$, by the estimate $\int_\Omega(\vec v(t,\cdot){\cdot}\nabla u)\,z\,\mathrm{d}x \le \|\vec v(t,\cdot)\|_{L^{2^*2/(2^*2-2^*-2)}(\Omega)}\|\nabla u\|_{L^2(\Omega;\mathbb{R}^n)}\|z\|_{L^{2^*}(\Omega)}$, we can rely on the conventional concept of a weak solution $u\in W^{1,2,2}(I;W^{1,2}(\Omega),W^{1,2}(\Omega)^*)$ as in Exercise 8.72.

8.8.4 Navier-Stokes equation $\frac{\partial}{\partial t}u{+}(u{\cdot}\nabla)u{-}\Delta u{+}\nabla\pi{=}g,\ \mathrm{div}\,u{=}0$

Another important example of a semilinear equation, or rather a system of equations, is the evolution version of the *Navier-Stokes equation*, cf. Remark 6.15,

$$\frac{\partial u}{\partial t} + (u{\cdot}\nabla)\,u - \Delta u + \nabla\pi = g, \quad \mathrm{div}\,u = 0, \quad u(0,\cdot) = u_0, \tag{8.205}$$

[66] Hint: Cf. Exercise 8.92 below.

for the velocity field $u : Q \to \mathbb{R}^n$ and the pressure $\pi : Q \to \mathbb{R}$ which is, in fact, a multiplicator to the constraint $\operatorname{div} u = 0$. This system is a model for a flow of a viscous incompressible fluid; the viscosity and the mass density is put equal to 1 in (8.205). Considering zero Dirichlet boundary conditions, we pose the problem into function spaces by putting $V := W^{1,2}_{0,\mathrm{div}}(\Omega;\mathbb{R}^n) = \{v \in W^{1,2}_0(\Omega;\mathbb{R}^n);\ \operatorname{div} u = 0\}$, cf. (6.29), endowed with the norm $\|v\|_V := \|\nabla v\|_{L^2(\Omega;\mathbb{R}^{n\times n})}$ and, to ensure $V \subset H$ densely, we define H as a closure of V in $L^2(\Omega;\mathbb{R}^n)$ in the L^2-norm. Then, in accord with Definition 8.1 (and Table 2. on p.215), $u \in W^{1,2,2}(I;V,V^*)$ is considered as a weak solution to (8.205) if $u(0,\cdot) = u_0$ and if

$$\Big\langle \frac{\partial u}{\partial t}, v\Big\rangle_{V^*\times V} + \int_\Omega (u{\cdot}\nabla)\, u \cdot v + \nabla u : \nabla v - gv \,\mathrm{d}x = 0 \tag{8.206}$$

for any $v \in V$ and for a.a. $t \in I$. Naturally, here the mapping $A : V \to V^*$ is defined by $\langle A(u), v\rangle := \int_\Omega (u{\cdot}\nabla)\, u \cdot v + \nabla u : \nabla v \,\mathrm{d}x$; this definition is correct for $n \le 3$.[67] Coercivity of A can be obtained by using (6.36): indeed it holds simply that $\langle A(v), v\rangle = \int_\Omega \nabla v : \nabla v + ((v\cdot\nabla)\, v)\cdot v \,\mathrm{d}x = \|\nabla v\|^2_{L^2(\Omega;\mathbb{R}^{n\times n})}$.

Example 8.76 (Pseudomonotone approach). For $n = 2$, we can estimate

$$\begin{aligned}\sup_{\|v\|_V\le 1} \Big| \int_\Omega ((u\cdot\nabla)\, u)\cdot v \,\mathrm{d}x\Big| &= \sup_{\|v\|_V\le 1} \int_\Omega ((u\cdot\nabla)\, v)\cdot u \,\mathrm{d}x \\ &\le \sup_{\|v\|_V\le 1} \big\|u\big\|^2_{L^4(\Omega;\mathbb{R}^n)}\big\|\nabla v\big\|_{L^2(\Omega;\mathbb{R}^{n\times n})} \le C^2_{\mathrm{GN}}\big\|u\big\|_{L^2(\Omega;\mathbb{R}^n)}\big\|\nabla u\big\|_{L^2(\Omega;\mathbb{R}^{n\times n})}\end{aligned} \tag{8.207}$$

where the Hölder inequality and the Gagliardo-Nirenberg inequality like (1.40) has been used. Hence A satisfies the growth condition (8.13) with $p = q = 2$ and $\mathfrak{C}(\zeta) = \max(1,\zeta)$. Hence we have guaranteed existence of a weak solution if $u_0 \in H$ and $g \in L^2(I; L^{2^{*\prime}}(\Omega;\mathbb{R}^n))$.

Example 8.77 (Weak-continuity approach). We consider the physically relevant case $n = 3$ (which covers $n = 2$ too), and will verify the condition (8.13) with $Z = V$. We have the embedding $W^{1,2}(\Omega) \subset L^6(\Omega)$; let us denote by N its norm, and estimate by the Hölder inequality and an interpolation:

$$\begin{aligned}\|A(u)\|_{V^*} &\le \sup_{\|\nabla v\|_{L^2(\Omega;\mathbb{R}^{n\times n})}\le 1} \int_\Omega \nabla u\cdot\nabla v + (u\cdot\nabla u)v \,\mathrm{d}x \\ &\le \sup_{\|\nabla v\|_{L^2(\Omega;\mathbb{R}^{n\times n})}\le 1} \big\|\nabla u\big\|_{L^2(\Omega;\mathbb{R}^{n\times n})}\big\|\nabla v\big\|_{L^2(\Omega;\mathbb{R}^{n\times n})} \\ &\qquad\qquad + \big\|u\big\|_{L^3(\Omega;\mathbb{R}^n)}\big\|\nabla u\big\|_{L^2(\Omega;\mathbb{R}^{n\times n})}\big\|v\big\|_{L^6(\Omega;\mathbb{R}^n)}\end{aligned}$$

[67] Since $4 > 2^*$ for $n \le 3$, this follows from the Hölder inequality $|\int_\Omega (u{\cdot}\nabla)\, u\cdot v \,\mathrm{d}x| \le \|u\|_{L^4(\Omega;\mathbb{R}^n)}\|\nabla u\|_{L^2(\Omega;\mathbb{R}^{n\times n})}\|v\|_{L^4(\Omega;\mathbb{R}^n)}$. In fact, using Gagliardo-Nirenberg inequality, the borderline case $n = 4$ can be covered, too.

$$\leq \|\nabla u\|_{L^2(\Omega;\mathbb{R}^{n\times n})} + N\|u\|_{L^6(\Omega;\mathbb{R}^n)}^{1/2}\|u\|_{L^2(\Omega;\mathbb{R}^n)}^{1/2}\|\nabla u\|_{L^2(\Omega;\mathbb{R}^{n\times n})}$$
$$\leq \|u\|_V + N^{3/2}\|u\|_H^{1/2}\|u\|_V^{3/2} \leq \max(1, N^{3/2}\|u\|_H^{1/2})(1+\|u\|_V^{3/2}).$$

Hencefore, we get (8.13) with $p = 2$, $q = 4$ and $\mathfrak{C}(r) = \max(1, N^{3/2}r^{1/2})$. As now $q' = 4/3$, the estimate (8.19a) yields $\frac{\partial}{\partial t}u \in L^{4/3}(I; V^*)$, which, however, is not in duality with $L^2(I; V) \ni u$ and the concept of the very weak solution and weak continuity is indeed urgent.[68]

Remark 8.78 (Uniqueness). We consider $n = 2$. Taking two weak solutions u_1 and u_2, subtracting the respective identities (8.206), testing it by $v = u_{12} := u_1 - u_2$, and using subsequently (6.36), the Hölder inequality, the Gagliardo-Nirenberg inequality like (1.40), one obtains

$$\frac{1}{2}\frac{\mathrm{d}}{\mathrm{d}t}\|u_{12}\|_{L^2(\Omega;\mathbb{R}^n)}^2 + \|\nabla u_{12}\|_{L^2(\Omega;\mathbb{R}^{n\times n})}^2 = \int_\Omega \big((u_1\cdot\nabla)u_1 - (u_2\cdot\nabla)u_2\big)\cdot u_{12}\mathrm{d}x$$
$$= \int_\Omega (u_{12}\cdot\nabla)u_2 \cdot u_{12} + (u_2\cdot\nabla)u_{12} \cdot u_{12}\,\mathrm{d}x$$
$$= \int_\Omega (u_{12}\cdot\nabla)u_2 \cdot u_{12}\,\mathrm{d}x \leq \|\nabla u_2\|_{L^2(\Omega;\mathbb{R}^{n\times n})}\|u_{12}\|_{L^4(\Omega;\mathbb{R}^n)}^2$$
$$\leq C_{\rm GN}^2\|\nabla u_2\|_{L^2(\Omega;\mathbb{R}^{n\times n})}\|u_{12}\|_{L^2(\Omega;\mathbb{R}^n)}\|\nabla u_{12}\|_{L^2(\Omega;\mathbb{R}^{n\times n})}$$
$$\leq \frac{1}{2}C_{\rm GN}^4\|\nabla u_2\|_{L^2(\Omega;\mathbb{R}^{n\times n})}^2\|u_{12}\|_{L^2(\Omega;\mathbb{R}^n)}^2 + \frac{1}{2}\|\nabla u_{12}\|_{L^2(\Omega;\mathbb{R}^{n\times n})}^2$$

with $C_{\rm GN}$ the constant from the Gagliardo-Nirenberg inequality $\|v\|_{L^4(\Omega)} \leq C_{\rm GN}\|v\|_{L^2(\Omega)}^{1/2}\|\nabla v\|_{L^2(\Omega;\mathbb{R}^n)}^{1/2}$. Then absorbing the last term in the left-hand side and using the Gronwall inequality when realizing that $t \mapsto \|\nabla u_2(t,\cdot)\|_{L^2(\Omega;\mathbb{R}^{n\times n})}^2 \in L^1(I)$ and $u_{12}(0,\cdot) = 0$, one obtains $u_{12}(t,\cdot) = 0$ for a.a. $t \in I$.

This technique does not work if $n = 3$ and surprisingly, in this physically relevant case, the uniqueness is a mysteriously difficult problem.[69]

Exercise 8.79. Derive the weak formulation (8.206).[70]

Exercise 8.80 (*Darcy-Brinkman system*). Modify the analysis of (8.205) in this section by adding another, lower-order dissipative term, namely

$$\frac{\partial u}{\partial t} + (u\cdot\nabla)\,u - \Delta u + u + \nabla\pi = g, \quad \mathrm{div}\,u = 0, \quad u(0,\cdot) = u_0; \tag{8.208}$$

[68] Note that the idea of putting $Z = V \cap W^{1,\infty}(\Omega;\mathbb{R}^n)$ would lead to $\frac{\partial}{\partial t}u \in L^2(I; Z^*)$ which is again not in duality with $L^2(I; V)$.

[69] This question, intimately related to regularity for (8.205), was identified by the Clay Mathematical Institute as one out of 7 most challenging mathematical "Millennium problems", and at the time of publishing even the 2nd edition of this book was still waiting for its (affirmative or not) answer, together with a $ 1 million award.

[70] Hint: Test (8.205) by $v \in V$, integrate it over Ω, and use Green's formula and the orthogonality $\int_\Omega(\nabla\pi)v\,\mathrm{d}x = -\int_\Omega \pi\,\mathrm{div}\,v\,\mathrm{d}x = 0$.

this additional term bears an interpretation as a reaction force to the flow through static porous media with a constant porosity.

8.8.5 Some more exercises

Exercise 8.81. Consider the parabolic problem:

$$\frac{\partial u}{\partial t} - \operatorname{div}\big(|\nabla u|^{p-2}\nabla u\big) + c(\nabla u) = g, \quad u(0) = u_0, \quad u|_\Sigma = 0, \tag{8.209}$$

with the continuous nonlinearity $c : \mathbb{R}^n \to \mathbb{R}$ having the growth restricted by

$$|c(s)| \le C\big(1 + |s|^{p-1-\delta}\big) \tag{8.210}$$

with some $\delta > 0$, and show the basic a-priori estimates of an approximate solution obtained by the Galerkin method if $u_0 \in L^2(\Omega)$ and $g \in L^{p'}(I; L^{p^{*\prime}}(\Omega))$.[71] Show the existence of a weak solution to (8.209) by convergence of Galerkin's solutions u_k by using the d-monotonicity of the p-Laplacean to prove first the strong convergence of ∇u_k, similarly as in Exercise 2.85.[72]

---

[71] Hint: Test the identity $\int_\Omega \big(\frac{\partial}{\partial t}u_k + c(\nabla u_k) - g\big)v + |\nabla u_k|^{p-2}\nabla u_k \cdot \nabla v \,\mathrm{d}x = 0$ by $v := u_k(t,\cdot)$:

$$\begin{aligned}\frac{1}{2}\frac{\mathrm{d}}{\mathrm{d}t}\|u_k\|^2_{L^2(\Omega)} + \|\nabla u_k\|^p_{L^p(\Omega)} &= \int_\Omega \big(g - c(\nabla u_k)\big)u_k \,\mathrm{d}x \\ &\le \varepsilon\|u\|^p_{L^{p^*}(\Omega)} + 2^{p'-1}C_\varepsilon\Big(\|c(\nabla u_k)\|^{p'}_{L^{p^{*\prime}}(\Omega)} + \|g\|^{p'}_{L^{p^{*\prime}}(\Omega)}\Big),\end{aligned}$$

from which the a-priori estimate of u_k in $L^\infty(I; L^2(\Omega)) \cap L^p(I; W^{1,p}(\Omega))$ follows by Gronwall's inequality and by using (8.210), so that

$$\|c(\nabla u_k)\|^{p'}_{L^{p^{*\prime}}(\Omega)} \le \int_\Omega C^{p'}\big(1 + |\nabla u_k|^{p-1-\delta}\big)^{p'} \mathrm{d}x \le C_{\varepsilon,\delta} + \varepsilon\|\nabla u_k\|^p_{L^p(\Omega;\mathbb{R}^n)}.$$

The dual estimate of $\frac{\partial}{\partial t}u_k$ in $L^{p'}(I; W^{1,p}_0(\Omega)^*_{\text{lcs}})$ can then be obtained standardly.

[72] Hint: Take a subsequence $u_k \rightharpoonup u$ in $W^{1,p,p'}(I; W^{1,p}_0(\Omega), W^{1,p}_0(\Omega)^*_{\text{lcs}})$. Use the norm $\|v\|_{W^{1,p}_0(\Omega)} := \|\nabla v\|_{L^p(\Omega;\mathbb{R}^n)}$ and, by (2.141), estimate

$$\begin{aligned}&\Big(\|u_k\|^{p-1}_{L^p(I;W^{1,p}_0(\Omega))} - \|v\|^{p-1}_{L^p(I;W^{1,p}_0(\Omega))}\Big)\Big(\|u_k\|_{L^p(I;W^{1,p}_0(\Omega))} - \|v\|_{L^p(I;W^{1,p}_0(\Omega))}\Big) \\ &\le \int_Q \big(|\nabla u_k|^{p-2}\nabla u_k - |\nabla v|^{p-2}\nabla v\big)\big(\nabla u_k - \nabla v\big)\,\mathrm{d}x\mathrm{d}t \\ &= \int_Q |\nabla u_k|^{p-2}\nabla u_k\cdot(\nabla u_k - \nabla v_k) + |\nabla u_k|^{p-2}\nabla u_k\cdot(\nabla v_k - \nabla v) - |\nabla v|^{p-2}\nabla v\cdot(\nabla u_k - \nabla v)\,\mathrm{d}x\mathrm{d}t \\ &= \int_Q \Big(g - c(\nabla u_k) - \frac{\partial u_k}{\partial t}\Big)(u_k - v_k) + |\nabla u_k|^{p-2}\nabla u_k\cdot(\nabla v_k - \nabla v) - |\nabla v|^{p-2}\nabla v\cdot(\nabla u_k - \nabla v)\,\mathrm{d}x\mathrm{d}t\end{aligned}$$

with $v_k(t,\cdot) \in V_k$. Assume $v_k \to v$ in $L^p(I; W^{1,p}(\Omega)))$. For $v = u$, $u_k - v_k \to u - u = 0$ in $L^p(Q)$ because of the compact embedding $W^{1,p}_0(\Omega) \Subset L^p(\Omega)$ and Aubin-Lions' Lemma 7.7, and then $\int_\Omega c(\nabla u_k)(u_k - v_k)\mathrm{d}x \to 0$ because $\{c(\nabla u_k)\}_{k\in\mathbb{N}}$ is bounded in $L^{p'}(Q)$ thanks to (8.210). Use

$$\limsup_{k\to\infty} \int_Q -\frac{\partial u_k}{\partial t}u_k \,\mathrm{d}x\mathrm{d}t = \limsup_{k\to\infty} \int_\Omega \frac{u_{0k}^2 - u_k(T)^2}{2}\mathrm{d}x \le \int_\Omega \frac{u_0^2 - u(T)^2}{2}\mathrm{d}x = \int_0^T -\Big\langle\frac{\partial u}{\partial t}, u\Big\rangle \mathrm{d}t$$

Exercise 8.82. Consider again the parabolic problem (8.209) with c satisfying

$$|c(s)| \le C\big(1+|s|^{p/2}\big) \tag{8.211}$$

and show existence of a weak solution to (8.209) if $p > 2n/(n+2)$, $u_0 \in W_0^{1,p}(\Omega)$ and $g \in L^2(Q)$ in a simpler way than in Exercise 8.81 by using the $L^2(Q)$-estimate on $\frac{\partial}{\partial t}u$ and convergence of Galerkin's approximations weakly in $W^{1,\infty,2}(I;W_0^{1,p}(\Omega),L^2(\Omega))$ and strongly in $L^p(I;W_0^{1,p}(\Omega))$.[73]

Exercise 8.83. Show how the coercivity works in the above concrete cases. Check the coercivity (8.95) or (8.60d).[74]

Exercise 8.84. Prove a-priori estimates and convergence of Galerkin approximants for the equation $\frac{\partial}{\partial t}u-\mathrm{div}(|\nabla u|^{p-2}\nabla u+|u|^{\mu}\nabla u) = g$, with $p > 2$ and some $\mu \ge 0$.[75]

Exercise 8.85 (*Singular perturbations* by a *biharmonic-term*). Consider

$$\frac{\partial u}{\partial t} - \mathrm{div}(|\nabla u|^{p-2}\nabla u) + \varepsilon\Delta^2 u = g, \qquad u(0,\cdot) = u_0, \quad u|_\Sigma = \frac{\partial u}{\partial \nu}\Big|_\Sigma = 0. \tag{8.212}$$

Denoting u_ε the weak solution to (8.212), execute the test by u_ε and prove a-priori

---

because $u_k(T) \rightharpoonup u(T)$ weakly in $L^2(\Omega)$, cf. (8.104). Push the other terms to zero, too. Conclude that $u_k \to u$ in $L^p(I;W_0^{1,p}(\Omega))$. Finally, pass to the limit directly in the Galerkin identity, which gives the integral identity $\int_0^T \big(\langle\frac{\partial}{\partial t}u,v\rangle + \int_\Omega |\nabla u|^{p-2}\nabla u\cdot\nabla v + c(\nabla u)v - gv\,\mathrm{d}x\big)\,\mathrm{d}t = 0$. Choosing $v(t,x) = \zeta(t)z(x)$ with $\zeta \in C_0(I)$ and z ranging a dense subset of $W^{1,p}(\Omega)$, prove the pointwise (for.a.a. t) equation of the type (8.126).

[73] Hint: A further test of the Galerkin identity by $\frac{\partial}{\partial t}u_k$ gives

$$\Big\|\frac{\partial u_k}{\partial t}\Big\|_{L^2(\Omega)}^2 + \frac{1}{p}\frac{\mathrm{d}}{\mathrm{d}t}\|\nabla u_k\|_{L^p(\Omega)}^p = \int_\Omega \big(g - c(\nabla u_k)\big)\frac{\partial u_k}{\partial t}\mathrm{d}x \le \|c(\nabla u_k) - g\|_{L^2(\Omega)}^2 + \Big\|\frac{\partial u_k}{\partial t}\Big\|_{L^2(\Omega)}^2,$$

from which the a-priori estimate of u_k in $W^{1,2}(I;L^2(\Omega)) \cap L^\infty(I;W^{1,p}(\Omega)))$ follows by Gronwall's inequality by using (8.211), so that $\|c(\nabla u_k)\|_{L^2(\Omega)}^2 \le 2C^2(\mathrm{meas}_n(\Omega) + \|\nabla u_k\|_{L^p(\Omega)}^p)$. The convergence of u_k to some u solving (8.209) can be made similarly as in Exercise 8.81 but we can make directly the limit passage $\lim_{k\to\infty}\int_Q(\frac{\partial}{\partial t}u_k)u_k\mathrm{d}x\mathrm{d}t = \int_Q(\frac{\partial}{\partial t}u)u\mathrm{d}x\mathrm{d}t$ because $\frac{\partial}{\partial t}u_k \rightharpoonup \frac{\partial}{\partial t}u$ weakly in $L^2(Q)$ and $u_k \rightharpoonup u$ weakly* in $W^{1,\infty,2}(I;W_0^{1,p}(\Omega),L^2(\Omega))$, hence by Aubin-Lions' Lemma 7.7 strongly in $L^2(Q)$ provided $p > 2n/(n+2)$ so that $W_0^{1,p}(\Omega) \Subset L^2(\Omega)$.

[74] Hint: Note that, e.g. for the case (8.180),

$$\langle A(v),v\rangle = \int_\Omega |\nabla v|^p + |v|^{q_1} + |v|^\mu v\,\mathrm{d}x + b\int_\Gamma |v|^{q_2}\mathrm{d}S \ge c\|v\|_{W^{1,p}(\Omega)}^q - \|v\|_{L^{\mu+1}(\Omega)}^{\mu+1} - C$$

where we used an equivalent norm on $W^{1,p}(\Omega)$. In particular, the semi-coercivity (8.95) holds for $\mu \le 1$ or $p > \mu+1$, $q_2 \ge p$ (but $q_2 \le p^{*\prime}$), $q_1 \ge 1$ (but $q_1 - 1 \le p^{\#\prime}$), and $b > 0$. Alternatively, $q_2 \ge 1$ is sufficient if $q_1 \ge p$. The weaker coercivity (8.60d) holds even for $q_2 > 1$ and $q_1 \ge 1$ or vice versa $q_2 \ge 1$ and $q_1 > 1$.

[75] Hint: Use the test by u_k to get $\{u_k\}_{k\in\mathbb{N}}$ bounded in $L^p(I;W^{1,p}(\Omega)) \cap L^\infty(I;L^2(\Omega))$, then estimate $\frac{\partial}{\partial t}u_k$, and show convergence, e.g., by Minty's trick for Δ_p combined with $\limsup_{k\to\infty}\int_Q \nabla u_k\cdot\nabla(u_k - v)\,\mathrm{d}x\mathrm{d}t \le \int_Q \nabla u\cdot\nabla(u-v)\,\mathrm{d}x\mathrm{d}t$.

estimates:

$$\big\|u_\varepsilon\big\|_{L^2(I;W_0^{2,2}(\Omega))} \le C/\sqrt{\varepsilon}, \qquad \big\|u_\varepsilon\big\|_{L^p(I;W_0^{1,p}(\Omega))\cap L^\infty(I;L^2(\Omega))} \le C, \tag{8.213a}$$

$$\Big\|\frac{\partial u_\varepsilon}{\partial t}\Big\|_{L^1(I;W^{-2,2}(\Omega)+W^{-1,p'}(\Omega))} \le C, \quad \Big\|\frac{\partial u_\varepsilon}{\partial t}+\varepsilon\Delta^2 u_\varepsilon\Big\|_{L^{p'}(I;W^{-1,p'}(\Omega))} \le C, \tag{8.213b}$$

assuming that $g \in L^{p'}(I;L^{p^{*\prime}}(\Omega))$ and $u_0 \in L^2(\Omega)$. Selecting a subsequence $u_\varepsilon \to u$ converging weakly* in $L^p(I;W_0^{1,p}(\Omega)) \cap L^\infty(I;L^2(\Omega))$ for $\varepsilon \to 0$, show the convergence[76]

$$\frac{\partial u_\varepsilon}{\partial t}+\varepsilon\Delta^2 u_\varepsilon \to \frac{\partial u}{\partial t} \qquad \text{weakly in } \ L^{p'}(I;W^{-1,p'}(\Omega)). \tag{8.214}$$

Also show that[77]

$$u_\varepsilon(T) \to u(T) \qquad \text{weakly in } \ L^2(\Omega). \tag{8.215}$$

Then show the convergence of u_ε to the weak solution of the initial-boundary-value problem $\frac{\partial}{\partial t}u - \mathrm{div}(|\nabla u|^{p-2}\nabla u) = g$, $u(0,\cdot) = u_0$, and $u|_\Sigma = 0$.[78] Assuming $g \in L^2(Q)$ and $u_0 \in W_0^{1,p}(\Omega)$, and regularizing the initial condition $u_\varepsilon(0) = u_{0\varepsilon} \in W_0^{2,2}(\Omega) \cap W^{1,p}(\Omega)$ so that $u_{0\varepsilon} \to u_0$ in $W_0^{1,p}(\Omega)$ while $\|u_{0\varepsilon}\|_{W^{2,2}(\Omega)} = \mathscr{O}(1/\sqrt{\varepsilon})$, execute the test by $\frac{\partial u_\varepsilon}{\partial t}$ to prove the a-priori estimates:

$$\big\|u_\varepsilon\big\|_{L^\infty(I;W_0^{1,p}(\Omega))} \le C, \quad \big\|u_\varepsilon\big\|_{L^\infty(I;W_0^{2,2}(\Omega))} \le \frac{C}{\sqrt{\varepsilon}}, \quad \Big\|\frac{\partial u_\varepsilon}{\partial t}\Big\|_{L^2(Q)} \le C. \tag{8.216}$$

Modify this example by considering other boundary conditions or also some other terms as $\mathrm{div}(a(u))$ or a quasilinear regularizing term as in Example 2.46.

Remark 8.86 (Strong convergence of the singular perturbations)**.** Strong convergence in Example 8.85 based on (8.213) is expectable but rather technical, however. We will use $\int_0^T \langle \frac{\partial u_\varepsilon}{\partial t} - \frac{\partial u}{\partial t}, u_\varepsilon - u\rangle \mathrm{d}t \ge 0$; here again it is important that $\frac{\partial u}{\partial t}$ belongs

---

[76] Hint: By (8.213b), choose $\frac{\partial u_\varepsilon}{\partial t}+\varepsilon\Delta^2 u_\varepsilon \to \dot u \in L^{p'}(I;W^{-1,p'}(\Omega))$ and, for any $z \in \mathscr{D}(Q)$, use

$$\begin{aligned}\langle \dot u, z\rangle &= \lim_{\varepsilon\to 0} \langle \frac{\partial u_\varepsilon}{\partial t} + \varepsilon\Delta^2 u_\varepsilon, z\rangle \\ &= \lim_{\varepsilon\to 0}\int_Q \frac{\partial u_\varepsilon}{\partial t} z + \varepsilon\nabla^2 v_\varepsilon{:}\nabla^2 z \,\mathrm{d}x\mathrm{d}t = -\lim_{\varepsilon\to 0}\int_Q u_\varepsilon\frac{\partial z}{\partial t} - \varepsilon\nabla^2 u_\varepsilon{:}\nabla^2 z\,\mathrm{d}x\mathrm{d}t = -\int_Q u\frac{\partial z}{\partial t}\,\mathrm{d}x\mathrm{d}t\end{aligned}$$

because, thanks to (8.213a), $\|\varepsilon\nabla^2 u_\varepsilon\|_{L^2(Q;\mathbb{R}^{n\times n})} = \varepsilon\mathscr{O}(1/\sqrt{\varepsilon}) = \mathscr{O}(\sqrt{\varepsilon}) \to 0$. Thus $\dot u$ is shown to be the distributional derivative of u.

[77] Hint: Use (8.214) and the first estimate in (8.213a) to show the $W^{-1,p'}$-weak convergence:

$$u_\varepsilon(T) = u_0 + \int_0^T \frac{\partial u_\varepsilon}{\partial t}\mathrm{d}t = v_0 + \int_0^T \frac{\partial u_\varepsilon}{\partial t}+\varepsilon\Delta^2 u_\varepsilon\,\mathrm{d}t - \varepsilon\int_0^T \Delta^2 u_\varepsilon\,\mathrm{d}t \to u_0 + \int_0^T \frac{\partial u}{\partial t}\,\mathrm{d}t = u(T)$$

and then the second estimate in (8.213a) implies (8.215).

[78] Hint: Use Minty's trick based on the monotonicity of the operator $\frac{\partial}{\partial t} - \Delta_p + \varepsilon\Delta^2$ similarly like in Exercise 2.100.

to $L^{p'}(I;W_0^{1,p}(\Omega)^*)$ and is thus in duality to u. By the d-monotonicity (2.141) of $-\Delta_p$, we get, for any $\tilde u \in L^2(I;W_0^{2,2}(\Omega)) \cap L^p(I;W^{1,p}(\Omega))$:

$$
\begin{aligned}
&\Big(\big\|\nabla u_\varepsilon\big\|_{L^p(Q;\mathbb{R}^n)}^{p-1} - \big\|\nabla u\big\|_{L^p(Q;\mathbb{R}^n)}^{p-1}\Big)\Big(\big\|\nabla u_\varepsilon\big\|_{L^p(Q;\mathbb{R}^n)} - \big\|\nabla u\big\|_{L^p(Q;\mathbb{R}^n)}\Big)\\
&\quad\le \int_Q \big(|\nabla u_\varepsilon|^{p-2}\nabla u_\varepsilon - |\nabla u|^{p-2}\nabla u\big)\cdot\nabla(u_\varepsilon - u)\,\mathrm{d}x\mathrm{d}t\\
&\quad\le \int_0^T\bigg(\Big\langle\frac{\partial u_\varepsilon}{\partial t} - \frac{\partial u}{\partial t}, u_\varepsilon - u\Big\rangle + \int_\Omega \big(|\nabla u_\varepsilon|^{p-2}\nabla u_\varepsilon - |\nabla u|^{p-2}\nabla u\big)\cdot\nabla(u_\varepsilon - u)\,\mathrm{d}x\bigg)\mathrm{d}t\\
&\quad= \int_Q\bigg(\frac{\partial u_\varepsilon}{\partial t}\cdot u_\varepsilon + |\nabla u_\varepsilon|^{p-2}\nabla u_\varepsilon\cdot\nabla u_\varepsilon\bigg) - \bigg(\frac{\partial u_\varepsilon}{\partial t}\cdot u + |\nabla u_\varepsilon|^{p-2}\nabla u_\varepsilon\cdot\nabla u\bigg)\mathrm{d}x\mathrm{d}t\\
&\qquad\qquad - \int_0^T\bigg(\Big\langle\frac{\partial u}{\partial t}, u_\varepsilon - u\Big\rangle + \int_\Omega |\nabla u|^{p-2}\nabla u\cdot\nabla(u_\varepsilon - u)\,\mathrm{d}x\bigg)\mathrm{d}t\\
&\quad= \int_Q\bigg(-\varepsilon|\nabla^2 u_\varepsilon|^2 + g u_\varepsilon + \varepsilon\nabla^2 u_\varepsilon : \nabla^2\tilde u - g\tilde u\\
&\qquad\qquad - \frac{\partial u_\varepsilon}{\partial t}(u - \tilde u) - |\nabla u_\varepsilon|^{p-2}\nabla u_\varepsilon\cdot\nabla(u - \tilde u)\bigg)\mathrm{d}x\mathrm{d}t\\
&\qquad\qquad - \int_0^T\bigg(\Big\langle\frac{\partial u}{\partial t}, u_\varepsilon - u\Big\rangle + \int_\Omega |\nabla u|^{p-2}\nabla u\cdot\nabla(u_\varepsilon - u)\,\mathrm{d}x\bigg)\mathrm{d}t\\
&\quad\le \int_Q \varepsilon\nabla^2 u_\varepsilon : \nabla^2\tilde u + g(u_\varepsilon - \tilde u) - \frac{\partial u_\varepsilon}{\partial t}(u - \tilde u) - |\nabla u_\varepsilon|^{p-2}\nabla u_\varepsilon\cdot\nabla(u - \tilde u)\,\mathrm{d}x\mathrm{d}t\\
&\qquad\qquad - \int_0^T\bigg(\Big\langle\frac{\partial u}{\partial t}, u_\varepsilon - u\Big\rangle + \int_\Omega |\nabla u|^{p-2}\nabla u\cdot\nabla(u_\varepsilon - u)\,\mathrm{d}x\bigg)\mathrm{d}t\\
&\quad\to \int_Q g(u - \tilde u) - \chi\cdot\nabla(u - \tilde u)\,\mathrm{d}x\mathrm{d}t - \int_0^T\Big\langle\frac{\partial u}{\partial t}, u - \tilde u\Big\rangle\mathrm{d}t \qquad \text{for } \varepsilon \to 0, \qquad (8.217)
\end{aligned}
$$

where χ is a weak limit (of a subsequence) of $|\nabla u_\varepsilon|^{p-2}\nabla u_\varepsilon$ in $L^{p'}(Q;\mathbb{R}^n)$. For the convergence (8.217), we have used (8.214) and (8.215), which allows for

$$
\begin{aligned}
&\lim_{\varepsilon\to 0}\int_Q \varepsilon\nabla^2 u_\varepsilon : \nabla^2\tilde u - \frac{\partial u_\varepsilon}{\partial t}(u - \tilde u)\,\mathrm{d}x\mathrm{d}t\\
&\quad= \lim_{\varepsilon\to 0}\int_0^T \Big\langle \varepsilon\Delta^2 u_\varepsilon + \frac{\partial u_\varepsilon}{\partial t}, \tilde u\Big\rangle + \Big\langle\frac{\partial u}{\partial t}, u_\varepsilon\Big\rangle\,\mathrm{d}t - \int_\Omega u_\varepsilon(T)u(T) - |u_0|^2\,\mathrm{d}x\\
&\quad= \int_0^T\Big\langle\frac{\partial u}{\partial t}, \tilde u\Big\rangle + \Big\langle\frac{\partial u}{\partial t}, u\Big\rangle\mathrm{d}t - \big\|u(T)\big\|_{L^2(\Omega)}^2 + \big\|u_0\big\|_{L^2(\Omega)}^2 = \int_0^T\Big\langle\frac{\partial u}{\partial t}, \tilde u - u\Big\rangle\mathrm{d}t.
\end{aligned}
$$

Eventually, pushing $\tilde u \to u$ in $L^p(I;W_0^{1,p}(\Omega))$ makes the final expression in (8.217) arbitrarily closed to 0, which yields $\|\nabla u_\varepsilon\|_{L^p(Q;\mathbb{R}^n)} \to \|\nabla u\|_{L^p(Q;\mathbb{R}^n)}$ and then $u_\varepsilon \to u$ in $L^p(I;W_0^{1,p}(\Omega))$ by the uniform convexity of the space $L^p(Q;\mathbb{R}^n)$.

Exercise 8.87 (*Conservation law regularized by* Δ). Consider

$$\frac{\partial u}{\partial t} + \mathrm{div}(F(u)) - \varepsilon\Delta u = g, \qquad u|_{t=0} = u_0, \qquad u|_\Sigma = 0, \tag{8.218}$$

where $F:\mathbb{R}\to\mathbb{R}^n$ has at most linear growth, i.e. $|F(r)| \le C_1 + C_2|r|$, and $\varepsilon > 0$. Make the basic estimates.[79] Assuming also that F is Lipschitz continuous, make an estimate of u in $W^{1,2}(I;L^2(\Omega))\cap L^\infty(I;W_0^{1,2}(\Omega))$.[80] Prove further a bound for u in $W^{1,\infty}(I;L^2(\Omega))\cap W^{1,2}(I;W_0^{1,2}(\Omega))$.[81] For estimation of the term $\mathrm{div}(F(u))$ on the left-hand side, see Exercise 9.27 below. For a special case $n = 1 = \varepsilon$ and $F(r) = \frac12 r^2$, consider the so-called (regularized) *Burgers equation*

$$\frac{\partial u}{\partial t} + u\frac{\partial u}{\partial x} - \frac{\partial^2 u}{\partial x^2} = g \text{ in } Q := (0,T)\times(0,1), \;\; u|_{x=0,1} = 0, \;\; u|_{t=0} = u_0. \tag{8.219}$$

Assuming $0 \le g \le K$ and $u_0 \in W^{1,2}(\Omega)$, $u_0 \ge 0$, prove $u \in L^\infty(I;W^{1,2}(\Omega))\cap W^{1,2}(I;L^2(\Omega))\cap L^2(I;W^{2,2}(\Omega))$ and $0 \le u(t,x) \le tK + \|u_0\|_{L^\infty(0,1)}$.[82] Using this regularity, prove also that the solution is unique.[83]

---

[79] Hint: Test by u gives

$$\frac12\frac{\mathrm{d}}{\mathrm{d}t}\|u\|^2_{L^2(\Omega)} + \varepsilon\|\nabla u\|^2_{L^2(\Omega;\mathbb{R}^n)} \le \Big(C_1\sqrt{\mathrm{meas}_n(\Omega)} + C_2\|u\|_{L^2(\Omega)}\Big)\|\nabla u\|_{L^2(\Omega;\mathbb{R}^n)},$$

so that by Young's and Gronwall's inequalities one obtains u bounded in $L^\infty(I;L^2(\Omega))\cap L^2(I;W_0^{1,2}(\Omega))$. Then the "dual" estimate of $\frac{\partial}{\partial t}u$ in $L^2(I;W^{-1,2}(\Omega))$ follows standardly.

[80] Hint: Test by $\frac{\partial}{\partial t}u$ gives

$$\Big\|\frac{\partial u}{\partial t}\Big\|^2_{L^2(\Omega)} + \frac{1}{\varepsilon^2}\frac{\mathrm{d}}{\mathrm{d}t}\|\nabla u\|^2_{L^2(\Omega;\mathbb{R}^n)} \le \sup_{r\in\mathbb{R}}|F'(r)|\,\|\nabla u\|_{L^2(\Omega;\mathbb{R}^n)}\Big\|\frac{\partial u}{\partial t}\Big\|_{L^2(\Omega)}.$$

Then use the Young and the Gronwall inequalities.

[81] Hint: Differentiating (8.218) in time and testing by $\frac{\partial}{\partial t}u$ gives:

$$\frac12\frac{\mathrm{d}}{\mathrm{d}t}\Big\|\frac{\partial u}{\partial t}\Big\|^2_{L^2(\Omega)} + \varepsilon\Big\|\nabla\frac{\partial u}{\partial t}\Big\|^2_{L^2(\Omega;\mathbb{R}^n)} \le \sup_{r\in\mathbb{R}}|F'(r)|\,\Big\|\frac{\partial u}{\partial t}\Big\|_{L^2(\Omega)}\Big\|\frac{\partial}{\partial t}\nabla u\Big\|_{L^2(\Omega;\mathbb{R}^n)}.$$

Then use again the Young and the Gronwall inequalities.

[82] Hint: First, test (8.219) by u to get $u \in L^2(I;W^{1,2}(\Omega))\cap L^\infty(I;L^2(\Omega))$; note that $\int_0^1 u^2\frac{\partial}{\partial x}u\,\mathrm{d}x = \frac13\int_0^1\frac{\partial}{\partial x}u^3\,\mathrm{d}x = 0$. Then test (8.219) by u^- to get $u \ge 0$. Then put $w = u - Kt$ to get $\frac{\partial}{\partial t}w + (w+Kt)\frac{\partial}{\partial x}w - \frac{\partial^2}{\partial x^2}w = g - K \le 0$ and test it by $(w - \|u_0\|_{L^\infty(0,1)})^+$ to get $w \le \|u_0\|_{L^\infty(0,1)}$. Eventually, test (8.219) by $\frac{\partial}{\partial t}u$ and use $u \in L^\infty(Q)$ to estimate

$$\Big\|\frac{\partial u}{\partial t}\Big\|^2_{L^2(0,1)} + \frac12\frac{\mathrm{d}}{\mathrm{d}t}\Big\|\frac{\partial u}{\partial x}\Big\|^2_{L^2(0,1)} = -\int_0^1 u\frac{\partial u}{\partial x}\frac{\partial u}{\partial t}\mathrm{d}x \le \|u\|_{L^\infty(Q)}\Big\|\frac{\partial u}{\partial x}\Big\|_{L^2(0,1)}\Big\|\frac{\partial u}{\partial t}\Big\|_{L^2(0,1)}$$

to get $u \in L^\infty(I;W^{1,2}(\Omega))\cap W^{1,2}(I;L^2(\Omega))$. Then from (8.219) read $u \in L^2(I;W^{2,2}(\Omega))$.

[83] Hint: Denoting $u_{12} := u_1 - u_2$, test the difference of (8.219) for u_1 and u_2 by u_{12}:

$$\begin{aligned}\frac12\frac{\mathrm{d}}{\mathrm{d}t}\|u_{12}\|^2_{L^2(0,1)} + \Big\|\frac{\partial u_{12}}{\partial x}\Big\|^2_{L^2(0,1)} &= \int_0^1\Big(u_2\frac{\partial u_2}{\partial x} - u_1\frac{\partial u_1}{\partial x}\Big)u_{12}\,\mathrm{d}x\\ &\le \Big\|\frac{\partial u_1}{\partial x}\Big\|_{L^\infty(0,1)}\|u_{12}\|^2_{L^2(0,1)} + \frac12\|u_2\|^2_{L^\infty(0,1)}\|u_{12}\|^2_{L^2(0,1)} + \frac12\Big\|\frac{\partial u_1}{\partial x}\Big\|^2_{L^2(0,1)}.\end{aligned}$$

Exercise 8.88 (*Allen-Cahn equation* [15][84]). Consider the initial-boundary-value problem for the semilinear equation

$$\frac{\partial u}{\partial t} - \Delta u + (u^2-1)u = 0, \qquad u(0,\cdot) = u_0, \quad \frac{\partial u}{\partial \nu}\Big|_\Sigma = 0. \tag{8.220}$$

Prove existence of a solution by Rothe's or Galerkin's method, deriving a-priori estimates of the type $\|u\|_{L^2(I;W^{1,2}(\Omega))\cap L^4(Q)}$ or $\|u\|_{L^\infty(I;W^{1,2}(\Omega)\cap L^4(\Omega))}$.

Exercise 8.89 (*Cahn-Hilliard equation* [89][85]). Consider the initial-boundary-value problem for the semilinear 4th-order equation

$$\frac{\partial u}{\partial t} - \Delta\beta(u) + \Delta^2 u = g, \qquad u(0,\cdot) = u_0, \quad u|_\Sigma = \frac{\partial u}{\partial \nu}\Big|_\Sigma = 0. \tag{8.221}$$

Prove existence of a solution by Rothe's or Galerkin's method, deriving a-priori estimates of u in $W^{1,2,2}(I;W_0^{2,2}(\Omega),W^{-2,2}(\Omega))\cap L^\infty(I;L^2(\Omega))$, assuming $\beta:\mathbb{R}\to\mathbb{R}$ smooth of the form $\beta=\beta_1+\beta_2$ with β_1 nondecreasing and β_2 Lipschitz continuous and $g\in L^2(I;L^{2^{**\prime}}(\Omega))$.[86]

Exercise 8.90 (*Viscous Cahn-Hilliard equation* [189][87]). Consider the semilinear 4th-order *pseudoparabolic equation*

$$\frac{\partial(u-\Delta u)}{\partial t} - \Delta\beta(u) + \Delta^2 u = g, \qquad u(0,\cdot) = u_0, \quad u|_\Sigma = \frac{\partial u}{\partial \nu}\Big|_\Sigma = 0 \tag{8.222}$$

with β qualified as in Exercise 8.89, and modify the a-priori estimates therein.

Exercise 8.91 (*Non-Newtonean fluids*[88]). Analogously to (6.26a), consider

$$\frac{\partial u}{\partial t} - \operatorname{div}\sigma\big(e(u)\big) + (u\cdot\nabla)u + \nabla\pi = g, \quad \operatorname{div} u = 0, \tag{8.223}$$

[84] Up to a suitable scaling, this equation is related to Ginzburg-Landau phase transition theory; for more details see e.g. Alikakos, Bates [14], Caginalp [87], Cahn, Hilliard [88], Hoffmann, Tang [205, Chap.2], Ohta, Mimura, and Kobayashi [317]. Let us remark that the full Ginzburg-Landau system is related to superconductivity and received great attention in physics, being reflected also by Nobel prizes to L.D. Landau in 1962 and (1/3) to V.L. Ginzburg in 2003.

[85] This equation has been proposed to model isothermal phase separation in binary alloys or mixtures. There is an extensive spool of related references, e.g. Artstein and Slemrod [20] or Elliott and Zheng [136], Novic-Cohen [315, 316], or von Wahl [422].

[86] Hint: Test (8.221) by u, use $\beta_1'\ge 0$ and Gagliardo-Nirenberg's inequality (Theorem 1.24 with $q=p=r=k=2$, $\beta=1$, $\lambda=1/2$) to estimate

$$\begin{aligned}\frac{1}{2}\frac{\mathrm{d}}{\mathrm{d}t}\|u\|^2_{L^2(\Omega)} + \|\Delta u\|^2_{L^2(\Omega)} + \int_\Omega \beta_1'(u)|\nabla u|^2\mathrm{d}x &= \int_\Omega gu - \beta_2'(u)|\nabla u|^2\mathrm{d}x\\ &\le \|g\|_{L^{2^{**\prime}}(\Omega)}\|u\|_{L^{2^{**}}(\Omega)} + \sup\beta_2'(\cdot)\|\nabla u\|^2_{L^2(\Omega;\mathbb{R}^n)}\\ &\le \|g\|_{L^{2^{**\prime}}(\Omega)}\|u\|_{L^{2^{**}}(\Omega)} + C_{\rm GN}\sup|\beta_2'(\cdot)|\,\|u\|_{L^2(\Omega)}\|\nabla^2 u\|_{L^2(\Omega;\mathbb{R}^{n\times n})},\end{aligned}$$

and then use still $\int_\Omega|\Delta u|^2\mathrm{d}x = \int_\Omega|\nabla^2 u|^2\mathrm{d}x$ under the considered boundary conditions, cf. Example 2.46, and eventually Gronwall's and Young's inequalities.

[87] This equation has been proposed by Grinfeld and Novick-Cohen [189] to model phase separation in glass and polymer systems.

[88] See Ladyzhenskaya [248] or Málek et al. [268, Sect.5.4.1]. Cf. also [81, 123] for $p>2n/(n+2)$.

with $u|_\Sigma = 0$, $u(0,\cdot) = u_0 \in L^2(\Omega;\mathbb{R}^n)$, $e(u)$ as in (6.26a), and $\sigma(e) = |e|^{p-2}e$; hence (6.28a,b) holds. Testing a Galerkin approximation of (8.223) by the approximate solution itself, prove existence of a weak solution if p is large enough.[89]

Exercise 8.92 (*Semi-implicit time discretization*). Consider $p = 2$ and the semilinear parabolic problem (8.167) with the linearization $\big[B(w,u)\big](v) := \int_\Omega \nabla u\cdot\nabla v + |w|^{q_1-2}uv\,\mathrm{d}x + \int_\Gamma |w|^{q_2-2}uv\,\mathrm{d}S$ and the semi-implicit formula (8.55), which leads to

$$\Big\langle\frac{\partial u_\tau}{\partial t},v\Big\rangle+\int_\Omega\nabla\bar u_\tau{\cdot}\nabla v+|\bar u_\tau^{\rm R}|^{q_1-2}\bar u_\tau v-\bar g_\tau v\,\mathrm{d}x=\int_\Gamma\bar h_\tau v-|\bar u_\tau^{\rm R}|^{q_2-2}\bar u_\tau v\,\mathrm{d}S \quad (8.224)$$

for a.a. $t\in I$ and all $v\in W^{1,2}(\Omega)$ with $\bar u_\tau^{\rm R}$ defined in (8.202). Make the basic a-priori estimate[90] and prove the convergence for $\tau\to 0$.[91]

8.9 Global monotonicity approach, periodic problems

Sometimes, other methods can be applied on the abstract level, too. Let us mention a "global approach" which can solve the Cauchy problem directly on $W^{1,p,p'}(I;V,V^*)$ provided $V\subset H$ and which can straightforwardly be adapted for

[89] Hint: Using the identity $\int_\Omega (u\cdot\nabla)u\cdot u\,\mathrm{d}x = 0$, cf. (6.36), the suggested test gives bounds of u in $L^\infty(I;L^2(\Omega;\mathbb{R}^n))\cap L^p(I;W^{1,p}_{0,\mathrm{div}}(\Omega;\mathbb{R}^n))$, for $W^{1,p}_{0,\mathrm{div}}(\Omega;\mathbb{R}^n)$ see (6.29). For the dual estimate of $\frac{\partial}{\partial t}u$ in $L^{p'}(I;W^{1,p}_{0,\mathrm{div}}(\Omega;\mathbb{R}^n)^*)$, use Green's Theorem 1.31 for the convective term:

$$\int_Q(u\cdot\nabla)u\cdot v\,\mathrm{d}x\mathrm{d}t=-\int_Q(u\cdot\nabla)v\cdot u\,\mathrm{d}x\mathrm{d}t\le C\|u\|^2_{L^{p^\circledast}(Q;\mathbb{R}^n)}\|\nabla v\|_{L^p(Q;\mathbb{R}^{n\times n})},$$

which needs $2/p^\circledast + 1/p \le 1$. In view of (8.131), identify that $p\ge 11/5$ (resp. $p\ge 2$) is needed for $n=3$ (resp. $n=2$). Make it more rigorous by using seminorms arisen in Galerkin's method. Alternatively, without using Green's theorem, prove an estimate of $\frac{\partial}{\partial t}u$ in $(L^p(I;W^{1,p}_{0,\mathrm{div}}(\Omega;\mathbb{R}^n))\cap L^\infty(I;L^2(\Omega)))^*$ to be used as suggested in Remark 8.12.

[90] Hint: Testing by u_τ^k gives

$$\frac12\|u_\tau^k\|^2_{L^2(\Omega)}-\frac12\|u_\tau^{k-1}\|^2_{L^2(\Omega)}+\tau\|\nabla u_\tau^k\|^2_{L^2(\Omega;\mathbb{R}^n)}\le\int_\Omega g_\tau^k u_\tau^k-|u_\tau^{k-1}|^{q_1-2}|u_\tau^k|^2\,\mathrm{d}x$$
$$+\int_\Gamma h_\tau^k u_\tau^k-|u_\tau^{k-1}|^{q_2-2}|u_\tau^k|^2\,\mathrm{d}S\le\|g_\tau^k\|_{L^{p^*\prime}(\Omega)}\|u_\tau^k\|_{L^{p^*}(\Omega)}+\|h_\tau^k\|_{L^{p^\#\prime}(\Gamma)}\|u_\tau^k\|_{L^{p^\#}(\Gamma)}.$$

Then proceed as in (8.27) to get the bound in $L^\infty(I;L^2(\Omega))\cap L^2(I;W^{1,2}(\Omega))$. Further, the strategy of Example 8.63(2) leads to the bound of $\frac{\partial}{\partial t}u_\tau$ in $L^2(I;W^{1,2}(\Omega)^*)$ and also of $\frac{\partial}{\partial t}\bar u_\tau$ in $\mathscr{M}(I;W^{1,2}(\Omega)^*)$.

[91] Hint: By Corollary 7.9 with the interpolation (8.128), realize that, for a subsequence, $\bar u_\tau\to u$ in $L^{2^\circledast-\varepsilon}(Q)$, $2^\circledast = 2+4/n$, $\varepsilon>0$; cf. also (8.152). Since $\bar u_\tau^{\rm R}$ inherits all a-priori estimates as $\bar u_\tau$, also $\bar u_\tau^{\rm R}\to u$ in $L^{2^\circledast-\varepsilon}(Q)$. Realize that these limits must indeed coincide with each other because $\|\bar u_\tau^{\rm R}-\bar u_\tau\|_{L^2(I;W^{1,2}(\Omega)^*)}=\mathscr{O}(\tau)$ just by a modification of (8.40) with the boundedness of $\{\frac{\mathrm{d}}{\mathrm{d}t}u_\tau\}_{0<\tau\le\tau_0}$ in $L^2(I;W^{1,2}(\Omega)^*)$. The strategy for the traces $\bar u_\tau|_\Sigma$ and $\bar u_\tau^{\rm R}|_\Sigma$ is as in (8.153). Then make the limit passage directly in (8.224) integrated over I.

periodic problems of the form

$$\frac{\mathrm{d}u}{\mathrm{d}t} + A\big(t, u(t)\big) = f(t) \ \text{ for a.a. } t\in I, \quad u(0) = u(T)\ . \tag{8.225}$$

Obviously, considering $A : \mathbb{R} \times V \to V^*$ and $f : \mathbb{R} \to V^*$ periodic with the period T, a solution u to $\frac{\mathrm{d}}{\mathrm{d}t}u + \mathscr{A}(u) = f$ having the same given period can be just constructed by a periodic prolongation of the solution $u : I \to V$ to (8.225); this is why we refer to (8.225) as the periodic problem and $u(0) = u(T)$ as a *periodic condition.*

Considering a mapping $L : \mathrm{dom}(L) \to L^{p'}(I; V^*)$, $\mathrm{dom}(L) \subset L^p(I; V)$, the following property of L will play an important role: for any $w \in L^{p'}(I; V^*)$ and $u \in L^p(I; V)$:

$$\Big(\forall v\in\mathrm{dom}(L):\ \big\langle w{-}L(v), u{-}v\big\rangle \ge 0\Big) \quad\Rightarrow\quad \begin{cases} u\in\mathrm{dom}(L), \\ w = L(u). \end{cases} \tag{8.226}$$

A monotone mapping L satisfies (8.226) if[92] and only if[93] it is *maximal monotone.* The base of the direct method is the following observation:

Lemma 8.93 (Maximal monotonicity of $\frac{\mathrm{d}}{\mathrm{d}t}$). *Let $L : u \mapsto \frac{\mathrm{d}}{\mathrm{d}t}u : \mathrm{dom}(L) \to L^{p'}(I; V^*)$ and either*

$$\mathrm{dom}(L) := \big\{u\in W^{1,p,p'}(I; V, V^*);\ \ u(0) = u_0\big\} \tag{8.227}$$

for $u_0 \in H$ fixed, or

$$\mathrm{dom}(L) := \big\{u\in W^{1,p,p'}(I; V, V^*);\ \ u(0) = u(T)\big\}. \tag{8.228}$$

Then L is monotone, radially continuous, and satisfies (8.226).

Proof. [94] The monotonicity of L follows from the fact that, for any $u, v \in \mathrm{dom}(L)$, by using (7.22), we have

$$\begin{aligned} \big\langle L(u){-}L(v), u{-}v\big\rangle &= \int_0^T \Big\langle \frac{\mathrm{d}(u{-}v)}{\mathrm{d}t}, u{-}v\Big\rangle\,\mathrm{d}t \\ &= \frac{1}{2}\big\|u(T){-}v(T)\big\|_H^2 - \frac{1}{2}\big\|u(0){-}v(0)\big\|_H^2 \ \begin{cases} \ge 0 & \text{in case (8.227)}, \\ = 0 & \text{in case (8.228)}, \end{cases} \end{aligned}$$

because $\|u(0) - v(0)\|_H = \|u_0 - u_0\|_H = 0$ in case (8.227) and $\|u(0) - v(0)\|_H = \|u(T) - v(T)\|_H$ in case (8.228).

[92] Supposing the contrary (i.e. (8.226) does not hold for some (u, w)), we can derive that $\mathrm{Graph}(L) \cup \{(u, w)\}$ would be a graph of a monotone mapping larger than $\mathrm{Graph}(L)$, i.e. L is not maximal monotone.

[93] Realize that, supposing $w \neq L(u)$, $\mathrm{Graph}(L) \cup \{(u, w)\}$ would be a graph of a monotone operator, contradicting the fact that L is maximal.

[94] See, e.g., Barbu [37, p.167] (only the case (8.227)), Gajewski et al. [168, Sect. VI.1.2], Zeidler [427, Sect. 32.3b] for $u_0 = 0$.

The radial continuity now means that $\langle L(u+\varepsilon v), v\rangle = \int_0^T \langle \frac{\mathrm{d}}{\mathrm{d}t}(u+\varepsilon v), v\rangle\,\mathrm{d}t \to \int_0^T \langle \frac{\mathrm{d}}{\mathrm{d}t}u, v\rangle\,\mathrm{d}t = \langle L(u), v\rangle$ for all v such that $u+\varepsilon v \in \mathrm{dom}(L)$ for some (and thus all) $\varepsilon \neq 0$, which is obvious.

For (8.227), let us first assume $u_0 \in V$ while, for (8.228), u_0 can be considered arbitrary from V, e.g. $u_0 = 0$. Then let us take $z \in V$ and $\varphi \in C^1(I)$ such that $\varphi(0) = \varphi(T) = 0$, and put $v(t) = \varphi(t)z + u_0$. Thus $v \in \mathrm{dom}(L)$. Obviously, $\frac{\mathrm{d}}{\mathrm{d}t}v = \varphi' z$. Using the premise in (8.226), we have

$$\begin{aligned} 0 \le \big\langle w - L(v), u - v\big\rangle &= \big\langle w, u\big\rangle + \big\langle L(v), v\big\rangle - \big\langle w, v\big\rangle - \big\langle L(v), u\big\rangle \\ &= \big\langle w, u\big\rangle + \frac{1}{2}\|v(T)\|_H^2 - \frac{1}{2}\|v(0)\|_H^2 - \int_0^T \big\langle w\varphi + \varphi' u, z\big\rangle\,\mathrm{d}t - \big\langle w, u_0\big\rangle \\ &= \big\langle w, u\big\rangle - \Big\langle \int_0^T (w\varphi + \varphi' u)\,\mathrm{d}t, z\Big\rangle - \big\langle w, u_0\big\rangle, \end{aligned} \tag{8.229}$$

where $v(T) = u_0 = v(0)$ has been used and where $\langle w, u_0\rangle$ means $\int_0^T \langle w(t), u_0\rangle \mathrm{d}t$. Note also that the first integral in (8.229) is the Lebesgue one while the second is a Bochner one. Since $z \in V$ is arbitrary, it must hold that $\int_0^T \varphi'(t)u(t) + \varphi(t)w(t)\,\mathrm{d}t = 0$. As φ is arbitrary, it must hold that $\frac{\mathrm{d}}{\mathrm{d}t}u = w$ in the sense of distributions.

Moreover, since $w \in L^{p'}(I; V^*)$, we have $u \in W^{1,p,p'}(I; V, V^*)$. To prove $u \in \mathrm{dom}(L)$ we have to show $u(0) = u_0$ or $u(0) = u(T)$, respectively.

Using the premise in (8.226) with $w = \frac{\mathrm{d}}{\mathrm{d}t}u$ and the by-part formula (7.22), we have

$$0 \le \Big\langle \frac{\mathrm{d}u}{\mathrm{d}t} - \frac{\mathrm{d}v}{\mathrm{d}t}, u - v\Big\rangle = \frac{1}{2}\big\|u(T) - v(T)\big\|_H^2 - \frac{1}{2}\big\|u(0) - v(0)\big\|_H^2. \tag{8.230}$$

In case (8.227) if $u_0 \in V$ is considered, we can set $v(t) = ((T-t)u_0 + tu_T^\varepsilon)/T$ with $u_T^\varepsilon \in V$ chosen in such a way that $u_T^\varepsilon \to u(T)$ in H; note that $u(T)$ has a good sense only in H due to Lemma 7.3 but not in V in general. From (8.230), we then have

$$0 \le \big\|u(T) - u_T^\varepsilon\big\|_H^2 - \big\|u(0) - u_0\big\|_H^2 \to -\big\|u(0) - u_0\big\|_H^2 \tag{8.231}$$

hence $u(0) = u_0$. In case (8.228), from (8.230) we get

$$\begin{aligned} 0 &\le \frac{1}{2}\big\|u(T)\big\|_H^2 - \frac{1}{2}\big\|u(0)\big\|_H^2 - \big\langle u(T), v(T)\big\rangle + \big\langle u(0), v(0)\big\rangle + \frac{1}{2}\big\|v(T)\big\|_H^2 - \frac{1}{2}\big\|v(0)\big\|_H^2 \\ &= \frac{1}{2}\big\|u(T)\big\|_H^2 - \frac{1}{2}\big\|u(0)\big\|_H^2 + \big(v(0), u(0) - u(T)\big) \end{aligned} \tag{8.232}$$

because $v(0) = v(T)$. As $v(0)$ is arbitrary, we get $u(0) = u(T)$.

If $u_0 \in H\backslash V$, we must replace the constant u_0 which then does not belong to $L^p(I; V)$ by some nonconstant $\tilde{u}_0 \in W^{1,p,p'}(I; V, V^*)$ and then to choose $v(t) = \varphi(t)z + \tilde{u}_0(t)$. Assuming $\tilde{u}_0(0) = u_0$, we have still $v \in \mathrm{dom}(L)$. Assuming still $\tilde{u}_0(0) = \tilde{u}_0(T)$, it causes only the additional term $\int_0^T \langle \frac{\mathrm{d}}{\mathrm{d}t}\tilde{u}_0, u\rangle\,\mathrm{d}t$ emerging in

(8.229). By choosing again z and then also φ arbitrary, one can argue exactly in the same way as before.[95] $\square$

Remark 8.94 (Other conditions). Lemma 8.93 explains why the initial or the periodic conditions are natural. E.g., if one would choose $\mathrm{dom}(L) := \{u \in W^{1,p,p'}(I;V,V^*);\ u(T) = u_T\}$, then L would not be monotone; i.e. prescribing a terminal condition $u(T) = u_T$ does not yield a well-posed problem. In case $\mathrm{dom}(L) := \{u \in W^{1,p,p'}(I;V,V^*);\ u(0) = u_0,\ u(T) = u_T\}$, L would be monotone but not maximal monotone; i.e. prescribing both the terminal and the initial conditions does not yield a well-posed problem, either. On the other hand, it is an easy exercise to prove a modification of Lemma 8.93 for the so-called *anti-periodic condition* $u(0) + u(T) = 0$.

Let us now prove an abstract result, abbreviating $\mathscr{V} = L^p(I;V)$. Let us also assume that $\mathscr{V}$ can be approximated by finite-dimensional subspaces $\mathscr{V}_k$ such that, for $u_0 = 0$, $\mathscr{V}_k \subset \mathrm{dom}(L)$, $\mathscr{V}_k \subset \mathscr{V}_{k+1}$, and $\bigcup_{k\in\mathbb{N}} \mathscr{V}_k$ is dense in $\mathrm{dom}(L)$ with respect to the norm $\|u\|_{\mathrm{dom}(L)} = \|u\|_{\mathscr{V}} + \|Lu\|_{\mathscr{V}^*}$; note that this space is separable so that such a chain of subspaces does exist.[96] Note also that we, in fact, assumed L linear and that, as $\mathrm{dom}(L) \neq \mathscr{V}$, we cannot use directly the Browder-Minty Theorem 2.18 for $L + \mathscr{A}$.

Lemma 8.95 (Surjectivity of $L + \mathscr{A}$). *Let $\mathscr{A} : \mathscr{V} \to \mathscr{V}^*$ be radially continuous and monotone, and $L : \mathrm{dom}(L) \to \mathscr{V}^*$ be affine and radially continuous[97] and satisfy (8.226). Moreover, let $\mathscr{V}$ admit the approximation by finite-dimensional subspaces in the above sense, and let $L + \mathscr{A}$ be coercive with respect to the norm of $\mathscr{V}$. Then $L + \mathscr{A}$ is surjective. Moreover, if $\mathscr{A}$ is strictly monotone, then $(L + \mathscr{A})^{-1} : \mathscr{V}^* \to \mathrm{dom}(L)$ does exist.*

---

[95] Such an approximation can be made as follows: after a possible re-normalization of the reflexive space V so that both V and V^* are strictly convex (by Asplund's theorem), we take $\tilde{u}_0$ on $I/2 = [0, T/2]$ as the solution to the initial-value problem $\frac{\mathrm{d}}{\mathrm{d}t}\tilde{u}_0 + J_p(\tilde{u}_0) = 0$ and $\tilde{u}_0(0) = u_0$ with $J_p : V \to V^*$ induced by the potential $\frac{1}{p}\|\cdot\|_V^p$; as for J_p cf. Example 8.103 below. By using the test with $\tilde{u}_0$, one has the estimate $\|\tilde{u}_0\|_{L^p(I/2;V)} \le C$. Then one can estimate

$$\begin{aligned}\left\|\frac{\mathrm{d}\tilde{u}_0}{\mathrm{d}t}\right\|_{L^{p'}(I/2;V^*)} &= \sup_{\|v\|_{L^p(I/2;V)}\le 1} \int_0^{T/2} \big\langle J_p(\tilde{u}_0), v\big\rangle \,\mathrm{d}t \\ &\le \sup_{\|v\|_{L^p(I/2;V)}\le 1} \int_0^{T/2} \big\|\tilde{u}_0\big\|_V^{p-1} \big\|v\big\|_V \,\mathrm{d}t \le \big\|\tilde{u}_0\big\|_{L^p(I/2;V)}^{p-1} \le C.\end{aligned}$$

On $[T/2, T]$, we define $\tilde{u}_0(t) := \tilde{u}_0(T-t)$ to have $\tilde{u}_0(0) = u_0 = \tilde{u}_0(T)$ and $\tilde{u}_0 \in W^{1,p,p'}(I;V,V^*)$.

[96] As we consider $\mathscr{V} = L^p(I;V)$ and $L = \frac{\mathrm{d}}{\mathrm{d}t}$, we get the norm on $\mathrm{dom}(L)$ identical with that induced from $W^{1,p,p'}(I;V,V^*)$. An example of $\mathscr{V}_k$ can be $\mathrm{span}\{\varphi v;\ v \in V_k,\ \varphi \in C([0,T])$ a polynom of the degree $\le k$, $\varphi(0) = 0\}$ with V_k from (2.7) in case of the initial-value problem. For the periodic problem, $\varphi(0) = 0$ is to be replaced by $\varphi(0) = \varphi(T)$.

[97] In fact, the assertion holds even without the requirement of the affinity and the radial-continuity assumptions about L.

Proof. [98] Take $w_0 \in \mathrm{dom}(L)$.[99] Then, for $\tilde{L}(u) := L(u+w_0) - L(w_0)$, $\tilde{L}$ is linear and $\mathrm{dom}(\tilde{L}) = \mathrm{dom}(L) - w_0$ is a linear subspace. We put still $\tilde{\mathscr{A}}(u) := \mathscr{A}(u+w_0)$; note that again $\tilde{\mathscr{A}}$ is monotone and radially continuous and $\tilde{L} + \tilde{\mathscr{A}}$ is coercive. The equation $L(u) + \mathscr{A}(u) = f$ is equivalent with $\tilde{L}\tilde{u} + \tilde{\mathscr{A}}(\tilde{u}) = f - L(w_0)$, their solutions being related to each other by $\tilde{u} + w_0 = u$. Thus we can assume that L is a linear operator without any loss of generality.

Consider $\mathscr{V}_k$ a finite-dimensional subspace of $\mathscr{V}$ as assumed, and endow $\mathscr{V}_k$ by the norm of $\mathscr{V}$. Denoting $I_k : \mathscr{V}_k \to \mathscr{V}$ the canonical inclusion, $I_k^* : \mathscr{V}^* \to \mathscr{V}_k^*$ and the norm of I_k and I_k^* is at most 1. We will show that

$$\exists u_k \in \mathscr{V}_k \ \ \forall v \in \mathscr{V}_k : \qquad \big\langle Lu_k + \mathscr{A}(u_k), v \big\rangle = \langle f, v \rangle. \tag{8.233}$$

Let us consider the mapping $B_k : u \mapsto I_k^*(Lu + \mathscr{A}(u)) : \mathscr{V}_k \to \mathscr{V}_k^*$, which is radially continuous, monotone, and also coercive due to the estimate

$$\begin{aligned} \frac{\langle B_k(u), u \rangle}{\|u\|_{\mathscr{V}}} &= \frac{\langle I_k^*(Lu + \mathscr{A}(u)), u \rangle}{\|u\|_{\mathscr{V}}} = \frac{\langle Lu + \mathscr{A}(u), I_k u \rangle}{\|u\|_{\mathscr{V}}} \\ &= \frac{\langle Lu + \mathscr{A}(u), u \rangle}{\|u\|_{\mathscr{V}}} \geq a\big(\|u\|_{\mathscr{V}}\big) \to +\infty \end{aligned} \tag{8.234}$$

for $\|u\|_{\mathscr{V}} \to \infty$, $u \in \mathscr{V}_k$. By the Browder-Minty Theorem 2.18, the equation $B_k(u) = I_k^* f$ has a solution u_k. Such u_k satisfies also (8.233) and, from (8.234),

$$\begin{aligned} a\big(\|u_k\|_{\mathscr{V}}\big) &\leq \frac{\langle B_k(u_k), u_k \rangle}{\|u_k\|_{\mathscr{V}}} \leq \big\|B_k(u_k)\big\|_{\mathscr{V}_k^*} = \big\|I_k^* f\big\|_{\mathscr{V}_k^*} \\ &\leq \big\|I_k^*\big\|_{\mathscr{L}(\mathscr{V}^*, \mathscr{V}_k^*)} \|f\|_{\mathscr{V}^*} = \big\|I_k\big\|_{\mathscr{L}(\mathscr{V}_k, \mathscr{V})} \|f\|_{\mathscr{V}^*} \leq \|f\|_{\mathscr{V}^*} \end{aligned} \tag{8.235}$$

hence $\{u_k\}_{k\in\mathbb{N}}$ is bounded in $\mathscr{V}$, and then also, by the monotonicity of L and by (8.233) and (8.235),

$$\begin{aligned} \big\langle \mathscr{A}(u_k), u_k \big\rangle &\leq \big\langle Lu_k, u_k \big\rangle + \big\langle \mathscr{A}(u_k), u_k \big\rangle = \big\langle I_k^* f, u_k \big\rangle = \big\langle f, I_k u_k \big\rangle \\ &= \big\langle f, u_k \big\rangle \leq \big\|f\big\|_{\mathscr{V}^*} \big\|u_k\big\|_{\mathscr{V}} \leq \big\|f\big\|_{\mathscr{V}^*} a^{-1}\big(\|f\|_{\mathscr{V}^*}\big). \end{aligned} \tag{8.236}$$

To conclude that $\{\mathscr{A}(u_k)\}_{k\in\mathbb{N}}$ is bounded in $\mathscr{V}^*$, as in (2.42), for any $\varepsilon > 0$, we estimate

$$\big\|\mathscr{A}(u_k)\big\|_{\mathscr{V}^*} \leq \frac{1}{\varepsilon} \sup_{\|v\|_{\mathscr{V}} \leq \varepsilon} \Big(\big\langle \mathscr{A}(u_k), u_k \big\rangle + \big\langle \mathscr{A}(v), v \big\rangle - \big\langle \mathscr{A}(v), u_k \big\rangle \Big). \tag{8.237}$$

Now we use that $\{\langle \mathscr{A}(u_k), u_k \rangle\}_{k\in\mathbb{N}}$ is bounded due to (8.236), $\{\langle \mathscr{A}(v), v \rangle; \|v\|_{\mathscr{V}} \leq \varepsilon\}$ is bounded if $\varepsilon > 0$ is small because $\mathscr{A}$ is locally bounded around the origin

[98] See Gajewski [168, Section III.2.2]. Alternatively, one can approximate the operator $L + \mathscr{A}$ by adding the duality mapping J.

[99] If $u_0 \in V$, we can take simply $w_0(t) = u_0$. If $u_0 \in H \setminus V$, we can take $w_0 \in W^{1,p,p'}(I; V, V^*)$, e.g., a solution of the initial-value problem $\frac{\mathrm{d}}{\mathrm{d}t} w_0 + J_p(w_0) \ni 0$ with $w_0(0) = u_0$ like in the proof of Lemma 8.93.

due to Lemma 2.15, and eventually $\{\langle \mathscr{A}(v), u_k\rangle\}_{k\in\mathbb{N}}$ is bounded by (8.235) if ε is small.

In view of the above a-priori estimates, we can consider some $(u,\chi) \in \mathscr{V}\times\mathscr{V}^*$ being the limit of a subsequence such that $u_k \rightharpoonup u$ and $\mathscr{A}(u_k) \rightharpoonup \chi$. Furthermore, take $z \in \mathscr{V}$, $v \in \mathrm{dom}(L)$ and $v_k \in \mathscr{V}_k$ such that $v_k \to v$ with respect to the norm $\|\cdot\|_{\mathrm{dom}(L)}$. Then, by (8.233), the monotonicity of L (written between v_k and u_k), and the monotonicity of $\mathscr{A}$ (written between z and u_k), we have

$$\begin{aligned} 0 &\le \big\langle Lv_k - Lu_k, v_k - u_k\big\rangle + \big\langle \mathscr{A}(z) - \mathscr{A}(u_k), z - u_k\big\rangle \\ &= \big\langle Lv_k - Lu_k - \mathscr{A}(u_k), v_k - u_k\big\rangle + \big\langle \mathscr{A}(z), z - u_k\big\rangle - \big\langle \mathscr{A}(u_k), z - v_k\big\rangle \\ &= \big\langle Lv_k - f, v_k - u_k\big\rangle + \big\langle \mathscr{A}(z), z - u_k\big\rangle - \big\langle \mathscr{A}(u_k), z - v_k\big\rangle. \end{aligned} \tag{8.238}$$

Now we can pass to the limit with $k \to \infty$. Note that L is continuous with respect to the norm $\|\cdot\|_{\mathrm{dom}(L)}$ so that $\langle Lv_k, v_k\rangle \to \langle Lv, v\rangle$. Thus we come to

$$\begin{aligned} 0 &\le \big\langle Lv - f, v - u\big\rangle + \big\langle \mathscr{A}(z), z - u\big\rangle - \big\langle \chi, z - v\big\rangle \\ &= \big\langle Lv - f + \chi, v - u\big\rangle + \big\langle \mathscr{A}(z) - \chi, z - u\big\rangle, \end{aligned} \tag{8.239}$$

which now holds for any $v \in \mathrm{dom}(L)$ and any $z \in \mathscr{V}$. Choosing $z := u$ in (8.239), we obtain

$$\big\langle Lv - f + \chi, v - u\big\rangle \ge 0 \tag{8.240}$$

for any $v \in \mathrm{dom}(L)$. This implies $u \in \mathrm{dom}(L)$ and $Lu = f - \chi$ because L satisfies (8.226). Knowing $u \in \mathrm{dom}(L)$, we can also choose $v := u$ in (8.239), which gives

$$\big\langle \mathscr{A}(z) - \chi, z - u\big\rangle \ge 0 \tag{8.241}$$

for any $z \in \mathscr{V}$. Since $\mathscr{A}$ is monotone and radially continuous, by the Minty trick (see Lemma 2.13) we obtain $\mathscr{A}(u) = \chi$. Altogether, $Lu + \mathscr{A}(u) = (f - \chi) + \chi = f$.

If $\mathscr{A}$ is strictly monotone, so is $L + \mathscr{A}$ and thus the solution to the equation $Lu + \mathscr{A}(u) = f$ is unique, which means that $(L + \mathscr{A})^{-1}$ is single-valued. □

Theorem 8.96 (Existence). *Let the Carathéodory mapping $A : I\times V \to V^*$ satisfy the growth condition (8.80) with $\mathfrak{C}$ constant and $A(t,\cdot)$ be radially continuous, monotone and semi-coercive in the sense of (8.95) with $Z := V$ but with $|\cdot|_V := \|\cdot\|_V$ and, in case (8.228), with $c_2 = 0$. Then both the Cauchy problem (8.1) and the periodic problem (8.225) have solutions. Moreover, if $A(t,\cdot)$ is strictly monotone for a.a. $t \in I$, these solutions are unique.*

Proof. First, as in the proof of Lemma 8.95, we can consider L linear without loss of generality. Since A is a Carathéodory mapping satisfying the growth condition (8.80) with $\mathfrak{C}$ constant, $\mathscr{A}$ maps $L^p(I;V)$ into $L^{p'}(I;V^*)$ and $\mathscr{A}$ is radially continuous, cf. Example 8.52.

Directly from (8.95) with $c_2 = 0$ and $|\cdot|_V := \|\cdot\|_V$, we get the coercivity of $L+\mathscr{A}$ with respect to the norm of $\mathscr{V}$ on $\mathrm{dom}(L)$ from (8.228) simply by integration over I:

$$\int_0^T\Big\langle\frac{\mathrm{d}u}{\mathrm{d}t}+A(t,u),u\Big\rangle\,\mathrm{d}t \geq \int_0^T \frac{1}{2}\frac{\mathrm{d}}{\mathrm{d}t}\|u\|_H^2 + c_0\|u(t)\|_V^p - c_1(t)\|u(t)\|_V\,\mathrm{d}t$$
$$\geq \frac{1}{2}\|u(T)\|_H^2 - \frac{1}{2}\|u(0)\|_H^2 + \int_0^T (c_0-\varepsilon)\|u(t)\|_V^p - C_\varepsilon c_1^{p'}(t)\,\mathrm{d}t$$
$$\geq (c_0-\varepsilon)\|u\|_{L^p(I;V)}^p - C_\varepsilon\|c_1\|_{L^{p'}(I)}^{p'}. \tag{8.242}$$

In case of (8.227), $c_2>0$ can be admitted by modifying (8.242) by estimating

$$\int_0^{t_1}\Big\langle\frac{\mathrm{d}u}{\mathrm{d}t}+A(t,u),u\Big\rangle\,\mathrm{d}t \geq \int_0^{t_1} \frac{1}{2}\frac{\mathrm{d}}{\mathrm{d}t}\|u\|_H^2 + c_0\|u\|_V^p - c_1\|u\|_V - c_2\|u\|_H^2\,\mathrm{d}t$$
$$\geq \frac{1}{2}\|u(t_1)\|_H^2 - \frac{1}{2}\|u_0\|_H^2 + \int_0^{t_1} (c_0-\varepsilon)\|u\|_V^p - C_\varepsilon c_1^{p'} - c_2\|u\|_H^2\,\mathrm{d}t$$

for any $t_1\in I$, from which one obtains

$$(c_0-\varepsilon)\|u\|_{L^p(I;V)}^p \leq \Big(\frac{1}{2}\|u_0\|_H^2 + \int_0^T\Big\langle\frac{\mathrm{d}u}{\mathrm{d}t}+A(t,u),u\Big\rangle + C_\varepsilon c_1^{p'}\Big)\mathrm{e}^{\int_0^T c_2\,\mathrm{d}t}$$

by using the Gronwall inequality (1.66), so that

$$\int_0^T\Big\langle\frac{\mathrm{d}u}{\mathrm{d}t}+A(t,u),u\Big\rangle\,\mathrm{d}t \geq (c_0-\varepsilon)\mathrm{e}^{-\|c_2\|_{L^1(I)}}\big\|u\big\|_{L^p(I;V)}^p - \frac{1}{2}\big\|u_0\big\|_H^2 - C_\varepsilon\big\|c_1\big\|_{L^{p'}(I)}^{p'},$$

which shows the coercivity $L+\mathscr{A}$ on $\mathrm{dom}(L)$ from (8.227).

Then we use Lemmas 8.93 and 8.95 for $\mathscr{V} = L^p(I;V)$ and L defined in Lemma 8.93; note that then $\mathrm{dom}(L)=W^{1,p,p'}(I;V,V^*)$. □

Remark 8.97. If $p<2$, the coercivity of $L+\mathscr{A}$ on $\mathrm{dom}(L)$ from (8.228) obviously fails for (8.95) with $c_2>0$. Also, the uniqueness for (8.225) fails if $A(t,\cdot)$ is merely monotone, while for (8.1) the monotonicity of $A(t,\cdot)$ is sufficient for the uniqueness, as we saw in Theorem 8.34. This is because $L+\mathscr{A}$ is then strictly monotone on $\mathrm{dom}(L)$ from (8.227) but not on $\mathrm{dom}(L)$ from (8.228).

Exercise 8.98 (Limit passage for $\frac{\partial u}{\partial t}-\Delta_p u$)**.** Modify Example 8.66 by exploiting monotonicity of $\frac{\partial u}{\partial t}-\Delta_p u$ instead of the monotonicity of $-\Delta_p u$ only.

8.10 Problems with a convex potential: direct method

For $\phi: V\to\mathbb{R}$ convex, let us define the *conjugate function* $\phi^*: V^*\to\mathbb{R}$ by

$$\phi^*(v^*) := \sup_{v\in V}\langle v^*,v\rangle - \phi(v). \tag{8.243}$$

The transformation $\phi\mapsto\phi^*$ is called the Legendre transformation in the smooth case, or the *Legendre-Fenchel transformation* in the general case.

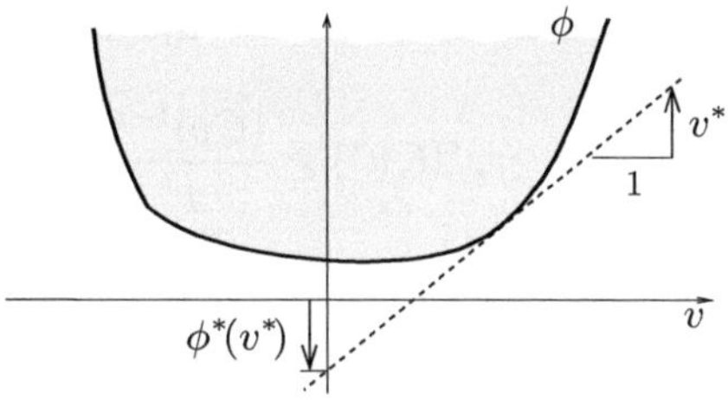

Figure 18. An illustration of a convex ϕ and the value of its conjugate ϕ^* at a given v^*.

Always, ϕ^* is convex, lower semicontinuous, and $\phi^*(v^*) + \phi(v) \geq \langle v^*, v\rangle$, which is called *Fenchel's inequality* [148]. Also $\phi^{**} = \phi$ if and only if ϕ is lower semicontinuous; recall that V is here considered as reflexive. If ϕ is lower semicontinuous and proper (i.e. $\phi > -\infty$ and $\phi \not\equiv +\infty$), then[100]

$$v^* \in \partial\phi(v) \quad\Leftrightarrow\quad v \in \partial\phi^*(v^*) \quad\Leftrightarrow\quad \phi^*(v^*) + \phi(v) = \langle v^*, v\rangle. \tag{8.244}$$

For $f \in V^*$, we have $[\phi - f]^*(v^*) = \phi^*(v^* + f)$ because

$$\begin{aligned}[\phi - f]^*(v^*) &= \sup_{v\in V}\langle v^*, v\rangle - \big(\phi(v) - \langle f, v\rangle\big) \\ &= \sup_{v\in V}\langle v^* + f, v\rangle - \phi(v) = \phi^*(v^* + f).\end{aligned} \tag{8.245}$$

For any constant c, one has $[\phi + c]^* = \phi^* - c$. Also, $[c\phi]^* = c\phi^*(\cdot/c)$ provided c is positive because

$$\big[c\phi\big]^*(v^*) = \sup_{v\in V}\langle v^*, v\rangle - c\phi(v) = c\Big(\sup_{v\in V}\Big\langle \frac{v^*}{c}, v\Big\rangle - \phi(v)\Big) = c\phi^*\Big(\frac{v^*}{c}\Big). \tag{8.246}$$

Furthermore, $\phi_1^* \leq \phi_2^*$ if $\phi_1 \geq \phi_2$. Moreover, ϕ smooth implies ϕ^* strictly convex.[101]

Suppose now that $A(t,u) = \varphi_u'(t,u)$ for some potential $\varphi : I \times V \to \mathbb{R}$ such that $\varphi(t,\cdot) : V \to \mathbb{R}$ is Gâteaux differentiable for a.a. $t \in I$ with $\varphi_u'(t,\cdot) : V \to V^*$ denoting its differential. Again, we consider the situation $V \subset H \subset V^*$ for a Hilbert space H. Let us define $\Phi : W^{1,p,p'}(I;V,V^*) \to \mathbb{R}$ by

$$\Phi(u) := \frac{1}{2}\|u(T)\|_H^2 + \int_0^T \varphi(t,u(t)) + \varphi^*\Big(t, f(t) - \frac{\mathrm{d}u}{\mathrm{d}t}\Big) - \langle f(t), u(t)\rangle\,\mathrm{d}t, \tag{8.247}$$

where $\varphi^*(t,\cdot)$ is conjugate to $\varphi(t,\cdot)$. Let us mention that Φ is well-defined provided φ is a convex Carathéodory integrand[102] satisfying

$$c\|u\|_V^p \leq \varphi(t,u) \leq C\big(1 + \|u\|_V^p\big), \tag{8.248}$$

[100]The inclusion $v^* \in \partial\phi(v)$ is equivalent to $\langle v^*, u\rangle - \phi(u) \leq \langle v^*, v\rangle - \phi(v)$ holding for any $u \in V$, which is equivalent to $\phi^*(v^*) = \sup_{u\in U}\langle v^*, u\rangle - \phi(u) = \langle v^*, v\rangle - \phi(v)$, from which already the equivalence of the first and the third statements follows. If ϕ is lower semicontinuous, then $\phi^{**} = \phi$, so that the equivalence with the second statement follows by symmetry.

[101]Suppose a contrary, i.e. Graph(ϕ^*) contains a segment, then $\partial\phi(u)$ is not a singleton for u, which contradicts Gâteaux differentiability of φ, cf. Exercise 5.34.

[102]This means that $\varphi(t,\cdot) : V \to \mathbb{R}$ is convex and continuous while $\varphi(\cdot, v) : I \to \mathbb{R}$ is measurable.

which implies the analogous estimates for the conjugate φ^*, namely

$$\frac{(cp)^{1-p'}}{p'}\|u^*\|_{V^*}^{p'} \geq \varphi^*(t,u^*) \geq \frac{(Cp)^{1-p'}}{p'}\|u^*\|_{V^*}^{p'} - C; \tag{8.249}$$

cf. Example 8.103. Note that (8.248) ensures that $\mathscr{N}_\varphi : L^p(I;V) \to L^1(I)$ while (8.249) ensures that $\varphi^*(t,\cdot)$ has at most p'-growth so that $\mathscr{N}_{\varphi^*} : L^{p'}(I;V^*) \to L^1(I)$; the needed fact that φ^* is a Carathéodory integrand can be proved from separability of V and from (8.249).[103]

Theorem 8.99 (BREZIS-EKELAND VARIATIONAL PRINCIPLE [68]). *Let φ be a Carathéodory function satisfying* (8.248) *and $\varphi(t,\cdot)$ be convex and continuously differentiable.*[104] *Then:*

(i) *If $u \in W^{1,p,p'}(I;V,V^*)$ solves the Cauchy problem* (8.1), *then u minimizes Φ over* $\mathrm{dom}(L)$ *from* (8.227) *and, moreover, $\Phi(u) = \frac{1}{2}\|u_0\|_H^2$.*

(ii) *Conversely, if $\Phi(u) = \frac{1}{2}\|u_0\|_H^2$ for some $u \in \mathrm{dom}(L)$ from* (8.227), *then u minimizes Φ over* $\mathrm{dom}(L)$ *and solves the Cauchy problem* (8.1).

Proof. By (8.245) we have $\varphi^*(t, f+\cdot) = [\varphi(t,\cdot) - f]^*$, and therefore, by using the Fenchel inequality and (7.22), we have always

$$\begin{aligned}\Phi(u) &= \frac{1}{2}\|u(T)\|_H^2 + \int_0^T \varphi(t,u(t)) - \langle f(t),u(t)\rangle + \varphi^*\Big(t, f(t) - \frac{\mathrm{d}u}{\mathrm{d}t}\Big)\,\mathrm{d}t \\ &\geq \int_0^T -\Big\langle \frac{\mathrm{d}u}{\mathrm{d}t}, u(t)\Big\rangle\,\mathrm{d}t + \frac{1}{2}\|u(T)\|_H^2 = \frac{1}{2}\|u_0\|_H^2 \end{aligned}\tag{8.250}$$

for any $u \in \mathrm{dom}(L)$. If u solves the Cauchy problem (8.1), i.e. $\frac{\mathrm{d}}{\mathrm{d}t}u + \varphi'_u(t,u(t)) = f(t)$ or, in other words, $-\frac{\mathrm{d}}{\mathrm{d}t}u = (\varphi(t,u) - \langle f(t),u\rangle)'_u$, then, by using (8.245) and (8.244),

$$\begin{aligned}&[\varphi - f(t)]\,(u(t)) + [\varphi - f(t)]^*\Big(-\frac{\mathrm{d}u}{\mathrm{d}t}\Big) \\ &\quad = \varphi\big(t,u(t)\big) - \big\langle f(t),u(t)\big\rangle + \varphi^*\Big(t, f(t) - \frac{\mathrm{d}u}{\mathrm{d}t}\Big) = -\Big\langle \frac{\mathrm{d}u}{\mathrm{d}t}, u(t)\Big\rangle. \end{aligned}\tag{8.251}$$

Hence this u attains the minimum of Φ on $\mathrm{dom}(L)$, proving thus (i).

[103] For a countable dense set $\{v_k\}_{k\in\mathbb{N}} \subset V$, we have $\varphi^*(t,v^*) = \sup_{k\in\mathbb{N}}\langle v^*,v_k\rangle - \varphi(t,v_k)$ and then $\varphi^*(\cdot,v^*)$, being a supremum of a countable collection of measurable functions $\{\langle v^*,v_k\rangle - \varphi(\cdot,v_k)\}_{k\in\mathbb{N}}$, is itself measurable. Moreover, (8.249) makes the convex functional $\varphi^*(t,\cdot)$ locally bounded from above on the Banach space V^*, hence it must be continuous.

[104] This guarantees, in particular, that A is a Carathéodory mapping: the continuity of $A(t,\cdot) = \varphi'(t,\cdot)$ is just assumed while the measurability of $A(\cdot,u) = \varphi'(\cdot,u)$ follows from the measurability of both $\varphi(\cdot,u+\varepsilon v)$ and $\varphi(\cdot,u)$ for any $u,v \in V$, hence by Lebesgue's Theorem 1.14 $\langle A(\cdot,u),v\rangle = \mathrm{D}\varphi(\cdot,u;v) = \lim_{\varepsilon\to 0}\frac{1}{\varepsilon}\varphi(\cdot,u+\varepsilon v) - \frac{1}{\varepsilon}\varphi(\cdot,u)$ is Lebesgue measurable, too, and eventually $A(\cdot,u)$ itself is Bochner measurable by Pettis' Theorem 1.34 by exploiting again the (generally assumed) separability of V.

Conversely, suppose that $\Phi(u) = \frac{1}{2}\|u_0\|_H^2$. Note that, in view of (8.250), $u \in \mathrm{dom}(L)$ then also minimizes Φ on $\mathrm{dom}(L)$. Moreover, by (8.250),

$$0 = \Phi(u) - \frac{1}{2}\|u_0\|_H^2 = \int_0^T \varphi(t, u(t)) + \varphi^*\Big(t, f(t) - \frac{\mathrm{d}u}{\mathrm{d}t}\Big) - \Big\langle f(t) - \frac{\mathrm{d}u}{\mathrm{d}t}, u(t)\Big\rangle \,\mathrm{d}t \geq 0;$$

the last inequality goes from the Fenchel inequality. Thus, for a.a. t, it holds that $\varphi(t, u(t)) - \langle f(t), u(t)\rangle + \varphi^*(t, f(t) - \frac{\mathrm{d}}{\mathrm{d}t}u) + \langle \frac{\mathrm{d}}{\mathrm{d}t}u, u(t)\rangle = 0$. By (8.244), this is equivalent with $\frac{\mathrm{d}}{\mathrm{d}t}u = f(t) - \varphi'_u(t, u(t))$, so that $u \in \mathrm{dom}(L)$ solves the Cauchy problem (8.1). □

Corollary 8.100. *Let the assumptions of Theorem* 8.99 *be fulfilled. Then the solution to* (8.1) *is unique.*

Proof. As φ is smooth, φ^* is strictly convex, and thus also Φ is strictly convex because L is injective on $\mathrm{dom}(L)$ from (8.227). Thus Φ can have only one minimizer on the affine manifold $\mathrm{dom}(L)$. By Theorem 8.99(i), it gives uniqueness of the solution to (8.1). □

Remark 8.101 (*Periodic problems*). Modification for periodic problem (8.225) uses: $\Phi(u) := \int_0^T \varphi(t, u(t)) + \varphi^*(t, f(t) - \frac{\mathrm{d}}{\mathrm{d}t}u) - \langle f(t), u(t)\rangle \mathrm{d}t$ and $\mathrm{dom}(L)$ from (8.228). The minimum of Φ on $\mathrm{dom}(L)$ is 0. Modification of Corollary 8.100 for periodic problems requires $\varphi(t,\cdot)$ strictly convex because L is not injective on $\mathrm{dom}(L)$ from (8.228) so that strict convexity of $\varphi^*(t,\cdot)$ does not ensure strict convexity of Φ.

Note that Theorem 8.99(i) stated the existence of a minimizer of Φ on $\mathrm{dom}(L)$ by means of an a-priori knowledge that the solution to the Cauchy problem (8.1) does exist. We can however proceed in the opposite way, which gives us another (so-called direct) method to prove existence of a solution to (8.1). Note that Theorem 8.99(ii) does not imply this existence result because $\min_{u\in\mathrm{dom}(L)} \Phi(u) = \frac{1}{2}\|u_0\|_H^2$ is not obvious unless we know that the solution to (8.1) exists.

Theorem 8.102 (Direct method). *Let the assumptions of Theorem* 8.99 *be fulfilled and let also* $\varphi^*(t,\cdot)$ *be smooth. Then:*

(i) *Φ attains its minimum on* $\mathrm{dom}(L)$.

(ii) *Moreover, this (unique) minimizer represents the solution to the Cauchy problem* (8.1).

Proof. (i) Φ is convex, continuous, $W^{1,p,p'}(I;V,V^*)$ is reflexive, and by Lemma 7.3 the mapping $u \mapsto u(0) : W^{1,p,p'}(I;V,V^*) \to H$ is continuous so that $\mathrm{dom}(L)$ is closed in $W^{1,p,p'}(I;V,V^*)$. Moreover, by (8.248) and (8.249), Φ is coercive on $\mathrm{dom}(L)$: indeed, due to the lower bound

$$\Phi(u) \geq \int_0^T \Big[c\|u\|_V^p - \langle f, u\rangle + \frac{(Cp)^{1-p'}}{p'}\Big\|f - \frac{\mathrm{d}u}{\mathrm{d}t}\Big\|_{V^*}^{p'} - C\Big] \mathrm{d}t, \tag{8.252}$$

obviously $\Phi(u) \to +\infty$ for $\|u\|_{L^p(I;V)} + \|\frac{\mathrm{d}}{\mathrm{d}t}u\|_{L^{p'}(I;V^*)} \to \infty$. Then the existence of a minimizer follows by the direct method.

(ii) We must calculate Φ' and then use simply $\langle\Phi'(u), v\rangle = 0$ for any v belonging to the tangent cone to $\mathrm{dom}(L)$ at u, i.e for any $v \in W^{1,p,p'}(I;V,V^*)$ with $v(0) = 0$.

Without loss of generality, we can consider $u_0 = 0$; cf. the proof of Lemma 8.95. Then $\mathrm{dom}(L)$ is a linear subspace and, if endowed by the topology of $W^{1,p,p'}(I;V,V^*)$, L is continuous and injective, and therefore L^{-1} does exist on $\mathrm{Range}(L)$.

Denote $\varphi_T(u) = \int_0^T \varphi(t,u(t))\,\mathrm{d}t$ and similarly $\varphi_T^*(\xi) = \int_0^T \varphi^*(t,\xi(t))\,\mathrm{d}t$. Note that $\varphi_T^* : L^{p'}(I;V^*) \to \mathbb{R}$ is conjugate to $\varphi_T : L^p(I;V) \to \mathbb{R}$.[105] Using $L : u \mapsto \frac{\mathrm{d}}{\mathrm{d}t}u : W^{1,p,p'}(I;V,V^*) \to L^{p'}(I;V^*)$ and (8.247), we can write

$$\Phi(u) = [\varphi_T - f](u) + \varphi_T^*(f - L(u)) + \frac{1}{2}\|u(T)\|_H^2. \tag{8.253}$$

Using the first equivalence in (8.244) and realizing that $\partial\varphi_T = \{\varphi_T'\}$ and $\partial\varphi_T^* = \{[\varphi_T^*]'\}$, we obtain $[\varphi_T^*]' = [\varphi_T']^{-1}$. In particular, denoting $w := [\varphi_T^*]'(f - L(u))$, we have $w = [\varphi_T']^{-1}(f - L(u))$, so that

$$\varphi_T'(w) = f - L(u). \tag{8.254}$$

We can then calculate[106]

$$\begin{aligned}\Phi'(u) &= \varphi_T'(u) - f - L^*\big([\varphi_T^*]'(f - L(u))\big) + u(T)\cdot\delta_T \\ &= \varphi_T'(u) - f + L(w) + (u(T) - w(T))\cdot\delta_T \end{aligned} \tag{8.255}$$

where $\delta_T : W^{1,p,p'}(I;V,V^*) \to H : u \mapsto u(T)$ and where we used also the identity $L^*(w) = -L(w) + w(T)\cdot\delta_T$ which follows from the by-part formula for $w \in W^{1,p,p'}(I;V,V^*) \subset L^{p'}(I;V^*)^*$ if $v(0) = 0$ is taken into account:

$$\begin{aligned}\big\langle L(w), v\big\rangle &= \Big\langle \frac{\mathrm{d}w}{\mathrm{d}t}, v\Big\rangle = -\Big\langle w, \frac{\mathrm{d}v}{\mathrm{d}t}\Big\rangle + \big\langle w(T), v(T)\big\rangle - \big\langle w(0), v(0)\big\rangle \\ &= \big\langle w, L(v)\big\rangle + \big\langle w(T), v(T)\big\rangle = \big\langle L^*(w), v\big\rangle + \big\langle w(T), v(T)\big\rangle. \end{aligned} \tag{8.256}$$

From $\langle\Phi'(u), v\rangle = 0$ with v vanishing on $[0, T-\varepsilon]$ and such that $v(T) = u(T) - w(T)$, passing also $\varepsilon \to 0$, we obtain from (8.255) that $0 = \langle v(T), u(T) - w(T)\rangle = \|u(T) - w(T)\|_H^2$, i.e.

$$u(T) = w(T). \tag{8.257}$$

[105] This follows from the identity $[\varphi_T]^*(\xi) = \sup_{u\in L^p(I;V)} \int_0^T \langle\xi(t), u(t)\rangle - \varphi(t,u(t))\,\mathrm{d}t = \int_0^T \sup_{u\in V}[\langle\xi(t), u\rangle - \varphi(t,u)]\mathrm{d}t = \varphi_T^*(\xi)$ which can be proved by a measurable-selection technique.

[106] We also use the formula $\partial\Phi(A(u)) = A^*\partial\Phi(u)$ which holds provided $0 \in \mathrm{int}(\mathrm{Range}(A) - \mathrm{Dom}(\Phi))$.

Furthermore, taking v with a compact support in $(0,T)$, we get from $\langle\Phi'(u),v\rangle = 0$ with Φ' in (8.255) that

$$\varphi_T'(u) - f + L(w) = 0. \tag{8.258}$$

Subtracting (8.254) and (8.258), testing it by $u-w$, and using monotonicity of φ_T', we get

$$0 = \big\langle L(u)-L(w), u-w\big\rangle + \big\langle \varphi_T'(w)-\varphi_T'(u), u-w\big\rangle \le \frac{1}{2}\frac{\mathrm{d}}{\mathrm{d}t}\big\|u-w\big\|_H^2. \tag{8.259}$$

Using (8.257) and the Gronwall inequality backward, we get $u=w$. Putting this into (8.258) (or alternatively into (8.254)), we get $\frac{\mathrm{d}}{\mathrm{d}t}u + \varphi_u'(t,u(t)) = f(t)$; here we use also that[107] $[\varphi_T'(u)](t) = \varphi_u'(t,u(t))$. As $u \in \mathrm{dom}(L)$, the initial condition $u(0)=u_0$ is satisfied, too. $\square$

Example 8.103. For $\phi(v) = \frac{1}{p}\|v\|_V^p$, the conjugate function is $\phi^*(v^*) = \frac{1}{p'}\|v^*\|_{V^*}^{p'}$, which follows by the Hölder inequality and which explains why $p' := p/(p-1)$ has been called a *conjugate exponent.* This also implies $(c\|\cdot\|_V^p)^* = cp\frac{1}{p'}\|\frac{\cdot}{cp}\|_{V^*}^{p'} = \frac{(cp)^{1-p'}}{p'}\|\cdot\|_{V^*}^{p'}$.

Example 8.104 (Parabolic evolution by p-Laplacean[108]). Considering $V = W_0^{1,p}(\Omega)$, $\varphi(t,u) = \int_\Omega \frac{1}{p}|\nabla u|^p \mathrm{d}x$ and $\langle f(t),u\rangle = \int_\Omega g(t,x)u(x)\,\mathrm{d}x$ corresponds, in the variant of the Cauchy problem, to the initial-boundary-value problem

$$\left.\begin{aligned} \frac{\partial u}{\partial t} - \mathrm{div}\big(|\nabla u|^{p-2}\nabla u\big) &= g && \text{in } Q,\\ u &= 0 && \text{on } \Sigma,\\ u(0,\cdot) &= u_0 && \text{in } \Omega; \end{aligned}\right\} \tag{8.260}$$

cf. Example 4.23. Let us abbreviate $\Delta_p : W_0^{1,p}(\Omega) \to W^{-1,p'}(\Omega)$ the p-Laplacean, this means $\Delta_p u := \mathrm{div}(|\nabla u|^{p-2}\nabla u)$. One can notice that $\varphi(u) = \frac{1}{p}\|u\|_{W_0^{1,p}(\Omega)}^p$ provided $\|u\|_{W_0^{1,p}(\Omega)} := \|\nabla u\|_{L^p(\Omega;\mathbb{R}^n)}$. Then $\varphi^*(\xi) = \frac{1}{p'}\|\xi\|_{W^{-1,p'}(\Omega)}^{p'}$. Moreover, one can see[109] that $\Delta_p u = -J_p(u)$ where $J_p : V \to V^*$ is the duality mapping with respect to the p-power defined by the formulae $\langle J_p(u),u\rangle = \|J_p(u)\|_{V^*}\|u\|_V$ and $\|J_p(u)\|_{V^*} = \|u\|_V^{p-1}$. Hence, $\|\xi\|_{V*} = \|J_p^{-1}(\xi)\|_V^{p-1}$ implies here $\|\xi\|_{W^{-1,p'}(\Omega)}^{p'} = \|\Delta_p^{-1}\xi\|_{W_0^{1,p}(\Omega)}^p$ so that

$$\varphi^*(\xi) = \frac{1}{p'}\big\|\xi\big\|_{W^{-1,p'}(\Omega)}^{p'} = \frac{1}{p'}\big\|\Delta_p^{-1}\xi\big\|_{W_0^{1,p}(\Omega)}^p = \frac{1}{p'}\big\|\nabla\,\Delta_p^{-1}\xi\big\|_{L^p(\Omega;\mathbb{R}^n)}^p. \tag{8.261}$$

---

107 See e.g. [209, Theorem II.9.24].

108 For the linear case (i.e. $p=2$) see Brezis and Ekeland [68] or also Aubin [26].

109 Cf. Proposition 3.14 which, however, must be modified. Note that, for $p=2$, $J_p = J$ with J the standard duality mapping (3.1).

It yields the following explicit form of the functional Φ:

$$\Phi(u) = \int_0^T \bigg(\int_\Omega \frac{1}{p}|\nabla u|^p + \frac{1}{p'}\Big|\nabla \Delta_p^{-1}\Big(g - \frac{\partial u}{\partial t}\Big)\Big|^p - gu \,\mathrm{d}x + \Big\langle \frac{\partial u}{\partial t}, u \Big\rangle \bigg) \mathrm{d}t. \quad (8.262)$$

We can observe that the integrand in (8.262) is nonlocal in space; some nonlocality (in space or in time) is actually inevitable as shown by Adler [5] who proved that there is no local variational principle yielding (8.260) as its Euler-Lagrange equation.

8.11 Bibliographical remarks

Further reading concerning evolution by pseudomonotone mappings can include monographs by Brezis [65], Gajewski et al. [168], Lions [261], Růžička [376, Sect. 3.3.5-6], Showalter [383, Chap.III], Zeidler [427, Vol.II B]. Beside Rothe's original artical [356], also e.g. [304] and special monographs by Kačur [219] and Rektorys [347] are devoted to Rothe's method. Moreover, various semi-implicit modifications of the basic Rothe method leading to efficient numerical schemes has been devised in [214, 221, 222].

Galerkin's method is in a special monograph Thomée [405], and also in Zeidler [427, Sect.30].

Quasilinear parabolic equations are thoroughly exposed in Ladyzhenskaya, Solonikov, Uraltseva [249], Liebermann [259, Chap.13], Lions [261, Chap.II.1], and Taylor [402, Chap.15]. Semilinear equations received special attentions in Henri [200], Pao [324] and Robinson [351]. Monotone parabolic equations are also in Wloka [424]. The weak solution we derived in Theorems 8.13 and 8.31 on an abstract level for weakly continuous mappings can in concrete cases be derived for quasilinear equations, too; cf. [268, Sect.5.3]. *Fully nonlinear equations* of the type $\frac{\partial}{\partial t}u + a(\Delta u) = g$ (also not mentioned in here) are, e.g., in Dong [126, Chap.9,10], or Liebermann [259, Chap.14–15].

Regularity theory for parabolic equations is exposed, e.g., in Bensoussan and Frehse [50], Kačur [219, Chap.3], Ladyzhenskaya et al. [249], Lions and Magenes [262], and Taylor [402, Chap.15]. For further advanced topics in parabolic problems see DiBenedetto [120], Galaktionov [169], and Zheng [429]. In the context of their optimal control, we refer e.g. to Fattorini [144, Part II] or Tröltzsch [410].

The Navier-Stokes equations have been thoroughly exposed by Constantin and Foias [106], Feistauer [147], Lions [261, Chap.I.6], Sohr [391], Taylor [402, Chap.17], and Temam [403]. Generalization for non-Newtonean fluids is in Ladyzhenskaya [248] and Málek et al. [268].

For time periodic problems we refer to Vejvoda et al. [415]. The method used for the existence Theorem 8.96 applies also to the general case when $\mathscr{A}$ is pseudomonotone, see Brezis [65] or also Zeidler [427, Section 32.4]. See also Lions [261, Section 7.2.2] for $\mathscr{A}$ being the mapping of type (M). The anti-periodic problems have been addresed e.g. in [8, 198].

The Brezis-Ekeland principle in the weaker version as in Theorem 8.99, invented in [68] and independently by Nayroles [301] and thus sometimes addressed as a Brezis-Ekeland-Nayroles principle, can be found even for nonsmooth problems in Aubin and Cellina [28, Section 3.4] or Aubin [26] for V a Hilbert space, $f = 0$, and autonomous systems. The improvement as a direct method, i.e. Theorem 8.102, is from [362]. Newer investigations are by Ghoussoub [185] and Tzou [186], Stefanelli [396], and Visintin [419, 420], cf. also (11.54) below. Other variational principles for parabolic equations had been surveyed by Hlaváček [203].

Chapter 9

Evolution governed by accretive mappings

Now we replace the weak compactness and monotonicity method by the norm topology technique and a completeness argument. Although, in comparison with the former technique, this method is not the basic one, it widens in a worthwhile way the range of the monotone-mapping approach presented in Chapter 8.

Again we consider the Cauchy problem (8.4) but now with $A : \mathrm{dom}(A) \to X$ an m-accretive mapping (or, more generally, $A+\lambda\mathbf{I}$ m-accretive for some $\lambda \geq 0$), X a Banach space whose norm will be denoted by $\|\cdot\|$ as in Chap. 3, $\mathrm{dom}(A)$ dense[1] in X, $f \in L^1(I;X)$, $u_0 \in X$.

9.1 Strong solutions

Let us agree to call $u \in W^{1,1}(I;X) \equiv W^{1,\infty,1}(I;X,X)$ a *strong solution* to the initial-value problem (8.4) if $\{A(u(t))\}_{t\geq 0}$ is bounded in X, hence in particular $u(t) \in \mathrm{dom}(A)$ for all $t \in I$, and (8.4) is valid a.e. on $I := [0,T]$, as well as the initial condition in (8.4) holds, and the distributional derivative $\frac{\mathrm{d}}{\mathrm{d}t}u \in L^1(I;X)$ is also the *weak derivative*, i.e. w-$\lim_{\varepsilon\to 0}\frac{1}{\varepsilon}u(t+\varepsilon) - \frac{1}{\varepsilon}u(t)$ for a.a. $t \in I$.[2] The following assertion will be found useful:

Lemma 9.1 (Chain rule). *Let $\Phi : X \to \mathbb{R}$ be convex and locally Lipschitz continuous, and $u \in W^{1,1}(I;X)$ have also the weak derivative. Then $\Phi \circ u : I \to \mathbb{R}$ is a.e. differentiable and $\frac{\mathrm{d}}{\mathrm{d}t}\Phi(u(t)) = \langle f, \frac{\mathrm{d}}{\mathrm{d}t}u(t)\rangle$ with any $f \in \partial\Phi(u(t))$ holds for a.a. $t \in I$.*

[1] In fact, if $\mathrm{cl}(\mathrm{dom}(A)) \neq X$, we must require $u_0 \in \mathrm{cl}(\mathrm{dom}(A))$ in Lemma 9.4 and Theorem 9.5.

[2] In general, $u \in W^{1,1}(I;X)$ need not have the weak derivative, but if it has, then it coincides with the distributional derivative $\frac{\mathrm{d}}{\mathrm{d}t}u$.

Proof. As $u \in W^{1,1}(I;X)$ is bounded and absolutely continuous and Φ locally Lipschitz, $\Phi \circ u$ is absolutely continuous, and hence a.e. differentiable. Consider $t \in I$ at which $\Phi \circ u$ as well as u have (weak) derivatives. As Φ is convex, $\Phi(u(t+\varepsilon)) \geq \Phi(u(t)) + \langle f, u(t+\varepsilon) - u(t)\rangle$ for any $f \in \partial\Phi(u(t))$ and any $\varepsilon \in [-t, T-t]$, cf. (5.2) for $v := u(t+\varepsilon)$. In particular, for $\varepsilon > 0$, we obtain

$$\frac{\Phi(u(t+\varepsilon)) - \Phi(u(t))}{\varepsilon} \geq \Big\langle f, \frac{u(t+\varepsilon) - u(t)}{\varepsilon}\Big\rangle \quad \text{and} \tag{9.1a}$$

$$\frac{\Phi(u(t)) - \Phi(u(t-\varepsilon))}{\varepsilon} \leq \Big\langle f, \frac{u(t) - u(t-\varepsilon)}{\varepsilon}\Big\rangle. \tag{9.1b}$$

Passing to the limit in (9.1a,b), we obtain respectively $\frac{\mathrm{d}}{\mathrm{d}t}\Phi(u(t)) \geq \langle f, \frac{\mathrm{d}}{\mathrm{d}t}u(t)\rangle$ and $\frac{\mathrm{d}}{\mathrm{d}t}\Phi(u(t)) \leq \langle f, \frac{\mathrm{d}}{\mathrm{d}t}u(t)\rangle$. □

The following assertion justifies the definition of the strong solution:

Proposition 9.2 (Uniqueness, continuous dependence). *Let A_λ, defined by*

$$A_\lambda := A + \lambda \mathbf{I} \tag{9.2}$$

with $\mathbf{I} : X \to X$ denoting the identity, be accretive, and X^* be separable. Then the strong solution, if it exists, is unique. Moreover,

$$\|u_1 - u_2\|_{C(I;X)} \leq \mathrm{e}^{\max(\lambda,0)T}\big(\|f_1 - f_2\|_{L^1(I;X)} + \|u_{01} - u_{02}\|\big) \tag{9.3}$$

with u_i being the unique strong solutions to $\frac{\mathrm{d}}{\mathrm{d}t}u_i + A(u_i) = f_i$, $u_i(0) = u_{0i}$, $i = 1, 2$.

Proof. Our strategy is to test the difference $\frac{\mathrm{d}}{\mathrm{d}t}(u_1 - u_2) + A(u_1) - A(u_2) = f_1 - f_2$ by $J(u_1 - u_2)$. Using Lemma 9.1 for $\Phi = \frac{1}{2}\|\cdot\|^2$ and that $\partial\Phi = J$, cf. Example 5.2, we obtain $\langle j^*, \frac{\mathrm{d}}{\mathrm{d}t}(u_1 - u_2)\rangle = \frac{1}{2}\frac{\mathrm{d}}{\mathrm{d}t}\|u_1 - u_2\|^2$ for any $j^* \in J(u_1 - u_2)$ a.e. on I. As we have the liberty in taking $j^* \in J(u_1 - u_2)$ arbitrarily, we select it so that $\langle j^*, A_\lambda(u_1) - A_\lambda(u_2)\rangle \geq 0$ a.e. on I, using the accretivity of A_λ. Then, for a.a. $t \in I$, we have

$$\begin{aligned}
\|u_1 - u_2\|\frac{\mathrm{d}}{\mathrm{d}t}\|u_1 - u_2\| &= \frac{1}{2}\frac{\mathrm{d}}{\mathrm{d}t}\|u_1 - u_2\|^2 \\
&\leq \frac{1}{2}\frac{\mathrm{d}}{\mathrm{d}t}\|u_1 - u_2\|^2 + \langle j^*, A_\lambda(u_1) - A_\lambda(u_2)\rangle \\
&= \Big\langle j^*, \frac{\mathrm{d}(u_1 - u_2)}{\mathrm{d}t} + A_\lambda(u_1) - A_\lambda(u_2)\Big\rangle = \langle j^*, f_1 - f_2 + \lambda(u_1 - u_2)\rangle \\
&\leq \|j^*\|_*\big(\|f_1 - f_2\| + \lambda\|u_1 - u_2\|\big) = \|u_1 - u_2\|\big(\|f_1 - f_2\| + \lambda\|u_1 - u_2\|\big),
\end{aligned} \tag{9.4}$$

then we divide[3] it by $\|u_1 - u_2\|$, which gives $\frac{\mathrm{d}}{\mathrm{d}t}\|u_1 - u_2\| \leq \|f_1 - f_2\| + \lambda\|u_1 - u_2\|$, and use the Gronwall inequality (1.66). This gives $\|u_1(t) - u_2(t)\| \leq \mathrm{e}^{\lambda t}(\|u_{01} - u_{02}\| + \int_0^t \|f_1(\vartheta) - f_2(\vartheta)\|\mathrm{e}^{-\lambda\vartheta}\mathrm{d}\vartheta)$. The uniqueness comes as a side product. □

[3] This step is legal. Indeed, assume that, at some $t > 0$, it holds that $\|u_1 - u_2\| = 0$ and simultaneously $\frac{\mathrm{d}}{\mathrm{d}t}\|u_1 - u_2\| > \|f_1 - f_2\| + \lambda\|u_1 - u_2\| \geq 0$, which would however imply existence of $\varepsilon > 0$ such that $\|u_1 - u_2\|(t - \varepsilon) = \|u_1 - u_2\|(t) + o(\varepsilon) < 0$, a contradition.

The existence of a solution will be proved by the *Rothe method*, based again on the recursive formula (8.5) combined with (8.58). The Rothe functions u_τ and $\bar{u}_\tau$ are again defined by (8.6) and (8.7), respectively.

Lemma 9.3 (EXISTENCE OF ROTHE'S SEQUENCE). *Let A_λ be m-accretive, $f \in L^1(I;X)$, $u_0 \in X$. Then u_τ does exist provided $\tau < 1/\lambda$ (or τ arbitrary if $\lambda \le 0$).*

Proof. We have

$$\big[\mathbf{I} + \tau A\big](u_\tau^1) = u_0 + \int_0^\tau f(t)\,\mathrm{d}t \in X, \tag{9.5}$$

which has at least one solution $u_\tau^1 \in \mathrm{dom}(A)$ because $\mathbf{I} + \tau A = \mathbf{I} + \tau A_\lambda - \tau\lambda\mathbf{I} = (1-\tau\lambda)[\mathbf{I}+\frac{\tau}{1-\tau\lambda}A_\lambda]$ and A_λ is m-accretive and thus $[\mathbf{I}+\lambda_1 A_\lambda]$ is surjective for any $\lambda_1 := \tau/(1-\tau\lambda)$ positive, in particular for any $\tau > 0$ sufficiently small, obviously $\tau < 1/\lambda$ (if $\lambda > 0$), cf. Definition 3.4 and (9.2).

Recursively, we obtain u_τ^2, u_τ^3, etc. □

Lemma 9.4 (A-PRIORI ESTIMATES). *Let A_λ be m-accretive, $f \in L^1(I;X)$, and $u_0 \in X$. Then, for $\tau < 1/\lambda$, it holds that*

$$\|u_\tau\|_{C(I;X)} \le C, \quad \|\bar{u}_\tau\|_{L^\infty(I;X)} \le C. \tag{9.6}$$

If, in addition, $f \in W^{1,1}(I;X)$ and $u_0 \in \mathrm{dom}(A)$, then also

$$\left\|\frac{\mathrm{d}u_\tau}{\mathrm{d}t}\right\|_{L^\infty(I;X)} \le C\big(\|f\|_{W^{1,1}(I;X)} + \|A(u_0)\|\big). \tag{9.7}$$

Proof. First, by making a transformation as in the proof of Lemma 8.95, one can assume $\mathrm{dom}(A) \ni 0$.

The identity (8.5) with (8.58) can be rewritten into the form

$$u_\tau^k = \Big[\frac{1}{\tau}\mathbf{I} + A_\lambda\Big]^{-1}\Big(\frac{1}{\tau}u_\tau^{k-1} + f_\tau^k + \lambda u_\tau^k\Big), \quad \text{where } f_\tau^k := \frac{1}{\tau}\int_{(k-1)\tau}^{k\tau} f(t)\,\mathrm{d}t. \tag{9.8}$$

As A_λ is accretive, $[\mathbf{I} + \tau A_\lambda]^{-1}$ is non-expansive (see Lemma 3.7) and therefore $[\frac{1}{\tau}\mathbf{I} + A_\lambda]^{-1}$ is Lipschitz continuous with the constant τ. Denoting $v_\tau := [\frac{1}{\tau}\mathbf{I} + A_\lambda]^{-1}(0)$, we obtain from (9.8) that $\|u_\tau^k - v_\tau\| \le \tau\|(\frac{1}{\tau}u_\tau^{k-1} + f_\tau^k + \lambda u_\tau^k) - 0\|$, from which we obtain

$$\|u_\tau^k\| \le \|u_\tau^{k-1}\| + \tau\big(\|f_\tau^k\| + \lambda\|u_\tau^k\|\big) + \|v_\tau\|. \tag{9.9}$$

Then we get the estimates (9.6) by using the discrete Gronwall inequality (1.70)[4] because $\tau\lambda < 1$ and because $\|v_\tau\| = \mathscr{O}(\tau)$; indeed,

$$\|v_\tau\| = \|v_\tau - 0\| \le \Big\|\big(v_\tau + \tau A_\lambda(v_\tau)\big) - \big(0 + \tau A_\lambda(0)\big)\Big\| = \tau\|A_\lambda(0)\| = \mathscr{O}(\tau) \tag{9.10}$$

[4]Note that the condition $\tau < 1/a$ in (1.70) reads here just as $\tau < 1/\lambda$.

because $[\mathbf{I}+\tau A_\lambda]^{-1}$ is non-expansive and because $v_\tau+\tau A_\lambda(v_\tau)=0$.

Furthermore, assuming $f\in W^{1,1}(I;X)$ and $u_0\in\mathrm{dom}(A)$, we apply $J(u_\tau^k-u_\tau^{k-1})$ to (8.5). We get, by using accretivity of A_λ,[5]

$$\begin{aligned}\frac{1}{\tau}\|u_\tau^k-u_\tau^{k-1}\|^2 &\le \Big\langle J(u_\tau^k-u_\tau^{k-1}),\frac{u_\tau^k-u_\tau^{k-1}}{\tau}\Big\rangle\\ &\qquad+\Big\langle J(u_\tau^k-u_\tau^{k-1}),A_\lambda(u_\tau^k)-A_\lambda(u_\tau^{k-1})\Big\rangle\\ &=\big\langle J(u_\tau^k-u_\tau^{k-1}),f_\tau^k-A(u_\tau^{k-1})+\lambda(u_\tau^k-u_\tau^{k-1})\big\rangle\\ &=\big\langle J(u_\tau^k-u_\tau^{k-1}),f_\tau^{k-1}-A(u_\tau^{k-1})+(f_\tau^k-f_\tau^{k-1})+\lambda(u_\tau^k-u_\tau^{k-1})\big\rangle\\ &=\Big\langle J(u_\tau^k-u_\tau^{k-1}),\frac{u_\tau^{k-1}-u_\tau^{k-2}}{\tau}+(f_\tau^k-f_\tau^{k-1})+\lambda(u_\tau^k-u_\tau^{k-1})\Big\rangle\\ &\le\big\|J(u_\tau^k-u_\tau^{k-1})\big\|_*\Big(\Big\|\frac{u_\tau^{k-1}-u_\tau^{k-2}}{\tau}\Big\|+\big\|f_\tau^k-f_\tau^{k-1}\big\|+\lambda\big\|u_\tau^k-u_\tau^{k-1}\big\|\Big).\end{aligned}\tag{9.11}$$

Dividing it by $\|u_\tau^k-u_\tau^{k-1}\|=\|J(u_\tau^k-u_\tau^{k-1})\|_*$, for $k\ge 2$ we get

$$\Big\|\frac{u_\tau^k-u_\tau^{k-1}}{\tau}\Big\|\le\Big\|\frac{u_\tau^{k-1}-u_\tau^{k-2}}{\tau}\Big\|+\tau\Big\|\frac{f_\tau^k-f_\tau^{k-1}}{\tau}\Big\|+\lambda\tau\Big\|\frac{u_\tau^k-u_\tau^{k-1}}{\tau}\Big\|,\tag{9.12}$$

which can be treated by Gronwall's inequality (1.70) if $\lambda\tau<1$, by using also summability of the second right-hand-side term in (9.12) uniformly in τ if $f\in W^{1,1}(I;X)$.[6] For $k=1$, similarly as in (9.11), we obtain $\|u_\tau^1-u_\tau^0\|^2\le\tau\langle J(u^1-u^0), f_\tau^1-A(u_\tau^0)+\lambda(u_\tau^1-u_\tau^0)\rangle\le\tau\|u_\tau^1-u_\tau^0\|\,\|f_\tau^1-A(u_\tau^0)+\lambda(u_\tau^1-u_\tau^0)\|$, from which further

$$\begin{aligned}\Big\|\frac{u_\tau^1-u_\tau^0}{\tau}\Big\| &\le \big\|f_\tau^1-A(u_\tau^0)+\lambda(u_\tau^1-u_\tau^0)\big\|\\ &\le \big\|f(0)\big\|+\big\|A(u_0)\big\|+\tau\Big(\big\|f\big\|_{W^{1,1}(I;X)}+\lambda\Big\|\frac{u_\tau^1-u_\tau^0}{\tau}\Big\|\Big).\end{aligned}\tag{9.13}$$

From this, exploiting also $u_0\in\mathrm{dom}(A)$, (9.7) follows. □

Theorem 9.5 (Existence of strong solutions, Kato [225, 226]). *Let X^* be uniformly convex,[7] A_λ be m-accretive for some $\lambda\ge 0$, $f\in W^{1,1}(I;X)$, and $u_0\in\mathrm{dom}(A)$. Then there is $u\in W^{1,\infty}(I;X)$ such that $u_\tau\to u$ in $C(I;X)$ and this u is a strong solution to (8.4).*

Proof. We show that $\{u_\tau\}_{\tau>0}$ is a Cauchy sequence in $C(I;X)$. For $\tau,\sigma>0$, we have $\frac{\mathrm{d}}{\mathrm{d}t}u_\tau+A(\bar u_\tau)=\bar f_\tau$ and $\frac{\mathrm{d}}{\mathrm{d}t}u_\sigma+A(\bar u_\sigma)=\bar f_\sigma$. Subtracting them, testing

[5] If J is set-valued, we must again select a suitable element from J to ensure non-negativity of the second left-hand-side term, while the required identity $\|u_\tau^k-u_\tau^{k-1}\|^2=\langle J(u_\tau^k-u_\tau^{k-1}),u_\tau^k-u_\tau^{k-1}\rangle$ holds always.

[6] Assuming $f\in W^{1,1}(I;X)$, one must modify (8.75)–(8.76) to bound $\sum_{k=1}^{T/\tau}\|f_\tau^k-f_\tau^{k-1}\|$ independently of τ.

[7] A counterexample by Webb (cf. [118, Sect.14.3, Example 6]) shows that this assumption is indeed essential.

it by $J(\bar{u}_\tau - \bar{u}_\sigma)$, and using also accretivity of A_λ and monotonicity[8] of J (see Lemma 3.2(ii)), we obtain

$$\begin{aligned}\frac{1}{2}\frac{\mathrm{d}}{\mathrm{d}t}\big\|u_\tau - u_\sigma\big\|^2 &\le \Big\langle J(u_\tau - u_\sigma), \frac{\mathrm{d}}{\mathrm{d}t}(u_\tau - u_\sigma)\Big\rangle + \big\langle J(\bar{u}_\tau - \bar{u}_\sigma), A_\lambda(\bar{u}_\tau) - A_\lambda(\bar{u}_\sigma)\big\rangle \\ &= \big\langle J(\bar{u}_\tau - \bar{u}_\sigma), \bar{f}_\tau - \bar{f}_\sigma\big\rangle + \Big\langle J(u_\tau - u_\sigma) - J(\bar{u}_\tau - \bar{u}_\sigma), \frac{\mathrm{d}}{\mathrm{d}t}(u_\tau - u_\sigma)\Big\rangle \\ &\quad + \lambda\big\langle J(\bar{u}_\tau - \bar{u}_\sigma), \bar{u}_\tau - \bar{u}_\sigma\big\rangle \le \big\|J(\bar{u}_\tau - \bar{u}_\sigma)\big\|_* \big\|\bar{f}_\tau - \bar{f}_\sigma\big\| + \lambda\big\|\bar{u}_\tau - \bar{u}_\sigma\big\|^2 \\ &\le \frac{1}{2}\big\|\bar{u}_\tau - \bar{u}_\sigma\big\|^2 + \frac{1}{2}\big\|\bar{f}_\tau - \bar{f}_\sigma\big\|^2 + \lambda\big\|\bar{u}_\tau - \bar{u}_\sigma\big\|^2,\end{aligned}$$

and the term $\|\bar{u}_\tau - \bar{u}_\sigma\|$ can further be estimated by

$$\big\|\bar{u}_\tau - \bar{u}_\sigma\big\| \le \big\|u_\tau - u_\sigma\big\| + \big\|\bar{u}_\tau - u_\tau\big\| + \big\|\bar{u}_\sigma - u_\sigma\big\| = \big\|u_\tau - u_\sigma\big\| + \mathscr{O}(\tau) + \mathscr{O}(\sigma) \tag{9.14}$$

due to the estimate

$$\big\|\bar{u}_\tau - u_\tau\big\|_{L^\infty(I;X)} \le \tau\Big\|\frac{\mathrm{d}}{\mathrm{d}t}u_\tau\Big\|_{L^\infty(I;X)} \le \tau C \tag{9.15}$$

where (9.7) was used. We eventually obtain

$$\frac{\mathrm{d}}{\mathrm{d}t}\big\|u_\tau - u_\sigma\big\|^2 \le \big\|u_\tau - u_\sigma\big\|^2 + \big\|\bar{f}_\tau - \bar{f}_\sigma\big\|^2 + 2\lambda\big\|u_\tau - u_\sigma\big\|^2 + \mathscr{O}(\max(\tau,\sigma)^2). \tag{9.16}$$

Using $\|\bar{f}_\tau - \bar{f}_\sigma\| \to 0$ in $L^2(I)$ for $\tau, \sigma \to 0$, by the Gronwall inequality, we get $\|u_\tau - u_\sigma\|_{C(I;X)} \to 0$ if $\tau, \sigma \to 0$; more precisely, $\|u_\tau - u_\sigma\|_{C(I;X)} = \mathscr{O}(\max(\tau,\sigma))$. Since $C(I;X)$ is complete, the limit of the sequence $\{u_\tau\}_{\tau>0}$ exists.

Take $t \in I$. Thanks to (9.7) and $f \in W^{1,1}(I;X) \subset L^\infty(I;X)$, $A(\bar{u}_\tau(t))$ is bounded so that we can assume that $A(\bar{u}_\tau(t))$ converges weakly in X to some $w(t)$; here we used reflexivity of X^* (and hence also of X) by Milman-Pettis' theorem. Simultaneously, $\bar{u}_\tau(t) \to u(t)$ because $u_\tau(t) \to u(t)$ and $u_\tau(t) - \bar{u}_\tau(t) \to 0$, cf. (9.15). Now we have to show that the graph of the m-accretive mapping A_λ is (norm×weak)-closed. Exploiting the assumed uniform convexity of X^*, hence continuity of J, cf. Lemma 3.2(iii), we have $J(v - \bar{u}_\tau(t)) \to J(v - u(t))$, and thus

$$\big\langle J(v - u(t)), A_\lambda(v) - w_\lambda\big\rangle = \lim_{\tau\to 0}\big\langle J(v - \bar{u}_\tau(t)), A_\lambda(v) - A_\lambda(\bar{u}_\tau(t))\big\rangle \ge 0 \tag{9.17}$$

for any $v \in \mathrm{dom}(A_\lambda)$ and for $w_\lambda := w(t) + \lambda u(t)$. As $\mathbf{I} + A_\lambda$ is surjective, we may choose $v \in V$ so that $v + A_\lambda(v) = u(t) + w_\lambda$, and then from (9.17) we get $0 \le \langle J(v - u(t)), A_\lambda(v) - w_\lambda\rangle = \langle J(v - u(t)), u(t) - v\rangle = -\|u(t) - v\|^2$. Hence $u(t) = v \in \mathrm{dom}(A_\lambda)$ and, from $v + A_\lambda(v) = u(t) + w_\lambda$, we further obtain $u(t) + A_\lambda(u(t)) = u(t) + w_\lambda$, i.e. $w_\lambda = A_\lambda(u(t))$ and thus also $w(t) = A(u(t))$.

Moreover, $\bar{f}_\tau \to f$ in $L^1(I;X)$ and also $\frac{\mathrm{d}}{\mathrm{d}t}u_\tau \rightharpoonup \frac{\mathrm{d}}{\mathrm{d}t}u$ weakly in $L^2(I;X)$. Altogether, we can pass to the limit in the equation $\frac{\mathrm{d}}{\mathrm{d}t}u_\tau + A(\bar{u}_\tau) = \bar{f}_\tau$, obtaining

[8] By monotonicity of J, it holds that $\langle J(u_\tau - u_\sigma) - J(\bar{u}_\tau - \bar{u}_\sigma), \frac{\mathrm{d}}{\mathrm{d}t}(u_\tau - u_\sigma)\rangle \le 0$.

(8.4). Since $u_\tau(0) = u_0$ and $u_\tau \to u$ in $C(I;X)$, the initial condition $u(0) = u_0$ is satisfied, too. By Milman-Pettis' theorem, the uniformly convex X^* is reflexive, and hence so is X. Therefore the distributional derivative $\frac{\mathrm{d}}{\mathrm{d}t}u$ is also the weak derivative; note that u, having the distributional derivative $\frac{\mathrm{d}}{\mathrm{d}t}u$ in $L^1(I;X)$, is also absolutely continuous due to the estimate $\|u(t) - u(s)\| \le \int_s^t \|\frac{\mathrm{d}}{\mathrm{d}\vartheta}u\| \,\mathrm{d}\vartheta$,[9] so that, by Komura's Theorem 1.39, $\frac{\mathrm{d}}{\mathrm{d}t}u$ is even the strong derivative. Eventually, $\{A(u(t))\}_{t\in I} = \{f(t) - \frac{\mathrm{d}}{\mathrm{d}t}u\}_{t\in I}$ is bounded in X, as required. □

9.2 Integral solutions

We define $u \in C(I;X)$ the *integral solution* (of type λ) if $u(0) = u_0$ and

$$\forall v\in\mathrm{dom}(A),\quad 0\le s\le t\le T:\ \ \frac{1}{2}\|u(t)-v\|^2 \le \frac{1}{2}\|u(s)-v\|^2 + \int_s^t \big\langle f(\vartheta)-A(v),u(\vartheta)-v\big\rangle_{\mathrm{s}} + \lambda\big\|u(\vartheta)-v\big\|^2\mathrm{d}\vartheta\,; \tag{9.18}$$

where λ refers to the accretivity of $A_\lambda := A+\lambda\mathbf{I}$ and where $\langle u,v\rangle_{\mathrm{s}} := \sup\big\langle u, J(v)\big\rangle$ is the semi-inner product, cf. (3.7). Note that integral solutions need not range over $\mathrm{dom}(A)$ and their time derivative need not exist, in contrast with the strong solutions.

Proposition 9.6 (Consistency of definition (9.18))**.** *Let $A_\lambda := A+\lambda\mathbf{I}$ be accretive (for a sufficiently large λ). Any strong solution is the integral solution.*

Proof. Using accretivity of A_λ and the properties (3.1) of the duality mapping J, we get the following calculations:[10]

$$\begin{aligned}\frac{1}{2}\|u(t)-v\|^2 - \frac{1}{2}\|u(s)-v\|^2 &= \int_s^t \frac{\mathrm{d}}{\mathrm{d}\vartheta}\frac{1}{2}\|u(\vartheta)-v\|^2\mathrm{d}\vartheta\\ &= \int_s^t\Big\langle J(u(\vartheta){-}v),\frac{\mathrm{d}u}{\mathrm{d}\vartheta}\Big\rangle\,\mathrm{d}\vartheta = \int_s^t\Big\langle J(u(\vartheta){-}v),\frac{\mathrm{d}u}{\mathrm{d}\vartheta} - f(\vartheta)+A(v)\Big\rangle\\ &\qquad + \big\langle J(u(\vartheta)-v), f(\vartheta)-A(v)\big\rangle\,\mathrm{d}\vartheta\\ &\le \int_s^t\big\langle J(u(\vartheta){-}v),-A(u(\vartheta))+A(v)\big\rangle + \big\langle f(\vartheta)-A(v),u(\vartheta)-v\big\rangle_{\mathrm{s}}\mathrm{d}\vartheta\\ &\le \int_s^t\big\langle J(u(\vartheta){-}v),-A_\lambda(u(\vartheta))+A_\lambda(v)\big\rangle\end{aligned}$$

[9]This can be seen from (7.2) used for $\varphi(\vartheta) := \int_s^t \varrho_\varepsilon(\vartheta-\theta)\mathrm{d}\theta$ with ϱ_ε from (7.11), and by passing to the limit with $\varepsilon\to 0$, which gives $u(t)-u(s) = \int_s^t \frac{\mathrm{d}}{\mathrm{d}\vartheta}u\,\mathrm{d}\vartheta$ at all Lebesgue points s and t of u, cf. Theorem 1.35,

[10]If J is set-valued, we must choose a suitable $j^*\in J(u(\vartheta)-v)$ in order to guarantee the non-positiveness of the first right-hand side term and use that the identity $\frac{\mathrm{d}}{\mathrm{d}t}\frac{1}{2}\|u(\vartheta)-v\|^2 = \langle j^*,\frac{\mathrm{d}u}{\mathrm{d}t}\rangle$ holds a.e. for whatever choice $j^*\in J(u(\vartheta)-v)$ we made, cf. Lemma 9.1.

$$+\lambda \left\langle J(u(\vartheta)-v), u(\vartheta)-v\right\rangle + \left\langle f(\vartheta)-A(v), u(\vartheta)-v\right\rangle_{\mathrm{s}} \mathrm{d}\vartheta$$

$$\leq \int_s^t \lambda \|u(\vartheta)-v\|^2 + \left\langle f(\vartheta)-A(v), u(\vartheta)-v\right\rangle_{\mathrm{s}} \mathrm{d}\vartheta \,. \tag{9.19}$$

□

Hence, Theorem 9.5 yields an integral solution if the data f and u_0 are regular enough and X is reflexive with X^* uniformly convex. We put the regularity of f and u_0 off, and later in Theorem 9.9 we get rid also of the reflexivity of X. Even more important and here more difficult, is to prove uniqueness that shows *selectivity* of the definition of integral-solutions, which is not self-evident at all.

Theorem 9.7 (Existence and uniqueness). *Let X^* be uniformly convex, A_λ be m-accretive, $f \in L^1(I;X)$, and $u_0 \in \mathrm{cl}\,\mathrm{dom}(A)$, in particular just $u_0 \in X$ if A is densely defined. Then* (8.4) *has a unique integral solution.*

Proof. Take $f_\varepsilon \in W^{1,1}(I;X)$ and $u_{0\varepsilon} \in \mathrm{dom}(A)$ such that $f_\varepsilon \to f$ in $L^1(I;X)$ and $u_{0\varepsilon} \to u_0$ in X. Denote $u_\varepsilon \in C(I;X)$ the strong solution to the problem

$$\frac{\mathrm{d}u_\varepsilon}{\mathrm{d}t} + A\big(u_\varepsilon(t)\big) = f_\varepsilon(t)\,, \qquad u_\varepsilon(0) = u_{0\varepsilon}\,, \tag{9.20}$$

obtained in Theorem 9.5, so that by (9.19) for any $v \in \mathrm{dom}(A)$ and any $0 \leq s \leq t \leq T$ it holds that

$$\frac{1}{2}\|u_\varepsilon(t)-v\|^2 \leq \frac{1}{2}\|u_\varepsilon(s)-v\|^2 + \int_s^t \left\langle f_\varepsilon(\vartheta)-A(v), u_\varepsilon(\vartheta)-v\right\rangle_{\mathrm{s}} + \lambda\|u_\varepsilon(\vartheta)-v\|^2 \mathrm{d}\vartheta. \tag{9.21}$$

By (9.3), $\{u_\varepsilon\}_{\varepsilon>0}$ is a Cauchy sequence in $C(I;X)$ which is complete, so that there is some $u \in C(I;X)$ such that $u_\varepsilon \to u$ in $C(I;X)$.

Since $u_\varepsilon(0) = u_{0\varepsilon} \to u_0$ and simultaneously $u_\varepsilon(0) \to u(0)$, we can see that $u(0) = u_0$. Moreover, passing to the limit in (9.21) and using the continuity of $\langle\cdot,\cdot\rangle_{\mathrm{s}}$,[11] we can see that u is an integral solution to (8.4).

For uniqueness of the integral solution, let us consider, besides the just obtained integral solution u, some other integral solution, say $\tilde{u}$. Take u_ε the strong solution corresponding to f_ε and $u_{0\varepsilon}$ as above. As $u_\varepsilon(\sigma) \in \mathrm{dom}(A)$ for arbitrary

[11]For the limit passage in the integral, we use Lebesgue's Theorem 1.14 and realize that, by Lemma 3.2(iii), J is continuous and thus so is $(u,v) \mapsto \langle u, J(v)\rangle = \langle u,v\rangle_{\mathrm{s}}$, and that $\{f_\varepsilon\}_{\varepsilon>0}$ can have an integrable majorant, and $\langle\cdot,\cdot\rangle_{\mathrm{s}}$ has a linear growth in the left-hand argument, while for the right-hand argument we have L^∞-a-priori estimates (9.6) valid for f's in $L^1(I;X)$ and u_0's in X.

$\sigma \in I$, we thus can put a test element $v := u_\varepsilon(\sigma)$ into (9.21), obtaining

$$\begin{aligned}\frac{1}{2}\|\tilde{u}(t) - u_\varepsilon(\sigma)\|^2 - \frac{1}{2}\|\tilde{u}(s) - u_\varepsilon(\sigma)\|^2 \\ \leq \int_s^t \big\langle f(\vartheta) - A(u_\varepsilon(\sigma)),\ \tilde{u}(\vartheta) - u_\varepsilon(\sigma)\big\rangle_{\mathrm{s}} + \lambda\|\tilde{u}(\vartheta) - u_\varepsilon(\sigma)\|^2 \mathrm{d}\vartheta \\ \leq \int_s^t \big\langle f(\vartheta) - f_\varepsilon(\sigma), \tilde{u}(\vartheta) - u_\varepsilon(\sigma)\big\rangle_{\mathrm{s}} \mathrm{d}\vartheta \\ + \lambda \int_s^t \|\tilde{u}(\vartheta) - u_\varepsilon(\sigma)\|^2 \,\mathrm{d}\vartheta + \int_s^t \Big\langle \frac{\mathrm{d}u_\varepsilon}{\mathrm{d}t}(\sigma), \tilde{u}(\vartheta) - u_\varepsilon(\sigma)\Big\rangle_{\mathrm{s}} \mathrm{d}\vartheta, \end{aligned} \tag{9.22}$$

where we used $A(u_\varepsilon(\sigma)) = f_\varepsilon(\sigma) - \frac{\mathrm{d}}{\mathrm{d}t}u_\varepsilon(\sigma)$. Let us apply $\int_a^b \mathrm{d}\sigma$ to (9.22). By using Fubini's theorem, we can re-write the last integral as

$$\begin{aligned}\int_a^b \Big[\int_s^t \Big\langle \frac{\mathrm{d}u_\varepsilon}{\mathrm{d}t}(\sigma), \tilde{u}(\vartheta) - u_\varepsilon(\sigma)\Big\rangle_{\mathrm{s}} \mathrm{d}\vartheta\Big] \,\mathrm{d}\sigma \\ = \int_s^t \Big[\int_a^b \Big\langle \frac{\mathrm{d}u_\varepsilon}{\mathrm{d}t}(\sigma), \tilde{u}(\vartheta) - u_\varepsilon(\sigma)\Big\rangle_{\mathrm{s}} \mathrm{d}\sigma\Big] \,\mathrm{d}\vartheta \\ = \frac{1}{2}\int_s^t \|\tilde{u}(\vartheta) - u_\varepsilon(a)\|^2 - \|\tilde{u}(\vartheta) - u_\varepsilon(b)\|^2 \,\mathrm{d}\vartheta. \end{aligned} \tag{9.23}$$

Abbreviating $\varphi_\varepsilon(t,\sigma) := \frac{1}{2}\|\tilde{u}(t) - u_\varepsilon(\sigma)\|^2$ and $\psi_\varepsilon(\vartheta,\sigma) := \langle f(\vartheta) - f_\varepsilon(\sigma),\ \tilde{u}(\vartheta) - u_\varepsilon(\sigma)\rangle_{\mathrm{s}}$, we get

$$\begin{aligned}\int_a^b \varphi_\varepsilon(t,\sigma) - \varphi_\varepsilon(s,\sigma) \,\mathrm{d}\sigma \leq \int_a^b \int_s^t \psi_\varepsilon(\sigma,\vartheta) + 2\lambda\varphi_\varepsilon(\vartheta,\sigma) \,\mathrm{d}\vartheta\mathrm{d}\sigma \\ + \int_s^t \varphi_\varepsilon(\vartheta,a) - \varphi_\varepsilon(\vartheta,b) \,\mathrm{d}\vartheta. \end{aligned} \tag{9.24}$$

Pass to the limit with $\varepsilon \to 0$, using $u_\varepsilon \to u$ in $C(I;X)$ (here the first part of this theorem is exploited) and $f_\varepsilon \to f$ in $L^1(I;X)$ (with an integrable majorant), so that in particular

$$\begin{aligned}\limsup_{\varepsilon\to 0} \int_a^b \psi_\varepsilon(\vartheta,\sigma) \,\mathrm{d}\sigma = \limsup_{\varepsilon\to 0} \int_a^b \big\langle f(\vartheta) - f_\varepsilon(\sigma), \tilde{u}(\vartheta) - u_\varepsilon(\sigma)\big\rangle_{\mathrm{s}} \mathrm{d}\sigma \\ \leq \int_a^b \big\langle f(\vartheta) - f(\sigma), \tilde{u}(\vartheta) - u(\sigma)\big\rangle_{\mathrm{s}} \mathrm{d}\sigma =: \int_a^b \psi(\vartheta,\sigma) \,\mathrm{d}\sigma; \end{aligned} \tag{9.25}$$

again we used the upper semicontinuity of $\langle\cdot,\cdot\rangle_{\mathrm{s}}$. Denoting naturally $\varphi(\vartheta,\sigma) := \frac{1}{2}\|\tilde{u}(\vartheta) - u(\sigma)\|^2$, we have eventually (9.24) without the subscript ε.

Now we need still to smooth φ and ψ for a moment. We can do it by convolution with a kernel like (7.11), cf. Figure 16(middle). For simplicity, here we use

$\frac{1}{\delta}\chi_{[-\delta/2,\delta/2]}$ with $\delta > 0$, prolong $\tilde{u}$ and u for $t < 0$ by continuity and f for $t < 0$ by zero, and define $\varphi_\delta(\vartheta, \sigma) := \frac{1}{\delta^2}\int_{-\delta/2}^{\delta/2}\int_{-\delta/2}^{\delta/2}\varphi(\vartheta - \xi, \sigma - \zeta)\mathrm{d}\zeta\mathrm{d}\xi$ and similarly $\psi_\delta(\vartheta, \sigma) := \frac{1}{\delta^2}\int_{-\delta/2}^{\delta/2}\int_{-\delta/2}^{\delta/2}\psi(\vartheta - \xi, \sigma - \zeta)\mathrm{d}\zeta\mathrm{d}\xi$. Using (9.24) (without ε, of course) we obtain

$$\begin{aligned}
&\int_a^b \varphi_\delta(t,\sigma) - \varphi_\delta(s,\sigma)\,\mathrm{d}\sigma + \int_s^t \varphi_\delta(\vartheta,b) - \varphi_\delta(\vartheta,a)\,\mathrm{d}\vartheta \\
&\quad = \frac{1}{\delta^2}\int_{-\delta/2}^{\delta/2}\int_{-\delta/2}^{\delta/2}\bigg(\int_a^b \varphi(t{-}\xi,\sigma{-}\zeta) - \varphi(s{-}\xi,\sigma{-}\zeta)\,\mathrm{d}\sigma \\
&\qquad\qquad + \int_s^t \varphi(\vartheta{-}\xi,b{-}\zeta) - \varphi(\vartheta{-}\xi,a{-}\zeta)\,\mathrm{d}\vartheta\bigg)\mathrm{d}\xi\mathrm{d}\zeta \\
&\quad = \frac{1}{\delta^2}\int_{-\delta/2}^{\delta/2}\int_{-\delta/2}^{\delta/2}\bigg(\int_{a-\zeta}^{b-\zeta} \varphi(t{-}\xi,\sigma) - \varphi(s{-}\xi,\sigma)\,\mathrm{d}\sigma \\
&\qquad\qquad + \int_{s-\xi}^{t-\xi} \varphi(\vartheta,b{-}\zeta) - \varphi(\vartheta,a{-}\zeta)\,\mathrm{d}\vartheta\bigg)\mathrm{d}\xi\mathrm{d}\zeta \\
&\quad \le \frac{1}{\delta^2}\int_{-\delta/2}^{\delta/2}\int_{-\delta/2}^{\delta/2}\bigg(\int_{a-\zeta}^{b-\zeta}\int_{s-\xi}^{t-\xi} \psi(\vartheta,\sigma) + 2\lambda\varphi(\vartheta,\sigma)\mathrm{d}\vartheta\mathrm{d}\sigma\bigg)\mathrm{d}\xi\mathrm{d}\zeta \\
&\quad = \frac{1}{\delta^2}\int_{-\delta/2}^{\delta/2}\int_{-\delta/2}^{\delta/2}\bigg(\int_a^b\int_s^t \psi(\vartheta{-}\xi,\sigma{-}\zeta){+}2\lambda\varphi(\vartheta{-}\xi,\sigma{-}\zeta)\mathrm{d}\vartheta\mathrm{d}\sigma\bigg)\mathrm{d}\xi\mathrm{d}\zeta \\
&\quad = \int_a^b\int_s^t \psi_\delta(\vartheta,\sigma) + 2\lambda\varphi_\delta(\vartheta,\sigma)\,\mathrm{d}\vartheta\mathrm{d}\sigma
\end{aligned} \tag{9.26}$$

for any $\delta/2 \le a \le b \le T - \delta/2$ and $\delta/2 \le s \le t \le T - \delta/2$. Thus we get (9.24) with δ instead of ε. As now φ_δ and ψ_δ are absolutely continuous, we can deduce

$$\frac{\partial}{\partial\vartheta}\varphi_\delta(\vartheta,\sigma) + \frac{\partial}{\partial\sigma}\varphi_\delta(\vartheta,\sigma) \le \psi_\delta(\vartheta,\sigma) + 2\lambda\varphi_\delta(\vartheta,\sigma)\ ; \tag{9.27}$$

to see it, just apply $\int_s^t \mathrm{d}\vartheta \int_a^b \mathrm{d}\sigma$ to (9.27) to get (9.26). Putting $\vartheta - \sigma$ and denoting $\widehat{\varphi}_\delta(\vartheta) := \varphi_\delta(\vartheta, \vartheta)$ and $\widehat{\psi}_\delta(\vartheta) := \psi_\delta(\vartheta, \vartheta)$, we obtain

$$\frac{\mathrm{d}}{\mathrm{d}\vartheta}\widehat{\varphi}_\delta(\vartheta) \le \widehat{\psi}_\delta(\vartheta) + 2\lambda\widehat{\varphi}_\delta(\vartheta). \tag{9.28}$$

From Gronwall's inequality, we get

$$\widehat{\varphi}_\delta(t) \le \Big(\widehat{\varphi}_\delta(0) + \int_0^t \big|\widehat{\psi}_\delta(\vartheta)\big|\mathrm{d}\vartheta\Big)\mathrm{e}^{2\lambda^+ t} \tag{9.29}$$

for any $t \in I$. Now we can pass $\delta \to 0$. Obviously, $\widehat{\varphi}_\delta(t) = \frac{1}{\delta^2}\int_{-\delta/2}^{\delta/2}\int_{-\delta/2}^{\delta/2}\varphi(t{-}\xi, t{-}\zeta)\mathrm{d}\zeta\mathrm{d}\xi \to \varphi(t,t) = \frac{1}{2}\|\tilde{u}(t) - u(t)\|^2$ for each $t \in I$. In particular, $\widehat{\varphi}_\delta(0) \to \frac{1}{2}\|\tilde{u}(0) -$

$u(0)\|^2 = \frac{1}{2}\|u_0 - u_0\|^2 = 0$. Moreover, using $|\psi(\vartheta,\sigma)| = |\langle f(\vartheta) - f(\sigma), \tilde{u}(\vartheta) - u(\sigma)\rangle_{\rm s}| \le C\|f(\vartheta) - f(\sigma)\|$ with $C = \|\tilde{u}\|_{C(I;X)} + \|u\|_{C(I;X)}$, we obtain

$$
\begin{aligned}
\int_0^t \big|\widehat{\psi}_\delta(\vartheta)\big| \,{\rm d}\vartheta &= \int_0^t \frac{1}{\delta^2}\bigg|\int_{-\delta/2}^{\delta/2}\int_{-\delta/2}^{\delta/2} \psi(\vartheta-\xi,\vartheta-\zeta)\,{\rm d}\zeta{\rm d}\xi\bigg|\,{\rm d}\vartheta \\
&\le \frac{C}{\delta^2}\int_0^t\int_{-\delta/2}^{\delta/2}\int_{-\delta/2}^{\delta/2}\big\|f(\vartheta{-}\xi) - f(\vartheta{-}\zeta)\big\|\,{\rm d}\zeta{\rm d}\xi{\rm d}\vartheta \\
&\le \int_0^t\frac{C}{\delta^2}\int_{-\delta/2}^{\delta/2}\int_{-\delta/2}^{\delta/2}\big\|f(\vartheta{-}\xi) - f(\vartheta)\big\| + \big\|f(\vartheta) - f(\vartheta{-}\zeta)\big\|\,{\rm d}\zeta{\rm d}\xi{\rm d}\vartheta \\
&\le C\int_0^t\frac{1}{\delta}\int_{-\delta/2}^{\delta/2}\big\|f(\vartheta{-}\xi) - f(\vartheta)\big\|\,{\rm d}\xi{\rm d}\vartheta \\
&\qquad\qquad + C\int_0^t\frac{1}{\delta}\int_{-\delta/2}^{\delta/2}\big\|f(\vartheta{-}\zeta) - f(\vartheta)\big\|\,{\rm d}\zeta{\rm d}\vartheta.
\end{aligned}
\tag{9.30}
$$

The last two terms then converge to zero, cf. Theorems 1.14 and 1.35. Hence from (9.29) we get in the limit that $\varphi(t,t) = 0$, so $\tilde{u}(t) = u(t)$ for any $t \in I$. □

Having the uniqueness of the integral solution, the stability (9.3) follows just by the limit passage by strong solutions corresponding to regularized data $(f_{i\varepsilon}, u_{0i\varepsilon}) \to (f_i, u_{0i})$, $i = 1, 2$. This yields:

Corollary 9.8 (Stability). *Let the conditions of Theorem 9.7 hold. Then the estimate* (9.3) *holds for u_i being the unique integral solution corresponding to the data $(f_i, u_{0i}) \in L^1(I;X) \times \mathrm{cl}\,\mathrm{dom}(A)$, $i = 1, 2$.*

The requirement of the uniform convexity of X^* (hence, in particular, reflexivity of X) we used in Theorem 9.7 can be restrictive in some applications but it can be weakened. We do it in the next assertion, proving thus existence of the integral solution by the Rothe method combined with a regularization of data.

Theorem 9.9 (Existence: the nonreflexive case). *Let $f \in L^1(I;X)$, $u_0 \in \mathrm{cl}\,\mathrm{dom}(A)$, and A_λ be m-accretive for some λ. Then (8.4) has an integral solution.*

Proof. We take the regularization $f_\varepsilon \in W^{1,1}(I;X)$ and $u_{0\varepsilon} \in \mathrm{dom}(A)$ as in the proof of Theorem 9.7, i.e. $f_\varepsilon \to f$ in $L^1(I;X)$ and $u_{0\varepsilon} \to u_0$ in X, but here we additionally assume

$$
\big\|f_\varepsilon\big\|_{W^{1,1}(I;X)} = \mathscr{O}\Big(\frac{1}{\varepsilon}\Big) \qquad\text{and}\qquad \big\|A(u_{0\varepsilon})\big\| = \mathscr{O}\Big(\frac{1}{\varepsilon}\Big). \tag{9.31}
$$

Denote $u_{\varepsilon\tau} \in C(I;X)$ the Rothe solution corresponding to f_ε and $u_{0\varepsilon}$ with a time step $\tau > 0$, i.e. it holds that $\frac{\rm d}{{\rm d}t}u_{\varepsilon\tau} + A(\bar{u}_{\varepsilon\tau}) = \overline{(f_\varepsilon)}_\tau$, and $u_{\varepsilon\tau}(0) = u_{0\varepsilon}$; here we used m-accretivity of A_λ. Then, combining (9.7) and (9.15),

$$
\big\|\bar{u}_{\varepsilon\tau} - u_{\varepsilon\tau}\big\| = \tau\mathscr{O}\Big(\frac{1}{\varepsilon}\Big) = \mathscr{O}\Big(\frac{\tau}{\varepsilon}\Big). \tag{9.32}
$$

Let us define $J_1 : X \to X^*$ by $\langle J_1(v), v\rangle = \|J_1(v)\|_*\|v\|$ and $\|J_1(v)\|_* = 1$ for $v \neq 0$ while $\|J_1(v)\|_* = 0$ for $v = 0$; cf. Example 8.104 for J_p. Then we test the difference of $\frac{\mathrm{d}}{\mathrm{d}t}u_{\tau\varepsilon} + A(\bar{u}_{\tau\varepsilon}) = \overline{(f_\varepsilon)}_\tau$ and $\frac{\mathrm{d}}{\mathrm{d}t}u_{\sigma\delta} + A(\bar{u}_{\sigma\delta}) = \overline{(f_\delta)}_\sigma$ by $J_1(\bar{u}_{\tau\varepsilon} - \bar{u}_{\sigma\delta})$. Realizing that J_1 has a potential $\|\cdot\|$ and applying Lemma 9.1, we get $\frac{\mathrm{d}}{\mathrm{d}t}\|u_{\tau\varepsilon} - u_{\sigma\delta}\| \leq \|\overline{(f_\varepsilon)}_\tau - \overline{(f_\delta)}_\sigma\| + \lambda\|\bar{u}_{\tau\varepsilon} - \bar{u}_{\sigma\delta}\|$ and by (9.32), like in (9.16), we get $\frac{\mathrm{d}}{\mathrm{d}t}\|u_{\tau\varepsilon} - u_{\sigma\delta}\| \leq \|\overline{(f_\varepsilon)}_\tau - \overline{(f_\delta)}_\sigma\| + \lambda\|u_{\tau\varepsilon} - u_{\sigma\delta}\| + \mathscr{O}(\max(\frac{\tau}{\varepsilon}, \frac{\sigma}{\delta}))$. By Gronwall's inequality, we can see that both $\{u_{\tau\varepsilon}\}_{\tau,\varepsilon>0,\,\tau=o(\varepsilon)}$ and $\{\bar{u}_{\tau\varepsilon}\}_{\tau,\varepsilon>0,\,\tau=o(\varepsilon)}$ are Cauchy in $C(I;X)$ with the same limit in this complete space, say u.

As J has a potential $\frac{1}{2}\|\cdot\|^2$, cf. Example 5.2, we have, as in (9.19), the estimate

$$\begin{aligned}\frac{1}{2}\|u_{\varepsilon\tau}^k - v\|^2 - \frac{1}{2}\|u_{\varepsilon\tau}^{k-1} - v\|^2 &\leq \big\langle j^*, u_{\varepsilon\tau}^k - u_{\varepsilon\tau}^{k-1}\big\rangle \\ &= \big\langle j^*, u_{\varepsilon\tau}^k - u_{\varepsilon\tau}^{k-1} - \tau f_{\varepsilon\tau}^k + \tau A(v) + \tau(f_{\varepsilon\tau}^k - A(v))\big\rangle \\ &= \big\langle j^*, -\tau A(u_{\varepsilon\tau}^k) + \tau A(v) + \tau(f_{\varepsilon\tau}^k - A(v))\big\rangle \\ &= \tau\big\langle j^*, A_\lambda(v) - A_\lambda(u_{\varepsilon\tau}^k) + f_{\varepsilon\tau}^k - A(v) + \lambda(u_{\varepsilon\tau}^k - v)\big\rangle \\ &\leq \tau\big\langle f_{\varepsilon\tau}^k - A(v), u_{\varepsilon\tau}^k - v\big\rangle_{\mathrm{s}} + \tau\lambda\|u_{\varepsilon\tau}^k - v\|^2,\end{aligned} \tag{9.33}$$

where the former inequality holds for any $j^* \in J(u_{\varepsilon\tau}^k - v)$ while for the last one we must select $j^* \in J(u_{\varepsilon\tau}^k - v)$ suitably so that $\langle j^*, A_\lambda(u_{\varepsilon\tau}^k) - A_\lambda(v)\rangle \geq 0$.

Summing it between two arbitrary time levels, we get

$$\frac{\|u_{\varepsilon\tau}(t) - v\|^2}{2} \leq \frac{\|u_{\varepsilon\tau}(s) - v\|^2}{2} + \int_s^t \big\langle\overline{(f_\varepsilon)}_\tau(\vartheta) - A(v), \bar{u}_{\varepsilon\tau}(\vartheta) - v\big\rangle_{\mathrm{s}} + \lambda\big\|\bar{u}_{\varepsilon\tau}(\vartheta) - v\big\|^2 \mathrm{d}\vartheta$$

with $t, s \in \{k\tau;\ k = 0, \ldots, T/\tau\}$. Fixing t and s, let us now make the limit passage with $\tau = T2^{-k}$, $k \to \infty$, $\varepsilon \to 0$, $\tau = o(\varepsilon)$. The above inequality turns then into (9.18) using again the upper semicontinuity of $\langle\cdot,\cdot\rangle_{\mathrm{s}}$ and, thanks to $\tau = o(\varepsilon)$, also using Lemma 8.7.[12] Altogether, we can see that u satisfies (9.18) for t and s from a dense subset of I. Then, by continuity, (9.18) holds for any $t, s \in I$.

Moreover, since $u_{\varepsilon\tau}(0) = u_{0\varepsilon} \to u_0$ and also $u_{\varepsilon\tau}(0) \to u(0)$, we can see that $u(0) = u_0$. Hence u is an integral solution to (8.4). $\square$

Remark 9.10 (*Periodic problems*[13]). If, in addition, A_λ is accretive for some $\lambda < 0$, then $u_0 \mapsto u(T)$ is a contraction on X, namely

$$\|u_1(T) - u_2(T)\| \leq \mathrm{e}^{\lambda T}\|u_{01} - u_{02}\|, \tag{9.34}$$

[12] For the limit passage in the integral, we note that $\{\overline{(f_\varepsilon)}_\tau\}_{\varepsilon=o(\tau)>0}$ can have an integrable majorant, $\langle\cdot,\cdot\rangle_{\mathrm{s}}$ has a linear growth in the left-hand argument, while for the right-hand argument $\bar{u}_{\varepsilon\tau}(\cdot) - v$ we have L^∞-a-priori estimates (9.6) valid for f's in $L^1(I;X)$ and u_0's in X. Then one is to use the upper semicontinuity of $\langle\cdot,\cdot\rangle_{\mathrm{s}}$, cf. Exercise 3.34 on p.112, and Fatou's Theorem 1.15.

[13] See Barbu [37, Sect.III.2.2], Brezis [66, Sect.III.6], Crandall and Pazy [110], Showalter [383, Sect.IV,Prop.7.3], or Straškraba and Vejvoda [399], or Vainberg [414, Sect.VIII.26.4].

cf. Corollary 9.8.[14] Having proved this contraction, one can use the Banach fixed point Theorem 1.12 to prove existence of a unique periodic integral solution, i.e. $u \in C(I;X)$ satisfying (9.18) and the *periodic condition* $u(T) = u(0)$. Trivially, also the mapping $u_0 \mapsto -u(T)$ is a contraction on X and thus one can prove existence of a unique anti-periodic integral solution, i.e. $u \in C(I;X)$ satisfying (9.18) and the *anti-periodic condition* $u(T) = -u(0)$.

Example 9.11 (Connection with the monotone-mapping approach)**.** Consider $A_1 : V \to V^*$ monotone, radially continuous, and bounded, $A_2 : H \to H$ Lipschitz continuous with a Lipschitz constant ℓ, and $A_1 + A_2 : V \to V^*$ coercive, $V \subset H \cong H^* \subset V^*$, V being embedded into H densely and compactly. We define X and $A : \mathrm{dom}(A) \to X$ by

$$X := H, \quad \mathrm{dom}(A) := \big\{v \in V;\ A_1(v) \in H\big\}, \quad A := \big(A_1 + A_2\big)\big|_{\mathrm{dom}(A)}. \tag{9.35}$$

Then A_λ is accretive for $\lambda \geq \ell$ because $J : X = H \cong H^* = X^*$ is the identity and

$$\begin{aligned}\big\langle J(u-v), A(u)-A(v) + \lambda(u-v)\big\rangle_{H^*\times H} &= \big(u-v, A_1(u)-A_1(v)\big) + \lambda\|u-v\|_H^2 \\ +\big(u-v, A_2(u)-A_2(v)\big) &\geq \big\langle A_1(u)-A_1(v), u-v\big\rangle_{V^*\times V} + (\lambda-\ell)\|u-v\|_H^2 \geq 0\end{aligned} \tag{9.36}$$

for any $u, v \in \mathrm{dom}(A)$ and for $\lambda \geq \ell$. Moreover, as $A_2 : V \to V^*$ is totally continuous, it is pseudomonotone as well as A_1, so that $\mathbf{I} + A_1 + A_2$ is also pseudomonotone. As $A_1 + A_2$ is coercive, $\mathbf{I} + A_1 + A_2$ is coercive, too. Thus, for any $f \in H$, the equation $u + A_1(u) + A_2(u) = f$ has a solution $u \in V$. Moreover, $A_1(u) = f - u - A_2(u) \in H$ so that $u \in \mathrm{dom}(A)$. Hence $\mathbf{I} + A : \mathrm{dom}(A) \to H$ is surjective, so that A_λ is m-accretive. This approach gives additional information about solutions to $\frac{\mathrm{d}}{\mathrm{d}t}u + A(u) = f$, $u(0) = u_0$, in comparison with Theorems 8.16 and 8.18. E.g. if $f \in W^{1,1}(I;H)$ and $u_0 \in V$ such that $A(u_0) \in H$, then $\frac{\mathrm{d}}{\mathrm{d}t}u$ exists everywhere even as a weak derivative.

Remark 9.12 (Lipschitz perturbation of accretive mappings)**.** The calculation (9.36) applies for a general case if $A = A_1 + A_2$ with $A_1 : \mathrm{dom}(A) \to X$ m-accretive and $A_2 : X \to X$ Lipschitz continuous. Then $A_\lambda := A + \lambda\mathbf{I}$ is m-accretive for $\lambda \geq \ell$, and then Theorems 9.5, 9.7, and 9.9 obviously extend to this case.

9.3 Excursion to nonlinear semigroups

The concept of evolution governed by autonomous (=time independent) accretive mappings is intimately related to the theory of semigroups, which is an extensively

[14]This can be seen from the estimate (9.3) for $f_1 = f_2$. If X^* is not uniformly convex, we can use (9.3) obtained in the limit for those integral solutions which arose by the regularization/time-discretization procedure in the proof of Theorem 9.9, even without having formally proved their uniqueness in this case.

developed area with lots of nice results, although its application is often rather limited. Here we only present a very minimal excursion into it.

A one-parametric collection $\{S_t\}_{t\geq 0}$ of mappings $S_t : X \to X$ is called a *C^0-semigroup* if $S_t \circ S_s = S_{t+s}$, $S_0 = \mathbf{I}$, and $(t,u) \mapsto S_t(u) : [0,+\infty) \times X \to X$ is separately continuous[15]. If, moreover, $\|S_t(u) - S_t(v)\| \leq e^{\lambda t}\|u-v\|$, then $\{S_t\}_{t\geq 0}$ is called a *C^0-semigroup of the type* λ. In particular, if $\lambda = 0$, we speak about a *non-expansive C^0-semigroup.*[16]

A natural temptation is to describe behaviour of S_t for all $t > 0$ by some object (called a generator) known at the origin $t = 0$. This idea reflects an everlasting ambition of mankind to forecast the future from the present. For this, we define the so-called (weak) generator $A_{\rm w}$ as

$$A_{\rm w}(u) = \operatorname*{w-lim}_{t\searrow 0} \frac{S_t(u) - u}{t} \tag{9.37}$$

with $\mathrm{dom}(A_{\rm w}) = \{u \in X, \ \text{ the limit in (9.37) exists}\}$. The relation of non-expansive semigroups with accretive mappings is very intimate:

Proposition 9.13. *If $\{S_t\}_{t\geq 0}$ is a non-expansive semigroup, then $A_{\rm w}$ is dissipative.*

Proof. Take $u, v \in X$ and an (even arbitrary) element $j^* \in J(u-v)$. Then

$$\begin{aligned}\Big\langle j^*, \frac{S_t(u) - u}{t} - \frac{S_t(v) - v}{t}\Big\rangle &= \frac{1}{t}\Big(\langle j^*, S_t(u) - S_t(v)\rangle - \|u - v\|^2\Big) \\ &\leq \frac{1}{t}\Big(\|S_t(u) - S_t(v)\| - \|u - v\|\Big)\|u - v\| \leq 0 \end{aligned}\tag{9.38}$$

provided S_t is non-expansive. Considering $u, v \in \mathrm{dom}(A_{\rm w})$, we can pass to the limit to obtain $\langle j^*, A_{\rm w}(u) - A_{\rm w}(v)\rangle \leq 0$. $\square$

The relation between the semigroup and its generators substantially depends on qualification of X. In general, there even exist non-expansive C^0-semigroups possessing no generator, i.e. $\mathrm{dom}(A_{\rm w}) = \emptyset$; such an example is due to Crandall and Liggett [107].

Using (and expanding) our previous results, we obtain a way to generate a C^0-semigroup by means of an m-dissipative generator.

Proposition 9.14. *Let X and X^* be uniformly convex and $A : \mathrm{dom}(A) \to X$ be m-accretive with $\mathrm{dom}(A)$ dense in X, and let $S_t(u_0) := u(t)$ with $u \in C([0,t];X)$ being a unique integral solution to the problem $\frac{\mathrm{d}}{\mathrm{d}t}u + A(u) = 0$ with the initial condition $u(0) = u_0$. Then:*

[15] This means that both $t \mapsto S_t(u)$ and $S_t(\cdot)$ are continuous. Equivalently, continuity of $t \mapsto S_t(u)$ is guaranteed by $\lim_{t\searrow 0} S_t = \mathbf{I}$ pointwise.

[16] In literature, from historical reasons, such semigroups are also called, not completely correctly, semigroups of contractions (as if λ were negative).

(i) $\{S_t\}_{t\geq 0}$ *is a non-expansive* C^0*-semigroup whose generator is* $-A$.

(ii) *The mapping* $t\mapsto A(u(t))$ *is weakly continuous provided* $u_0\in\mathrm{dom}(A)$.

(iii) *Then also the weak derivative* $\frac{\mathrm{d}}{\mathrm{d}t}u(t)$ *exists for all* $t\geq 0$ *and* $\frac{\mathrm{d}}{\mathrm{d}t}u(t)+A(u(t))=0$.

Proof. It is obvious that $\{S_t\}_{t\geq 0}$ is a C^0-semigroup. The non-expansiveness of S_t follows from (9.3), cf. Corollary 9.8, i.e. here $\|S_t(u_{01})-S_t(u_{02})\| := \|u_1(t)-u_2(t)\|\leq\|u_{01}-u_{02}\|$.

As $\{A(u(t))\}_{t\geq 0}$ is bounded in X and, by Milman-Pettis' theorem, X is reflexive, we can assume that $A(u(t_k))\rightharpoonup\xi$ in X for (a suitable sequence) $t_k\to t$ with $t\geq 0$. Simultaneously, $u(t_k)\to u(t)$. As A is m-accretive (hence also maximally accretive) and X^* is uniformly convex, A has a (norm$\times$weak)-closed graph (see the proof of Theorem 9.5), and thus $A(u(t))=\xi$ and $\text{w-lim}_{s\to t}A(u(s))=A(u(t))$, proving thus (ii).

Then, passing to the limit in $\frac{1}{\varepsilon}u(t+\varepsilon)-\frac{1}{\varepsilon}u(t) = -\frac{1}{\varepsilon}\int_t^{t+\varepsilon}A(u(s))\,\mathrm{d}s$ gives $\frac{\mathrm{d}}{\mathrm{d}t}u(t)=\text{w-lim}_{\varepsilon\to 0}-\frac{1}{\varepsilon}\int_t^{t+\varepsilon}A(u(s))\,\mathrm{d}s=-A(u(t))$, proving thus (iii).

For $u_0\in\mathrm{dom}(A)$, we thus have $\lim_{t\searrow 0}\frac{1}{\varepsilon}S_t(u_0)-\frac{1}{\varepsilon}u_0=-A(u_0)$. But simultaneously, by definition (9.37), it is equal to $A_{\mathrm{w}}(u_0)$. Thus $\mathrm{dom}(A)\subset\mathrm{dom}(-A_{\mathrm{w}})$. By Proposition 9.13 and Exercise 3.37, $-A_{\mathrm{w}}$ is accretive and A is maximally accretive, hence $\mathrm{dom}(A)=\mathrm{dom}(-A_{\mathrm{w}})$. □

The above assertion can be generalized for $A_\lambda := A+\lambda\mathbf{I}$ m-accretive and then $\{S_t\}_{t\geq 0}$ a C^0-semigroup of type λ. Also, it conversely holds that a non-expansive C^0-semigroup on X, uniformly convex together with its dual, yields $u(\cdot):t\mapsto S_t(u_0)$ weakly differentiable everywhere, $A_{\mathrm{w}}(u(\cdot))$ weakly continuous, and $\frac{\mathrm{d}}{\mathrm{d}t}u(t)=A_{\mathrm{w}}(u(t))$ for all $t\geq 0$;[17] let us remark that its generator A_{w}, which is dissipative by Proposition 9.13, need not be m-dissipative.

For X^* uniformly convex and A m-accretive, we have, in fact, proved[18] that the so-called *Crandall-Liggett formula* [107]

$$S_t(u)=\lim_{k\to\infty}\left[\mathbf{I}+\frac{t}{k}A\right]^{-k}(u) \tag{9.39}$$

generates a non-expansive C^0-semigroup. In fact, by sophisticated combinatorial arguments, it can be proved for a general Banach space X that the limit in (9.39) is even uniform with respect to t ranging over bounded intervals I.

[17] If X is a Hilbert space, any maximally accretive mapping is m-accretive, and thus we can prove it simply by taking a maximally accretive extension A of $-A_{\mathrm{w}}$ (which does exists by a standard Zorn-lemma argument) and by applying Proposition 9.14 when realizing that $u(t)=S_t(u_0)$ for any $u_0\in\mathrm{dom}(A_w)\subset\mathrm{dom}(-A)$. In a general uniformly convex Banach space we refer, e.g., to Barbu [37, Theorem III.1.2].

[18] Realize that $[\mathbf{I}+\frac{t}{k}A]^{-k}(u_0)=u_\tau^k$ with u_τ^k from (8.5) with $f_\tau^k\equiv 0$ and $\tau=t/k$, and then the convergence $\lim_{k\to\infty,\tau\to 0,k\tau=t}u_\tau^k=\lim_{\tau\to 0}u_\tau(t)=u(t)=S_t(u_0)$ has been obtained in the proof of Theorem 9.7 provided $u_0\in\mathrm{dom}(A)$ and k's forming an ever-refining sequence of partitions of $[0,t]$, while for a general $u_0\in X$ this proof must be modified so that the convergence $u_{0\tau}\to u_0$ is employed first. The (even Lipschitz) continuity of $S_t(\cdot)$ follows from the estimate (9.3).

In the Hilbert case, we can use, in particular, Example 9.11. In this situation, the equation $\frac{\mathrm{d}}{\mathrm{d}t}u+A_1(u)+A_2(u)=0$ induces a C^0-semigroup on H of type λ with λ referring to the Lipschitz constant of A_2. Again $f\in L^1(I;H)$ is allowed similarly as in Theorem 8.13, which completes the previous results.[19] In a general Hilbert case, the relation between non-expansive C^0-semigroups and their generators is very intimate: the generator is always densely defined, and there is a one-to-one correspondence between m-accretive mappings and generators of non-expansive C^0-semigroups.

For the linear operator A (and again a Banach space X), the consequence of Proposition 9.14 is the following:

Corollary 9.15. *Let X and X^* be uniformly convex, $A_0:\mathrm{dom}(A_0)\to X$, with $\mathrm{dom}(A_0)$ dense in X, be linear, let $\langle A_0(v),v\rangle_{\mathrm{s}}\le 0$ and let $A_0-\mathbf{I}$ be surjective. Then A_0 generates a non-expansive C^0-semigroup.*

Proof. We take $A=-A_0$. Then A is m-accretive and $S_t(v):=u(t)$ with u being the unique integral solution to $\frac{\mathrm{d}}{\mathrm{d}t}u+A(u(t))=0$, $u(0)=v$. □

In fact, the above assertion holds for a general Banach space (even also as a converse implication), which is known as the *Lumer-Phillips theorem* [266].

We will still consider a special "semilinear" (but partly non-autonomous) situation, namely that $A(t,v):=A_1v+A_2(t,v)$ with $-A_1$ being a linear generator of a non-expansive C^0-semigroup $\{S_t\}_{t\ge0}\subset\mathscr{L}(X,X)$ and $A_2:I\times X\to X$ a Carathéodory mapping qualified later. We call $u\in C(I;X)$ a *mild solution* to the Cauchy problem (8.1) if the following Volterra-type integral equation

$$u(t)=S_tu_0+\int_0^t S_{t-s}\big(f(s)-A_2(s,u(s))\big)\,\mathrm{d}s \tag{9.40}$$

holds for any $t\in I$. Existence and uniqueness of a mild solution can be shown quite easily:

Proposition 9.16 (Existence and uniqueness). *Let $A(t,v):=A_1v+A_2(t,v)$ with $-A_1$ a linear generator of a non-expansive C^0-semigroup $\{S_t\}_{t\ge0}$, and the Carathéodory mapping $A_2:I\times X\to X$ satisfy $A_2(\cdot,0)\in L^1(0,T;X)$ and $\|A_2(t,v_1)-A_2(t,v_2)\|\le\ell(t)\|v_1-v_2\|$ for some $\ell\in L^1(0,T)$ and $v_1,v_2\in X$, and let $f\in L^1(I;X)$ and $u_0\in X$. Then there is just one mild solution $u\in C(I;X)$ to (8.1).*

Proof. Uniqueness follows simply by subtracting (9.40) written for two solutions u_1 and u_2, which gives $u_{12}(t):=u_1(t)-u_2(t)=\int_0^t S_{t-s}(A_2(s,u_2(s))-$

[19] In fact, one can still prove that, if $A_1(u_0)\in H$, then there is $u\in W^{1,\infty}(I;H)$ a strong solution to $\frac{\mathrm{d}}{\mathrm{d}t}u+A_1(u)+A_2(u)=f$, $u(0)=u_0$, and even $\frac{\mathrm{d}^+}{\mathrm{d}t}u+A_1(u)+A_2(u)=f$ holds everywhere on I where $\frac{\mathrm{d}^+}{\mathrm{d}t}$ denotes the right derivative, cf. e.g. [37, Theoem III.2.5.].

$A_2(s,u_1(s)))\,\mathrm{d}s$, hence

$$\begin{aligned}\big\|u_{12}(t)\big\| &\le \int_0^t \big\|S_{t-s}\big\|_{\mathscr{L}(X,X)}\big\|A_2(s,u_2(s))-A_2(s,u_1(s))\big\|\mathrm{d}s\\ &\le \int_0^t \ell(s)\big\|u_{12}(s)\big\|\,\mathrm{d}s \end{aligned}\tag{9.41}$$

because S_{t-s} is non-expansive, from which $u_{12}=0$ follows by Gronwall's inequality.

For the existence we use Banach's fixed point Theorem 1.12. Considering $\bar u_1,\bar u_2\in C(I;X)$, we calculate $u_1,u_2\in C(I;X)$ as $u_i(t):=S_tu_0+\int_0^t S_{t-s}\big(f(s)-A_2(s,\bar u_i(s))\big)\,\mathrm{d}s$, $i=1,2$. Like (9.41), we can estimate $\|u_1(t)-u_2(t)\|\le\int_0^t\ell(t)\|\bar u_1(s)-\bar u_2(s)\|\mathrm{d}s$ so that $\|u_1-u_2\|_{C(0,t_1;X)}\le\big(\int_0^{t_1}\ell(s)\mathrm{d}s\big)\|\bar u_1-\bar u_2\|_{C(0,t_1;X)}$. Hence the mapping $\bar u\mapsto u$ is a contraction on $C(0,t_1;X)$ if $t_1>0$ is so small that $\int_0^{t_1}\ell(s)\,\mathrm{d}s<1$, and has thus a fixed point $u\in C(0,t_1;X)$, being obviously a mild solution on $[0,t_1]$. Then, starting from $u(t_1)$ instead of u_0, we get a mild solution on $[t_1,t_2]$ with $t_2>0$ so small that $\int_{t_1}^{t_2}\ell(s)\,\mathrm{d}s<1$. Altogether, for $t\in[t_1,t_2]$, it holds that

$$\begin{aligned}u(t)&=S_{t-t_1}u(t_1)+\int_{t_1}^t S_{t-s}\big(f(s)-A_2(s,u(s))\big)\,\mathrm{d}s\\ &=S_{t-t_1}\Big(S_{t_1}u_0+\int_0^{t_1}S_{t_1-s}\big(f(s)-A_2(s,u(s))\big)\,\mathrm{d}s\Big)+\int_{t_1}^t S_{t-s}\big(f(s)-A_2(s,u(s))\big)\,\mathrm{d}s\\ &=S_tu_0+\int_0^t S_{t-s}\big(f(s)-A_2(s,u(s))\big)\,\mathrm{d}s,\end{aligned}$$

so we have obtained a mild solution on $[0,t_2]$. As $\ell\in L^1(0,T)$, we can continue such prolongation until some $t_k\ge T$. □

Proposition 9.17. *Let A, f and u_0 be qualified as in Proposition 9.16. Then:*

(i) (CONSISTENCY.) *Any strong solution to $\frac{\mathrm{d}}{\mathrm{d}t}u+A_1u+A_2(t,u)=f$, $u(0)=u_0$, with $-A_1$ a generator of a linear C^0-semigroup $\{S_t\}_{t\ge0}$ is a mild solution.*

(ii) (SELECTIVITY I.) *If the mild solution is weakly differentiable, then it is the strong solution, too.*

(iii) (SELECTIVITY II.) *If X^* is uniformly convex, $f\in W^{1,1}(I;X)$, $u_0\in\mathrm{dom}(A_1)$, and A_2 time independent, then the mild solution is also the strong solution.*

Proof. Obviously, $\frac1\varepsilon S_{t+\varepsilon}v-\frac1\varepsilon S_tv=(\frac1\varepsilon S_\varepsilon-\frac1\varepsilon\mathbf{I})S_tv=S_t(\frac1\varepsilon S_\varepsilon-\frac1\varepsilon\mathbf{I})v$, hence in the limit $\frac{\mathrm{d}}{\mathrm{d}t}S_tv=A_{\mathrm{w}}S_tv=S_tA_{\mathrm{w}}v=-A_1S_tv$, where $\frac{\mathrm{d}}{\mathrm{d}t}$ denotes the weak derivative and $A_{\mathrm{w}}=-A_1$ is the generator of $\{S_t\}_{t\ge0}$.

Then, as to (i), it holds that $\frac{\mathrm{d}}{\mathrm{d}s}S_{t-s}u(s)=A_1S_{t-s}u(s)+S_{t-s}\frac{\mathrm{d}}{\mathrm{d}s}u(s)=A_1S_{t-s}u(s)+S_{t-s}(-A_1u(s)+f(s)-A_2(s,u(s)))=S_{t-s}(f(s)-A_2(s,u(s)))$, which gives (9.40) after the integration over $[0,t]$.

As to (ii), differentiating (9.40), one obtains

$$\begin{aligned}\frac{du}{dt} &= \frac{dS_t}{dt}u_0 + \int_0^t \frac{dS_{t-s}}{dt}\big(f(s) - A_2(s,u(s))\big)\,ds + f(t) - A_2(t,u(t))\\ &= A_w S_t u_0 + \int_0^t A_w S_{t-s}\big(f(s){-}A_2(s,u(s))\big)\,ds + f(t) - A_2(t,u(t))\\ &= A_w u(t) + f(t) - A_2(t,u(t)).\end{aligned}$$

As to (iii), by Theorem 9.5, our problem possesses a strong solution u. By this Proposition 9.17(i), u is also a mild solution. By Proposition 9.16, there is no other mild solution. □

Remark 9.18 (*Non-autonomous systems*[20]). For the general time-dependent A as used in (8.1), instead of a one-parametric C^0-semigroup of type λ, it is natural to consider a two-parametric collection of mappings $\{U_{t,s} : X \to X\}_{0\le s\le t}$ such that $U_{t,t} = \mathbf{I}$, $U_{t,\vartheta} \circ U_{\vartheta,s} = U_{t,s}$ for any $0 \le s \le \vartheta \le t$, and $\|U_{t,s}(u) - U_{t,s}(v)\| \le e^{\lambda(t-s)}\|u - v\|$ and $(t,s) \mapsto U_{t,s}(v)$ is continuous for any $u, v \in X$. If $\mathbf{I} + \lambda A(t,\cdot)$ are accretive for all $t \ge 0$, $\{U_{t,s}\}_{0\le s\le t}$ can be generated by the Crandall-Liggett formula (9.39) naturally generalized as

$$U_{t,s}(u) := \lim_{k\to\infty} \prod_{i=1}^{k} \Big[\mathbf{I} + \frac{t-s}{k}A\big(s + i\frac{t-s}{k}, \cdot\big)\Big]^{-1}(u). \tag{9.42}$$

For the autonomous case when $A(t,\cdot) \equiv A$, we can put simply $S_t := U_{0,t}$ to obtain the previous situation; note that then also $U_{t,s} \equiv S_{t-s}$ and (9.42) just coincides with (9.39).

9.4 Applications to initial-boundary-value problems

Example 9.19 (*Nonlinear heat transfer*). We consider heat transfer in a homogeneous isotropic but temperature-dependent medium (8.197) moving by a forced advection with the given velocity field $\vec{v}$:

$$\mathfrak{c}(\theta)\Big(\frac{\partial\theta}{\partial t} + \vec{v}\cdot\nabla\theta\Big) - \mathrm{div}\big(\kappa(\theta)\nabla\theta\big) + c_0(x,\theta) = g. \tag{9.43}$$

We consider the time-independent Neumann boundary condition $\kappa(\theta)\frac{\partial\theta}{\partial\nu} = h$ and the initial condition $\theta|_{t=0} = \theta_0$. We first apply the *enthalpy transformation* and also the *Kirchhoff transformation* (8.198), which turns the outlined problem into the form

$$\left.\begin{aligned}\frac{\partial u}{\partial t} + \vec{v}\cdot\nabla u - \Delta\beta(u) + \gamma(x,u) &= g &&\text{in } Q,\\ \frac{\partial}{\partial\nu}\beta(u) &= h &&\text{on } \Sigma,\\ u(0,\cdot) &= u_0 &&\text{on } \Omega,\end{aligned}\right\} \tag{9.44}$$

[20] First relevant papers are by Browder [75], Crandall, Pazy [110], and Kato [225].

with $\beta(r) := [\widehat{\kappa} \circ \widehat{\mathfrak{c}}^{-1}](r)$, cf. Example 8.71, and $\gamma(x,r) := c_0(x, \widehat{\mathfrak{c}}^{-1}(r))$ and $u_0 \in \widehat{\mathfrak{c}}(\theta_0)$. The m-accretive mapping approach, cf. also Remark 3.25, requires $\vec{v} \in W^{1,\infty}(\Omega;\mathbb{R}^n)$ such that $\operatorname{div}\vec{v} \leq 0$ and $(\vec{v}|_\Sigma)\cdot\nu = 0$ and then can be based on the setting:

$$X := L^1(\Omega), \qquad A(u) := \vec{v}\cdot\nabla u - \Delta\beta(u) + \gamma(x,u), \tag{9.45a}$$

$$\operatorname{dom}(A) := \Big\{u \in L^1(\Omega);\ \ \Delta\beta(u) - \vec{v}\cdot\nabla u \in L^1(\Omega),\ \ \frac{\partial}{\partial\nu}\beta(u) = h\Big\}; \tag{9.45b}$$

of course, $\Delta\beta(u) - \vec{v}\cdot\nabla u =: f$ is meant in the sense of distributions, i.e. $\langle f, z\rangle = \beta(u)\Delta z + u(\vec{v}\cdot\nabla z) - u\,(\operatorname{div}\vec{v})\,z$ for any $z \in \mathscr{D}(\Omega)$. If $\gamma(x,\cdot)$ is Lipschitz continuous with the Lipschitz constant ℓ, then A_λ is accretive for $\lambda \geq \ell$, which follows from $\int_\Omega (\gamma(u) - \gamma(v) + \lambda(u-v))\,\mathrm{sign}(u-v)\,\mathrm{d}x \geq \int_\Omega(-\ell|u-v| + \lambda|u-v|)\,\mathrm{d}x \geq 0$. Then Theorem 9.9 provides existence of an integral solution for physically natural data qualification, i.e.

$$g \in L^1(I; L^1(\Omega)) \cong L^1(Q) \quad \text{ and } \quad w_0 \in L^1(\Omega), \tag{9.46}$$

so that the heat sources have a finite energy (without any further restrictions), while for h we required $h \in L^{2^{\#'}}(\Gamma)$ in Proposition 3.22.[21]

In fact, we need $\beta = \widehat{\kappa}\circ\widehat{\mathfrak{c}}^{-1}$ to be Lipschitz continuous and increasing, and we do not need $\widehat{\mathfrak{c}}$ to be strictly increasing, and thus $\mathfrak{c}(\cdot) \geq \varepsilon > 0$ need not be upper-bounded. Even more, $\widehat{\mathfrak{c}}$ can be a monotone set-valued mapping which corresponds to Dirac distributions in $\mathfrak{c}$. This is an enthalpy formulation of the *Stefan problem*. Then $u_0 \in \widehat{\mathfrak{c}}(\theta_0)$ is to be determined because the initial temperature θ_0 need not bear enough information if $\widehat{\mathfrak{c}}$ jumps at θ_0; each such a jump is related to the respective latent heat of the particular phase transformations.

Example 9.20 (*Scalar conservation law*[22]). In view of Section 3.2.4, we consider the initial-boundary-value problem

$$\left.\begin{aligned} \frac{\partial u}{\partial t} + \frac{\partial}{\partial x}F(u) &= g && \text{in } Q := (0,T)\times(0,1),\\ u(\cdot,0) &= u_{\mathrm{D}} && \text{on } (0,T),\\ u(0,\cdot) &= u_0 && \text{on } \Omega := (0,1). \end{aligned}\right\} \tag{9.47}$$

The m-accretive mapping approach requires F strongly monotone, cf. Proposition 3.26, and then we obtain an integral solution $u \in C(I; L^1(0,1))$ if $g \in L^1(Q)$ and $u_0 \in L^1(0,1)$. A special case $F(r) = \frac{1}{2}r^2$ leads to the so-called *Burgers equation*

$$\frac{\partial u}{\partial t} + u\frac{\partial u}{\partial x} = g \text{ in } Q := (0,T)\times(0,1), \quad u|_{x=0} = u_{\mathrm{D}}, \quad u|_{t=0} = u_0. \tag{9.48}$$

[21] For a general time-dependent boundary heat flux $h \in L^1(\Sigma)$ see [361] and also Sections 12.1 and 12.7–12.9 below.

[22] See Zeidler [427, Vol.III, Sect.57.6].

The above theory, however, does not apply directly since F is now not strongly monotone. Assuming $u_0 \geq 0$, $u_{\rm D} \geq 0$, we can expect $u \geq 0$, cf. also Exercise 8.87, and then modify $F(r) = \frac{1}{2}|r|r$ which is strictly (but not strongly) monotone. Then we modify dom(A) from (3.39b) for $\{u \in L^\infty(0,1);\ u(0) = u_{\rm D},\ \frac{\rm d}{{\rm d}x}(|u|u) \in L^1(0,1)$ in the weak sense$\}$, without requiring $u \in W^{1,1}(0,1)$.

Example 9.21 (*Conservation law on* $\mathbb{R}^n$)**.** [23] In view of Remark 3.27, we can consider also the initial-value problem

$$\left.\begin{array}{rcll}\dfrac{\partial u}{\partial t} + \displaystyle\sum_{i=1}^n \frac{\partial}{\partial x_i} F_i(u) &=& g & \text{in } Q := (0,T)\times\mathbb{R}^n,\\ u(0,\cdot) &=& u_0 & \text{on } \mathbb{R}^n.\end{array}\right\} \tag{9.49}$$

Assuming $F \in C^1(\mathbb{R};\mathbb{R}^n)$ and $\limsup_{|u|\to 0} |F(u)|/|u| < \infty$, the mapping A defined as the closure in $L^1(\mathbb{R}^n) \times L^1(\mathbb{R}^n)$ of the mapping $u \mapsto {\rm div}(F(u)) : C_0^1(\mathbb{R}^n) \to C_0(\mathbb{R}^n)$ is m-accretive on $L^1(\Omega)$, and then we obtain an integral solution $u \in C(I; L^1(\mathbb{R}^n))$ if $g \in L^1(Q)$ and $u_0 \in L^1(\mathbb{R}^n)$.

Example 9.22 (*Hamilton-Jacobi equation*)**.** In view of Remark 3.28, the (one-dimensional) Hamilton-Jacobi equation

$$\left.\begin{array}{rcll}\dfrac{\partial u}{\partial t} + F\big(\dfrac{\partial u}{\partial x}\big) &=& g & \text{in } Q := (0,T)\times(0,1),\\ u|_\Sigma &=& 0 & \text{on } \Sigma := (0,T)\times\{0,1\},\\ u(0,\cdot) &=& u_0 & \text{on } \Omega := (0,1),\end{array}\right\} \tag{9.50}$$

with $F{:}\mathbb{R}\to\mathbb{R}$ increasing, has an integral solution $u \in C(\bar{Q})$ if $g \in L^1(I; C([0,1]))$ and $u_0 \in C([0,1])$.

Example 9.23 (Nonlinear test I)**.** Consider again the quasilinear boundary-value problem (8.167). Being inspired by Section 3.2.2 (i.e. accretivity of Δ_p in $L^q(\Omega)$ and the concrete form of the duality mapping J in $L^q(\Omega)$, see Propositions 3.16 and 3.13), we can test (8.167) by $|u|^{q-2}u$, $q \geq 1$, as we did in (9.9).[24] The particular terms can be estimated as

$$\int_\Omega \frac{\partial u}{\partial t}|u|^{q-2}u\,{\rm d}x = \frac{1}{q}\int_\Omega \frac{\partial |u|^q}{\partial t}\,{\rm d}x = \frac{1}{q}\frac{\rm d}{{\rm d}t}\|u\|_{L^q(\Omega)}^q, \tag{9.51a}$$

$$\int_\Omega -{\rm div}\big(|\nabla u|^{p-2}\nabla u\big)|u|^{q-2}u\,{\rm d}x$$
$$= \int_\Omega |\nabla u|^{p-2}\nabla u\cdot\nabla\big(|u|^{q-2}u\big)\,{\rm d}x - \int_\Gamma |\nabla u|^{p-2}\frac{\partial u}{\partial\nu}|u|^{q-2}u\,{\rm d}S$$

[23] See Barbu [38, Sections 2.3.2 and 4.3.4], Dafermos [114, Chap.VI], or Miyadera [287, Chap.7]. Other techniques are exposed in Málek at al. [268, Chap.2].

[24] The calculations (9.51) are only formal unless we have proved regularity of u in advance. In the context of results we have proved, a rigorous derivation is to be made by time discretization.

$$= (q-1)\int_\Omega |\nabla u|^p|u|^{q-2}\,dx + \int_\Gamma |u|^{q_2+q-2} - h|u|^{q-2}u\,dS, \tag{9.51b}$$

$$\int_\Omega |u|^{q_1-2}u|u|^{q-2}u\,dx = \|u\|^{q_1+q-2}_{L^{q_1+q-2}(\Omega)}, \tag{9.51c}$$

cf. Section 3.2.2 for (9.51b). Altogether, abbreviating $p_i = q_i+q-2$ for $i = 1, 2$ and realizing that $|\nabla u|^p|u|^{q-2} = (p/(p+q-2))^p|\nabla|u|^{(p+q-2)/p}|^p$, we get

$$\begin{aligned}
&\frac{1}{q}\frac{d}{dt}\|u\|^q_{L^q(\Omega)} + \frac{(q-1)\,p^p}{(p+q-2)^p}\big\|\nabla|u|^{(p+q-2)/p}\big\|^p_{L^p(\Omega)} + \|u\|^{p_1}_{L^{p_1}(\Omega)} + \|u\|^{p_2}_{L^{p_2}(\Gamma)}\\
&= \int_\Omega g|u|^{q-2}u\,dx + \int_\Gamma h|u|^{q-2}u\,dS\\
&\le \|g\|_{L^q(\Omega)}\big\|\,|u|^{q-1}\big\|_{L^{q'}(\Omega)} + \|h\|_{L^{p_2/(q_2-1)}(\Gamma)}\big\|\,|u|^{q-1}\big\|_{L^{p_2/(q-1)}(\Gamma)}\\
&\le \|g\|_{L^q(\Omega)}\big(1+\|u\|^q_{L^q(\Omega)}\big) + \frac{q_2-1}{p_2}\|h\|^{p_2/(q_2-1)}_{L^{p_2/(q_2-1)}(\Gamma)} + \frac{q-1}{p_2}\|u\|^{p_2}_{L^{p_2}(\Gamma)}.
\end{aligned} \tag{9.52}$$

We assume $g \in L^1(I; L^q(\Omega))$, $h \in L^{p_2/(q_2-1)}(\Sigma)$, and $u_0 \in L^q(\Omega)$, and use Gronwall's inequality to get the estimate of u in $L^\infty(I; L^q(\Omega)) \cap L^{q_1+q-1}(Q)$ and of $|u|^{(p+q-2)/p}$ in $L^p(I; W^{1,p}(\Omega))$. If $p = 2$, from the term $(q-1)|u|^{q-2}|\nabla u|^2$ in (9.51b) we obtain through (1.46) still an estimate of u in $L^q(I; W^{2/q-\epsilon,q}(\Omega))$ if $q \ge 2$.

Having the estimate $\int_Q (q-1)|\nabla u|^p|u|^{q-2}dxdt \le C_1$ and $\sup_t \|u(t,\cdot)\|_{L^q(\Omega)} \le C_2$, we can derive the estimate of ∇u in a suitable Lebesgue space even for an arbitrary $p > 1$. Take $r \ge 1$. By Hölder inequality, assuming $r < p$,

$$\begin{aligned}
\int_Q |\nabla u|^r dxdt &= \int_Q |\nabla u|^r|u|^{(q-2)r/p}|u|^{(2-q)r/p}dxdt\\
&\le \Big(\int_Q |\nabla u|^p|u|^{q-2}dxdt\Big)^{r/p}\Big(\int_Q |u|^{(2-q)r/(p-r)}dxdt\Big)^{(p-r)/p}\\
&= \Big(\frac{C_1}{q-1}\Big)^{r/p}\Big(\int_0^T \|u(t,\cdot)\|^{(2-q)r/(p-r)}_{L^{(2-q)r/(p-r)}(\Omega)}dt\Big)^{(p-r)/p}.
\end{aligned} \tag{9.53}$$

By the Gagliardo-Nirenberg inequality (1.39) with the equivalent norm on $W^{1,r}(\Omega)$ chosen as $\|u\|_{L^q(\Omega)} + \|\nabla u\|_{L^r(\Omega;\mathbb{R}^n)}$, one obtains:

$$\begin{aligned}
\|u(t,\cdot)\|_{L^{(2-q)r/(p-r)}(\Omega)} &\le C_{\rm GN}\big(\|u(t,\cdot)\|_{L^q(\Omega)} + \|\nabla u(t,\cdot)\|_{L^r(\Omega;\mathbb{R}^n)}\big)^\lambda\|u(t,\cdot)\|^{1-\lambda}_{L^q(\Omega)}\\
&\le C_{\rm GN}C_2^{1-\lambda}\big(C_2 + \|\nabla u(t,\cdot)\|_{L^r(\Omega;\mathbb{R}^n)}\big)^\lambda
\end{aligned} \tag{9.54}$$

for

$$\frac{p-r}{(2-q)r} \ge \lambda\Big(\frac{1}{r} - \frac{1}{n}\Big) + (1-\lambda)\frac{1}{q}. \tag{9.55}$$

We raise (9.54) to the power $(2-q)r/(p-r)$ power, use it in (9.53), and choose $\lambda := (p-r)/(2-q)$:

$$
\begin{aligned}
&\Big(\int_0^T \|u(t,\cdot)\|_{L^{(2-q)r/(p-r)}(\Omega)}^{(2-q)r/(p-r)}\mathrm{d}t\Big)^{(p-r)/p} \\
&\le \Big(\int_0^T C_{\mathrm{GN}}^{\frac{p-r}{2-q}} C_2^{\frac{(1-\lambda)(p-r)}{2-q}} \big(C_2+\|\nabla u(t,\cdot)\|_{L^r(\Omega;\mathbb{R}^n)}\big)^{\frac{\lambda(2-q)r}{p-r}}\mathrm{d}t\Big)^{\frac{p-r}{p}} \\
&\le \Big(\int_0^T C_{\mathrm{GN}}^{\frac{p-r}{2-q}} C_2^{1-\frac{\lambda(p-r)}{2-q}} 2^{r-1}\big(C_2^r+\|\nabla u(t,\cdot)\|_{L^r(\Omega;\mathbb{R}^n)}^r\big)\mathrm{d}t\Big)^{\frac{p-r}{p}} \\
&= C_3 + C_4\Big(\int_Q |\nabla u|^r\mathrm{d}x\mathrm{d}t\Big)^{(p-r)/p}
\end{aligned}
\tag{9.56}
$$

for suitable C_3 and C_4. Substituting this choice of $\lambda := (p-r)/(2-q)$ into (9.55), one gets, after some algebra,

$$
r \le \frac{qp+np+qn-2n}{q+n}, \qquad \text{and also} \quad 0 \le \frac{p-r}{2-q} \le 1 \quad \text{and} \quad r<p. \tag{9.57}
$$

Since always $(p-r)/p<1$, by linking (9.53) with (9.56), we get the bound of ∇u in $L^r(Q;\mathbb{R}^n)$.

Example 9.24 (Nonlinear test II). Considering again the problem (8.167), we can differentiate it in time and test it by $|\frac{\partial}{\partial t}u|^{q-2}\frac{\partial}{\partial t}u$, $q \ge 1$, as we did in (9.11).[25] The particular terms arising on the left-hand side can be treated as

$$
\int_\Omega \frac{\partial^2 u}{\partial t^2}\Big|\frac{\partial u}{\partial t}\Big|^{q-2}\frac{\partial u}{\partial t}\,\mathrm{d}x = \frac{1}{q}\frac{\mathrm{d}}{\mathrm{d}t}\int_\Omega\Big|\frac{\partial u}{\partial t}\Big|^q\mathrm{d}x = \frac{1}{q}\frac{\mathrm{d}}{\mathrm{d}t}\Big\|\frac{\partial u}{\partial t}\Big\|_{L^q(\Omega)}^q, \tag{9.58a}
$$

$$
\begin{aligned}
\int_\Omega -\frac{\partial}{\partial t}\mathrm{div}\big(|\nabla u|^{p-2}\nabla u\big)\Big|\frac{\partial u}{\partial t}\Big|^{q-2}\frac{\partial u}{\partial t}\,\mathrm{d}x &= \int_\Omega \frac{\partial}{\partial t}\big(|\nabla u|^{p-2}\nabla u\big) \\
\cdot\nabla\Big(\Big|\frac{\partial u}{\partial t}\Big|^{q-2}\frac{\partial u}{\partial t}\Big)\mathrm{d}x &- \int_\Gamma \frac{\partial}{\partial t}\Big(|\nabla u|^{p-2}\frac{\partial u}{\partial \nu}\Big)\Big|\frac{\partial u}{\partial t}\Big|^{q-2}\frac{\partial u}{\partial t}\,\mathrm{d}S \\
&\ge \int_\Omega (q-1)\frac{8p-8}{p^2}\Big(\frac{\partial|\nabla u|^{p/2}}{\partial t}\Big)^2\Big|\frac{\partial u}{\partial t}\Big|^{q-2}\mathrm{d}x \\
&\quad + \int_\Gamma (q_2-1)|u|^{q_2-2}\Big|\frac{\partial u}{\partial t}\Big|^q - \frac{\partial h}{\partial t}\Big|\frac{\partial u}{\partial t}\Big|^{q-2}\frac{\partial u}{\partial t}\,\mathrm{d}S,
\end{aligned}
\tag{9.58b}
$$

$$
\int_\Omega \frac{\partial(|u|^{q_1-1}u)}{\partial t}\Big|\frac{\partial u}{\partial t}\Big|^{q-2}\frac{\partial u}{\partial t}\,\mathrm{d}x = \int_\Omega (q_1-1)|u|^{q_1-2}\Big|\frac{\partial u}{\partial t}\Big|^q\mathrm{d}x \ge 0 \tag{9.58c}
$$

cf. the calculations in (8.174). Like (8.174), it requires $p>1$. Assuming h constant in time so that $\frac{\partial}{\partial t}h$ simply vanishes, we obtain

$$
\frac{1}{q}\frac{\mathrm{d}}{\mathrm{d}t}\Big\|\frac{\partial u}{\partial t}\Big\|_{L^q(\Omega)}^q \le \int_\Omega \frac{\partial g}{\partial t}\Big|\frac{\partial u}{\partial t}\Big|^{q-2}\frac{\partial u}{\partial t}\mathrm{d}x \le \Big\|\frac{\partial g}{\partial t}\Big\|_{L^q(\Omega)}\Big(1+\Big\|\frac{\partial u}{\partial t}\Big\|_{L^q(\Omega)}^q\Big). \tag{9.59}
$$

[25] Again, the calculations in this example are only formal unless we have proved regularity of u in advance. A rigorous derivation would have to be made by time discretization.

By Gronwall's inequality, we get $u \in W^{1,\infty}(I;L^q(\Omega))$ provided $g \in W^{1,1}(I;L^q(\Omega))$ and $\mathrm{div}(|\nabla u_0|^{p-2}\nabla u_0) - |u_0|^{q_1-2}u_0 = \frac{\partial}{\partial t}u|_{t=0} \in L^q(\Omega)$.

If $p = 2$, we obtain a term $(q-1)|\frac{\partial}{\partial t}u|^{q-2}|\frac{\partial}{\partial t}\nabla u|^2$ in (9.58b), cf. (8.174). Then, from (1.46), we obtain still an estimate of $\frac{\partial}{\partial t}u$ in $L^q(I;W^{2/q-\epsilon,q}(\Omega))$ if $q \geq 2$.

Moreover, if q_1=2, the respective term, i.e. now $\int_\Omega |\frac{\partial}{\partial t}u|^q \mathrm{d}x \geq 0$, gives also the estimate $W^{1,q}(I;L^q(\Omega))$. This estimate is weaker in comparison with the usual $W^{1,\infty}(I;L^q(\Omega))$-estimate (9.7) but, contrary to it, is uniform with respect to $q \to \infty$ if $g \in W^{1,\infty}(I;L^\infty(\Omega))$, $u_0 \in L^\infty(\Omega)$, and $\Delta_p u_0 \in L^\infty(\Omega)$. Thus we get a uniform estimate $L^\infty(I;L^q(\Omega))$, from which the boundedness of u in $L^\infty(Q)$ follows. The idea of passing to L^∞-bounds by increasing q is called *Moser's trick* [295] but it is usually organized in a much more sophisticated way.

Remark 9.25 (Nonlinear test I revisited: case $q = 1$). Note that the L^r-estimate of ∇u obtained in Example 9.23 does not work for $q = 1$ because of the factor $1/(q-1)$ in (9.53). Boccardo and Gallouët [56, Formula (2.7)] give the bound for $\int_Q |\nabla u|^p/(1+|u|)^{1+\varepsilon_0}\,\mathrm{d}x\mathrm{d}t$, which then gives, essentially by the same procedure as above, the exponent $r < (p+np-n)/(n+1)$. In fact, instead of the test by $|u|^{q-2}u$ for $q = 1$, one can rather use the test by $\mathrm{sign}_\varepsilon(u)$ from (3.23) like in (3.2.2) or by $\chi(u) := 1 - \frac{1}{(1+u)^\varepsilon}$, $\varepsilon > 0$, proposed by Feireisl and Málek [145]. For $p = 2$, cf. (12.17) below.

qualification of					quality of u
g	h	u_0	p	q	
$L^1(I;L^q(\Omega))$	$L^{\frac{p_2}{q_2-1}}(\Sigma)$	$L^q(\Omega)$	>1	≥ 1	$L^\infty(I;L^q(\Omega)) \cap L^{p_1}(Q)$
			$=2$	≥ 2	$L^q(I;W^{2/q-\epsilon,q}(\Omega))$
$W^{1,1}(I;L^q(\Omega))$	constant in time	$\Delta_p u_0 \in L^q(\Omega)$	>1	≥ 1	$W^{1,\infty}(I;L^q(\Omega))$
		$u_0 \in L^{q_1 q-q}(\Omega)$	$=2$	≥ 2	$W^{1,q}(I;W^{2/q-\epsilon,q}(\Omega))$

Table 5. Summary of Examples 9.23–9.24; $p_1 := q_1+q-2$ *and* $p_2 := q_2+q-2$.

Exercise 9.26 (*Heat equation with advection:* a nonlinear test). Consider again the heat equation in the enthalpy formulation $\frac{\partial}{\partial t}u + v\cdot\nabla u - \Delta\beta(u)+\gamma(u) = g$, cf. (9.44), and assume $\mathrm{div}\,v \leq 0$ and $v|_\Gamma \cdot \nu \geq 0$ as in Exercise 2.91, and test it by $|u|^{q-2}u$, $q \geq 1$, to obtain the estimate of u in $L^\infty(I;L^q(\Omega))$. For β strongly monotone and $q \geq 2$, from (1.46) derive also an estimate in $L^q(I;W^{2/q-\epsilon,q}(\Omega))$.[26]

[26] Hint: Use Example 9.23 but modify (9.51b) to

$$\int_\Omega \Delta\beta(u)|u|^{q-2}u\,\mathrm{d}x = (1-q)\int_\Omega \beta'(u)|u|^{q-2}|\nabla u|^2\mathrm{d}x + \int_\Gamma (h - b(u))|u|^{q-2}u\,\mathrm{d}S \leq \int_\Gamma h|u|^{q-2}u\,\mathrm{d}S$$

because $\beta'(r) \geq 0$ and $b(r)r \geq 0$ in (8.199). Moreover, treat the advective term as in (3.31).

Exercise 9.27 (*Conservation law regularized*). For $\varepsilon > 0$ fixed, as in Exercise 8.87, consider again

$$\frac{\partial u}{\partial t} + \operatorname{div}\big(F(u)\big) - \varepsilon\Delta u = g, \qquad u|_{t=0} = u_0, \qquad u|_\Sigma = 0, \tag{9.60}$$

test it by $|u|^{q-2}u$, $q \geq 1$, and prove an a-priori estimate in $L^\infty(I; L^q(\Omega))$ and, if $q \geq 2$, also in $L^q(I; W^{2/q-\epsilon,q}(\Omega))$.[27]

Remark 9.28 (*Positivity of temperature*). The estimates in Exercise 9.26 can be interpolated as in Example 9.23 with Remark 9.25 to obtain the estimate of ∇u in $L^r(Q;\mathbb{R}^n)$ for $1 \leq r < \frac{n+2}{n+1}$. Often, the meaning of the solution u to (9.44) is the absolute temperature and then the desired property would be its positivity. Indeed, such positivity can be proved by a comparison argument similarly like in [146, Sect. 4.2.1]. Assuming $\inf u_0 > 0$, $h \geq 0$, and $\gamma(u) - g \leq C|u|^{\omega-2}u$ for some $\omega \geq 2$ and $C \geq 0$, we can compare u with some v spatially constant, i.e. $v(t,x) = y(t)$, such that y solves the initial-value problem for the ordinary differential equations $\frac{\mathrm{d}y}{\mathrm{d}t} + |y|^{\omega-2}y = 0$ with $y(0) = y_0 := \inf u_0 > 0$. This problem has just one solution which is positive.[28] Then v solves the initial-boundary-value problem

$$\left.\begin{aligned} \frac{\partial v}{\partial t} + \vec{v}\cdot\nabla v - \Delta\beta(v) + C|v|^{\omega-2}v &= 0 && \text{in } Q,\\ \frac{\partial}{\partial\nu}\beta(v) &= 0 && \text{on } \Sigma,\\ u(0,\cdot) &= y_0 := \inf u_0 && \text{on } \Omega. \end{aligned}\right\} \tag{9.61}$$

The comparison between (9.44) and (9.61) can formally[29] be performed by subtracting the respective weak formulations and testing them by $(u-v)^-$. Important

[27] Hint: Denote by $T_q : \mathbb{R} \to \mathbb{R}$ the inverse mapping to $r \mapsto |r|^{q-2}r$, i.e. $T_q(r) = |r|^{(2-q)/(q-1)}r$, and by $F_{i,q} : \mathbb{R} \to \mathbb{R}$ the primitive function to $F_i \circ T_q : \mathbb{R} \to \mathbb{R}$ such that $F_{i,q}(0) = 0$, $i = 1, \ldots, n$. Then the F-term, if tested as suggested, vanishes; indeed, by using Green's formula twice,

$$\begin{aligned} \int_\Omega \operatorname{div}(F(u))|u|^{q-2}u\,\mathrm{d}x &= -\int_\Omega F(u)\cdot\nabla(|u|^{q-2}u)\,\mathrm{d}x = -\int_\Omega [F\circ T_q](v)\cdot\nabla v\,\mathrm{d}x \\ &= -\int_\Omega \sum_{i=1}^n [F_i\circ T_q](v)\frac{\partial v}{\partial x_i}\,\mathrm{d}x = -\int_\Omega \sum_{i=1}^n \frac{\partial F_{i,q}(v)}{\partial x_i}\,\mathrm{d}x = -\int_\Omega \operatorname{div}(F_{1,q},\ldots,F_{n,q})(v)\,\mathrm{d}x \\ &= \int_\Omega (F_{1,q},\ldots,F_{n,q})(v)\cdot(\nabla 1)\,\mathrm{d}x - \int_\Gamma \sum_{i=1}^n F_{i,q}(v)\nu_i\,\mathrm{d}S = -\int_\Gamma \sum_{i=1}^n F_{i,q}(v)\nu_i\,\mathrm{d}S = 0, \end{aligned}$$

where we used the substitution $v := |u|^{q-2}u = T_q^{-1}(u)$. The remaining terms have the positive sign as in Example 9.23 for $p = 2$. If $q \geq 2$, use (1.46) to get the fractional-derivative estimate. Note that the growth of F can be superlinear: the condition (2.55a), requiring $\mathscr{N}_F{:}L^{(p^*-\epsilon)'}(\Omega){\to}L^{p'}(\Omega;\mathbb{R}^n)$, implies $|F(r)| \leq C{+}|r|^{(p^*-\epsilon)'/p'}$.

[28] For $\omega = 2$, one has $y(t) = \mathrm{e}^{-Ct}y_0$ while, for $\omega > 2$, one can find $y(t) = (y_0{+}C(\omega{-}2)t)^{1/(2-\omega)}$.

[29] If the assumptions on the data do not give sufficient integrability for such a test, one has to make it in a suitable approximation and then refine it to the limit.

phenomena are that the advective term $\int_Q \vec{v}\cdot\nabla(u-v)(u-v)^- \mathrm{d}x\mathrm{d}t$ is non-negative by the argument as in Exercise 2.91 and similarly the $\Delta\beta$-term gives

$$\begin{aligned}(\nabla\beta(u)-\nabla\beta(v))\cdot\nabla(u-v)^- &= \nabla\beta(u)\cdot\nabla(u-v)^- \\ &= \chi_{\{u\le v\}}\nabla\beta(u)\cdot\nabla u = \chi_{\{u\le v\}}\beta'(u)|\nabla u|^2 \ge 0 \quad (9.62)\end{aligned}$$

a.e. on Q, which can be shown by a smoothening like in the proof of Propositions 3.16 or 3.22; here $\chi_{\{u\le v\}}$ denotes the characteristic function of the set $\{(t,x)\in Q;\ u(t,x)\le v(t,x)\}$. As a result, this comparison gives $u\ge v>0$ on Q.

9.5 Applications to some systems

The accretivity technique works, in a limited extent, even for systems of equations on special occasions.

An interesting example[30] is the abstract initial-value problem (the *Cauchy problem*) for the 2nd-order *doubly nonlinear* equation

$$\frac{\mathrm{d}^2u}{\mathrm{d}t^2} + A\Big(\frac{\mathrm{d}u}{\mathrm{d}t}\Big) + B(u) = f(t), \qquad u(0)=u_0\,, \qquad \frac{\mathrm{d}u}{\mathrm{d}t}(0) = v_0, \tag{9.63}$$

which can equivalently be written as the system of two 1st-order equations:

$$\frac{\mathrm{d}u}{\mathrm{d}t} - v = 0, \qquad u(0)=u_0, \tag{9.64a}$$

$$\frac{\mathrm{d}v}{\mathrm{d}t} + A(v) + B(u) = f, \qquad v(0)=v_0. \tag{9.64b}$$

Assumptions we make are the following, cf. also Theorem 11.33(ii) below:

$$\begin{aligned}
&A: V\to V^* \ \text{ monotone and radially continuous,} && (9.65a)\\
&B = B_1 + B_2 \text{ with}\\
&B_1: V\to V^* \ \text{ linear, continuous, symmetric, i.e. } B_1^*=B_1, \text{ and}\\
&\qquad\qquad \langle B_1u,u\rangle \ge c_0\|u\|_V^2 - c_1\|u\|_H^2, \quad c_0>0,\ c_1\ge 0,\\
&B_2: H\to H \ \text{ Lipschitz continuous } (\ell\text{=the Lipschitz constant}), && (9.65b)\\
&V \text{ and } H \text{ Hilbert spaces.} && (9.65c)
\end{aligned}$$

[30] Cf. Barbu [38, Sect.4.3.5] where A is admitted set-valued, describing thus a 2nd-order evolution variational inequality. For $A=0$ and $B=\mathrm{div}(\mathbb{A}(x)\nabla u)$ see e.g. Renardy and Rogers [349, Sect.11.3.2].

If $A \equiv 0$, (9.63) is a semilinear *hyperbolic equation.* We put

$$X := V \times H, \tag{9.66a}$$

$$\mathrm{dom}(C) := \big\{(u,v) \in X;\ \ A(v) + B_1 u \in H,\ \ v \in V\big\}, \ \text{ and} \tag{9.66b}$$

$$C(u,v) := \big(\lambda u - v\,,\, \lambda v + A(v) + B(u)\big)\,, \quad \text{ where} \tag{9.66c}$$

$$\lambda \geq \sup_{\substack{u \in V\\ v\in H}} \frac{(c_1+\ell)(u,v)}{\langle B_1 u, u\rangle + c_1\|u\|_H^2 + \|v\|_H^2}, \quad \lambda > \sqrt{c_1+\ell} - 1; \tag{9.66d}$$

note that, due to (9.65b), the denominator in (9.66d) is lower bounded by $c_0\|u\|_V^2 + \|v\|_H^2$. Then, putting $w := (u,v)$, the system (9.64) can be rewritten as

$$\frac{\mathrm{d}w}{\mathrm{d}t} + C(w) - \lambda w = (0,f)\,, \qquad w(0) = (u_0,v_0)\,. \tag{9.67}$$

We endow X with an inner product

$$\Big\langle (u_1,v_1),(u_2,v_2)\Big\rangle_{X\times X} := \Big\langle B_1 u_1, u_2\Big\rangle_{V^*\times V} + c_1\Big\langle u_1,u_2\Big\rangle_{H\times H} + \Big\langle v_1,v_2\Big\rangle_{H\times H} \tag{9.68}$$

and identify X^* with X itself. Then the duality mapping J is the identity and one can show $C : \mathrm{dom}(C) \to X$ is accretive, i.e. monotone with respect to the product (9.68), cf. Remark 3.10; indeed for any $w_1, w_2 \in \mathrm{dom}(C)$ one has

$$\begin{aligned}
\big\langle C(w_1) - C(w_2), J(w_{12})\big\rangle_{X\times X^*} &= \big(C(w_1) - C(w_2), w_{12}\big)_{X\times X}\\
&= \big\langle B_1(\lambda u_{12} - v_{12}), u_{12}\big\rangle_{V^*\times V} + c_1\big(\lambda u_{12} - v_{12}, u_{12}\big)_{X\times X}\\
&\quad + \big(\lambda v_{12} + A(v_1) - A(v_2) + B_1 u_{12} + B_2(u_1) - B_2(u_2), v_{12}\big)_{X\times X}\\
&= \big\langle A(v_1) - A(v_2), v_{12}\big\rangle_{V^*\times V} + \big(B_2(u_1) - B_2(u_2), v_{12}\big)_{X\times X}\\
&\quad - c_1\big(v_{12}, u_{12}\big)_{X\times X} + \lambda\Big(\big\langle B_1 u_{12}, u_{12}\big\rangle_{V^*\times V} + c_1\big\|u_{12}\big\|_H^2 + \big\|v_{12}\big\|_H^2\Big) \geq 0
\end{aligned} \tag{9.69}$$

where we again abbreviated $w_{12} := w_1 - w_2$, $u_{12} := u_1 - u_2$ and $v_{12} := v_1 - v_2$, and use both monotonicity of A and that λ is large enough, cf. (9.66d). Moreover, C is maximal monotone, which means that for any $(f_0, f_1) \in V \times H =: X$ there is $w \equiv (u,v) \in \mathrm{dom}(C)$ such that $w + C(w) = (f_0, f_1)$, i.e.

$$u + \lambda u - v = f_0, \tag{9.70a}$$

$$v + \lambda v + A(v) + B_1 u + B_2(u) = f_1. \tag{9.70b}$$

From (9.70a) one gets $u = (v{+}f_0)/(1{+}\lambda)$. Putting this into (9.70b) yields

$$(1{+}\lambda)v + \frac{B_1 v}{1{+}\lambda} + A(v) + B_2\Big(\frac{v{+}f_0}{1{+}\lambda}\Big) = f_1 - \frac{B_1 f_0}{1{+}\lambda} \in V^*. \tag{9.71}$$

Since, due to (9.66d), we have $1{+}\lambda > (c_1{+}\ell)/(1{+}\lambda)$, the mapping $v \mapsto (1{+}\lambda)v + B_1 v/(1{+}\lambda) + B_2((v{+}f_0)/(1{+}\lambda))$ is monotone, coercive, and bounded as a mapping

$V \to V^*$. By Proposition 2.20, (9.71) has a solution $v \in V$. Then also $u = (v + f_0)/(1+\lambda) \in V$ and thus, by (9.70b), $A(v) + B_1u = f - (1+\lambda)v \in H$ so that altogether $(u,v) \in \mathrm{dom}(C)$. Therefore, C is also m-accretive.

Assuming, in addition, that $V_0 := \{u \in V;\ B_1u \in H\}$ is dense in V, we have $\mathrm{dom}(C)$ dense in X.[31] Then Theorem 9.7[32] directly implies the existence of an integral solution $(u,v) \in C(I;X)$ to (9.64), i.e. also a solution $u \in C(I;V) \cap C^1(I;H)$ to (9.63), provided $u_0 \in V$, $v_0 \in H$, $f \in L^1(I;H)$.

Example 9.29. For $V := W_0^{1,2}(\Omega)$, $H := L^2(\Omega)$, $a : \mathbb{R}^n \to \mathbb{R}^n$ monotone, $|a(s)| \le C(1+|s|)$, and $c : \mathbb{R} \to \mathbb{R}$ Lipschitz continuous, the equation

$$\frac{\partial^2 u}{\partial t^2} - \mathrm{div}\, a\Big(\nabla \frac{\partial u}{\partial t}\Big) - \Delta u + c(u) = g \tag{9.72}$$

with the zero Dirichlet boundary conditions in the weak formulation satisfies all the above requirements; note that obviously $V_0 = \{u \in W_0^{1,2}(\Omega);\ \Delta u \in L^2(\Omega)\} \supset W^{2,2}(\Omega) \cap W_0^{1,2}(\Omega)$ is dense in $W_0^{1,2}(\Omega)$.

Example 9.30 (*Linearized thermo-visco-elasticity*[33]). More sophisticated usage of the transformation (9.63)→(9.64) is for a system that arises by a linearization of the full thermo-visco-elasticity system, cf. (12.4)-(12.5) below[34]:

$$\frac{\partial^2 u}{\partial t^2} - \mu_{\mathrm{v}}\Delta \frac{\partial u}{\partial t} - \mu\Delta u + \alpha\nabla\theta = g, \tag{9.73a}$$

$$\frac{\partial \theta}{\partial t} - \kappa\Delta\theta + \alpha\theta_0 \mathrm{div}\frac{\partial u}{\partial t} = h, \tag{9.73b}$$

under the initial conditions $u = u_0$, $\frac{\partial}{\partial t}u = v_0$, $\theta = \theta_0$, where θ_0 is a constant, and some boundary conditions, say $u|_\Sigma = u_{\mathrm{D}}$ with u_{D} constant in time and $\frac{\partial}{\partial\nu}\theta = 0$. Here the following notation is used:

$u : Q \to \mathbb{R}^n$ is an unknown displacement,
$\theta : Q \to R$ is an unknown temperature,
$\mu_{\mathrm{v}} \ge 0$ (resp. $\mu > 0$) a coefficient related to viscosity (resp. elasticity) response,
$\kappa > 0$ a coefficient expressing heat conductivity, and
α a coefficient expressing thermal expansion,

[31] Indeed, as V is assumed dense in H, we can first approximate any $v \in H$ by some $\tilde v \in V$. Then, for $f \in H$ arbitrary, take $u \in V$ such that $A(\tilde v) + B_1u = f \in H$, hence $B_1u = f - A(\tilde v) \in V^*$, and taking $\tilde u \in V$ such that $B_1u_1 = A(\tilde v)$, we have $B_1z = f \in H$ for $z = \tilde u + u$, hence u ranges over $V_0 - \tilde u$ which is, by the assumption, dense in V.

[32] Note that, e.g., Theorem 8.30 for evolution via monotone mapping $\mathscr{A}_1 := C$ with lower-order terms $\mathscr{A}_2 := 0$ and $\mathscr{A}_3 := -\lambda\mathrm{I}$ cannot be used directly because we are now not in the situation that $C : X \to X^*$ with X compactly embedded into a pivot Hilbert space.

[33] A special case $n = 1$ and $\mu = \theta_0 = 1$ is in Zheng [429, Sect.2.7] and $\mu_{\mathrm{v}} = 0$ is in Jiang and Racke [217, Sect.7.2].

[34] For simplicity, we consider here $\lambda = \lambda_{\mathrm{v}} = \gamma = 0$, $\mathfrak{c} = \varrho = 1$, and α in place of $\alpha(3\lambda+2\mu)$. The linearization uses the natural assumption that the temperature varies only very slightly around θ_0 and the process is very slow so that the contribution of the terms quadratic in the velocity $\frac{\partial}{\partial t}u$ in (12.5) is only small and can be well neglected.

g, h are mechanical loading and heat sources, respectively.
Denoting $v := \frac{\partial}{\partial t}u/\sqrt{\mu}$ and $z = \theta/\sqrt{\theta_0\mu}$ and dividing (9.73a) by $\sqrt{\mu}$ and (9.73b) by $\sqrt{\theta_0\mu}$, the system (9.73) transforms into

$$\frac{\partial}{\partial t}\begin{pmatrix} u \\ v \\ z \end{pmatrix} + \begin{pmatrix} 0 & -\sqrt{\mu} & 0 \\ -\sqrt{\mu}\Delta & -\mu_{\rm v}\Delta & \alpha\sqrt{\theta_0}\nabla \\ 0 & \alpha\sqrt{\theta_0}\,{\rm div} & -\kappa\Delta \end{pmatrix}\begin{pmatrix} u \\ v \\ z \end{pmatrix} = \begin{pmatrix} 0 \\ g/\sqrt{\mu} \\ h/\sqrt{\theta_0\mu} \end{pmatrix}. \tag{9.74}$$

The setting $X := W^{1,2}(\Omega;\mathbb{R}^n) \times L^2(\Omega;\mathbb{R}^n) \times L^2(\Omega)$ with the inner product defined by $((u_1, v_1, z_1), (u_2, v_2, z_2)) := \int_\Omega \nabla u_1{:}\nabla u_2 + v_1{\cdot}v_2 + z_1 z_2\,{\rm d}x$ and ${\rm dom}(A) := \{(u, v, z)\in X;\ A(u, v, z)\in X,\ u|_\Gamma = u_{\rm D},\ v|_\Gamma = 0,\ \frac{\partial}{\partial\nu}z = 0\}$ with A defined by the matrix in (9.74) makes A accretive; indeed,

$$\begin{aligned}\big(A(u, v, z), (u, v, z)\big) &= \int_\Omega \Big(- \sqrt{\mu}\nabla v : \nabla u - \sqrt{\mu}\Delta u\cdot v - \mu_{\rm v}\Delta v\cdot v \\ &\qquad + \alpha\sqrt{\theta_0}\,\nabla z\cdot v + \alpha\sqrt{\theta_0}\,{\rm div}(v)\,z - \kappa\Delta z\,z\Big)\,{\rm d}x \\ &= \mu_{\rm v}\big\|\nabla v\big\|^2_{L^2(\Omega;\mathbb{R}^{n\times n})} + \kappa\big\|\nabla z\big\|^2_{L^2(\Omega;\mathbb{R}^n)} \ge 0. \end{aligned} \tag{9.75}$$

The m-accretivity follows by Lax-Milgram Theorem 2.19. Then, by Theorem 9.7, we obtain a unique integral solution $u \in C(I;W^{1,2}(\Omega;\mathbb{R}^n))\cap C^1(I;L^2(\Omega;\mathbb{R}^n))$ and $\theta \in C(I;L^2(\Omega))$ provided $g \in L^1(I;L^2(\Omega;\mathbb{R}^n))$, $h \in L^1(I;L^2(\Omega))$.

Example 9.31 (Generalized standard materials [9, 194, 197]). Other usage of the transformation (9.63)→(9.64) is for a model of Halphen and Nguen's [197] *generalized standard materials*[35], i.e. isothermal model of materials with internal parameters $z \in \mathbb{R}^m$. At small strains, it is governed by the following system of the equilibrium equation for u and of the evolution inclusion for z:

$$\varrho\frac{\partial^2 u}{\partial t^2} - {\rm div}\,\sigma = g, \tag{9.76a}$$

$$\sigma = \sigma_{\rm v}\Big(\frac{\partial w}{\partial t}\Big) + \psi'_{\rm e}(w, z), \qquad w = e(u) := \frac{1}{2}(\nabla u)^\top + \frac{1}{2}\nabla u, \tag{9.76b}$$

$$\gamma\Big(\frac{\partial z}{\partial t}\Big) + \varrho\,\psi'_z(w, z) \ni 0, \tag{9.76c}$$

where $\psi : \mathbb{R}^{n\times n}_{\rm sym} \times \mathbb{R}^m \to \mathbb{R}$ with $\mathbb{R}^{n\times n}_{\rm sym}$ the set of $n \times n$ symmetric matrices is quadratic positive definite. To be more specific, let us consider

$$\varrho\psi(e, z) := \frac{1}{2}(e - \mathbb{B}z)^\top\mathbb{D}(e - \mathbb{B}z) + \frac{1}{2}z^\top\mathbb{L}z \tag{9.77}$$

with $\mathbb{D} \in \mathbb{R}^{n\times n\times n\times n}_{\rm sym}$, $\mathbb{B} \in \mathbb{R}^{n\times n\times m}$, and $\mathbb{L} \in \mathbb{R}^{m\times m}_{\rm sym}$. The meaning of the variables and the constants is:

[35] This covers many special cases, among them so-called Prandtl-Reuss or Maxwell materials, see Alber [9, Chapter 3] for these and many more cases. For the accretive formulation of the Prandtl-Reuss plasticity or of the plasticity with hardening, see also [384].

$u: Q \to \mathbb{R}^n$ is the displacement,
$w: Q \to \mathbb{R}^{n\times n}_{\rm sym}$ the strain (as a function of (t, x)),
$z: Q \to \mathbb{R}^m$ the internal parameters ,
$e: \mathbb{R}^{n\times n} \to \mathbb{R}^{n\times n}_{\rm sym}$ the small-strain tensor, cf. (6.22),
$\sigma_{\rm v}: \mathbb{R}^{n\times n}_{\rm sym} \to \mathbb{R}^{n\times n}_{\rm sym}$ a monotone viscous-stress tensor,
$\sigma: Q \to \mathbb{R}^{n\times n}_{\rm sym}$ the total stress tensor, here $\sigma = \sigma_{\rm v}(\frac{\partial}{\partial t}w) + \mathbb{D}(w - \mathbb{B}z)$,
$\gamma: \mathbb{R}^m \rightrightarrows \mathbb{R}^m$ a maximal strictly monotone (possibly set-valued) mapping,[36]
$\varrho > 0$ mass density,
$g: Q \to \mathbb{R}^n$ an external force.

We have to specify initial conditions for u, $\frac{\partial}{\partial t}u$, and z, and boundary conditions for u; as to the latter point, let us consider zero Dirichlet conditions for simplicity. Let us assume

$$\sigma_{\rm v}, \gamma^{-1} \text{ continuous with at most linear growth,} \tag{9.78a}$$

$$\exists \varepsilon > 0\ \forall e \in \mathbb{R}^{n\times n}_{\rm sym}: \quad \sigma_{\rm v}(e){:}e \ge \varepsilon|e|^2, \tag{9.78b}$$

$$c(0) \ni 0. \tag{9.78c}$$

Note that γ^{-1} indeed does exist since we assumed γ strictly monotone. Denoting $v := \frac{\partial}{\partial t}u$ as in (9.64a), the system (9.76) can be written as the first-order system in terms of (v, w, z) as $\frac{\partial}{\partial t}(v, w, z) + C(v, w, z) = (g, 0, 0)$ with C defined by

$$C(v, w, z) := -\Big(\frac{\mathrm{div}\big(\sigma_{\rm v}(e(v)) + \mathbb{D}(w - \mathbb{B}z)\big)}{\varrho}\,,\ e(v)\,,\ \gamma^{-1}\big(-\mathbb{B}^\top\mathbb{D}(w - \mathbb{B}z) - \mathbb{L}z\big)\Big).$$

We set $X := L^2(\Omega;\mathbb{R}^n) \times L^2(\Omega;\mathbb{R}^{n\times n}) \times L^2(\Omega;\mathbb{R}^m)$ with the norm $\|(v, w, z)\|_X = (\int_\Omega \varrho|v|^2 + 2\varrho\psi(e, z)\,\mathrm{d}x)^{1/2}$, which makes X a Hilbert space, and $\mathrm{dom}(C) := \{(v, w, z) \in X;\ v \in W^{1,2}_0(\Omega, \mathbb{R}^n),\ A(v, w, z) \in X\}$. This makes C accretive: indeed, as J is the identity, for any $(v_1, w_1, z_1), (v_2, w_2, z_2) \in \mathrm{dom}(C)$, we have the estimate

$$\begin{aligned}
\big\langle C(v_1, w_1, z_1) - C(v_1, w_1, z_1), J(v_{12}, w_{12}, z_{12})\big\rangle_{X\times X^*} \\
= \int_\Omega -\mathrm{div}\big(\sigma_{\rm v}(e(v_1)) - \sigma_{\rm v}(e(v_2)) + \mathbb{D}(w_{12} - \mathbb{B}z_{12})\big)\cdot v_{12} \\
- \big(e(v_{12}) - \mathbb{B}\xi_{12}\big)^\top\mathbb{D}(w_{12} - \mathbb{B}z_{12}) - \xi_{12}^\top\mathbb{L}z_{12}\,\mathrm{d}x \\
= \int_\Omega \Big(\sigma_{\rm v}(e(v_1)) - \sigma_{\rm v}(e(v_2)) \pm \mathbb{D}(w_{12} - \mathbb{B}z_{12})\Big){:}\nabla v_{12} \\
- \xi_{12}^\top\big(\mathbb{B}^\top\mathbb{D}(w_{12} - \mathbb{B}z_{12}) + \mathbb{L}z_{12}\big)\,\mathrm{d}x \ge 0
\end{aligned} \tag{9.79}$$

where "$\pm$" indicates the terms that cancel each other and where we abbreviated $\xi_i := \gamma^{-1}(-\mathbb{B}^\top\mathbb{D}(w_i - \mathbb{B}z_i) - \mathbb{L}z_i)$ for $i = 1, 2$ and, as before, $v_{12} := v_1 - v_2$, $w_{12} := w_1 - w_2$, $\xi_{12} := \xi_1 - \xi_2$, etc. We used also that $\sigma_{\rm v}(\cdot)$ is assumed monotone

[36] When $\gamma(0)$ is not a singleton, this allows for modelling activated processes in evolution of z.

and that $\sigma:\nabla v = \sigma:e(v)$ because σ is symmetric. The last term in (9.79) is indeed non-negative as $\gamma(\cdot)$ is monotone. To prove that C is m-accretive, we show, for any $(g, g_1, g_2) \in X$, existence of some $(v, w, z) \in \mathrm{dom}(C)$ such that $(v, w, z) + C(v, w, z) = (g, g_1, g_2)$. Considering $V := W_0^{1,2}(\Omega; \mathbb{R}^n) \times L^2(\Omega; \mathbb{R}^{n\times n}) \times L^2(\Omega; \mathbb{R}^m)$ and now $C : V \to V^*$ in the weak formulation, the existence of $(v, w, z) \in V$ follows by Browder-Minty theorem 2.18; the radial continuity of C follows by (9.78a) while its coercivity follows by (9.78b,c) if (9.76) is used for $(v_2, w_2, z_2) := (0, 0, 0)$. Moreover, since also $C(v, w, z) = (g - v, g_1 - w, g_2 - z) \in X$, we have $(v, w, z) \in \mathrm{dom}(C)$. In particular, Theorem 9.7 then gives us existence of a unique integral solution to (9.76) provided still $g \in L^1(I; L^2(\Omega; \mathbb{R}^n))$, $u(0, \cdot) \in W^{1,2}(\Omega; \mathbb{R}^n)$ so that $w(0, \cdot) \in L^2(\Omega; \mathbb{R}^{n\times n}_{\mathrm{sym}})$, $v(0, \cdot) = \frac{\partial}{\partial t} u(0, \cdot) \in L^2(\Omega; \mathbb{R}^n)$, and $z(0, \cdot) \in L^2(\Omega; \mathbb{R}^m)$. For the more difficult non-dissipative case $\sigma_{\mathrm{v}} = 0$ we refer to Alber [9, Chap 4].

Example 9.32 (*Phase-field system*[37]). Solidification processes can be described by the system

$$\frac{\partial u}{\partial t} = \Delta u + \zeta \frac{\partial v}{\partial t} + g, \qquad u|_{t=0} = u_0, \tag{9.80a}$$

$$\xi \frac{\partial v}{\partial t} = \xi \Delta v - \frac{1}{\xi} c(v) - u, \qquad v|_{t=0} = v_0, \tag{9.80b}$$

for the unknown u and v having the meaning of a temperature and an order parameter, respectively, and with fixed $\zeta > 0$ and (small) $\xi > 0$. Considering zero Dirichlet boundary conditions, we define $X := L^2(\Omega)^2$, $\mathrm{dom}(A) = \{z \in W_0^{1,2}(\Omega);\ \Delta z \in L^2(\Omega)\}^2$, and $A := A_1 + A_2$ with $A_1(u, v) := (\frac{\zeta}{\xi} u - \Delta(u+\zeta v), -\Delta v)$ and $A_2(u, v) := (\frac{\zeta}{\xi^2} c(v), \frac{1}{\xi^2} c(v) + \frac{1}{\xi} u)$. Obviously, $\frac{\partial}{\partial t}(u, v) + A(u, v) = (g, 0)$ is just (9.80), namely (9.80b) multiplied by ζ/ξ is added to (9.80a) and (9.80b) is divided by ξ. Considering $c : \mathbb{R} \to \mathbb{R}$ Lipschitz continuous and the Hilbert space X endowed with the inner product $((u_1, v_1), (u_2, v_2)) := \int_\Omega u_1 u_2 + \zeta^2 v_1 v_2 \, \mathrm{d}x$ and identified with its own dual, the linear operator A_1 is accretive,

$$\begin{aligned}\big\langle A_1(u, v), J(u, v)\big\rangle_{X\times X^*} &= \big(A_1(u, v), (u, v)\big)_{X\times X} = \int_\Omega \Big(\frac{\zeta}{\xi} u - \Delta(u+\zeta v)\Big) u \\ &\quad - \zeta^2 \Delta v\, v \, \mathrm{d}x = \int_\Omega \frac{\zeta}{\xi} u^2 + |\nabla u|^2 + \zeta \nabla u \cdot \nabla v + \zeta^2 |\nabla v|^2 \, \mathrm{d}x \\ &\geq \frac{\zeta}{\xi} \big\| u \big\|^2_{L^2(\Omega)} + \frac{1}{2} \big\| \nabla u \big\|^2_{L^2(\Omega;\mathbb{R}^n)} + \frac{\zeta^2}{2} \big\| \nabla v \big\|^2_{L^2(\Omega;\mathbb{R}^n)} \geq 0, \end{aligned} \tag{9.81}$$

while A_2 is Lipschitz continuous. By Lax-Milgram Theorem 2.19, A_1 can be shown m-accretive. Besides, $\mathrm{dom}(A) \supset W^{2,2}(\Omega)^2 \cap W_0^{1,2}(\Omega)^2$ shows $\mathrm{dom}(A)$ dense in X. Then, by Theorem 9.7, we obtain a unique integral solution $(u, v) \in C(I; L^2(\Omega)^2)$ provided $g \in L^1(I; L^2(\Omega))$ and $u_0, v_0 \in L^2(\Omega)$.

[37] For more details see Sect. 12.5 below.

Example 9.33. Modification of Example 9.20 leads naturally to a system of m equations for $u = (u_1, \ldots, u_m) : (0,T) \times (0,1) \to \mathbb{R}^m$:

$$\left.\begin{aligned} \frac{\partial u_i}{\partial t} + \frac{\partial}{\partial x} F_i(u_i) + G_i(u_1, \ldots, u_m) &= 0 && \text{in } Q := (0,T) \times (0,1), \\ u_i(\cdot, 0) &= 0 && \text{on } (0,T) \times \{0\}, \\ u_i(0, \cdot) &= u_0 && \text{on } \Omega := (0,1), \quad i = 1, \ldots, m. \end{aligned}\right\} \tag{9.82}$$

To apply the m-accretive mapping approach, we assume F_i strongly monotone and $G : \mathbb{R}^n \to \mathbb{R}^n$ Lipschitz continuous, and then put:

$$X := L^1(0,1;\mathbb{R}^m), \qquad A(u) := \Big(\frac{\partial}{\partial x} F_1(u_1), \ldots, \frac{\partial}{\partial x} F_m(u_m)\Big) + G(u), \tag{9.83a}$$

$$\mathrm{dom}(A) := \Big\{ u \in W^{1,1}(0,1;\mathbb{R}^m);\ \frac{\partial}{\partial x} F(u) \in L^1(0,1;\mathbb{R}^m),\ u(0) = 0 \Big\}. \tag{9.83b}$$

Choosing λ greater than the Lipschitz constant of G, the mapping $A_\lambda : A + \lambda \mathbf{I}$ will be accretive.

Remark 9.34 (*Carleman's system*[38]). Some other systems that give rise to accretive mappings exist, as e.g. the following two-dimensional hyperbolic system

$$\frac{\partial u}{\partial t} + \frac{\partial u}{\partial x_1} + u^2 - v^2 = 0, \tag{9.84a}$$

$$\frac{\partial v}{\partial t} + \frac{\partial v}{\partial x_2} + v^2 - u^2 = 0, \tag{9.84b}$$

considered on $\Omega := (\mathbb{R}^+)^2$ together with the boundary conditions $u(t,0,x_2) = v(t,x_1,0) = 0$ and the initial conditions $u(0,x) = u_0 \geq 0$ and $v(0,x) = v_0 \geq 0$. The setting $\mathrm{dom}(A) := \{(u,v) \in L^1((\mathbb{R}^+)^2)^2 \cap L^\infty((\mathbb{R}^+)^2)^2;\ \frac{\partial}{\partial x_1} u + u^2 - v^2 \in L^1((\mathbb{R}^+)^2),\ \frac{\partial}{\partial x_2} v + v^2 - u^2 \in L^1((\mathbb{R}^+)^2),\ u,v \geq 0,\ u(t,0,x_2) = v(t,x_1,0) = 0\}$ makes the underlying mapping A accretive.

9.6 Bibliographical remarks

In general, see the monographs mentioned in Sect. 3.4. Some more detailed comments are as follows.

The notion of integral solution has been introduced by Bénilan and Brézis [48] for X a Hilbert space and then by Bénilan [45, 46] for X a general Banach space. See also Barbu [37, Sect.III.2.1], Deimling [118, Sect.14.3], Hu and Papageorgiou [209, Part I, Sect.III.8]. An equivalent definition involving the inequality

$$\mathrm{e}^{-2\lambda t} \big\| u(t) - v \big\|^2 \leq \mathrm{e}^{-2\lambda s} \big\| u(s) - v \big\|^2 + 2 \int_s^t \mathrm{e}^{-2\lambda \vartheta} \big\langle f(\vartheta) - A(v), u(\vartheta) - v \big\rangle_{\mathrm{s}} \, \mathrm{d}\vartheta \tag{9.85}$$

[38] See Miyadera [287, Examples 2.3, 4.10, 6.2].

has been used by Miyadera [287, Sect.5.1]. Besides, an alternative definition

$$\big\|u(t)-v\big\| \le \big\|u(s)-v\big\| + \int_s^t \big\langle f(\vartheta)-A(v), u(\vartheta)-v\big\rangle_+ + \lambda\big\|u(\vartheta)-v\big\|\,\mathrm{d}\vartheta \qquad (9.86)$$

with $\langle u, v\rangle_+ = \inf_{\varepsilon>0} \frac{1}{\varepsilon}\|u+\varepsilon v\| - \frac{1}{\varepsilon}\|u\|$ can be used, see Showalter [383, Chap.IV.8] or Zeidler [427, Chap.57]; it holds that $\langle u, J(v)\rangle \le \langle u, v\rangle_+\|v\|$. This definition makes some estimates easier, e.g. it shows that (9.3) holds also for integral solutions. A combination of (9.85) and (9.86) is in Barbu [38, Sect.4.1.1].

There is an alternative technique to prove existence of an integral solution to (8.4) based on a regularization of A: instead of the Rothe approximation and Theorem 9.5, it is possible to use the solution of $\frac{\mathrm{d}}{\mathrm{d}t}u + [\mathscr{Y}_\varepsilon(A)](u) = f$ where $\mathscr{Y}_\varepsilon(A) := \varepsilon^{-1}J(\mathbf{I} - (\mathbf{I} + \varepsilon J^{-1}A)^{-1})$ is the Yosida approximation of A; for $X = \mathbb{R}^n$ cf. (2.164b) and for X general see Remark 5.18. For this approach see Barbu [37, Sections 3.1-2] and [38, Sections 4.1.2], Miyadera [287, Chap.3], Yosida [425, Sect.XIV.6-7], or Zeidler [427, Sect.31.1].

Uniqueness in nonreflexive case (not proved here) can be found in Barbu [38, Sections 4.1.1] (by a smoothing method) or Showalter [383, Sect.IV.8] (by Rothe's method). This is related to the Crandall-Liggett formula for the general Banach space, see, e.g. Barbu [37, Sect.III.1.2] or Pavel [326, Sect.3.2].

An accretive approach to the *Klein-Gordon equation*, cf. Exercise 11.41, is in Barbu [38, Sect. 4.3.5], Cazenave and Haraux [90], or Kobayashi and Oharu [234].

For non-expansive semigroups see Barbu [37], Belleni-Morante and McBridge [40, Chap.5], Cioranescu [95, Chap.VI], Crandall and Pazy [109], Hu and Papageorgiou [209, Part I, Sect.III.8], Ito and Kappel [212, Chap.5], Miyadera [287, Chap.3-4], Pavel [326, Chap.II], Pazy [330, Chap.6], Renardy and Rogers [349, Chap.11], or Zeidler [427, Sect.57.5]. In case of X being a Hilbert case, see in particular Barbu [37, Sect.4.1], Brezis [66], Zeidler [427, Sect.31.1] or Zheng [429, Chap.II]. Semilinear parabolic equations treated on the base of the convolution (9.40) and their mild solution are in Cezenave and Haraux [90], Fattorini [144, Chap.5], Henri [200], Miklavčič [284, Chap.5–6], Pazy [330, Chap.8], or Zheng [429]. Let us remark that the *mild solution* has sometimes alternatively the meaning of a limit of the Rothe sequence $\bar{u}_\tau$ in $C(I;X)$; cf. Barbu [38, Sect.4.1.1]. A semigroup approach to Navier-Stokes equations is in Kobayashi and Oharu [234], Miklavčič [284, Sect.6.5] or Sohr [391].

Chapter 10

Evolution governed by certain set-valued mappings

Each of the above presented techniques bears a generalization for the case of set-valued mappings. Now, as in Chapter 5, without narrowing substantially possible applications, we will restrict ourselves to the monotonicity method for an initial-value problem for the quite special type of inclusions:

$$\frac{\mathrm{d}u}{\mathrm{d}t} + \partial\Phi\big(u(t)\big) + A\big(t, u(t)\big) \ni f(t), \qquad u(0) = u_0, \tag{10.1}$$

with $\Phi : V \to \mathbb{R} \cup \{+\infty\}$ a convex potential and $A : I \times V \to V^*$ a Carathéodory mapping such that $\mathscr{A}$ is pseudomonotone.

As before, we will first deal with the problem on an abstract level. The peculiarity is connected with a possible presence of an indicator function in Φ so that no growth condition can be assumed on $\partial\Phi$ and thus no "dual" estimate on the time derivative of the solution is at our disposal. So, one must either rely on a regularity or confine oneself to a weak solution which does not involve any time derivative of the solution itself.

10.1 Abstract problems: strong solutions

As a *strong solution* to (10.1), we will understand $u \in W^{1,\infty,2}(I; V, H)$ such that (10.1) holds a.e., in particular $u(t) \in \mathrm{Dom}(\Phi)$ for a.a. $t \in I$.

In view of the definition (5.2) of $\partial\Phi$, i.e. $\partial\Phi(u(t)) := \{\xi \in V^*;\ \forall v \in V : \langle \xi, v - u(t)\rangle + \Phi(u(t)) \le \Phi(v)\}$, we can write (10.1) in the equivalent form:

$$\left\langle \frac{\mathrm{d}u}{\mathrm{d}t} + A\big(t, u(t)\big), v - u(t) \right\rangle + \Phi(v) - \Phi\big(u(t)\big) \ge \big\langle f(t), v - u(t)\big\rangle \tag{10.2}$$

for any $v \in V$. Note that for $v \notin \mathrm{Dom}(\Phi)$ this inequality is trivial. A typical example is: $\Phi = \varphi + \delta_K$ with $\varphi : V \to \mathbb{R}$ and $K \subset V$ convex. Then (10.2) turns into the variational inequality for a.a. $t \in I$:

$$\text{Find } u(t) \in K: \quad \forall v \in K: \\ \Big\langle \frac{\mathrm{d}u}{\mathrm{d}t} + A(t, u(t)), v - u(t) \Big\rangle + \varphi(v) - \varphi\big(u(t)\big) \geq \big\langle f(t), v - u(t) \big\rangle. \tag{10.3}$$

For the special case $\varphi = 0$, (10.3) can equally be written in the form

$$\frac{\mathrm{d}u}{\mathrm{d}t} + A\big(t, u(t)\big) \in f(t) - N_K\big(u(t)\big). \tag{10.4}$$

Thus we have arrived back at (10.1) for a special case $\partial\Phi = \partial\delta_K = N_K$.

Lemma 10.1 (Uniqueness). *If A satisfies* (8.114), *i.e.* $\langle A(t,u) - A(t,v), u - v \rangle \geq -c(t)\|u - v\|_H^2$ *with* $c \in L^1(I)$, *then* (10.1) *has at most one strong solution.*

Proof. Take $u_1, u_2 \in W^{1,\infty,2}(I; V, H)$ two strong solutions to (10.1). Put $u := u_1$ and $v := u_2$ into (10.2):

$$\Big\langle \frac{\mathrm{d}u_1}{\mathrm{d}t} + A(t, u_1), u_2 - u_1 \Big\rangle + \Phi(u_2) - \Phi(u_1) \geq \big\langle f, u_2 - u_1 \big\rangle \tag{10.5}$$

for a.a. $t \in I$, and analogously for $u := u_2$ and $v := u_1$ we have

$$\Big\langle \frac{\mathrm{d}u_2}{\mathrm{d}t} + A(t, u_2), u_1 - u_2 \Big\rangle + \Phi(u_1) - \Phi(u_2) \geq \big\langle f, u_1 - u_2 \big\rangle \tag{10.6}$$

for a.a. $t \in I$. Adding (10.5) and (10.6) and abbreviating $u_{12} = u_1 - u_2$, one gets[1]

$$0 \leq -\Big\langle \frac{\mathrm{d}u_{12}}{\mathrm{d}t}, u_{12} \Big\rangle - \big\langle A(t, u_1) - A(t, u_2), u_{12} \big\rangle \leq -\frac{1}{2}\frac{\mathrm{d}}{\mathrm{d}t}\big\|u_{12}\big\|_H^2 + c(t)\big\|u_{12}\big\|_H^2. \tag{10.7}$$

By the Gronwall inequality and by $u_{12}(0) = 0$, one gets $u_{12} = 0$. □

Here, we demonstrate a usage of a *regularization* method in order to get a sequence of parabolic equations (which we already know how to solve from Chapter 8):

$$\frac{\mathrm{d}u_\varepsilon}{\mathrm{d}t} + \Phi'_\varepsilon\big(u_\varepsilon(t)\big) + A\big(t, u_\varepsilon(t)\big) = f(t)\,, \qquad u_\varepsilon(0) = u_0, \tag{10.8}$$

[1] As we assume $\frac{\mathrm{d}}{\mathrm{d}t}u \in L^2(I; H)$, we do not have $\frac{\mathrm{d}}{\mathrm{d}t}u \in L^p(I; V^*)$ guaranteed if $p < 2$. However, we certainly have $u \in L^\infty(I; H)$, cf. Lemma 7.1, and thus the first duality in (10.7) can be understood as the inner product in $L^2(I; H)$ and then Lemma 7.3 can be used for $p = 2$ and $V = H$.

depending on a regularization parameter ε. Let us assume that $\Phi_\varepsilon : V \to \mathbb{R}$ is convex and smooth and satisfies

$$\forall v \in W^{1,\infty,2}(I;V,H): \qquad \limsup_{\varepsilon\to 0} \int_0^T \Phi_\varepsilon\big(v(t)\big)\,\mathrm{d}t \le \int_0^T \Phi\big(v(t)\big)\,\mathrm{d}t, \quad (10.9\mathrm{a})$$

$$u_\varepsilon \overset{*}{\rightharpoonup} u \text{ in } W^{1,\infty,2}(I;V,H) \;\Rightarrow\; \liminf_{\varepsilon\to 0} \int_0^T \Phi_\varepsilon\big(u_\varepsilon(t)\big)\,\mathrm{d}t \ge \int_0^T \Phi\big(u(t)\big)\,\mathrm{d}t; \quad (10.9\mathrm{b})$$

cf. also (5.38).

Theorem 10.2 (A-PRIORI ESTIMATES AND CONVERGENCE). *Let A be semi-coercive in the sense (8.95) with $Z = V$ and satisfy the growth condition (8.80), and let (10.9) with $\Phi_\varepsilon \ge 0$ be fulfilled, $A(t,\cdot) : V \to V^*$ be pseudomonotone with $A(t,v) = A_1(v) + A_2(t,v)$ such that*

$$A_1 = \varphi' \quad \textit{with some } \varphi : V \to \mathbb{R},\ \ \varphi(v) \ge c_0|v|_V^p \textit{ for some } c_0 \ge 0, \qquad (10.10\mathrm{a})$$

$$\|A_2(t,v)\|_H \le \gamma(t) + C\|v\|_V^{p/2} \quad \textit{with some } \gamma \in L^2(I),\ C \in \mathbb{R}, \qquad (10.10\mathrm{b})$$

$u_0 \in V \cap \mathrm{Dom}(\Phi)$, $f \in L^2(I;H)$. Then $\|u_\varepsilon\|_{W^{1,\infty,2}(I;V,H)} \le C$ and, for a subsequence, $u_\varepsilon \overset{*}{\rightharpoonup} u$ in $W^{1,\infty,2}(I;V,H)$ and u is the strong solution to (10.1).

Proof. Let us note that the solution u_ε to (10.8) does exist.[2] The a-priori estimate results from testing (10.8) by $\frac{\mathrm{d}}{\mathrm{d}t}u_\varepsilon$:

$$\begin{aligned}\Big\|\frac{\mathrm{d}u_\varepsilon}{\mathrm{d}t}\Big\|_H^2 + \frac{\mathrm{d}}{\mathrm{d}t}\Phi_\varepsilon(u_\varepsilon) + \frac{\mathrm{d}}{\mathrm{d}t}\varphi(u_\varepsilon) &= \Big\langle f - A_2(t,u_\varepsilon), \frac{\mathrm{d}u_\varepsilon}{\mathrm{d}t}\Big\rangle \\ &\le \frac12\big\|f - A_2(t,u_\varepsilon)\big\|_H^2 + \frac12\Big\|\frac{\mathrm{d}u_\varepsilon}{\mathrm{d}t}\Big\|_H^2 \\ &\le \big\|f\big\|_H^2 + 2\gamma^2 + 2C^2\big\|u_\varepsilon\big\|_V^p + \frac12\Big\|\frac{\mathrm{d}u_\varepsilon}{\mathrm{d}t}\Big\|_H^2.\end{aligned} \qquad (10.11)$$

Then, by using the strategy (8.63)–(8.64) and the Gronwall inequality, one obtains the estimate[3]

$$\Big\|\frac{\mathrm{d}u_\varepsilon}{\mathrm{d}t}\Big\|_{L^2(I;H)} \le C, \qquad \|u_\varepsilon\|_{L^\infty(I;V)} \le C. \qquad (10.12)$$

Now, take a subsequence $u_\varepsilon \overset{*}{\rightharpoonup} u$ in $W^{1,\infty,2}(I;V,H)$. As Φ_ε is convex, (10.8) implies

$$\Big\langle \frac{\mathrm{d}u_\varepsilon}{\mathrm{d}t} + A(t,u_\varepsilon(t)), v - u_\varepsilon(t)\Big\rangle + \Phi_\varepsilon(v) \ge \big\langle f(t), v - u_\varepsilon(t)\big\rangle + \Phi_\varepsilon\big(u_\varepsilon(t)\big). \quad (10.13)$$

Now, we consider $v = v(t)$ with $v \in L^p(I;V) \cap L^\infty(I;H)$ and integrate (10.13) over I. Then we can use the usual "parabolic" trick

$$\begin{aligned}\limsup_{\varepsilon\to 0}\Big\langle \frac{\mathrm{d}u_\varepsilon}{\mathrm{d}t}, v - u_\varepsilon\Big\rangle &= \limsup_{\varepsilon\to 0}\Big(\Big\langle \frac{\mathrm{d}u_\varepsilon}{\mathrm{d}t}, v\Big\rangle - \frac12\|u_\varepsilon(T)\|_H^2 + \frac12\|u_0\|_H^2\Big) \\ &\le \Big\langle \frac{\mathrm{d}u}{\mathrm{d}t}, v\Big\rangle - \frac12\|u(T)\|_H^2 + \frac12\|u_0\|_H^2 = \Big\langle \frac{\mathrm{d}u}{\mathrm{d}t}, v - u\Big\rangle,\end{aligned} \qquad (10.14)$$

[2] It follows by methods of Chapter 8 by the a-priori estimates derived in (10.11).

[3] Note that $\Phi_\varepsilon(u_0) + \varphi(u_0) \le C < +\infty$, which follows by (10.9a) from the assumption $u_0 \in \mathrm{Dom}(\Phi) \cap V$, and also $\Phi_\varepsilon(u_\varepsilon(T)) + \varphi(u_\varepsilon(T)) \ge \inf_{v\in V,\delta>0} \Phi_\delta(v) + \varphi(v) \ge 0$.

relying on the fact that $u_\varepsilon(T) \rightharpoonup u(T)$ in H because the mapping $u \mapsto u(T): W^{1,2}(I;H) \to H$ is weakly continuous; the by-part integration formula (7.22) is now backed up by Lemma 7.3 with $W^{1,2,2}(I;H,H)$ instead of $W^{1,p,p'}(I;V,V^*)$. Furthermore, we can use the test $v:= u$ because $u \in W^{1,\infty,2}(I;V,H)$, which gives

$$\liminf_{\varepsilon\to 0}\int_0^T \big\langle A(t,u_\varepsilon), u-u_\varepsilon\big\rangle\,dt \geq \liminf_{\varepsilon\to 0}\int_0^T\Big\langle f-\frac{du_\varepsilon}{dt}, u-u_\varepsilon\Big\rangle$$
$$-\Phi_\varepsilon(u)+\Phi_\varepsilon(u_\varepsilon)\,dt \geq \liminf_{\varepsilon\to 0}\int_0^T\Big\langle f-\frac{du_\varepsilon}{dt}, u-u_\varepsilon\Big\rangle\,dt$$
$$-\limsup_{\varepsilon\to 0}\int_0^T\Phi_\varepsilon(u)\,dt+\liminf_{\varepsilon\to 0}\int_0^T\Phi_\varepsilon(u_\varepsilon)\,dt \geq \int_0^T -\Phi(u)+\Phi(u)\,dt = 0;$$

note that we used both (10.9) and (10.14). Using Lemma 8.29, the obtained pseudomonotonicity of $\mathscr{A}$ yields

$$\limsup_{\varepsilon\to 0}\int_0^T \big\langle A(t,u_\varepsilon), v-u_\varepsilon\big\rangle\,dt \leq \int_0^T \big\langle A(t,u), v-u\big\rangle\,dt. \tag{10.15}$$

Altogether, using (10.14), (10.15), (10.9a), and (10.9b), we can pass with $\varepsilon \to 0$ directly in (10.13) integrated over $[0,T]$, which gives

$$0 \leq \int_0^T\Big\langle\frac{du}{dt}-f, v-u\Big\rangle+\big\langle A(t,u), v-u\big\rangle+\Phi\big(v(t)\big)-\Phi\big(u(t)\big)\,dt. \tag{10.16}$$

From this, we get $\langle\frac{d}{dt}u-f, v-u(t)\rangle+\langle A(t,u(t)), v-u(t)\rangle+\Phi(v)-\Phi(u(t))\geq 0$ for any $v\in V$ and for a.a. $t\in I$.[4]

Moreover, $u(0)=u_0$; realize that certainly $u_\varepsilon \rightharpoonup u$ in $C(I;H)$ and $u_\varepsilon(0)=u_0$. Hence, u is the strong solution to (10.1). □

Remark 10.3. In fact, the proof of Theorem 10.2 requires verification of (10.9b) only for a limit u of any subsequence of $\{u_\varepsilon\}_{\varepsilon>0}$. For example, if K is closed in V, let us consider[5]

$$\Phi := \delta_K\,, \qquad \Phi_\varepsilon(v) := \frac{1}{\varepsilon}\inf_{w\in K}\|w-v\|_V^p\,. \tag{10.17}$$

Since (10.11) implies $\sup_{t\in[0,T]}\Phi_\varepsilon(u_\varepsilon(t))\leq C$ (as $\inf\varphi > -\infty$ is assumed), in the limit one has $u(t)\in K$ for a.a. t. Then $\int_0^T\Phi(u(t))\,dt = 0$.[6] Since $\Phi_\varepsilon\geq 0$, (10.9b) is

[4] Assume the contrary, choose a suitable $v=v(t)$ in a measurable (and also integrable) way.

[5] For $p=2$, Φ'_ε from (10.17) is the Yosida approximation of $\partial\Phi=\partial\delta_K=N_K(\cdot)$.

[6] As $v\mapsto \mathrm{dist}_V(v,K)^p := \inf_{w\in K}\|w-v\|_V^p$ is certainly continuous on V with a p-growth, by Theorem 1.43, the corresponding Nemytskiĭ mapping $L^p(I;V)\to L^1(I)$ is continuous and hence the mapping $v\mapsto\int_0^T\mathrm{dist}_V(v(t),K)^p dt$ is a continuous functional on $L^p(I;V)$ which is convex, hence weakly lower-semicontinuous. As we have $u_\varepsilon \rightharpoonup u$ weakly in $L^p(I;V)$ and $\|\mathrm{dist}_V(u_\varepsilon(\cdot),K)\|_{L^\infty(I)} = \mathscr{O}(\sqrt[p]{\varepsilon})$ thanks to (10.11), we get in the limit

$$\int_0^T \mathrm{dist}_V\big(u(t),K\big)^p dt \leq \liminf_{\varepsilon\to 0}\int_0^T \mathrm{dist}_V\big(u_\varepsilon(t),K\big)^p dt = \liminf_{\varepsilon\to 0} T\,\mathscr{O}\big(\sqrt[p]{\varepsilon}\big)^p = 0.$$

satisfied for this u. The condition (10.9a) is satisfied because, for $v(\cdot) \notin K$ on some subset of I with a positive measure, we have trivially $\limsup_{\varepsilon\to 0}\int_0^T \Phi_\varepsilon(v)\,dt \le +\infty = \int_0^T \Phi(v)\,dt$ while in the opposite case, i.e. for $v(\cdot) \in K$ a.e. on I, we have $\lim_{\varepsilon\to 0}\int_0^T \Phi_\varepsilon(v)\,dt = \lim_{\varepsilon\to 0} 0 = 0 = \int_0^T \Phi(v)\,dt$.

If K is closed in H, modification of (10.17) by replacing V with H is possible, as well. In this case, the natural option is to consider[7]

$$\Phi := \delta_K\,, \qquad \Phi_\varepsilon(v) := \frac{1}{\varepsilon}\inf_{w\in K}\|w-v\|_H^2\,. \tag{10.18}$$

Example 10.4. One can apply also the *Rothe method*, which leads to a sequence of variational problems at each time level:

$$\frac{u_\tau^k - u_\tau^{k-1}}{\tau} + \partial\Phi(u_\tau^k) + A_\tau^k(u_\tau^k) \ni f_\tau^k\,, \qquad u_\tau^0 = u_0\,, \tag{10.19}$$

with $A_\tau^k(u) := \frac{1}{\tau}\int_{(k-1)\tau}^{k\tau} A(t,u)\,dt$ and $f_\tau^k := \frac{1}{\tau}\int_{(k-1)\tau}^{k\tau} f(t)\,dt$ as in (8.81). Existence of the Rothe approximate solutions can then be shown by Corollary 5.19.

10.2 Abstract problems: weak solutions

As in Definition 8.2, we say now that $u \in L^p(I;V)$ is a *weak solution* to the initial-value problem (10.1) if

$$\int_0^T \Big\langle \frac{dv}{dt} + A(t,u(t)) - f(t), v(t)-u(t)\Big\rangle_{V^*\times V} + \Phi\big(v(t)\big) - \Phi\big(u(t)\big)\,dt \ge -\frac{1}{2}\big\|v(0)-u_0\big\|_H^2 \tag{10.20}$$

for all $v \in W^{1,p,p'}(I;V,V^*)$. As in each definition, questions about consistency and *selectivity* immediately arise, cf. Sect. 2.4.1. Let us make clear the former one, the latter being justified by Proposition 10.8 below.

Lemma 10.5. *Any strong solution $u \in W^{1,\infty,2}(I;V,H)$ to (10.1) is also the weak solution in the sense (10.20).*

Proof. If $v = v(t)$ with $v \in W^{1,p,p'}(I;V,V^*)$, we have after integration of (10.2)

[7] In this case, we use $\|\mathrm{dist}_H(u_\varepsilon(t),K)\|_{L^\infty(I)} = \mathscr{O}(\sqrt{\varepsilon})$. This Φ_ε is Gâteaux differentiable.

over I that

$$\int_0^T \Big\langle \frac{\mathrm{d}v}{\mathrm{d}t} + A\big(t,u(t)\big) - f(t), v(t) - u(t)\Big\rangle + \Phi\big(v(t)\big) - \Phi\big(u(t)\big)\,\mathrm{d}t$$
$$= \int_0^T \Big\langle \frac{\mathrm{d}u}{\mathrm{d}t} + A\big(t,u(t)\big) - f(t), v(t) - u(t)\Big\rangle + \Phi\big(v(t)\big) - \Phi\big(u(t)\big)$$
$$+ \Big\langle \frac{\mathrm{d}v}{\mathrm{d}t} - \frac{\mathrm{d}u}{\mathrm{d}t}, v(t) - u(t)\Big\rangle\,\mathrm{d}t \geq \frac{1}{2}\big\|v(T) - u(T)\big\|_H^2 - \frac{1}{2}\big\|v(0) - u_0\big\|_H^2. \tag{10.21}$$

This already gives (10.20). □

We also suppose a certain consistency of the operator $L = \frac{\mathrm{d}}{\mathrm{d}t}$ and the "constraints" involved implicitly in Φ:

$\forall u \in L^p(I;V) \cap L^\infty(I;H) \quad \forall u_0 \in H \quad \exists$ a sequence $\{u_\delta\}_{\delta>0} \subset W^{1,p,p'}(I;V,V^*)$:

$$\limsup_{\delta\to0} \int_0^T \Phi(u_\delta)\,\mathrm{d}t \leq \int_0^T \Phi(u)\,\mathrm{d}t, \tag{10.22a}$$
$$u = \lim_{\delta\to0} u_\delta \qquad \text{in } L^p(I;V), \tag{10.22b}$$
$$\limsup_{\delta\to0} \int_0^T \Big\langle \frac{\mathrm{d}u_\delta}{\mathrm{d}t}, u_\delta - u\Big\rangle\,\mathrm{d}t \leq 0, \tag{10.22c}$$
$$u_0 = \lim_{\delta\to0} u_\delta(0) \qquad \text{in } H. \tag{10.22d}$$

Theorem 10.6 (A-PRIORI ESTIMATES AND CONVERGENCE).[8] *Let A satisfy the growth condition (8.80) and the semicoercivity (8.95) with $Z = V$ and $\mathscr{A} : L^p(I;V) \cap L^\infty(I;H) \to L^{p'}(I;V^*)$ be pseudomonotone, let Φ'_ε satisfy $\langle \Phi'_\varepsilon(v), v\rangle \geq 0$ and the growth condition*

$$\forall \varepsilon > 0\ \exists \mathfrak{C}_\varepsilon : \mathbb{R}\to\mathbb{R} \text{ increasing } \forall v \in V :\ \big\|\Phi'_\varepsilon(v)\big\|_{V^*} \leq \mathfrak{C}_\varepsilon(\|v\|_H)\big(1+\|v\|_V^{p-1}\big), \tag{10.23}$$

moreover $f \in L^{p'}(I;V^*)$, $u_0 \in H$, and let (10.9) be strengthened to

$$\forall v \in W^{1,p,p'}(I;V,V^*) : \qquad \limsup_{\varepsilon\to0} \int_0^T \Phi_\varepsilon\big(v(t)\big)\,\mathrm{d}t \leq \int_0^T \Phi\big(v(t)\big)\,\mathrm{d}t, \tag{10.24a}$$
$$u_\varepsilon \overset{*}{\rightharpoonup} u \text{ in } L^p(I;V) \cap L^\infty(I;H) \Rightarrow \liminf_{\varepsilon\to0} \int_0^T \Phi_\varepsilon\big(u_\varepsilon(t)\big)\,\mathrm{d}t \geq \int_0^T \Phi\big(u(t)\big)\,\mathrm{d}t, \tag{10.24b}$$

and eventually (10.22) be fulfilled. Then the sequence of weak solutions $\{u_\varepsilon\}_{\varepsilon>0}$ to (10.8) satisfies $\|u_\varepsilon\|_{L^\infty(I;H)\cap L^p(I;V)} \leq C$ and $u_\varepsilon \rightharpoonup u$ (subsequences) in $L^p(I;V)$ with u being the weak solution to (10.1).

[8]This assertion is essentially due to Brézis [65], see also Lions [261, Ch.II, Sect.9.3] or Showalter [383, Ch.III, Thm.7.1]. For A linear, see also Duvaut and Lions [130, p.51].

Proof. As shown in Chapter 8, the approximate solution $u_\varepsilon \in W^{1,p,p'}(I;V,V^*)$ does exist. Using a test by u_ε and by incorporating (8.21), we have the estimate:

$$\begin{aligned}
\frac{1}{2}\frac{\mathrm{d}}{\mathrm{d}t}\|u_\varepsilon\|_H^2 + c_0|u_\varepsilon|_V^p &\le \frac{1}{2}\frac{\mathrm{d}}{\mathrm{d}t}\|u_\varepsilon\|_H^2 + \big\langle \Phi_\varepsilon'(u_\varepsilon)+A(t,u_\varepsilon), u_\varepsilon\big\rangle + c_1|u_\varepsilon|_V + c_2\|u_\varepsilon\|_H^2\\
&= \langle f, u_\varepsilon\rangle + c_1|u_\varepsilon|_V + c_2\|u_\varepsilon\|_H^2\\
&\le C_{\rm P}\|f\|_{V^*}\big(|u_\varepsilon|_V + \|u_\varepsilon\|_H\big) + c_1|u_\varepsilon|_V + c_2\|u_\varepsilon\|_H^2\\
&\le \zeta(C_{\rm P}+c_1)|u_\varepsilon|_V^p + c_1C_\zeta + C_{\rm P}C_\zeta\|f\|_{V^*}^{p'} + \big(c_2+C_{\rm P}\|f\|_{V^*}\big)\Big(\frac{1}{4}+\|u_\varepsilon\|_H^2\Big)
\end{aligned}\tag{10.25}$$

with $\zeta > 0$ which is to be chosen small enough, namely $\zeta < c_0/(C_{\rm P}+c_1)$, and with $C_{\rm P}$ from (8.9) and C_ζ depending on p and on ζ like in (1.22). By Gronwall's inequality and by (8.9), this gives $\{u_\varepsilon\}_{\varepsilon>0}$ bounded in $L^\infty(I;H)\cap L^p(I;V)$. For $\varepsilon > 0$ fixed, we can get also the estimate of $\frac{\mathrm{d}}{\mathrm{d}t}u_\varepsilon$ in $L^{p'}(I;V^*)$ because we assumed a growth condition of the type (8.80) for $A + \Phi_\varepsilon'$, although not uniformly with respect to $\varepsilon > 0$, cf. (10.23). Thus we can use the by-part formula (7.22) and, by testing (10.8) by $v - u_\varepsilon$, can write

$$\begin{aligned}
0 &= \int_0^T \Big\langle \frac{\mathrm{d}v}{\mathrm{d}t}+A(t,u_\varepsilon)-f, v-u_\varepsilon\Big\rangle + \langle \Phi_\varepsilon'(u_\varepsilon), v-u_\varepsilon\rangle + \Big\langle \frac{\mathrm{d}u_\varepsilon}{\mathrm{d}t}-\frac{\mathrm{d}v}{\mathrm{d}t}, v-u_\varepsilon\Big\rangle \mathrm{d}t\\
&\le \int_0^T \Big\langle \frac{\mathrm{d}v}{\mathrm{d}t}+A(t,u_\varepsilon)-f, v-u_\varepsilon\Big\rangle + \Phi_\varepsilon(v) - \Phi_\varepsilon(u_\varepsilon)\,\mathrm{d}t + \frac{1}{2}\|v(0)-u_0\|_H^2
\end{aligned}\tag{10.26}$$

for any $v\in W^{1,p,p'}(I;V,V^*)$. The inequality in (10.26) arose from $\langle \Phi_\varepsilon'(u_\varepsilon), v-u_\varepsilon\rangle \le \Phi_\varepsilon(v) - \Phi_\varepsilon(u_\varepsilon)$ (due to convexity of Φ_ε) and from the obvious inequality $0 \le \frac{1}{2}\|v(T) - u_\varepsilon(T)\|_H^2$.

Choosing a subsequence, we have $u_\varepsilon \overset{*}{\rightharpoonup} u$ in $L^p(I;V)\cap L^\infty(I;H)$. We are now to prove (10.15). We cannot put $v := u$ because we do not have the needed information $\frac{\mathrm{d}}{\mathrm{d}t}u \in L^{p'}(I;V^*)$, hence we must employ the regularization u_δ of u from (10.22). Then, since $u_\delta \in W^{1,p,p'}(I;V,V^*)$, we can use (10.26) for $v = u_\delta$, which gives

$$\begin{aligned}
\liminf_{\varepsilon\to0}\int_0^T \langle A(t,u_\varepsilon), u-u_\varepsilon\rangle \mathrm{d}t &= \liminf_{\varepsilon\to0}\int_0^T \big\langle A(t,u_\varepsilon), u_\delta-u_\varepsilon\big\rangle + \big\langle A(t,u_\varepsilon), u-u_\delta\big\rangle \mathrm{d}t\\
&\ge \liminf_{\varepsilon\to0}\bigg(\int_0^T \Big\langle f - \frac{\mathrm{d}u_\delta}{\mathrm{d}t}, u_\delta - u_\varepsilon\Big\rangle - \Phi_\varepsilon(u_\delta) + \Phi_\varepsilon(u_\varepsilon)\,\mathrm{d}t\\
&\qquad\qquad - \frac{1}{2}\big\|u_\delta(0)-u_0\big\|_H^2 - \big\|\mathscr{A}(u_\varepsilon)\big\|_{L^{p'}(I;V^*)}\big\|u-u_\delta\big\|_{L^p(I;V)}\bigg)\\
&\ge \int_0^T \Phi(u) - \Phi(u_\delta) + \Big\langle \frac{\mathrm{d}u_\delta}{\mathrm{d}t} - f, u - u_\delta\Big\rangle \mathrm{d}t \;-\; \frac{1}{2}\big\|u_\delta(0) - u_0\big\|_H^2\\
&\qquad\qquad - \limsup_{\varepsilon\to0}\big\|\mathscr{A}(u_\varepsilon)\big\|_{L^{p'}(I;V^*)}\big\|u - u_\delta\big\|_{L^p(I;V)}
\end{aligned}\tag{10.27}$$

where (10.24) has been used. Then, passing with $\delta \to 0$, by (10.22) and the boundedness of $\{\|\mathscr{A}(u_\varepsilon)\|_{L^{p'}(I;V^*)}\}_{\varepsilon>0}$ by (8.80), we can push the right-hand side of (10.27) to zero, hence we eventually get $\liminf_{\varepsilon\to 0}\int_0^T\langle A(t,u_\varepsilon),u-u_\varepsilon\rangle\mathrm{d}t \geq 0$. From this, (10.15) follows because $\mathscr{A}$ is assumed pseudomonotone from $L^p(I;V)\cap L^\infty(I;H)$ to its (unspecified) dual.

Then, we can estimate from above the limit superior of the right-hand side of (10.26), which will itself be non-negative, too:

$$\begin{aligned} 0 \leq\ & \lim_{\varepsilon\to 0}\int_0^T\Big\langle\frac{\mathrm{d}v}{\mathrm{d}t}-f,v-u_\varepsilon\Big\rangle\,\mathrm{d}t+\limsup_{\varepsilon\to 0}\int_0^T\big\langle A(t,u_\varepsilon),v-u_\varepsilon\big\rangle\,\mathrm{d}t\\ &+\limsup_{\varepsilon\to 0}\int_0^T\Phi_\varepsilon(v)\,\mathrm{d}t-\liminf_{\varepsilon\to 0}\int_0^T\Phi_\varepsilon(u_\varepsilon)\,\mathrm{d}t+\frac{1}{2}\big\|v(0)-u_0\big\|_H^2\\ \leq\ & \int_0^T\Big\langle\frac{\mathrm{d}v}{\mathrm{d}t}-f+A(t,u),v-u\Big\rangle\,\mathrm{d}t+\int_0^T\Phi(v)-\Phi(u)\,\mathrm{d}t+\frac{1}{2}\big\|v(0)-u_0\big\|_H^2; \end{aligned} \tag{10.28}$$

note that we used (10.15) and (10.24). □

Remark 10.7. We did not have any information about $\frac{\mathrm{d}}{\mathrm{d}t}u$ in the preceding Theorem 10.6, which is why we could not expect any pseudomonotonicity of $\mathscr{A}$ inherited from pseudomonotonicity of $A(t,\cdot)$ as in Lemmas 8.8 or 8.29. The assumed pseudomonotonicity of $\mathscr{A}$ can be then obtained as, e.g., Example 8.52 or Remark 8.45.

Proposition 10.8 (Uniqueness of the weak solution). *Let $A(t,\cdot)$ be strictly monotone for a.a. $t\in I$ and Φ admit the approximation property* (10.22); *then there is at most one weak solution to* (10.1) *in the class $L^\infty(I;H)$.*

Proof. [9] Take $u_1,u_2\in L^p(I;V)\cap L^\infty(I;H)$ two weak solutions, i.e. both u_1 and u_2 satisfy (10.20). Let us sum (10.20) for u_1 and u_2, and test it by $v\equiv u_\delta := \mathfrak{R}_\delta(u,u_0)$ with $u:=\frac{1}{2}u_1+\frac{1}{2}u_2$ and with $u_\delta=\mathfrak{R}_\delta(u,u_0)$ denoting a regularization procedure with the properties (10.22). This gives:

$$\begin{aligned} \frac{1}{2}\big\langle\mathscr{A}(u_1)-\mathscr{A}(u_2),u_2-u_1\big\rangle &= \lim_{\delta\to 0}\Big(\big\langle\mathscr{A}(u_1),u_\delta-u_1\big\rangle+\big\langle\mathscr{A}(u_2),u_\delta-u_2\big\rangle\Big)\\ &\geq \liminf_{\delta\to 0}\bigg(\int_0^T\Phi(u_1)+\Phi(u_2)-2\Phi(u_\delta)\,\mathrm{d}t\\ &\qquad\qquad+\Big\langle f-\frac{\mathrm{d}u_\delta}{\mathrm{d}t},2u_\delta-u_1+u_2\Big\rangle-\|u_\delta(0)-u_0\|_H^2\bigg)\\ &\geq 2\liminf_{\delta\to 0}\int_0^T\Phi(u)-\Phi(u_\delta)\,\mathrm{d}t+2\lim_{\delta\to 0}\Big\langle f,u_\delta-u\Big\rangle\\ &\qquad\qquad-2\limsup_{\delta\to 0}\Big\langle\frac{\mathrm{d}u_\delta}{\mathrm{d}t},u_\delta-u\Big\rangle-\lim_{\delta\to 0}\|u_\delta(0)-u_0\|_H^2. \end{aligned} \tag{10.29}$$

[9] See Lions [261, Chap.II,Sect.9.4] or Showalter [383, Chap.III, Prop.7.1] for $p\geq 2$.

Using (10.22) successively to the particular terms we push them to zero for $\delta \to 0$. Altogether, this means $\langle \mathscr{A}(u_1) - \mathscr{A}(u_2), u_1 - u_2 \rangle \le 0$, which gives $u_1 = u_2$ by the assumed strict monotonicity of $A(t, \cdot)$. □

Example 10.9 (The regularization procedure (10.22)). Let us illustrate (10.22) for a special case

$$\Phi(u) = \varphi(u) + \delta_K(u) \tag{10.30}$$

with $K \subset H$ convex and closed in H and $\varphi : V \to \mathbb{R}$ continuous and satisfying $0 \le \varphi(v) \le C(1 + \|v\|_V^p)$. Then, assuming also $u_0 \in K = \mathrm{cl}_H(K \cap V)$, we can use the construction (7.19), here with δ in place of ε and with the approximation $u_{0\delta} \to u_0$ in H with some $\{u_{0\delta}\}_{\delta>0} \subset K \cap V$. Obviously, we get $u_\delta \in W^{1,\infty,\infty}(I; V, H) \subset W^{1,p,p'}(I; V, V^*)$ with the properties (10.22b-d), cf. (7.18a-c). The Nemytskiĭ mapping $\mathscr{N}_\varphi : L^p(I; V) \to L^1(I)$ is continuous. By (10.22b), $\int_0^T \varphi(u_\delta(t))\,\mathrm{d}t \to \int_0^T \varphi(u(t))\,\mathrm{d}t$. Moreover, the convolutory integral (7.19) remains valued in K if $u(t) \in K$ for a.a. $t \in I$ and $u_{0\delta} \in K$ so that, in particular, $\bar{u}_\delta$ from the proof of Lemma 7.4 is valued in K; here we used the convexity of K and closedness of K in H. Hence, in this case (10.22a) obviously holds because $\int_0^T \delta_K(u_\delta(t))\,\mathrm{d}t = 0 = \int_0^T \delta_K(u(t))\,\mathrm{d}t$. If $u(t) \notin K$ for t from a set in I with a positive Lebesgue measure, then the right-hand integral in (10.22a) equals $+\infty$ and therefore (10.22a) holds in this case, too.

Example 10.10 (The regularization procedure (10.24)). For $K \subset H$ closed and $\Phi = \delta_K$, the regularization (10.18) can now be shown to satisfy (10.24) similarly as we did in Remark 10.3.

10.3 Examples of unilateral parabolic problems

We illustrate the above abstract theory on the evolution variant of the obstacle problem (5.18) in a special form (i.e. p-Laplacean with zero boundary condition).

Example 10.11 (An obstacle problem: very weak solution). We consider, for $w \in W^{1,p}(\Omega) \cap L^2(\Omega)$ independent of time, the following complementarity problem:

$$\left.\begin{array}{ll} \left.\begin{array}{ll} \dfrac{\partial u}{\partial t} - \mathrm{div}\big(|\nabla u|^{p-2}\nabla u\big) \ge g\,, & u \ge w, \\ \Big(\dfrac{\partial u}{\partial t} - \mathrm{div}\big(|\nabla u|^{p-2}\nabla u\big) - g\Big)(u - w) = 0, & \end{array}\right\} & \text{in } Q, \\ \dfrac{\partial u}{\partial \nu} \ge 0, \qquad u \ge w, \qquad \dfrac{\partial u}{\partial \nu}(u - w) = 0 & \text{on } \Sigma, \\ u(0, \cdot) = u_0 & \text{on } \Omega. \end{array}\right\} \tag{10.31}$$

The weak formulation results as in (10.20) in a *parabolic variational inequality*: we seek $u(t, \cdot) \ge w$ for a.a. $t \in I$ such that

$$\int_Q \Big(\frac{\partial v}{\partial t} - g\Big)(v - u) + |\nabla u|^{p-2}\nabla u \cdot \nabla(v - u)\,\mathrm{d}x\mathrm{d}t \ge -\frac{1}{2}\int_\Omega \big(v(0, \cdot) - u_0\big)^2 \mathrm{d}x \tag{10.32}$$

holds for any $v \in W^{1,p,p'}(I;W^{1,p}(\Omega),L^{p^{*\prime}}(\Omega))$, $v(t,\cdot) \geq w$ for a.a. $t \in I$. The *regularization* using a quadratic-penalty method arises as in (10.8) with (10.18):

$$\left.\begin{aligned} \frac{\partial u_\varepsilon}{\partial t} - \mathrm{div}\big(|\nabla u_\varepsilon|^{p-2}\nabla u_\varepsilon\big) + \frac{1}{\varepsilon}(u_\varepsilon - w)^- &= g && \text{in } Q,\\ \frac{\partial u}{\partial \nu} &= 0 && \text{on } \Sigma,\\ u(0,\cdot) &= u_0 && \text{on } \Omega, \end{aligned}\right\} \tag{10.33}$$

where $v^- := \min(0,v)$. Suppose: $u_0 \in L^2(\Omega)$, $u_0 \geq w$, and $g \in L^{p'}(I;L^{p^{*\prime}}(\Omega))$. The a-priori estimate can be obtained by multiplication of the equation in (10.33) by $u_\varepsilon - w$, integration over Ω, and by using Green's Theorem 1.31:

$$\begin{aligned} &\frac{1}{2}\frac{\mathrm{d}}{\mathrm{d}t}\|u_\varepsilon\|^2_{L^2(\Omega)} + \|\nabla u_\varepsilon\|^p_{L^p(\Omega;\mathbb{R}^n)} + \frac{1}{\varepsilon}\|(u_\varepsilon - w)^-\|^2_{L^2(\Omega)}\\ &\quad = \int_\Omega g(u_\varepsilon - w) + \frac{\partial u_\varepsilon}{\partial t}w + |\nabla u_\varepsilon|^{p-2}\nabla u_\varepsilon \nabla w \,\mathrm{d}x\\ &\quad \leq N\|g\|_{L^{p^{*\prime}}(\Omega)}\|u_\varepsilon - w\|_{W^{1,p}(\Omega)} + \int_\Omega \frac{\partial u_\varepsilon}{\partial t} w \,\mathrm{d}x + \big\||\nabla u_\varepsilon|^{p-1}\big\|_{L^{p'}(\Omega)}\|\nabla w\|_{L^p(\Omega;\mathbb{R}^n)}\\ &\quad \leq C_{\mathrm{P}}N\|g\|_{L^{p^{*\prime}}(\Omega)}\Big(\|\nabla u_\varepsilon - \nabla w\|_{L^p(\Omega;\mathbb{R}^n)} + \|u_\varepsilon - w\|_{L^2(\Omega)}\Big)\\ &\qquad + \int_\Omega \frac{\partial u_\varepsilon}{\partial t} w \,\mathrm{d}x + \frac{1}{p'}\|\nabla u_\varepsilon\|^p_{L^p(\Omega;\mathbb{R}^n)} + \frac{1}{p}\|\nabla w\|^p_{L^p(\Omega;\mathbb{R}^n)} \end{aligned} \tag{10.34}$$

where N is the norm of the embedding $W^{1,p}(\Omega) \subset L^{p^*}(\Omega)$ and C_{P} is the constant from the Poincaré inequality (1.55). After absorbing the last-but-one term in the left-hand side following the strategy (8.168), and making the integration over $[0,t]$, we can use Gronwall's inequality to estimate:

$$\begin{aligned} \|u_\varepsilon(t)\|^2_{L^2(\Omega)} - \|u_0\|^2_{L^2(\Omega)} &= \int_0^t \frac{\mathrm{d}}{\mathrm{d}t}\|u_\varepsilon(\vartheta)\|^2_{L^2(\Omega)}\mathrm{d}\vartheta \leq 2\int_0^t \bigg(\|u_\varepsilon(\vartheta,\cdot)\|^2_{L^2(\Omega)}\\ &+ \int_\Omega \frac{\partial u_\varepsilon}{\partial t}w\,\mathrm{d}x\bigg)\mathrm{d}\vartheta + K\Big(\|w\|^2_{L^2(\Omega)} + \|\nabla w\|^p_{L^p(\Omega;\mathbb{R}^n)}\Big) + K\|g\|^{p'}_{L^{p'}(0,t;L^{p^{*\prime}}(\Omega))} \end{aligned}$$

with some constant K together with the estimate

$$\int_0^t\int_\Omega \frac{\partial u_\varepsilon}{\partial t} w\mathrm{d}x\mathrm{d}\vartheta = \int_\Omega \big(u_\varepsilon(t) - u_0\big)w\mathrm{d}x \leq \big\|u_\varepsilon(t)\big\|^2_{L^2(\Omega)} + \big\|u_0\big\|^2_{L^2(\Omega)} + \frac{1}{2}\|w\|^2_{L^2(\Omega)}.$$

Altogether, this gives the estimates

$$\big\|u_\varepsilon\big\|_{L^\infty(I;L^2(\Omega))} \leq C, \tag{10.35a}$$

$$\big\|\nabla u_\varepsilon\big\|_{L^p(Q;\mathbb{R}^n)} \leq C, \tag{10.35b}$$

$$\big\|(u_\varepsilon - w)^-\big\|_{L^2(Q)} \leq \sqrt{\varepsilon}C. \tag{10.35c}$$

Note that the usual dual estimate $\|\frac{\partial}{\partial t}u_\varepsilon\|_{L^{p'}(I;W^{1,p}(\Omega)^*)} \le \varepsilon^{-1/2}C \to \infty$ cannot be used, hence one cannot expect convergence to a weak solution.

However, the convergence to the very weak solution in the sense (10.32) can then be proved by the methods we used for the weak solution of the abstract problem in Theorem 10.6 combined with direct treatment of pseudomonotonicity of $-\Delta_p$ by the Minty trick. Let us test the regularized equation (10.33) by $v - u_\varepsilon$ with $v \ge w$, and apply Green's Theorem 1.31. Realizing that $(u_\varepsilon - w)^-(v - u_\varepsilon) \le 0$ whenever $v \ge w$, it yields

$$\int_Q \Big(\frac{\partial u_\varepsilon}{\partial t} - g\Big)(v - u_\varepsilon) + |\nabla u_\varepsilon|^{p-2}\nabla u_\varepsilon \cdot \nabla(v - u_\varepsilon)\,\mathrm{d}x\mathrm{d}t$$
$$= -\frac{1}{\varepsilon}\int_Q (u_\varepsilon - w)^-(v - u_\varepsilon)\,\mathrm{d}x\mathrm{d}t \ge 0. \tag{10.36}$$

Assuming $v \in W^{1,p,p'}(I; W^{1,p}(\Omega), L^{p'}(\Omega))$, we can make the by-part integration:

$$\int_Q \frac{\partial u_\varepsilon}{\partial t}(v - u_\varepsilon)\,\mathrm{d}x\mathrm{d}t = \int_Q \frac{\partial v}{\partial t}(v - u_\varepsilon)\,\mathrm{d}x\mathrm{d}t + \int_Q \Big(\frac{\partial u_\varepsilon}{\partial t} - \frac{\partial v}{\partial t}\Big)(v - u_\varepsilon)\,\mathrm{d}x\mathrm{d}t$$
$$= \int_Q \frac{\partial v}{\partial t}(v - u_\varepsilon)\,\mathrm{d}x\mathrm{d}t - \frac{1}{2}\|u_\varepsilon(T,\cdot) - v(T,\cdot)\|^2_{L^2(\Omega)} + \frac{1}{2}\|u_0 - v(0,\cdot)\|^2_{L^2(\Omega)}$$
$$\le \int_Q \frac{\partial v}{\partial t}(v - u_\varepsilon)\,\mathrm{d}x\mathrm{d}t + \frac{1}{2}\|u_0 - v(0,\cdot)\|^2_{L^2(\Omega)}$$

to obtain

$$\int_Q \Big(\frac{\partial v}{\partial t} - g\Big)(v - u_\varepsilon) + |\nabla u_\varepsilon|^{p-2}\nabla u_\varepsilon \cdot \nabla(v - u_\varepsilon)\,\mathrm{d}x\mathrm{d}t \ge -\frac{1}{2}\|u_0 - v(0,\cdot)\|^2_{L^2(\Omega)}. \tag{10.37}$$

Now we apply the regularization procedure (7.19) which results in $u_\delta(t,x) := \delta^{-1}\int_0^{+\infty} \mathrm{e}^{-s/\delta}u(t-s,x)\,\mathrm{d}s$ if $u(t,x)$ is prolonged for $t < 0$ suitably as in Lemma 7.4. By the technique (10.27), we obtain $\liminf_{\varepsilon\to 0}\int_Q |\nabla u_\varepsilon|^{p-2}\nabla u_\varepsilon \cdot \nabla(u - u_\varepsilon)\,\mathrm{d}x\mathrm{d}t \ge 0$. By the monotonicity, boundedness, and radial continuity of the p-Laplacean as a mapping $L^p(I; W^{1,p}(\Omega)) \to L^{p'}(I; W^{1,p}(\Omega)^*)$, we have also $\limsup_{\varepsilon\to 0}\int_Q |\nabla u_\varepsilon|^{p-2}\nabla u_\varepsilon \cdot \nabla(v - u_\varepsilon)\,\mathrm{d}x\mathrm{d}t \le \int_Q |\nabla u|^{p-2}\nabla u \cdot \nabla(v - u)\,\mathrm{d}x\mathrm{d}t$, cf. Lemma 2.9. Now we can estimate from above the limit superior in (10.37), which just gives (10.32). Of course, we used also $u \ge 0$ implied by (10.35c).

Example 10.12 (An obstacle problem: weak solution)**.** Consider again the problem (10.31) and assume now that $g \in L^2(Q)$ and $u_0 \in W^{1,p}(\Omega)$, $u_0 \ge w$ a.e. in Ω, and $p > \max(1, 2n/(n+2))$ so that $W^{1,p}(\Omega) \Subset L^2(\Omega)$. The weak formulation of the problem (10.31) requires $u(t,\cdot) \ge w$ to satisfy, for any $v \ge w$ and for a.a. $t \in I$, the inequality:

$$\int_\Omega \Big(\frac{\partial u}{\partial t} - g(t,x)\Big)\big(v(x) - u(t,x)\big)$$
$$+ \big|\nabla u(t,x)\big|^{p-2}\nabla u(t,x) \cdot \big(\nabla v(x) - \nabla u(t,x)\big)\,\mathrm{d}x \ge 0\,. \tag{10.38}$$

The needed a-priori estimate (10.12) for the approximate solution u_ε can be obtained by multiplication of the equation in (10.33) by $\frac{\partial}{\partial t}u_\varepsilon$, integration over Ω, and usage of Green's Theorem 1.31 with the boundary condition in (10.33):

$$\begin{aligned}\Big\|\frac{\partial u_\varepsilon}{\partial t}\Big\|^2_{L^2(\Omega)} + \frac{1}{p}\frac{\partial}{\partial t}\big\|\nabla u_\varepsilon\big\|^p_{L^p(\Omega;\mathbb{R}^n)} + \frac{1}{2\varepsilon}\frac{\partial}{\partial t}\big\|(u_\varepsilon - w)^-\big\|^2_{L^2(\Omega)} \\ = \int_\Omega g(t,\cdot)\frac{\partial u_\varepsilon}{\partial t}\,\mathrm{d}x \le \frac{1}{2}\|g(t,\cdot)\|^2_{L^2(\Omega)} + \frac{1}{2}\Big\|\frac{\partial u_\varepsilon}{\partial t}\Big\|^2_{L^2(\Omega)}.\end{aligned} \tag{10.39}$$

The last term is to be absorbed in the first one. This gives the estimates

$$\Big\|\frac{\partial u_\varepsilon}{\partial t}\Big\|_{L^2(Q)} \le C, \tag{10.40a}$$

$$\big\|\nabla u_\varepsilon\big\|_{L^\infty(I;L^p(\Omega;\mathbb{R}^n))} \le C, \tag{10.40b}$$

$$\big\|(u_\varepsilon - w)^-\big\|_{L^\infty(I;L^2(\Omega))} \le \sqrt{\varepsilon}C. \tag{10.40c}$$

In view of these estimates, we can suppose that (up to a subsequence) $u_\varepsilon \to u$ in $L^2(Q)$,[10] and also $u_\varepsilon \rightharpoonup u$ in $L^p(I;W^{1,p}(\Omega))$ and in $W^{1,2}(I;L^2(\Omega))$.

Moreover, the continuity of the Nemytskiĭ mapping $v \mapsto (v-w)^- : L^2(Q) \to L^2(Q)$, the convergence of $u_\varepsilon \to u$ in $L^2(Q)$, and (10.40c) yields also

$$\big\|(u-w)^-\big\|_{L^2(Q)} = \lim_{\varepsilon\to 0}\big\|(u_\varepsilon - w)^-\big\|_{L^2(Q)} \le \lim_{\varepsilon\to 0}\sqrt{\varepsilon}C = 0, \tag{10.41}$$

i.e. $u \ge w$ for a.a. $(t,x) \in Q$. We want to make a limit passage in (10.36). For $p = 2$, we can use the concavity of the functional[11] $u \mapsto \int_Q |\nabla u|^{p-2}\nabla u\cdot\nabla(v-u)\,\mathrm{d}x\mathrm{d}t = \int_Q \nabla u\cdot\nabla(v-u)\,\mathrm{d}x\mathrm{d}t$ which, by taking into account its continuity, implies its weak upper semicontinuity. In general, for $p \ne 2$, we can use the Minty trick (see Lemma 2.13) quite similarly as we did in the steady-state problem, cf. (5.76)–(5.77). For any $v \ge w$, by monotonicity of the p-Laplacean and by (10.36), we have

$$\begin{aligned}0 &\ge \limsup_{\varepsilon\to 0}\int_Q \big(|\nabla u_\varepsilon|^{p-2}\nabla u_\varepsilon - |\nabla v|^{p-2}\nabla v\big)\cdot\nabla(v-u_\varepsilon)\,\mathrm{d}x\mathrm{d}t \\ &\ge \lim_{\varepsilon\to 0}\int_Q \Big(g - \frac{\partial u_\varepsilon}{\partial t}\Big)(v-u_\varepsilon) - |\nabla v|^{p-2}\nabla v\cdot\nabla(v-u_\varepsilon)\,\mathrm{d}x\mathrm{d}t \\ &= \int_Q \Big(g - \frac{\partial u}{\partial t}\Big)(v-u) - |\nabla v|^{p-2}\nabla v\cdot\nabla(v-u)\,\mathrm{d}x\mathrm{d}t\end{aligned} \tag{10.42}$$

[10] Here we use Aubin-Lions Lemma 7.7, which gives $u_\varepsilon \to u$ in $L^\gamma(I;L^{p^*-\epsilon}(\Omega))$ with $\gamma < +\infty$ arbitrary, which is embedded into $L^2(Q)$ if $p > \max(1, 2n/(n+2))$.

[11] Unfortunately, the function $a \mapsto |a|^{p-2}a(b-a)$ is indeed not concave if $p \ne 2$.

as $u_\varepsilon \to u$ in $L^2(Q)$.[12] Let us put $v := (1-\delta)u + \delta z$ for $z \geq w$ a.e.; note that $v \geq w$ for any $\delta \in [0,1]$. After dividing it by δ, this gives

$$\int_Q \Big(g - \frac{\partial u}{\partial t}\Big)(z-u) - \big|\nabla u + \delta\nabla(z-u)\big|^{p-2}\big(\nabla u + \delta\nabla(z-u)\big)\cdot\nabla(z-u)\,\mathrm{d}x\mathrm{d}t \leq 0.$$

Passing to the limit with $\delta \to 0$ gives

$$\int_Q \Big(\frac{\partial}{\partial t}u - g\Big)(z-u) + |\nabla u|^{p-2}\nabla u\cdot\nabla(z-u)\,\mathrm{d}x\mathrm{d}t \geq 0,$$

which further gives the point-wise inequality (10.38), cf. Example 8.49.

Alternatively, for $p \in (1,+\infty)$ arbitrary, we can use the fact that the elliptic part has a potential and transforms the problem into the form

$$\int_Q \frac{\partial u_\varepsilon}{\partial t}(v-u_\varepsilon) + \frac{1}{p}|\nabla v|^p \mathrm{d}x\mathrm{d}t \geq \int_Q \frac{1}{p}|\nabla u_\varepsilon|^p + g(v-u_\varepsilon)\,\mathrm{d}x\mathrm{d}t, \tag{10.43}$$

and then use the weak lower semicontinuity of $u \mapsto \int_Q |\nabla u|^p \mathrm{d}x\mathrm{d}t : L^p(Q) \to \mathbb{R}$. This gives in the limit $\int_Q \frac{\partial u}{\partial t}(v-u) + \frac{1}{p}|\nabla v|^p\,\mathrm{d}x\mathrm{d}t \geq \int_Q \frac{1}{p}|\nabla u|^p + g(v-u)\,\mathrm{d}x\mathrm{d}t$, from which already $\int_Q \frac{\partial u}{\partial t}(v-u) + |\nabla u|^{p-2}\nabla u\cdot\nabla(v-u)\,\mathrm{d}x\mathrm{d}t \geq \int_Q g(v-u)\,\mathrm{d}x\mathrm{d}t$ follows because the convex functional $u \mapsto \frac{1}{p}\int_\Omega |\nabla u|^p\,\mathrm{d}x$ is just the potential of the mapping $A: W^{1,p}(\Omega) \to W^{1,p}(\Omega)^*$ defined as $\langle A(u), v\rangle = \int_\Omega |\nabla u|^{p-2}\nabla u\cdot\nabla v\,\mathrm{d}x$.

Strong convergence in $L^p(I;W^{1,p}(\Omega))$ can be proved by putting $v := u$ into (10.42), which gives

$$\begin{aligned}\int_Q \Big(\big|\nabla u_\varepsilon\big|^{p-2}\nabla u_\varepsilon - |\nabla u|^{p-2}\nabla u\Big)\cdot\nabla(u_\varepsilon - u)\,\mathrm{d}x\mathrm{d}t \\ \leq \int_Q \Big(g - \frac{\partial u_\varepsilon}{\partial t}\Big)(u_\varepsilon - u) - |\nabla u|^{p-2}\nabla u\cdot\nabla(u_\varepsilon - u)\,\mathrm{d}x\mathrm{d}t \to 0.\end{aligned} \tag{10.44}$$

Then, by the d-monotonicity of the p-Laplacean (cf. Example 2.83) and the uniform convexity of $L^p(I;W^{1,p}(\Omega))$, we get strong convergence in this space.

Exercise 10.13. Augment (10.31) by a lower-order term, say $c(u)$, or $c(\nabla u)$, or $\mathrm{div}(a_0(u))$ with $a_0 : \mathbb{R} \to \mathbb{R}^n$, and modify Example 10.12 accordingly.

Exercise 10.14. Modify Example 10.12 by considering the unilateral complementarity condition only on Σ as we did in the steady-state case in (5.95).

---

[12]Otherwise, we could alternatively use $u_\varepsilon(T) \rightharpoonup u(T)$ in $L^2(\Omega)$ and estimate only "lim inf" by

$$\begin{aligned}\liminf_{\varepsilon\to 0}\int_Q \frac{\partial u_\varepsilon}{\partial t}u_\varepsilon\mathrm{d}x\mathrm{d}t &= \liminf_{\varepsilon\to 0}\frac{1}{2}\|u_\varepsilon(T)\|^2_{L^2(\Omega)} - \frac{1}{2}\|u_0\|^2_{L^2(\Omega)} \\ &\geq \frac{1}{2}\|u(T)\|^2_{L^2(\Omega)} - \frac{1}{2}\|u_0\|^2_{L^2(\Omega)} = \int_Q \frac{\partial u}{\partial t}u\,\mathrm{d}x\mathrm{d}t.\end{aligned}$$

Exercise 10.15. Modify (10.42) by using, instead of the monotonicity of the p-Laplacean $-\Delta_p$, the monotonicity of $\frac{\partial}{\partial t} - \Delta_p$ or of $-\Delta_p + \frac{1}{\varepsilon}(\cdot - w)^-$ or of $\frac{\partial}{\partial t} - \Delta_p + \frac{1}{\varepsilon}(\cdot - w)^-$.

Example 10.16 (*Continuous casting: one-phase Stefan problem*)**.** In Sect. 5.6.2 we derived the following variational inequality to be satisfied for any $v \geq 0$:

$$\int_\Omega \sum_{i=1}^3 \kappa_i \frac{\partial u}{\partial x_i} \frac{\partial (v-u)}{\partial x_i} + cv_3 \frac{\partial u}{\partial x_3}(v-u)\,\mathrm{d}x + \int_\Gamma b(x)u(v-u)\,\mathrm{d}S \geq -\ell v_3 \int_\Omega (v-u)\,\mathrm{d}S + \int_\Gamma h(v-u)\,\mathrm{d}S \tag{10.45}$$

for $\kappa_1 = \kappa_2 = \kappa_3 = 1$. Here we neglect the diffusion flux in the x_3-variable (i.e. we put $\kappa_3 = 0$ while holding $\kappa_1 = \kappa_2 = 1$) and denote by $t = x_3/v_3$ the "residential" time (then $v_3 \frac{\partial}{\partial x_3} u = \frac{\partial}{\partial t} u$) and L and Ω_2 the length and the cross-section of the casted workpiece. Thus, for $T = L/v_3$, $I = [0,T]$, $\Omega = \Omega_2 \times I$ on Figure 14 on p. 167. Now we put $Q = I \times \Omega_2$ and $\Sigma = I \times \Gamma_2$ with $\Gamma_2 := \partial\Omega_2$ and also use the notation $u(t, x_1, x_2)$ (resp. $\mathrm{d}x\mathrm{d}t$) instead of $u(x_1, x_2, x_3)$ (resp. $\mathrm{d}x$). Thus (10.45) turns into

$$\int_Q \nabla u \cdot \nabla(v-u) + c\frac{\partial u}{\partial t}(v-u)\,\mathrm{d}t\mathrm{d}x + \int_\Sigma b(x)u(v-u)\,\mathrm{d}S\mathrm{d}t \geq -\ell v_3 \int_Q (v-u)\,\mathrm{d}t\mathrm{d}x + \int_\Sigma h(v-u)\,\mathrm{d}S\mathrm{d}t; \tag{10.46}$$

of course, now $\nabla = (\frac{\partial}{\partial x_1}, \frac{\partial}{\partial x_2})$. On the top side we have prescribed the Dirichlet boundary condition (cf. again Figure 14) which now turns into the initial condition, while on the bottom side of Γ we now do not prescribe any condition at all. Thus we arrived at the parabolic variational inequality:

$$\left.\begin{array}{l} \left.\begin{array}{l} \frac{\partial u}{\partial t} - \mathrm{div}(\kappa \nabla u) + \ell v_3 \geq 0, \qquad u \geq 0, \\ \Big(\frac{\partial u}{\partial t} - \mathrm{div}(\kappa\nabla u) + \ell v_3\Big)u = 0, \end{array}\right\} \quad \text{in } Q, \\ \frac{\partial u}{\partial \nu} + bu \geq 0, \quad u \geq 0, \quad \Big(\frac{\partial u}{\partial \nu} + bu\Big)u = 0 \qquad \text{on } \Sigma, \\ u(0,\cdot) = u_0 \qquad\qquad\qquad\qquad\qquad\qquad \text{on } \Omega_2. \end{array}\right\} \tag{10.47}$$

The Baiocchi transformation (5.129) of temperature θ adapted for moving boundary problems is called the *Duvaut transformation*[13]:

$$u(t, x_1, x_2) := -\int_0^t \theta(\vartheta, x_1, x_2)\mathrm{d}\vartheta. \tag{10.48}$$

[13]See Duvaut [129]. The new variable is called freezing index; see also Baiocchi [29], Crank [111, Sect.6.4.5], Duvaut-Lions [130, Appendix 3], Rodrigues [354, Sect.2.11].

Exercise 10.17 (*Elliptic regularization*[14]). Denote u_ε the solution to (10.45) with $\kappa_1 = \kappa_2 = 1$ and $\kappa_3 = \varepsilon$ with $\varepsilon > 0$ and show that $u_\varepsilon \to u$ in a suitable topology for $\varepsilon \searrow 0$, where u solves (10.46). Note that the boundary condition on the bottom part of Γ does not influence the limit.

10.4 Bibliographical remarks

Evolution variational inequalities and unilateral parabolic problems are addressed by Barbu [37, Sect.IV.3], Elliott, Ockendon [135], Glowinski, Lions, Trémolières [182], Lions [261, Chap.2,Sect.9 and Chap.3,Sect.6], Naumann [299], Showalter [383], and Zeidler [427, Chap.55]. A fundamental paper is by Brézis [65, Chap.II]. Applications to mechanics, in particular to contact problems, is in Duvaut, Lions [130], Eck, Jarušek, Krbec [132], Frémond [154], Hlaváček, Nečas [308], Kikuchi and Oden [231].

Variational inequalities in the context of their optimal control are in Barbu [38, Chap.5] or Tiba [404]. For application of Rothe's method as in Example 10.4 we refer to Kačur [219, Section 5.2].

[14]See Lions [260].

Chapter 11

Doubly-nonlinear problems

In this chapter we touch upon some selected problems not mentioned so far. This concerns situations with the time-derivative $\frac{\mathrm{d}}{\mathrm{d}t}u$ appearing nonlinearly (Section 11.1) or acting on a nonlinearity (Section 11.2), in the former case also in combination with the second time-derivative $\frac{\mathrm{d}^2}{\mathrm{d}t^2}u$ involved linearly (Section 11.3).

11.1 Inclusions of the type $\partial\Psi(\frac{\mathrm{d}}{\mathrm{d}t}u) + \partial\Phi(u) \ni f$

First, we begin with the initial-value problem for the inclusion

$$\partial\Psi\Big(\frac{\mathrm{d}u}{\mathrm{d}t}\Big) + A\big(u(t)\big) \ni f(t)\,, \qquad u(0) = u_0. \tag{11.1}$$

As both $\partial\Psi$ and A can be nonlinear and even set-valued (e.g. $A = \partial\Phi$), we speak about (a special case of) the so-called doubly nonlinear problem. Again, we pose the problem in the framework of Gelfand's triple $V \subset H \subset V^*$ with compact and dense embeddings.

11.1.1 Potential Ψ valued in $\mathbb{R} \cup \{+\infty\}$.

The first option will simultaneously be an illustration of a technique, not yet mentioned, based on the test of a differentiated-in-time inclusion by $\frac{\mathrm{d}^2}{\mathrm{d}t^2}u$. For this, we consider $A : V \to V^*$ in a special form

$$\begin{aligned} A = A_1 + A_2, \quad & A_1 : V \to V^* \text{ linear, } A_1^* = A_1, \text{ and } \big\langle A_1 v, v\big\rangle \geq c_0 |v|_V^2, \\ & A_2 : H \to H \text{ Lipschitz continuous,} \end{aligned} \tag{11.2}$$

and $\Psi : H \to \mathbb{R} \cup \{+\infty\}$ uniformly convex on H in the sense

$$\forall \xi_1 \in \partial\Psi(v_1),\ \ \xi_2 \in \partial\Psi(v_2): \qquad \big\langle \xi_1 - \xi_2, v_1 - v_2\big\rangle \geq c_1 \big\|v_1 - v_2\big\|_H^2 \tag{11.3}$$

with some c_0 and c_1 positive, and $|\cdot|_V$ again a seminorm on V satisfying the abstract Poincaré-type inequality (8.9). The requirement (11.3) implies that one can write $\Psi(v) = \Psi_0(v) + \frac{1}{2}c_1\|v\|_H^2$ with some Ψ_0 convex, hence also one has

$$\partial\Psi(v) = \partial\Psi_0(v) + c_1 v. \tag{11.4}$$

On the other hand, we did not impose any growth restriction on Ψ so that, in particular, Ψ can take values $+\infty$. By (11.2), A_1 has a quadratic potential, namely $A_1 = \Phi'$ with $\Phi(v) = \frac{1}{2}\langle A_1 v, v\rangle$, hence as a special case of (11.1) we consider the inclusion

$$\partial\Psi\big(\frac{\mathrm{d}u}{\mathrm{d}t}\big) + \Phi'\big(u(t)\big) + A_2\big(u(t)\big) \ni f(t)\ , \quad u(0) = u_0, \tag{11.5}$$

with Φ quadratic, and thus smooth, so that $\partial\Phi(v) = \{\Phi'(v)\}$.

We will call $u \in W^{1,2}(I;V)$ a strong solution to (11.1) if $u(0) = u_0$ and the inclusion in (11.1) is satisfied for a.a. $t \in I$. Equivalently, it means

$$\forall v\in V\ \forall(\text{a.a.})\,t\in I:\ \Big\langle A\big(u(t)\big), v-\frac{\mathrm{d}u}{\mathrm{d}t}\Big\rangle + \Psi(v) - \Psi\big(\frac{\mathrm{d}u}{\mathrm{d}t}\big) \geq \Big\langle f(t), v-\frac{\mathrm{d}u}{\mathrm{d}t}\Big\rangle. \tag{11.6}$$

We will analyze it via the *Rothe method*, which is now based on the recursive formula:

$$\partial\Psi\Big(\frac{u_\tau^k - u_\tau^{k-1}}{\tau}\Big) + A\big(u_\tau^k\big) \ni f_\tau^k\ , \qquad u_\tau^0 = u_0,\ \ f_\tau^k := \frac{1}{\tau}\int_{(k-1)\tau}^{k\tau} f(t)\,\mathrm{d}t. \tag{11.7}$$

This determines recursively the Rothe solutions u_τ and $\bar{u}_\tau$. As we will need also an analog of the 2nd-order time derivative, we have to introduce the piecewise affine interpolation $[\frac{\mathrm{d}}{\mathrm{d}t}u_\tau]^{\mathrm{i}}$ of the piecewise constant time derivative $\frac{\mathrm{d}}{\mathrm{d}t}u_\tau$; cf. Figure 17 on p. 216. This interpolated derivative is defined only on the interval $[\tau, T]$, and its derivative is obviously piecewise constant and imitates the second-order derivative of u_τ by the following symmetric second-order difference formula:

$$\frac{\mathrm{d}}{\mathrm{d}t}\Big[\frac{\mathrm{d}u_\tau}{\mathrm{d}t}\Big]^{\mathrm{i}}\bigg|_{[k\tau,(k+1)\tau]} = \frac{u_\tau^{k+1} - 2u_\tau^k + u_\tau^{k-1}}{\tau}\,, \qquad k = 1,\ldots,T/\tau - 1. \tag{11.8}$$

Proposition 11.1. *Let $A : V \to V^*$ satisfy* (11.2) *and $\Psi : H \to \mathbb{R}\cup\{+\infty\}$ be uniformly convex on H in the sense of* (11.3)*, lower semicontinuous on V, proper and $\partial\Psi(0) \ni 0$, and $V \Subset H$. Moreover, let $f \in W^{1,2}(I;H)$ and $u_0 \in V$ be a steady state with respect to $f(0)$ in the sense that $A(u_0) = f(0)$.*

(i) *Then the Rothe functions $u_\tau \in C(I;V)$ and $\bar{u}_\tau \in L^\infty(I;V)$ do exist and we have the estimates*

$$\big\|u_\tau\big\|_{W^{1,\infty}(I;V)} \leq C, \qquad \Big\|\frac{\mathrm{d}}{\mathrm{d}t}\Big[\frac{\mathrm{d}u_\tau}{\mathrm{d}t}\Big]^{\mathrm{i}}\Big\|_{L^2(I;H)} \leq C \tag{11.9}$$

for τ sufficiently small, where $[\cdot]^{\mathrm{i}}$ denotes the piecewise affine interpolation defined on the whole interval I by considering formally $u_\tau^k = u_0$ for $k = -1$; hence $\frac{\mathrm{d}}{\mathrm{d}t}[\frac{\mathrm{d}}{\mathrm{d}t}u_\tau]^{\mathrm{i}}|_{[0,\tau]} := (u_\tau^1 - u_0)/\tau^{-2}$.

(ii) *There is a subsequence such that $u_\tau \to u$ weakly* in $W^{1,\infty}(I;V)$, and every such u is a strong solution to (11.1).*

To make a-priori estimates, let us first outline the procedure heuristically: assume, for a moment, $\Psi \in C^2(V)$ and $A_2 \in C^1(H,H)$, differentiate $\Phi'(\frac{\mathrm{d}}{\mathrm{d}t}u) + A_1u + A_2(u) = f$ in time and test it by $\frac{\mathrm{d}^2}{\mathrm{d}t^2}u$, and use symmetry of A_1 (so that $\langle A_1\frac{\mathrm{d}}{\mathrm{d}t}u, \frac{\mathrm{d}^2}{\mathrm{d}t^2}u\rangle = \frac{1}{2}\frac{\mathrm{d}}{\mathrm{d}t}\langle A_1\frac{\mathrm{d}}{\mathrm{d}t}u, \frac{\mathrm{d}}{\mathrm{d}t}u\rangle$):

$$\begin{aligned}\Psi''\Big(\frac{\mathrm{d}u}{\mathrm{d}t}\Big)\Big(\frac{\mathrm{d}^2u}{\mathrm{d}t^2},\frac{\mathrm{d}^2u}{\mathrm{d}t^2}\Big) &+ \frac{1}{2}\frac{\mathrm{d}}{\mathrm{d}t}\Big\langle A_1\frac{\mathrm{d}u}{\mathrm{d}t},\frac{\mathrm{d}u}{\mathrm{d}t}\Big\rangle \\ &= \Big\langle \frac{\mathrm{d}}{\mathrm{d}t}\partial\Psi(\frac{\mathrm{d}u}{\mathrm{d}t}),\frac{\mathrm{d}^2u}{\mathrm{d}t^2}\Big\rangle + \Big\langle A_1\frac{\mathrm{d}u}{\mathrm{d}t},\frac{\mathrm{d}^2u}{\mathrm{d}t^2}\Big\rangle \\ &= \Big\langle \frac{\mathrm{d}f}{\mathrm{d}t} - A_2'(u)\frac{\mathrm{d}u}{\mathrm{d}t},\frac{\mathrm{d}^2u}{\mathrm{d}t^2}\Big\rangle \le \frac{1}{c_1}\Big\|\frac{\mathrm{d}f}{\mathrm{d}t}\Big\|_H^2 + \frac{\ell^2}{c_1}\Big\|\frac{\mathrm{d}u}{\mathrm{d}t}\Big\|_H^2 + \frac{c_1}{2}\Big\|\frac{\mathrm{d}^2u}{\mathrm{d}t^2}\Big\|_H^2, \end{aligned} \tag{11.10}$$

where $\ell := \|A_2'(\cdot)\|_{C^0(H,\mathscr{L}(H,H))}$ is the Lipschitz constant of $A_2 : H \to H$. By (11.3), $\Psi''(\cdot)(\xi,\xi) \ge c_1\|\xi\|_H^2$, hence the first term can be estimated from below $c_1\|\frac{\mathrm{d}^2}{\mathrm{d}t^2}u\|_H^2$ while the last term is to be absorbed in it. We integrate it over $(0,t)$ and use $\langle A_1\frac{\mathrm{d}}{\mathrm{d}t}u, \frac{\mathrm{d}}{\mathrm{d}t}u\rangle \ge c_0|\frac{\mathrm{d}}{\mathrm{d}t}u|_V^2$. We obtain

$$\begin{aligned}\frac{c_0}{2}\Big|\frac{\mathrm{d}u}{\mathrm{d}t}(t)\Big|_V^2 &+ \frac{c_1}{2}\int_0^t\Big\|\frac{\mathrm{d}^2u}{\mathrm{d}t^2}\Big\|_H^2\mathrm{d}t \\ &\le \int_0^t\Big(\frac{1}{c_1}\Big\|\frac{\mathrm{d}f}{\mathrm{d}t}\Big\|_H^2 + \frac{\ell^2}{c_1}\Big\|\frac{\mathrm{d}u}{\mathrm{d}t}\Big\|_H^2\Big)\mathrm{d}t + \frac{1}{2}\Big\langle A_1\frac{\mathrm{d}u}{\mathrm{d}t}(0),\frac{\mathrm{d}u}{\mathrm{d}t}(0)\Big\rangle. \end{aligned} \tag{11.11}$$

Further, we denote $U(t) := \int_0^t\|\frac{\mathrm{d}^2}{\mathrm{d}\vartheta^2}u\|_H^2\mathrm{d}\vartheta$ so that $\frac{\mathrm{d}}{\mathrm{d}t}U = \|\frac{\mathrm{d}^2}{\mathrm{d}t^2}u\|_H^2$ and use

$$\begin{aligned}\Big\|\frac{\mathrm{d}u}{\mathrm{d}t}(t)\Big\|_H^2 &= \Big\|\frac{\mathrm{d}u}{\mathrm{d}t}(0) + \int_0^t\frac{\mathrm{d}^2u}{\mathrm{d}\vartheta^2}\mathrm{d}\vartheta\Big\|_H^2 \\ &\le 2\Big\|\int_0^t\frac{\mathrm{d}^2u}{\mathrm{d}\vartheta^2}\mathrm{d}\vartheta\Big\|_H^2 + 2\Big\|\frac{\mathrm{d}u}{\mathrm{d}t}(0)\Big\|_H^2 \le 2TU(t) + 2\Big\|\frac{\mathrm{d}u}{\mathrm{d}t}(0)\Big\|_H^2 \end{aligned} \tag{11.12}$$

to substitute it into (11.11) to get

$$\begin{aligned}\frac{c_0}{2}\Big|\frac{\mathrm{d}u}{\mathrm{d}t}(t)\Big|_V^2 + \frac{c_1}{2}U(t) &\le \int_0^t\Big(\frac{1}{c_1}\Big\|\frac{\mathrm{d}f}{\mathrm{d}t}\Big\|_H^2 + \frac{\ell^2}{c_1}2TU(\vartheta)\Big)\mathrm{d}\vartheta \\ &\quad + 2T\Big\|\frac{\mathrm{d}u}{\mathrm{d}t}(0)\Big\|_H^2 + \frac{1}{2}\Big\langle A_1\frac{\mathrm{d}u}{\mathrm{d}t}(0),\frac{\mathrm{d}u}{\mathrm{d}t}(0)\Big\rangle, \end{aligned} \tag{11.13}$$

from which a bound for $U(t)$ uniform in $t \in I$ follows by the Gronwall inequality. For $t = T$ it implies a bound for u in $W^{2,2}(I;H)$ and, putting it again into (11.11) and using (8.9), also in $W^{1,\infty}(I;V)$. For using Gronwall's inequality, $\langle A_1\frac{\mathrm{d}}{\mathrm{d}t}u(0),\frac{\mathrm{d}}{\mathrm{d}t}u(0)\rangle$ must be finite, for which we need the imposed qualification of u_0 with respect to $f(0)$ because $\frac{\mathrm{d}}{\mathrm{d}t}u(0)\in[\partial\Psi]^{-1}(f(0)-A(u_0))=[\partial\Psi]^{-1}(0)$. In view of the assumption $\partial\Psi(0)\ni 0$, we can see that $\frac{\mathrm{d}}{\mathrm{d}t}u(0)=0$.

Proof of Proposition 11.1. Seeking $u=u_\tau^k$ satisfying the inclusion (11.7) is equivalent to seeking u solving $\partial\varphi(u)+A_2(u)\ni f_\tau^k$ where

$$\varphi(v):=\tau\Psi\Big(\frac{v-u_\tau^{k-1}}{\tau}\Big)+\frac{1}{2}\langle A_1v,v\rangle. \tag{11.14}$$

The existence of such u can be shown by Corollary 5.19; the coercivity follows from (11.2) and (11.3) by the estimate[1] $c_0|v|_V^2+(c_1/\tau)\|v-u_\tau^{k-1}\|_H^2\le\langle\partial\varphi(v),v\rangle\le\langle f_\tau^k-A_2(v),v\rangle\le\|f_\tau^k\|_H\|v\|_H+C(1+\|v\|_H^2)$ while the pseudomonotonicity of A_2 is due to its continuity and the compactness of $V\Subset H$.

Now, following the strategy (11.10), we are to prove a-priori estimates. In terms of the time-difference

$$\delta_\tau^k:=\frac{u_\tau^k-u_\tau^{k-1}}{\tau}, \tag{11.15}$$

we modify (11.7) by using (11.4), i.e. $c_1\delta_\tau^k+\partial\Psi_0(\delta_\tau^k)+A_1u_\tau^k+A_2(u_\tau^k)\ni f_\tau^k$, and write it for k and $k{+}1$ in the form:

$$\big\langle c_1\delta_\tau^k+A_1u_\tau^k+A_2(u_\tau^k)-f_\tau^k\,,\,v-\delta_\tau^k\big\rangle+\Psi_0(v)\ge\Psi_0\big(\delta_\tau^k\big), \tag{11.16a}$$
$$\big\langle c_1\delta_\tau^{k+1}+A_1u_\tau^{k+1}+A_2(u_\tau^{k+1})-f_\tau^{k+1},v-\delta_\tau^{k+1}\big\rangle+\Psi_0(v)\ge\Psi_0\big(\delta_\tau^{k+1}\big). \tag{11.16b}$$

As we defined formally $u_\tau^{-1}=u_0$, we have $\delta_\tau^0=0$. As $A_1u_0+A_2(u_0)=f(0)$ and Ψ is minimized at 0 (due to its convexity and the assumption $\partial\Psi(0)\ni 0$), the inequality (11.16a) holds for $k=0$ as well; one can imagine f extended continuously for $t<0$ by a constant, hence $f_\tau^0:=f(0)$. We put $v:=\delta_\tau^{k+1}$ into (11.16a) and $v:=\delta_\tau^k$ into (11.16b). Adding them for the suggested substitutions and using the formula (8.25)[2] yields

$$\begin{aligned}c_1\big\|\delta_\tau^{k+1}-\delta_\tau^k\big\|_H^2&+\frac{\tau}{2}\big\langle A_1\delta_\tau^{k+1},\delta_\tau^{k+1}\big\rangle-\frac{\tau}{2}\big\langle A_1\delta_\tau^k,\delta_\tau^k\big\rangle\\&\le\big\langle c_1\delta_\tau^{k+1}-c_1\delta_\tau^k,\delta_\tau^{k+1}-\delta_\tau^k\big\rangle+\big\langle A_1u_\tau^{k+1}-A_1u_\tau^k,\delta_\tau^{k+1}-\delta_\tau^k\big\rangle\\&\le\big\langle f_\tau^{k+1}-f_\tau^k,\delta_\tau^{k+1}-\delta_\tau^k\big\rangle-\big\langle A_2(u_\tau^{k+1})-A_2(u_\tau^k),\delta_\tau^{k+1}-\delta_\tau^k\big\rangle\\&\le\frac{1}{c_1}\big\|f_\tau^{k+1}-f_\tau^k\big\|_H^2+\tau^2\frac{\ell^2}{c_1}\big\|\delta_\tau^{k+1}\big\|_H^2+\frac{c_1}{2}\big\|\delta_\tau^{k+1}-\delta_\tau^k\big\|_H^2.\end{aligned} \tag{11.17}$$

The last term is to be absorbed in the first left-hand-side term. Then, to imitate

[1] We use $\partial\Psi(0)\ni 0$ with (11.3) for $\xi_2=0$, $v_2=0$, $v_1=(v-u_\tau^{k-1})/\tau$.

[2] It is here $\tau\langle A_1\delta_\tau^{k+1},\delta_\tau^{k+1}-\delta_\tau^k\rangle\ge\frac{\tau}{2}\langle A_1\delta_\tau^{k+1},\delta_\tau^{k+1}\rangle-\frac{\tau}{2}\langle A_1\delta_\tau^k,\delta_\tau^k\rangle$.

(11.11), we sum it for $k=0,\dots,l$, $l\le T/\tau$, which, after multiplying by $1/\tau$, gives

$$\frac{1}{2}\big\langle A_1\delta_\tau^{l+1},\delta_\tau^{l+1}\big\rangle+\tau\sum_{k=0}^{l}\frac{c_1}{2}\Big\|\frac{\delta_\tau^{k+1}-\delta_\tau^k}{\tau}\Big\|_H^2$$
$$\le\tau\sum_{k=0}^{l}\bigg(\frac{1}{c_1}\Big\|\frac{f_\tau^{k+1}-f_\tau^k}{\tau}\Big\|_H^2+\frac{\ell^2}{c_1}\big\|\delta_\tau^{k+1}\big\|_H^2\bigg)+\frac{1}{2}\big\langle A_1\delta_\tau^0,\delta_\tau^0\big\rangle. \tag{11.18}$$

Then, after using the discrete analog of (11.12), we use the discrete Gronwall inequality (1.70) provided τ is sufficiently small. The boundedness of the term $\tau^{-1}\sum_{k=1}^{l}\|f_\tau^{k+1}-f_\tau^k\|_H^2$ follows from the assumption $f\in W^{1,2}(I;H)$ as in (8.75)-(8.76). Note that $\langle A_1\delta_\tau^k,\delta_\tau^k\rangle$ is certainly bounded (as it even vanishes) for $k=0$ because $\delta_\tau^0=0$. Altogether, we get the estimates (11.9). Also,

$$\Big\|\Big[\frac{du_\tau}{dt}\Big]^{\mathrm{i}}-\frac{du_\tau}{dt}\Big\|_{L^2(I;H)}=\frac{\tau}{\sqrt{3}}\Big\|\frac{d}{dt}\Big[\frac{du_\tau}{dt}\Big]^{\mathrm{i}}\Big\|_{L^2(I;H)}=\mathscr{O}(\tau), \tag{11.19}$$

cf. (8.50).

Then the convergence (in terms of a subsequence) $[\frac{d}{dt}u_\tau]^{\mathrm{i}}\to\frac{d}{dt}u$ in $L^2(I;H)$, which follows by Aubin-Lions Lemma 7.7, implies also $\frac{d}{dt}u_\tau\to\frac{d}{dt}u$ in $L^2(I;H)$. Also, we have

$$\liminf_{\tau\to0}\int_0^T\Big\langle A_1\bar u_\tau,\frac{du_\tau}{dt}\Big\rangle\,dt=\liminf_{\tau\to0}\frac{1}{2}\big\langle A_1u_\tau(T),u_\tau(T)\big\rangle$$
$$-\frac{1}{2}\big\langle A_1u_0,u_0\big\rangle+\lim_{\tau\to0}\int_0^T\Big\langle A_1(\bar u_\tau-u_\tau),\frac{du_\tau}{dt}\Big\rangle\,dt$$
$$\ge\frac{1}{2}\big\langle A_1u(T),u(T)\big\rangle-\frac{1}{2}\big\langle A_1u_0,u_0\big\rangle=\int_0^T\Big\langle A_1u,\frac{du}{dt}\Big\rangle\,dt \tag{11.20}$$

where we used also[3] $u_\tau(T)\rightharpoonup u(T)$ in V and $\int_0^T\langle A_1(\bar u_\tau-u_\tau),\frac{d}{dt}u_\tau\rangle dt=\mathscr{O}(\tau)$ because $\|\bar u_\tau-u_\tau\|_{L^2(I;V)}\le\tau\|\frac{d}{dt}u_\tau\|_{L^2(I;V)}=\mathscr{O}(\tau)$. Then we can make the limit passage in the equivalent form of (11.7), namely

$$\int_0^T\Big\langle A_1\bar u_\tau+A_2(\bar u_\tau)-\bar f_\tau,v-\frac{du_\tau}{dt}\Big\rangle+\Psi(v)-\Psi\Big(\frac{du_\tau}{dt}\Big)\,dt\ge0 \tag{11.21}$$

by using

$$\liminf_{\tau\to0}\int_0^T\Psi\Big(\frac{du_\tau}{dt}\Big)\,dt\ge\int_0^T\Psi\Big(\frac{du}{dt}\Big)\,dt \tag{11.22}$$

because $v\mapsto\int_0^T\Psi(v(t))\,dt:L^\infty(I;V)\to\bar{\mathbb{R}}$ is weakly* lower semicontinuous and $\frac{d}{dt}u_\tau\to\frac{d}{dt}u$ weakly* in $L^\infty(I;V)$ due to the estimate in (11.9). Eventually,

[3] By $\frac{d}{dt}u_\tau\rightharpoonup\frac{d}{dt}u$ in $L^2(I;V)$ it follows that $u_\tau(T)=u_0+\int_0^T\frac{d}{dt}u_\tau dt\rightharpoonup u_0+\int_0^T\frac{d}{dt}udt=u(T)$.

$A_2(\bar{u}_\tau) \to A_2(u)$ in $L^2(I;H)$, which yields $\lim_{\tau\to 0}\int_0^T (A_2(\bar{u}_\tau), v - \frac{\mathrm{d}}{\mathrm{d}t}u_\tau)\mathrm{d}t = \int_0^T (A_2(u), v - \frac{\mathrm{d}}{\mathrm{d}t}u)\,\mathrm{d}t$. Altogether, (11.21) in the limit results in

$$\int_0^T \Big\langle A_1u + A_2(u) - f, v - \frac{\mathrm{d}u}{\mathrm{d}t}\Big\rangle + \Psi(v) - \Psi\big(\frac{\mathrm{d}u}{\mathrm{d}t}\big)\,\mathrm{d}t \geq 0 \tag{11.23}$$

from which the pointwise variant (11.6) follows. □

Remark 11.2 (Alternative estimation if $A_2 = 0$). The strategy (11.10)–(11.13) can be modified if $A_2 = 0$ by estimating

$$\begin{aligned}\int_0^t \Big\langle \frac{\mathrm{d}f}{\mathrm{d}t}, \frac{\mathrm{d}^2u}{\mathrm{d}t^2}\Big\rangle \,\mathrm{d}t &= \Big\langle \frac{\mathrm{d}f}{\mathrm{d}t}(t), \frac{\mathrm{d}u}{\mathrm{d}t}(t)\Big\rangle - \int_0^t \Big\langle \frac{\mathrm{d}^2f}{\mathrm{d}t^2}, \frac{\mathrm{d}u}{\mathrm{d}t}\Big\rangle \,\mathrm{d}t - \Big\langle \frac{\mathrm{d}f}{\mathrm{d}t}(0), \frac{\mathrm{d}u}{\mathrm{d}t}(0)\Big\rangle\\
&\leq C_{\mathrm{P}}\Big\|\frac{\mathrm{d}f}{\mathrm{d}t}(t)\Big\|_{V^*}\Big(\Big|\frac{\mathrm{d}u}{\mathrm{d}t}(t)\Big|_V + \Big\|\frac{\mathrm{d}u}{\mathrm{d}t}(t)\Big\|_H\Big)\\
&\quad + \int_0^t C_{\mathrm{P}}\Big\|\frac{\mathrm{d}^2f}{\mathrm{d}t^2}\Big\|_{V^*}\Big(\Big|\frac{\mathrm{d}u}{\mathrm{d}t}\Big|_V + \Big\|\frac{\mathrm{d}u}{\mathrm{d}t}\Big\|_H\Big)\,\mathrm{d}t + \Big\|\frac{\mathrm{d}f}{\mathrm{d}t}(0)\Big\|_{V^*}\Big\|\frac{\mathrm{d}u}{\mathrm{d}t}(0)\Big\|_V\\
&\leq C\Big\|\frac{\mathrm{d}f}{\mathrm{d}t}(t)\Big\|_{V^*}^2 + \frac{c_0}{4}\Big|\frac{\mathrm{d}u}{\mathrm{d}t}(t)\Big|_V^2 + \frac{c_1}{4}U(t)\\
&\quad + \int_0^t C\Big\|\frac{\mathrm{d}^2f}{\mathrm{d}t^2}\Big\|_{V^*}\Big(1 + \Big|\frac{\mathrm{d}u}{\mathrm{d}t}\Big|_V^2 + U\Big)\,\mathrm{d}t + \Big\|\frac{\mathrm{d}f}{\mathrm{d}t}(0)\Big\|_{V^*}^2 + C\end{aligned}$$

with C depending on C_{P}, c_0, c_1, and on $\|\frac{\mathrm{d}u}{\mathrm{d}t}(0)\|_V$. Usage of Gronwall's inequality then needs $f \in W^{2,1}(I;V^*)$. Like in Remark 8.23, the Rothe method now needs either usage of the finer version of the discrete Gronwall inequality like in Remark 8.15 or a controlled smoothening of f like in (8.15) to be involved in (11.7). Combining it with the previous strategy (11.10)–(11.13), we can thus allow for $f \in W^{1,2}(I;H) + W^{2,1}(I;V^*)$.

Example 11.3 (*Pseudoparabolic equations*). Equations with the time-derivative involved in a (possibly nonlinear) differential operator are sometimes called *pseudoparabolic*. An example is the problem with the regularized q-Laplacean:

$$\left.\begin{aligned} -\mathrm{div}\Big(\big(\varepsilon + |\nabla\frac{\partial u}{\partial t}|^{q-2}\big)\nabla\frac{\partial u}{\partial t}\Big) - \Delta u + c(u) &= g, && \text{in } Q,\\ u &= 0, && \text{on } \Sigma,\\ u(0,\cdot) &= u_0 && \text{on } \Omega,\end{aligned}\right\} \tag{11.24}$$

with $c : \mathbb{R} \to \mathbb{R}$ Lipschitz continuous and $1 < q \leq 2$. This problem fits with the above presented theory for $V = W_0^{1,2}(\Omega)$, $H = L^2(\Omega)$, $\Psi(v) = \int_\Omega \frac{1}{q}|\nabla v|^q + \frac{\varepsilon}{2}|\nabla v|^2\mathrm{d}x$, $A_1 = -\Delta$, and $A_2 = \mathscr{N}_c$; then c_1 in (11.3) is $\varepsilon C_{\mathrm{P}}^{-2}$ with C_{P} from the Poincaré inequality $\|v\|_{L^2(\Omega)} \leq C_{\mathrm{P}}\|\nabla v\|_{L^2(\Omega;\mathbb{R}^n)}$.

Example 11.4 (*Parabolic variational inequalities of "type II"*[4]). The above presented abstract theory is fitted to a unilateral constraint acting on the time derivative. An example is the following complementarity problem:

$$\left.\begin{array}{ll}\left.\begin{array}{r}\dfrac{\partial u}{\partial t} - \Delta u + c(u) \geq g, \quad \dfrac{\partial u}{\partial t} \geq 0 \\ \left(\dfrac{\partial u}{\partial t} - \Delta u + c(u) - g\right)\dfrac{\partial u}{\partial t} = 0\end{array}\right\} & \text{in } Q, \\ u = 0 & \text{on } \Sigma, \\ u(0,\cdot) = u_0 & \text{on } \Omega,\end{array}\right\} \tag{11.25}$$

with $c : \mathbb{R} \to \mathbb{R}$ again Lipschitz continuous. Variational inequality can be obtained by using the Green formula: find $u \in W^{1,\infty}(I; L^2(\Omega)) \cap L^\infty(I; W_0^{1,2}(\Omega))$ with $\frac{\partial}{\partial t}u \geq 0$ a.e. in Q such that

$$\begin{aligned}&\forall v \in W_0^{1,2}(\Omega), \quad v \geq 0 \ \text{ a.e. on } \Omega : \\ &\int_\Omega \left(\frac{\partial u}{\partial t} + c(u) - g\right)\left(v - \frac{\partial u}{\partial t}\right) + \nabla u \cdot \nabla\left(v - \frac{\partial u}{\partial t}\right) \mathrm{d}x \geq 0.\end{aligned} \tag{11.26}$$

This fits with the above abstract scheme with $\Psi : L^2(\Omega) \to [0, +\infty]$ defined as

$$\Psi(v) := \begin{cases} \frac{1}{2}\|v\|^2_{L^2(\Omega)} & \text{if } v \geq 0 \ \text{ a.e. in } \Omega \\ +\infty & \text{otherwise,}\end{cases} \tag{11.27a}$$

$$A_1(v) = -\Delta v, \qquad A_2(v) = c(v). \tag{11.27b}$$

Example 11.5 (*Boundary inequalities*). A unilateral constraint on $\frac{\partial}{\partial t}u$ can be realized on the boundary Γ. An example is the following complementarity problem:

$$\left.\begin{array}{ll}\dfrac{\partial u}{\partial t} - \Delta u = g, & \text{in } Q, \\ \dfrac{\partial u}{\partial \nu} + bu \geq 0, \quad \dfrac{\partial u}{\partial t} \geq 0, \quad \left(\dfrac{\partial u}{\partial \nu} + bu\right)\dfrac{\partial u}{\partial t} = 0 & \text{on } \Sigma, \\ u(0,\cdot) = u_0 & \text{on } \Omega.\end{array}\right\} \tag{11.28}$$

Variational inequality results by the Green formula: find $u \in W^{1,2}(I; W^{1,2}(\Omega))$ with $\frac{\partial}{\partial t}u|_\Gamma \geq 0$ such that, for all $v \in W^{1,2}(\Omega)$ with $v|_\Gamma \geq 0$ it holds that

$$\int_\Omega \left(\frac{\partial u}{\partial t} - g\right)\left(v - \frac{\partial u}{\partial t}\right) + \nabla u \cdot \nabla\left(v - \frac{\partial u}{\partial t}\right)\mathrm{d}x + \int_\Gamma bu\left(v - \frac{\partial u}{\partial t}\right)\mathrm{d}S \geq 0; \tag{11.29}$$

we assume $u \in W^{1,2}(I; W^{1,2}(\Omega))$ to give a good sense of $\nabla\frac{\partial}{\partial t}u$ and of $\frac{\partial}{\partial t}u|_\Gamma$.

[4]See Duvant, Lions [130, Chap.II], Glowinski, Lions, Trémoliers [182, Chap.6, Sect. 5] or Kačur [219, Sect.5.3] for more information.

This fits with the above abstract scheme with $V = W^{1,p}(\Omega)$, $H = L^2(\Omega)$, $\Psi : W^{1,2}(\Omega) \to [0,+\infty]$ defined as

$$\Psi(v) := \begin{cases} \frac{1}{2}\|v\|^2_{L^2(\Omega)} & \text{if } v|_\Gamma \geq 0 \text{ a.e. on } \Gamma, \\ +\infty & \text{otherwise.} \end{cases} \tag{11.30}$$

If still $g \in W^{1,2}(I; L^2(\Omega))$ and u_0 qualifies appropriately, by Proposition 11.1, (11.29) has a solution.

11.1.2 Potential Φ valued in $\mathbb{R} \cup \{+\infty\}$

Strengthening the assumptions on Ψ, we can allow the leading part of A, i.e. A_1 in (11.2), nonlinear and even set-valued, and also the very restrictive assumption on the initial condition we made in Proposition 11.1 can thus be put off. Thus, we get another special case of the doubly nonlinear problem. We confine ourselves to a case when $A = A_1 + A_2$ with A_1 having a convex, possibly nonsmooth ($\mathbb{R}\cup\{+\infty\}$)-valued potential, let us denote it by Φ, i.e. $A_1 = \partial\Phi$, and $A_2 : H \to H$. Thus, we have in mind the inclusion (11.5) with (possibly nonsmooth) $\Phi : V \to \mathbb{R} \cup \{+\infty\}$ and $\Psi : H \to \mathbb{R}$ being convex, coercive, and bounded in the sense

$$\big\langle \partial\Psi(v), v \big\rangle \geq\ c_0\|v\|^q_H - c_1\|v\|_H, \tag{11.31a}$$

$$\big\|\partial\Psi(v)\big\|_H \leq\ C\big(1 + \|v\|^{q-1}_H\big), \tag{11.31b}$$

$$\big\langle \partial\Phi(v), v \big\rangle \geq\ c_2|v|^p_V - c_3|v|_V, \tag{11.31c}$$

$$\big\|A_2(v)\big\|_H \leq\ C\big(1 + \|v\|^{q-1}_H\big); \tag{11.31d}$$

for some c_0, c_2 positive, $|\cdot|_V$ a seminorm on V satisfying the abstract Poincaré-type inequality (8.9), and $p, q > 1$; in fact, the concrete value of p will not play any role, cf. the estimates (11.35).

By a weak solution to (11.5) we will understand $u \in W^{1,\infty,q}(I; V, H)$ such that $u(0) = u_0 \in H$ and, for some $w, z \in L^{q'}(I; H)$, the following system of two inequalities and one "merging" equality holds:

$$w + z + \mathscr{A}_2(u) = f, \tag{11.32a}$$

$$\int_0^T \Big\langle w - \xi, \frac{\mathrm{d}u}{\mathrm{d}t} - v \Big\rangle_{H\times H} \mathrm{d}t \geq 0 \quad \forall v \in L^q(I;H),\ \xi \in L^{q'}(I;H),\ \xi \in \partial\Psi(v), \tag{11.32b}$$

$$\int_0^T \big\langle z - \xi, u - v \big\rangle_{V^*\times V} \mathrm{d}t \geq 0 \qquad \forall v \in L^q(I;V),\ \xi \in L^{q'}(I;V^*),\ \xi \in \partial\Phi(v). \tag{11.32c}$$

The philosophy behind this definition can be seen, without going into details, from the fact that (11.32b) means $w \in \partial\Psi(\frac{\mathrm{d}}{\mathrm{d}t}u)$ due to the maximal monotonicity of $\partial\Psi$, cf. Theorem 5.3(ii), and similarly (11.32c) means $z \in \partial\Phi(u)$. Hence (11.32a) expresses just the inclusion (11.5).

We will analyze it again by Rothe's method, which consists in the following recursive formula:

$$\partial\Psi\Big(\frac{u_\tau^k-u_\tau^{k-1}}{\tau}\Big)+\partial\Phi(u_\tau^k)+A_2(u_\tau^k)\ni f_\tau^k\,,\quad u_\tau^0=u_0, \tag{11.33}$$

and f_τ^k from (8.58). This determines recursively the Rothe solutions u_τ and $\bar u_\tau$, and (11.33) for $k=1,\dots,T/\tau$ then means that, for some $\bar w_\tau$ and $\bar z_\tau$ piecewise constant on the considered partition of I, the following identity and inequalities hold:

$$\bar w_\tau+\bar z_\tau+\mathscr{A}_2(\bar u_\tau)=\bar f_\tau, \tag{11.34a}$$

$$\int_0^T\Big\langle\bar w_\tau-\xi,\frac{\mathrm{d}u_\tau}{\mathrm{d}t}-v\Big\rangle_{H\times H}\mathrm{d}t\ge0\quad\forall v,\xi\in L^\infty(I;H),\ \xi\in\partial\Psi(v), \tag{11.34b}$$

$$\int_0^T\langle\bar z_\tau-\xi,\bar u_\tau-v\rangle_{V^*\times V}\mathrm{d}t\ge0\qquad\forall v\in L^\infty(I;V),\ \forall\xi\in L^\infty(I;V^*),\ \xi\in\partial\Phi(v). \tag{11.34c}$$

In fact, it suffices to require (11.34b,c) to hold for the specified ξ and v piecewise constant on the considered partition of I only.

Proposition 11.6 (Colli and Visintin[5]). *Let $V\Subset H$, the convex, lower semicontinuous functionals $\Phi:V\to\mathbb{R}\cup\{+\infty\}$ and $\Psi:H\to\mathbb{R}$ satisfy* (11.31a-c)*, $A_2:H\to H$ be continuous and satisfy* (11.31d)*, $f\in L^{q'}(I;H)$ and $u_0\in\mathrm{dom}(\Phi)$; in particular, $u_0\in V$ due to* (11.31c)*. Then:*

(i) *For $\tau>0$ sufficiently small, $u_\tau\in C(I;V)$ and $\bar u_\tau\in L^\infty(I;V)$ do exist and the following a-priori estimates hold:*

$$\big\|u_\tau\big\|_{W^{1,\infty,q}(I;V,H)}\le C, \tag{11.35a}$$

$$\big\|\bar w_\tau\big\|_{L^{q'}(I;H)}\le C, \tag{11.35b}$$

$$\big\|\bar z_\tau\big\|_{L^{q'}(I;H)}\le C. \tag{11.35c}$$

(ii) *There is a subsequence and some $(u,w,z)\in W^{1,\infty,q}(I;V;H)\times L^{q'}(I;H)^2$ such that $(u_\tau,\bar w_\tau,\bar z_\tau)$ converges weakly* to (u,w,z) and any (u,w,z) obtained in this way satisfies* (11.32).

Proof. Existence of the Rothe sequence follows by Corollary 5.19. For this, we define the convex functional $\varphi:V\to\bar{\mathbb{R}}$ by $v\mapsto\Phi(v)+\tau\Psi\big((v-u_\tau^{k-1})/\tau\big)$. Then, any solution to $\partial\varphi(u)+A_2(u)\ni f_\tau^k$ solves also (11.33).[6] The needed pseudomonotonicity of $A_2:V\to V^*$ follows from the compactness of $V\Subset H$ and the continuity of

[5] See the original work by Colli and Visintin [104] and Colli [101], or also the monograph [418, Sect.III.2] for more details in the special L^2-case $\partial\Phi(u):=-\mathrm{div}(a(\nabla u))$.

[6] Here we use Exercise 5.31.

$A_2 : H \to H$, while the coercivity follows from the estimate

$$\begin{aligned}\tfrac{1}{2}\tau^{1-q}c_0\|u\|_H^q - \tilde{c}_1\|u\|_H + c_2|u|_V^p - c_3|u|_V &\le \big\langle \partial\varphi(u), u\big\rangle + C_\tau \\ &\le \big\langle f_\tau^k - A_2(u), u\big\rangle + C_\tau \le \|f_\tau^k - A_2(u)\|_H^{q'} + \|u\|_H^q + C_\tau \\ &\le 2^{q'-1}\|f_\tau^k\|_H^{q'} + 4^{q'-1}C^{q'}\big(1 + \|u\|_H^q\big) + \|u\|_H^q + C_\tau, \end{aligned} \tag{11.36}$$

where C_τ is a constant depending on τ and u_τ^{k-1}. The last two terms with $\|u\|_H^2$ can be absorbed in the left-hand side if τ is small enough.

Let us first outline the a-priori estimate heuristically: test (11.5) by $\frac{\mathrm{d}}{\mathrm{d}t}u$ and estimate:

$$\begin{aligned}c_0\Big\|\frac{\mathrm{d}u}{\mathrm{d}t}\Big\|_H^q - c_1\Big\|\frac{\mathrm{d}u}{\mathrm{d}t}\Big\|_H + \frac{\mathrm{d}}{\mathrm{d}t}\Phi(u) &\le \Big\langle \partial\Psi\big(\frac{\mathrm{d}u}{\mathrm{d}t}\big), \frac{\mathrm{d}u}{\mathrm{d}t}\Big\rangle + \Big\langle \partial\Phi(u), \frac{\mathrm{d}u}{\mathrm{d}t}\Big\rangle \\ &\le \Big\langle f - A_2(u), \frac{\mathrm{d}u}{\mathrm{d}t}\Big\rangle \le 2^{q'-1}C_\varepsilon\Big(\|f\|_H^{q'} + \|A_2(u)\|_H^{q'}\Big) + \varepsilon\Big\|\frac{\mathrm{d}u}{\mathrm{d}t}\Big\|_H^q \\ &\le 2^{q'-1}C_\varepsilon\Big(\|f\|_H^{q'} + 2^{q'-1}C^{q'}\big(1 + \|u\|_H^q\big)\Big) + \varepsilon\Big\|\frac{\mathrm{d}u}{\mathrm{d}t}\Big\|_H^q \end{aligned} \tag{11.37}$$

where c_0 and c_1 come from (11.31a), C from (11.31d), and where C_ε is from (1.22) with q instead of p. The last term is to be absorbed in the first left-hand-side term provided $\varepsilon < c_0$ is chosen. Similarly, the c_1-term is to be handled "at the right-hand side" by Young inequality, too. As in (8.64), we denote $U(t) := \int_0^t \|\frac{\mathrm{d}}{\mathrm{d}\vartheta}u\|_H^q \mathrm{d}\vartheta$ so that $\frac{\mathrm{d}}{\mathrm{d}t}U = \|\frac{\mathrm{d}}{\mathrm{d}t}u\|_H^q$ and, by using also

$$\begin{aligned}\|u(t)\|_H^q = \Big\|u_0 + \int_0^t \frac{\mathrm{d}u}{\mathrm{d}\vartheta}\mathrm{d}\vartheta\Big\|_H^q &\le 2^{q-1}\Big\|\int_0^t \frac{\mathrm{d}u}{\mathrm{d}\vartheta}\mathrm{d}\vartheta\Big\|_H^q + 2^{q-1}\big\|u_0\big\|_H^q \\ &\le 2^{q-1}t^{q-1}U(t) + 2^{q-1}\big\|u_0\big\|_H^q, \end{aligned} \tag{11.38}$$

the estimate (11.37) yields

$$\frac{\mathrm{d}}{\mathrm{d}t}\Big(U + \Phi(u)\Big) \le C\big(\|f(t)\|_H^{q'} + U(t)\big) \tag{11.39}$$

with some C large enough. Then, by Gronwall's inequality, we get $U(t) + \Phi(u(t))$ bounded independently of t. Then, using also the semi-coercivity[7] of Φ and (11.38), we get $\|u(t)\|_H$ and $|u(t)|_V$ bounded independently of t, which bounds u in $L^\infty(I;V)$ through the Poincaré-type inequality (8.9). Eventually, $U(T) < +\infty$ bounds u in $W^{1,q}(I;H)$.

In the discrete scheme, we test (11.33) by $u_\tau^k - u_\tau^{k-1}$: More precisely, we test $w_\tau^k + z_\tau^k = f_\tau^k - A_2(u_\tau^k)$ by δ_τ^k, where $w_\tau^k \in \partial\Psi(\delta_\tau^k)$ with δ_τ^k the time difference (11.15) and $z_\tau^k \in \partial\Phi(u_\tau^k)$. The last inclusion means $\Phi(v) \ge \Phi(u_\tau^k) + \langle z_\tau^k, v - u_\tau^k\rangle$. Using $v = u_\tau^{k-1}$ and copying the strategy (11.37), we obtain

[7] The semi-coercivity of Φ means here $\Phi(v) \ge \tilde{c}_2|v|_V^p - \tilde{c}_3|v|_V$ with some $\tilde{c}_2 > 0$ and follows from (11.31c) and the assumed convexity of Φ by Theorem 4.4(i).

$$
\begin{aligned}
c_0\|\delta_\tau^k\|_H^q + \frac{\Phi(u_\tau^k) - \Phi(u_\tau^{k-1})}{\tau} &\le \langle w_\tau^k, \delta_\tau^k\rangle + \langle z_\tau^k, \delta_\tau^k\rangle + c_1\|\delta_\tau^k\|_H \\
&\le \langle f_\tau^k - A_2(u_\tau^k), \delta_\tau^k\rangle + c_1\|\delta_\tau^k\|_H \\
&\le 4^{q'-1}C_\varepsilon\big(\|f_\tau^k\|_H^{q'} + C^{q'} + C^{q'}\|u_\tau^k\|_H^q\big) + \varepsilon\|\delta_\tau^k\|_H^q + c_1\big(C_\varepsilon + \varepsilon\|\delta_\tau^k\|_H^q\big). \quad (11.40)
\end{aligned}
$$

Taking $\varepsilon < c_0/(1+c_1)$ and applying the discrete Gronwall inequality (1.70), like (11.38)–(11.39), gives estimate (11.35a).

As $\bar{w}_\tau \in \partial\Psi(\frac{\mathrm{d}}{\mathrm{d}t}u_\tau)$, by (11.31b) we have $\|\bar{w}_\tau(t)\|_H \le C(1 + \|\frac{\mathrm{d}}{\mathrm{d}t}u_\tau\|_H^{q-1})$, which gives also the estimate (11.35b). Moreover, from $\bar{z}_\tau = \bar{f}_\tau - \bar{w}_\tau - A_2(\bar{u}_\tau)$ we get also (11.35c).

As for the limit passage in (11.34), let us choose a subsequence such that:

$$u_\tau \rightharpoonup u \qquad \text{weakly* in } W^{1,\infty,q}(I;V,H), \tag{11.41}$$

$$\bar{w}_\tau \rightharpoonup w \qquad \text{weakly in } L^{q'}(I;H), \tag{11.42}$$

$$\bar{z}_\tau \rightharpoonup z \qquad \text{weakly in } L^{q'}(I;H). \tag{11.43}$$

From (11.41) and the Aubin-Lions Lemma 7.7, it also follows $u_\tau \to u$ in $L^q(I;H)$. Moreover, we have $\bar{f}_\tau \to f$ in $L^{q'}(I;H)$, cf. Remark 8.15. As we have

$$\|u_\tau - \bar{u}_\tau\|_{L^q(I;H)} = \frac{\tau}{\sqrt[q]{q+1}}\left\|\frac{\mathrm{d}u_\tau}{\mathrm{d}t}\right\|_{L^q(I;H)} = \mathscr{O}(\tau), \tag{11.44}$$

cf. (8.50), we have also $\bar{u}_\tau \to u$ in $L^q(I;H)$, and thus the convergence $\int_0^T\langle\bar{z}_\tau, \bar{u}_\tau\rangle\mathrm{d}t \to \int_0^T\langle z, u\rangle\mathrm{d}t$ in (11.34c) is obvious. Hence $z \in \partial\Phi(u)$ is proved. Moreover, $\mathscr{A}_2(\bar{u}_\tau) \to \mathscr{A}_2(u)$ in $L^{q'}(I;H)$ due to the continuity of the Nemytskiĭ mapping $\mathscr{A}_2 : L^q(I;H) \to L^{q'}(I;H)$, using the continuity of $A_2 : H \to H$ and the growth condition (11.31d); cf. Theorem 1.43. The limit passage in (11.34a) is then obvious. As for (11.34b), we use

$$
\begin{aligned}
\limsup_{\tau\to0}\int_0^T\Big\langle\bar{w}_\tau, \frac{\mathrm{d}u_\tau}{\mathrm{d}t}\Big\rangle\mathrm{d}t &= \limsup_{\tau\to0}\int_0^T\Big\langle\bar{f}_\tau - \mathscr{A}_2(\bar{u}_\tau) - \bar{z}_\tau, \frac{\mathrm{d}u_\tau}{\mathrm{d}t}\Big\rangle\mathrm{d}t \\
&\le \lim_{\tau\to0}\int_0^T\Big\langle\bar{f}_\tau - \mathscr{A}_2(\bar{u}_\tau), \frac{\mathrm{d}u_\tau}{\mathrm{d}t}\Big\rangle\mathrm{d}t - \liminf_{\tau\to0}\big(\Phi(u_\tau(T)) - \Phi(u_0)\big) \\
&\le \int_0^T\Big\langle f - \mathscr{A}_2(u), \frac{\mathrm{d}u}{\mathrm{d}t}\Big\rangle\mathrm{d}t - \Phi(u(T)) + \Phi(u_0) \\
&= \int_0^T\Big\langle f - \mathscr{A}_2(u), \frac{\mathrm{d}u}{\mathrm{d}t}\Big\rangle - \frac{\mathrm{d}}{\mathrm{d}t}\Phi(u(t))\,\mathrm{d}t \\
&= \int_0^T\Big\langle f - \mathscr{A}_2(u) - z, \frac{\mathrm{d}u}{\mathrm{d}t}\Big\rangle\mathrm{d}t = \int_0^T\Big\langle w, \frac{\mathrm{d}u}{\mathrm{d}t}\Big\rangle\mathrm{d}t, \qquad (11.45)
\end{aligned}
$$

where the first inequality is based on

$$\int_0^T\Big\langle \bar{z}_\tau, \frac{\mathrm{d}u_\tau}{\mathrm{d}t}\Big\rangle \mathrm{d}t = \sum_{k=1}^{T/\tau}\big\langle z_\tau^k, u_\tau^k - u_\tau^{k-1}\big\rangle \geq \sum_{k=1}^{T/\tau}\Phi(u_\tau^k) - \Phi(u_\tau^{k-1}) = \Phi\big(u_\tau(T)\big) - \Phi(u_0),$$

which follows from convexity of Φ and from that $z_\tau^k \in \partial\Phi(u_\tau^k)$, and the second inequality (11.45) is based on the convergence[8] $u_\tau(T) \rightharpoonup u(T)$ and the weak lower semi-continuity of $\Phi : V \to \bar{\mathbb{R}}\cup\{+\infty\}$. For the last line in (11.45), we used also the *chain rule*, namely that $\frac{\mathrm{d}}{\mathrm{d}t}\Phi(u) = \langle z, \frac{\mathrm{d}}{\mathrm{d}t}u\rangle$ for any $z \in \partial\Phi(u)$. This needs rather subtle arguments:[9] For any $\varepsilon > 0$ we can consider a finite partition $0 \leq t_0^\varepsilon < t_1^\varepsilon < \cdots < t_{k_\varepsilon}^\varepsilon \leq T$ such that $\lim_{\varepsilon\to 0}\max_{i=1,\ldots,k_\varepsilon}(t_i^\varepsilon - t_{i-1}^\varepsilon) = 0$ and z is defined at all t_i^ε and both z_ε and $z_\varepsilon^{\mathrm{R}}$, defined by $z_\varepsilon|_{(t_{i-1}^\varepsilon, t_i^\varepsilon)} = z(t_i^\varepsilon)$ and $z_\varepsilon^{\mathrm{R}}|_{(t_{i-1}^\varepsilon, t_i^\varepsilon)} = z(t_{i-1}^\varepsilon)$, converge to z weakly in $L^{q'}(I;H)$ for $\varepsilon \to 0$. As $z(t_i^\varepsilon) \in \partial\Phi(u(t_i^\varepsilon))$ for $i = 1,\ldots,k_\varepsilon$, we have

$$\big\langle u(t_i^\varepsilon) - u(t_{i-1}^\varepsilon), z(t_{i-1}^\varepsilon)\big\rangle \leq \Phi\big(u(t_i^\varepsilon)\big) - \Phi\big(u(t_{i-1}^\varepsilon)\big) \leq \big\langle u(t_i^\varepsilon) - u(t_{i-1}^\varepsilon), z(t_i^\varepsilon)\big\rangle. \tag{11.46}$$

Then, summing it up for $i = 1,\ldots,k_\varepsilon$, we obtain

$$\begin{aligned}\int_0^T\Big\langle\frac{\mathrm{d}u}{\mathrm{d}t}, z_\varepsilon^{\mathrm{R}}\Big\rangle_{H\times H}\mathrm{d}t &= \sum_{i=1}^{k_\varepsilon}\tau_i^\varepsilon\Big\langle\frac{u(t_i^\varepsilon) - u(t_{i-1}^\varepsilon)}{\tau_i^\varepsilon}, z(t_{i-1}^\varepsilon)\Big\rangle_{V\times V^*} \\ &\leq \Phi\big(u(t_{k_\varepsilon}^\varepsilon)\big) - \Phi\big(u(t_0^\varepsilon)\big) \\ &\leq \sum_{i=1}^{k_\varepsilon}\tau_i^\varepsilon\Big\langle\frac{u(t_i^\varepsilon) - u(t_{i-1}^\varepsilon)}{\tau_i^\varepsilon}, z(t_i^\varepsilon)\Big\rangle_{V\times V^*} = \int_0^T\Big\langle\frac{\mathrm{d}u}{\mathrm{d}t}, z_\varepsilon\Big\rangle_{H\times H}\mathrm{d}t \end{aligned}\tag{11.47}$$

with $\tau_i^\varepsilon := t_i^\varepsilon - t_{i-1}^\varepsilon$. As before, we have $\liminf_{\varepsilon\to 0}\Phi(u(t_{k_\varepsilon}^\varepsilon)) \geq \Phi(u(T))$ because $t_{k_\varepsilon}^\varepsilon \to T$ and $u(t_{k_\varepsilon}^\varepsilon) \rightharpoonup u(T)$ in V. Then, passing $\varepsilon \to 0$ in (11.47), we arrive at the last equality in (11.45).[10] □

[8] This follows from the boundedness of $\{u_\tau(T)\}_{\tau>0}$ in V and of $\{\frac{\mathrm{d}}{\mathrm{d}t}u_\tau\}_{\tau>0}$ in $L^q(I;H)$.

[9] In a different way, namely by using Yosida's regularization of Φ, this chain rule was proved in [396, Prop. 2.2] even for H a reflexive Banach space similarly as used also in (11.78) below. Note that for a special case $\Phi = \frac{1}{2}\|\cdot\|^2$, we stated such a chain rule already in Lemma 9.1 and then used it in (9.19).

[10] Cf. also Visintin [418, Prop.XI.4.11]. More in detail, both z_ε and $z_\varepsilon^{\mathrm{R}}$ are to be understood as defined equal to zero on $I \setminus [t_0^\varepsilon, t_{k_\varepsilon}^\varepsilon]$ in (11.79). This is a classical result that we can rely on $\lim_{\varepsilon\to 0}\sum_{i=1}^{k_\varepsilon}(t_i^\varepsilon - t_{i-1}^\varepsilon)z(t_i^\varepsilon) = \int_0^T z(t)\,\mathrm{d}t$ for suitable partitions and, as this holds for a.a. partitions, we may also require symmetrically that $\lim_{\varepsilon\to 0}\sum_{i=1}^{k_\varepsilon}(t_i^\varepsilon - t_{i-1}^\varepsilon)z(t_{i-1}^\varepsilon) = \int_0^T z(t)\,\mathrm{d}t$. Here we need rather this result for the Stieltjes-type integral $\lim_{\varepsilon\to 0}\sum_{i=1}^{k_\varepsilon}\int_{t_{i-1}^\varepsilon}^{t_i^\varepsilon}\big\langle\dot{u}(t), z(t_i^\varepsilon)\big\rangle\,\mathrm{d}t = \int_0^T\big\langle\dot{u}(t), z(t)\big\rangle\,\mathrm{d}t$ for some $\dot{u} \in L^q(I;H)$ fixed, cf. also [116, Lemma 4.12]. Using it for $\dot{u} = \frac{\mathrm{d}}{\mathrm{d}t}u$, we obtain also the convergence in the two integrals in (11.47). In addition, these partitions can be assumed nested, and we can pick up one common point inside I, and investigate the limit passage in (11.79) separately on the right-hand and the left-hand half-intervals. In the former option, it is like if t_0^ε would be fixed in (11.79) and then we can see that even $\lim_{\varepsilon\to 0}\Phi(u(t_{k_\varepsilon}^\varepsilon)) = \Phi(u(T))$. The analogous argument for the left-hand half-interval then yields $\lim_{\varepsilon\to 0}\Phi(u(t_0^\varepsilon)) = \Phi(u(0))$. Let us remark that the technique of replacement of Lebesgue integral by suitable Riemann sums dates back to Hahn [195] in 1914.

Remark 11.7. Realize that we used only monotonicity of $\partial\Psi$ (not potentiality) for the basic a-priori estimates. Hencefore, the generalization for a maximal monotone mapping in place of $\partial\Psi$ is possible. Anyhow, the potentiality of $\partial\Psi$ allows for alternative formulation of (11.32b), namely

$$\int_0^T \Psi(v) + \Big\langle w, \frac{\mathrm{d}u}{\mathrm{d}t} - v\Big\rangle_{H\times H} \mathrm{d}t \geq \int_0^T \Psi\Big(\frac{\mathrm{d}u}{\mathrm{d}t}\Big)\,\mathrm{d}t \tag{11.48a}$$

to hold for any $v \in L^q(I;H)$. Similarly, (11.32c) bears an alternative formulation

$$\int_0^T \Phi(v) + \langle z, u-v\rangle_{V^*\times V} \mathrm{d}t \geq \int_0^T \Phi(u)\,\mathrm{d}t \tag{11.48b}$$

to be valid for any $v \in L^q(I;V)$.

Remark 11.8 (Energy balance)**.** The estimate (11.37) has, in concrete motivated cases, a "physical" interpretation. If Φ is a stored energy and Ψ a (pseudo)potential of dissipative forces, then $\langle\partial\Psi(\frac{\mathrm{d}}{\mathrm{d}t}u), \frac{\mathrm{d}}{\mathrm{d}t}u\rangle + \frac{\mathrm{d}}{\mathrm{d}t}\Phi(u) = \langle f, \frac{\mathrm{d}}{\mathrm{d}t}u\rangle$ expresses the balance between the dissipation rate, the rate of change of stored energy, and the power of external loading f. Disregarding the non-potential term A_2, this balance (as an inequality) is just the core of (11.37).

Proposition 11.9 (Dynamic minimization of Φ)**.** *Let $f{=}0$ and $A_2{=}0$, and*[11]

$$\exists c,\alpha{>}0\ \ \forall\varepsilon{>}0\ \ \forall v{\in}H:\ \Big(\xi{\in}\partial\Psi(v):\ \|\xi\|_H \geq \varepsilon\Big) \ \ \Rightarrow\ \ \langle\xi, v\rangle \geq c\varepsilon^\alpha. \tag{11.49}$$

Considering $I{=}[0,+\infty)$ as in Remark 8.22, it holds that $\lim_{t\to+\infty}\Phi(u(t)) = \min\Phi$.

Proof. [12] Testing (11.5) by $\frac{\mathrm{d}}{\mathrm{d}t}u$ and integrating it over $[t_1,t_2]$ yields an energy estimate:

$$\Phi\big(u(t_2)\big) - \Phi\big(u(t_1)\big) + \int_{t_1}^{t_2} \inf\Big\langle\partial\Psi\big(\frac{\mathrm{d}u}{\mathrm{d}t}(\vartheta)\big), \frac{\mathrm{d}u}{\mathrm{d}t}(\vartheta)\Big\rangle\,\mathrm{d}\vartheta \leq 0; \tag{11.50}$$

it is to be proved by smoothing of Ψ and a limit passage. Thus[13]

$$\lim_{\substack{t_1\to+\infty\\ t_2\to+\infty}} \int_{t_1}^{t_2} \inf\Big\langle\partial\Psi\big(\frac{\mathrm{d}u}{\mathrm{d}t}\big), \frac{\mathrm{d}u}{\mathrm{d}t}\Big\rangle\,\mathrm{d}t \leq \lim_{\substack{t_1\to+\infty\\ t_2\to+\infty}} \Big(\Phi\big(u(t_2)\big) - \Phi\big(u(t_1)\big)\Big) = 0, \tag{11.51}$$

so $\{\int_0^t \inf\langle\partial\Psi(\frac{\mathrm{d}}{\mathrm{d}t}u(\vartheta)), \frac{\mathrm{d}}{\mathrm{d}t}u(\vartheta)\rangle\mathrm{d}\vartheta\}_{t>0}$ is Cauchy, hence the limit, denoted by definition $\int_0^{+\infty} \inf\langle\partial\Psi(\frac{\mathrm{d}}{\mathrm{d}t}u(\vartheta)), \frac{\mathrm{d}}{\mathrm{d}t}u(\vartheta)\rangle\mathrm{d}\vartheta$, does exist and is finite. Put

$$I_\varepsilon := \Big\{t \in [0,+\infty);\ \sup\Big\|\partial\Psi\big(\frac{\mathrm{d}u}{\mathrm{d}t}\big)\Big\|_H < \varepsilon\Big\}. \tag{11.52}$$

[11] The condition (11.49) is satisfied, e.g., for $\Psi(v) = \|v\|_H^q$ with $q > 1$. Then $\alpha = q'$. In particular, it holds for the "linear" evolution $\frac{\mathrm{d}}{\mathrm{d}t}u + \partial\Phi(u) \ni 0$ where $q = 2$. On the other hand, it does not hold for $\Psi(v) = \|v\|_H$.

[12] Cf. Aubin and Cellina [28, Chap.3, Sect.4] for the special case $\Psi(v) = \frac{1}{2}\|v\|_H^2$.

[13] Here we use that $t \mapsto \Phi(u(t))$ is bounded from below and, due to (11.50), nondecreasing.

Then the measure of I_ε must be infinite, otherwise by (11.49) we would have $\int_0^{+\infty} \inf\langle\partial\Psi(\frac{\mathrm{d}u}{\mathrm{d}t}(\vartheta)), \frac{\mathrm{d}u}{\mathrm{d}t}(\vartheta)\rangle \mathrm{d}\vartheta \geq c\varepsilon^\alpha \mathrm{meas}(\mathbb{R}^+ \setminus I_\varepsilon) = +\infty$, a contradiction. Hence, for any $t \in I_\varepsilon$, we can take $\xi \in \partial\Psi(\frac{\mathrm{d}u}{\mathrm{d}t})$ such that $-\xi \in \partial\Phi(u(t))$, and then we have

$$\begin{aligned}\inf_{\vartheta>0} \Phi\big(u(\vartheta)\big) &\leq \Phi\big(u(t)\big) \leq \Phi(v) + \big\langle -\xi, u(t) - v\big\rangle \\ &\leq \Phi(v) + \big\|\xi\big\|_H \big\|u(t) - v\big\|_H \leq \Phi(v) + \varepsilon\big\|u(t) - v\big\|_H \end{aligned} \tag{11.53}$$

for $v \in V \subset H$ arbitrary. Thus $\inf_{\vartheta>0} \Phi(u(\vartheta)) \leq \Phi(v)$. □

Remark 11.10 (*Stefanelli's variational principle* [396]). Considering a fully potential situation (i.e. $A_2 \equiv 0$) and following the idea of the Brezis-Ekeland principle from Sect. 8.10, one can apply two Fenchel inequalities to (11.32) written as $w \in \partial\Psi(\frac{\mathrm{d}}{\mathrm{d}t}u)$ and $f - w \in \partial\Phi(u)$. Summing it up, this would give the functional $\mathfrak{F}(u,w) := \int_0^T \Psi(\frac{\mathrm{d}u}{\mathrm{d}t}) + \Psi^*(w) - \langle w, \frac{\mathrm{d}u}{\mathrm{d}t}\rangle + \Phi(u) + \Phi^*(f - w) - \langle f - w, u\rangle \,\mathrm{d}t$ to be minimized on $W^{1,\infty,q}(I;V,H) \times L^{q'}(I;H)$ subject to the constraint $u(0) = u_0$. Obviously, $\mathfrak{F}(u,w) \geq 0$. Moreover, $\mathfrak{F}(u,w) = 0$ means exactly that $w \in \partial\Psi(\frac{\mathrm{d}}{\mathrm{d}t}u)$ and $f - w \in \partial\Phi(u)$ hold a.e. on I. Knowing existence of such (u,w) from Proposition 11.6, we can claim existence of such a minimizer. In contrast to (8.247), the potential $\mathfrak{F}$ need not be convex. Also its coercivity and, because of the term $\langle w, \frac{\mathrm{d}u}{\mathrm{d}t}\rangle$, weak lower semicontinuity are not obvious. Inspired by (11.45), the latter drawback can be avoided by substituting $\int_0^T \langle w, \frac{\mathrm{d}u}{\mathrm{d}t}\rangle \,\mathrm{d}t = \int_0^T \langle f, \frac{\mathrm{d}u}{\mathrm{d}t}\rangle \,\mathrm{d}t - \Phi(u(T)) + \Phi(u_0)$. Yet, it relies on $w \in f - \partial\Phi(u)$ which is not guaranteed here and which thus might violate even the non-negativity of $\mathfrak{F}$. Therefore, in [396], the following modification has been devised:

$$\begin{aligned}\mathfrak{F}(u,w) := &\bigg(\int_0^T \Psi\Big(\frac{\mathrm{d}u}{\mathrm{d}t}\Big) + \Psi^*(w) - \Big\langle f, \frac{\mathrm{d}u}{\mathrm{d}t}\Big\rangle \,\mathrm{d}t + \Phi(u(T)) - \Phi(u_0)\bigg)^+ \\ &+ \int_0^T \Phi(u) + \Phi^*(f - w) - \langle f - w, u\rangle \,\mathrm{d}t. \end{aligned} \tag{11.54}$$

Now again $\mathfrak{F}(u,w) \geq 0$ and $\mathfrak{F}(u,w) = 0$ means exactly that $w \in \partial\Psi(\frac{\mathrm{d}}{\mathrm{d}t}u)$ and $f - w \in \partial\Phi(u)$ hold a.e. on I. Like the Brezis-Ekeland principle, $\mathfrak{F}$ does not involve any derivatives of Φ and Ψ and is especially designed for nonsmooth problems. It may serve for the *direct method* of finding solutions to (11.1) as well as various methods of asymptotic analysis, cf. [396].

Exercise 11.11. Modify the proof of Proposition 11.6(ii) for the alternative formulation (11.48).

Exercise 11.12 (*Decoupling by semi-implicit Rothe method*). Consider the system of two inclusions governed by $\Phi(u,v)$, and $\Psi(\dot{u},\dot{v}) = \Psi_1(\dot{u}) + \Psi_2(\dot{v})$, i.e.

$$\partial\Psi_1\Big(\frac{\mathrm{d}u}{\mathrm{d}t}\Big) + \partial_u\Phi(u,v) = f_1 \quad \text{and} \quad \partial\Psi_2\Big(\frac{\mathrm{d}v}{\mathrm{d}t}\Big) + \partial_v\Phi(u,v) = f_2, \tag{11.55a}$$

and, instead of a fully implicit discretisation, the following *semi-implicit formula:*

$$\partial\Psi_1\Big(\frac{u_\tau^k-u_\tau^{k-1}}{\tau}\Big)+\partial_u\Phi(u_\tau^k,v_\tau^{k-1})=f_{1,\tau}^k, \tag{11.56a}$$

$$\partial\Psi_2\Big(\frac{v_\tau^k-v_\tau^{k-1}}{\tau}\Big)+\partial_v\Phi(u_\tau^k,v_\tau^k)=f_{2,\tau}^k, \tag{11.56b}$$

and modify the proof of Proposition 11.6, assuming that Φ is only separately convex, i.e. $\Phi(\cdot,v)$ and $\Phi(u,\cdot)$ are convex.[14]

Exercise 11.13 (*Evolutionary quasivariational inequality*). Consider the u-dependent $\Psi=\Psi(u,v)$ and, instead of (11.1), the inclusion

$$\partial_v\Psi\Big(u,\frac{\mathrm{d}u}{\mathrm{d}t}\Big)+\partial\Phi\big(u(t)\big)\ni f(t)\,,\qquad u(0)=u_0. \tag{11.57}$$

Modify the definition (11.48)[15] and the proof of Proposition 11.6(ii) for suitably qualified Ψ.

Exercise 11.14. Like in Exercise 8.62, assuming Φ smooth for simplicity, modify the estimation scenario (11.37) for A_2 having the (possibly nonpolynomial) growth $\|A_2(u)\|_H\le C(1+\|u\|_H^{q-1}+\Phi(u)^{1/q'-\epsilon})$ instead of (11.31d). Assuming $\epsilon>0$, modify also the coercivity estimate (11.36). Finish the proof of Proposition 11.6 by assuming $A_2:V_1\to H$ continuous for some $V_1\Supset V$.

Example 11.15 (Quasistatic *friction*). Referring to Exercise 11.13, consider $V=\{v\in W^{1,2}(\Omega);\ v=0 \text{ on } \Gamma_{\rm D}\}$ and the functionals $\Psi:V\times V\to\mathbb{R}$ and $\Phi:V\to\mathbb{R}$ given by

$$\Psi(u,v)=\int_\Omega\frac12|\nabla v|^2\,\mathrm{d}x+\int_\Gamma\mu b(u^+)|v|\,\mathrm{d}S, \tag{11.58a}$$

$$\Phi(u)=\int_\Omega\frac12|\nabla u|^2\,\mathrm{d}x+\int_{\Gamma_{\rm N}}\widehat{b}(u^+)\,\mathrm{d}S \tag{11.58b}$$

with $\widehat{b}(r)=\frac{1}{\alpha}|r|^\alpha$ and $b(r)=|r|^{\alpha-2}r$ with $1\le\alpha<2^\#$. The test by $\frac{\partial u}{\partial t}$ gives the a-priori estimate $u\in W^{1,2}(I;W^{1,2}(\Omega))$, which also gives compactness of the traces of $\frac{\partial u}{\partial t}$ on $\Sigma_{\rm N}$ in $L^2(I;L^\alpha(\Gamma_{\rm N}))$ by combination of the Aubin-Lions theorem with the trace operator, cf. p. 254. This allows for easy limit passage in the terms as $\int_{\Sigma_{\rm N}}\mu b(u^+)|\frac{\partial u}{\partial t}|\,\mathrm{d}S\mathrm{d}t$ and $\int_{\Sigma_{\rm N}}\mu b(u^+)|v|\,\mathrm{d}S\mathrm{d}t$ occurring in the weak formulation. Such a problem represents a scalar variant of a (regularized) unilateral frictional contact problem on $\Gamma_{\rm N}$ with a friction coefficient $\mu>0$.

---

[14] Hint: To get u_τ^k, minimize $u\mapsto\Phi(u,v_\tau^{k-1})+\tau\Psi_1((u-u_\tau^{k-1})\tau)-\langle f_{1,\tau}^k,u\rangle$. Then minimize $v\mapsto\Phi(u_\tau^k,v)+\tau\Psi_2((v-v_\tau^{k-1})\tau)-\langle f_{2,\tau}^k,v\rangle$ get v_τ^k. For a-priori estimates, use Remark 8.25.

[15] Hint: instead of (11.48a), consider $\int_0^T\Psi(u,v)+\langle w,\frac{\mathrm{d}u}{\mathrm{d}t}-v\rangle\mathrm{d}t\ge\int_0^T\Psi(u,\frac{\mathrm{d}u}{\mathrm{d}t})\,\mathrm{d}t$ for any v.

11.1.3 Uniqueness and continuous dependence on data

The doubly-nonlinear structure of (11.1) makes uniqueness of the solution not fully automatic.[16] We present two techniques to address this nontrivial task.

Proposition 11.16. *Let the assumptions of Proposition* 11.1 *be fulfilled. Having two sets of data* $(f,u_0)=(f_i,u_{0i})$ *and the corresponding strong solutions* u_i, $i=1,2$, *the following estimate*

$$\big\|u_1-u_2\big\|_{W^{1,2}(I;H)\cap L^\infty(I;V)}\le C\Big(\big\|f_1-f_2\big\|_{L^2(I;H)}+\sqrt{\langle A_1(u_{01}-u_{02}),u_{01}-u_{02}\rangle}\,\Big)$$

holds. In particular, it implies uniqueness of the strong solution.

Proof. We write (11.6) modified by using (11.4) for $u:=u_1$ and put $v:=\frac{\mathrm{d}}{\mathrm{d}t}u_2$, and also for $u:=u_2$ and put $v:=\frac{\mathrm{d}}{\mathrm{d}t}u_1$. This gives

$$\Big\langle c_1\frac{\mathrm{d}u_1}{\mathrm{d}t}+A_1u_1(t)+A_2(u_1),\frac{\mathrm{d}u_2}{\mathrm{d}t}-\frac{\mathrm{d}u_1}{\mathrm{d}t}\Big\rangle+\Psi_0\big(\frac{\mathrm{d}u_2}{\mathrm{d}t}\big)-\Psi_0\big(\frac{\mathrm{d}u_1}{\mathrm{d}t}\big)\ge\Big\langle f_1(t),\frac{\mathrm{d}u_2}{\mathrm{d}t}-\frac{\mathrm{d}u_1}{\mathrm{d}t}\Big\rangle,\tag{11.59}$$

$$\Big\langle c_1\frac{\mathrm{d}u_2}{\mathrm{d}t}+A_1u_2(t)+A_2(u_2),\frac{\mathrm{d}u_1}{\mathrm{d}t}-\frac{\mathrm{d}u_2}{\mathrm{d}t}\Big\rangle+\Psi_0\big(\frac{\mathrm{d}u_1}{\mathrm{d}t}\big)-\Psi_0\big(\frac{\mathrm{d}u_2}{\mathrm{d}t}\big)\ge\Big\langle f_2(t),\frac{\mathrm{d}u_1}{\mathrm{d}t}-\frac{\mathrm{d}u_2}{\mathrm{d}t}\Big\rangle.\tag{11.60}$$

Summing (11.59) with (11.60), one gets

$$\begin{aligned}c_1\Big\|\frac{\mathrm{d}u_1}{\mathrm{d}t}-\frac{\mathrm{d}u_2}{\mathrm{d}t}\Big\|_H^2&+\frac12\frac{\mathrm{d}}{\mathrm{d}t}\langle A_1(u_1-u_2),u_1-u_2\rangle\\&\le\Big\langle f_1-f_2,\frac{\mathrm{d}u_1}{\mathrm{d}t}-\frac{\mathrm{d}u_2}{\mathrm{d}t}\Big\rangle-\Big\langle A_2(u_1)-A_2(u_2),\frac{\mathrm{d}u_1}{\mathrm{d}t}-\frac{\mathrm{d}u_2}{\mathrm{d}t}\Big\rangle\\&\le\frac{1}{c_1}\|f_1-f_2\|_H^2+\frac{\ell^2N^2}{c_1}\|u_1-u_2\|_V^2+\frac{c_1}{2}\Big\|\frac{\mathrm{d}u_1}{\mathrm{d}t}-\frac{\mathrm{d}u_2}{\mathrm{d}t}\Big\|_H^2,\end{aligned}\tag{11.61}$$

from which the claimed estimate follows by Gronwall's inequality as in (8.63); here ℓ stands for the Lipschitz constant of A_2 and N for the norm of the embedding $V\subset H$. In particular, for $f_1{=}f_2$ and $u_{01}{=}u_{02}$, one gets $u_1{=}u_2$, i.e. the uniqueness. □

The nonlinear leading part needs finer technique and additional assumptions.

Proposition 11.17 (Mielke and Theil[17]). *If, in addition to the assumption of Proposition* 11.6, $A_2=0$ *and* Φ *is uniformly convex and smooth enough so that* Φ' *is strongly monotone and satisfies the Taylor expansion formula*

$$\big\|\Phi''(u_1)(u_2-u_1)+\Phi'(u_1)-\Phi'(u_2)\big\|_{V^*}\le C\big\|u_1-u_2\big\|_V^2,\tag{11.62}$$

then the solution to (11.5) *is unique in the class* $W^{1,1}(I;V)$.

[16] For a counterexample see Brokate, Krejčí and Schnabel [70].

[17] See Mielke and Theil [283, Theorem 7.4] for a bit modified case. Later works are by Mielke [278, Theorem 3.4] and by Brokate, Krejčí and Schnabel [70].

Proof. Take $u_1, u_2 \in W^{1,1}(I;V)$ two solutions to (11.5). We have $\partial\Psi(\frac{\mathrm{d}}{\mathrm{d}t}u_1) + \Phi'(u_1) \ni f$, which means equivalently $\langle f - \Phi'(u_1) - \xi, \frac{\mathrm{d}}{\mathrm{d}t}u_1 - v\rangle \geq 0$ for any $\xi \in \partial\Psi(v)$. As $f - \Phi'(u_2) \in \partial\Psi(\frac{\mathrm{d}}{\mathrm{d}t}u_2)$, we can substitute $\xi := f - \Phi'(u_2)$ and $v := \frac{\mathrm{d}}{\mathrm{d}t}u_2$, which gives

$$\Big\langle \Phi'(u_2) - \Phi'(u_1), \frac{\mathrm{d}u_1}{\mathrm{d}t} - \frac{\mathrm{d}u_2}{\mathrm{d}t}\Big\rangle \geq 0. \tag{11.63}$$

Furthermore, put

$$\alpha(t) := \big\langle \Phi'(u_1) - \Phi'(u_2), u_1 - u_2\big\rangle, \quad r_i := \Phi''(u_i)(u_i - u_{3-i}) + \Phi'(u_{3-i}) - \Phi'(u_i)$$

for $i = 1, 2$. Then

$$\begin{aligned}\frac{\mathrm{d}\alpha}{\mathrm{d}t} &= \Big\langle \Phi''(u_1)\frac{\mathrm{d}u_1}{\mathrm{d}t} - \Phi''(u_2)\frac{\mathrm{d}u_2}{\mathrm{d}t}, u_1 - u_2\Big\rangle + \Big\langle \Phi'(u_1) - \Phi'(u_2), \frac{\mathrm{d}u_1}{\mathrm{d}t} - \frac{\mathrm{d}u_2}{\mathrm{d}t}\Big\rangle \\ &= \sum_{i=1,2} \Big\langle \Phi''(u_i)(u_i - u_{3-i}) + \Phi'(u_i) - \Phi'(u_{3-i}), \frac{\mathrm{d}u_i}{\mathrm{d}t}\Big\rangle \\ &= \sum_{i=1,2} \Big\langle r_i + 2\Phi'(u_i) - 2\Phi'(u_{3-i}), \frac{\mathrm{d}u_i}{\mathrm{d}t}\Big\rangle.\end{aligned}$$

By (11.62), by strong monotonicity $\langle \Phi'(u_1) - \Phi'(u_2), u_1 - u_2\rangle \geq c\big\|u_1 - u_2\big\|_V^2$ for some $c > 0$, and by (11.63), we can estimate

$$\begin{aligned}\frac{\mathrm{d}\alpha}{\mathrm{d}t} &\leq C\|u_1 - u_2\|_V^2\bigg(\Big\|\frac{\mathrm{d}u_1}{\mathrm{d}t}\Big\|_V + \Big\|\frac{\mathrm{d}u_2}{\mathrm{d}t}\Big\|_V\bigg) \\ &\quad + 2\Big\langle \Phi'(u_1) - \Phi'(u_2), \frac{\mathrm{d}u_1}{\mathrm{d}t} - \frac{\mathrm{d}u_2}{\mathrm{d}t}\Big\rangle \leq \frac{C}{c}\bigg(\Big\|\frac{\mathrm{d}u_1}{\mathrm{d}t}\Big\|_V + \Big\|\frac{\mathrm{d}u_2}{\mathrm{d}t}\Big\|_V\bigg)\alpha(t)\end{aligned} \tag{11.64}$$

and, from $\alpha(0) = 0$, we get $\alpha \equiv 0$ by Gronwall's inequality. Hence $u_1 = u_2$. □

Remark 11.18. The assumption (11.62) requires $C^{2,1}$-smoothness of Φ. Then, the constant C in (11.62) can be taken as $\frac{1}{2}\|\Phi''\|_{C^{0,1}(V;\mathscr{L}(V;V^*))}$. For the linear leading part, Φ is quadratic and (11.62) is trivially fulfilled with $C \equiv 0$.

11.2 Inclusions of the type $\frac{\mathrm{d}}{\mathrm{d}t}E(u) + \partial\Phi(u) \ni f$

Some physically motivated problems lead to double nonlinearity of a structure other than (11.1), namely

$$\frac{\mathrm{d}E(u)}{\mathrm{d}t} + A\big(u(t)\big) \ni f(t), \qquad u(0) = u_0. \tag{11.65}$$

We again consider it posed in the Gelfand triple $V \Subset H \cong H^* \Subset V^*$. Moreover, we will consider another Banach space V_1 such that $V \subset V_1 \subset H$ (hence $H \subset V_1^* \subset V^*$) and $E : V_1 \to V_1^*$ monotone (or possibly even maximal monotone set-valued

$E : V_1 \rightrightarrows V_1^*$), and $A := A_1 + A_2$ with $A_1 := \partial\Phi$, $\Phi : V \to \mathbb{R} \cup \{+\infty\}$ proper convex, and $A_2 : V \to V^*$. The strong solution is then understood as a couple $(u, w) \in L^p(I; V) \times W^{1,\infty,p'}(I; V_1^*, V^*)$ such that $w(0) \in E(u_0)$ and

$$\forall v \in L^p(I; V) : \quad \int_0^T \Phi(v) + \Big\langle \frac{\mathrm{d}w}{\mathrm{d}t} + A_2(u) - f, v - u \Big\rangle_{V^* \times V} - \Phi(u)\,\mathrm{d}t \ge 0, \tag{11.66a}$$

$$\forall \xi \in L^{q'}(I; V_1^*)\ \forall v \in L^q(I; V_1),\ \xi \in E(v) : \quad \int_0^T \langle w - \xi, u - v \rangle_{V_1^* \times V_1} \mathrm{d}t \ge 0 \tag{11.66b}$$

with some $q > 1$. As $E : V_1 \rightrightarrows V_1^*$ is maximal monotone, (11.66b) means $w(t) \in E(u(t))$ while, as Φ is convex, (11.66a) means $f(t) - \frac{\mathrm{d}}{\mathrm{d}t}w - A_2(u(t)) \in \partial\Phi(u(t))$ for a.a. $t \in I$. Hence (11.66) indeed corresponds to (11.65).

11.2.1 The case $E := \partial\Psi$.

Let us consider the set-valued case $E = \partial\Psi$ with $\Psi : V_1 \to \mathbb{R}$ convex. We apply the naturally modified *Rothe method* which now seeks recursively the couple $(u_\tau^k, w_\tau^k) \in V \times V_1^*$ such that

$$\frac{w_\tau^k - w_\tau^{k-1}}{\tau} + A(u_\tau^k) \ni f_\tau^k\,, \qquad w_\tau^k \in E(u_\tau^k), \tag{11.67}$$

for $k = 1, \ldots, T/\tau$, and with $w_\tau^0 \in E(u_0)$, $u_0 \in V_1$.

Lemma 11.19. *Assume* $\Psi : V_1 \to \mathbb{R}$ *and* $A = A_1 + A_2$ *with* $A_1 = \partial\Phi$, $\Phi : V \to \mathbb{R}$ *convex continuous, and* $A_2 : V \to V^*$ *pseudomonotone. Moreover, if* $f \in L^{p'}(I; V^*)$ *and* $\Psi^*(E(u_0)) < +\infty$,[18] *and if*

$$\exists c_0 > 0,\ c_1, c_2 \in \mathbb{R} : \qquad \langle A(v), v \rangle \ge c_0 \|v\|_V^p - c_1 \|v\|_V - c_2 \|E(v)\|_{V_1^*}^{q'}, \tag{11.68a}$$

$$\exists c_3 \in \mathbb{R} : \qquad \|E(v)\|_{V_1^*} \le c_3 \big(1 + \|v\|_{V_1}^{q-1}\big)\,, \tag{11.68b}$$

$$\exists \mathfrak{C} : \mathbb{R} \to \mathbb{R}\ \textit{increasing} : \qquad \big\|\partial\Phi(v)\big\|_{V^*} \le \mathfrak{C}\big(\|E(v)\|_{V_1^*}\big)\big(1 + \|v\|_V^{p-1}\big), \tag{11.68c}$$

$$\big\|A_2(v)\big\|_{V^*} \le \mathfrak{C}\big(\|E(v)\|_{V_1^*}\big)\big(1 + \|v\|_V^{p-1}\big) \tag{11.68d}$$

for some $q > 1$, *and if* $\tau \le \tau_0 < (q-1)/(q2^{q-1}c_2c_3^{q-1})$, *then there exists the Rothe's sequence* $\{u_\tau\}_{\tau>0}$ *and* $\{w_\tau\}_{\tau>0}$ *and*

$$\big\|w_\tau\big\|_{L^\infty(I;V_1^*)} \le C, \qquad \big\|u_\tau\big\|_{L^p(I;V)} \le C, \qquad \Big\|\frac{\mathrm{d}w_\tau}{\mathrm{d}t}\Big\|_{L^{p'}(I;V^*)} \le C. \tag{11.69}$$

Proof. Define $B(v) := E(v) + \tau A(v)$. Then one can take $u_\tau^k = v$ with v solving $B(v) \ni \tau f_\tau^k + w_\tau^{k-1}$, which does exist by Corollary 5.19,[19] and $w_\tau^k \in E(u_\tau^k)$.

---

[18] Legendre-Fenchel's conjugate Ψ^* to Ψ is defined $\Psi^*(v^*) := \sup_{v \in V} \langle v^*, v \rangle - \Psi(v)$, cf. (8.243).

[19] We use $B(v) := \partial\Psi(v) + \tau\partial\Phi(v) + \tau A_2(v) \supset \partial[\Psi + \tau\Phi](v) + \tau A_2(v)$, cf. Example 5.31. Note also that, by (11.68b) and (11.73), it holds that $\langle w, \partial\Psi^*(w)\rangle \ge \varepsilon\|w\|_{V_1^*}^{q'}$ for some $\varepsilon > 0$. For $E(u) = w$, i.e. $u = \partial\Psi^{-1}(w) = \partial\Psi^*(w)$, we have $\langle E(u), u\rangle \ge \varepsilon\|E(u)\|_{V_1^*}^{q'}$, and therefore, by (11.68a) and for $\tau > 0$ small enough, the mapping B is coercive as required in Corollary 5.19.

Let us first derive the a-priori estimate heuristically, assuming Ψ^* and Ψ smooth. Using

$$\frac{d}{dt}\Psi^*(w)=\Big\langle[\Psi^*]'(w),\frac{dw}{dt}\Big\rangle_{V\times V^*}=\Big\langle[\Psi']^{-1}(w),\frac{dw}{dt}\Big\rangle_{V\times V^*}=\Big\langle u,\frac{dw}{dt}\Big\rangle_{V\times V^*} \tag{11.70}$$

with $w = \Psi'(u)$, cf. (8.244), and testing (11.65) by u, we obtain

$$\frac{d}{dt}\Psi^*(w) + \langle A(u), u\rangle = \Big\langle \frac{dw}{dt} + A(u), u\Big\rangle = \langle f, u\rangle. \tag{11.71}$$

By (11.68a) and by Young's inequality $\langle f, u\rangle \le C_\varepsilon\|f\|_{V^*}^{p'} + \varepsilon\|u\|_V^p$, it further gives

$$\frac{d}{dt}\Psi^*(w) + c_0\|u\|_V^p \le c_1\|u\|_V + C_\varepsilon\|f\|_{V^*}^{p'} + \varepsilon\|u\|_V^p + c_2\|w\|_{V_1^*}^{q'} \tag{11.72}$$

and, taking $\varepsilon = c_0/2$, we can make an integration over $[0, t]$ and estimate

$$\frac{(q-1)\|w(t)\|_{V_1^*}^{q'}}{q2^{q-1}c_3^{q-1}} \le \Psi^*\big(w(t)\big) + \frac{c_3}{q'} \le \Psi^*\big(E(u_0)\big) + C + \int_0^t c_2\|w(\vartheta)\|_{V_1^*}^{q'}d\vartheta \tag{11.73}$$

with some C depending on q, c_0, c_1, c_3, and $\|f\|_{L^{p'}(I;V^*)}$, where we used the lower bound for Ψ^* obtained from $\Psi(v) \le c_3(\frac{1}{q'} + \frac{2}{q}\|v\|_{V_1}^q)$ as (8.248)–(8.249).[20] By Gronwall's inequality, it yields the estimate of w in $L^\infty(I; V_1^*)$. After integration (11.72) over I, we get also the estimate of u in $L^p(I; V)$. Further, the dual estimate of $\frac{d}{dt}w$ in $L^{p'}(I; V^*)$ follows from

$$\begin{aligned}\sup_{\|v\|_{L^p(I;V)}\le 1}\Big\langle\frac{dw}{dt}, v\Big\rangle &= \sup_{\|v\|_{L^p(I;V)}\le 1}\big\langle f-\mathscr{A}(u), v\big\rangle \le \big\|f-\mathscr{A}(u)\big\|_{L^{p'}(I;V^*)}\\ &\le \|f\|_{L^{p'}(I;V^*)} + 2\mathfrak{C}\big(\|w\|_{L^\infty(I;V_1^*)}\big)\Big(\int_0^T\big(1+\|u(t)\|_V^{p-1}\big)^{p'}dt\Big)^{1/p'}\\ &\le \|f\|_{L^{p'}(I;V^*)} + 4\mathfrak{C}\big(\|w\|_{L^\infty(I;V_1^*)}\big)\big(T^{1/p} + \|u\|_{L^p(I;V)}\big)^{p-1}.\end{aligned} \tag{11.74}$$

Rigorously, one must proceed by testing (11.67) by u_τ^k. The difference analog of (11.70) reads simply as

$$\frac{\Psi^*(w_\tau^k) - \Psi^*(w_\tau^{k-1})}{\tau} \le \Big\langle\frac{w_\tau^k - w_\tau^{k-1}}{\tau}, u_\tau^k\Big\rangle, \tag{11.75}$$

provided $u_\tau^k \in \partial\Psi^*(w_\tau^k)$, which just follows from the definition of the subdifferential, cf. (5.2), or equivalently provided $w_\tau^k \in \partial\Psi(u_\tau^k)$, cf. (8.244). Then the difference analog of (11.71)–(11.72) and the discrete Gronwall inequality instead of (11.73) provided $\tau \le \tau_0$ sufficiently small as specified are simple, as well as the analog to (11.74). □

[20] The upper bound for $\Psi(v)$ follows from (11.68b) when one uses the formula (4.6) with $\Phi(0) = 0$, as we can without loss of generality, which gives $\Psi(v) = \int_0^1\langle E(tv), v\rangle dt \le \int_0^1 c_3(1 + \|tv\|_{V_1})\|v\|_{V_1}dt = c_3(\|v\|_{V_1} + \|v\|_{V_1}^q/q) \le c_3(1/q' + 2\|v\|_{V_1}^q/q)$.

Proposition 11.20. *Let, in addition to the assumptions of Lemma* 11.19, *also* V_1 *be separable,* $A_2 : V \to V^*$ *be monotone and radially continuous, and* $V \Subset V_1$. *Then* (11.65) *possesses a strong solution. Moreover, any weak* limit* (u, w) *of (a subsequence of)* $\{(\bar{u}_\tau, w_\tau)\}_{\tau>0}$ *in* $L^p(I;V) \times W^{1,\infty,p'}(I;V_1^*,V^*)$ *solves* (11.65).

Proof. In view of (11.67), we have

$$\int_0^T \Phi(v) + \Big\langle \frac{\mathrm{d}w_\tau}{\mathrm{d}t} + A_2(\bar{u}_\tau) - \bar{f}_\tau, v - \bar{u}_\tau \Big\rangle_{V^*\times V} - \Phi(\bar{u}_\tau)\,\mathrm{d}t \geq 0, \tag{11.76a}$$

$$\int_0^T \big\langle \bar{w}_\tau - \xi, \bar{u}_\tau - v \big\rangle_{V_1^*\times V_1} \,\mathrm{d}t \geq 0, \tag{11.76b}$$

provided $\xi(t) \in E(v(t))$ for a.a. $t \in I$. To obtain (11.66b), we select a subsequence converging as indicated, cf. the a-priori estimates (11.69), and pass to the limit in (11.76b) by proving

$$\lim_{\tau\to 0} \int_0^T \big\langle \bar{w}_\tau, \bar{u}_\tau \big\rangle_{V_1^*\times V_1} \mathrm{d}t = \lim_{\tau\to 0} \int_0^T \big\langle \bar{w}_\tau, \bar{u}_\tau \big\rangle_{V^*\times V} \mathrm{d}t = \int_0^T \big\langle w, u \big\rangle_{V_1^*\times V_1} \mathrm{d}t. \tag{11.77}$$

We know that $\bar{u}_\tau \rightharpoonup u$ in $L^p(I;V)$ and, by Aubin-Lions' compact-embedding lemma, $W^{1,\infty,p'}(I;V_1^*,V^*) \Subset L^{p'}(I;V^*)$, also $w_\tau \to w$ in $L^{p'}(I;V^*)$. As $\frac{\mathrm{d}}{\mathrm{d}t}w_\tau$ is bounded in $L^{p'}(I;V^*)$, we have $\|w_\tau - \bar{w}_\tau\|_{L^{p'}(I;V^*)} = \tau \|\frac{\mathrm{d}}{\mathrm{d}t}w_\tau\|_{L^{p'}(I;V^*)} / \sqrt[p']{p'+1} = \mathscr{O}(\tau)$, cf. (8.50), and therefore also $\bar{w}_\tau \to w$ in $L^{p'}(I;V^*)$. In this way, (11.77) is proved. To get (11.66a), we must make a limit passage in (11.76a). We employ

$$\begin{aligned}
\liminf_{\tau\to 0} \int_0^T \Big\langle \frac{\mathrm{d}w_\tau}{\mathrm{d}t}, \bar{u}_\tau \Big\rangle_{V^*\times V} \mathrm{d}t &= \liminf_{\tau\to 0} \tau \sum_{k=1}^{T/\tau} \Big\langle \frac{w_\tau^k - w_\tau^{k-1}}{\tau}, u_\tau^k \Big\rangle_{V_1^*\times V_1} \\
&\geq \liminf_{\tau\to 0} \tau \sum_{k=1}^{T/\tau} \frac{\Psi^*(w_\tau^k) - \Psi^*(w_\tau^{k-1})}{\tau} \\
&= \liminf_{\tau\to 0} \Psi^*\big(w_\tau(T)\big) - \Psi^*\big(w_0\big) \\
&\geq \Psi^*\big(w(T)\big) - \Psi^*\big(w_0\big) = \int_0^T \Big\langle \frac{\mathrm{d}w}{\mathrm{d}t}, u \Big\rangle_{V^*\times V} \mathrm{d}t,
\end{aligned} \tag{11.78}$$

where (11.75) and (11.70) have been used together with the fact that $w_\tau(T) \rightharpoonup w(T)$ in V_1^*.[21] Moreover, the last equality in (11.78) represents a *chain rule* for Φ^* which holds because $u(t) \in E^{-1}(w(t))$ for a.a. $t \in I$ has been already proved by limiting (11.76b) and because Ψ^* is a potential of E^{-1}. To prove this equality, for any $\varepsilon > 0$ we can consider a finite partition $0 \leq t_0^\varepsilon < t_1^\varepsilon < \cdots < t_{k_\varepsilon}^\varepsilon \leq T$ like we did for (11.45), here guaranteeing that u is defined at all t_i^ε and both u_ε and $u_\varepsilon^{\mathrm{R}}$,

[21] Note that $\{w_\tau(T)\}_{\tau>0}$ is bounded in V_1^* due to (11.69) and, by the weak continuity of $W^{1,p'}(I;V^*) \to V^* : w \mapsto w(T)$, its weak limit in V^* (and thus in V_1^*, too) is just $w(T)$.

defined by $u_\varepsilon|_{(t_{i-1}^\varepsilon,t_i^\varepsilon)} = u(t_i^\varepsilon)$ and $u_\varepsilon^{\mathrm{R}}|_{(t_{i-1}^\varepsilon,t_i^\varepsilon)} = u(t_{i-1}^\varepsilon)$, converge to u weakly in $L^p(I;V)$ for $\varepsilon \to 0$. As $u(t_i^\varepsilon) \in \partial\Psi^*(w(t_i^\varepsilon))$ for $i = 1,\ldots,k_\varepsilon$, like (11.46), we have

$$\big\langle w(t_i^\varepsilon)-w(t_{i-1}^\varepsilon), u(t_{i-1}^\varepsilon)\big\rangle \le \Psi^*\big(w(t_i^\varepsilon)\big)-\Psi^*\big(w(t_{i-1}^\varepsilon)\big) \le \big\langle w(t_i^\varepsilon)-w(t_{i-1}^\varepsilon), u(t_i^\varepsilon)\big\rangle.$$

Then, summing it up for $i = 1,\ldots,k_\varepsilon$, we obtain

$$\begin{aligned}\int_0^T\Big\langle\frac{\mathrm{d}w}{\mathrm{d}t}, u_\varepsilon^{\mathrm{R}}\Big\rangle_{V^*\times V}\mathrm{d}t &= \sum_{i=1}^{k_\varepsilon}\tau_i^\varepsilon\Big\langle\frac{w(t_i^\varepsilon)-w(t_{i-1}^\varepsilon)}{\tau_i^\varepsilon}, u(t_{i-1}^\varepsilon)\Big\rangle_{V^*\times V}\\ &\le \Psi^*\big(w(t_{k_\varepsilon}^\varepsilon)\big) - \Psi^*\big(w(t_0^\varepsilon)\big)\\ &\le \sum_{i=1}^{k_\varepsilon}\tau_i^\varepsilon\Big\langle\frac{w(t_i^\varepsilon)-w(t_{i-1}^\varepsilon)}{\tau_i^\varepsilon}, u(t_i^\varepsilon)\Big\rangle_{V^*\times V} = \int_0^T\Big\langle\frac{\mathrm{d}w}{\mathrm{d}t}, u_\varepsilon\Big\rangle_{V^*\times V}\mathrm{d}t \qquad (11.79)\end{aligned}$$

with $\tau_i^\varepsilon := t_i^\varepsilon - t_{i-1}^\varepsilon$. As before, we have $\liminf_{\varepsilon\to0}\Psi^*(w(t_{k_\varepsilon}^\varepsilon)) \ge \Psi^*(w(T))$ because $t_{k_\varepsilon}^\varepsilon \to T$ and $w(t_{k_\varepsilon}^\varepsilon) \rightharpoonup w(T)$ in V_1^*. Then, passing $\varepsilon \to 0$ in (11.79), we arrive at the last equality in (11.78) similarly as we did for proving (11.45).[22] Using the monotonicity of A_2, (11.76a), the convexity of Φ, and (11.78), we obtain

$$\begin{aligned}0 &\le \limsup_{\tau\to0}\int_0^T\big\langle A_2(\bar u_\tau), \bar u_\tau - v\big\rangle - \big\langle A_2(v), \bar u_\tau - v\big\rangle\,\mathrm{d}t\\ &\le \limsup_{\tau\to0}\int_0^T\Phi(v) - \Phi(\bar u_\tau) + \Big\langle\frac{\mathrm{d}w_\tau}{\mathrm{d}t} - \bar f_\tau, v - \bar u_\tau\Big\rangle - \big\langle A_2(v), \bar u_\tau - v\big\rangle\,\mathrm{d}t\\ &\le \int_0^T\Phi(v) - \Phi(u) + \Big\langle\frac{\mathrm{d}w}{\mathrm{d}t} - f + A_2(v), v - u\Big\rangle\,\mathrm{d}t. \qquad (11.80)\end{aligned}$$

Then (11.66a) follows by Minty's trick, i.e. by putting $v := (1-\varepsilon)u + \varepsilon z$, $\varepsilon \in (0,1]$, using the convexity of Φ for $\Phi(v) - \Phi(u) \le \varepsilon(\Phi(z) - \Phi(u))$, dividing it by ε, and passing to the limit with $\varepsilon \searrow 0$ by using (11.68b,d) and Lebesgue dominated-convergence Theorem 1.14 as in (8.165).

Finally, $E(u_0) \ni w_\tau(0) \rightharpoonup w(0)$ in V_1^* and the convexity and closedness of $E(u_0)$ implies $w(0) \in E(u_0)$. □

Exercise 11.21. Like in (11.48b) relying on potentiality of $E = \partial\Psi$, one can modify (11.66b) to the variational inequality

$$\int_0^T\Psi(v) + \langle w, u - v\rangle_{V_1^*\times V_1}\,\mathrm{d}t \ge \int_0^T\Psi(u)\,\mathrm{d}t \qquad (11.81)$$

[22] Note that we cannot directly use that $\frac{\mathrm{d}}{\mathrm{d}t}\Psi^*(w) = \langle u, \frac{\mathrm{d}}{\mathrm{d}t}w\rangle_{V\times V^*}$ for any $u \in \partial\Psi^*(w) \cap V$ like Lemma 9.1 with $\Psi^* : V_1 \to \mathbb{R}$ instead of $\Phi : V \to \mathbb{R}$ together with reflexivity of V^* (so that by Komura's Theorem 1.39 $\frac{\mathrm{d}}{\mathrm{d}t}w$ is also the strong derivative) because we could assume Ψ^* locally Lipschitz continuous on V_1^* but hardly on V^* where w is valued as an absolutely continuous mapping.

to hold for any $v \in L^q(I;V_1)$. Modify the proof of Proposition 11.20 for this alternative definition.[23]

Exercise 11.22. Considering an abstract Poincaré inequality (8.9) modified as $\|v\|_V \le C_{\rm P}(|v|_V + \|v\|_{V_1^*})$, weaken the semi-coercivity assumption (11.68) as $\langle A(v), v\rangle \ge c_0|v|_V^p - c_1|v|_V - c_2\|E(v)\|_{V_1^*}^{q'}$ and modify the estimate (11.72) accordingly.

Remark 11.23. A general drawback of the technique based on the test by u is that it does not yield any information about $\frac{{\rm d}u}{{\rm d}t}$, which prevented us from using Papageorgiou's Lemma 8.8 and a possibly non-monotone A_2. Note that, formally, $\frac{{\rm d}u}{{\rm d}t} = \frac{\rm d}{{\rm d}t}E^{-1}(w) = [E^{-1}]'(w)\frac{{\rm d}w}{{\rm d}t}$ but $[E^{-1}]'(w) \in \mathscr{L}(V_1^*, V_1)$ while $\frac{{\rm d}w}{{\rm d}t}$ is valued only in V^* but not in V_1^*. In some specific cases, $[E^{-1}]'(w)$ admits an extension on V^*, which may open a way to estimate $\frac{{\rm d}u}{{\rm d}t}$; compare $E = $identity on H used in Chapters 8 and 10.

11.2.2 The case E non-potential

An alternative approach to analysis of (11.65) is based on testing by $\frac{\rm d}{{\rm d}t}u$. It allows us to consider Φ unbounded and valued in $\mathbb{R}\cup\{+\infty\}$ and to replace the assumption on potentiality of E by its uniform monotonicity. In addition, assuming E smooth with $E' : V_1 \to V_1^*$ bounded, one can use the *semi-implicit Rothe method*:

$$E'(u_\tau^{k-1})\frac{u_\tau^k - u_\tau^{k-1}}{\tau} + A(u_\tau^k) \ni f_\tau^k \tag{11.82}$$

for $k = 1, \dots, T/\tau$, $u_\tau^0 = u_0$. This, linearizing partly the problem, can lead to advantageous numerical strategies after applying additionally the Galerkin method.

Lemma 11.24. *Let $A = A_1 + A_2$ with $A_1 = \partial\Phi$ with $\Phi : V \to \mathbb{R}\cup\{+\infty\}$ convex lower semicontinuous, and let $f \in L^2(I;V_1^*)$, $u_0 \in {\rm dom}(\Phi)$, and*

$$\exists c_1 > 0 \quad \forall u, v \in V_1 : \qquad \big\langle E'(u)v, v\big\rangle_{V_1^* \times V_1} \ge c_1\|v\|_{V_1}^2, \tag{11.83a}$$

$$\exists c_0 > 0 \quad \forall v \in V : \qquad \Phi(v) \ge c_0|v|_V^p, \tag{11.83b}$$

$$\exists c_2 \in \mathbb{R} \quad \forall v \in V : \qquad \big\|A_2(v)\big\|_{V_1^*} \le c_2\big(1 + \|v\|_{V_1}\big) \tag{11.83c}$$

with some $p > 0$ (whose value is not reflected in (11.84)) and with $|\cdot|_V$ referring to a seminorm satisfying a modified Poincaré inequality (8.9), namely $\|v\|_V \le C_{\rm P}(|v|_V + \|v\|_{V_1})$. Then, for $\tau > 0$ sufficiently small, the following a-priori estimates hold:

$$\big\|u_\tau\big\|_{L^\infty(I;V)} \le C, \qquad \Big\|\frac{{\rm d}u_\tau}{{\rm d}t}\Big\|_{L^2(I;V_1)} \le C. \tag{11.84}$$

[23] Hint: Instead of (11.76b), use $\int_0^T \Psi(v) + \langle \bar w_\tau, \bar u_\tau - v\rangle_{V_1^* \times V_1} \,{\rm d}t \ge \int_0^T \Psi(\bar u_\tau)\,{\rm d}t$, and make the limit passage by weak lower semicontinuity of $u \mapsto \int_0^T \Psi(u)\,{\rm d}t$ and again by (11.77).

Proof. Let us first proceed heuristically: testing (11.65) by $\frac{\mathrm{d}}{\mathrm{d}t}u$, using that $\langle\frac{\mathrm{d}}{\mathrm{d}t}E(u), \frac{\mathrm{d}}{\mathrm{d}t}u\rangle = \langle E'(u)\frac{\mathrm{d}}{\mathrm{d}t}u, \frac{\mathrm{d}}{\mathrm{d}t}u\rangle \geq c_1\|\frac{\mathrm{d}}{\mathrm{d}t}u\|_{V_1}^2$, and integrating it over $[0,t]$ gives

$$\begin{aligned}\Phi(u(t)) - \Phi(u_0) + c_1U(t) &\leq \int_0^t \frac{\mathrm{d}}{\mathrm{d}\vartheta}\Phi(u) + \Big\langle E'(u(\vartheta))\frac{\mathrm{d}u}{\mathrm{d}\vartheta}, \frac{\mathrm{d}u}{\mathrm{d}\vartheta}\Big\rangle_{V_1^*\times V_1}\mathrm{d}\vartheta \\ &= \int_0^t \Big\langle\frac{\mathrm{d}E(u)}{\mathrm{d}\vartheta} + \Phi'(u(\vartheta)), \frac{\mathrm{d}u}{\mathrm{d}\vartheta}\Big\rangle\,\mathrm{d}\vartheta = \int_0^t \Big\langle f(\vartheta) - A_2(u(\vartheta)), \frac{\mathrm{d}u}{\mathrm{d}\vartheta}\Big\rangle\,\mathrm{d}\vartheta \\ &\leq \int_0^t \frac{c_1}{2}\Big\|\frac{\mathrm{d}u}{\mathrm{d}\vartheta}\Big\|_{V_1}^2 + \frac{1}{c_1}\|f(\vartheta)\|_{V_1^*}^2 + 2\frac{c_2^2}{c_1}(1 + \|u(\vartheta)\|_{V_1}^2)\,\mathrm{d}\vartheta \\ &\leq \frac{c_1}{2}U(t) + \frac{1}{c_1}\|f\|_{L^2([0,t];V_1^*)}^2 + 2T\frac{c_2^2}{c_1}\Big(1 + 2\|u_0\|_{V_1}^2 + 2\int_0^t U(\vartheta)\,\mathrm{d}\vartheta\Big),\end{aligned} \tag{11.85}$$

where the last estimate follows as (8.63) and, as before, $U(t) := \int_0^t \|\frac{\mathrm{d}}{\mathrm{d}\vartheta}u\|_{V_1}^2\mathrm{d}\vartheta$. By Gronwall's inequality (1.66) and by (11.83b) together with the abstract Poincaré-type inequality (8.9), it yields the estimate of u in $L^\infty(I;V)$ and of $\frac{\mathrm{d}}{\mathrm{d}t}u$ in $L^2(I;V_1)$.

Rigorously, the a-priori estimate (11.84) can be obtained by testing (11.82) by $(u_\tau^k - u_\tau^{k-1})/\tau$:

$$\begin{aligned}c_1\Big\|\frac{u_\tau^k - u_\tau^{k-1}}{\tau}\Big\|_{V_1}^2 + \frac{\Phi(u_\tau^k) - \Phi(u_\tau^{k-1})}{\tau} &\leq \frac{\Phi(u_\tau^k) - \Phi(u_\tau^{k-1})}{\tau} \\ + \Big\langle E'(u_\tau^{k-1})\Big(\frac{u_\tau^k - u_\tau^{k-1}}{\tau}\Big), \frac{u_\tau^k - u_\tau^{k-1}}{\tau}\Big\rangle &\leq \Big\langle f_\tau^k - A_2(u_\tau^k), \frac{u_\tau^k - u_\tau^{k-1}}{\tau}\Big\rangle\end{aligned} \tag{11.86}$$

and by continuing the strategy (11.85) by the discrete Gronwall inequality. □

Proposition 11.25. *Let, in addition to the assumptions of Lemma 11.24, also $E' : V_1 \to \mathscr{L}(V_1, V_1^*)$ be continuous and bounded in the sense $\|E'(v)\|_{\mathscr{L}(V_1,V_1^*)} \leq C(1+\|v\|_{V_1}^q)$ for some $q > 1$, $A_2 : V_1 \to V_1^*$ be continuous and $V \Subset V_1$. Then (11.65) possesses a strong solution. Moreover, any (u, w), with $w = E(u)$ and u a weak* limit of (a subsequence of) $\{u_\tau\}_{\tau>0}$ in $W^{1,\infty,2}(I;V,V_1)$, solves (11.65).*

Proof. Choosing a convergent subsequence $u_\tau \overset{*}{\rightharpoonup} u$ in $W^{1,\infty,2}(I;V,V_1)$, we make a limit passage in

$$E'(\bar{u}_\tau^{\mathrm{R}})\frac{\mathrm{d}u_\tau}{\mathrm{d}t} + \partial\Phi(\bar{u}_\tau) + A_2(\bar{u}_\tau) \ni \bar{f}_\tau, \tag{11.87}$$

with the "retarded" Rothe function $\bar{u}_\tau^{\mathrm{R}}$ as defined in (8.202). Using $V \Subset V_1$ and Aubin-Lions' lemma, we have $u_\tau \to u$ in $L^{q_1}(I;V_1)$ for any $q_1 < +\infty$. Then, as $\|u_\tau - \bar{u}_\tau^{\mathrm{R}}\|_{L^2(I;V_1)} = \tau\|\frac{\mathrm{d}}{\mathrm{d}t}u_\tau\|_{L^2(I;V_1)}/\sqrt{3} = \mathscr{O}(\tau)$, cf. (8.50), we have $\bar{u}_\tau^{\mathrm{R}} \to u$ in $L^2(I;V_1)$ and, by the interpolation between $L^2(I;V_1)$ and $L^\infty(I;V_1)$, also $\bar{u}_\tau^{\mathrm{R}} \to u$ in $L^{q_1}(I;V_1)$. By the same arguments, also $\bar{u}_\tau \to u$ in $L^{q_1}(I;V_1)$. By continuity of the Nemytskiĭ mapping induced by E' as $L^{q_1}(I;V_1) \to L^{q_1/q}(I;\mathscr{L}(V_1,V_1^*))$, and by $\frac{\mathrm{d}}{\mathrm{d}t}u_\tau \rightharpoonup \frac{\mathrm{d}}{\mathrm{d}t}u$ in $L^2(I;V_1)$, we can see that $E'(\bar{u}_\tau^{\mathrm{R}})\frac{\mathrm{d}}{\mathrm{d}t}u_\tau$ converges to $E'(u)\frac{\mathrm{d}}{\mathrm{d}t}u$ weakly in $L^{2q_1/(2q+q_1)}(I;V_1^*)$. Then $\langle E'(\bar{u}_\tau^{\mathrm{R}})\frac{\mathrm{d}}{\mathrm{d}t}u_\tau, \bar{u}_\tau\rangle \to \langle E'(u)\frac{\mathrm{d}}{\mathrm{d}t}u, u\rangle$ provided we choose $q_1 \geq 2q + 2$. By (11.83c), we have the Nemytskiĭ mapping induced by

A_2 continuous as $L^{q_1}(I;V_1) \to L^{q_1}(I;V_1^*)$, hence $A_2(\bar{u}_\tau) \to A_2(u)$ in $L^{q_1}(I;V_1^*)$. Then, using also the convexity of Φ, we can pass to the limit in (11.87) written in the form (11.66a), i.e. we can make limit superior in

$$\int_0^T \Phi(v) + \Big\langle E'(\bar{u}_\tau^{\mathrm{R}})\frac{\mathrm{d}u_\tau}{\mathrm{d}t} + A_2(\bar{u}_\tau) - \bar{f}_\tau, v - \bar{u}_\tau \Big\rangle_{V_1^*\times V_1} - \Phi(\bar{u}_\tau)\,\mathrm{d}t \geq 0. \tag{11.88}$$

This gives just (11.66a) when realizing $E'(u)\frac{\mathrm{d}}{\mathrm{d}t}u = \frac{\mathrm{d}}{\mathrm{d}t}E(u)$ and putting $w = E(u)$. Then (11.66b) follows, too. Eventually, the limit passage in the initial condition $u_0 = u_\tau(0) \rightharpoonup u(0)$ yields $u(0) = u_0$. $\square$

Remark 11.26. Like in Remark 8.23, a more general $f \in L^2(I;V_1^*) + W^{1,1}(I;V^*)$ can be considered. Also, like in Exercise 8.62, a modification of the growth condition (11.83c) as $\|A_2(v)\|_{V_1^*} \leq c_2(1 + \|v\|_{V_1} + \Phi(v)^{1/2})$ can also be considered.

Example 11.27 (*Heat equation*). The nonlinear heat equation in the form (8.199) can be transformed by considering $\beta(u)$ as the unknown function itself into the form $\frac{\partial}{\partial t}(\beta^{-1}(u)) - \Delta u = g$ with the boundary condition $\frac{\partial}{\partial\nu}u + (b_1 + b_2|\widehat{\kappa}^{-1}\beta^{-1}(u)|^3)\widehat{\kappa}^{-1}(\beta^{-1}(u)) = h$. Assuming $\beta : \mathbb{R} \to \mathbb{R}$ increasing and satisfying $|\beta^{-1}(r)| \leq C(1 + |r|^{q-1})$ for some $q \geq 2$, we can apply the approach from Sections 11.2.1-2 with $V := W^{1,2}(\Omega)$, $V_1 := L^q(\Omega)$, $H := L^2(\Omega)$. Even (11.68b) allows for $\beta : \mathbb{R} \to \mathbb{R}$ to be only non-decreasing, which causes the problem to be elliptic on regions where the values of the solution u range over the interval(s) where β is constant. Thus one can treat it as a so-called *parabolic/elliptic equation*, here arising from the phenomenon that the heat capacity $\mathfrak{c}$ in (8.197) decays to zero for some values of temperature. Then one speaks about a *degenerate heat equation*.

Example 11.28 (*Landau-Lifschitz-Gilbert equation* [179, 251]). In fact, this section 11.2.2 applies to the inclusions of the type $B(u)\frac{\mathrm{d}u}{\mathrm{d}t} + A(u) \ni f$ even in the cases when $B(u)\frac{\mathrm{d}u}{\mathrm{d}t}$ cannot be reduced to $\frac{\mathrm{d}}{\mathrm{d}t}E(u)$ for any E. An example of such a structure is a *pseudoparabolic* equation

$$\alpha\,\frac{\partial u}{\partial t} - \beta(|u|)u \times \frac{\partial u}{\partial t} - \mu\,\Delta u + c(u) = g, \tag{11.89}$$

where u is $\mathbb{R}^3$-valued, $\times : \mathbb{R}^3\times\mathbb{R}^3 \to \mathbb{R}^3$ is the vector product of two 3-dimensional vectors, $\alpha, \mu > 0$, $\beta : \mathbb{R} \to \mathbb{R}$ is continuous, and $c : \mathbb{R}^3 \to \mathbb{R}^3$ is continuous with a coercive potential. Such an equation describes evolution of a magnetization vector u in magnetic materials.[24] Considering further an initial-boundary-value problem

[24]The minima of the potential of c determines directions of easy magnetisation, the so-called gyromagnetic β-term causes non-dissipative precession of the magnetisation vector, while the α-term determines attenuation of this precession, the μ-term is the so-called exchange energy which is of quantum origin, and g is the external magnetisation.

for (11.89), the *semi-implicit discretisation* of the type (11.82) yields

$$\alpha\,\frac{u_\tau^k - u_\tau^{k-1}}{\tau} - \beta(|u_\tau^{k-1}|)u_\tau^{k-1}\times\frac{u_\tau^k - u_\tau^{k-1}}{\tau} - \mu\,\Delta u_\tau^k + c(u_\tau^k) = g_\tau^k. \tag{11.90}$$

Exercise 11.29. Prove existence of a weak solution $u_\tau^k \in W^{1,2}(\Omega;\mathbb{R}^3)$ to (11.90) with a suitable boundary condition (e.g. $\frac{\partial}{\partial\nu}u_\tau^k = 0$ on Γ), then prove a-priori bounds for u_τ in $L^\infty(I;W^{1,2}(\Omega;\mathbb{R}^3))\cap W^{1,2}(I;L^2(\Omega;\mathbb{R}^3))$ provided $g\in L^2(Q;\mathbb{R}^3)$ and the initial and boundary data are qualified appropriately, and eventually prove convergence for $\tau\to 0$ to a weak solution to (11.89).[25]

11.2.3 Uniqueness

The notion of monotonicity of A can be generalized to be suitable for the doubly-nonlinear structure (11.65) with $E : V_1 \to V_1^*$ invertible such that $E^{-1}(E(V)) \subset V$: the mapping $A : V \to V^*$ is called E-*monotone* (in Gajewski's sense [164]) if

$$\forall u,v\in V,\ \ z = E^{-1}\Big(\frac{E(u)+E(v)}{2}\Big):\quad \big\langle A(u), u-z\big\rangle + \big\langle A(v), v-z\big\rangle \ge 0. \tag{11.91}$$

If, in addition, strict inequality in (11.91) implies $u = v$, then A is called strictly E-monotone. Obviously, if E is linear, then (strict) E-monotonicity is just the conventional (strict) monotonicity.

In case that $E = \Psi'$ and A is E-monotone, the function

$$\varrho(u,v) := \Psi^*\big(E(u)\big) + \Psi^*\big(E(v)\big) - 2\Psi^*\Big(\frac{E(u)+E(v)}{2}\Big), \tag{11.92}$$

introduced by Gajewski [164], can measure a "distance" of two solutions u_1 and u_2 corresponding to two initial conditions u_{01} and u_{02} in the sense that, for all $t\in I$, the mapping $u_0 \mapsto u(t)$ is non-expansive:

$$\varrho\big(u_1(t), u_2(t)\big) \le \varrho\big(u_{01}, u_{02}\big). \tag{11.93}$$

Indeed, as in the last equality in (11.78) with t instead of T, we have

$$\Psi^*\big(E(u(t))\big) - \Psi^*\big(E(u(0))\big) = \int_0^t \Big\langle u(\vartheta), \frac{\mathrm{d}E(u(\vartheta))}{\mathrm{d}\vartheta}\Big\rangle\,\mathrm{d}\vartheta. \tag{11.94}$$

Then, using subsequently the definition (11.92) and the formula (11.94) with $z(t) = E^{-1}(\frac12 E(u_1(t)) + \frac12 E(u_2(t)))$ and assuming a special case $f\equiv 0$, we obtain

$$\varrho\big(u_1(t), u_2(t)\big) - \varrho\big(u_{01}, u_{02}\big) = \int_0^t \Big\langle u_1(\vartheta), \frac{\mathrm{d}E(u_1(\vartheta))}{\mathrm{d}\vartheta}\Big\rangle$$

[25] Hint: For existence of $u^k\in W^{1,2}(\Omega;\mathbb{R}^3)$, use coercivity on the particular time levels; realize the cancellation $(u_\tau^{k-1}\times u^k)\cdot u^k = 0$. For a-priori estimates, execute the test by $u_\tau^k - u_\tau^{k-1}$ and realize the cancellation $(u_\tau^{k-1}\times(u_\tau^k - u_\tau^{k-1}))\cdot(u_\tau^k - u_\tau^{k-1}) = 0$. For convergence use $u_\tau\to u$ in $L^2(Q;\mathbb{R}^3)$ by the Rellich-Kondrachov Theorem 1.21 and also $u_\tau^{\mathrm{R}} - u_\tau \to 0$ in $L^2(Q;\mathbb{R}^3)$.

$$+\Big\langle u_2(\vartheta), \frac{\mathrm{d}E(u_2(\vartheta))}{\mathrm{d}\vartheta}\Big\rangle - 2\Big\langle \frac{\mathrm{d}}{\mathrm{d}\vartheta}\frac{E(u_1(\vartheta))+E(u_2(\vartheta))}{2}, z(\vartheta)\Big\rangle\,\mathrm{d}\vartheta$$
$$= \int_0^t \Big\langle u_1(\vartheta)-z(\vartheta), \frac{\mathrm{d}E(u_1(\vartheta))}{\mathrm{d}\vartheta}\Big\rangle + \Big\langle u_2(\vartheta)-z(\vartheta), \frac{\mathrm{d}E(u_2(\vartheta))}{\mathrm{d}\vartheta}\Big\rangle\,\mathrm{d}\vartheta$$
$$= -\int_0^t \big\langle A(u_1(\vartheta)), u_1(\vartheta)-z(\vartheta)\big\rangle + \big\langle A(u_2(\vartheta)), u_2(\vartheta)-z(\vartheta)\big\rangle\,\mathrm{d}\vartheta \le 0. \tag{11.95}$$

Proposition 11.30 (Gajewski, Gröger, Nečas[26]). *Let $E=\Psi'$ be invertible, Ψ be convex, and let:*

(i) *A be strictly E-monotone and $f=0$, or*

(ii) *$A=\partial\Phi$ with both Φ and $\Phi\circ E^{-1}$ convex, E be strongly monotone in the sense $\langle E(u)-E(v), u-v\rangle_{V_1^*\times V_1} \ge m\|u-v\|_{V_1}^2$ with $m>0$ and Lipschitz continuous, $[E^{-1}]': V_1^* \to \mathscr{L}(V_1^*, V_1)$ be Lipschitz continuous, and $f\in L^1(I;V_1^*)$.*

Then the inclusion (11.65) *admits at most one strong solution.*

Proof. In case (i), (11.95) implies $\varrho\big(u_1(t), u_2(t)\big)=0$, thus $u_1(t)=u_2(t)$ for a.a. $t\in I$.

As to (ii), abbreviating $w=E(u)$, (11.65) can be written as the system of two inclusions:

$$\frac{\mathrm{d}w}{\mathrm{d}t} + \partial\Phi(u) \ni f, \tag{11.96a}$$
$$\frac{\mathrm{d}u}{\mathrm{d}t} + [E^{-1}]'(w)\partial\Phi\big(E^{-1}(w)\big) \ni [E^{-1}]'(w)f. \tag{11.96b}$$

Note that $\frac{\mathrm{d}}{\mathrm{d}t}w = \frac{\mathrm{d}}{\mathrm{d}t}(E(u)) = E'(u)\frac{\mathrm{d}}{\mathrm{d}t}u$. Let us consider two solutions (u_1,w_1) and (u_2,w_2), write (11.96) for them, subtract the particular inclusions in (11.96) tested by u_1-u_2 and w_1-w_2, respectively, and sum it up. Altogether, using also the assumed monotonicity of A and of $\partial[\Phi\circ E^{-1}]$,[27] we get

$$\frac{\mathrm{d}}{\mathrm{d}t}\big\langle w_1-w_2, u_1-u_2\big\rangle_{V_1^*\times V_1} = \Big\langle \frac{\mathrm{d}w_1}{\mathrm{d}t} - \frac{\mathrm{d}w_2}{\mathrm{d}t}, u_1-u_2\Big\rangle_{V^*\times V}$$
$$+\Big\langle \frac{\mathrm{d}u_1}{\mathrm{d}t} - \frac{\mathrm{d}u_2}{\mathrm{d}t}, w_1-w_2\Big\rangle_{V_1^*\times V_1} \le \big\langle\big([E^{-1}]'(w_1)-[E^{-1}]'(w_2)\big)f, w_1-w_2\big\rangle_{V_1^*\times V_1}$$
$$\le L\big\|w_1-w_2\big\|_{V_1^*}^2\big\|f\big\|_{V_1^*} \le LM^2\big\|u_1-u_2\big\|_{V_1}^2\big\|f\big\|_{V_1^*}$$

where L and M are the Lipschitz constants of $[E^{-1}]'$ and of E, respectively. Integrating over I, one obtains $m\|u_1(t)-u_2(t)\|_{V_1}^2 \le \langle w_1(t)-w_2(t), u_1(t)-u_2(t)\rangle \le LM^2\int_0^t \big\|u_1(\vartheta)-u_2(\vartheta)\big\|_{V_1}^2\big\|f(\vartheta)\big\|_{V_1^*}\mathrm{d}\vartheta$, from which $u_1(t)=u_2(t)$ a.e. on I follows by Gronwall's inequality. □

[26] The case (i) is basically due to Gajewski [164] while the case (ii) has earlier been investigated by Gröger and Nečas [192] in a narrower setting $V=V_1=H$.

[27] The convexity of $\Phi\circ E^{-1}$ implies E-monotonicity of $\partial\Phi$.

11.3 2nd-order equations

We will now treat the abstract 2nd-order *doubly-nonlinear Cauchy problem* (9.63) in a non-autonomous variant, i.e.:

$$\frac{\mathrm{d}^2 u}{\mathrm{d}t^2} + A\Big(t, \frac{\mathrm{d}u}{\mathrm{d}t}\Big) + B\big(t, u(t)\big) = f(t)\ , \quad u(0) = u_0\ , \quad \frac{\mathrm{d}u}{\mathrm{d}t}(0) = v_0. \tag{11.97}$$

Enough dissipation (i.e. A uniformly monotone on V) causes that (11.97) represents, in fact, a parabolic problem in terms of "velocity" $\frac{\mathrm{d}}{\mathrm{d}t}u$.

By the *strong solution* we will understand $u \in W^{2,\infty,p,p'}(I; V, V, V^*)$, cf. the notation (7.4), such that (11.97) holds for a.a. $t \in I$.

Considering finite-dimensional subspaces V_k of $Z \subset V$ satisfying (2.7) with Z from (8.95), we apply the Galerkin method with $u_{0k} \in V_k$ approximating u_0 in V and $v_{0k} \in V_k$ approximating v_0 in H. Existence of a solution u_k of the resulting initial-value problem for the system of ordinary-differential equations can be proved by the usual prolongation technique combined with the following a-priori estimates.

Lemma 11.31 (A-PRIORI ESTIMATES). *Let $A : I \times V \to V^*$ be semi-coercive in the sense of* (8.95), *let $B : V \to V^*$ be time-independent and have a potential $\Phi : V \to \mathbb{R}^+$, and let $f \in L^{p'}(I; V^*)$, $u_{0k}, v_{0k} \in V_k$, $\lim_{k\to\infty} u_{0k} = u_0$ in V and $\lim_{k\to\infty} v_{0k} = v_0$ in H. Then, with C independent of k,*

$$\Big\|\frac{\mathrm{d}u_k}{\mathrm{d}t}\Big\|_{L^\infty(I;H)\cap L^p(I;V)} \le C \quad \& \quad \|u_k\|_{L^\infty(I;V)} \le C. \tag{11.98}$$

Moreover, if A satisfies the growth conditions (8.80), *i.e.*

$$\|A(t,u)\|_{V^*} \le \mathfrak{C}(\|u\|_H)(\gamma_A(t) + \|u\|_V^{p-1}),$$

and B satisfies $\|B(u)\|_{V^*} \le \mathfrak{C}(\|u\|_V)$ with $\gamma_A \in L^{p'}(I)$ and $\mathfrak{C}$ increasing, then

$$\Big|\frac{\mathrm{d}^2 u_k}{\mathrm{d}t^2}\Big|_{p',l} \le C \tag{11.99}$$

for any $k \ge l$; for the seminorm $|\cdot|_{p',l}$ see (8.94).

Proof. For (11.98), let us test the equation $\frac{\mathrm{d}^2}{\mathrm{d}t^2}u_k + \mathscr{A}(\frac{\mathrm{d}}{\mathrm{d}t}u_k) + \mathscr{B}(u_k) = f_k$ by $\frac{\mathrm{d}}{\mathrm{d}t}u_k$, and use $\langle \frac{\mathrm{d}^2}{\mathrm{d}t^2}u_k, \frac{\mathrm{d}}{\mathrm{d}t}u_k\rangle = \frac{1}{2}\frac{\mathrm{d}}{\mathrm{d}t}\|\frac{\mathrm{d}}{\mathrm{d}t}u_k\|_H^2$, cf. also (7.23). Using also a strategy like that used in (8.21), we get

$$\begin{aligned}\frac{1}{2}\frac{\mathrm{d}}{\mathrm{d}t}\Big\|\frac{\mathrm{d}u_k}{\mathrm{d}t}\Big\|_H^2 &+ c_0\Big|\frac{\mathrm{d}u_k}{\mathrm{d}t}\Big|_V^p + \frac{\mathrm{d}}{\mathrm{d}t}\Phi(u_k) \le c_1(t)\Big|\frac{\mathrm{d}u_k}{\mathrm{d}t}\Big|_V + c_2(t)\Big\|\frac{\mathrm{d}u_k}{\mathrm{d}t}\Big\|_H^2 \\ &+ \Big\langle f, \frac{\mathrm{d}u_k}{\mathrm{d}t}\Big\rangle \le C_\varepsilon c_1^{p'} + \varepsilon(t)\Big|\frac{\mathrm{d}u_k}{\mathrm{d}t}\Big|_V^p + c_2(t)\Big\|\frac{\mathrm{d}u_k}{\mathrm{d}t}\Big\|_H^2 \\ &+ C_\varepsilon C_{\mathrm{P}}\|f\|_{V^*}^{p'} + \varepsilon C_{\mathrm{P}}\Big|\frac{\mathrm{d}u_k}{\mathrm{d}t}\Big|_V^p + C_{\mathrm{P}}\|f\|_{V^*}\Big(\frac{1}{2} + \frac{1}{2}\Big\|\frac{\mathrm{d}u_k}{\mathrm{d}t}\Big\|_H^2\Big).\end{aligned} \tag{11.100}$$

For $\varepsilon > 0$ small enough, we get (11.98) by the Gronwall inequality. As $\{u_{0k}\}_{k\in\mathbb{N}} \subset V$ is bounded and $\{\frac{\mathrm{d}}{\mathrm{d}t}u_k\}_{k\in\mathbb{N}} \subset L^p(I;V)$ is bounded, by Lemma 7.1 with $V_1 = V_2 = V$ we get even $\{u_k\}_{k\in\mathbb{N}} \in C(I;V)$ bounded. The dual estimate (11.99) can be obtained from

$$\begin{aligned}
\left|\frac{\mathrm{d}^2u_k}{\mathrm{d}t^2}\right|_{p',l} &:= \sup_{\substack{\|z\|_{L^p(I;V)}\le 1\\ z(t)\in V_l \text{ for a.a. } t\in I}} \left\langle\frac{\mathrm{d}^2u_k}{\mathrm{d}t^2}, z\right\rangle = \left\langle f - \mathscr{A}\Big(\frac{\mathrm{d}u_k}{\mathrm{d}t}\Big) - \mathscr{B}(u_k), z\right\rangle \\
&\le \sup_{\|z\|_{L^p(I;V)}\le 1} \left\| f + \mathscr{A}\Big(\frac{\mathrm{d}u_k}{\mathrm{d}t}\Big) + \mathscr{B}(u_k)\right\|_{L^{p'}(I;V^*)} \|z\|^p_{L^p(I;V)} \\
&\le \|f\|_{L^{p'}(I;V^*)} + \mathfrak{C}\Big(\Big\|\frac{\mathrm{d}u_k}{\mathrm{d}t}\Big\|_{L^\infty(I;H)}\Big) \\
&\quad\times\left(\|\gamma_A\|_{L^{p'}(I)} + \left\|\frac{\mathrm{d}u_k}{\mathrm{d}t}\right\|^{p-1}_{L^p(I;V)}\right) + T^{1/p'}\mathfrak{C}\big(\|u_k\|_{L^\infty(I;V)}\big). \qquad (11.101)
\end{aligned}$$

□

Lemma 11.32 (Other estimates). *Let*

$$\begin{aligned}
A = A_1 + A_2 \quad & \textit{with } A_1 \textit{ time-independent}, \\
& A_1 = \Psi' \textit{ for a convex potential } \Psi : V \to \mathbb{R}, \\
& \Psi(v) \ge c_0\|v\|_V^p, \quad c_0 > 0, \\
& \|A_2(t,v)\|_H \le C_1(t) + C_2\|v\|_V^{p/2}, \quad C_1 \in L^2(I), \qquad (11.102) \\
B = B_1 + B_2 \quad & \textit{with } B_1 : V \to V^* \textit{ smooth (and time-independent)}, \\
& \big\|B_1(u)\big\|_{V^*} \le C_3\big(1 + \|u\|_V^{p-1}\big), \\
& \big\langle [B_1'(u)](v), v\big\rangle \le C_4\big(1 + \|u\|_V^{p-2}\big)\|v\|_V^2, \\
& \big\|B_2(t,u)\big\|_H \le C_5(t) + C_6\|u\|_V^{p/2}, \quad C_5 \in L^2(I), \qquad (11.103)
\end{aligned}$$

with $p \ge 2$ *if* $B_1 \ne 0$, *and let* $f \in L^2(I;H)$, $u_{0k}, v_{0k} \in V_k$ *and now both* $\lim_{k\to\infty} u_{0k} = u_0$ *and* $\lim_{k\to\infty} v_{0k} = v_0$ *in* V. *Then*

$$\left\|\frac{\mathrm{d}u_k}{\mathrm{d}t}\right\|_{L^\infty(I;V)} \le C \quad \& \quad \left\|\frac{\mathrm{d}^2u_k}{\mathrm{d}t^2}\right\|_{L^2(I;H)} \le C. \qquad (11.104)$$

Proof. For (11.104), we test the Galerkin equation by $\frac{\mathrm{d}^2}{\mathrm{d}t^2}u_k$. Using (11.102), one gets

$$\begin{aligned}
\left\|\frac{\mathrm{d}^2u_k}{\mathrm{d}t^2}\right\|_H^2 + \frac{\mathrm{d}}{\mathrm{d}t}\Psi\Big(\frac{\mathrm{d}u_k}{\mathrm{d}t}\Big) &= \left\langle f(t) - A_2\Big(t, \frac{\mathrm{d}u_k}{\mathrm{d}t}\Big) - B(t,u_k), \frac{\mathrm{d}^2u_k}{\mathrm{d}t^2}\right\rangle \\
&\le 2\big\|f(t)\big\|_H^2 + 2\left\|A_2\Big(t, \frac{\mathrm{d}u_k}{\mathrm{d}t}\Big)\right\|_H^2 - \left\langle B_1(u_k), \frac{\mathrm{d}^2u_k}{\mathrm{d}t^2}\right\rangle \\
&\qquad + 2\big\|B_2(t,u_k)\big\|_H^2 + \frac{1}{2}\left\|\frac{\mathrm{d}^2u_k}{\mathrm{d}t^2}\right\|_H^2
\end{aligned}$$

$$\le 2\big\|f(t)\big\|_H^2 + 4C_1^2(t) + 4C_2^2\Big\|\frac{\mathrm{d}u_k}{\mathrm{d}t}\Big\|_V^p - \Big\langle B_1(u_k), \frac{\mathrm{d}^2u_k}{\mathrm{d}t^2}\Big\rangle + 4C_5^2(t) + 4C_6^2\big\|u_k\big\|_V^p + \frac{1}{2}\Big\|\frac{\mathrm{d}^2u_k}{\mathrm{d}t^2}\Big\|_H^2. \quad (11.105)$$

Absorbing the last term in the left-hand side and integrating this estimate over $[0,t]$ and using the coercivity of Ψ assumed in (11.102) and the by-part formula $\int_0^t\langle B_1(u_k), \frac{\mathrm{d}^2}{\mathrm{d}\vartheta^2}u_k\rangle\mathrm{d}\vartheta = \langle B_1(u_k(t)), \frac{\mathrm{d}}{\mathrm{d}t}u_k(t)\rangle - \langle B_1(u_{0k}), v_{0k}\rangle - \int_0^t\langle[B_1'(u_k)](\frac{\mathrm{d}}{\mathrm{d}\vartheta}u_k), \frac{\mathrm{d}}{\mathrm{d}\vartheta}u_k\rangle\mathrm{d}\vartheta$, we get

$$\begin{aligned} c_0\Big\|\frac{\mathrm{d}u_k}{\mathrm{d}t}\Big\|_V^p + \int_0^t\Big\|\frac{\mathrm{d}^2u_k}{\mathrm{d}\vartheta^2}\Big\|_H^2\mathrm{d}\vartheta &\le \Psi\Big(\frac{\mathrm{d}u_k}{\mathrm{d}t}(t)\Big) + \int_0^t\Big\|\frac{\mathrm{d}^2u_k}{\mathrm{d}\vartheta^2}\Big\|_H^2\mathrm{d}\vartheta \\ &\le \Psi(v_{0k}) + \int_0^t\bigg(2\big\|f(\vartheta)\big\|_H^2 + 4C_1^2(\vartheta) + 4C_2^2\Big\|\frac{\mathrm{d}u_k}{\mathrm{d}\vartheta}\Big\|_V^p \\ &\quad + \Big\langle[B_1'(u_k)]\Big(\frac{\mathrm{d}u_k}{\mathrm{d}\vartheta}\Big), \frac{\mathrm{d}u_k}{\mathrm{d}\vartheta}\Big\rangle + 4C_5^2(\vartheta) + 4C_6^2\big\|u_k(\vartheta)\big\|_V^p\bigg)\mathrm{d}\vartheta \\ &\quad - \Big\langle B_1\big(u_k(t)\big), \frac{\mathrm{d}u_k}{\mathrm{d}t}(t)\Big\rangle + \big\langle B_1(u_{0k}), v_{0k}\big\rangle \\ &\le \Psi(v_{0k}) + 2\|f\|_{L^2(I;H)}^2 + 4\|C_1\|_{L^2(I)}^2 + 4\|C_5\|_{L^2(I)}^2 \\ &\quad + \int_0^t\bigg(4C_2^2\Big\|\frac{\mathrm{d}u_k}{\mathrm{d}\vartheta}\Big\|_V^p + C_4 + 2C_4\Big\|\frac{\mathrm{d}u_k}{\mathrm{d}\vartheta}\Big\|_V^p + (C_4+4C_6^2)\big\|u_k(\vartheta)\big\|_V^p\bigg)\mathrm{d}\vartheta \\ &\quad + C_\varepsilon C_3^{p'}\big(1+\|u_k(t)\|_V^{p-1}\big)^{p'} + \varepsilon\Big\|\frac{\mathrm{d}u_k}{\mathrm{d}t}(t)\Big\|_V^p + \big\langle B_1(u_{0k}), v_{0k}\big\rangle; \quad (11.106)\end{aligned}$$

note that $p\ge2$ was needed to apply Hölder's inequality if $B_1\ne0$. For $\varepsilon<c_0$ we can absorb the last-but-one term in the left-hand side. As in (11.38), we can estimate

$$\big\|u_k(t)\big\|_V^p \le (2t)^{p-1}U_k(t) + 2^{p-1}\big\|u_{0k}\big\|_V^p, \quad \text{with } U_k(t) := \int_0^t\Big\|\frac{\mathrm{d}u_k}{\mathrm{d}\vartheta}\Big\|_V^p\mathrm{d}\vartheta. \quad (11.107)$$

Hence the estimate (11.106) exhibits the structure

$$\Big\|\frac{\mathrm{d}u_k}{\mathrm{d}t}\Big\|_V^p + \int_0^t\Big\|\frac{\mathrm{d}^2u_k}{\mathrm{d}\vartheta^2}\Big\|_H^2\mathrm{d}\vartheta \le C + C\int_0^t\bigg(\Big\|\frac{\mathrm{d}u_k}{\mathrm{d}\vartheta}\Big\|_V^p + U_k(\vartheta)\bigg)\mathrm{d}\vartheta + CU_k(t) \quad (11.108)$$

with a sufficiently large constant C. Adding $(C+1)U_k(t) = (C+1)\int_0^t\|\frac{\mathrm{d}}{\mathrm{d}\vartheta}u_k\|_V^p\mathrm{d}\vartheta$, we obtain

$$\Big\|\frac{\mathrm{d}u_k}{\mathrm{d}t}\Big\|_V^p + \int_0^t\Big\|\frac{\mathrm{d}^2u_k}{\mathrm{d}\vartheta^2}\Big\|_H^2\mathrm{d}\vartheta + U_k(t) \le C + \int_0^t\bigg((2C+1)\Big\|\frac{\mathrm{d}u_k}{\mathrm{d}\vartheta}\Big\|_V^p + CU_k(\vartheta)\bigg)\mathrm{d}\vartheta,$$

which eventually allows us to use Gronwall's inequality to conclude the bound for $\|\frac{\mathrm{d}}{\mathrm{d}t}u_k(t)\|_V$ uniformly for $t\in I$ as well as the second estimate in (11.104). $\square$

Theorem 11.33 (CONVERGENCE). *Suppose $V \Subset H$ and the assumptions of Lemma 11.31 hold (so that the a-priori estimates (11.98)–(11.99) are at our disposal), and one of the following situations holds:*

(i) $A = A_1 + A_2$, *with A_1 is linear, time-independent, symmetric (i.e. $A_1^* = A_1$), and positive semi-definite, $\mathscr{A}_2$ continuous as a mapping $L^q(I;H) \to L^{p'}(I;V^*)$ with some $q < +\infty$, and $B : V \to V^*$ is semi-coercive and pseudomonotone.*

(ii) $A(t,\cdot) : V \to V^*$ *is pseudomonotone for a.a. $t \in I$ and $B = B_1 + B_2$ with $B_1 : V \to V^*$ linear, monotone and symmetric in the sense $B_1^* = B_1$, $\mathscr{B}_2 : W^{1,p}(I;V) \to L^{p'}(I;V^*)$ totally continuous, and, for simplicity, $u_0 \in V_1$ (hence $u_0 \in V_k$ for any $k > 1$, too).*

Then u_k converges (as a subsequence) to a strong solution to (11.97).

Proof. Take $z \in W^{1,p,p'}(I;V,V^*)$ and a sequence $\{z_k\}_{k\in\mathbb{N}}$, $z_k \in W^{1,\infty}(I;V_k)$, such that $z_k \to z$ in $W^{1,p,p'}(I;V,V^*_{\rm lcs})$; by a density argument it does exist.[28] We use z_k as a test function for (11.97). By the by-part integration, we obtain

$$\int_0^T -\Big\langle \frac{\mathrm{d}u_k}{\mathrm{d}t}, \frac{\mathrm{d}z_k}{\mathrm{d}t}\Big\rangle + \Big\langle \mathscr{A}\Big(\frac{\mathrm{d}u_k}{\mathrm{d}t}\Big), z_k\Big\rangle + \big\langle \mathscr{B}(u_k), z_k\big\rangle - \big\langle f, z_k\big\rangle \,\mathrm{d}t \\ + \Big\langle \frac{\mathrm{d}u_k}{\mathrm{d}t}(T), z_k(T)\Big\rangle = \big\langle v_{0k}, z_k(0)\big\rangle. \tag{11.109}$$

Let us now choose a subsequence such that

$$u_\tau \rightharpoonup u \ \text{ in } W^{1,p}(I;V) \quad \text{and} \quad u_\tau \overset{*}{\rightharpoonup} u \ \text{ in } \ W^{1,\infty}(I;H). \tag{11.110}$$

Then also

$$u_k(T) = u_0 + \int_0^T \frac{\mathrm{d}u_k}{\mathrm{d}t}\,\mathrm{d}t \rightharpoonup u_0 + \int_0^T \frac{\mathrm{d}u}{\mathrm{d}t}\,\mathrm{d}t = u(T) \quad \text{in } V, \tag{11.111}$$

hence $u_k(T) \to u(T)$ in $H \Supset V$. By (11.99) and by the interpolated Aubin-Lions Lemma 7.8[29], we then have $\{\frac{\mathrm{d}}{\mathrm{d}t}u_k\}_{k\in\mathbb{N}}$ relatively compact in $L^q(I;H)$ with any $q < +\infty$. Therefore,

$$\frac{\mathrm{d}u_k}{\mathrm{d}t} \to \frac{\mathrm{d}u}{\mathrm{d}t} \quad \text{in } L^q(I;H). \tag{11.112}$$

In particular, $\frac{\mathrm{d}}{\mathrm{d}t}u_k \to \frac{\mathrm{d}}{\mathrm{d}t}u$ in $L^2(I;H)$. Moreover, by the $L^\infty(I;H)$-estimate of $\{\frac{\mathrm{d}}{\mathrm{d}t}u_k\}_{k\in\mathbb{N}}$, choosing (for a moment only) a subsequence, $\frac{\mathrm{d}}{\mathrm{d}t}u_k(T)$ converges weakly in H. By (11.99), $\frac{\mathrm{d}}{\mathrm{d}t}u_k(T) = \int_0^T \frac{\mathrm{d}^2}{\mathrm{d}t^2}u_k \mathrm{d}t + v_{0k} \to \int_0^T \frac{\mathrm{d}^2}{\mathrm{d}t^2}u \mathrm{d}t + v_0$ in $V^*_{\rm lcs}$, hence

$$\frac{\mathrm{d}u_\tau}{\mathrm{d}t}(T) \rightharpoonup \frac{\mathrm{d}u}{\mathrm{d}t}(T) \quad \text{in } H. \tag{11.113}$$

[28] Cf. the proof of Lemma 8.28 with the arguments in the proof of Theorem 8.31.

[29] We use Lemma 7.8 here with $V_1 = V$, $V_2 = V_4 = H$, and $V_3 = V^*_{\rm lcs}$.

Putting $z_k - u_k$ instead of z_k into (11.109), one gets

$$\liminf_{k\to\infty} \big\langle \mathscr{B}(u_k), z_k - u_k \big\rangle = \liminf_{k\to\infty} \Big(\int_0^T \langle f, z_k - u_k\rangle + \Big\langle \frac{\mathrm{d}u_k}{\mathrm{d}t}, \frac{\mathrm{d}z_k}{\mathrm{d}t} - \frac{\mathrm{d}u_k}{\mathrm{d}t} \Big\rangle \,\mathrm{d}t$$

$$- \int_0^T \Big\langle A_1 \frac{\mathrm{d}u_k}{\mathrm{d}t}, z_k - u_k \Big\rangle \,\mathrm{d}t - \Big\langle \mathscr{A}_2\Big(\frac{\mathrm{d}u_k}{\mathrm{d}t}\Big), z_k - u_k \Big\rangle$$

$$- \Big\langle \frac{\mathrm{d}u_k}{\mathrm{d}t}(T), z_k(T) - u_k(T) \Big\rangle + \langle v_{0k}, z_k(0) - u_{0k}\rangle \Big)$$

$$= \int_0^T \langle f, z{-}u\rangle + \Big\langle \frac{\mathrm{d}u}{\mathrm{d}t}, \frac{\mathrm{d}z}{\mathrm{d}t} - \frac{\mathrm{d}u}{\mathrm{d}t} \Big\rangle \,\mathrm{d}t - \limsup_{k\to\infty} \int_0^T \Big\langle A_1 \frac{\mathrm{d}u_k}{\mathrm{d}t}, z_k{-}u_k \Big\rangle \,\mathrm{d}t$$

$$- \Big\langle \mathscr{A}_2\Big(\frac{\mathrm{d}u}{\mathrm{d}t}\Big), z - u \Big\rangle - \Big\langle \frac{\mathrm{d}u}{\mathrm{d}t}(T), z(T) - u(T) \Big\rangle + \langle v_0, z(0) - u_0\rangle$$

because $\frac{\mathrm{d}}{\mathrm{d}t}u_k \to \frac{\mathrm{d}}{\mathrm{d}t}u$ in $L^2(I;H)$, $\frac{\mathrm{d}}{\mathrm{d}t}z_k \to \frac{\mathrm{d}}{\mathrm{d}t}z$ in $L^2(I;H)$, and we also used (11.111) together with $V \Subset H$ and (11.113). The term with A_2 uses $\frac{\mathrm{d}}{\mathrm{d}t}u_k \to \frac{\mathrm{d}}{\mathrm{d}t}u$ in $L^q(I;H)$ and the assumption that $\mathscr{A}_2 : L^q(I;H) \to L^{p'}(I;V^*)$ is continuous.

By (11.111) and by the weak upper semi-continuity of $z \mapsto -\langle A_1 z, z\rangle : V \to \mathbb{R}$, one gets

$$\limsup_{k\to\infty} \Big\langle \mathscr{A}_1 \frac{\mathrm{d}u_k}{\mathrm{d}t}, z_k - u_k \Big\rangle$$

$$= \lim_{k\to\infty} \int_0^T \Big\langle A_1 \frac{\mathrm{d}u_k}{\mathrm{d}t}, z_k \Big\rangle \,\mathrm{d}t - \liminf_{k\to\infty} \frac{\langle A_1 u_k(T), u_k(T)\rangle}{2} + \frac{\langle A_1 u_0, u_0\rangle}{2}$$

$$\leq \int_0^T \Big\langle A_1 \frac{\mathrm{d}u}{\mathrm{d}t}, z \Big\rangle \,\mathrm{d}t - \frac{\langle A_1 u(T), u(T)\rangle}{2} + \frac{\langle A_1 u_0, u_0\rangle}{2} = \Big\langle \mathscr{A}_1 \frac{\mathrm{d}u}{\mathrm{d}t}, z{-}u \Big\rangle. \tag{11.114}$$

Then

$$\liminf_{k\to\infty} \big\langle \mathscr{B}(u_k), z_k{-}u_k \big\rangle \geq \int_0^T \langle f, z{-}u\rangle + \Big\langle \frac{\mathrm{d}u}{\mathrm{d}t}, \frac{\mathrm{d}z}{\mathrm{d}t} - \frac{\mathrm{d}u}{\mathrm{d}t} \Big\rangle \mathrm{d}t$$

$$- \Big\langle \mathscr{A}\Big(\frac{\mathrm{d}u}{\mathrm{d}t}\Big), z{-}u \Big\rangle - \Big\langle \frac{\mathrm{d}u}{\mathrm{d}t}(T), z(T) \Big\rangle + \big\langle v_0, z(0) \big\rangle. \tag{11.115}$$

We have $\{\mathscr{B}(u_k)\}_{k\in\mathbb{N}}$ bounded in $L^{p'}(I;V^*)$ (cf. the assumptions in Lemma 11.31) and $z_k \to z$ in $L^p(I;V)$, so that

$$\liminf_{k\to\infty} \big\langle \mathscr{B}(u_k), z{-}u_k \big\rangle = \lim_{k\to\infty} \big\langle \mathscr{B}(u_k), z{-}z_k \big\rangle$$

$$+ \liminf_{k\to\infty} \big\langle \mathscr{B}(u_k), z_k{-}u_k \big\rangle = \liminf_{k\to\infty} \big\langle \mathscr{B}(u_k), z_k{-}u_k \big\rangle. \tag{11.116}$$

In particular, for $z := u$ we have $\limsup_{k\to\infty} \langle \mathscr{B}(u_k), u_k - u\rangle \leq 0$ and, in view of Lemma 8.29, we can use the pseudomonotonicity of $\mathscr{B}$ to conclude that, for any $z \in L^p(I;V)$, $\liminf_{k\to\infty}\langle\mathscr{B}(u_k), u_k - z\rangle \geq \langle\mathscr{B}(u), u - z\rangle$. Joining it with (11.115) and (11.116), one gets $\langle\mathscr{B}(u), u-z\rangle \leq \int_0^T \langle f, u-z\rangle - \langle\frac{\mathrm{d}}{\mathrm{d}t}u, \frac{\mathrm{d}}{\mathrm{d}t}z - \frac{\mathrm{d}}{\mathrm{d}t}u\rangle\mathrm{d}t - \langle\mathscr{A}(\frac{\mathrm{d}}{\mathrm{d}t}u), u-z\rangle - \langle\frac{\mathrm{d}}{\mathrm{d}t}u(T), z(T)-u(T)\rangle + \langle v_0, z(0)-u(0)\rangle$. As it holds for any z, we can conclude that

$$\big\langle\mathscr{B}(u), z\big\rangle = \int_0^T \Big\langle f - A\big(t, \frac{\mathrm{d}u}{\mathrm{d}t}\big), z\Big\rangle - \Big\langle\frac{\mathrm{d}u}{\mathrm{d}t}, \frac{\mathrm{d}z}{\mathrm{d}t}\Big\rangle\mathrm{d}t - \Big\langle\frac{\mathrm{d}u}{\mathrm{d}t}(T), z(T)\Big\rangle + \big\langle v_0, z(0)\big\rangle. \tag{11.117}$$

Moreover, the initial conditions $\frac{\mathrm{d}}{\mathrm{d}t}u(0) = v_0$ and $u(0) = u_0$ are satisfied by the continuity arguments. As $z \in W^{1,p,p'}(I;V,V^*)$ we can use the formula (7.15) for $z(T) = 0 = z(0)$, which enables us to rewrite (11.117) into the form (11.97).

In the case (ii), we use the pseudomonotonicity of the mapping $\mathscr{A}+\mathscr{B}\circ L^{-1}$ where $[L^{-1}v](t) = \int_0^t v(\vartheta)\,\mathrm{d}\vartheta + u_0$ is the inverse mapping to $L = \frac{\mathrm{d}}{\mathrm{d}t} : \mathrm{dom}(L) \to L^p(I;V)$ with

$$\mathrm{dom}(L) = \big\{u \in W^{1,p}(I;V);\quad u(0) = u_0\big\}, \tag{11.118}$$

cf. (8.227). Note that u_k is the Galerkin approximation to (11.97) if and only if $v_k = L^{-1}u_k$ and $\langle\frac{\mathrm{d}}{\mathrm{d}t}v_k + [\mathscr{A}+\mathscr{B}\circ L^{-1}](v_k) - f, z_k\rangle = 0$ for any $z_k \in L^p(I;V_k)$ and $v_k(0) = v_{0k}$; here $u_0 \in V_k$ has been employed. By Lemma 8.29, $\mathscr{A}$ is pseudomonotone on $W^{1,p,p'}(I;V,V^*_{\mathrm{lcs}}) \cap L^\infty(I;H)$. The mapping $v \mapsto \mathscr{B}_1(L^{-1}v) : L^p(I;V) \to C(I;V) \subset L^{p'}(I;V^*)$ is monotone: indeed, for $v_1, v_2 \in L^p(I;V)$ we again abbreviate $v_{12} = v_1 - v_2$ and then, using the symmetry and monotonicity of B_1, we have

$$\begin{aligned}
\big\langle\mathscr{B}_1(L^{-1}v_1) - \mathscr{B}_1(L^{-1}v_2), v_1 - v_2\big\rangle &= \int_0^T \Big\langle B_1\Big(\int_0^t v_{12}(\vartheta)\mathrm{d}\vartheta\Big), v_{12}(t)\Big\rangle\mathrm{d}t \\
&= \int_0^T \Big\langle B_1\Big(\int_0^t v_{12}(\vartheta)\,\mathrm{d}\vartheta\Big), \frac{\mathrm{d}}{\mathrm{d}t}\int_0^t v_{12}(\vartheta)\,\mathrm{d}\vartheta\Big\rangle\,\mathrm{d}t \\
&= \frac{1}{2}\int_0^T \frac{\mathrm{d}}{\mathrm{d}t}\Big\langle B_1\Big(\int_0^t v_{12}(\vartheta)\,\mathrm{d}\vartheta\Big), \int_0^t v_{12}(\vartheta)\,\mathrm{d}\vartheta\Big\rangle\,\mathrm{d}t \\
&= \frac{1}{2}\Big\langle B_1\Big(\int_0^T v_{12}(\vartheta)\,\mathrm{d}\vartheta\Big), \int_0^T v_{12}(\vartheta)\,\mathrm{d}\vartheta\Big\rangle - \frac{1}{2}\langle B_1 0, 0\rangle \geq 0.
\end{aligned} \tag{11.119}$$

Moreover, $\mathscr{B}_1 \circ L^{-1} : L^p(I;V) \to L^{p'}(I;V^*)$ is bounded as both $L^{-1} : L^p(I;V) \to L^\infty(I;V)$ and $B_1 : V \to V^*$ are bounded. Hence, it is also radially continuous. By Lemma 2.9, $\mathscr{B}_1 \circ L^{-1}$ is pseudomonotone. Also, $L^{-1} : L^p(I;V) \to W^{1,p}(I;V)$ is weakly continuous, and $\mathscr{B}_2 : W^{1,p}(I;V) \to L^{p'}(I;V)$ is assumed totally continuous, $\mathscr{B}_2 \circ L^{-1} : L^p(I;V^*) \to L^{p'}(I;V^*)$ is totally continuous. Altogether, due to Lemma 2.11(i) and Corollary 2.12, $\mathscr{A} + \mathscr{B}_1 \circ L^{-1} + \mathscr{B}_2 \circ L^{-1}$ is pseudomonotone. Then we can employ Theorem 8.30 with $\mathscr{A}+\mathscr{B}\circ L^{-1}$ in place of $\mathscr{A}$ to get v solving $\frac{\mathrm{d}}{\mathrm{d}t}v + \mathscr{A}(v) + \mathscr{B}(L^{-1}v) = f$ and $v(0) = v_0$. Then it suffices to put $u = L^{-1}v$. □

Remark 11.34 (Energy balance). Testing the equation (11.97) by $\frac{\mathrm{d}}{\mathrm{d}t}u$, which leads to the a-priori estimate (11.100), has in concrete motivated cases a "physical" interpretation. If Φ is a potential of B (cf. Lemma 11.31) then, integrating over $[0,t]$, this test leads to

$$\underbrace{\frac{1}{2}\Big\|\frac{\mathrm{d}u}{\mathrm{d}t}(t)\Big\|_H^2+\Phi\big(u(t)\big)}_{\text{total energy at time } t} \;+\; \underbrace{\int_0^t\Big\langle A\Big(\frac{\mathrm{d}u}{\mathrm{d}\vartheta}\Big),\frac{\mathrm{d}u}{\mathrm{d}\vartheta}\Big\rangle\,\mathrm{d}\vartheta}_{\text{dissipated energy}}$$

$$= \underbrace{\frac{1}{2}\big\|v_0\big\|_H^2+\Phi\big(u_0\big)}_{\text{total energy at time } 0} \;+\; \underbrace{\int_0^t\Big\langle f(\vartheta),\frac{\mathrm{d}u}{\mathrm{d}\vartheta}\Big\rangle\,\mathrm{d}\vartheta}_{\text{work of external forces}} \tag{11.120}$$

which just expresses the balance of "mechanical" energy. Here, the total energy means the sum of the "kinetic" energy $\|\frac{\mathrm{d}}{\mathrm{d}t}u\|_H^2$ and the "stored" energy $\Phi(u)$, cf. also (12.11) below.

Proposition 11.35 (UNIQUENESS[30]). *Let A be "weakly monotone" in the sense of (8.114) and one of the following situations takes place:*

(i) *B is linear of the form $B(t,u) = B_1u$ with $B_1 : V \to V^*$ monotone and symmetric, i.e. $B_1^* = B_1$.*

(ii) *B is Lipschitz continuous on H, i.e. $\|B(t,u)-B(t,v)\|_H \le \ell(t)\|u-v\|_H$ with $\ell \in L^2(I)$.*

(iii) *$\|B(t,u)-B(t,v)\|_{V^*}^{p'/2} \le \ell(t)\|u-v\|_H$ with $\ell \in L^2(I)$ and $\langle A(t,u)-A(t,v),u-v\rangle \ge c_0\|u-v\|_V^p - c_1(t)\|u-v\|_V - c_2(t)\|u-v\|_H^2$ with $c_0 > 0$, $c_1 \in L^{p'}(I)$, and $c_2 \in L^1(I)$.*

Then (11.97) possesses at most one (strong) solution.

Proof. We take two solutions u_1 and u_2, subtract (11.97) for $u = u_1$ and $u = u_2$, and test it by $v = \frac{\mathrm{d}}{\mathrm{d}t}u_{12}$ where $u_{12} := u_1 - u_2$. We thus get

$$\frac{1}{2}\frac{\mathrm{d}}{\mathrm{d}t}\Big(\Big\|\frac{du_{12}}{dt}\Big\|_H^2+\big\langle B_1(u_{12}),u_{12}\big\rangle\Big) = \frac{1}{2}\frac{\mathrm{d}}{\mathrm{d}t}\Big\|\frac{du_{12}}{dt}\Big\|_H^2+\Big\langle B(t,u_1)-B(t,u_2),\frac{du_{12}}{dt}\Big\rangle$$
$$= -\Big\langle A\big(t,\frac{du_1}{dt}\big)-A\big(t,\frac{du_2}{dt}\big),\frac{du_{12}}{dt}\Big\rangle \le c(t)\Big\|\frac{du_{12}}{dt}\Big\|_H^2.$$

By the Gronwall inequality and by $u_{12}(0) = 0$ and $\frac{\mathrm{d}}{\mathrm{d}t}u_{12}(0) = 0$, one gets $u_1 = u_2$.

In the case (ii), we can estimate

$$\frac{1}{2}\frac{\mathrm{d}}{\mathrm{d}t}\Big\|\frac{du_{12}}{dt}\Big\|_H^2 = -\Big\langle A\big(t,\frac{\mathrm{d}}{\mathrm{d}t}u_1\big)-A\big(t,\frac{\mathrm{d}}{\mathrm{d}t}u_2\big),\frac{du_{12}}{dt}\Big\rangle$$
$$+\Big\langle B(t,u_2)-B(t,u_1),\frac{du_{12}}{dt}\Big\rangle$$
$$\le c(t)\Big\|\frac{du_{12}}{dt}\Big\|_H^2+\big\|B(t,u_1)-B(t,u_2)\big\|_H^2+\Big\|\frac{du_{12}}{dt}\Big\|_H^2 \tag{11.121}$$
$$\le c(t)\Big\|\frac{du_{12}}{dt}\Big\|_H^2+\ell(t)^2\big\|u_{12}\big\|_H^2+\Big\|\frac{du_{12}}{dt}\Big\|_H^2$$

[30] For the case (iii), cf. also Zeidler [427, Chap.33].

where $c(\cdot)$ comes from (8.114). Abbreviating still $\frac{\mathrm{d}}{\mathrm{d}t}u_{12} = v_{12}$, we get

$$\frac{1}{2}\frac{\mathrm{d}}{\mathrm{d}t}\|v_{12}\|_H^2 \le c(t)\|v_{12}\|_H^2 + \ell(t)^2\|u_{12}\|_H^2 + \|v_{12}\|_H^2. \tag{11.122}$$

Multiplying $\frac{\mathrm{d}}{\mathrm{d}t}u_{12} = v_{12}$ by u_{12}, one gets

$$\frac{1}{2}\frac{\mathrm{d}}{\mathrm{d}t}\|u_{12}\|_H^2 = \langle v_{12}, u_{12}\rangle \le \frac{1}{2}\|v_{12}\|_H^2 + \frac{1}{2}\|u_{12}\|_H^2. \tag{11.123}$$

Now we apply the Gronwall inequality to the system (11.122)–(11.123) together with $u_{12}(0)=0$ and $v_{12}(0)=0$, which gives, in particular, that $u_{12}(t)=0$ for all t.

In the case (iii), we get analogously

$$\begin{aligned}\frac{1}{2}\frac{\mathrm{d}}{\mathrm{d}t}\|v_{12}\|_H^2 + c_0\|v_{12}\|_V^p &\le c_1(t)\|v_{12}\|_V + c_2(t)\|v_{12}\|_H^2 \\ &\quad + \langle B(t,u_2) - B(t,u_1), v_{12}\rangle \le C_\varepsilon c_1(t)^{p'} + c_2(t)\|v_{12}\|_H^2 \\ &\quad + C_\varepsilon\|B(t,u_1) - B(t,u_2)\|_{V^*}^{p'} + \varepsilon\|v_{12}\|_V^p. \end{aligned} \tag{11.124}$$

We choose $\varepsilon < c_0$ to absorb the last term and also the estimate $\|B(t,u_1) - B(t,u_2)\|_{V^*}^{p'} \le \ell(t)^2\|u_{12}\|_H^2$. Then we apply again the Gronwall inequality to the system (11.123)–(11.124). □

Remark 11.36. Velocity $\frac{\partial}{\partial t}u$ is indeed a natural test function, as already claimed in Remark 11.34, while u itself is not a suitable test function here. This is related to troubles typically arising in variational inequalities with obstacles like $u \ge 0$, i.e. $B = \partial\Phi$ with $\Phi = \delta_K$, $K := \{v \ge 0\}$. E.g., after a penalization, one can consider $\frac{\partial^2}{\partial t^2}u_\varepsilon - \Delta\frac{\partial}{\partial t}u_\varepsilon + \frac{1}{\varepsilon}u_\varepsilon^- = g$. By testing by $\frac{\partial}{\partial t}u_\varepsilon$, one gets $\|\frac{\partial}{\partial t}u_\varepsilon\|_{L^\infty(I;L^2(\Omega))\cap L^2(I;W^{1,2}(\Omega))} = \mathscr{O}(1)$ and $\|u_\varepsilon^-\|_{L^\infty(I;L^2(\Omega))} = \mathscr{O}(\sqrt{\varepsilon})$. But, we must test by $v - u_\varepsilon$ to prove convergence, and we get after the by-part integration the term $\int_Q |\frac{\partial}{\partial t}u_\varepsilon|^2$ with a "bad" sign. Note that we do not have the "dual estimate" to $\frac{\partial^2}{\partial t^2}u_\varepsilon$ uniform in ε. Also, a test by $\frac{\partial^2}{\partial t^2}u_\varepsilon$ yields the penalty term $\varepsilon^{-1}u_\varepsilon^-\frac{\partial^2}{\partial t^2}u_\varepsilon$ which cannot be estimated "on the left-hand side".

Remark 11.37 (*Rothe method*)**.** The semidiscretization in time is also here applicable to (11.97): we define $u_\tau^k \in V$, $k = 1,\dots,K$, by the following recursive formula:

$$\frac{u_\tau^k - 2u_\tau^{k-1} + u_\tau^{k-2}}{\tau^2} + A_\tau^k\Big(\frac{u_\tau^k - u_\tau^{k-1}}{\tau}\Big) + B_\tau^k(u_\tau^k) = f_\tau^k\,, \tag{11.125a}$$

$$u_\tau^0 = u_0, \qquad u_\tau^{-1} = u_0 - \tau v_0, \tag{11.125b}$$

where again $f_\tau^k := \frac{1}{\tau}\int_{(k-1)\tau}^{k\tau} f(t)\,\mathrm{d}t$ and $A_\tau^k(u) := \frac{1}{\tau}\int_{(k-1)\tau}^{k\tau} A(t,u)\,\mathrm{d}t$ and $B_\tau^k(u) := \frac{1}{\tau}\int_{(k-1)\tau}^{k\tau} B(t,u)\,\mathrm{d}t$. Existence of the Rothe sequence then needs, as in Lemma 8.5,

A_τ^k and B_τ^k to be pseudomonotone and semi-coercive for $k = 1, \dots, T/\tau$. Various modifications of the above procedure are as usual. Instead of (11.99) or (11.104), we need here to estimate $\frac{\mathrm{d}}{\mathrm{d}t}[\frac{\mathrm{d}}{\mathrm{d}t}u_\tau]^{\mathrm{i}}$ in $L^{p'}(I;V^*)$ or in $L^2(I;H)$, where $[\cdot]^{\mathrm{i}}$ denotes the piecewise-linear interpolation operator, cf. Figure 17 on p.216, defined now on the whole interval $I = [0,T]$ because, thanks to (11.125b), we defined the Rothe sequence $\{u_\tau^k\}$ even for $k = -1$. By using the discrete *by-part summation* formula

$$\sum_{k=1}^{T/\tau} \langle u^k - 2u^{k-1} + u^{k-2}, z^k \rangle = \langle u^{T/\tau} - u^{T/\tau-1}, z^{T/\tau} \rangle - \langle u^0 - u^{-1}, z^1 \rangle - \sum_{k=2}^{T/\tau} \langle u^{k-1} - u^{k-2}, z^k - z^{k-1} \rangle \tag{11.126}$$

and considering a test function $z \in W^{1,p,p'}(I;V,V^*)$, one obtains an analog of (11.109), namely

$$\int_0^T \Big\langle \mathscr{A}\Big(\frac{\mathrm{d}u_\tau}{\mathrm{d}t}\Big), \bar{z}_\tau \Big\rangle + \langle \mathscr{B}(\bar{u}_\tau), \bar{z}_\tau \rangle - \langle \bar{f}_\tau, \bar{z}_\tau \rangle \,\mathrm{d}t - \int_\tau^T \Big\langle \frac{\mathrm{d}u_\tau}{\mathrm{d}t}(\cdot - \tau), \frac{\mathrm{d}z_\tau}{\mathrm{d}t} \Big\rangle \,\mathrm{d}t + \Big\langle \frac{\mathrm{d}u_\tau}{\mathrm{d}t}(T), z_\tau(T) \Big\rangle = \langle v_0, z_\tau(\tau) \rangle \tag{11.127}$$

where z_τ and $\bar{z}_\tau$ are respectively the piecewise constant and piecewise affine interpolants of the values $\{z(k\tau)\}_{k=1}^{T/\tau}$. Alternatively, without such interpolants, one can use an analog of (11.109) as

$$\int_0^T \Big\langle \mathscr{A}\Big(\frac{\mathrm{d}u_\tau}{\mathrm{d}t}\Big), z \Big\rangle + \langle \mathscr{B}(\bar{u}_\tau), z \rangle - \langle \bar{f}_\tau, z \rangle - \Big\langle \Big[\frac{\mathrm{d}u_\tau}{\mathrm{d}t}\Big]^{\mathrm{i}}, \frac{\mathrm{d}z}{\mathrm{d}t} \Big\rangle \,\mathrm{d}t + \Big\langle \frac{\mathrm{d}u_\tau}{\mathrm{d}t}(T), z(T) \Big\rangle = \langle v_0, z(0) \rangle \tag{11.128}$$

and prove the convergence of the piece-wise affine interpolant $[\frac{\mathrm{d}u_\tau}{\mathrm{d}t}]^{\mathrm{i}}$ towards $\frac{\mathrm{d}u}{\mathrm{d}t}$.

11.4 Exercises

Exercise 11.38 (Penalty-function method for type-II parabolic inequalities). Consider the complementarity problem (11.25). Show the connection between (11.25) and (11.26) analogously as done in Proposition 5.9. The L^2-type penalty-function method leads to the initial-boundary-value problem

$$\left.\begin{aligned} \frac{\partial u_\varepsilon}{\partial t} + \frac{1}{\varepsilon}\Big(\frac{\partial u_\varepsilon}{\partial t}\Big)^- - \Delta u_\varepsilon + c(u_\varepsilon) &= g && \text{in } Q, \\ u_\varepsilon &= 0 && \text{on } \Sigma, \\ u_\varepsilon(0,\cdot) &= u_0 && \text{on } \Omega. \end{aligned}\right\} \tag{11.129}$$

As (11.37) above, test (11.129) by $\frac{\partial}{\partial t}u_\varepsilon$ and formulate assumptions on u_0 and on $c(\cdot)$ to obtain the estimates

$$\big\|u_\varepsilon\big\|_{L^\infty(I;W_0^{1,2}(\Omega))} \le C, \quad \Big\|\frac{\partial u_\varepsilon}{\partial t}\Big\|_{L^2(Q)} \le C, \quad \Big\|\Big(\frac{\partial u_\varepsilon}{\partial t}\Big)^-\Big\|_{L^2(Q)} \le C\sqrt{\varepsilon}. \tag{11.130}$$

Test (11.129) by $v - \frac{\partial}{\partial t}u_\varepsilon$ and show the convergence of a selected subsequence $\{u_\varepsilon\}_{\varepsilon>0}$ to the solution of (11.26).[31]

Exercise 11.39 (*Landau-Lifschitz-Gilbert equation* modified). Consider the modification of (11.89):

$$\partial\psi\Big(\frac{\partial u}{\partial t}\Big) - \beta(|u|)u \times \frac{\partial u}{\partial t} - \mu\,\Delta u + c(u) = g \tag{11.131}$$

with $\psi : \mathbb{R}^3 \to \mathbb{R}$ a convex (possibly nonsmooth) function; this modification describes pinning effects of dry-friction type [374]. Using the orthogonality $(\beta(|u|)u\times\frac{\partial u}{\partial t})\cdot\frac{\partial u}{\partial t} = 0$, one can design the definition of the weak solution in the spirit of (11.81) as

$$\int_Q \psi(v) + \mu\nabla u{:}\nabla\Big(v-\frac{\partial u}{\partial t}\Big) + (c(u)-g)\Big(v-\frac{\partial u}{\partial t}\Big) \\ + \Big(\beta(|u|)u\times\frac{\partial u}{\partial t}\Big)\cdot v\,\mathrm{d}x\mathrm{d}t \ge \int_Q \psi\Big(\frac{\partial u}{\partial t}\Big)\mathrm{d}x\mathrm{d}t. \tag{11.132}$$

---

[31] Hint: For $v \ge 0$, it holds that

$$\int_Q \Big(\frac{\partial u_\varepsilon}{\partial t}+c(u_\varepsilon)-g\Big)\Big(v-\frac{\partial u_\varepsilon}{\partial t}\Big) + \nabla u_\varepsilon\cdot\nabla\Big(v-\frac{\partial u_\varepsilon}{\partial t}\Big)\mathrm{d}x\mathrm{d}t = \frac{1}{\varepsilon}\int_Q \Big(\frac{\partial u_\varepsilon}{\partial t}\Big)^-\Big(\frac{\partial u_\varepsilon}{\partial t}-v\Big)\mathrm{d}x\mathrm{d}t \ge 0$$

so that, formally, we have

$$\begin{aligned}
0 &\le \limsup_{\varepsilon\to 0}\int_Q \Big(\frac{\partial u_\varepsilon}{\partial t}+c(u_\varepsilon)-g\Big)\Big(v-\frac{\partial u_\varepsilon}{\partial t}\Big) + \nabla u_\varepsilon\cdot\nabla\Big(v-\frac{\partial u_\varepsilon}{\partial t}\Big)\mathrm{d}x\mathrm{d}t\\
&= -\liminf_{\varepsilon\to 0}\Big(\Big\|\frac{\partial u_\varepsilon}{\partial t}\Big\|^2_{L^2(Q)} + \frac{1}{2}\|\nabla u_\varepsilon(T,\cdot)\|^2_{L^2(\Omega;\mathbb{R}^n)}\Big)\\
&\quad + \lim_{\varepsilon\to 0}\int_Q (c(u_\varepsilon)-g)\Big(v-\frac{\partial u_\varepsilon}{\partial t}\Big) + \frac{\partial u_\varepsilon}{\partial t}v + \nabla u_\varepsilon\cdot\nabla v\mathrm{d}x\mathrm{d}t + \frac{1}{2}\|\nabla u_0\|^2_{L^2(\Omega;\mathbb{R}^n)}\\
&\le \int_Q \Big(\frac{\partial u}{\partial t}+c(u)-g\Big)\Big(v-\frac{\partial u}{\partial t}\Big) + \nabla u_\varepsilon\cdot\nabla\Big(v-\frac{\partial u}{\partial t}\Big)\mathrm{d}x\mathrm{d}t
\end{aligned}$$

where also $\nabla u_\varepsilon(T,\cdot) \rightharpoonup \nabla u(T,\cdot)$ weakly in $L^2(\Omega;\mathbb{R}^n)$ has been used. Note that, as we do not have $\nabla\frac{\partial}{\partial t}u_\varepsilon \in L^2(Q)$ guaranteed by our a-priori estimates (11.130), the term $\int_Q \nabla u_\varepsilon\cdot\nabla\frac{\partial}{\partial t}u_\varepsilon\mathrm{d}x\mathrm{d}t$ gets a meaning only if put equal to $\frac{1}{2}\|\nabla u_\varepsilon(T,\cdot)\|^2_{L^2(\Omega;\mathbb{R}^n)} - \frac{1}{2}\|u_0\|^2_{L^2(\Omega;\mathbb{R}^n)}$, which can be justified either by using Galerkin's approximation or by a limit of mollified u_ε.

Finally, as $\xi \mapsto \|\xi^-\|^2_{L^2(Q)}$ is a convex continuous functional on $L^2(Q)$, it is weakly lower semicontinuous, and by $\partial u_\varepsilon/\partial t \rightharpoonup \partial u/\partial t$ and by the last estimate in (11.130) we have

$$\Big\|\Big(\frac{\partial u}{\partial t}\Big)^-\Big\|^2_{L^2(Q)} \le \liminf_{\varepsilon\to 0}\Big\|\Big(\frac{\partial u_\varepsilon}{\partial t}\Big)^-\Big\|^2_{L^2(Q)} \le \lim_{\varepsilon\to 0} C^2\varepsilon = 0$$

so that $\frac{\partial}{\partial t}u \ge 0$ a.e. in Q.

Modify the semi-implicit discretisation (11.90) and prove existence of its weak solution, show a-priori estimates and make a limit passage in a discrete analog of (11.132).

Exercise 11.40. Consider the initial-boundary-value problem:

$$\left.\begin{aligned} &\frac{\partial^2 u}{\partial t^2} - \Delta\frac{\partial u}{\partial t} + \left|\frac{\partial u}{\partial t}\right|^{p-2}\frac{\partial u}{\partial t} - \mathrm{div}\big(|\nabla u|^{q-2}\nabla u\big) = g && \text{in } Q,\\ &u(0,\cdot) = u_0\ , \quad \frac{\partial u}{\partial t}(0,\cdot) = v_0, && \text{in } \Omega,\\ &u|_\Sigma = 0 && \text{on } \Sigma. \end{aligned}\right\} \tag{11.133}$$

Apply the Galerkin method, denote the approximate solution by u_k. Qualify u_0, v_0 and g appropriately and prove a-priori estimates of u_k in $L^\infty(I;W^{1,q}(\Omega)) \cap W^{1,2}(I;W^{1,2}(\Omega)\cap L^p(\Omega))\cap W^{1,\infty}(I;L^2(\Omega))\cap W^{2,2}(I;W^{-1,2}(\Omega))$.[32] Assume $p < q^*$ and prove convergence by using monotonicity and the Minty trick.[33]

[32] Hint: Test the equation in (11.133) by $\frac{\partial}{\partial t}u_k$, obtaining

$$\int_\Omega \frac{1}{2}\frac{\partial}{\partial t}\left|\frac{\partial u_k}{\partial t}\right|^2 + \left|\nabla\frac{\partial u_k}{\partial t}\right|^2 + \left|\frac{\partial u_k}{\partial t}\right|^p + \frac{1}{q}\frac{\partial}{\partial t}|\nabla u_k|^q \mathrm{d}x$$
$$= \int_\Omega g(t,\cdot)\frac{\partial u_k}{\partial t}\mathrm{d}x \le C_\varepsilon\|g(t,\cdot)\|^2_{W^{1,2}(\Omega)^*} + \varepsilon\left\|\frac{\partial u_k}{\partial t}\right\|^2_{W^{1,2}(\Omega)}$$

from which the estimate follows by Gronwall's inequality assuming $u_0 \in W^{1,q}(\Omega)$, $v_0 \in L^2(\Omega)$, and $g \in L^2(I;L^{2^*}(\Omega))$. For the "dual" estimate of $\frac{\partial^2}{\partial t^2}u_k$ use the strategy (11.101).

[33] Hint: Use monotonicity of the q-Laplacean and (11.133) to write

$$0 \le \int_Q |\nabla u_k|^{q-2}\nabla u_k - |\nabla z|^{q-2}\nabla z\cdot\nabla(u_k - z)\,\mathrm{d}x\mathrm{d}t$$
$$= \int_Q \Big(g - \frac{\partial^2 u_k}{\partial t^2} - \left|\frac{\partial u_k}{\partial t}\right|^{p-2}\frac{\partial u_k}{\partial t}\Big)(u_k - z) - \Big(\nabla\frac{\partial u_k}{\partial t} + |\nabla z|^{q-2}\nabla z\Big)\cdot\nabla(u_k - z)\,\mathrm{d}x\mathrm{d}t.$$

Realize that, by Aubin-Lions' lemma, $\frac{\partial}{\partial t}u_k \to \frac{\partial}{\partial t}u$ (as a subsequence) strongly in $L^2(Q)$ due to the compact embedding $L^2(I;W^{1,2}(\Omega))\cap W^{1,2}(I;W^{-1,2}(\Omega)) \Subset L^2(Q)$. Also $u_k(T,\cdot) \rightharpoonup u(T,\cdot)$ weakly in $W^{1,2}(\Omega)$, hence strongly in $L^2(\Omega)$, and $\frac{\partial}{\partial t}u_k(T,\cdot) \rightharpoonup \frac{\partial}{\partial t}u(T,\cdot)$ weakly in $L^2(\Omega)$. Then estimate the limit superior:

$$\limsup_{k\to\infty}\Big(-\int_Q \frac{\partial^2 u_k}{\partial t^2}u_k + \nabla\frac{\partial u_k}{\partial t}\cdot\nabla u_k\mathrm{d}x\mathrm{d}t\Big) = \limsup_{k\to\infty}\Big(\int_Q\left|\frac{\partial u_k}{\partial t}\right|^2\mathrm{d}x\mathrm{d}t$$
$$+ \int_\Omega v_0u_0 - \frac{\partial u_k}{\partial t}(T,\cdot)u_k(T,\cdot) + \frac{1}{2}|u_0|^2 - \frac{1}{2}|\nabla u_k(T,\cdot)|^2\mathrm{d}x\Big)$$
$$\le \int_Q\left|\frac{\partial u}{\partial t}\right|^2\mathrm{d}x\mathrm{d}t + \int_\Omega v_0u_0 - \frac{\partial u}{\partial t}(T,\cdot)u(T,\cdot) + \frac{1}{2}|u_0|^2 - \frac{1}{2}|\nabla u(T,\cdot)|^2\mathrm{d}x$$
$$= -\int_0^T\Big(\Big\langle\frac{\partial^2 u}{\partial t^2}, u\Big\rangle + \int_\Omega\nabla\frac{\partial u}{\partial t}\cdot\nabla u\,\mathrm{d}x\Big)\mathrm{d}t.$$

The limit passage in the lower-order term $|\frac{\partial}{\partial t}u_k|^{p-2}\frac{\partial}{\partial t}u_k$ can be made by compactness. If $p \le 2$, then $\frac{\partial}{\partial t}u_k \to \frac{\partial}{\partial t}u$ in $L^p(Q)$ while, if $p > 2$, then at least $L^{p-\epsilon}(Q)$ because $\{\frac{\partial}{\partial t}u_k\}_{k\in\mathbb{N}}$ is bounded in $L^p(Q)$. If $p < q^*$, make the limit passage in $\int_Q|\frac{\partial}{\partial t}u_k|^{p-2}(\frac{\partial}{\partial t}u_k)u_k\mathrm{d}x\mathrm{d}t$ when

Exercise 11.41 (Klein-Gordon equation, generalized[34]). Consider the initial-Dirichlet-boundary-value problem for the semilinear *hyperbolic equation*

$$\frac{\partial^2 u}{\partial t^2} - \Delta u + |u|^{q-2}u = g, \quad u(0,\cdot) = u_0, \quad \frac{\partial u}{\partial t}(0,\cdot) = v_0, \quad u|_\Sigma = 0. \qquad (11.134)$$

The variant $q = 3$ is called the *Klein-Gordon equation*, having applications in quantum physics. For $q>1$, derive a-priori estimates of u in $W^{1,\infty}(I;L^2(\Omega)) \cap L^\infty(I;W_0^{1,2}(\Omega)\cap L^q(\Omega))$ by testing it by $\frac{\partial}{\partial t}u$. Prove convergence of the Galerkin approximations u_k by weak continuity.[35] If $q \geq 2$ is small enough, prove uniqueness by using the test function $v = \frac{\partial}{\partial t}u_1 - u_2$, with u_1, u_2 being two weak solutions.[36]

Exercise 11.42 (Viscous regularization of Klein-Gordon equation). Consider the initial-boundary-value problem:

$$\left.\begin{array}{ll} \dfrac{\partial^2 u}{\partial t^2} - \mu\mathrm{div}\Big(\Big|\dfrac{\partial u}{\partial t}\Big|^{p-2}\nabla\dfrac{\partial u}{\partial t}\Big) - \Delta u + c(u) = g & \text{in } Q,\\ u(0,\cdot) = u_0\ , \quad \dfrac{\partial u}{\partial t}(0,\cdot) = v_0, & \text{in } \Omega,\\ u|_\Sigma = 0 & \text{on } \Sigma, \end{array}\right\} \qquad (11.135)$$

with $\mu > 0$. Apply the Galerkin method, denote the approximate solution by u_k, and prove a-priori estimates for u_k in $L^\infty(I;W^{1,2}(\Omega)) \cap W^{1,p}(I;W^{1,p}(\Omega)) \cap W^{2,p'}(I;W^{\max(2,p)}(\Omega))$ and specify qualifications on u_0, v_0, g, and $c(\cdot)$.[37] Prove

realizing boundedness of $\{u_k\}_{k\in\mathbb{N}}$ in $L^\infty(I;W^{1,q}(\Omega)) \subset L^{q^*}(Q)$. Finally, put $z = u + \delta w$ and finish the proof by Minty's trick.

[34] Cf. Barbu [38, Sect.4.3.5], Jerome [215], or Lions [261, Sect.I.1].

[35] Hint: For $q < p^*+1$, use the Aubin-Lions Lemma 7.7 to get compactness in $L^{q-1}(Q)$ which allows for a limit passage through the term $\int_Q |u|^{q-2}uv\,\mathrm{d}x\mathrm{d}t$ if $v \in L^\infty(Q)$. For $q \geq p^*+1$, interpolate between $W^{1,2}(\Omega)$ and $L^q(\Omega)$ and get again compactness in $L^{q-1}(Q)$ but now by Lemma 7.8.

[36] Hint: Abbreviating $u_{12} = u_1 - u_2$, realize that $r \mapsto |r|^{q-2}r$ is Lipschitz continuous on $[r_1, r_2]$ with the Lipschitz constant $(q-1)\max(|r_1|^{q-2}, |r_2|^{q-2})$; this test gives

$$\frac{1}{2}\frac{\mathrm{d}}{\mathrm{d}t}\Big(\Big\|\frac{\partial u_{12}}{\partial t}\Big\|^2_{L^2(\Omega)} + \|\nabla u_{12}\|^2_{L^2(\Omega;\mathbb{R}^n)}\Big) = \int_\Omega \big(|u_2|^{q-1}u_2 - |u_1|^{q-1}u_1\big)\frac{\partial u_{12}}{\partial t}\,\mathrm{d}x$$
$$\leq \frac{q-1}{2}\max\big(\|\,|u_1|^{q-2}\|_{L^\alpha(\Omega)}, \|\,|u_2|^{q-2}\|_{L^\alpha(\Omega)}\big)\Big(\|u_{12}\|^2_{L^{2^*}(\Omega)} + \Big\|\frac{\partial u_{12}}{\partial t}\Big\|^2_{L^2(\Omega)}\Big)$$

with α so that $\alpha^{-1} + (2^*)^{-1} + 2^{-1} = 1$. Exploiting that $u_1, u_2 \in L^\infty(I;L^{2^*}(\Omega))$ and assuming q so small that $\alpha \geq 2^*/(q-2)$, proceed by Gronwall inequality.

[37] Hint: Test the equation in (11.135) by $\frac{\partial}{\partial t}u_k$, obtaining

$$\int_\Omega \frac{1}{2}\frac{\partial}{\partial t}\Big|\frac{\partial u_k}{\partial t}\Big|^2 + \mu\Big|\nabla\frac{\partial u_k}{\partial t}\Big|^p + \frac{1}{2}\frac{\partial}{\partial t}|\nabla u_k|^2\mathrm{d}x$$
$$= \int_\Omega \big(g(t,\cdot) - c(u_k)\big)\frac{\partial u_k}{\partial t}\mathrm{d}x \leq C_\varepsilon\|g(t,\cdot) - c(u_k)\|^{p'}_{W^{1,p}(\Omega)^*} + \varepsilon\Big\|\frac{\partial u_k}{\partial t}\Big\|^p_{W^{1,p}(\Omega)},$$

from which the claimed estimates follow by Gronwall's inequality if $u_0 \in W^{1,2}(\Omega)$, $v_0 \in L^2(\Omega)$, $g \in L^{p'}(I;L^{p^{*\prime}}(\Omega))$, and $c(\cdot)$ has an at most $2/p'$-growth. For $p < 2$, an at most linear growth of convergence by monotonicity and the Minty trick.[38] Eventually, denoting u_μ the solution to (11.135), prove that u_μ approaches the solution to the Klein-Gordon equation (11.134) if $c(r) = |r|^{q-2}r$ when $\mu \to 0$.[39]

Exercise 11.43 (Martensitic transformation in *shape-memory alloys*[40]). Consider

$$\left.\begin{aligned}
&\frac{\partial^2 u}{\partial t^2} - \mu\Delta\frac{\partial u}{\partial t} - \mathrm{div}\big(\sigma(\nabla u)\big) + \lambda\Delta^2 u = g && \text{in } Q,\\
&u(0,\cdot) = u_0\ , \qquad \frac{\partial u}{\partial t}(0,\cdot) = v_0, && \text{in } \Omega,\\
&u = 0\ , \qquad \frac{\partial u}{\partial \nu} = 0, && \text{on } \Sigma.
\end{aligned}\right\} \tag{11.136}$$

Assume $\mu, \lambda > 0$, $g \in L^2(I; L^{2^{**}{}'}(\Omega))$, $u_0 \in W_0^{2,2}(\Omega)$, $v_0 \in L^2(\Omega)$, and at most linear growth of $\sigma : \mathbb{R}^n \to \mathbb{R}^n$, consider Galerkin's approximation, and derive the a-priori estimates in $W^{1,\infty}(I; L^2(\Omega)) \cap W^{1,2}(I; W^{1,2}(\Omega)) \cap L^\infty(I; W_0^{2,2}(\Omega))$ by testing (11.136) by $\frac{\partial}{\partial t}u$.[41] Then estimate still $\frac{\partial^2}{\partial t^2}u$ in $L^2(I; W^{-2,2}(\Omega))$ and prove convergence of Galerkin's approximants. Make also the limit passage in (11.136)

$c(\cdot)$ can be allowed if $\int_\Omega c(u_k)\frac{\partial}{\partial t}u_k \mathrm{d}x \le \frac{1}{2}\|c(u_k)\|^2_{L^2(\Omega)} + \frac{1}{2}\|\frac{\partial}{\partial t}u_k\|^2_{L^2(\Omega)}$ is used. Alternatively, one can impose a condition $c(r)r \ge 0$ and estimate $c(u_k)\frac{\partial}{\partial t}u_k = \frac{\partial}{\partial t}\int_0^{u_k} c(\xi)\mathrm{d}\xi$ on the left-hand side. The "dual" estimate of $\frac{\partial^2}{\partial t^2}u_k$ then follows by the strategy (11.101).

[38] Hint: By the monotonicity of the p-Laplacean and by (11.135),

$$\begin{aligned}
0 &\le \int_Q \mu\Big(|\nabla\frac{\partial u_k}{\partial t}|^{p-2}\nabla\frac{\partial u_k}{\partial t} - |\nabla\frac{\partial z}{\partial t}|^{p-2}\nabla\frac{\partial z}{\partial t}\Big)\cdot\nabla\frac{\partial(u_k - z)}{\partial t}\mathrm{d}x\mathrm{d}t\\
&= \int_Q \Big(g - c(u) - \frac{\partial^2 u_k}{\partial t^2}\Big)\frac{\partial(u_k - z)}{\partial t} - \Big(\nabla u_k + |\nabla\frac{\partial z}{\partial t}|^{p-2}\nabla\frac{\partial z}{\partial t}\Big)\cdot\nabla\frac{\partial(u_k - z)}{\partial t}\mathrm{d}x\mathrm{d}t.
\end{aligned}$$

By using $\frac{\partial}{\partial t}u_k(T,\cdot) \rightharpoonup \frac{\partial}{\partial t}u(T,\cdot)$ weakly in $L^2(\Omega)$ and $u_k(T,\cdot) \rightharpoonup u(T,\cdot)$ weakly in $W^{1,2}(\Omega)$, estimate the limit superior as $\limsup_{k\to\infty}\int_Q -\frac{\partial^2}{\partial t^2}u_k\frac{\partial}{\partial t}u_k - \nabla u_k\cdot\nabla\frac{\partial}{\partial t}u_k \mathrm{d}x\mathrm{d}t = \limsup_{k\to\infty}\frac{1}{2}\int_\Omega |v_0|^2 - |\frac{\partial}{\partial t}u_k(T,\cdot)|^2 + |\nabla u_0|^2 - |\nabla u(T,\cdot)|^2\mathrm{d}x \le \frac{1}{2}\int_\Omega |v_0|^2 - \big|\frac{\partial u}{\partial t}(T,\cdot)|^2 + |\nabla u_0|^2 - |\nabla u(T,\cdot)|^2\mathrm{d}x = -\int_Q \frac{\partial^2 u}{\partial t^2}\frac{\partial u}{\partial t} + \nabla u\cdot\nabla\frac{\partial u}{\partial t}\mathrm{d}x\mathrm{d}t$.

[39] Hint: Realize that $\|u_\mu\|_{W^{1,p}(I;W^{1,p}(\Omega))} = \mathscr{O}(\mu^{-1/p})$ hence the term $\int_Q \mu|\frac{\partial}{\partial t}u|^{p-2}\nabla\frac{\partial}{\partial t}u\cdot\nabla v\,\mathrm{d}x\mathrm{d}t = \mathscr{O}(\mu^{1-1/p})$ with v fixed vanishes for $\mu \searrow 0$.

[40] In the vectorial variant, $u(t,\cdot) : \Omega \to \mathbb{R}^n$ is the "displacement", cf. Example 6.7, and (11.136) describes isothermal vibrations of a "viscous" solid whose stress response $\sigma : \mathbb{R}^{n\times n} \to \mathbb{R}^{n\times n}$ need not be monotone and need not have any quasiconvex (cf. Remark 6.5) potential, and which has some capilarity-like behaviour with $\lambda > 0$ possibly small. The multi-well potential of σ may describe various phases (called martensite or austenite) in so-called shape-memory alloys and then (11.136) is a very simple model for a solid-solid phase transformation, cf. [343] for a critical discussion. For mathematical treatment of this capilarity case see e.g. Abeyaratne and Knowles [1] or Hoffmann and Zochowski [206], cf. also Brokate and Sprekels [71, Chap.5].

[41] Hint: This test gives

$$\begin{aligned}
\frac{1}{2}\frac{\mathrm{d}}{\mathrm{d}t}\Big(\Big\|\frac{\partial u}{\partial t}\Big\|^2_{L^2(\Omega)} + \lambda\|\nabla^2 u\|^2_{L^2(\Omega;\mathbb{R}^{n\times n})}\Big) + \mu\Big\|\frac{\partial\nabla u}{\partial t}\Big\|^2_{L^2(\Omega;\mathbb{R}^n)} &= \int_\Omega gu - \sigma(\nabla u)\cdot\frac{\partial\nabla u}{\partial t}\mathrm{d}x\\
&\le \frac{1}{2}\|g\|^2_{L^{2^{**}{}'}(\Omega)} + \frac{1}{2}\|u\|^2_{L^{2^{**}}(\Omega)} + \frac{1}{2\mu}\|\sigma(\nabla u)\|^2_{L^2(\Omega;\mathbb{R}^n)} + \frac{\mu}{2}\Big\|\frac{\partial\nabla u}{\partial t}\Big\|^2_{L^2(\Omega;\mathbb{R}^n)}
\end{aligned}$$

with $\mu \searrow 0$, showing existence of a solution to the semilinear *hyperbolic equation* $\frac{\partial^2}{\partial t^2}u - \mathrm{div}(\sigma(\nabla u)) + \lambda\Delta^2 u = g$.[42]

Exercise 11.44. Modify Exercise 11.42 by replacing the term $c(u)$ by $c(\nabla u)$ or $\mathrm{div}\big(a_0(u)\big)$ with $a_0 : \mathbb{R} \to \mathbb{R}^n$.

Exercise 11.45. Show that $B := -\Delta_p$ formulated weakly with Dirichlet boundary conditions, i.e. $V = W_0^{1,p}(\Omega)$, satisfies (11.103) provided $p \geq 2$.[43]

Exercise 11.46 (*Rothe method*)**.** Applying the Rothe method to the system (9.64), one obtains the system

$$\frac{v_\tau^k - v_\tau^{k-1}}{\tau} + A_\tau^k(v_\tau^k) + B_\tau^k(u_\tau^k) = f_\tau^k\,, \qquad u_\tau^0 = u_0, \tag{11.137a}$$

$$\frac{u_\tau^k - u_\tau^{k-1}}{\tau} - v_\tau^k = 0\,, \qquad v_\tau^0 = v_0, \tag{11.137b}$$

to be solved recurrently for $k = 1, \ldots, T/\tau$. Compare (11.137) with (11.125).[44] Consider varying time step τ_k (which might be useful for efficient numerical implementation) and modify (11.125) correspondingly and perform an energy-type estimate by testing (11.137a) by v_τ^k.[45]

11.5 Bibliographical remarks

Doubly nonlinear problems from Sections 11.1–11.2 have, in concrete cases, been thoroughly exposed in Visintin [418, Sect.III.1] and, under the name *pseudoparabolic equations*, in Gajewski, Gröger and Zacharias [168, Chap.V].

In particular, the structure in Section 11.1 has been investigated by Colli and Visintin [101, 104]. Also, differently from Remark 11.10, if Φ is smooth, a variational principle can be devised by ususing the functional[46]

$$u \mapsto \int_0^T \Psi\Big(\frac{\mathrm{d}u}{\mathrm{d}t}\Big) + \Psi^*(f - \Phi'(u)) + \Big\langle f, \frac{\mathrm{d}u}{\mathrm{d}t}\Big\rangle\,\mathrm{d}t + \Phi(u(T)). \tag{11.138}$$

and continue by estimation of $\|\nabla u\|^2_{L^2(\Omega;\mathbb{R}^n)} \leq 2\|\nabla u_0\|^2_{L^2(\Omega;\mathbb{R}^n)} + 2t\int_0^t \|\frac{\partial}{\partial t}\nabla u\|^2_{L^2(\Omega;\mathbb{R}^n)}$ and by Gronwall's inequality.

[42]Hint: Denoting u_μ the weak solution to (11.136), realize that $\|\frac{\partial}{\partial t}u_\mu\|_{L^2(I;W^{1,2}(\Omega))} = \mathscr{O}(1/\sqrt{\mu})$ while the other estimates are independent of $\mu > 0$, so that the term $\int_Q \mu\nabla\frac{\partial}{\partial t}u_\mu \cdot \nabla v\,\mathrm{d}x\mathrm{d}t = \mathscr{O}(\sqrt{\mu})$ and vanishes in the weak formulation if $\mu \to 0$.

[43]Hint: Paraphrase (8.174) to show $\langle B'(u)v, v\rangle = \int_\Omega |\nabla u|^{p-2}|\nabla v|^2 + (p-2)|\nabla u|^{p-4}(\nabla v\cdot\nabla u)^2$ and then use Hölder's inequality to show (11.103).

[44]Hint: Substituting v_τ^k from (11.137b) into (11.137a) gives exactly (11.125a) but (11.137b) is not considered for $k = 0$ so the formal value u_τ^{-1} from (11.125b) is not needed now.

[45]Hint: The second-order difference $(u_\tau^k - 2u_\tau^{k-1} + u_\tau^{k-2})/\tau^2$ turns into a non-symmetric difference $u_\tau^k/\tau_k^2 - u_\tau^{k-1}(\tau_k + \tau_{k-1})/(\tau_k^2\tau_{k-1}) + u_\tau^{k-2}/(\tau_k\tau_{k-1})$.

[46]The construction of (11.54) is by Fenchel identity applied to (Ψ, Ψ^*) at the argument $(\frac{\mathrm{d}u}{\mathrm{d}t}, f - \Phi'(u))$, which yields the equivalence between $\partial\Psi(\frac{\mathrm{d}u}{\mathrm{d}t}) = f - \Phi'(u)$ and that the functional $\int_0^T \Psi(\frac{\mathrm{d}u}{\mathrm{d}t}) + \Psi^*(f - \Phi'(u)) - \langle\frac{\mathrm{d}u}{\mathrm{d}t}, f - \Phi'(u)\rangle\,\mathrm{d}t = \int_0^T \Psi(\frac{\mathrm{d}u}{\mathrm{d}t}) + \Psi^*(f - \Phi'(u)) + \langle\frac{\mathrm{d}u}{\mathrm{d}t}, f - \Phi'(u)\rangle\,\mathrm{d}t + \Phi(u(T)) - \Phi(u_0)$ takes its minimal value, i.e. 0.

For a special quadratic Φ arising in plasticity, see [397] while a general Φ was considered in [398] with Ψ homogeneous degree-1 in both cases, i.e. $\Psi(av) = a\Psi(v)$ for any $a \geq 0$. A more general Ψ was considered in [280]. In fact, a lot of physical applications are based on (11.5) with Ψ homogeneous degree-1. This is related to so-called *rate-independent systems* and requires $q = 1$ in (11.31a)–(11.31b) so that the results presented in Section 11.1.2 do not cover this case. Instead of the system of two variational inequalities (11.32), a mathematically more suitable definition of the solution has been proposed by Mielke and Theil [282] and works merely with energetics of the process $u : [0,T] \to V$. This process is called an *energetic solution* if, besides the initial condition $u(0) = u_0$, it satisfies the stability and the energy inequality in the sense:

$$\forall t{\in}I\ \ \forall v{\in}V: \ \Phi(u(t)) - \langle f(t), u(t)\rangle \leq \Phi(v) - \langle f(t), v\rangle + \Psi\big(u(t){-}v\big), \tag{11.139a}$$

$$\forall t{\geq}s: \quad \Phi(u(t)) - \langle f(t), u(t)\rangle + \mathrm{Var}_\Psi(u; s, t) \leq \Phi(u(s)) - \langle f(s), u(s)\rangle - \int_s^t \Big\langle \frac{\partial f}{\partial \vartheta}, u(\vartheta)\Big\rangle \,\mathrm{d}\vartheta \tag{11.139b}$$

where $\mathrm{Var}_\Psi(u; s, t)$ denotes the total variation of Ψ along the process u during the interval $[s, t]$ defined by $\mathrm{Var}_\Psi(u; s, t) := \sup \sum_{i=1}^{j} \Psi(u(t_{i-1}) - u(t_i))$ with the supremum taken over all $j \in \mathbb{N}$ and over all partitions of $[0, t]$ in the form $0 = t_0 < t_1 < \cdots < t_{j-1} < t_j = t$. Note that this definition does not involve explicitly time derivative $\frac{\mathrm{d}}{\mathrm{d}t}u$ which indeed need not exist in an conventional sense. Cf. Mielke [278, 279] for a survey of the related theory and applications, and for other concepts of solution. Sometimes, a generalization for $\Psi = \Psi(u, \frac{\mathrm{d}}{\mathrm{d}t}u)$ is useful. The special case of a homogeneous degree-1 potential $\Psi(u, \cdot) := [\delta_{K(u)}]^*$ with a convex set $K(u) \subset V$ then leads to the inclusion $\partial_{\mathrm{d}u/\mathrm{d}t}\Psi(u, \frac{\mathrm{d}}{\mathrm{d}t}u) + \Phi'(u) = [\partial\delta_{K(u)}]^{-1}(\frac{\mathrm{d}}{\mathrm{d}t}u) + \Phi'(u) \ni f$, i.e. $\frac{\mathrm{d}}{\mathrm{d}t}u \in N_{K(u)}(f{-}\Phi'(u))$. Processes u governed by such inclusions are called the *sweeping processes*, see e.g. Krejčí at al. [239, 240, 241] and Kunze and Monteiro Marques [246].

The doubly nonlinear structure in Sections 11.2 first occurred probably in Grange and Mignot [187], and was investigated in particular by Aizicovici and Hokkanen [7], Alt and Luckhaus [11] (both even with a possible degeneracy of the parabolic term), DiBenedetto and Showalter [121], Gajewski [164] (with application to semiconductors), Gröger and Nečas [192], Otto [319, 320], Showalter [382], and Stefanelli [395]. A thorough exposition is in the monographs by Hokkanen and Morosanu [207, Chap.10], and Hu and Papageorgiou [209, Part II, Sect.II.5].

The 2nd-order evolution has been addressed by Gajewski et al. [168, Chap.VII], Lions [261, Chap.II.6 and III.6], and Zeidler [427, Chap.33 and 56]. The structure of Theorem 11.33(i) even with B set-valued, arising from a concrete unilateral problem, has been addressed by Jarušek et al. [213]. For both A and B potential and nonlinear in the highest derivatives, e.g., $\frac{\partial^2}{\partial t^2}u - \Delta_p\frac{\partial}{\partial t}u - \Delta_q u = g$, see Bulíček, Málek, and Rajagopal [84] who assumed $p \geq 2$ and $q \leq 2$, improving thus former results by Friedman and Nečas [156]. A similar problem is also in Biazutti [52].

Chapter 12

Systems of equations: particular examples

Just as in steady-state problems, no abstract theory exists universally for a broader class of systems of nonlinear equations.[1] Thus, as in Chapter 6, we confine ourselves to some illustrative examples having straightforward physical motivation and using the previously exposed techniques in a nontrivial manner.

12.1 Thermo-visco-elasticity

The first example is indeed rather nontrivial, illustrating on a rather simple thermo-mechanical system how the previous L^1-type estimates for the heat equation can be executed jointly with the estimates of the remaining part of the system.

We assume a body occupying the domain $\Omega \subset \mathbb{R}^n$, $n \leq 3$, made from isotropic linearly-responding elastic and heat conductive material described in terms of the small strains. Let us briefly derive a thermodynamically consistent system. The departure point is the specific Helmholtz free energy considered here as:

$$\psi(e, \theta) := \frac{\lambda_{\rm e}}{2}\big(\mathrm{Tr}(e)\big)^2 + \mu_{\rm e}|e|^2 - \alpha \mathrm{Tr}(e)\theta - \psi_0(\theta) \tag{12.1}$$

and the dissipation rate:[2]

$$\xi(\dot{e}) := \lambda_{\rm v}\big(\mathrm{Tr}(\dot{e})\big)^2 + 2\mu_{\rm v}|\dot{e}|^2, \qquad \dot{e} = \frac{\partial e}{\partial t}, \tag{12.2}$$

where (and in following formulae):

[1] Some of the previous results, however, can be adopted for systems of a special form simply by considering u vector-valued, see e.g. Ladyzhenskaya et al. [249, Chap.VII].

[2] Up to the factor $\frac{1}{2}$ related to 2-homogeneity, ξ is also a "quasipotential" of dissipative forces.

$u : \Omega \to \mathbb{R}^n$ is the displacement, cf. Example 6.7,
$\theta : \Omega \to \mathbb{R}$ temperature,
$e = e(u) = \frac{1}{2}(\nabla u)^\top + \frac{1}{2}\nabla u$ the small-strain tensor, cf. Example 6.8,
μ_e, λ_e are Lamé constants related to elastic response, cf. (6.23),
$\alpha = \alpha_0(n\lambda_e + 2\mu_e)$ with α_0 the thermal dilatation coefficient,
$\psi_0(\theta)$ the thermal part of the free energy,
$\mathfrak{c}(\theta) := \theta\psi_0''(\theta)$ is heat capacity,
ϱ mass density,
μ_v, λ_v are Lamé constants related to viscous response,
κ heat conduction coefficient,
g the prescribed bulk force,
h the prescribed surface force,
f the external prescribed heat flux through the boundary Γ.

Of course, "Tr(·)" in (12.1) stands for a trace of a square matrix. The particular terms in (12.1) are related respectively to the elastic stored energy, temperature dilatation, and a contribution of chaotic vibrations of the atomic grid (=heat). The quadratic form of ξ in (12.2) is related to linear viscosity.

Moreover, as standard in thermodynamics, we define the specific entropy by so-called Gibbs' relation $s = -\psi_\theta'$ $(\equiv -\frac{\partial}{\partial\theta}\psi)$ and the specific internal energy w by

$$w := \psi + \theta s = \frac{\lambda_e}{2}(\operatorname{div} u)^2 + \mu_e\big|e(u)\big|^2 + \vartheta \quad \text{where } \vartheta = \theta\psi_0'(\theta) - \psi(\theta). \tag{12.3}$$

The elastic and viscous stress tensors are defined as ψ_e' and $\frac{1}{2}\xi'$. The equilibrium equation balances the total stress $\sigma = \psi_e' + \frac{1}{2}\xi'$ with the inertial forces and outer loading g:

$$\varrho\frac{\partial^2 u}{\partial t^2} - \operatorname{div}\sigma = g \quad \text{with} \quad \sigma = \operatorname{div}\Big(\lambda_e u + \lambda_v\frac{\partial u}{\partial t}\Big)\mathbb{I} + 2e\Big(\mu_e u + \mu_v\frac{\partial u}{\partial t}\Big) - \alpha\theta\mathbb{I}, \tag{12.4}$$

where $\mathbb{I} \in \mathbb{R}^{n\times n}$ is the identity matrix. The heat equation then can be obtained from the energy balance requiring that the kinetic energy and the internal energy in a closed system is preserved, cf. (12.11) below. Defining still the heat flux $-\kappa(\theta)\nabla\theta$ (isotropic nonlinear medium), we complete (12.4) by the heat equation

$$\begin{aligned}\theta\frac{\partial s}{\partial t} &= \mathfrak{c}(\theta)\frac{\partial\theta}{\partial t} + \theta\,\alpha\operatorname{div}\frac{\partial u}{\partial t} = \operatorname{div}(\kappa(\theta)\nabla\theta) + \xi\Big(e\big(\frac{\partial u}{\partial t}\big)\Big)\\ &= \operatorname{div}(\kappa(\theta)\nabla\theta) + \lambda_v\big(\operatorname{div}\frac{\partial u}{\partial t}\big)^2 + 2\mu_v\Big|e\big(\frac{\partial u}{\partial t}\big)\Big|^2\end{aligned} \tag{12.5}$$

and finally we choose some boundary conditions, e.g. an unsupported mechanically and thermally loaded body, and initial conditions:

$$\nu\cdot\sigma = h, \qquad \kappa(\theta)\frac{\partial\theta}{\partial\nu} = f \qquad\qquad \text{on } \Sigma, \tag{12.6a}$$

$$u(0,\cdot) = u_0, \qquad \frac{\partial u}{\partial t}(0,\cdot) = v_0, \qquad \theta(0,\cdot) = \theta_0 \qquad \text{on } \Omega. \tag{12.6b}$$

The important fact is that the above procedure satisfies the 2nd thermodynamical law.[3]

We employ the *enthalpy transformation*, cf. Examples 8.71 and 11.27. We denote $\vartheta = \widehat{\mathfrak{c}}(\theta) := \int_0^\theta \mathfrak{c}(\varrho)\mathrm{d}\varrho$ and abbreviate $\gamma(\vartheta) := \widehat{\mathfrak{c}}^{-1}(\vartheta)$ for $\vartheta \geq 0$ and, formally, $\gamma(\vartheta) = 0$ for $\vartheta < 0$. As in Example 8.71, we define $\beta := \widehat{\kappa}\circ\gamma$ with $\widehat{\kappa}$ a primitive of κ. Note that ϑ has the same meaning as in (12.3) up to the constant $\psi_0(0)$. By such substitution, the resulted system then takes the form

$$\varrho\frac{\partial^2 u}{\partial t^2} - \mathrm{div}\Big(\mathrm{div}\big(\lambda_{\mathrm{e}}u+\lambda_{\mathrm{v}}\frac{\partial u}{\partial t}\big)\mathbb{I}+2e\big(\mu_{\mathrm{e}}u+\mu_{\mathrm{v}}\frac{\partial u}{\partial t}\big) - \alpha\gamma(\vartheta)\mathbb{I}\Big) = g, \tag{12.7a}$$

$$\frac{\partial\vartheta}{\partial t} - \Delta\beta(\vartheta) = \lambda_{\mathrm{v}}\big(\mathrm{div}\frac{\partial u}{\partial t}\big)^2 + 2\mu_{\mathrm{v}}\Big|e\big(\frac{\partial u}{\partial t}\big)\Big|^2 - \alpha\gamma(\vartheta)\mathrm{div}\frac{\partial u}{\partial t}, \tag{12.7b}$$

with the boundary and initial conditions

$$\left.\begin{array}{l}\Big(\mathrm{div}\big(\lambda_{\mathrm{e}}u+\lambda_{\mathrm{v}}\frac{\partial u}{\partial t}\big)\mathbb{I}+2e\big(\mu_{\mathrm{e}}u+\mu_{\mathrm{v}}\frac{\partial u}{\partial t}\big) - \alpha\gamma(\vartheta)\mathbb{I}\Big)\cdot\nu = h,\\ \dfrac{\partial\beta(\vartheta)}{\partial\nu} = f\end{array}\right\}\quad\text{on }\Sigma, \tag{12.8a}$$

$$u(0,\cdot) = u_0, \qquad \frac{\partial u}{\partial t}(0,\cdot) = v_0, \qquad \vartheta(0,\cdot) = \gamma^{-1}(\theta_0) \qquad \text{on }\Omega. \tag{12.8b}$$

The energy balance can be obtained formally by multiplication of (12.4) by $\frac{\partial}{\partial t}u$ and (12.5) by 1, and by using Green's formula for (12.4) and (12.5):

$$\frac{\mathrm{d}}{\mathrm{d}t}\Big(\frac{\varrho}{2}\Big\|\frac{\partial u}{\partial t}\Big\|^2_{L^2(\Omega)} + \int_\Omega \varphi(e(u))\,\mathrm{d}x\Big) + \int_\Omega\Big(\xi\big(e(\frac{\partial u}{\partial t})\big) - \alpha\gamma(\vartheta)\mathrm{div}\frac{\partial u}{\partial t}\Big)\mathrm{d}x = \int_\Omega g\cdot\frac{\partial u}{\partial t}\,\mathrm{d}x + \int_\Gamma h\cdot\frac{\partial u}{\partial t}\,\mathrm{d}S, \tag{12.9}$$

$$\frac{\mathrm{d}}{\mathrm{d}t}\int_\Omega\vartheta\mathrm{d}x - \int_\Omega\Big(\xi\big(e(\frac{\partial u}{\partial t})\big) - \alpha\gamma(\vartheta)\mathrm{div}\frac{\partial u}{\partial t}\Big)\mathrm{d}x = \int_\Gamma f\,\mathrm{d}S, \tag{12.10}$$

where $\varphi(e) := \psi(e,0)$. Summing (12.9) with (12.10), we get the total-energy balance:

$$\frac{\mathrm{d}}{\mathrm{d}t}\Big(\underbrace{\frac{\varrho}{2}\Big\|\frac{\partial u}{\partial t}\Big\|^2_{L^2(\Omega)}}_{\text{kinetic energy}} + \underbrace{\int_\Omega w(t,x)\,\mathrm{d}x}_{\text{internal energy}}\Big) := \frac{\mathrm{d}}{\mathrm{d}t}\int_\Omega\frac{\varrho}{2}\Big|\frac{\partial u}{\partial t}\Big|^2 + \varphi(e(u)) + \vartheta\,\mathrm{d}x$$
$$= \underbrace{\int_\Omega g\cdot\frac{\partial u}{\partial t}\,\mathrm{d}x + \int_\Gamma h\cdot\frac{\partial u}{\partial t}\,\mathrm{d}S}_{\text{power of external forces}} + \underbrace{\int_\Gamma f\,\mathrm{d}S}_{\text{power of external heat flux}}. \tag{12.11}$$

[3]Indeed, dividing (12.5) by θ, the Clausius-Duhem inequality reads as:
$$\frac{\mathrm{d}}{\mathrm{d}t}\int_\Omega s(t,x)\,\mathrm{d}x = \int_\Omega\frac{\xi(e(\frac{\partial u}{\partial t}))+\mathrm{div}(\kappa(\theta)\nabla\theta)}{\theta}\,\mathrm{d}x = \int_\Omega\frac{\xi(e(\frac{\partial}{\partial t}u))}{\theta} + \kappa(\theta)\frac{|\nabla\theta|^2}{\theta^2}\mathrm{d}x + \int_\Gamma\frac{f}{\theta}\mathrm{d}S \geq 0$$
provided $\theta > 0$ and $f \geq 0$. For the positivity of temperature θ, cf. Remark 12.10 below.

To prove existence of a weak solution of this system, we apply the Rothe method with a suitable *regularization* to compensate the growth of the non-monotone terms on the right-hand side of the heat equation (12.7b). More specifically, we consider

$$\varrho\frac{u_\tau^k-2u_\tau^{k-1}+u_\tau^{k-2}}{\tau^2}-\mathrm{div}\Big(\mathrm{div}\Big(\lambda_\mathrm{e}u_\tau^k+\lambda_\mathrm{v}\frac{u_\tau^k-u_\tau^{k-1}}{\tau}\Big)\mathbb{I}-\alpha\gamma(\vartheta_\tau^k)\mathbb{I}$$
$$+2e\Big(\mu_\mathrm{e}u_\tau^k+\mu_\mathrm{v}\frac{u_\tau^k-u_\tau^{k-1}}{\tau}\Big)\Big)=g_\tau^k+\tau\mathrm{div}\big(|e(u_\tau^k)|^{\eta-2}e(u_\tau^k)\big), \tag{12.12a}$$

$$\frac{\vartheta_\tau^k-\vartheta_\tau^{k-1}}{\tau}-\Delta\beta(\vartheta_\tau^k)=\lambda_\mathrm{v}\Big(\mathrm{div}\frac{u_\tau^k-u_\tau^{k-1}}{\tau}\Big)^2$$
$$+2\mu_\mathrm{v}\Big|e\Big(\frac{u_\tau^k-u_\tau^{k-1}}{\tau}\Big)\Big|^2-\alpha\gamma(\vartheta_\tau^k)\mathrm{div}\frac{u_\tau^k-u_\tau^{k-1}}{\tau}, \tag{12.12b}$$

with the corresponding boundary and the regularized initial conditions

$$\Big(\mathrm{div}\Big(\lambda_\mathrm{e}u_\tau^k+\lambda_\mathrm{v}\frac{u_\tau^k-u_\tau^{k-1}}{\tau}\Big)\mathbb{I}-\alpha\gamma(\vartheta_\tau^k)\mathbb{I}+\tau|e(u_\tau^k)|^{\eta-2}e(u_\tau^k)$$
$$+2e\Big(\mu_\mathrm{e}u_\tau^k+\mu_\mathrm{v}\frac{u_\tau^k-u_\tau^{k-1}}{\tau}\Big)\Big)\cdot\nu=0,\qquad \frac{\partial\beta(\vartheta_\tau^k)}{\partial\nu}=f_\tau^k\qquad\text{on }\Gamma, \tag{12.13a}$$
$$u_\tau^0=u_{0\tau},\qquad u_\tau^{-1}=u_\tau^0-\tau v_0,\qquad \vartheta_\tau^0=\gamma^{-1}(\theta_0)\qquad\text{on }\Omega. \tag{12.13b}$$

In terms of the original data, we assume:

$$\exists c_{\min}>0,\ \omega>2n/(n{+}2)\ \forall\theta\in\mathbb{R}^+:\ \ \mathfrak{c}(\theta)\ge c_{\min}(1{+}\theta)^{\omega-1}, \tag{12.14a}$$
$$\exists\,\beta_0>0\ \forall\theta\in\mathbb{R}^+:\ \ \kappa(\theta)\ge\beta_0\mathfrak{c}(\theta), \tag{12.14b}$$
$$\varrho>0,\quad \mu_\mathrm{e}>0,\quad \mu_\mathrm{v}>0,\quad \lambda_\mathrm{e}>-2\mu_\mathrm{e}/n,\quad \lambda_\mathrm{v}>-2\mu_\mathrm{v}/n, \tag{12.14c}$$
$$g\in L^2(I;L^{2^{*\prime}}(\Omega;\mathbb{R}^n)),\quad h\in L^2(I;L^{2^{\#\prime}}(\Gamma;\mathbb{R}^n)), \tag{12.14d}$$
$$u_0\in W^{1,2}(\Omega;\mathbb{R}^n),\quad v_0\in L^2(\Omega;\mathbb{R}^n), \tag{12.14e}$$
$$\theta_0\ge0,\quad f\ge0,\quad \widehat{\mathfrak{c}}(\theta_0)\in L^1(\Omega),\quad f\in L^1(\Sigma). \tag{12.14f}$$

Note that (12.14a) requires a slight growth of the heat capacity and conductivity if $n\ge2$ and that it bounds the growth of γ as $|\gamma(\vartheta)|\le C(1{+}\vartheta^{1/\omega})$. Also note that, due to (12.14b), it holds that $\beta'(\vartheta)=[\kappa/\mathfrak{c}](\gamma(\vartheta))\ge\beta_0$.

Lemma 12.1 (Existence of Rothe's solution). *Let* (12.14) *hold and* $\eta>4$, *and let* $u_{0\tau}\in W^{1,\eta}(\Omega;\mathbb{R}^n)$. *Then the boundary-value problem* (12.12)–(12.13) *has a solution* $(u_\tau^k,\vartheta_\tau^k)\in W^{1,\eta}(\Omega;\mathbb{R}^n)\times W^{1,2}(\Omega)$ *such that* $\vartheta_\tau^k\ge0$ *for any* $k=1,...,T/\tau$.

Lemma 12.2 (Energy estimates). *Let* (12.14) *hold and* $\eta>4$, *and let also* $\|u_{0\tau}\|_{W^{1,\eta}(\Omega;\mathbb{R}^n)}=\mathscr{O}(\tau^{-1/\eta})$. *Then:*

$$\|u_\tau\|_{W^{1,\infty}(I;L^2(\Omega;\mathbb{R}^n))\cap L^\infty(I;W^{1,2}(\Omega;\mathbb{R}^n))} \le C, \tag{12.15a}$$
$$\|\bar\vartheta_\tau\|_{L^\infty(I;L^1(\Omega))} \le C, \tag{12.15b}$$
$$\|e(u_\tau)\|_{L^\infty(I;L^\eta(\Omega;\mathbb{R}^{n\times n}))} \le C\tau^{-1/\eta}. \tag{12.15c}$$

Proposition 12.3 (Further estimates). *Under the assumption of Lemma 12.2, it also holds that*

$$\|u_\tau\|_{W^{1,2}(I;W^{1,2}(\Omega;\mathbb{R}^n))} \le C, \tag{12.16a}$$
$$\|\nabla\bar\vartheta_\tau\|_{L^r(Q;\mathbb{R}^n)} \le C_r \qquad \text{with } r < \frac{n+2}{n+1}, \tag{12.16b}$$
$$\|\bar\vartheta_\tau\|_{L^q(Q)} \le C_q \qquad \text{with } q < \frac{n+2}{n}, \tag{12.16c}$$
$$\Big\|\frac{\partial}{\partial t}\Big[\frac{\partial u_\tau}{\partial t}\Big]^{\mathrm{i}}\Big\|_{L^2(I;W^{1,2}(\Omega;\mathbb{R}^n)^*)+L^{\eta'}(I;W^{1,\eta}(\Omega;\mathbb{R}^n)^*)} \le C, \tag{12.16d}$$
$$\Big\|\frac{\partial}{\partial t}\Big[\frac{\partial u_\tau}{\partial t}\Big]^{\mathrm{i}} - \tau\mathrm{div}\big(|e(\bar u_\tau)|^{\eta-2}e(\bar u_\tau)\big)\Big\|_{L^2(I;W^{1,2}(\Omega;\mathbb{R}^n)^*)} \le C, \tag{12.16e}$$
$$\Big\|\frac{\partial\vartheta_\tau}{\partial t}\Big\|_{L^1(I;W^{3,2}(\Omega)^*)} \le C. \tag{12.16f}$$

Proof. We use the test of (12.12b) by $\chi(\vartheta_\tau^k) := 1-(1+\vartheta_\tau^k)^{-\varepsilon}$ as suggested in Remark 9.25 for $p=2$. After summation for $k=1,...,T/\tau$, we obtain

$$\begin{aligned}
L &:= \beta_0\varepsilon\int_Q \frac{|\nabla\bar\vartheta_\tau|^2}{(1+\bar\vartheta_\tau)^{1+\varepsilon}}\,\mathrm{d}x\mathrm{d}t = \beta_0\int_Q \chi'(\bar\vartheta_\tau)|\nabla\bar\vartheta_\tau|^2\,\mathrm{d}x\mathrm{d}t\\
&\le \int_Q \chi'(\bar\vartheta_\tau)\beta'(\bar\vartheta_\tau)\nabla\bar\vartheta_\tau\cdot\nabla\bar\vartheta_\tau\,\mathrm{d}x\mathrm{d}t = \int_Q \nabla\beta(\bar\vartheta_\tau)\cdot\nabla\chi(\bar\vartheta_\tau)\,\mathrm{d}x\mathrm{d}t\\
&\le \int_Q \nabla\beta(\bar\vartheta_\tau)\cdot\nabla\chi(\bar\vartheta_\tau)\,\mathrm{d}x\mathrm{d}t + \int_\Omega \widehat\chi(\bar\vartheta_\tau(T,\cdot))\,\mathrm{d}x\\
&\le \int_\Omega \widehat\chi(\vartheta_0)\,\mathrm{d}x + \int_Q \bar r_\tau\chi(\bar\vartheta_\tau)\,\mathrm{d}x\mathrm{d}t + \int_\Sigma \bar f_\tau\chi(\bar\vartheta_\tau)\,\mathrm{d}S\mathrm{d}t\\
&\le \|\gamma(\theta_0)\|_{L^1(\Omega)} + \|\bar r_\tau\|_{L^1(Q)} + \|\bar f_\tau\|_{L^1(\Sigma)} =: C_0 + \|\bar r_\tau\|_{L^1(Q)},
\end{aligned} \tag{12.17}$$

where $\widehat\chi$ is a primitive of χ such that $\widehat\chi(0)=0$ and where we abbreviated the heat sources $\bar r_\tau := \xi(e(\frac{\partial}{\partial t}u_\tau)) - \alpha\gamma(\bar\vartheta_\tau)\mathrm{div}\frac{\partial}{\partial t}u_\tau$. The inequality on the 4th line of (12.17) uses monotonicity of χ hence convexity of $\widehat\chi$ so that the "*discrete chain rule*" holds:

$$\frac{\widehat\chi(\vartheta_\tau^k) - \widehat\chi(\vartheta_\tau^{k-1})}{\tau} \le \frac{\vartheta_\tau^k - \vartheta_\tau^{k-1}}{\tau}\chi(\vartheta_\tau^k).$$

Further, we test (12.12b) by $u_\tau^k - u_\tau^{k-1}$, sum it for $k=1,\dots,T/\tau$, and add to (12.17) with a sufficiently big weight to see the dissipation $\xi(e(\frac{\partial}{\partial t}u_\tau))$ on the left-hand side. Thus, we obtain

$$\xi\Big(e\big(\frac{\partial u_\tau}{\partial t}\big)\Big) + L \le C_1 + C_1\Big\|\alpha\gamma(\bar\vartheta_\tau)\mathrm{div}\frac{\partial u_\tau}{\partial t}\Big\|_{L^1(Q)} \tag{12.18}$$

with some C_1. By calculations like[4] (9.53) with $p = 2$ and $q = 1-\varepsilon$, for $1 \le r < 2$, we obtain

$$\int_Q |\nabla\bar\vartheta_\tau|^r \mathrm{d}x\mathrm{d}t \le \Big(\frac{L}{\beta_0\varepsilon}\Big)^{r/2}\Big(\int_0^T \big\|1+\bar\vartheta_\tau(t,\cdot)\big\|^{(1+\varepsilon)r/(2-r)}_{L^{(1+\varepsilon)r/(2-r)}(\Omega)}\mathrm{d}t\Big)^{(2-r)/2}. \tag{12.19}$$

Then we use (9.54) to obtain

$$\begin{aligned}
&\big\|1+\bar\vartheta_\tau(t,\cdot)\big\|_{L^{(1+\varepsilon)r/(2-r)}(\Omega)}\\
&\quad\le C_{\rm GN}\Big(\big\|1+\bar\vartheta_\tau(t,\cdot)\big\|_{L^1(\Omega)} + \big\|\nabla\bar\vartheta_\tau(t,\cdot)\big\|_{L^r(\Omega;\mathbb{R}^n)}\Big)^\lambda \big\|1+\bar\vartheta_\tau(t,\cdot)\big\|^{1-\lambda}_{L^1(\Omega)}\\
&\quad\le C_{\rm GN}\big(\mathrm{meas}_n(\Omega)+C\big)^{1-\lambda}\Big(\mathrm{meas}_n(\Omega) + C + \big\|\nabla\bar\vartheta_\tau(t,\cdot)\big\|_{L^r(\Omega;\mathbb{R}^n)}\Big)^\lambda
\end{aligned} \tag{12.20}$$

with C from (12.15b), provided (9.55) holds, i.e. provided

$$\frac{2-r}{(1+\varepsilon)r} \ge \lambda\Big(\frac{1}{r} - \frac{1}{n}\Big) + 1 - \lambda \qquad \text{with} \qquad 0 < \lambda \le 1. \tag{12.21}$$

We raise (12.20) to the power $(1+\varepsilon)r/(2-r)$, use it in (12.19), and choose $\lambda := (2-r)/(1+\varepsilon)$. As in (9.56), we obtain

$$\Big(\int_0^T \big\|1+\bar\vartheta_\tau(t,\cdot)\big\|^{(1+\varepsilon)r/(2-r)}_{L^{(1+\varepsilon)r/(2-r)}(\Omega)}\mathrm{d}t\Big)^{\frac{2-r}{2}} \le C_2 + C_2\Big(\int_Q |\nabla\bar\vartheta_\tau|^r \mathrm{d}x\mathrm{d}t\Big)^{\frac{2-r}{2}}. \tag{12.22}$$

Merging (12.19) with (12.17) and with (12.22) gives an estimate of the type $\|\nabla\bar\vartheta_\tau\|^r_{L^r(Q;\mathbb{R}^n)}/(1 + \|\nabla\bar\vartheta_\tau\|^{r(1-r/2)}_{L^r(Q;\mathbb{R}^n)}) \le C(1 + \|\bar r_\tau\|_{L^1(Q)})^{r/2}$, i.e.,

$$\|\nabla\bar\vartheta_\tau\|^r_{L^r(Q;\mathbb{R}^n)} - C_3 \le \left(\frac{\|\nabla\bar\vartheta_\tau\|^r_{L^r(Q;\mathbb{R}^n)}}{1 + \|\nabla\bar\vartheta_\tau\|^{r(1-r/2)}_{L^r(Q;\mathbb{R}^n)}}\right)^{2/r} \le C_3\big(1 + \|\bar r_\tau\|_{L^1(Q)}\big) \tag{12.23}$$

for C_3 large enough. Substituting our choice of $\lambda := (2-r)/(1+\varepsilon)$ into (12.21)[5], one gets, after some algebra, the condition

$$r \le \frac{2 + n - \varepsilon n}{1 + n}\,. \tag{12.24}$$

[4]We use $1+\bar\vartheta_\tau$ instead of u in (9.53). In fact, $C_1/(q-1)$ in (9.53) is now $L/(\beta_0\varepsilon)$ so that we can indeed consider $q < 1$.

[5]Note that $0 < \lambda < 1$ needed in (12.21) is automatically ensured by $1 \le r < 2$ and $\varepsilon > 0$.

For $0 < \omega \leq 2$, using that $\gamma(\cdot)$ has a growth at most $1/\omega$, we can estimate the last term in (12.18) as

$$\begin{aligned}\Big|\int_Q \alpha\gamma(\bar\vartheta_\tau)\mathrm{div}\frac{\partial u_\tau}{\partial t}\,\mathrm{d}x\mathrm{d}t\Big| &\leq C\big\|\gamma(\bar\vartheta_\tau)\big\|^2_{L^2(Q)} + \delta\Big\|\frac{\partial e(u_\tau)}{\partial t}\Big\|^2_{L^2(Q;\mathbb{R}^{n\times n})}\\ &\leq C_\delta + C_\delta\big\|\bar\vartheta_\tau\big\|^{2/\omega}_{L^{2/\omega}(Q)} + \delta\Big\|\frac{\partial e(u_\tau)}{\partial t}\Big\|^2_{L^2(Q;\mathbb{R}^{n\times n})}\end{aligned} \tag{12.25}$$

and absorb the last term in the left-hand side of (12.18) by choosing $\delta < 2C_2\varepsilon_1$. For $\omega \geq 2$, we can simply make this estimate like in the case $\omega = 2$. Further, by the Gagliardo-Nirenberg inequality, we estimate:

$$\big\|\bar\vartheta_\tau\big\|_{L^{2/\omega}(\Omega)} \leq K_{\rm GN}\big\|\bar\vartheta_\tau\big\|^{1-\mu}_{L^1(\Omega)}\Big(\big\|\bar\vartheta_\tau\big\|_{L^1(\Omega)} + \big\|\nabla\bar\vartheta_\tau\big\|_{L^r(\Omega;\mathbb{R}^n)}\Big)^\mu \tag{12.26}$$

for

$$\frac{\omega}{2} \geq \mu\Big(\frac{1}{r} - \frac{1}{n}\Big) + 1 - \mu, \qquad 0 \leq \mu < 1, \tag{12.27}$$

and, by (12.19), (12.22), and the first estimate in (12.23), we have $\|\nabla\bar\vartheta_\tau\|_{L^r(\Omega;\mathbb{R}^d)} \leq C_3 + C_2^{2/r}L$ so that we can estimate

$$\begin{aligned}C_\delta\big\|\bar\vartheta_\tau\big\|^{2/\omega}_{L^{2/\omega}(\Omega)} &\leq C_\delta K_{\rm GN}^{2/\omega}C_3^{2(1-\mu)/\omega}\big(C_3 + \big\|\nabla\bar\vartheta_\tau\big\|_{L^r(\Omega;\mathbb{R}^n)}\big)^{2\mu/\omega}\\ &\leq C'_\delta + \delta\big\|\nabla\bar\vartheta_\tau\big\|^r_{L^r(\Omega;\mathbb{R}^n)} \leq C'_\delta + \delta C_3 + \delta C_2^{2/r}L\end{aligned} \tag{12.28}$$

provided $2\mu/\omega < r$, i.e.,

$$\frac{\omega}{2} > \frac{\mu}{r}. \tag{12.29}$$

The last term in (12.28) is to be absorbed in the left-hand side of (12.18). The optimal choice of μ makes the right-hand sides of (12.27) and (12.29) mutually equal, which gives $\mu = n/(n{+}1)$. Note that always $0 < \mu < 1$, as required for (12.26). Taking into account that $r < (n{+}2)/(n{+}1)$, from (12.27) (or, equally, from (12.29)) we obtain

$$\omega > \frac{2n}{n+2}. \tag{12.30}$$

It eventually gives the estimates (12.16a) and (12.16b).

The estimate (12.16c) arises by a suitable interpolation between (12.15b) and (12.16b). The remaining "dual" estimates in (12.16) follows from the already obtained ones. □

Proposition 12.4 (Convergence for $\tau \to 0$). *Let (12.14) hold and $\eta > 4$, and let also $u_{0\tau} \to u_0$ in $W^{1,2}(\Omega;\mathbb{R}^n)$ and even $\|u_{0\tau}\|_{W^{1,\eta}(\Omega;\mathbb{R}^n)} = o(\tau^{-1/\eta})$. Then there is a subsequence such that*

$$u_\tau \to u \qquad \text{strongly in } W^{1,2}(I;W^{1,2}(\Omega;\mathbb{R}^n)), \tag{12.31a}$$

$$\bar\vartheta_\tau \to \vartheta \qquad \text{strongly in } L^q(Q) \text{ with any } 1 \le q < (n{+}2)/n, \tag{12.31b}$$

and any (u,ϑ) obtained by this way is a weak solution to the initial-boundary-value problem (12.7)–(12.8). In particular, (12.7)–(12.8) has a weak solution.

Proof. Choose a weakly converging subsequence in the topology of the estimates (12.16). By the interpolated Aubin-Lions' Lemma 7.8, combining (12.15b), (12.16b) and (12.16f), one obtains (12.31b).

Like in (11.109) with Remark 11.37, we use the by-part summation (11.126) to obtain the identity

$$\begin{aligned}
&\int_\Omega \varrho\frac{\partial u_\tau}{\partial t}(T)\cdot v_\tau(T)\,\mathrm{d}x - \int_\tau^T\!\!\int_\Omega \varrho\frac{\partial u_\tau}{\partial t}(\cdot-\tau)\cdot\frac{\partial v_\tau}{\partial t}\,\mathrm{d}x\mathrm{d}t\\
&\quad + \int_Q \mathrm{div}\Big(\lambda_\mathrm{e}\bar u_\tau + \lambda_\mathrm{v}\frac{\partial u_\tau}{\partial t} - \alpha\gamma(\bar\vartheta_\tau)\mathbb{I}\Big)\mathrm{div}\,\bar v_\tau\\
&\quad + 2e\Big(\mu_\mathrm{e}\bar u_\tau{+}\mu_\mathrm{v}\frac{\partial u_\tau}{\partial t} + \tau|e(\bar u_\tau)|^{\eta-2}e(\bar u_\tau)\Big){:}e(\bar v_\tau)\,\mathrm{d}x\mathrm{d}t\\
&\quad = \int_\Omega v_0\cdot v_\tau(\tau)\,\mathrm{d}x + \int_Q \bar g_\tau\cdot\bar v_\tau\,\mathrm{d}x\mathrm{d}t + \int_\Sigma \bar h_\tau\cdot\bar v_\tau\,\mathrm{d}x\mathrm{d}t,
\end{aligned}\tag{12.32}$$

where $\bar v_\tau$ and v_τ is the piecewise constant and the affine interpolants of the test function values $\{v(k\tau)\}_{k=1}^{T/\tau}$.

For v smooth, by (12.15c), one has

$$\Big|\int_Q \tau|e(\bar u_\tau)|^{\eta-2}e(\bar u_\tau){:}e(v)\,\mathrm{d}x\mathrm{d}t\Big| \le \tau\big\|e(\bar u_\tau)\big\|_{L^\eta(Q;\mathbb{R}^{n\times n})}^{\eta-1}\big\|e(v)\big\|_{L^\eta(Q;\mathbb{R}^{n\times n})} = \mathscr{O}(\tau^{1/\eta}) \tag{12.33}$$

and thus can see that the regularizing η-term disappears in the limit for $\tau \to 0$. Altogether, we proved the convergence of (12.32) to the weak formulation of the mechanical part, namely

$$\begin{aligned}
&\int_\Omega \varrho\frac{\partial u}{\partial t}(T)\cdot v(T)\,\mathrm{d}x - \int_Q \varrho\frac{\partial u}{\partial t}\cdot\frac{\partial v}{\partial t} + \mathrm{div}\Big(\lambda_\mathrm{e}u + \lambda_\mathrm{v}\frac{\partial u}{\partial t} - \alpha\gamma(\vartheta)\mathbb{I}\Big)\mathrm{div}\,v\\
&+ 2e\Big(\mu_\mathrm{e}\bar u_\tau{+}\mu_\mathrm{v}\frac{\partial u}{\partial t}\Big){:}e(v)\,\mathrm{d}x\mathrm{d}t = \int_\Omega v_0\cdot v(0)\,\mathrm{d}x + \int_Q g\cdot v\,\mathrm{d}x\mathrm{d}t + \int_\Sigma h\cdot v\,\mathrm{d}x\mathrm{d}t.
\end{aligned}\tag{12.34}$$

By (12.16e), it holds that

$$\varrho\frac{\partial}{\partial t}\Big[\frac{\partial u_\tau}{\partial t}\Big]^{\mathrm{i}} - \tau\mathrm{div}\big(|e(\bar u_\tau)|^{\eta-2}e(\bar u_\tau)\big) \to \zeta \quad \text{weakly in } L^2(I;W^{1,2}(\Omega;\mathbb{R}^n)^*) \tag{12.35}$$

and, like in Exercise 8.85, we can show that $\zeta = \varrho\frac{\partial^2 u}{\partial t^2}$.

It is essential that $\varrho\frac{\partial^2 u}{\partial t^2}$ is in duality with $\frac{\partial u}{\partial t}$, so that we can legally substitute $v = \frac{\partial u}{\partial t}$ into (12.34) to obtain mechanical-energy equality

$$\int_\Omega \frac{\varrho}{2}\Big|\frac{\partial u}{\partial t}(T)\Big|^2 + \varphi\big(e(u(T))\big)\,\mathrm{d}x + \int_Q \xi\Big(e\big(\frac{\partial u}{\partial t}\big)\Big)\,\mathrm{d}x\mathrm{d}t = \int_\Omega \frac{\varrho}{2}|v_0|^2 + \varphi(e(u_0))\,\mathrm{d}x + \int_Q g\cdot\frac{\partial u}{\partial t} + \alpha\gamma(\vartheta)\mathrm{div}\frac{\partial u}{\partial t}\,\mathrm{d}x\mathrm{d}t + \int_\Sigma h\cdot\frac{\partial u}{\partial t}\,\mathrm{d}S\mathrm{d}t \tag{12.36}$$

with $\varphi(e) = \psi(e,0)$ as in (12.9).

For the limit passage in the heat equation, we need to prove strong convergence of $e(\frac{\partial}{\partial t}u_\tau)$ in $L^2(Q;\mathbb{R}^{n\times n})$. We use

$$\begin{aligned}
\int_Q \xi\Big(e\big(\frac{\partial u}{\partial t}\big)\Big)\,\mathrm{d}x\mathrm{d}t &\le \liminf_{\tau\to0}\int_Q \xi\Big(e\big(\frac{\partial u_\tau}{\partial t}\big)\Big)\,\mathrm{d}x\mathrm{d}t \le \limsup_{\tau\to0}\int_Q \xi\Big(e\big(\frac{\partial u_\tau}{\partial t}\big)\Big)\,\mathrm{d}x\mathrm{d}t\\
&\le \limsup_{\tau\to0}\int_\Omega \frac{\varrho}{2}|v_0|^2 - \frac{\varrho}{2}\Big|\frac{\partial u_\tau}{\partial t}(T)\Big|^2 + \varphi(e(u_{0\tau})) - \varphi\big(e(u_\tau(T))\big) + \frac{\tau}{\eta}\big|e(u_{0\tau})\big|^\eta\\
&\qquad - \frac{\tau}{\eta}\big|e(u_\tau(T))\big|^\eta\mathrm{d}x + \int_Q g\cdot\frac{\partial u_\tau}{\partial t} + \alpha\gamma(\bar\vartheta_\tau)\mathrm{div}\frac{\partial u_\tau}{\partial t}\,\mathrm{d}x\mathrm{d}t + \int_\Sigma h\cdot\frac{\partial u_\tau}{\partial t}\,\mathrm{d}S\mathrm{d}t\\
&\le \int_\Omega \frac{\varrho}{2}|v_0|^2 - \frac{\varrho}{2}\Big|\frac{\partial u}{\partial t}(T)\Big|^2 + \varphi(e(u_0)) - \varphi\big(e(u(T))\big)\,\mathrm{d}x\\
&\qquad + \int_Q g\cdot\frac{\partial u}{\partial t} + \alpha\gamma(\vartheta)\mathrm{div}\frac{\partial u}{\partial t}\,\mathrm{d}x\mathrm{d}t + \int_\Sigma h\cdot\frac{\partial u}{\partial t}\,\mathrm{d}S\mathrm{d}t = \int_Q \xi\Big(e\big(\frac{\partial u}{\partial t}\big)\Big)\,\mathrm{d}x\mathrm{d}t.
\end{aligned}$$

Note that the last equality is exactly (12.36). Thus $\lim_{\tau\to0}\int_Q \xi(e(\frac{\partial}{\partial t}u_\tau))\,\mathrm{d}x\mathrm{d}t = \int_Q \xi(e(\frac{\partial}{\partial t}u))\,\mathrm{d}x\mathrm{d}t$. By (12.14c), the quadratic form ξ is coercive and thus $(\int_Q \xi(e(\cdot))\,\mathrm{d}x\mathrm{d}t)^{1/2}$ is an equivalent norm on $L^2(Q;\mathbb{R}^{n\times n}_{\mathrm{sym}})$ which keeps it uniformly convex. Therefore $e(\frac{\partial}{\partial t}u_\tau) \to e(\frac{\partial}{\partial t}u)$ strongly in $L^2(Q;\mathbb{R}^{n\times n})$. Then the limit passage in the semi-linear heat equation is simple.

In particular, using Lemma 12.1 and the above arguments, some weak solution to (12.7)–(12.8) indeed exists because, due to the qualification (12.14e) of u_0, the regularization $u_{0\tau}$ with all above required properties always exists. $\square$

Exercise 12.5. Prove Lemma 12.1 by using coercivity[6] and pseudomonotonicity of the underlying mapping and then using Brézis' Theorem 2.6. Further prove non-negativity[7] of temperature, i.e. $\vartheta_\tau^k \ge 0$.

[6]Hint: test (12.12a) by u_τ^k and (12.12b) by ϑ_τ^k and perform the a-priori estimate. Estimate the nonmonotone terms as $\int_\Omega |e(u_\tau^k)|^2\vartheta_\tau^k\mathrm{d}x \le C_{\varepsilon,\eta} + \varepsilon\|e(u_\tau^k)\|^\eta_{L^\eta(\Omega;\mathbb{R}^{n\times n})} + \varepsilon\|\vartheta_\tau^k\|^2_{L^2(\Omega)}$ provided $\eta > 0$, and also $|\int_\Omega \gamma(\vartheta_\tau^k)\mathrm{div}(u_\tau^k)\vartheta_\tau^k\mathrm{d}x| \le C_{\varepsilon,\eta} + \varepsilon\|e(u_\tau^k)\|^\eta_{L^\eta(\Omega;\mathbb{R}^{n\times n})} + \varepsilon\|\vartheta_\tau^k\|^2_{L^2(\Omega)}$ when realizing that $\gamma(\cdot)$ as at most linear growth, and similarly for the term $\int_\Omega \gamma(\vartheta_\tau^k)\mathrm{div}\,u_\tau^k\mathrm{d}x$.

[7]Hint: test (12.12b) by $(\vartheta_\tau^k)^-$ and realize that $\gamma(\vartheta) = 0$ for $\vartheta \le 0$.

Exercise 12.6. Prove Lemma 12.2 by imitating (12.11).[8]

Exercise 12.7. Make the interpolation between (12.15b) and (12.16b) in such a way so as to obtain (12.16c).

Exercise 12.8. Prove the strong convergence of $e(\frac{\partial}{\partial t}u_\tau)$ in $L^2(Q;\mathbb{R}^{n\times n})$ by the direct estimate of $e(\frac{\partial}{\partial t}(u_\tau-u))$ similar to (8.217), using also Remark 8.11.

Exercise 12.9. Assuming $\omega > 2$ instead of $\omega > 2n/(n+2)$ in (12.14a), prove the estimates (12.15) and (12.16a) directly without the interpolation (12.17)–(12.28) by making the test of (12.12a) by $(u_\tau^k-u_\tau^{k-1})/\tau$ and of (12.12b) by $1/2$. [9]

Remark 12.10 (*Positivity of temperature*)**.** It is a rather delicate interplay between possible cooling via the adiabatic term and sufficient heating via the dissipative term which guarantees positivity of temperature in thermally coupled systems, cf. [102, 146]. Here one can adapt the procedure from Remark 9.28. One can estimate (12.7b) as

$$\begin{aligned}\frac{\partial\vartheta}{\partial t}-\Delta\beta(\vartheta)&=\lambda_{\rm v}\big(\mathrm{div}\frac{\partial u}{\partial t}\big)^2+2\mu_{\rm v}\Big|e\big(\frac{\partial u}{\partial t}\big)\Big|^2-\alpha\gamma(\vartheta)\mathrm{div}\frac{\partial u}{\partial t}\\&\ge\frac{\lambda_{\rm v}}{2}\big(\mathrm{div}\frac{\partial u}{\partial t}\big)^2+2\mu_{\rm v}\Big|e\big(\frac{\partial u}{\partial t}\big)\Big|^2-\frac{\alpha}{2}\Big|\gamma(\vartheta)\Big|^2\ge-\frac{\alpha}{2}C|\vartheta|^2,\qquad(12.37)\end{aligned}$$

where C comes from the growth restriction $|\gamma(\vartheta)|\le C\vartheta$ which is ensured by (12.14a) and the definition $\gamma=\widehat{\mathfrak{c}}^{-1}$. Assuming $\vartheta_0(\cdot)\ge\vartheta_{0,\min}>0$, we compare (12.37) with a solution to the Riccati ordinary-differential equation $\frac{\mathrm{d}}{\mathrm{d}t}\chi+\frac{\alpha}{2}C\chi^2=0$ which, for $\chi(0)=\vartheta_{0,\min}>0$, gives a sub-solution of the heat equation. This initial-value problem has the solution $\chi(t)=2/(\alpha Ct+2/\vartheta_{0,\min})\ge 2/(\alpha CT+2/\vartheta_{0,\min})>0$. Considering $\vartheta_{\rm sub}(t,x)=\chi(t)$, we can subtract (12.37) from $\frac{\partial}{\partial t}\vartheta_{\rm sub}+\frac{\alpha}{2}C\vartheta_{\rm sub}^2=0$, test by $(\vartheta-\vartheta_{\rm sub})^-\le 0$, integrate over Ω at each $t\in I$, and use Green's formula. Thus we obtain

$$\begin{aligned}\frac{1}{2}\frac{\mathrm{d}}{\mathrm{d}t}\int_\Omega\big((\vartheta-\vartheta_{\rm sub})^-\big)^2\mathrm{d}x&\le\int_\Omega(\vartheta-\vartheta_{\rm sub})^-\frac{\partial}{\partial t}(\vartheta-\vartheta_{\rm sub})\,\mathrm{d}x\\&\le\int_\Omega\frac{\alpha}{2}C\big(\vartheta^2-\vartheta_{\rm sub}^2\big)\big(\vartheta-\vartheta_{\rm sub}\big)^--\beta'(\vartheta)\nabla\vartheta\cdot\nabla\big(\vartheta-\vartheta_{\rm sub}\big)^-\mathrm{d}x\\&\qquad-\int_\Gamma f\big(\vartheta-\vartheta_{\rm sub}\big)^-\mathrm{d}S\le 0,\qquad(12.38)\end{aligned}$$

where the last inequality is also due to $\vartheta\ge 0$ proved previously, and also due to the assumption $f\ge 0$ a.e. in Σ, and eventually also due to

$$\begin{aligned}\beta'(\vartheta)\nabla\vartheta\cdot\nabla\big(\vartheta-\vartheta_{\rm sub}\big)^-&=\beta'(\vartheta)\nabla(\vartheta-\vartheta_{\rm sub})\cdot\nabla\big(\vartheta-\vartheta_{\rm sub}\big)^-\\&=\beta'(\vartheta)\nabla(\vartheta-\vartheta_{\rm sub})^-\cdot\nabla\big(\vartheta-\vartheta_{\rm sub}\big)^-\ge 0\qquad(12.39)\end{aligned}$$

[8] Hint: test (12.12a) by $u_\tau^k-u_\tau^{k-1}$ and (12.12b) by 1 and realize the cancellation effects of the heat sources like in (12.11). Use also that ϑ_τ is bounded from below, as already proved in Lemma 12.1.

[9] Hint: Like (12.25), estimate $|\int_\Omega\alpha\gamma(\bar\vartheta_\tau)\mathrm{div}\frac{\partial u_\tau}{\partial t}\,\mathrm{d}x|\le C_\delta+\delta\|\bar\vartheta_\tau\|_{L^1(\Omega)}+\delta\|e(\frac{\partial u_\tau}{\partial t})\|^2_{L^2(\Omega;\mathbb{R}^{n\times n})}$.

a.e. on Ω; cf. Proposition 1.28. Realizing the initial condition $(\vartheta(0){-}\vartheta_{\rm sub}(0))^- = (\vartheta_0{-}\vartheta_{0,\min}(0))^- = 0$, we easily conclude that $(\vartheta{-}\vartheta_{\rm sub})^- = 0$ a.e. on Q, whence

$$\vartheta(t,x) \ge \vartheta_{\rm sub}(t) \ge \frac{2}{\alpha CT{+}2/\vartheta_{0,\min}} > 0 \quad \text{for a.a. } (t,x){\in}Q. \tag{12.40}$$

Remark 12.11 (*Galerkin method*). Applying the Galerkin method to a thermally coupled system is not so straightforward because the "nonlinear tests" by ϑ^- and by $\chi(\vartheta)$ (used in Exercise 12.5 and in (12.17), respectively) cannot be executed on finite-dimensional subspaces. Anyhow, careful regularization in the mechanical part controlled independently of the Galerkin approximation helps. Here, as in (12.12), one can consider a Galerkin approximation of the system:

$$\varrho\frac{\partial^2 u_\varepsilon}{\partial t^2} - {\rm div}\Big({\rm div}\big(\lambda_{\rm e}u_\varepsilon{+}\lambda_{\rm v}\frac{\partial u_\varepsilon}{\partial t}\big)\mathbb{I}{+}2e\big(\mu_{\rm e}u_\varepsilon{+}\mu_{\rm v}\frac{\partial u_\varepsilon}{\partial t}\big) - \alpha\gamma(\vartheta_\varepsilon)\mathbb{I}\Big)$$
$$= g + \varepsilon\,{\rm div}\Big(\Big|e\big(\frac{\partial u_\varepsilon}{\partial t}\big)\Big|^{\eta-2} e\big(\frac{\partial u_\varepsilon}{\partial t}\big)\Big), \tag{12.41a}$$
$$\frac{\partial\vartheta_\varepsilon}{\partial t} - \Delta\beta(\vartheta_\varepsilon) = \lambda_{\rm v}\big({\rm div}\frac{\partial u_\varepsilon}{\partial t}\big)^2 + 2\mu_{\rm v}\Big|e\big(\frac{\partial u_\varepsilon}{\partial t}\big)\Big|^2 - \alpha\gamma(\vartheta_\varepsilon){\rm div}\frac{\partial u_\varepsilon}{\partial t} \tag{12.41b}$$

with the corresponding boundary conditions with a regularized heat flux $f_\varepsilon \in L^2(I;L^{2^{\#'}}(\Gamma))$ and with a regularized initial condition $\vartheta(0) = \vartheta_{0,\varepsilon} \in W^{1,2}(\Omega)$. One then proceeds in the following steps. First, testing the particular equations in (12.41) by u_ε and ϑ_ε (meant in its Galerkin approximation), we obtain

$$\frac{\rm d}{{\rm d}t}\Big(\frac{\varrho}{2}\Big\|\frac{\partial u_\varepsilon}{\partial t}\Big\|^2_{L^2(\Omega;\mathbb{R}^n)} + \int_\Omega \varphi(e(u_\varepsilon))\,{\rm d}x + \frac12\big\|\vartheta_\varepsilon\big\|^2_{L^2(\Omega)}\Big)$$
$$+ \int_\Omega \xi\Big(e\big(\frac{\partial u_\varepsilon}{\partial t}\big)\Big)\,{\rm d}x + \varepsilon\Big\|e\big(\frac{\partial u_\varepsilon}{\partial t}\big)\Big\|^\eta_{L^\eta(\Omega;\mathbb{R}^{n\times n})} + \beta_0\big\|\nabla\vartheta_\varepsilon\big\|^2_{L^2(\Omega;\mathbb{R}^n)}$$
$$= \int_\Omega \lambda_{\rm v}\big({\rm div}\frac{\partial u_\varepsilon}{\partial t}\big)^2\vartheta_\varepsilon + 2\mu_{\rm v}\Big|e\big(\frac{\partial u_\varepsilon}{\partial t}\big)\Big|^2\vartheta_\varepsilon + \alpha\gamma(\vartheta_\varepsilon)(1{-}\vartheta_\varepsilon){\rm div}\frac{\partial u_\varepsilon}{\partial t}$$
$$+ g\cdot\frac{\partial u_\varepsilon}{\partial t}\,{\rm d}x + \int_\Gamma h\cdot\frac{\partial u_\varepsilon}{\partial l} + f_\varepsilon\vartheta_\varepsilon\,{\rm d}S$$
$$\le C_{\varepsilon,\eta} + \frac{\varepsilon}{2}\Big\|e\big(\frac{\partial u_\varepsilon}{\partial t}\big)\Big\|^\eta_{L^\eta(\Omega;\mathbb{R}^{n\times n})} + \big\|\vartheta_\varepsilon\big\|^2_{L^2(\Omega)} + \int_\Omega g\cdot\frac{\partial u_\varepsilon}{\partial t}\,{\rm d}x + \int_\Gamma h\cdot\frac{\partial u_\varepsilon}{\partial t} + f_\varepsilon\vartheta_\varepsilon\,{\rm d}S \tag{12.42}$$

with $C_{\varepsilon,\eta} < +\infty$ depending on ε and η, and with $\varphi(e) = \psi(e,0)$ as in (12.9) and ξ from (12.2). The last inequality in (12.42) holds if η is sufficiently large, namely $\eta > \max(4, 2/(1{-}1/\omega))$. By the usual Hölder/Young-inequality treatment of the last integrals in (12.42) and by Gronwall's inequality, we get estimates of the Galerkin approximation of u_ε bounded in $W^{1,\infty}(I;L^2(\Omega;\mathbb{R}^n)) \cap W^{1,\eta}(I;W^{1,\eta}(\Omega;\mathbb{R}^n))$ and of ϑ_ε bounded in $L^\infty(I;L^2(\Omega)) \cap L^2(I;W^{1,2}(\Omega))$. From (12.41b), we then get an estimate of the time derivative of the Galerkin approximation of ϑ_ε. This allows us to pass to the limit in the Galerkin approximation to obtain a weak solution $(u_\varepsilon, \vartheta_\varepsilon)$ to (12.41). Then we can perform the "nonlinear tests" of (12.41b) by ϑ_ε^- and by $\chi(\vartheta_\varepsilon)$ to obtain estimates analogous to (12.16), and to converge $\varepsilon \to 0$ analogously as we did for $\tau \to 0$ before.

Let us still recall a variation of this procedure based on a separate Galerkin approximation of the thermal and the mechanical parts (again combined with some regularization) and make a limit passage successively: first in the heat part, then execute the needed nonlinear estimates (and suppress possible regularization), and eventually make a limit passage in the mechanical part, cf. [78, 368].

Remark 12.12 (*Semi-implicit Rothe method*)**.** The Rothe method (12.12) can be modified to make *decoupling* of the problem (like in Remark 8.25 but now little differently), which yields a variational structure at each time level (like in Exercise 8.74) and which also suggests (after an additional spatial discretisation) an efficient numerical strategy. One can even avoid the regularization by the term $\tau\mathrm{div}|e(u)|^{\eta-2}e(u)$ used in (12.12). Namely, we devise:

$$\varrho\frac{u_\tau^k-2u_\tau^{k-1}+u_\tau^{k-2}}{\tau^2} - \mathrm{div}\bigg(\mathrm{div}\Big(\lambda_{\mathrm{e}}u_\tau^k+\lambda_{\mathrm{v}}\frac{u_\tau^k-u_\tau^{k-1}}{\tau}\Big)\mathbb{I} - \alpha\gamma(\vartheta_\tau^{k-1})\mathbb{I} + 2e\Big(\mu_{\mathrm{e}}u_\tau^k+\mu_{\mathrm{v}}\frac{u_\tau^k-u_\tau^{k-1}}{\tau}\Big)\bigg) = g_\tau^k \tag{12.43a}$$

$$\frac{\vartheta_\tau^k-\vartheta_\tau^{k-1}}{\tau} - \mathrm{div}\big(\beta'(\vartheta_\tau^{k-1})\nabla\vartheta_\tau^k\big) = \lambda_{\mathrm{v}}\Big(\mathrm{div}\frac{u_\tau^k-u_\tau^{k-1}}{\tau}\Big)^2 + 2\mu_{\mathrm{v}}\Big|e\Big(\frac{u_\tau^k-u_\tau^{k-1}}{\tau}\Big)\Big|^2 - \alpha\gamma(\vartheta_\tau^k)\mathrm{div}\frac{u_\tau^k-u_\tau^{k-1}}{\tau}, \tag{12.43b}$$

with corresponding boundary conditions. Like in Exercise 12.5, one can prove $\vartheta_\tau^k \geq 0$. Yet, nothing is gratis, and here one must impose a quite strong assumption on the growth of γ, namely $\omega > 2$ in (12.14a) so that $|\gamma(\vartheta)| \leq C_\omega(1+|\vartheta|^{1/2})$. One can then perform a-priori estimates (12.15a,b) and (12.16a) by testing (12.43a) by $\frac{u_\tau^k-u_\tau^{k-1}}{\tau}$ as before in (12.9) but the heat-transfer part (12.43b) is tested by $\frac{1}{2}$, instead of 1 used previously for (12.10). The adiabatic terms do not cancel, yielding $\alpha(\gamma(\vartheta_\tau^{k-1})-\frac{1}{2}\gamma(\vartheta_\tau^k))\mathrm{div}\frac{u_\tau^k-u_\tau^{k-1}}{\tau}$.[10] Afterward the a-priori estimate (12.16), simplified by avoiding the regularization, and then convergence can be proved.

Remark 12.13. The thermo-visco-elastic system (12.4)–(12.5) has been treated by Dafermos, Hsiao, and Slemrod [113, 115, 388] for $n = 1$ and constant coefficients c

[10] This can be estimated by Hölder inequality

$$\int_\Omega \alpha(\gamma(\vartheta_\tau^{k-1})-\frac{1}{2}\gamma(\vartheta_\tau^k))\mathrm{div}\frac{u_\tau^k-u_\tau^{k-1}}{\tau}\,\mathrm{d}x \leq |\alpha|\big\|\gamma(\vartheta_\tau^{k-1})-\frac{1}{2}\gamma(\vartheta_\tau^k)\big\|_{L^2(\Omega)}\Big\|\mathrm{div}\frac{u_\tau^k-u_\tau^{k-1}}{\tau}\Big\|_{L^2(\Omega)}$$
$$\leq \frac{\alpha^2C_\omega^2}{\lambda_{\mathrm{v}}}\Big(\mathrm{meas}_n(\Omega) + \int_\Omega 2\vartheta_\tau^{k-1}+\vartheta_\tau^k\,\mathrm{d}x\Big) + \frac{\lambda_{\mathrm{v}}}{2}\Big\|\mathrm{div}\frac{u_\tau^k-u_\tau^{k-1}}{\tau}\Big\|_{L^2(\Omega)}^2$$

and then treated by the discrete Gronwall inequality.

and κ, while for the multidimensional situation only local-in-time results (together with uniqueness and regularity) have been proved by Bonetti and Bonfanti [61]. For $\mathfrak{c} = \mathfrak{c}(\theta)$ growing like $\theta^{1/2+\varepsilon}$, $\varepsilon > 0$, the existence of solutions for $n = 3$ was recently treated by Blanchard and Guibé [54] by a Schauder fixed point, and in [368] by using Galerkin approximation directly for (12.4)–(12.5) without enthalpy transformation. Regularity has been treated by Pawłow and Zajączkowski [328] in case $\mathfrak{c} = \mathfrak{c}(\theta)$ growing linearly. Considering the heat-transfer coefficient κ dependent on $\nabla\theta$, the multidimensional case was treated also by Nečas at al. [306, 309] and by Eck, Jarušek and Krbec [132, Sect.5.4.2.2]. For $\nu_{\rm v} = \mu_{\rm v} = 0$ see Jiang and Racke [217, Chap.7]. For some modified models see e.g. Eck and Jarušek [131].

12.2 Buoyancy-driven viscous flow

The evolution version of the *Oberbeck-Boussinesq model* from Sect. 6.2 for the Newtonean-fluid case looks as[11]

$$\frac{\partial u}{\partial t} + (u\cdot\nabla)\,u - \Delta u + \nabla\pi = g(1-\alpha\theta), \tag{12.44a}$$

$$\operatorname{div} u = 0, \tag{12.44b}$$

$$\frac{\partial\theta}{\partial t} + u\cdot\nabla\theta - \kappa\Delta\theta = 0, \tag{12.44c}$$

where the notation is as in Section 6.2; for simplicity, the viscosity coefficient and the mass density now equals 1. Still we consider the initial conditions and the boundary condition as no-slip for u and as Newton's condition for θ, i.e.:

$$u(0,\cdot) = u_0, \qquad \theta(0,\cdot) = \theta_0 \qquad \text{on } \Omega, \tag{12.45a}$$

$$u = 0\,, \qquad \kappa\frac{\partial\theta}{\partial\nu} + \beta\theta = h \qquad \text{on } \Sigma. \tag{12.45b}$$

As to the data g, h, u_0, and θ_0, we assume, having in mind $n = 3$, that

$$g\in L^\infty(I;L^3(\Omega;\mathbb{R}^n)),\ h\in L^2(I;L^{4/3}(\Gamma)),\ u_0\in L^2(\Omega;\mathbb{R}^n),\ \theta_0\in L^2(\Omega). \tag{12.46}$$

We are going to use Schauder's-type fixed-point technique. For (v,ϑ) given, we consider u being the very weak solution to the *Oseen equation*

$$\left.\begin{aligned} \frac{\partial u}{\partial t} + (v\cdot\nabla)u - \Delta u + \nabla\pi = g(1-\alpha\vartheta),\quad \operatorname{div} u &= 0, && \text{in } Q,\\ u|_\Sigma &= 0 && \text{on } \Sigma,\\ u(0,\cdot) &= u_0 && \text{on } \Omega, \end{aligned}\right\} \tag{12.47}$$

[11] See, e.g., Lions [261, Ch.I, Sect.9.2] or Straughan [392], or Rajagopal et al. [344]. For a more general model expanding the heat equation by an adiabatic and dissipative heat sources see e.g. [223, 310].

and then θ being the very weak solution to

$$\left.\begin{aligned}\frac{\partial\theta}{\partial t}+v\cdot\nabla\theta-\kappa\Delta\theta &= 0 && \text{in } Q,\\ \kappa\frac{\partial\theta}{\partial\nu}+\beta\theta &= h && \text{on } \Sigma,\\ \theta(0,\cdot) &= \theta_0 && \text{on } \Omega.\end{aligned}\right\} \tag{12.48}$$

Note that, for a given (v,ϑ), (12.47) and (12.48) are linear. We denote $W^{1,2}_{0,\mathrm{div}}(\Omega;\mathbb{R}^n)=\{v\in W^{1,2}_0(\Omega;\mathbb{R}^n);\ \mathrm{div}\,v=0\}$, cf. (6.29).

Lemma 12.14 (A-PRIORI ESTIMATES). *Let $n\le 3$ and (12.46) hold. Then there is a very weak solution (u,θ) to (12.47) and (12.48) satisfying, for some $C_1,\dots,C_4$,*

$$\big\|u\big\|_{L^2(I;W^{1,2}_{0,\mathrm{div}}(\Omega;\mathbb{R}^n))\cap L^\infty(I;L^2(\Omega;\mathbb{R}^n))}\le C_1\Big(1+\big\|\vartheta\big\|_{L^2(I;W^{1,2}(\Omega))}\Big), \tag{12.49a}$$

$$\Big\|\frac{\partial u}{\partial t}\Big\|_{L^{4/3}(I;W^{1,2}_{0,\mathrm{div}}(\Omega;\mathbb{R}^n)^*)}\le C_2\Big(1+\big\|\vartheta\big\|_{L^2(I;W^{1,2}(\Omega))} + \big\|v\big\|_{L^2(I;W^{1,2}_{0,\mathrm{div}}(\Omega;\mathbb{R}^n))\cap L^\infty(I;L^2(\Omega;\mathbb{R}^n))}\Big), \tag{12.49b}$$

$$\big\|\theta\big\|_{L^2(I;W^{1,2}(\Omega))\cap L^\infty(I;L^2(\Omega))}\le C_3, \tag{12.49c}$$

$$\Big\|\frac{\partial\theta}{\partial t}\Big\|_{L^{4/3}(I;W^{-1,2}(\Omega))}\le C_4\Big(1+\big\|v\big\|_{L^2(I;W^{1,2}_{0,\mathrm{div}}(\Omega;\mathbb{R}^n))\cap L^\infty(I;L^2(\Omega;\mathbb{R}^n))}\Big). \tag{12.49d}$$

Proof. Let us consider the Galerkin approximation of (12.47)–(12.48) and, after deriving the a-priori estimate, we can pass to the limit by the same strategy as in the proof of Lemma 12.15 below. We proceed only heuristically.[12] Test (12.47) by u and use Green's theorem:

$$\frac{1}{2}\frac{\mathrm{d}}{\mathrm{d}t}\|u\|^2_{L^2(\Omega;\mathbb{R}^n)}+\int_\Omega((v\cdot\nabla)\,u)\cdot u+|\nabla u|^2+\nabla\pi\cdot u\,\mathrm{d}x$$
$$=\int_\Omega g(1-\alpha\vartheta)\cdot u\,\mathrm{d}x\le C\|g\|_{L^3(\Omega;\mathbb{R}^n)}\Big(1+\|\vartheta\|_{L^6(\Omega)}\Big)\|u\|_{L^6(\Omega;\mathbb{R}^n)}. \tag{12.50}$$

Using $\int_\Omega((v\cdot\nabla)u)\cdot u\,\mathrm{d}x=0$ provided $\mathrm{div}\,v=0$, cf. (6.36), and $\int_\Omega\nabla\pi\cdot u\,\mathrm{d}x=-\int_\Omega\pi\,\mathrm{div}\,u\,\mathrm{d}x=0$, one obtains (12.49) by Young's inequality and integration over I.

Moreover, for v bounded in $L^2(I;W^{1,2}_{0,\mathrm{div}}(\Omega;\mathbb{R}^n))\cap L^\infty(I;L^2(\Omega;\mathbb{R}^n))$, we get

[12] More precisely, we can perform the estimates (12.50) and (12.53) only in Galerkin's approximations because we do not have the by-part formula at our disposal for very weak solutions themselves, and then these bounds are inherited in the limit very weak solution, too.

also the dual estimate (12.49b):[13]

$$\Big\|\frac{\partial u}{\partial t}\Big\|_{L^{4/3}(I;W^{1,2}_{0,\mathrm{div}}(\Omega;\mathbb{R}^n)^*)} := \sup_{\|z\|_{L^4(I;W^{1,2}_{0,\mathrm{div}}(\Omega;\mathbb{R}^n))}\le 1} \Big\langle \frac{\partial u}{\partial t}, z\Big\rangle$$

$$= \sup_{\|z\|_{L^4(I;W^{1,2}_{0,\mathrm{div}}(\Omega;\mathbb{R}^n))}\le 1} \int_Q \nabla u{:}\nabla z + (v\cdot\nabla)u\cdot z - g(1-\alpha\vartheta)z\,\mathrm{d}x\mathrm{d}t$$

$$\le \big\|\nabla u\big\|_{L^2(Q;\mathbb{R}^{n\times n})}\Big(\sqrt[4]{T}+N^{3/2}\big\|v\big\|^{1/2}_{L^2(I;W^{1,2}(\Omega;\mathbb{R}^n))}\big\|v\big\|^{1/2}_{L^\infty(I;L^2(\Omega;\mathbb{R}^n))}\Big)$$
$$+NC\|g\|_{L^\infty(I;L^3(\Omega;\mathbb{R}^n))}\Big(\sqrt{T}+\|\vartheta\|_{L^2(I;L^6(\Omega))}\Big) \tag{12.51}$$

where N denotes the norm of the embedding $W^{1,2}(\Omega)\subset L^6(\Omega)$ and where we used the Hölder inequality and the interpolation (1.63) for the convective term:[14]

$$\int_Q (v\cdot\nabla)u\cdot z\,\mathrm{d}x\mathrm{d}t \le \big\|v\big\|_{L^4(I;L^3(\Omega;\mathbb{R}^3))}\big\|\nabla u\big\|_{L^2(Q;\mathbb{R}^{3\times3})}\big\|z\big\|_{L^4(I;L^6(\Omega;\mathbb{R}^3))}$$
$$\le \big\|v\big\|^{1/2}_{L^2(I;L^6(\Omega;\mathbb{R}^3))}\big\|v\big\|^{1/2}_{L^\infty(I;L^2(\Omega;\mathbb{R}^3))}\big\|\nabla u\big\|_{L^2(Q;\mathbb{R}^{3\times3})}\big\|z\big\|_{L^4(I;L^6(\Omega;\mathbb{R}^3))}. \tag{12.52}$$

Using (12.49a) and (12.49b), the estimate (12.102c) follows.

As to (12.49c), we test (12.48) by θ and use Green's theorem:

$$\frac{1}{2}\frac{\mathrm{d}}{\mathrm{d}t}\|\theta\|^2_{L^2(\Omega)} + \int_\Omega (v\cdot\nabla\theta)\theta\,\mathrm{d}x + \kappa\int_\Omega |\nabla\theta|^2\mathrm{d}x + \beta\int_\Gamma \theta^2\mathrm{d}S$$
$$= \int_\Gamma h\theta\,\mathrm{d}x \le \|h\|_{L^{4/3}(\Gamma)}\|\theta\|_{L^4(\Gamma)}, \tag{12.53}$$

and then (12.49c) follows by the identity (6.33) and the Poincaré inequality (1.56).

The estimate (12.49d) follows similarly as (12.51):

$$\Big\|\frac{\partial\theta}{\partial t}\Big\|_{L^{4/3}(I;W^{1,2}_{0,\mathrm{div}}(\Omega;\mathbb{R}^n)^*)} \le \big\|\nabla\theta\big\|_{L^2(Q;\mathbb{R}^3)}\Big(\Big(\sqrt[4]{T}$$
$$+N^{3/2}\big\|v\big\|^{1/2}_{L^2(I;W^{1,2}(\Omega;\mathbb{R}^3))}\big\|v\big\|^{1/2}_{L^\infty(I;L^2(\Omega;\mathbb{R}^3))}\Big) + N_\Gamma\|h-\beta\theta\|_{L^2(I;L^4(\Gamma))}\Big) \tag{12.54}$$

where N_Γ denotes the norm of the trace operator $W^{1,2}(\Omega)\subset L^4(\Gamma)$. □

[13] Alternatively, with the same effect, we could use Green's formula and then the Hölder inequality and interpolation in (12.52):

$$\int_\Omega (v\cdot\nabla)u\cdot z\,\mathrm{d}x = -\int_\Omega (v\cdot\nabla)z\cdot u\,\mathrm{d}x \le \|v\|_{L^{8/3}(I;L^4(\Omega;\mathbb{R}^3))}\|\nabla z\|_{L^4(I;L^2(\Omega;\mathbb{R}^{3\times3}))}\|u\|_{L^{8/3}(I;L^4(\Omega;\mathbb{R}^3))}$$
$$\le \|v\|^{3/4}_{L^2(I;L^6(\Omega;\mathbb{R}^3))}\|v\|^{1/4}_{L^\infty(I;L^2(\Omega;\mathbb{R}^3))}\|\nabla z\|_{L^4(I;L^2(\Omega;\mathbb{R}^{3\times3}))}\|u\|^{3/4}_{L^2(I;L^6(\Omega;\mathbb{R}^3))}\|u\|^{1/4}_{L^\infty(I;L^2(\Omega;\mathbb{R}^3))}.$$

[14] We use Proposition 1.41 for $p_1=q_2=2$, $q_1=6$, $p_2=+\infty$, and $\lambda=1/2$; cf. also Example 8.77.

Let us now abbreviate $\mathscr{W}_1 := W^{1,2,4/3}(I; W^{1,2}_{0,\mathrm{div}}(\Omega;\mathbb{R}^n), W^{1,2}_{0,\mathrm{div}}(\Omega;\mathbb{R}^n)^*) \cap L^\infty(I; L^2(\Omega;\mathbb{R}^n))$ and $\mathscr{W}_2 := W^{1,2,4/3}(I; W^{1,2}(\Omega), W^{1,2}(\Omega)^*) \cap L^\infty(I; L^2(\Omega))$. We define the mapping

$$M : \mathscr{W}_1{\times}\mathscr{W}_2 \rightrightarrows \mathscr{W}_1{\times}\mathscr{W}_2 \tag{12.55}$$

as $M(v,\vartheta)$ being the set of very weak solutions (u,θ) to (12.47)–(12.48) satisfying the bounds (12.49).

Lemma 12.15 (Continuity). *Let $n \le 3$ and* (12.46) *hold. The set-valued mapping $M : \mathscr{W}_1{\times}\mathscr{W}_2 \rightrightarrows \mathscr{W}_1{\times}\mathscr{W}_2$, see* (12.55), *is weakly upper semi-continuous.*

Proof. Assume $(v_k,\theta_k) \overset{*}{\rightharpoonup} (v,\theta)$ in $\mathscr{W}_1 \times \mathscr{W}_2$. By the "interpolated" Aubin-Lions' Lemma 7.8, we have $v_k \to v$ in $L^{2^\circledast-\epsilon}(Q;\mathbb{R}^n)$, cf. (8.152); if $n \le 3$, we can consider $2^\circledast \ge 10/3$, cf. (8.131). As $2^\circledast > 2$, we certainly have $(v_k\cdot\nabla)u_k \rightharpoonup (v\cdot\nabla)u$ weakly in $L^1(Q;\mathbb{R}^n)$. Similarly, we have also $v_k\cdot\nabla\theta_k \rightharpoonup v\cdot\nabla\theta$ weakly in $L^1(Q)$. Hence we can make the limit passage just by weak continuity. □

Proposition 12.16 (Existence of a fixed point). *Let $n \le 3$ and* (12.46) *hold. The set-valued mapping M has a fixed point $(u,\theta) \in M(u,\theta)$ which is a very weak solution to* (12.44)–(12.45).

Proof. The closed bounded convex set

$$\begin{aligned}\Big\{(u,\theta) \in \mathscr{W}_1 \times \mathscr{W}_2;\ & \|u\|_{L^2(I;W^{1,2}_{0,\mathrm{div}}(\Omega;\mathbb{R}^n))\cap L^\infty(I;L^2(\Omega;\mathbb{R}^n))} \le C_1(1{+}C_3),\\ & \Big\|\frac{\partial u}{\partial t}\Big\|_{L^{4/3}(I;W^{1,2}_{0,\mathrm{div}}(\Omega;\mathbb{R}^n)^*)} \le C_2(1{+}C_1{+}C_3{+}C_1C_3),\\ & \|\theta\|_{L^2(I;W^{1,2}(\Omega))\cap L^\infty(I;L^2(\Omega))} \le C_3,\\ & \Big\|\frac{\partial\theta}{\partial t}\Big\|_{L^{4/3}(I;W^{-1,2}(\Omega))} \le C_4(1{+}C_1{+}C_1C_3)\Big\}\end{aligned} \tag{12.56}$$

is weakly* compact in $\mathscr{W}_1 \times \mathscr{W}_2$. Due to Lemma 12.14, M maps this set into itself. As (12.47)–(12.48) are linear and (12.49) are convex inequalities, the values of M are convex. By Lemma 12.14, the values of M are nonempty. Taking into account Lemma 12.15, we can use Kakutani fixed-point theorem 1.11 to get $(u,\theta) \in M(u,\theta)$. □

Exercise 12.17. Apply Galerkin's method to (12.44) and prove convergence of the approximate solutions and thus existence of the very weak solution to (12.44) without using Schauder's-type fixed-point theorem.

12.3 Predator-prey system

Let us deal with a special evolution variant of the *Lotka-Volterra system* (6.51):[15]

[15] This special model and its analysis is after [346].

$$\left.\begin{array}{ll}\left.\begin{array}{l}\dfrac{\partial u}{\partial t}-d_1\Delta u=u\big(a_1-b_1u-c_1v\big)\\[2mm]\dfrac{\partial v}{\partial t}-d_2\Delta v=v\big(a_2-c_2u\big)\end{array}\right\} & \text{in } Q,\\[2mm] u(t,\cdot)|_\Gamma=0,\quad v(t,\cdot)|_\Gamma=0 & \text{on } \Sigma,\\[2mm] u(0,\cdot)=u_0,\quad v(0,\cdot)=v_0 & \text{on } \Omega.\end{array}\right\} \tag{12.57}$$

We consider the constants corresponding to the *predator-prey* variant, i.e. $b_1 \geq 0$, $d_1, d_2 > 0$, $a_1, c_1 > 0$ and $a_2, c_2 < 0$, so that u and v represent a prey and a predator densities, respectively; then a_1 is the growth rate of the prey species in the absence of the predators while $-a_2$ is the death rate of the predators in the absence of the prey; for further interpretation see Section 6.3. The maximal concentration of preys in the absence of predators (i.e. the carrying capacity of the environment) is $a_1/b_1 =: \gamma_1$. Instead of using Schauder's fixed-point theorem as in Section 6.3, we can now be more constructive and use the *semi-implicit Rothe method*: to be more specific, we seek $u_\tau^k, v_\tau^k \in W_0^{1,2}(\Omega)$ satisfying

$$\frac{u_\tau^k - u_\tau^{k-1}}{\tau} - d_1\Delta u_\tau^k = u_\tau^{k-1} a_1\big(1 - b_1 u_\tau^k\big) - c_1 u_\tau^k v_\tau^{k-1}, \tag{12.58a}$$

$$\frac{v_\tau^k - v_\tau^{k-1}}{\tau} - d_2\Delta v_\tau^k = v_\tau^k\big(a_2 - c_2 u_\tau^k\big) \tag{12.58b}$$

for $k = 1, \ldots, T/\tau$, while for $k = 0$ we consider

$$u^0 = u_{0\tau}, \quad v^0 = v_{0\tau}, \tag{12.59}$$

with some $u_{0\tau}, v_{0\tau} \in W_0^{1,2}(\Omega)$ such that $u_{0\tau} \to u_0$, $v_{0\tau} \to v_0$ in $L^2(\Omega)$ and $\|u_{0\tau}\|_{W_0^{1,2}(\Omega)} = \mathscr{O}(1/\sqrt{\tau})$, and $\|v_{0\tau}\|_{W_0^{1,2}(\Omega)} = \mathscr{O}(1/\sqrt{\tau})$. Note that the boundary-value problems (12.58) are linear and (12.58b) is *decoupled* from (12.58a), which suggests an efficient numerical strategy after a further discretization by a Galerkin method.

Lemma 12.18 (A-PRIORI BOUNDS). *If $\tau \leq \tau_0 < 1/(2a_2 - 2c_2\gamma_1)^+$ with $\gamma_1 = a_1/b_1$, the elliptic problems in (12.58) have unique solutions $(u_\tau^k, v_\tau^k) \in W_0^{1,2}(\Omega)^2$ for all $k = 1, \ldots, T/\tau$, which satisfy $0 \leq u_\tau^k \leq \gamma_1$ and $0 \leq v_\tau^k$ provided $0 \leq u_0 \leq \gamma_1$ and $0 \leq v_0 \in L^2(\Omega)$. Furthermore, the following a-priori estimates hold:*

$$\big\|(u_\tau, v_\tau)\big\|_{L^2(I;W_0^{1,2}(\Omega))^2} \leq C, \quad \Big\|\Big(\frac{\partial u_\tau}{\partial t}, \frac{\partial v_\tau}{\partial t}\Big)\Big\|_{L^2(I;W^{-1,2}(\Omega))^2} \leq C. \tag{12.60}$$

Proof. We use an induction argument for the first part of the lemma. Let us suppose that $u_\tau^{k-1}, v_\tau^{k-1} \in L^\infty(\Omega)$ satisfy $0 \leq u_\tau^{k-1} \leq \gamma_1$ and $0 \leq v_\tau^{k-1}$. Then, we have to prove that u_τ^k and v_τ^k belong to $W_0^{1,2}(\Omega)$ and inherit these bounds.

The linear problem (12.58a) for u_τ^k is coercive on $W_0^{1,2}(\Omega)$ because $b_1, c_1, u_\tau^{k-1}, v_\tau^{k-1} \geq 0$, so that by Lax-Milgram's Theorem 2.19 it possesses a unique weak solution. Let us show that $u_\tau^k \geq 0$. Testing the weak formulation of (12.58a) by $(u_\tau^k)^-$, one gets

$$\frac{1}{\tau}\big\|(u_\tau^k)^-\big\|_{L^2(\Omega)}^2 + d_1\big\|\nabla(u_\tau^k)^-\big\|_{L^2(\Omega;\mathbb{R}^n)}^2 \leq \int_\Omega \Big(\frac{1}{\tau}+a_1\Big)u_\tau^{k-1}(u_\tau^k)^-\,\mathrm{d}x \leq 0. \quad (12.61)$$

Hence, we get $(u_\tau^k)^- = 0$. Let us further prove that $u_\tau^k \leq \gamma_1$. Testing the weak formulation of (12.58a) by $(u_\tau^k-\gamma_1)^+$, we obtain

$$\begin{aligned}\big\|(u_\tau^k-\gamma_1)^+\big\|_{L^2(\Omega)}^2 + \tau d_1\big\|\nabla(u_\tau^k-\gamma_1)^+\big\|_{L^2(\Omega;\mathbb{R}^n)}^2 = \int_\Omega (u_\tau^{k-1}-\gamma_1)(u_\tau^k-\gamma_1)^+ \\ -\tau\Big(a_1 y_\tau^{k-1}\frac{u_\tau^k-\gamma_1}{\gamma_1}(u_\tau^k-\gamma_1)^+ + c_1u_\tau^k(u_\tau^k-\gamma_1)^+v_\tau^{k-1}\Big)\mathrm{d}x \leq 0 \quad (12.62)\end{aligned}$$

since $0 \leq u_\tau^{k-1} \leq \gamma_1$ and $v_\tau^{k-1} \geq 0$. Hence, $u_\tau^k \leq \gamma_1$ a.e. in Ω.

Now, (12.58b) is a linear boundary-value problem for v_τ^k which is coercive on $W_0^{1,2}(\Omega)$ if the coefficient $\frac{1}{\tau} - a_2 + c_2u_\tau^k$ is non-negative. Taking into account $u^k \leq \gamma_1$, it needs the condition $\tau < 1/(a_2-c_2\gamma_1)^+$. Therefore, by Lax-Milgram's Theorem 2.19, it possesses a unique solution $v_\tau^k \in W_0^{1,2}(\Omega)$. Let us show that $v_\tau^k \geq 0$. Testing the weak formulation of (12.58b) by $(v_\tau^k)^-$, one gets

$$\Big(\frac{1}{\tau}-a_2+c_2\gamma_1\Big)\big\|(v_\tau^k)^-\big\|_{L^2(\Omega)}^2 + d_2\big\|\nabla(v_\tau^k)^-\big\|_{L^2(\Omega;\mathbb{R}^n)}^2 \leq \int_\Omega \frac{v_\tau^{k-1}}{\tau}(v_\tau^k)^-\,\mathrm{d}x \leq 0;$$

recall that $c_2 \leq 0$. Hence, if $\tau < 1/(a_2-c_2\gamma_1)^+$, we get $(v_\tau^k)^- = 0$.

Let us now prove the estimates (12.60). Testing (12.58a) by u_τ^k and using Young's inequality and non-negativity of v_τ^{k-1}, we have

$$\frac{1}{2\tau}\|u_\tau^k\|_{L^2(\Omega)}^2 + d_1\|\nabla u_\tau^k\|_{L^2(\Omega;\mathbb{R}^n)}^2 \leq \int_\Omega \Big(\frac{1}{\tau}+\frac{a_1}{2}\Big)(u_\tau^{k-1})^2 + \frac{a_1}{2}(u_\tau^k)^2\mathrm{d}x. \quad (12.63)$$

Summing it for $k = 1,\dots,T/\tau$ yields boundedness of $\bar{u}_\tau$ and, by using the technique of combination (8.18) with (8.38), also of u_τ in $L^2(I;H_0^1(\Omega))$.

Now, testing (12.58b) by v_τ^k and using Young inequality we also have

$$\begin{aligned}\frac{1}{\tau}\big\|v_\tau^k\big\|_{L^2(\Omega)}^2 + d_2\big\|\nabla v_\tau^k\big\|_{L^2(\Omega;\mathbb{R}^n)}^2 &\leq \int_\Omega \frac{1}{\tau}v_\tau^{k-1}v_\tau^k + (a_2-c_2u_\tau^k)(v_\tau^k)^2\,\mathrm{d}x \\ &\leq \frac{1}{2\tau}\big\|v_\tau^{k-1}\big\|_{L^2(\Omega)}^2 + \frac{1}{2\tau}\big\|v_\tau^k\big\|_{L^2(\Omega)}^2 + (a_2-c_2\gamma_1)\big\|v_\tau^k\big\|_{L^2(\Omega)}^2. \quad (12.64)\end{aligned}$$

By using the discrete Gronwall inequality (1.70), we get $\bar{v}_\tau$ and, by using again the technique of combination (8.18) with (8.38), also v_τ bounded in $L^2(I;W_0^{1,2}(\Omega))$ independently of τ provided $\tau \leq \tau_0 < 1/(2a_2-2c_2\gamma_1)^+$, as assumed.

Using the "retarded" function $\bar{u}_\tau^{\rm R}$ as defined in (8.202) and analogously for $\bar{v}_\tau^{\rm R}$, the scheme (12.58) can be written down in a "compact" form as

$$\frac{\partial u_\tau}{\partial t} - d_1\Delta\bar{u}_\tau = a_1\bar{u}_\tau^{\rm R}\Big(1-\frac{\bar{u}_\tau}{\gamma_1}\Big) - c_1\bar{u}_\tau\bar{v}_\tau^{\rm R}, \tag{12.65a}$$

$$\frac{\partial v_\tau}{\partial t} - d_2\Delta\bar{v}_\tau = a_2\bar{v}_\tau - c_2\bar{u}_\tau\bar{v}_\tau. \tag{12.65b}$$

In view of this, we can estimate

$$\Big\|\frac{\partial u_\tau}{\partial t}\Big\|^2_{L^2(I;W^{-1,2}(\Omega))} = \sup_{\|z\|_{L^2(I;W_0^{1,2}(\Omega))}\le 1}\int_Q d_1\nabla\bar{u}_\tau{\cdot}\nabla z + a_1\bar{u}_\tau^{\rm R}\Big(1{-}\frac{\bar{u}_\tau}{\gamma_1}\Big)z$$
$$-c_1\bar{u}_\tau\bar{v}_\tau^{\rm R} z\,{\rm d}x{\rm d}t \le d_1\big\|\nabla\bar{u}_\tau\big\|_{L^2(Q;\mathbb{R}^n)} + {\rm meas}_{n+1}(Q)^{1/2}\frac{a_1\gamma_1}{4} + c_1\gamma_1\big\|\bar{v}_\tau^{\rm R}\big\|_{L^2(Q)}$$

which bounds $\frac{\partial}{\partial t}u_\tau$. Similarly, from (12.65), one obtains

$$\Big\|\frac{\partial v_\tau}{\partial t}\Big\|^2_{L^2(I;W^{-1,2}(\Omega))} = \sup_{\|z\|_{L^2(I;W_0^{1,2}(\Omega))}\le 1}\int_Q d_1\nabla\bar{v}_\tau{\cdot}\nabla z + a_2\bar{v}_\tau z$$
$$-c_2\bar{u}_\tau\bar{v}_\tau z\,{\rm d}x{\rm d}t \le d_1\big\|\nabla\bar{v}_\tau\big\|_{L^2(Q;\mathbb{R}^n)} + \max(|a_2-c_2\gamma_1|,-a_2)\big\|\bar{v}_\tau\big\|_{L^2(Q)}. \qquad \square$$

Proposition 12.19 (Convergence, uniqueness). *For $\tau\searrow 0$, $(u_\tau,v_\tau)\rightharpoonup(u,v)$ weakly in $\mathscr{W}^2$ with $\mathscr{W} := W^{1,2,2}(I;W_0^{1,2}(\Omega),W^{-1,2}(\Omega))$ and (u,v) is the unique weak solution to* (12.57).

Proof. The mentioned converging (sub)sequence does exist thanks to (12.60). By Aubin-Lions' lemma, also $(u_\tau,v_\tau)\to(u,v)$ in $L^2(Q)^2$. By interpolation and by (8.50), (8.31), and (12.60), it holds that

$$\big\|u_\tau{-}\bar{u}_\tau\big\|_{L^2(Q)} \le C_0\big\|u_\tau{-}\bar{u}_\tau\big\|^{1/2}_{L^2(I;W_0^{1,2}(\Omega))}\big\|u_\tau{-}\bar{u}_\tau\big\|^{1/2}_{L^2(I;W^{-1,2}(\Omega))}$$
$$\le C_0\sqrt{2}\big\|u_\tau\big\|^{1/2}_{L^2(I;W_0^{1,2}(\Omega))}\sqrt{\frac{\tau}{3}}\Big\|\frac{\partial u_\tau}{\partial t}\Big\|^{1/2}_{L^2(I;W^{-1,2}(\Omega))} = \mathscr{O}(\sqrt{\tau}). \tag{12.66}$$

Hence $\bar{u}_\tau\to u$ strongly in $L^2(Q)$. By analogous arguments, also $\bar{v}_\tau\to v$ in $L^2(Q)$.

Now we are to make a limit passage in (12.65) just by weak continuity. Note that the nonlinearity $\mathbb{R}_+^2\times\mathbb{R}^2\to\mathbb{R}^2 : (r_1,r_1',r_2,r_2')\mapsto\big(a_1r_1'(1-r_1/\gamma_1)-c_1r_1r_2', a_2r_2-c_2r_1r_2\big)$ has at most a quadratic growth so that, by continuity of the respective Nemytskiĭ mapping $L^2(Q)^4\to L^1(Q)^2$, the right-hand sides of (12.65a,b) converge strongly in $L^1(Q)$ to $a_1u(1-u/\gamma_1)-c_1uv$ and a_2v-c_2uv, respectively. The limit passage in the left-hand-side terms in (12.65a,b) is obvious because they are linear. Note that the mapping $u\mapsto u(0,\cdot):\mathscr{W}\to L^2(\Omega)$ is weakly continuous which allows us to pass to the limit in the respective initial conditions.

To prove uniqueness of the solution to (12.57), we consider two weak solutions (u_1, v_1) and (u_2, v_2), subtract the corresponding equations and test them by $u_{12} := u_1 - u_2$ and $v_{12} := v_1 - v_2$, respectively. This gives

$$\begin{aligned}\frac{1}{2}\frac{\mathrm{d}}{\mathrm{d}t}\Big(\big\|u_{12}\big\|^2_{L^2(\Omega)} + \big\|v_{12}\big\|^2_{L^2(\Omega)}\Big) + d_1\big\|\nabla u_{12}\big\|^2_{L^2(\Omega;\mathbb{R}^n)} + d_2\big\|\nabla(v_{12})\big\|^2_{L^2(\Omega;\mathbb{R}^n)}\\ + \int_\Omega b_1(u_1+u_2)(u_{12})^2\mathrm{d}x - a_2\big\|v_{12}\big\|^2_{L^2(\Omega)}\\ = a_1\big\|u_{12}\big\|^2_{L^2(\Omega)} - \int_\Omega c_1(u_1v_1{-}u_2v_2)u_{12} - c_2(u_1v_1{-}u_2v_2)v_{12}\mathrm{d}x\\ \le a_1\big\|u_{12}\big\|^2_{L^2(\Omega)} + c_1\big\|u_1\big\|_{L^\infty(\Omega)}\big\|v_{12}\big\|_{L^2(\Omega)}\big\|u_{12}\big\|_{L^2(\Omega)}\\ + c_2\big\|u_1\big\|_{L^\infty(\Omega)}\big\|v_{12}\big\|^2_{L^2(\Omega)} + c_2\big\|u_{12}\big\|_{L^2(\Omega)}\big\|v_2\big\|_{L^\infty(\Omega)}\big\|v_{12}\big\|_{L^2(\Omega)},\end{aligned}$$

where we also used that u_1, u_2, v_1, and v_2 are non-negative and that u_1 and v_2 have upper bounds. Then, by Young's and Gronwall's inequalities, we get $u_{12} = 0$ and $v_{12} = 0$. Thus we showed the uniqueness and thus the convergence of the whole sequence $\{(u_\tau, v_\tau)\}_{\tau>0}$. □

Exercise 12.20. Having (12.60) at disposal, execute the test of (12.58) by $u_\tau^k{-}u_\tau^{k-1}$ and $v_\tau^k{-}v_\tau^{k-1}$ to prove the a-priori bound of u_τ and v_τ in $W^{1,2}(I;L^2(\Omega)) \cap L^\infty(I;W_0^{1,p}(\Omega))$.[16]

12.4 Semiconductors

Modelling of transient regimes of semiconductor devices conventionally relies on the evolution variant of Roosbroeck's *drift-diffusion* system (6.68), i.e.

$$\operatorname{div}(\varepsilon\nabla\phi) = n - p + c_{\mathrm{D}} \qquad \text{in } Q, \tag{12.67a}$$

$$\frac{\partial n}{\partial t} - \operatorname{div}(\nabla n - n\nabla\phi) = r(n,p) \qquad \text{in } Q, \tag{12.67b}$$

$$\frac{\partial p}{\partial t} - \operatorname{div}(\nabla p + p\nabla\phi) = r(n,p) \qquad \text{in } Q, \tag{12.67c}$$

where we use the conventional notation of Section 6.5 except the sign convention of r. We can see that the magnetic field is still neglected and the electric field ϕ, which varies much faster than the carrier concentrations n and p, is governed

[16]Hint: In terms of the interpolants, test (12.65) respectively by $\frac{\partial}{\partial t}u_\tau$ and $\frac{\partial}{\partial t}v_\tau$ to obtain $\int_0^t \|\frac{\partial}{\partial t}u_\tau\|^2_{L^2(\Omega)} + \|\frac{\partial}{\partial t}v_\tau\|^2_{L^2(\Omega)}\,\mathrm{d}t + \frac{1}{2}d_1\|\nabla u_\tau(t)\|^2_{L^2(\Omega;\mathbb{R}^n)} + \frac{1}{2}d_2\|\nabla v_\tau(t)\|^2_{L^2(\Omega;\mathbb{R}^n)} \le \int_0^t f_1\frac{\partial}{\partial t}u_\tau + f_2\frac{\partial}{\partial t}v_\tau\,\mathrm{d}t + \frac{1}{2}d_1\|\nabla u_0\|^2_{L^2(\Omega;\mathbb{R}^n)} + \frac{1}{2}d_2\|\nabla v_0\|^2_{L^2(\Omega;\mathbb{R}^n)}$ for $t = k\tau$, $k = 1, ..., T/\tau$, where f_1 and f_2 denote the right-hand sides of (12.65a) and (12.65b), respectively. Realize that, by (12.60), we have already estimated $f_1, f_2 \in L^\infty(I;L^{2^*}(\Omega))$.

by the quasistatic equation (12.67a) which therefore does not involve any time derivative of ϕ.

Of course, (12.67b,c) is to be completed by initial conditions

$$n(0,\cdot) = n_0, \qquad p(0,\cdot) = p_0, \tag{12.68}$$

and some boundary conditions; e.g. Dirichlet ones of Γ_D with $\mathrm{meas}_{n-1}(\Gamma_D) > 0$ (electrodes with time-varying voltage) and zero Neumann one on $\Gamma_N = \Gamma \setminus \Gamma_D$ (an isolated part), i.e.

$$\phi|_{\Sigma_D} = \phi_\Sigma|_{\Sigma_D},\ n|_{\Sigma_D} = n_\Sigma|_{\Sigma_D},\ p|_{\Sigma_D} = p_\Sigma|_{\Sigma_D} \quad \text{on } \Sigma_D := (0,T)\times\Gamma_D, \tag{12.69a}$$

$$\frac{\partial\phi}{\partial\nu} = \frac{\partial n}{\partial\nu} = \frac{\partial p}{\partial\nu} = 0 \quad \text{on } \Sigma_N := (0,T)\times\Gamma_N, \tag{12.69b}$$

with n_Σ and p_Σ constant in time, i.e. $n_\Sigma(t,\cdot) = n_\Gamma$ and $p_\Sigma(t,\cdot) = p_\Gamma$. Again, we made an exponential-type transformation but now slightly different than (6.70)[17], namely we introduce a new variable set (ϕ, u, v) related to (ϕ, n, p) by

$$n = \mathrm{e}^u\,, \qquad p = \mathrm{e}^v \tag{12.70}$$

and abbreviate

$$s(u,v) := r\big(\mathrm{e}^u, \mathrm{e}^v\big). \tag{12.71}$$

Obviously, (12.70) transforms the currents $j_n = \nabla n - n\nabla\phi = \mathrm{e}^u\nabla(u-\phi)$ and $j_p = -\nabla p - p\nabla\phi = -\mathrm{e}^v\nabla(v+\phi)$. Another elegant trick[18], proposed by Gajewski [164, 165], consists in time-differentiation of (12.67a), which leads, by using (12.67b,c) together with the fact that concentration of dopants $c_D = c_D(x)$ is time-independent, to the pseudoparabolic equation $\frac{\partial}{\partial t}(-\mathrm{div}(\varepsilon\nabla\phi)) = \frac{\partial}{\partial t}(p - n - c_D) = \mathrm{div}(j_n - j_p)$. Of course, now we need the initial condition for ϕ, namely $\phi(0,\cdot) = \phi_0$, with ϕ_0 satisfying

$$\mathrm{div}\big(\varepsilon\nabla\phi_0\big) = n_0 - p_0 + c_D \qquad \text{on } \Omega; \tag{12.72}$$

and the initial conditions (12.68) now transform to

$$u_0 = \ln(n_0)\,, \qquad v_0 = \ln(p_0). \tag{12.73}$$

Hence the system (12.67) transforms to

$$\frac{\partial}{\partial t}\big(\mathrm{div}(\varepsilon\nabla\phi)\big) + \mathrm{div}\big(\mathrm{e}^u\nabla(\phi-u) + \mathrm{e}^v\nabla(\phi+v)\big) = 0 \tag{12.74a}$$

$$\frac{\partial}{\partial t}\big(\mathrm{e}^u\big) - \mathrm{div}\big(\mathrm{e}^u\nabla(u-\phi)\big) = s(u,v), \tag{12.74b}$$

$$\frac{\partial}{\partial t}\big(\mathrm{e}^v\big) - \mathrm{div}\big(\mathrm{e}^v\nabla(v+\phi)\big) = s(u,v), \tag{12.74c}$$

[17] Realize that (6.70) would not result in a doubly-nonlinear structure like (11.65).

[18] For alternative analysis of (12.67) without differentiating (12.67a) in time see e.g. [166, 167] or [119] or [276, Sect.3.7].

while the boundary conditions (12.69) transform to

$$\phi(t,\cdot)|_{\Sigma_D} = \phi_\Sigma(t,\cdot)|_{\Sigma_D}, \quad u(t,\cdot)|_{\Gamma_D} = u_\Gamma|_{\Gamma_D}, \quad v(t,\cdot)|_{\Gamma_D} = v_\Gamma|_{\Gamma_D} \quad \text{on } \Gamma_D, \tag{12.75a}$$

$$\frac{\partial\phi}{\partial\nu} = \frac{\partial u}{\partial\nu} = \frac{\partial v}{\partial\nu} = 0 \qquad \text{on } \Sigma_N, \tag{12.75b}$$

where $u_\Gamma := \ln(n_\Gamma)$ and $v_\Gamma := \ln(p_\Gamma)$.

The weak solution to (12.67) with (12.68)-(12.69) is understood as $(\phi,u,v) \in L^2(I;W^{1,2}(\Omega))^3$ such that also $\frac{\partial}{\partial t}(\operatorname{div}(\varepsilon\nabla\phi)), \frac{\partial}{\partial t}\mathrm{e}^u, \frac{\partial}{\partial t}\mathrm{e}^v \in L^2(I;W_0^{1,2}(\Omega)^*)$, (12.75a) holds, $u(0,\cdot) = u_0$ and $v(0,\cdot) = v_0$ with u_0 and v_0 from (12.73), and the integral identity

$$\Big\langle\frac{\partial}{\partial t}(\operatorname{div}(\varepsilon\nabla\phi)), z_1\Big\rangle + \Big\langle\frac{\partial}{\partial t}(\mathrm{e}^u), z_2\Big\rangle + \Big\langle\frac{\partial}{\partial t}(\mathrm{e}^v), z_3\Big\rangle \\ = \int_\Omega \mathrm{e}^u\nabla(u-\phi)\cdot\nabla(z_1-z_2) - \mathrm{e}^v\nabla(\phi+v)\cdot\nabla(z_1+z_3) + s(u,v)(z_2+z_3)\,\mathrm{d}x \tag{12.76}$$

holds for a.a. $t \in I$ and all $z \in W^{1,2}(\Omega)^3$ such that $z|_{\Gamma_D} = 0$.

The analysis of the original model is complicated and we confine ourselves only to a modified model arising by truncation of the nonlinearity $\xi \mapsto \mathrm{e}^\xi$ and then also of $s(\cdot,\cdot)$, i.e. by replacing them by

$$e_l(\xi) := \mathrm{e}^{\min(l,\max(-l,\xi))}, \qquad s_l(u,v) := r(e_l(u), e_l(v))\ ; \tag{12.77}$$

here l is a positive constant. Hence $e_l(r) = \mathrm{e}^r$ for $r \in [-l,l]$. We will analyze it by the Galerkin approximation by using the subspaces V_k of $W^{1,2}_{\Gamma_D}(\Omega) := \{v \in W^{1,2}(\Omega);\ v|_{\Gamma_D} = 0\}$, and assuming, for simplicity, that $u_0, v_0, u_\Gamma, v_\Gamma, \phi_\Sigma(t,\cdot) \in V_k$ for any k, while ϕ_0 has to be approximated by a suitable $\phi_{0k} \in V_k$. Note that e_l modifies the original nonlinearity out of the interval $[-l,l]$ and makes, in particular, s_l bounded. Hence we get an approximate solution $(\phi_{kl}, u_{kl}, v_{kl})$.

Lemma 12.21 (A-PRIORI BOUNDS). *Let $u_\Gamma, v_\Gamma \in W^{1,2}(\Omega)$, $\phi_\Sigma \in L^\infty(I;W^{1,2}(\Omega))$. Then, for $l \in \mathbb{N}$ fixed, the approximate solution $(\phi_{kl}, u_{kl}, v_{kl})$ satisfies:*

$$\big\|\phi_{kl}\big\|_{L^\infty(I;W^{1,2}(\Omega))} \le C_l, \tag{12.78a}$$

$$\big\|u_{kl}\big\|_{L^2(I;W^{1,2}(\Omega))} \le C_l, \qquad \big\|v_{kl}\big\|_{L^2(I;W^{1,2}(\Omega))} \le C_l, \tag{12.78b}$$

$$\big\|e_l(u_{kl})\big\|_{L^\infty(Q)} \le C_l, \qquad \big\|e_l(v_{kl})\big\|_{L^\infty(Q)} \le C_l, \tag{12.78c}$$

$$\Big\|\frac{\partial}{\partial t}(\operatorname{div}(\varepsilon\nabla\phi))\Big\|_{L^2(I;W^{-1,2}(\Omega))} \le C_l, \tag{12.78d}$$

$$\Big\|\frac{\partial}{\partial t}e_l(u_{kl})\Big\|_{L^2(I;W^{-1,2}(\Omega))} \le C_l, \quad \Big\|\frac{\partial}{\partial t}e_l(v_{kl})\Big\|_{L^2(I;W^{-1,2}(\Omega))} \le C_l. \tag{12.78e}$$

Proof. Let us test (12.74) modified as outlined above by $([\phi_{kl}-\phi_\Sigma](t,\cdot), u_{kl}(t,\cdot)-u_\Gamma, v_{kl}(t,\cdot)-v_\Gamma)$ itself. This, after integration over $[0,t]$ and the by-part integration $\int_0^t \frac{\partial}{\partial t} e_l(u_{kl})u_\Gamma \mathrm{d}t = (e_l(u_{kl}(t,\cdot)) - e_l(u_0))u_\Gamma$, gives

$$\begin{aligned}
&\int_\Omega \Big[\frac{\varepsilon}{2}|\nabla\phi_{kl}|^2 + [\widehat{e}_l]^*\big(e_l(u_{kl})\big) + [\widehat{e}_l]^*\big(e_l(v_{kl})\big)\Big](t,\cdot)\,\mathrm{d}x \\
&\quad + \int_0^t\int_\Omega e_l(u_{kl})\big|\nabla(u_{kl}-\phi_{kl})\big|^2 + e_l(v_{kl})\big|\nabla(v_{kl}+\phi_{kl})\big|^2\,\mathrm{d}x\mathrm{d}t \\
&\quad = \int_\Omega \Big(\frac{\varepsilon}{2}|\nabla\phi_0|^2 + [\widehat{e}_l]^*\big(e_l(u_0)\big) + [\widehat{e}_l]^*\big(e_l(v_0)\big) + \varepsilon[\nabla\phi_{kl}\cdot\nabla\phi_\Sigma](t,\cdot) \\
&\quad - \varepsilon\nabla\phi_{0k}\cdot\nabla\phi_\Sigma(0,\cdot) + \big(e_l(u_{kl}(t,\cdot)) - e_l(u_0)\big)u_\Gamma + \big(e_l(v_{kl}(t,\cdot)) - e_l(v_0)\big)v_\Gamma\Big)\mathrm{d}x \\
&\quad + \int_0^t\int_\Omega s_l(u_{kl},v_{kl})(u_{kl}+v_{kl}) - \varepsilon\nabla\phi_{kl}\cdot\nabla\frac{\partial\phi_\Sigma}{\partial t} \\
&\quad + e_l(u_{kl})\nabla(\phi_{kl}-u_{kl})\cdot\nabla(\phi_\Sigma-u_\Gamma) + e_l(v_{kl})\nabla(\phi_{kl}+v_{kl})\cdot\nabla(\phi_\Sigma+v_\Gamma)\,\mathrm{d}x\mathrm{d}t
\end{aligned}$$

where $[\widehat{e}_l]^*(\cdot)$ is the Legendre-Fenchel conjugate to the primitive function $\widehat{e}_l : \xi \mapsto \int_0^\xi e_l(\zeta)\mathrm{d}\zeta$ to e_l, cf. the formula (11.70). We further use the estimates

$$\begin{aligned}
\int_\Omega e_l(u_{kl})\big|\nabla(u_{kl}-\phi_{kl})\big|^2\mathrm{d}x &\geq \int_\Omega e_l(u_{kl})\Big(\frac{1}{2}\big|\nabla u_{kl}\big|^2 - \big|\nabla\phi_{kl}\big|^2\Big)\mathrm{d}x \\
&\geq \int_\Omega \frac{\mathrm{e}^{-l}}{2}\big|\nabla u_{kl}\big|^2 - \mathrm{e}^l\big|\nabla\phi_{kl}\big|^2\mathrm{d}x,
\end{aligned} \tag{12.79}$$

which yields the term $\mathrm{e}^l|\nabla\phi_{kl}|^2$ to be treated "on the right-hand side" by Gronwall's inequality. Similarly,

$$\int_\Omega e_l(v_{kl})\big|\nabla(v_{kl}+\phi_{kl})\big|^2\mathrm{d}x \geq \int_\Omega \frac{\mathrm{e}^{-l}}{2}\big|\nabla v_{kl}\big|^2 - \mathrm{e}^l\big|\nabla\phi_{kl}\big|^2\mathrm{d}x. \tag{12.80}$$

Using the obvious estimate $\widehat{e}_l(\xi) \leq \mathrm{e}^l|\xi|$ and thus $[\widehat{e}_l]^*(\zeta) \geq \delta_{[-\mathrm{e}^l,\mathrm{e}^l]}$, by Gronwall's inequality we eventually obtain the estimates (12.78a-c). Furthermore, from the equations themselves we obtain dual estimates on the time derivatives (12.78d,e). □

Proposition 12.22 (Convergence). *The approximate Galerkin solution $(\phi_{kl}, u_{kl}, v_{kl})$ converges (as a subsequence) for $k \to \infty$ (l kept fixed) weakly in $L^2(I;W^{1,2}(\Omega))^3$ to a weak solution, let us denote it by (ϕ_l, u_l, v_l), of (12.74) with e^u and e^v replaced by $e_l(u)$ and $e_l(v)$, respectively.*

Proof. The declared convergence can be shown in parallel to Section 11.2.1. In particular, by Aubin-Lions' lemma we have $w_{kl} := e_l(u_{kl}) \to w_l$ in $L^2(Q)$; here we also use that $\nabla w_{kl} = e_l'(u_{kl})\nabla u_{kl}$ is bounded in $L^2(Q;\mathbb{R}^n)$ due to (12.78b). Also $u_{kl} \rightharpoonup u_l$ in $L^2(Q)$, thus $\int_Q w_{kl}u_{kl}\,\mathrm{d}x\mathrm{d}t \to \int_Q w_l u_l\,\mathrm{d}x\mathrm{d}t$ so that by maximal monotonicity of e_l one can conclude that $w_l := e_l(u_l)$. Similarly, we can also prove $e_l(v_{kl}) \to e_l(v_l)$ in $L^2(Q)$, and therefore also $s_l(u_{kl}, v_{kl}) = r(e_l(u_{kl}), e_l(v_{kl})) \to r(e_l(u_l), e_l(v_l)) = s_l(u_l, v_l)$. Then one can pass to the limit in the Galerkin identity

$$\lim_{k\to\infty} \int_Q \frac{\partial w_{kl}}{\partial t} z + e_l(u_{kl})\nabla(u_{kl} - \phi_{kl}) \cdot \nabla z - s_l(u_{kl}, v_{kl}) z \,\mathrm{d}x\mathrm{d}t$$
$$= \int_0^T \bigg(\Big\langle \frac{\partial w_l}{\partial t}, z \Big\rangle + \int_\Omega e_l(u_l)\nabla(u_l - \phi_l) \cdot \nabla z - s_l(u_l, v_l) z \,\mathrm{d}x \bigg) \mathrm{d}t. \tag{12.81}$$

Analogous limit passage can be made in the other equations; the term $\frac{\partial}{\partial t}(\mathrm{div}(\varepsilon\nabla\phi_{kl}))$ is linear hence the limit passage is possible by a weak convergence due to the estimate (12.78d). □

Remark 12.23 (Limit passage for $l \to +\infty$). The strategy to pass to the original system (12.74) is to show a-priori bounds for u_l and v_l in $L^\infty(Q)$ independent of l and then, if l is chosen bigger than these bounds, (ϕ_l, u_l, v_l) is the weak solution of the non-modified system (12.74). For this, rather nontrivial step, we refer to Gajewski [163] and Gajewski and Gröger [167].

Remark 12.24 (Index-2 differential-algebraic system). When applying Galerkin approximation to (12.67), we obtain a system of so-called differential-algebraic equations (DAEs), i.e. time derivative is involved only in some components (here corresponding to u and v), while the rest (here corresponding to ϕ-components) forms an algebraic system. If one can eliminate the algebraic part after differentiating it k-times, we say that the DAEs have the (differential) index $k+1$. Therefore, as Gajewski's transformation (12.74) shows, the original system (12.67) can be viewed (in its Galerkin approximation) as an index-2 DAE.

Remark 12.25 (*Nernst-Planck-Poisson system*). The special case $c_\mathrm{D} \equiv 0$ and $r \equiv 0$ is a basic model for electro-diffusion of ions in electrolytes, which was formulated by W. Nernst and M. Planck at the end of the 19th century.[19]

12.5 Phase-field model

To describe solidification/melting processes at the microscale, models of a so-called phase-field type can be used. A basic Caginalp's model [87][20] consists of the

[19] W.H. Nernst received the Nobel prize in chemistry for his work in thermochemistry in 1920 (while also M.K.E.L. Planck received Nobel's prize already in 1918 in physics but not directly related to the system (12.67) with $c_\mathrm{D} = r = 0$).

[20] For further study see Brokate and Sprekels [71, Sect.6.2], Elliott and Zheng [137], Kenmochi and Niezgódka [229], Zheng [428, Sect.4.1].

following system:

$$\left.\begin{array}{ll}\left.\begin{array}{l}\dfrac{\partial\theta}{\partial t}=\Delta\theta+\dfrac{\partial v}{\partial t}+g,\\[2mm] \dfrac{\partial v}{\partial t}=\Delta v-c(v)-\theta\end{array}\right\} & \text{in } Q,\\[2mm] \theta=0,\quad v=0 & \text{on } \Sigma,\\ \theta(0,\cdot)=\theta_0,\quad v(0,\cdot)=v_0 & \text{on } \Omega\end{array}\right\} \tag{12.82}$$

where θ plays the role of a temperature and the "order parameter" v distinguishes particular phases according to where the (typically nonconvex) potential $\widehat{c}$ of $c:\mathbb{R}\to\mathbb{R}$ attains its minima. This is an interesting system not only for its applications but also for "training" purposes because there are various ways to get a-priori estimates and then prove existence of a solution. Let us outline the a-priori estimates heuristically.

First option: Summing the equations in (12.82) gives $\frac{\partial}{\partial t}\theta=\Delta(\theta+v)-c(v)-\theta+g$, and then testing it by θ yields the estimate

$$\begin{aligned}\frac{1}{2}\frac{\mathrm{d}}{\mathrm{d}t}\big\|\theta\big\|_{L^2(\Omega)}^2+\big\|\nabla\theta\big\|_{L^2(\Omega;\mathbb{R}^n)}^2+\big\|\theta\big\|_{L^2(\Omega)}^2&=\int_\Omega\nabla\theta{\cdot}\nabla v+\big(g-c(v)\big)\theta\,\mathrm{d}x\\ &\le\frac{1}{2}\big\|\nabla\theta\big\|_{L^2(\Omega;\mathbb{R}^n)}^2+\frac{1}{2}\big\|\nabla v\big\|_{L^2(\Omega;\mathbb{R}^n)}^2+\frac{3}{2}\big\|\theta\big\|_{L^2(\Omega)}^2+\big\|g\big\|_{L^2(\Omega)}^2+\big\|c(v)\big\|_{L^2(\Omega)}^2.\end{aligned}\tag{12.83}$$

Then, assuming $c(v)v\ge -c_1-c_2v^2$, $c_1,c_2\ge 0$, and testing the second equation in (12.82) by v yields:

$$\begin{aligned}\frac{1}{2}\frac{\mathrm{d}}{\mathrm{d}t}\big\|v\big\|_{L^2(\Omega)}^2+\big\|\nabla v\big\|_{L^2(\Omega;\mathbb{R}^n)}^2&=\int_\Omega\big(\theta-c(v)\big)v\,\mathrm{d}x\\ &\le c_1\mathrm{meas}_n(\Omega)+\frac{1}{2}\big\|\theta\big\|_{L^2(\Omega)}^2+\Big(\frac{1}{2}+c_2\Big)\big\|v\big\|_{L^2(\Omega)}^2.\end{aligned}\tag{12.84}$$

Summing (12.83) and (12.84), by standard procedure via Gronwall's inequality one obtains the a-priori estimates

$$\big\|\theta\big\|_{L^\infty(I;L^2(\Omega))}\le C,\quad \big\|\nabla\theta\big\|_{L^2(Q;\mathbb{R}^n)}\le C,\tag{12.85}$$

$$\big\|v\big\|_{L^\infty(I;L^2(\Omega))}\le C,\quad \big\|\nabla v\big\|_{L^2(Q;\mathbb{R}^n)}\le C,\tag{12.86}$$

provided $\theta_0,v_0\in L^2(\Omega)$, $g\in L^2(Q)$, and $c(\cdot)$ has at most linear growth because of the last term in (12.83).[21] The "dual" estimates of $\frac{\partial}{\partial t}\theta$ and $\frac{\partial}{\partial t}v$ in $L^2(I;W^{1,2}(\Omega)^*)$ then follow standardly.

[21]Standardly, $\widehat{c}$ is of the type $\widehat{c}(r)=(r^2-1)^2$, which is not consistent with this approach, however.

Second option: Testing the first equation of (12.82) by θ and the second one by $\frac{\partial}{\partial t}v$ gives

$$\frac{1}{2}\frac{\mathrm{d}}{\mathrm{d}t}\big\|\theta\big\|^2_{L^2(\Omega)} + \big\|\nabla\theta\big\|^2_{L^2(\Omega;\mathbb{R}^n)} = \int_\Omega \frac{\partial v}{\partial t}\theta + g\theta \,\mathrm{d}x \le \int_\Omega \frac{\partial v}{\partial t}\theta \,\mathrm{d}x$$
$$+ \varepsilon\big\|\theta\big\|^2_{L^{2^*}(\Omega)} + C_\varepsilon\big\|g\big\|^2_{L^{2^{*\prime}}(\Omega)}, \tag{12.87}$$

$$\Big\|\frac{\partial v}{\partial t}\Big\|^2_{L^2(\Omega)} + \frac{1}{2}\frac{\mathrm{d}}{\mathrm{d}t}\big\|\nabla v\big\|^2_{L^2(\Omega)} + \frac{\mathrm{d}}{\mathrm{d}t}\int_\Omega \widehat{c}(v)\,\mathrm{d}x = -\int_\Omega \theta\,\frac{\partial v}{\partial t}\,\mathrm{d}x\,, \tag{12.88}$$

where $\widehat{c} : \mathbb{R} \to \mathbb{R}$ is the potential (i.e. the primitive function) of c. Supposing $\theta_0 \in L^2(\Omega)$, $v_0 \in W^{1,2}(\Omega)$, $g \in L^2(I; L^{2^{*\prime}}(\Omega))$, and $\widehat{c} \ge 0$, and summing (12.87) and (12.88), and using Gronwall's inequality eventually yields the estimates:

$$\big\|\theta\big\|_{L^\infty(I;L^2(\Omega))} \le C, \qquad \big\|\nabla\theta\big\|_{L^2(Q;\mathbb{R}^n)} \le C, \tag{12.89}$$

$$\big\|\nabla v\big\|_{L^\infty(I;L^2(\Omega;\mathbb{R}^n))} \le C, \qquad \Big\|\frac{\partial v}{\partial t}\Big\|_{L^2(Q)} \le C. \tag{12.90}$$

Coming back to the first equation of (12.82), one gets standardly the "dual" estimate of $\frac{\partial}{\partial t}\theta$ in $L^2(I; W^{1,2}(\Omega)^*)$.

Third option: Testing the second equation in (12.82) again by $\frac{\partial}{\partial t}v$ but the first one by $\frac{\partial}{\partial t}\theta$ gives, besides (12.88), the estimate

$$\Big\|\frac{\partial \theta}{\partial t}\Big\|^2_{L^2(\Omega)} + \frac{1}{2}\frac{\mathrm{d}}{\mathrm{d}t}\big\|\nabla\theta\big\|^2_{L^2(\Omega;\mathbb{R}^n)} = \int_\Omega \Big(\frac{\partial v}{\partial t} + g\Big)\frac{\partial \theta}{\partial t}\mathrm{d}x$$
$$\le \frac{1}{\sqrt{2}}\bigg(\Big\|\frac{\partial \theta}{\partial t}\Big\|^2_{L^2(\Omega)} + \Big\|\frac{\partial v}{\partial t}\Big\|^2_{L^2(\Omega)} + \big\|g\big\|^2_{L^2(\Omega)}\bigg). \tag{12.91}$$

Supposing $\theta_0, v_0 \in W^{1,2}(\Omega)$, and $g \in L^2(Q)$ and summing (12.88) and (12.91), and estimating $|\int_\Omega \theta\,\frac{\partial}{\partial t}v\,\mathrm{d}x| \le N^2\|\nabla\theta\|^2_{L^2(\Omega;\mathbb{R}^n)} + \frac{1}{4}\|\frac{\partial}{\partial t}v\|^2_{L^2(\Omega)}$ in (12.88) with N the norm of the embedding $W^{1,2}(\Omega) \subset L^2(\Omega)$ gives via Gronwall's inequality again the estimates (12.90) together with

$$\big\|\theta\big\|_{L^\infty(I;W^{1,2}(\Omega))} \le C, \qquad \Big\|\frac{\partial \theta}{\partial t}\Big\|_{L^2(Q)} \le C. \tag{12.92}$$

Exercise 12.26 (Scaling). Modify the above estimate for the scaling by ζ and ξ as in (9.80) similarly as done in Example 9.32.

Exercise 12.27 (Galerkin's method). Prove existence of a weak solution to (12.82) by convergence of the approximate solution constructed by the Galerkin method based on the estimates (12.83)–(12.84), or (12.87)–(12.88), or (12.88)–(12.91). Note that Aubin-Lions' Lemma 7.7 is used for the only nonlinear term c.[22]

[22] Hint in case of (12.87)–(12.88): Make a growth assumption $|c(r)| \le C(1 + |r|^{2^*-\epsilon})$, $\epsilon > 0$, so that, in view of (12.90) and Aubin-Lions' Lemma 7.7, Galerkin's sequence of v's is compact in $L^{2^*-\epsilon}(Q)$ and the limit passage through the nonlinear term c is possible as the Nemytskiĭ mapping $\mathscr{N}_c : L^{2^*-\epsilon}(Q) \to L^1(Q)$ is continuous.

Exercise 12.28 (*Semi-implicit Rothe method*). An advantageous modification of Rothe's method that would lead to a de-coupling of the system at each time level can be based on the semi-implicit formula:

$$\frac{\theta_\tau^k-\theta_\tau^{k-1}}{\tau}=\Delta\theta_\tau^k+\frac{v_\tau^k-v_\tau^{k-1}}{\tau}+g_\tau^k,\qquad \frac{v_\tau^k-v_\tau^{k-1}}{\tau}=\Delta v_\tau^k-c(v_\tau^k)+\theta_\tau^{k-1}. \tag{12.93}$$

Modify the a-priori estimates (12.88) and (12.91) and prove the convergence.[23]

Exercise 12.29 (*Penrose and Fife's generalization* [333]). Consider

$$\left.\begin{aligned}\frac{\partial\theta}{\partial t}&=\Delta\beta(\theta)+d(v)\frac{\partial v}{\partial t}+g,\\ \frac{\partial v}{\partial t}&=\Delta v-c(v)-d(v)\beta(\theta)\end{aligned}\right\} \tag{12.94}$$

with $\beta:\mathbb{R}\to\mathbb{R}$ increasing. Note that for $d(\cdot)=1$ and $\beta(r)=r$, we get (12.82). The physically justified option suggested by Penrose and Fife [333] uses $\beta(r)=-1/r$, which requires quite sophisticated techniques,[24] however. Introducing the new variable $u=\beta(\theta)$ and $e(\cdot)=\beta(\cdot)^{-1}$, (12.94) transforms to a doubly-nonlinear system $\frac{\partial}{\partial t}e(u)-\Delta u=d(v)\frac{\partial}{\partial t}v+g$ and $\frac{\partial}{\partial t}v-\Delta v=-c(v)-d(v)u$. Assume, for simplicity, $\inf e'(\cdot)\ge\varepsilon>0$ and qualify also $c(\cdot)$ and $d(\cdot)$ appropriately, and use the technique from Sect. 11.2.1 to get the a-priori estimate of u in $L^2(I;W^{1,2}(\Omega))\cap L^\infty(I;L^2(\Omega))$ and of v in $W^{1,\infty,2}(I;W^{1,2}(\Omega),L^2(\Omega))$ by testing these equations by u and $\frac{\partial}{\partial t}v$, respectively.[25] After deriving still a dual estimate for $\frac{\partial}{\partial t}e(u)$, prove convergence of, say, Rothe approximations as in Sect. 11.2.1.

---

[23] Hint: Modification of the a-priori estimates (12.88) and (12.91) can be based simply on $\theta_\tau^{k-1}=\theta_\tau^k-\tau\frac{\theta_\tau^k-\theta_\tau^{k-1}}{\tau}$ so that one can estimate $\|\theta_\tau^{k-1}\|_{L^2(\Omega)}^2\le2\|\theta_\tau^k\|_{L^2(\Omega)}^2+2\tau^2\|\frac{\theta_\tau^k-\theta_\tau^{k-1}}{\tau}\|_{L^2(\Omega)}^2$. In other words, $\|\bar\theta_\tau^{\rm R}\|_{L^2(\Omega)}^2\le2\|\bar\theta_\tau\|_{L^2(\Omega)}^2+2\tau^2\|\frac{\partial}{\partial t}\theta_\tau\|_{L^2(\Omega)}^2$ where the "retarded" Rothe function $\bar\theta_\tau^{\rm R}$ is as in (8.202). The additional term $\tau^2\|\frac{\partial}{\partial t}\theta_\tau\|_{L^2(\Omega)}^2$ can be absorbed if τ is small enough. The convergence of the scheme based on (12.93), i.e.

$$\frac{\partial\theta_\tau}{\partial t}=\Delta\bar\theta_\tau+\frac{\partial v_\tau}{\partial t}+\bar g_\tau,\qquad\frac{\partial v_\tau}{\partial t}=\Delta\bar v_\tau-c(\bar v_\tau)+\bar\theta_\tau^{\rm R},$$

can be proved as in Exercise 12.27 when using also $\|\bar\theta_\tau^{\rm R}-\bar\theta_\tau\|_{L^2(\Omega)}=\tau\|\frac{\partial}{\partial t}\theta_\tau\|_{L^2(\Omega)}=\mathscr{O}(\tau)$.

[24] See Brokate and Sprekels [71, Sect.6.3], Colli and Sprekels [103], Elliott and Zheng [137], or Zheng [428, Sect.4.1.2]. Asymptotic behaviour under scaling of the terms $\frac{\partial}{\partial t}v$ and Δv in (12.94) and relation to a modified Stefan problem is in [103].

[25] Hint: Use $\int_\Omega(\frac{\partial}{\partial t}e(u))u\,dx=\frac{d}{dt}\int_\Omega[\widehat e]^*(u)\,dt$ with $[\widehat e]^*$ the conjugate function of the primitive function of e satisfying $[\widehat e]^*(r)\le|r|^2/(2\varepsilon)$ because $\widehat e(r)\ge\varepsilon|r|^2/2$, cf. (8.248)–(8.249), and note that the arisen terms $\pm\int_\Omega u\,d(v)\frac{\partial}{\partial t}v\,dx$ cancel with each other, and eventually estimate

$$\frac{d}{dt}\Big([\widehat e]^*(u)+\frac12|\nabla v|^2+\widehat c(v)\Big)+|\nabla u|^2+\Big|\frac{\partial v}{\partial t}\Big|^2=\int_\Omega gu\,dx\le\|g\|_{L^{2^*\prime}(\Omega)}\|u\|_{L^{2^*}(\Omega)}.$$

Exercise 12.30 (Kenmochi-Niezgódka's modification [228, 230][26]). Consider a *Cahn-Hilliard equation* $\frac{\partial}{\partial t}v + \Delta(\Delta v - c(v) - \theta) = 0$ instead of $\frac{\partial}{\partial t}v = \Delta v - c(v) - \theta$ in (12.82) and derive the a-priori estimates by testing by θ and by $\Delta^{-1}\frac{\partial}{\partial t}v$.[27]

Exercise 12.31 (Beneš' generalizations [43, 44]). Augment the second equation in (12.82) to $\frac{\partial}{\partial t}v = \Delta v - c(v) - \theta + \psi(\theta)|\nabla v|$ with $\psi : \mathbb{R} \to \mathbb{R}$ continuous and bounded, and modify all above estimates[28] and prove convergence of the Galerkin approximation[29]. Consider further a given velocity field $\vec{v} : Q \to \mathbb{R}^n$ and augment (12.82) by the advection terms, i.e. $\frac{\partial}{\partial t}\theta + \vec{v}\cdot\nabla\theta$ and $\frac{\partial}{\partial t}v + \vec{v}\cdot\nabla v$ instead of $\frac{\partial}{\partial t}\theta$ and $\frac{\partial}{\partial t}v$, respectively, and modify all above estimates[30] and prove convergence of the Galerkin approximation[31].

12.6 Navier-Stokes-Nernst-Planck-Poisson-type system

An incompressible ionized mixture of L mutually reacting chemical constituents occurs in various biological or chemical applications. We accept a so-called Eckart–

[26] For such a sort of model see also Alt and Pawłow [12].

[27] Hint: Test the first equation in (12.82) by θ and the Cahn-Hilliard equation by $J^{-1}\frac{\partial}{\partial t}v$ for $J : W_0^{1,2}(\Omega) \to W^{-1,2}(\Omega)$ the duality mapping, i.e. by $J^{-1}\frac{\partial}{\partial t}v = -\Delta^{-1}\frac{\partial}{\partial t}v$, cf. (3.18). By using several times Green's formula, realize that $-\int_\Omega \frac{\partial}{\partial t}v(\Delta^{-1}\frac{\partial}{\partial t}v)\,\mathrm{d}x = \|\frac{\partial}{\partial t}v\|^2_{W^{-1,2}(\Omega)}$ because of the definition (3.1) and because J^{-1} itself is the duality mapping $W^{-1,2}(\Omega) \to W_0^{1,2}(\Omega)$, further that

$$-\int_\Omega \Delta^2 v\Big(\Delta^{-1}\frac{\partial v}{\partial t}\Big)\mathrm{d}x = -\int_\Omega \Delta v \Delta\Big(\Delta^{-1}\frac{\partial v}{\partial t}\Big)\mathrm{d}x = -\int_\Omega \Delta v\frac{\partial v}{\partial t}\,\mathrm{d}x = \frac{1}{2}\frac{\mathrm{d}}{\mathrm{d}t}\int_\Omega |\nabla v|^2\,\mathrm{d}x,$$

and $\int_\Omega \Delta c(v)(\Delta^{-1}\frac{\partial}{\partial t}v)\,\mathrm{d}x = \int_\Omega c(v)\frac{\partial}{\partial t}v\,\mathrm{d}x = \frac{\mathrm{d}}{\mathrm{d}t}\int_\Omega \hat{c}(v)\,\mathrm{d}x$ with $\hat{c}$ again the primitive function to c, and also $-\int_\Omega \Delta\theta(\Delta^{-1}\frac{\partial}{\partial t}v)\,\mathrm{d}x = -\int_\Omega \theta\frac{\partial}{\partial t}v\,\mathrm{d}x$ cancels with the corresponding term coming from the first equation in (12.82), and obtain the estimates of $\theta \in L^2(I;W_0^{1,2}(\Omega))\cap L^\infty(I;L^2(\Omega))$ and $v \in L^\infty(I;W_0^{1,2}(\Omega))\cap W^{1,2}(I;W^{-1,2}(\Omega))$. From the first equation in (12.82), obtain now also the estimate of $\frac{\partial}{\partial t}\theta$ in $L^2(I;W^{-1,2}(\Omega))$ and from the Cahn-Hilliard equation $\Delta^2 v = \Delta(c(v) + \theta) - \frac{\partial}{\partial t}v$ eventually the estimate of v in $L^2(I;W^{2,2}(\Omega))$ under a suitable qualification of $c(\cdot)$. Then show convergence, e.g., of Galerkin's approximants. For details of Galerkin's approximation we refer (up to some sign conventions) to [230].

[28] Hint: Estimate the term $\psi(\theta)|\nabla v|$ as $\int_\Omega \psi(\theta)|\nabla v|v\,\mathrm{d}x \le \max\psi(\mathbb{R})^2\|v\|^2_{L^2(\Omega)} + \frac{1}{4}\|\nabla v\|^2_{L^2(\Omega;\mathbb{R}^n)}$ for (12.84) or $\int_\Omega \psi(\theta)|\nabla v|\frac{\partial}{\partial t}v\,\mathrm{d}x \le \max\psi(\mathbb{R})^2\|\nabla v\|^2_{L^2(\Omega;\mathbb{R}^n)} + \frac{1}{4}\|\frac{\partial}{\partial t}v\|^2_{L^2(\Omega)}$ for (12.88).

[29] Hint: Show the strong convergence of θ by Aubin-Lions' Lemma 7.7 and of ∇v by uniform monotonicity of the Laplacean as in Exercise 8.81, and then pass to the limit in the term $\psi(\theta)|\nabla v|$ by continuity.

[30] Hint: Assume $\vec{v} \in L^\infty(Q;\mathbb{R}^n)$ and estimate the term $\vec{v}\cdot\nabla\theta$ as $\int_\Omega -(\vec{v}(t,\cdot)\cdot\nabla\theta)\,\theta\,\mathrm{d}x \le \frac{1}{4}\|\nabla\theta\|^2_{L^2(\Omega;\mathbb{R}^n)} + \|\vec{v}(t,\cdot)\|^2_{L^\infty(\Omega;\mathbb{R}^n)}\|\theta\|^2_{L^2(\Omega)}$ for (12.83) or as $\int_\Omega -(\vec{v}(t,\cdot)\cdot\nabla\theta)\frac{\partial}{\partial t}\theta\,\mathrm{d}x \le \frac{1}{4}\|\frac{\partial}{\partial t}\theta\|^2_{L^2(\Omega)} + \|\vec{v}(t,\cdot)\|^2_{L^\infty(\Omega;\mathbb{R}^n)}\|\nabla\theta\|^2_{L^2(\Omega;\mathbb{R}^n)}$ for (12.91), and analogously for the term $\vec{v}\cdot\nabla v$ for (12.84) and (12.88). Alternatively, assuming $\operatorname{div}\vec{v} = 0$ and $\vec{v}|_\Sigma = 0$, use the calculations (6.33) for (12.83) and (12.84) to cause these terms to vanish.

[31] Hint: Show the strong convergence of θ and v by Aubin-Lions' lemma and pass to the limit in the terms $\vec{v}\cdot\nabla\theta$ and $\vec{v}\cdot\nabla v$ by the weak convergence.

Prigogine's phenomenological concept [133, 340][32] balancing only the barycentric momentum. Under certain simplifications[33], it leads to a system of $n+L+1$ differential equations combining the Navier-Stokes, and the Nernst-Planck equation modified for moving media, and the Poisson equation for the electrostatic field:

$$\varrho\frac{\partial v}{\partial t} + \varrho(v\cdot\nabla)v - \operatorname{div}(\mu\nabla v) + \nabla\pi = \sum_{\ell=1}^{L} u_\ell f_\ell\,, \qquad \operatorname{div} v = 0\,, \qquad f_\ell = -e_\ell\nabla\phi, \tag{12.95a}$$

$$\frac{\partial u_\ell}{\partial t} + \operatorname{div}\big(j_\ell + u_\ell v\big) = r_\ell(u_1,\dots,u_L)\,,$$
$$j_\ell = -d_1\nabla u_\ell - d_2 u_\ell(e_\ell - q_{\rm tot})\nabla\phi, \qquad \ell = 1,\dots,L\,, \tag{12.95b}$$

$$-\operatorname{div}(\varepsilon\nabla\phi) = q_{\rm tot}, \qquad q_{\rm tot} = \sum_{\ell=1}^{L} e_\ell u_\ell\,, \tag{12.95c}$$

with the initial conditions

$$v(0,\cdot) = v_0\,, \qquad u_\ell(0,\cdot) = u_{\ell 0} \quad \text{on } \Omega. \tag{12.96}$$

The meaning of the variables is:

v barycenter velocity,
π pressure,
u_ℓ concentration of ℓ-constituent, presumably to satisfy $\sum_{\ell=1}^{L} u_\ell = 1$, $u_\ell \geq 0$,
ϕ electrostatic potential,
$q_{\rm tot}$ the total electric charge,

and of the data is:

$\mu > 0$ viscosity,
$\varrho > 0$ density,
e_ℓ valence (= charge) of the ℓ-constituent,
$\varepsilon > 0$ permitivity,
$r_\ell(u_1,\dots,u_L)$ production rate of the ℓ-constituent by chemical reactions,
f_ℓ body force acting on the ℓ-constituent: $f_\ell = -e_\ell\nabla\phi$, [34]
j_ℓ phenomenological flux of the ℓ-constituent given in (12.95b) with
$d_1, d_2 > 0$ diffusion and mobility coefficients, respectively.

[32] I. Prigogine received the Nobel prize in chemistry for non-equilibrium thermodynamics, particularly the theory of dissipative structures, in 1977. An alternative, Truesdell's description of mixtures, balances momenta of each constituent and counts for interactive forces among them. Both concepts, completed by energy balance, have points that are thermodynamically still not fully justified. For relations between (12.95) and Truesdel's model see Samohýl [378].

[33] In particular, we consider isothermal processes, a volume additivity hypothesis with each constituent incompressible, small electrical currents (i.e. magnetic field is neglected), and the diffusion and mobility coefficients as well as mass densities being the same for each constituent. Moreover, we neglect the electric field outside of the specimen Ω.

[34] This comes from Lorenz' force acting on a charge e_ℓ moving in the electromagnetic field (E,B), i.e. $f_\ell = e_\ell(E + v_\ell \times B)$ after simplification $E = -\nabla\phi$ and $B = 0$.

In other words, $j_\ell = -d_2(u_\ell\nabla\mu_\ell - q_{\rm tot}\nabla\phi)$ where

$$\mu_\ell = e_\ell\phi + \frac{d_1}{d_2}\ln(u_\ell), \tag{12.97}$$

plays the role of an electrochemical potential. The equations (12.95a-c) thus balance the barycentric momentum (under the incompressibility condition), mass of particular constituents, and electric induction $\varepsilon\nabla\phi$. The interpretation of $q_{\rm tot}\nabla\phi$ in j_ℓ is a "reaction force"[35] keeping the natural requirement $\sum_{\ell=1}^L j_\ell = 0$ satisfied, which eventually fixes also the constraint $\sum_{\ell=1}^L u_\ell = 1$. We still consider some boundary conditions, e.g. a closed container, which, in some simplified version, leads to:

$$v = 0, \qquad u_\ell = u_{\ell\Gamma}, \qquad \phi = 0 \quad \text{ on } \Sigma. \tag{12.98}$$

Besides, we naturally assume $r_\ell : \mathbb{R}^L \to \mathbb{R}$ continuous and the mass conservation in all chemical reactions and non-negative production rate of ℓth constituent if there its concentration vanishes, and the volume-additivity constraint holds for the initial and the boundary conditions, i.e.

$$\sum_{\ell=1}^L r_\ell(u_1,\dots,u_L) = 0\ , \qquad r_\ell(u_1,\dots,u_{\ell-1},0,u_{\ell+1},\dots,u_L) \geq 0, \tag{12.99a}$$

$$\sum_{\ell=1}^L u_{\ell 0} = 1, \qquad \sum_{\ell=1}^L u_{\ell\Gamma} = 1, \qquad u_{\ell 0} \geq 0, \qquad u_{\ell\Gamma} \geq 0. \tag{12.99b}$$

For analysis, we define a so-called retract $K : \mathscr{M} \to \{\xi\in\mathscr{M};\ \xi_\ell \geq 0,\ \ell = 1,\dots,L\}$, where $\mathscr{M}$ denotes the affine manifold $\{\xi\in\mathbb{R}^L;\ \sum_{\ell=1}^L \xi_\ell = 1\}$, by

$$K_\ell(\xi) := \frac{\xi_\ell^+}{\sum_{k=1}^L \xi_k^+}, \qquad \xi_\ell^+ := \max(\xi_\ell, 0). \tag{12.100}$$

Note that K is continuous and bounded on $\mathscr{M}$. Starting with $\bar u \equiv (\bar u_\ell)_{\ell=1,..,L}$ and $\bar v$ given such that $\sum_{\ell=1}^L \bar u_\ell = 1$, we solve subsequently the Poisson, the approximate Navier-Stokes[36], and finally the generalized Nernst-Planck equations, i.e.

$$-\operatorname{div}(\varepsilon\nabla\phi) = \bar q_{\rm tot}\ , \qquad \bar q_{\rm tot} = \sum_{\ell=1}^L e_\ell K_\ell(\bar u)\ , \tag{12.101a}$$

$$\varrho\frac{\partial v}{\partial t} + \varrho(\bar v\cdot\nabla)v - \operatorname{div}(\mu\nabla v) + \nabla\pi = \bar q_{\rm tot}\nabla\phi\ , \qquad \operatorname{div} v = 0\ , \tag{12.101b}$$

$$\begin{aligned}\frac{\partial u_\ell}{\partial t} - \operatorname{div}(d_1\nabla u_\ell) + \operatorname{div}(u_\ell\bar v) &= r_\ell\big(K(\bar u)\big)\\ &\quad - \operatorname{div}\big(d_2K_\ell(\bar u)(e_\ell - \bar q_{\rm tot})\nabla\phi\big), \quad \ell = 1,..,L.\end{aligned} \tag{12.101c}$$

[35] This force is usually small because $|q_{\rm tot}|$ is small in comparison with $\max_{\ell=1,\dots,L}|e_\ell|$. Often, even the electro-neutrality assumption $q_{\rm tot} = 0$ is postulated. For derivation of this force and clarification of specific simplifications see Samohýl [378].

[36] The equation (12.101b) is called an Oseen problem.

Let $W^{1,2}_{0,\mathrm{div}}(\Omega;\mathbb{R}^n) := \{v \in W^{1,2}_0(\Omega;\mathbb{R}^n);\ \mathrm{div}\, v = 0\}$. Recall the notion of a very weak solution to the Navier-Stokes equation (12.95a) defined in Section 8.8.4 and analogously to (12.101b). Also, we use this concept for (12.95b) and (12.101c). The "technical" difficulty is that the time-derivatives $\frac{\partial}{\partial t}v$ and $\frac{\partial}{\partial t}u_\ell$ are not in duality with v and u_ℓ themselves.

Lemma 12.32 (A-PRIORI BOUNDS). *Let* (12.99) *hold and let* $n \le 3$. *For any* $\bar v \in L^2(I;W^{1,2}_{0,\mathrm{div}}(\Omega;\mathbb{R}^n)) \cap L^\infty(I;L^2(\Omega;\mathbb{R}^n))$ *and* $\bar u \in L^2(Q;\mathbb{R}^L)$ *such that* $\bar u(\cdot) \in \mathscr{M}$ *a.e. in* Q, *the equations* (12.101) *have very weak solutions* (v,ϕ,u) *which satisfy:*

$$\big\|\nabla\phi\big\|_{L^\infty(I;L^2(\Omega;\mathbb{R}^n))} \le C_0, \tag{12.102a}$$

$$\big\|v\big\|_{L^2(I;W^{1,2}(\Omega;\mathbb{R}^n))\cap L^\infty(I;L^2(\Omega;\mathbb{R}^n))} \le C_1, \tag{12.102b}$$

$$\Big\|\frac{\partial v}{\partial t}\Big\|_{L^{4/3}(I;W^{1,2}_{0,\mathrm{div}}(\Omega;\mathbb{R}^n)^*)} \le C_2 + C_3\big\|\bar v\big\|_{L^2(I;W^{1,2}(\Omega;\mathbb{R}^3))\cap L^\infty(I;L^2(\Omega;\mathbb{R}^n))}, \tag{12.102c}$$

$$\big\|u_\ell\big\|_{L^2(I;W^{1,2}(\Omega))\cap L^\infty(I;L^2(\Omega))} \le C_4, \tag{12.102d}$$

$$\Big\|\frac{\partial u_\ell}{\partial t}\Big\|_{L^{4/3}(I;W^{1,2}(\Omega)^*)} \le C_5 + C_6\big\|\bar v\big\|_{L^2(I;W^{1,2}(\Omega;\mathbb{R}^3))\cap L^\infty(I;L^2(\Omega;\mathbb{R}^n))}, \tag{12.102e}$$

with the constants $C_0,\dots,C_6$ *independent of* $\bar u$ *and* $\bar v$. *Besides,* u *satisfies the constraint* $\bar u(\cdot) \in \mathscr{M}$ *a.e. in* Q *(but not necessarily* $u_\ell \ge 0$*).*

Proof. We consider $n = 3$, the case $n \le 2$ being thus covered, too. Existence of very weak solutions to the particular decoupled linear equations (12.101a), (12.101b), and (12.101c) can be proved by standard arguments, based on the bounds below.

The estimate (12.102a) is obvious if one tests (12.101a) by ϕ itself and realizes that the right-hand side of (12.101a) is a-priori bounded in $L^\infty(Q)$. The estimate (12.102b) for v can be obtained by testing the weak formulation of the approximate Navier-Stokes system (12.101b) by v; note that the term $\int_\Omega(\bar v\cdot\nabla)v\cdot v\,\mathrm{d}x$ vanishes, cf. also Section 8.8.4.[37]

The dual estimate (12.102c) for $\frac{\partial}{\partial t}v$ can then be obtained as in (12.51) by testing (12.101b) by a suitable z as follows:

$$\varrho\Big\|\frac{\partial v}{\partial t}\Big\|_{L^{4/3}(I;W^{1,2}_{0,\mathrm{div}}(\Omega;\mathbb{R}^n)^*)} \le \big\|\nabla v\big\|_{L^2(Q;\mathbb{R}^{3\times3})}\Big(\mu\sqrt[4]{T}+\varrho N^{3/2}\big\|\bar v\big\|^{1/2}_{L^2(I;W^{1,2}(\Omega))}$$
$$\times\big\|\bar v\big\|^{1/2}_{L^\infty(I;L^2(\Omega;\mathbb{R}^3))}\Big) + N\max_{\ell=1,..,L}|e_\ell|\,\big\|\nabla\phi\big\|_{L^{4/3}(I;L^{6/5}(\Omega))}. \tag{12.103}$$

Finally, the estimate (12.102e) for u_ℓ can be obtained by testing (12.101c) by

[37] More precisely, we can do it in Galerkin's approximations and then these bounds are inherited in the limit very weak solution, too.

u_ℓ.[38] The dual estimate for $\frac{\partial}{\partial t}u_\ell$ can again be obtained analogously as (12.103).[39]

Now, we have to prove that the constraint $\sum_{\ell=1}^L u_\ell = 1$ is satisfied. Let us abbreviate $\sigma(t,\cdot) := \sum_{\ell=1}^L u_\ell(t,\cdot)$. By summing (12.95b) for $\ell = 1,\dots,L$, one gets

$$\frac{\partial\sigma}{\partial t} = \sum_{\ell=1}^L r_\ell(K(\bar u)) + \mathrm{div}\Big(d_1\nabla\sigma - \bar v\sigma + \sum_{\ell=1}^L d_2 K_\ell(\bar u)\Big(e_\ell - \sum_{k=1}^L e_k K_k(\bar u)\Big)\nabla\phi\Big) = \mathrm{div}\big(d_1\nabla\sigma\big) - v\cdot\nabla\sigma \tag{12.104}$$

where (12.99a) has been used. Thus (12.104) results in the linear equation $\frac{\partial}{\partial t}\sigma + v\cdot\nabla\sigma - \mathrm{div}(d_1\nabla\sigma) = 0$. We assumed $\sigma|_{t=0} = \sum_{\ell=1}^L u_{\ell 0} = 1$ and $\sigma|_\Sigma = \sum_{\ell=1}^L u_{\ell\Gamma} = 1$ on Σ, cf. (12.96) and (12.98) with (12.99b), so that the unique solution to this equation is $\sigma(t,\cdot)\equiv 1$ for any $t > 0$.[40] □

Let us abbreviate

$$\mathscr{W}_1 := W^{1,2,4/3}\big(I; W^{1,2}(\Omega;\mathbb{R}^L), W^{1,2}(\Omega;\mathbb{R}^L)^*\big) \cap L^\infty(I;L^2(\Omega;\mathbb{R}^L)), \tag{12.105a}$$

$$\mathscr{W}_2 := W^{1,2,4/3}\big(I; W^{1,2}_{0,\mathrm{div}}(\Omega;\mathbb{R}^n), W^{1,2}_{0,\mathrm{div}}(\Omega;\mathbb{R}^n)^*\big) \cap L^\infty(I;L^2(\Omega;\mathbb{R}^n)). \tag{12.105b}$$

If $n \le 3$, Aubin-Lions' Lemma 7.7 gives the compact embeddings $\mathscr{W}_1 \Subset L^2(I;L^{6-\epsilon}(\Omega;\mathbb{R}^L))$ for any $\epsilon > 0$, and similarly $\mathscr{W}_2 \Subset L^2(I;L^{6-\epsilon}(\Omega;\mathbb{R}^n))$. Moreover, let us abbreviate $M : \mathscr{W}_1\times\mathscr{W}_2 \rightrightarrows \mathscr{W}_1\times\mathscr{W}_2$ defined by $(u,v)\in M(\bar u,\bar v)$ if u is a very weak solution to (12.101b) satisfying (12.102e) and v is a very weak solution (12.101b) satisfying (12.102b,c) with ϕ a weak solution to (12.101a).[41]

Lemma 12.33 (Continuity). *Let* (12.99a) *hold and let* $n \le 3$. *Then the set-valued mapping* M *is weakly* upper semicontinuous if restricted to* $\{(\bar u,\bar v)\in\mathscr{W}_1\times\mathscr{W}_2;\ \bar u(\cdot)\in\mathscr{M}$ *a.e. in* $Q\}$.

[38] Again, more precisely, we can do it in Galerkin's approximations and then these bounds are inherited in the limit very weak solution, too.

[39] The term $\mathrm{div}(u_\ell\bar v)$ in (12.101c) suggests the estimate

$$\int_Q u_\ell\bar v\cdot\nabla z\,\mathrm{d}x \le \|u_\ell\|_{L^2(I;L^6(\Omega))}\|\bar v\|^{1/2}_{L^2(I;L^6(\Omega;\mathbb{R}^n))}\|\bar v\|^{1/2}_{L^\infty(I;L^2(\Omega;\mathbb{R}^n))}\|\nabla z\|_{L^4(I;L^2(\Omega;\mathbb{R}^n))}$$

while the last term in (12.101c) can be estimated as

$$\int_Q d_2K_\ell(\bar u)(e_\ell - \bar q_{\mathrm{tot}})\nabla\phi\cdot\nabla z\,\mathrm{d}x\mathrm{d}t \le 2d_2\max_{\ell=1,..,L}|e_\ell|\,\|\nabla\phi\|_{L^2(Q;\mathbb{R}^n)}\|\nabla z\|_{L^2(Q;\mathbb{R}^n)}.$$

[40] Cf. Theorem 8.36.

[41] Note how carefully M is defined: not every weak solution necessarily satisfies the a-priori estimates because we cannot perform the desired tests. However, we can do it for the Galerkin solutions and then pass to the limit so that we can show that $M(\bar u,\bar v)$ is at least nonempty.

Proof. Taking a sequence of $\{(\bar{u}^k, \bar{v}^k)\}_{k\in\mathbb{N}}$ converging weakly to $(\bar{u}, \bar{v})$ in $\mathscr{W}_1 \times \mathscr{W}_2$, by Aubin-Lions' Lemma 7.7, $\bar{u}^k \to \bar{u}$ strongly in $L^2(Q;\mathbb{R}^L)$, hence $\phi^k \to \phi$ in $L^q(I;W^{1,2}(\Omega))$, and also $K_\ell(\bar{u}^k)\nabla\phi^k \to K_\ell(\bar{u})\nabla\phi$ in $L^q(I;L^2(\Omega;\mathbb{R}^3))$ with $q < +\infty$ arbitrary. Then the limit passage in (12.101b) is routine; obviously $\int_Q (\bar{v}^k\cdot\nabla)v^k\cdot z\,dxdt \to \int_Q (\bar{v}\cdot\nabla)v\cdot z\,dxdt$ at least for those test functions z which are also in $L^\infty(Q)$ because $\bar{v}^k \to \bar{v}$ strongly in $L^2(Q;\mathbb{R}^3)$ and $\nabla v^k \to \nabla v$ weakly $L^2(Q;\mathbb{R}^{3\times 3})$.[42] The limit passage in the very weak formulation of (12.101c) with $(u^k, v^k, \phi^k, \bar{u}^k)$ in place of $(u, v, \phi, \bar{u})$ easily follows by standard arguments using the a-priori estimates (12.102e). The a-priori estimates (12.102) themselves are preserved in the limit, too. □

Proposition 12.34 (Existence of a fixed point). *Let* (12.99) *hold and let* $n \le 3$. *The mapping* $(\bar{u}, \bar{v}) \mapsto (u, v)$ *has a fixed point* (u, v) *on the convex set*

$$\begin{aligned}\Big\{(u,v)\in \mathscr{W}_1\times\mathscr{W}_2:\ &\big\|u\big\|_{L^2(I;W^{1,2}(\Omega;\mathbb{R}^L))\cap L^\infty(I;L^2(\Omega;\mathbb{R}^L))} \le C_4,\\ &\Big\|\frac{\partial u}{\partial t}\Big\|_{L^{4/3}(I;W^{1,2}(\Omega;\mathbb{R}^L)^*)} \le C_5 + C_1C_3,\\ &\big\|v\big\|_{L^2(I;W^{1,2}(\Omega;\mathbb{R}^3))\cap L^\infty(I;L^2(\Omega;\mathbb{R}^3))} \le C_1,\\ &\Big\|\frac{\partial v}{\partial t}\Big\|_{L^{4/3}(I;W^{1,2}_{0,\mathrm{div}}(\Omega;\mathbb{R}^n)^*)} \le C_2 + C_1C_3,\ \ \sum_{\ell=1}^L u_\ell = 1\Big\}\end{aligned}\tag{12.106}$$

with $C_1, \dots, C_5$ *from* (12.102). *Moreover, every such fixed point satisfies also* $u_\ell \ge 0$ *for any* ℓ. *Thus, considering also* ϕ *related to this fixed point* (u, v), *the triple* (ϕ, v, u) *is a very weak solution to the system* (12.95).

Proof. The weak upper semi-continuity of M has been proved in Lemma 12.33. By a-priori estimates (12.102b-d) and by arguments such as (12.104), this mapping maps the convex set (12.106) into itself. Both $\mathscr{W}_1$ and $\mathscr{W}_2$ are compact if endowed with the weak* topologies. Thanks to the linearity of (12.101) and convexity of $\{(u, v)\}$ satisfying (12.102b-d) for $(\bar{u}, \bar{v})$ given, the set $M(\bar{u}, \bar{v})$ is convex. By Lemma 12.32, also $M(\bar{u}, \bar{v}) \ne \emptyset$. By Kakutani's Theorem 1.11, we obtain existence of a fixed point.

The constraint $\sum_{\ell=1}^L u_\ell = 1$ is, as proved in (12.104), satisfied and, at this fixed point, we have additionally $u_\ell(t,\cdot) \ge 0$ satisfied for any t. To see this, test (12.101c) with $u_\ell = \bar{u}_\ell$ by the negative part u_ℓ^- of u_ℓ. Realizing $K_\ell(u)\nabla u_\ell^- = 0$ because, for a.a. $(t, x) \in Q$, either $K_\ell(u(t, x)) = 0$ (if $u_\ell(t, x) \le 0$) or $\nabla u_\ell(t, x)^- = 0$ (if $u_\ell(t, x) > 0$), and $r_\ell(\cdot)u_\ell^- \ge 0$ because of (12.99a)[43], we obtain $u_\ell^- = 0$ a.e. in Q.

[42] Here we used density of $L^\infty(Q) \cap L^2(I; W^{1,2}_{0,\mathrm{div}}(\Omega;\mathbb{R}^n))$ in $L^2(I; W^{1,2}_{0,\mathrm{div}}(\Omega;\mathbb{R}^n))$.

[43] To be more precise, we can assume, for a moment, that r_ℓ is defined on the whole $\mathbb{R}^L$ in such a way that $r_\ell(u_1, \dots, u_L) \ge 0$ for $u_\ell < 0$. As we are just proving that $u_\ell \ge 0$, the values of r_ℓ for negative concentrations are eventually irrelevant.

The non-negativity of u_ℓ together with $\sum_{\ell=1}^{L} u_\ell = 1$ ensures that $u(t,x) \in \mathrm{Range}(K)$ for a.a. $(t,x) \in Q$ so that $u_\ell = K_\ell(u)$ and thus the triple (ϕ, v, u) is a very weak solution not only to (12.101) with $\bar{v} = v$ and $\bar{u} = u$ but even to the original system (12.95). □

Remark 12.35. By neglecting the Navier-Stokes part (12.95a) and considering a stationary medium, i.e. $v = 0$ and π constant, (12.95) reduces to the Nernst-Planck-Poisson system, see Remark 12.25. Conversely, one can consider extension of the incompressible model (12.95) for anisothermal situations, see [365, 366].[44]

Exercise 12.36. Prove the a-priori bounds (12.102b) and (12.102e) in detail.

Exercise 12.37. Perform the estimate (12.103) for $n = 4$ but in the norm $\mathscr{M}(I; W^{1,2}_{0,\mathrm{div}}(\Omega;\mathbb{R}^n)^*)$. [45]

Exercise 12.38. Prove the limit passage in (12.101c) in detail by using the a-priori estimates (12.102e).

Exercise 12.39 (Galerkin approach[46])**.** Apply Galerkin's method directly to (12.95), using the retract K as in (12.101). Modify the a-priori estimates (12.102) to this case, as well as (12.104), and then make a limit passage, proving thus the existence of the very weak solution to (12.95) without the fixed-point argument.

Exercise 12.40 (Highly viscous Stokes case)**.** Assuming the viscous term in (12.95a) is dominant, omit the convective term $\varrho(v\cdot\nabla)v$ in (12.95a) so that (12.95a) becomes the Stokes equation. Modify the analysis: use a weak solution instead of the very weak ones[47] and Schauder fixed-point theorem instead of the Kakutani one.

12.7 Thermistor with eddy currents

The evolution variant of the steady-state thermistor model (6.57) should certainly count with nonstationary heat transfer and augment (6.57a) by $\mathfrak{c}(\theta)\frac{\partial\theta}{\partial t}$, cf. also Exercises 12.45 and 12.46 below. Under higher frequencies, one should count also with a magnetic field induced by electric current which, when varying in time,

---

[44] This is to be done by making some parameters dependent on temperature θ, in particular the chemical reaction rates $r_\ell = r_\ell(u,\theta)$, and adding the heat equation

$$\mathfrak{c}\frac{\partial\theta}{\partial t} - \mathrm{div}\big(\kappa\nabla\theta + \mathfrak{c}\,v\,\theta\big) = \mu|\nabla v|^2 + \sum_{\ell=1}^{L}\Big(f_\ell j_\ell + h_\ell(\theta) r_\ell(u,\theta)\Big)$$

where $h_\ell(\theta)$ are specific enthalpies, κ the heat conductivity, and $\mathfrak{c}$ the specific heat capacity. Then one can show that the total energy, i.e. the sum of the kinetic, the electrostatic, the internal energies, and the (negative) total enthalpy, i.e. $\int_\Omega(\frac{1}{2}\varrho|v|^2 + \frac{1}{2}\varepsilon|\nabla\phi|^2 + \mathfrak{c}\,\theta - \sum_{\ell=1}^{L} h_\ell u_\ell)\,\mathrm{d}x$, is conserved in an isolated system.

[45] Hint: Use $z \in C(I; L^4(\Omega;\mathbb{R}^4))$ in (12.52).

[46] This more constructive approach is after [364].

[47] Hint: Derive the "dual" estimates (12.102c,d) with L^2-norms instead of $L^{4/3}$-norm.

influences backward the electric current itself. In a so-called *eddy-current approximation*, the resulting system is[48]

$$\mathfrak{c}(\theta)\frac{\partial\theta}{\partial t} - \mathrm{div}\big(\kappa(\theta)\nabla\theta\big) = \sigma(\theta)|e|^2 \qquad \text{in } Q\,, \tag{12.107a}$$

$$\mu_0\frac{\partial h}{\partial t} + \mathrm{curl}\, e = 0 \qquad \text{in } Q\,, \tag{12.107b}$$

$$\mathrm{curl}\, h = \sigma(\theta)e \qquad \text{in } Q\,, \tag{12.107c}$$

with the following interpretation:

θ is the temperature (possibly rescaled by the enthalpy transformation),
e is the intensity of electric field,
h is the intensity of magnetic field,
σ the electric conductivity (depending on θ),
$\mathfrak{c}$ heat capacity (considered $\equiv 1$ without loss of generality, cf. Example 8.71),
κ the heat conductivity (depending on θ), and
μ_0 the vacuum permeability.

The equations (12.107b,c) are the rest of the Maxwell system in electrically conductive medium with no magnetisation nor polarization after the mentioned eddy-current approximation. Note that one can eliminate h by applying a curl-operator to (12.107b) and time-differentiation to (12.107c) so that, for σ constant and taking the identity $\mathrm{curl}^2 h = \nabla\mathrm{div}\, h - \Delta h$ into account, one obtains formally $\mu_0\sigma\frac{\partial}{\partial t}h - \Delta h = -\nabla\mathrm{div}\, h$. This reveals a parabolic character of the eddy-current approximation of the (originally hyperbolic) full Maxwell system.[49]

The system (12.107) is to be completed by boundary conditions. This is a little delicate point here, cf. also [62, 373], and we consider a simple situation in which there is no electro-magnetic field outside Ω and set: [50]

$$\kappa(\theta)\frac{\partial\theta}{\partial\nu} = f_{\mathrm{e}} \qquad \text{and} \qquad \nu\times h = \widehat{\jmath}_{\mathrm{e}} \qquad \text{on } \Sigma, \tag{12.108a}$$

[48]Indeed, steady states of (12.107) give (6.57) because (12.107b) reduces to $\mathrm{curl}\, e = 0$ which means existence of a potential ϕ so that $e = \nabla\phi$, and then applying div-operator to (12.107c) yields (6.57b) because $\mathrm{div}(\sigma(\theta)\nabla\phi) = \mathrm{div}(\mathrm{curl}\, h - \sigma(\theta)e) = 0$.

[49]The full hyperbolic Maxwell system involves $\varepsilon_0\frac{\partial e}{\partial t} + \mathrm{curl}\, h = \sigma(\theta)e$ instead of (12.107c) with ε_0 denoting the vacuum permittivity. The eddy-current approximation neglects the so-called displacement current $\varepsilon_0\frac{\partial e}{\partial t}$ and is legitimate in highly conductive media like metals, cf. [18].

[50]The boundary condition (12.108a) allows for injection of the normal current j_{e} like in Exercise 6.30. Indeed, by using (12.107) with $\mathrm{div}\,\mathrm{curl}\, h \equiv 0$ and $\mathrm{curl}\nabla v \equiv 0$ and assuming h and Γ regular enough, we have

$$\int_\Gamma(\sigma(\theta)e\cdot\nu)v\,\mathrm{d}S = \int_\Gamma(\mathrm{curl}\, h\cdot\nu)v\,\mathrm{d}S = \int_\Omega \mathrm{curl}\, h\cdot\nabla v + (\mathrm{div}\,\mathrm{curl}\, h)v\,\mathrm{d}x$$
$$= \int_\Omega h\mathrm{curl}\nabla v\,\mathrm{d}x + \int_\Gamma(\nu\times h)\cdot\nabla v\,\mathrm{d}S = \int_\Gamma(\nu\times h)\cdot\nabla v\,\mathrm{d}S = \int_\Gamma \widehat{\jmath}_{\mathrm{e}}\cdot\nabla v\,\mathrm{d}S = -\int_\Gamma \mathrm{div}_{\mathrm{S}}(\widehat{\jmath}_{\mathrm{e}})v\,\mathrm{d}S$$

for any smooth v. The last equality is due to the surface Green formula (2.103) written for $a\cdot\nu = \widehat{\jmath}_{\mathrm{e}}$, i.e. $\int_\Gamma \widehat{\jmath}_{\mathrm{e}}\cdot\nabla v\,\mathrm{d}S = \int_\Gamma(\mathrm{div}_{\mathrm{S}}\nu)(\widehat{\jmath}_{\mathrm{e}}\cdot\nu) - \mathrm{div}_{\mathrm{S}}(\widehat{\jmath}_{\mathrm{e}})v\,\mathrm{d}S$, when counting also that $\widehat{\jmath}_{\mathrm{e}}\cdot\nu = 0$. Thus we can identify the normal electric current through Γ as $j_{\mathrm{e}} = \sigma(\theta)e\cdot\nu = -\mathrm{div}_{\mathrm{S}}(\widehat{\jmath}_{\mathrm{e}})$. For $v = 1$, we can also see that always $\int_\Gamma j_{\mathrm{e}}\,\mathrm{d}S = 0$, i.e. electric charge preservation.

where f_e is the external heat flux and $\widehat{\jmath}_e$ a tangential surface current prescribed on the boundary. Further we prescribe initial conditions

$$\theta(0,\cdot)=\theta_0 \qquad \text{and} \qquad h(0,\cdot)=h_0 \qquad\qquad \text{in } \Omega; \tag{12.108b}$$

note that, in view of (12.107c), we have prescribed also $e(0,\cdot)=\operatorname{curl} h_0/\sigma(\theta_0)$.

We cannot apply the transformation of the Joule heat $\sigma|e|^2$ like in (6.59) but, anyhow, the problem is still well fitted to usage of a fixed-point argument after a suitable decoupling. To this goal, we need to make the enthalpy transformation as we did in Example 8.71 and thus, up to a re-scaling of temperature and the corresponding modification of the nonlinearities σ and κ, we can assume $\mathfrak{c}$ constant (and equal 1 without any loss of generality). In analog to (6.61), we design the fixed point as

$$M := M_2\circ\big(M_1\times\mathrm{id}\big) : \vartheta \mapsto \theta \quad \text{with} \quad M_1 : \vartheta \mapsto e \ \text{ and } \ M_2 : (\vartheta, e) \mapsto \theta, \tag{12.109}$$

where, for ϑ given, e together with h forms the unique weak solution to the system

$$\mu_0\frac{\partial h}{\partial t}+\operatorname{curl} e=0, \qquad \operatorname{curl} h-\sigma(\vartheta)e=0 \qquad \text{in } Q, \tag{12.110a}$$

$$\nu\times h=\widehat{\jmath}_e \qquad \text{on } \Sigma, \tag{12.110b}$$

$$h(0,\cdot)=h_0 \qquad \text{in } \Omega, \tag{12.110c}$$

and where, for such (ϑ, e), then θ solves in a weak sense the initial-boundary-value problem

$$\frac{\partial\theta}{\partial t}-\operatorname{div}\big(\kappa(\vartheta)\nabla\theta\big)=\sigma(\vartheta)|e|^2 \qquad \text{in } Q\,, \tag{12.111a}$$

$$\kappa(\vartheta)\frac{\partial\theta}{\partial\nu}=f_e \qquad \text{on } \Sigma\,, \tag{12.111b}$$

$$\theta(0,\cdot)=\theta_0 \qquad \text{in } \Omega. \tag{12.111c}$$

Obviously, any fixed point θ of M completed with the corresponding (e,h) will solve (12.107)–(12.108) with $\mathfrak{c}=1$. Note that the enthalpy transformation and the suitable decoupling made the problem (12.111) linear so that the set of its solution is convex. This allows for usage of Kakutani's fixed-point Theorem 1.11 without investigating a rather delicate issue of uniqueness of a solution to the heat equation (12.111) with L^1-data. Thus we consider M_2 and thus also M as set-valued, and assume

$$\sigma,\kappa:\mathbb{R}\to\mathbb{R} \text{ continuous and bounded}, \quad \inf\sigma(\cdot)>0,\ \inf\kappa(\cdot)>0, \tag{12.112a}$$

$$h_0\in L^2(\Omega;\mathbb{R}^3), \quad \theta_0\in L^1(\Omega), \quad f_e\in L^1(\Sigma), \quad f_e\ge 0, \quad \theta_0\ge 0, \tag{12.112b}$$

$$\exists\, h_e\in W^{1,1}(I;L^2(\Omega;\mathbb{R}^3)):\ \widehat{\jmath}_e=\nu\times h_e|_\Sigma \ \ \&\ \operatorname{curl} h_e\in L^2(Q;\mathbb{R}^3). \tag{12.112c}$$

Note that, in fact, (12.112c) is a natural qualification of $\widehat{\jmath}_{\rm e}$. Then we derive the mentioned existence by the following two lemmas, based on technique from [372]. We denote

$$L^2_{\rm curl,0}(\Omega;\mathbb{R}^3) := \big\{v \in L^2(\Omega;\mathbb{R}^3);\ \operatorname{curl} v \in L^2(\Omega;\mathbb{R}^3),\ v\times\nu = 0 \text{ on } \Gamma\big\}. \quad (12.113)$$

The weak formulation of (12.110) can be obtained by a curl-variant of the Green formula[51] and by the by-part integration in time, which results in

$$\int_Q h\cdot\operatorname{curl} v + e\cdot\operatorname{curl} z - \sigma(\vartheta)e\cdot v - \mu_0 h\cdot\frac{\partial z}{\partial t}\,\mathrm{d}x\mathrm{d}t = \int_\Sigma \widehat{\jmath}_{\rm e}\cdot v\,\mathrm{d}S\mathrm{d}t + \int_\Omega \mu_0 h_0\cdot v(0,\cdot)\,\mathrm{d}x \tag{12.114}$$

for all $v, z : Q \to \mathbb{R}^3$ smooth with $v(T,\cdot) = 0$. The weak formulation of the original system (12.107b,c) with the boundary and initial conditions (12.108) is analogous, just using θ in place of ϑ.

Lemma 12.41 (The mapping M_1). *For any $\vartheta \in L^1(Q)$, the system* (12.110) *has a unique weak solution (h, e) with $h \in L^\infty(I; L^2(\Omega;\mathbb{R}^3))$ and $e \in L^2(Q;\mathbb{R}^3)$ with* $\operatorname{curl} h \in L^2(Q;\mathbb{R}^3)$, $h - h_{\rm e} \in L^2(I; L^2_{\rm curl,0}(\Omega;\mathbb{R}^3))$, *and* $\frac{\partial h}{\partial t} \in L^2(I; L^2_{\rm curl,0}(\Omega;\mathbb{R}^3)^*)$. *Moreover, the mapping $M_1 : \vartheta \mapsto e : L^1(Q) \to L^2(Q;\mathbb{R}^3)$ is continuous.*

Sketch of the proof. We use the variant of (12.114) on the time interval $[0, t]$ without by-part integration in time, i.e.,

$$\int_0^t \Big(\mu_0\Big\langle\frac{\partial h}{\partial t}, z\Big\rangle + \int_\Omega h\cdot\operatorname{curl} v + e\cdot\operatorname{curl} z - \sigma(\vartheta)e\cdot v\,\mathrm{d}x\Big)\mathrm{d}t = \int_0^t\!\int_\Gamma \widehat{\jmath}_{\rm e}\cdot v\,\mathrm{d}S\mathrm{d}t, \tag{12.115}$$

where $\langle\cdot,\cdot\rangle$ denotes the duality between $L^2_{\rm curl,0}(\Omega;\mathbb{R}^3)$ and its dual. We further extend the identity (12.115) by continuity to allow for the test by $v = -e$ and by $z = h - h_{\rm e} \in L^2(I; L^2_{\rm curl,0}(\Omega;\mathbb{R}^3))$ with $h_{\rm e}$ from (12.112c), and use the curl-cancellation.[52] Making by-part integration, one thus obtains

$$\begin{aligned}\int_\Omega \frac{1}{2}|h(t)|^2\mathrm{d}x + \int_0^t\!\int_\Omega \frac{\sigma(\vartheta)}{\mu_0}|e|^2\mathrm{d}x\mathrm{d}t = \int_0^t\!\int_\Omega h\cdot\frac{\partial h_{\rm e}}{\partial t}\,\mathrm{d}x\mathrm{d}t - \int_\Omega h_{\rm e}(t,\cdot)\cdot h(t,\cdot)\,\mathrm{d}x\\ + \int_\Omega \Big(\frac{1}{2}h_0 + h_{\rm e}(0,\cdot)\Big)\cdot h_0\,\mathrm{d}x.\end{aligned}$$

By Gronwall's inequality, using (12.112), we obtain a-priori estimates $h \in L^\infty(I; L^2(\Omega;\mathbb{R}^3))$ and $e \in L^2(Q;\mathbb{R}^3)$. In fact, one should execute this scenario

[51] This means $\int_\Omega \operatorname{curl} u\cdot v\,\mathrm{d}x = \int_\Omega u\cdot\operatorname{curl} v\,\mathrm{d}x + \int_\Gamma (u\times v)\cdot\nu\,\mathrm{d}S$. In particular, we use $\int_\Omega \operatorname{curl} h\cdot v\,\mathrm{d}x - \int_\Omega h\cdot\operatorname{curl} v\,\mathrm{d}x = \int_\Gamma (h\times v)\cdot\nu\,\mathrm{d}S = \int_\Gamma (\nu\times h)\cdot v\,\mathrm{d}S = \int_\Gamma \widehat{\jmath}_{\rm e}\cdot v\,\mathrm{d}S$.

[52] The curl-cancellation means $\int_\Omega e\cdot\operatorname{curl} h - h\cdot\operatorname{curl} e\,\mathrm{d}x = \int_\Gamma (\nu\times h)\cdot e\,\mathrm{d}S$, which equals zero if the homogeneous boundary conditions $\nu\times h = 0$ are considered.

rigorously rather for some approximate solution and then pass to the limit by weak convergence to the linear system (12.110).

We then have also a-priori estimates $\operatorname{curl} h = \sigma(\vartheta)e \in L^2(Q;\mathbb{R}^3)$ and $\frac{\partial h}{\partial t} \in L^2(I; L^2_{\mathrm{curl},0}(\Omega;\mathbb{R}^3)^*)$ because $\langle \frac{\partial h}{\partial t}, v\rangle = -\langle \operatorname{curl} e, v\rangle/\mu_0 = -\int_Q e \operatorname{curl} v \,\mathrm{d}x\mathrm{d}t/\mu_0$ is bounded provided v ranges over a unit ball in $L^2(I; L^2_{\mathrm{curl},0}(\Omega;\mathbb{R}^3))$. It is important that these estimates are uniform with respect to ϑ because of the assumption (12.112a) on $\sigma(\cdot)$. Also it is important that $\frac{\partial h}{\partial t}$ is in duality with $h - h_{\mathrm{e}}$, and one can thus perform the by-part integration formula in time, which yields uniqueness of the weak solution to the linear elliptic/parabolic problem (12.110).

For the claimed continuity, let us consider a sequence $\vartheta_k \to \vartheta$ and the corresponding solutions (h_k, e_k) to (12.110) with ϑ_k in place of ϑ, then subtract the corresponding equations and test them respectively by $h_k - h$ and $e_k - e$. Using again the curl-cancellation property and the by-part integration in time, one gets

$$\inf \sigma(\cdot) \int_Q \big|e_k - e\big|^2 \mathrm{d}x\mathrm{d}t \le \frac{\mu_0}{2}\int_\Omega \big|h_k(T) - h(T)\big|^2 \mathrm{d}x + \int_Q \sigma(\vartheta_k)\big|e_k - e\big|^2 \mathrm{d}x\mathrm{d}t$$
$$= \int_Q \big(\sigma(\vartheta) - \sigma(\vartheta_k)\big) e \cdot \big(e_k - e\big)\,\mathrm{d}x\mathrm{d}t \to 0; \qquad (12.116)$$

the last convergence is due to the a-priori bounds for e_k and e in $L^2(\Omega;\mathbb{R}^3)$. From the left-hand side of (12.116), one thus obtains the desired strong convergence. $\square$

Lemma 12.42 (The mapping M_2). *For any $\vartheta \in L^1(Q)$ and $e \in L^2(\Omega;\mathbb{R}^3)$, the problem (12.111) has a weak solution $\theta \in L^\infty(I; L^1(\Omega)) \cap L^r(I; W^{1,r}(\Omega))$ with $r < \frac{5}{4}$ and $\frac{\partial \theta}{\partial t} \in L^1(I; W^{1,\infty}(\Omega)^*)$ and the mapping $(\vartheta, e) \mapsto \{\theta$ is a solution to (12.111)$\}$: $L^1(Q) \times L^2(\Omega;\mathbb{R}^3) \rightrightarrows L^1(Q)$ is upper semi-continuous and has convex non-empty closed values and a relatively compact range if restricted on a ball in $L^1(Q)$ with a sufficiently large radius.*

Sketch of the proof. For existence of an integral solution $\theta \in C(I; L^1(\Omega))$ see Example 9.19. In fact, it is also a weak solution with $\nabla\theta$ estimated in $L^r(Q;\mathbb{R}^3)$ with $r < 5/4$, cf. Remark 9.25, and $\frac{\partial \theta}{\partial t}$ estimated in $L^1(I; W^{1,\infty}(\Omega)^*)$. As the problem (12.111) is linear, its solutions form a convex set. This set is also closed and depends continuously on ϑ and e, as claimed. For this, we use the compact embedding $L^r(I; W^{1,r}(\Omega)) \cap W^{1,1}(I; W^{1,\infty}(\Omega)^*) \Subset L^1(Q)$ due to Aubin-Lions' theorem as well as the continuity of the Nemytskiĭ mapping $(\vartheta, e) \mapsto \sigma(\vartheta)|e|^2$: $L^1(Q) \times L^2(Q;\mathbb{R}^3) \to L^1(Q)$. $\square$

For the existence of a fixed point of M from (12.109), it then suffices to use Kakutani's Theorem 1.11 on a sufficiently large ball in $L^1(Q)$ which M maps into itself. Such a ball always exists because simply the whole set $M_1(L^1(\Omega))$ is bounded in $L^2(Q;\mathbb{R}^3)$.

Exercise 12.43. Realize that $\sigma(\vartheta_k) \to \sigma(\vartheta)$ in $L^p(\Omega)$, $p < +\infty$, but not in $L^\infty(\Omega)$, and prove the convergence in (12.116).[53] Realize why the other variant of (12.116) as $\int_Q \sigma(\vartheta)|e_k - e|^2 \mathrm{d}x\mathrm{d}t = \int_Q (\sigma(\vartheta) - \sigma(\vartheta_k)) e_k \cdot (e_k - e)\,\mathrm{d}x\mathrm{d}t$ would not give the desired effect.

Exercise 12.44. Instead of the fixed-point argument (12.109), use the Rothe method with a regularization like in (12.12), namely

$$\frac{\theta_\tau^k - \theta_\tau^{k-1}}{\tau} - \mathrm{div}\big(\kappa(\theta_\tau^k)\nabla\theta_\tau^k\big) = \sigma(\theta_\tau^k)|e_\tau^k|^2, \tag{12.117a}$$

$$\mu_0 \frac{h_\tau^k - h_\tau^{k-1}}{\tau} + \mathrm{curl}\, e_\tau^k = 0, \tag{12.117b}$$

$$\mathrm{curl}\, h_\tau^k = \sigma(\theta_\tau^k) e_\tau^k + \tau |e_\tau^k|^{\eta-2} e_\tau^k \tag{12.117c}$$

on Ω with the boundary conditions $\kappa(\theta_\tau^k)\frac{\partial}{\partial\nu}\theta_\tau^k = h_{\mathrm{e},\tau}^k$ and $\nu \times h_\tau^k = \widehat{\jmath}_{\mathrm{e},\tau}^{\,k}$ on Γ, and with a sufficiently large η. Prove existence of a weak solution $(\theta_\tau^k, h_\tau^k, e_\tau^k) \in W^{1,2}(\Omega) \times L^2(\Omega;\mathbb{R}^3) \times L^\eta(\Omega;\mathbb{R}^3)$ such that $\theta_\tau^k \geq 0$, $\mathrm{curl}\, h_\tau^k \in L^{\eta'}(\Omega;\mathbb{R}^3)$, and $\mathrm{curl}\, e_\tau^k \in L^2(\Omega;\mathbb{R}^3)$.[54] Further prove the a-priori estimates from Lemmas 12.41 and 12.42 together with $\|\bar{e}_\tau\|_{L^\eta(Q;\mathbb{R}^3)} = \mathscr{O}(\tau^{-1/\eta})$,[55] and show convergence (in terms of subsequences) of Rothe's solutions towards a weak solution of the continuous problem.[56] Alternatively to (12.117), use a semi-implicit discretization like in Remark 12.12.

Exercise 12.45. Analyse the system which arises from (12.107) by neglecting $\mu_0\frac{\partial h}{\partial t}$, namely

$$\mathfrak{c}(\theta)\frac{\partial\theta}{\partial t} - \mathrm{div}\big(\kappa(\theta)\nabla\theta\big) = \sigma(\theta)|\nabla\phi|^2 \quad \text{on } Q\,, \tag{12.118a}$$

$$-\mathrm{div}\big(\sigma(\theta)\nabla\phi\big) = 0 \quad \text{on } Q\,, \tag{12.118b}$$

with the initial condition $\theta(0,\cdot) = \theta_0$ and the boundary conditions

$$\kappa(\theta)\frac{\partial\theta}{\partial\nu} = f_{\mathrm{e}} \quad \text{and} \quad \sigma(\theta)\frac{\partial\phi}{\partial\nu} = j_{\mathrm{e}} \quad \text{on } \Sigma. \tag{12.119}$$

[53] Hint: Consider the Nemytskiĭ mapping $\vartheta_k \mapsto (\sigma(\vartheta) - \sigma(\vartheta_k))e : L^1(\Omega) \to L^2(\Omega)$ instead of the mapping $\vartheta_k \mapsto \sigma(\vartheta_k) : L^1(\Omega) \to L^\infty(\Omega)$.

[54] Hint: Choose $\eta > 4$ so that, when making the test of the particular equations in (12.117) respectively by θ_τ^k, $h_\tau^k - h_{\mathrm{e},\tau}^k$, and e_τ^k, the left-hand-side terms $|\theta_\tau^k|^2/\tau + \tau|e_\tau^k|^\eta$ dominates the term $\sigma(\theta_\tau^k)|e_\tau^k|^2\theta_\tau^k$ arising from the right-hand side of (12.117a), so that the underlying pseudomonotone mapping is coercive and Brézis' Theorem 2.6 ensures existence of a solution to (12.117). To prove non-negativity of some θ_τ^k, test (12.117a) by $(\theta_\tau^k)^-$.

[55] Hint: First execute the test by 1, $h_\tau^k - h_{\mathrm{e},\tau}^k$, and e_τ^k. Then test separately (12.117a) by $\chi(\vartheta_\tau^k) := 1 - (1+\vartheta_\tau^k)^{-\varepsilon}$ as suggested in Remark 9.25 for $p = 2$ to get $\nabla\bar{\vartheta}_\tau$ estimated in $L^r(Q;\mathbb{R}^3)$, $r < 5/4$.

[56] Hint: The semilinear terms can be limited by weak convergence combined with Aubin-Lions theorem. The regularizing term $\tau|\bar{e}_\tau|^{\eta-2}\bar{e}_\tau$ can be shown to vanish in the limit. The strong convergence of the Joule heat in $L^1(Q)$ can modify (12.116) by using also (8.52)–(8.53).

Assume the zero-current $\int_\Gamma j_{\rm e}(t,\cdot)\,{\rm d}S = 0$ for all t with $j_{\rm e} \in L^1(I;L^{2^{\#}{}'}(\Gamma))$, use the enthalpy transformation to get $\mathfrak{c}(\cdot) = 1$ like in (12.111), and prove existence of a weak solution by modifying Lemmas 12.41 and 12.42.[57]

Exercise 12.46. Consider (12.118) alternatively with the boundary conditions

$$\kappa(\theta)\frac{\partial\theta}{\partial\nu} = f_{\rm e} \quad \text{on } \Sigma, \qquad \phi|_{\Sigma_{\rm D}} = \phi_{\rm D} \quad \text{on } \Sigma_{\rm D}, \qquad \frac{\partial\phi}{\partial\nu} = 0 \quad \text{on } \Sigma_{\rm N}, \tag{12.120}$$

and transform (12.118a) into $\frac{\partial\theta}{\partial t} - {\rm div}\big(\kappa(\theta)\nabla\theta\big) = {\rm div}\big(\sigma(\theta)\phi\nabla\phi\big)$ with the boundary condition $\kappa(\theta)\frac{\partial\theta}{\partial\nu} + \sigma(\theta)\phi\frac{\partial\phi}{\partial\nu} = f_{\rm e}$ like in Section 6.4, referring now to the rescaled temperature. Prove existence of a weak solution $\theta \in L^2(I;W^{1,2}(\Omega)) \cap W^{1,2}(I;W^{1,2}(\Omega)^*)$ and $\phi \in L^\infty(I;W^{1,2}(\Omega))\cap L^\infty(Q)$.[58] Prove uniqueness for small $f_{\rm e}$ and $j_{\rm e}$.[59]

12.8 Thermodynamics of magnetic materials

Magnetic materials may undergo transformation between ferromagnetic and paramagnetic states, depending mainly on temperature, and thus exhibit nontrivial coupling with heat transfer. The resulting system, devised and analysed in [336], couples the *Landau-Lifschitz-Gilbert equation* (see Exercise 11.28) with the heat equation. The unknown fields are $\mathbb{R}^3$-valued magnetisation u and temperature θ. The resulting system is

$$\alpha\,\frac{\partial u}{\partial t} - \beta(|u|)u \times \frac{\partial u}{\partial t} - \mu\,\Delta u + \varphi_0'(u) + \theta\varphi_1'(u) = h, \tag{12.121a}$$

$$\mathfrak{c}(\theta)\frac{\partial\theta}{\partial t} - {\rm div}(\kappa(\theta)\,\nabla\theta) = \alpha\Big|\frac{\partial u}{\partial t}\Big|^2 + \theta\varphi_1'(u)\cdot\frac{\partial u}{\partial t} \tag{12.121b}$$

with the following interpretation:

u magnetisation vector,
θ the absolute temperature,
h given outer magnetic field,
$\alpha > 0$ an attenuation constant,

[57] Hint: Design the fixed point via $M_1 : \vartheta \mapsto \phi$ and $M_2 : (\vartheta,\phi) \mapsto \theta$ like before. Derive the a-priori bound of $\phi \in L^1(I;W^{1,2}(\Omega))$ provided $\phi(t,\cdot)$ is shifted by a suitable constant, e.g. so that $\int_\Omega \phi(t,\cdot)\,{\rm d}x = 0$ for all t. Here, a version (1.58) of the Poincaré inequality is to be used. For continuity of the mapping M_2 use again (12.116) modified for $e = \nabla\phi$.

[58] Hint: Test the heat equation by θ to estimate $\frac12\frac{\rm d}{{\rm d}t}\|\theta\|^2_{L^2(\Omega)} + \min\kappa(\cdot)\|\nabla\theta\|^2_{L^2(\Omega;\mathbb{R}^n)} \le \max\sigma(\cdot)\|\phi\|_{L^\infty(\Omega)}\|\nabla\phi\|_{L^2(\Omega;\mathbb{R}^n)}\|\nabla\theta\|_{L^2(\Omega;\mathbb{R}^n)} + \|f_{\rm e}\|_{L^{2^{\#}{}'}(\Gamma)}\|\theta\|_{L^{2^{\#}}(\Gamma)}$ by taking the a-priori bound of $\phi \in W^{1,2}(\Omega) \cap L^\infty(\Omega)$ into account, provided ϕ is shifted by a suitable constant, e.g. so that $\int_\Omega \phi(t,\cdot)\,{\rm d}x = 0$ for all t. Here, a version (1.58) of the Poincaré inequality is to be used. For continuity of a fixed-point mapping or alternatively for convergence of Rothe's or Galerkin's approximate solutions, modify (12.116).

[59] Hint: Modify the procedure from Exercise 6.32.

$\beta = \beta(|u|) > 0$ the inverse gyromagnetic ratio,
$\mu > 0$ the exchange-energy constant,
$\kappa = \kappa(\theta) > 0$ heat conductivity,
$\mathfrak{c} = \mathfrak{c}(\theta) > 0$ heat capacity.

The thermodynamics of (12.121) can be derived from the free energy considered partly linearized, namely[60]

$$\psi(u,\theta) = \phi_0(u) + \theta\phi_1(u) + \phi_2(\theta). \tag{12.122}$$

This gives the heat capacity $\mathfrak{c}(\theta) = -\theta\phi_2''(\theta)$ and allows for varying between a multiwell $\psi(\cdot,\theta)$ for lower temperatures (corresponding to the so-called ferromagnetic phase) and a single-well $\psi(\cdot,\theta)$ for higher temperatures (corresponding to the so-called paramagnetic phase), cf. [336] for specific examples. Thus "continuous switching" of the multi/single-well character of $\psi(\cdot,\theta)$ model the ferro/paramagnetic transformation. In view of (12.127b,d) below, the slight growth of both $\mathfrak{c}(\theta)$ and $\kappa(\theta)$ like $\sim \theta^{1/5+\epsilon}$ is needed for the analysis, similarly as in (12.14a,b).

To facilitate its analysis, we consider it after the enthalpy transformation and define

$$\gamma(u,\vartheta) = \theta\varphi_1'(u), \quad \hat\kappa(\vartheta) = \frac{\kappa(\theta)}{\mathfrak{c}(\theta)} \quad \text{with} \quad \theta = \hat{\mathfrak{c}}^{-1}(\vartheta) \quad \text{where} \quad \hat{\mathfrak{c}}(\theta) = \int_0^\theta \mathfrak{c}(r)\,\mathrm{d}r. \tag{12.123}$$

In terms of the re-scaled temperature $\vartheta = \hat{\mathfrak{c}}(\theta)$, (12.121) takes the form:

$$\alpha\frac{\partial u}{\partial t} - \beta(|u|)u\times\frac{\partial u}{\partial t} - \mu\,\Delta u + \varphi_0'(u) + \gamma(u,\vartheta) = h, \tag{12.124a}$$

$$\frac{\partial\vartheta}{\partial t} - \mathrm{div}(\hat\kappa(\vartheta)\,\nabla\vartheta) = \alpha\Big|\frac{\partial u}{\partial t}\Big|^2 + \gamma(u,\vartheta)\cdot\frac{\partial u}{\partial t} \tag{12.124b}$$

on Q. We complete (12.124) by the initial conditions

$$u(0,\cdot) = u_0, \qquad \vartheta(0,\cdot) = \vartheta_0 \qquad \text{in } \Omega, \tag{12.125a}$$

and the boundary conditions (for simplicity linear in terms of ϑ):

$$\frac{\partial u}{\partial\nu} = 0, \qquad \hat\kappa(\vartheta)\frac{\partial\vartheta}{\partial\nu} + b\vartheta = b\vartheta_{\mathrm{e}} \qquad \text{on } \Sigma. \tag{12.125b}$$

It is illustrative to prove existence of a weak solution to the initial-boundary-value-problem (12.124) by a suitable *regularization*. For $\varepsilon > 0$, we add an attenu-

[60] Note that, in terms of ψ, the system (12.121) reads as $\alpha\frac{\partial u}{\partial t} - \beta(|u|)u\times\frac{\partial u}{\partial t} - \mu\,\Delta u + \psi_u'(u,\theta) = h$ and $\theta\frac{\partial}{\partial t}\psi_\theta'(u,\theta) + \mathrm{div}(\kappa(\theta)\nabla\theta) = -\alpha|\frac{\partial u}{\partial t}|^2$. The latter equation is, in fact, the entropy equation.

ation $\varepsilon|\frac{\partial u}{\partial t}|^{p-2}\frac{\partial u}{\partial t}$ with a sufficiently large p. Thus we consider the system

$$\Big(\alpha+\varepsilon\Big|\frac{\partial u_\varepsilon}{\partial t}\Big|^{p-2}\Big)\frac{\partial u_\varepsilon}{\partial t} - \beta(|u_\varepsilon|)u_\varepsilon\times\frac{\partial u_\varepsilon}{\partial t} - \mu\,\Delta u_\varepsilon + \varphi_0'(u_\varepsilon) + \gamma(u_\varepsilon,\vartheta_\varepsilon) = h, \tag{12.126a}$$

$$\frac{\partial\vartheta_\varepsilon}{\partial t} - \mathrm{div}\big(\hat\kappa(\vartheta_\varepsilon)\nabla\vartheta_\varepsilon\big) = \alpha\Big|\frac{\partial u_\varepsilon}{\partial t}\Big|^2 + \gamma(u_\varepsilon,\vartheta_\varepsilon)\cdot\frac{\partial u_\varepsilon}{\partial t} \tag{12.126b}$$

on Q with the partly regularized initial and boundary conditions

$$u_\varepsilon(0,\cdot) = u_0, \qquad \vartheta_\varepsilon + u_\varepsilon(0,\cdot) = \vartheta_{0,\varepsilon} \qquad \text{in } \Omega, \tag{12.126c}$$

$$\frac{\partial u_\varepsilon}{\partial\nu} = 0, \qquad \hat\kappa(\vartheta_\varepsilon)\frac{\partial\vartheta_\varepsilon}{\partial\nu} + b\vartheta_\varepsilon = b\vartheta_{\mathrm{e},\varepsilon} \qquad \text{on } \Sigma. \tag{12.126d}$$

We make the following assumptions:

$$\beta,\hat\kappa:\mathbb{R}^+\to\mathbb{R}^+,\ \varphi_0':\mathbb{R}^3\to\mathbb{R}^3,\ \gamma:\mathbb{R}^3\times\mathbb{R}^+\to\mathbb{R}^3 \text{ continuous}, \tag{12.127a}$$

$$\exists\,\epsilon>0,\ C\in\mathbb{R}\ \forall(u,\vartheta)\in\mathbb{R}^3\times\mathbb{R}^+:\quad |\gamma(u,\vartheta)| \le C(1+|\vartheta|^{5/6-\epsilon}), \tag{12.127b}$$

$$\varphi_0(u) \ge \epsilon|u|^2, \quad |\varphi_0'(u)| \le C(1+|u|^2), \tag{12.127c}$$

$$\hat\kappa(\vartheta)\ge\epsilon, \quad \hat\kappa(\vartheta)\le C, \quad |\beta(|u|)| \le C(1+|u|^2), \tag{12.127d}$$

$$b\in L^\infty(\Sigma), \quad b\ge 0, \quad \vartheta_{\mathrm{e}}\in L^1(\Sigma), \quad \vartheta_{\mathrm{e}}\ge 0, \quad h\in W^{1,1}(I;L^2(\Omega;\mathbb{R}^3)), \tag{12.127e}$$

$$u_0\in W^{1,2}(\Omega;\mathbb{R}^3), \quad \vartheta_0\in L^1(\Omega), \quad \vartheta_0\ge 0. \tag{12.127f}$$

Lemma 12.47. *Let $p\ge 12$ and $\varepsilon>0$ be fixed, and let $\vartheta_{\mathrm{e},\varepsilon}\in L^2(\Sigma)$ and $\vartheta_{0,\varepsilon}\in L^2(\Omega)$. The regularized system* (12.126) *possesses a weak solution $u_\varepsilon\in W^{1,p}(I;L^p(\Omega;\mathbb{R}^3))$ and $\vartheta_\varepsilon\in L^2(I;W^{1,2}(\Omega))\cap W^{1,2}(I;L^2(\Omega))$ such that $\vartheta_\varepsilon\ge 0$.*

Sketch of the proof. Realize the structure of (12.126) as the *pseudo-parabolic equation* similarly like in Exercise 11.28. Besides convergence of some approximate solutions (cf. Exercise 12.51 below), the essential point is to execute suitable a-priori estimates. For this, test (12.126a) by $\frac{\partial u_\varepsilon}{\partial t}$ and (12.126b) by ϑ_ε. Sum it up to obtain

$$\begin{aligned}
&\frac{\mathrm{d}}{\mathrm{d}t}\int_\Omega\frac{\mu}{2}\big|\nabla u_\varepsilon\big|^2 + \varphi_0(u_\varepsilon) + \frac12|\vartheta_\varepsilon|^2\,\mathrm{d}x\\
&\qquad\qquad + \int_\Omega\alpha\Big|\frac{\partial u_\varepsilon}{\partial t}\Big|^2 + \varepsilon\Big|\frac{\partial u_\varepsilon}{\partial t}\Big|^p + \hat\kappa(\vartheta_\varepsilon)|\nabla\vartheta_\varepsilon|^2\,\mathrm{d}x + \int_\Gamma b\,\vartheta_\varepsilon^2\,\mathrm{d}S\\
&= \int_\Omega\alpha\Big|\frac{\partial u_\varepsilon}{\partial t}\Big|^2\vartheta_\varepsilon + (\vartheta_\varepsilon-1)\gamma(u_\varepsilon,\vartheta_\varepsilon)\cdot\frac{\partial u_\varepsilon}{\partial t} + h\cdot\frac{\partial u_\varepsilon}{\partial t}\,\mathrm{d}x + \int_\Gamma b\,\vartheta_{\mathrm{e},\varepsilon}\vartheta_\varepsilon\,\mathrm{d}S\\
&\le \alpha\Big\|\frac{\partial u_\varepsilon}{\partial t}\Big\|_{L^4(\Omega;\mathbb{R}^3)}^2\big\|\vartheta_\varepsilon\big\|_{L^2(\Omega)} + C\Big(1+\big\|\vartheta_\varepsilon\big\|_{L^2(\Omega)}^{11/6-\epsilon}\Big)\Big\|\frac{\partial u_\varepsilon}{\partial t}\Big\|_{L^4(\Omega;\mathbb{R}^3)}\\
&\qquad + \big\|h\big\|_{L^2(\Omega;\mathbb{R}^3)}\Big\|\frac{\partial u_\varepsilon}{\partial t}\Big\|_{L^2(\Omega;\mathbb{R}^3)} + \big\|b\big\|_{L^\infty(\Gamma)}\big\|\vartheta_{\mathrm{e},\varepsilon}\big\|_{L^2(\Gamma)}\big\|\vartheta_\varepsilon\big\|_{L^2(\Gamma)}
\end{aligned} \tag{12.128}$$

with a sufficiently large C. Note that the orthogonality $(\beta(|u_\varepsilon|)u_\varepsilon\times\frac{\partial u_\varepsilon}{\partial t})\cdot\frac{\partial u_\varepsilon}{\partial t}=0$ has been used. By Young's and Gronwall's inequalities, one obtains the estimates $u_\varepsilon\in W^{1,p}(I;L^p(\Omega;\mathbb{R}^3))$ and $\vartheta_\varepsilon\in L^2(I;W^{1,2}(\Omega))\cap L^\infty(I;L^2(\Omega))$; here $p\geq 12$ and then the estimate $\|\vartheta_\varepsilon\|_{L^2(\Omega)}^{11/6-\epsilon}\|\frac{\partial u_\varepsilon}{\partial t}\|_{L^4(\Omega;\mathbb{R}^3)}\leq C_\delta+\delta\|\vartheta_\varepsilon\|_{L^2(\Omega)}^2+\delta\|\frac{\partial u_\varepsilon}{\partial t}\|_{L^{12}(\Omega;\mathbb{R}^3)}^{12}$ with a suitably small $\delta>0$ has been used.

Moreover, test (12.126b) by $\frac{\partial\vartheta_\varepsilon}{\partial t}$ to get $\frac{\partial\vartheta_\varepsilon}{\partial t}\in L^2(Q)$ when realizing that the right-hand side of (12.126b) is in $L^2(Q)$ since $\frac{\partial u_\varepsilon}{\partial t}\in L^{12}(Q;\mathbb{R}^3)$ and $\vartheta_\varepsilon\in L^{10/3}(Q)$ have already been proved.

Eventually, $\vartheta_\varepsilon\geq 0$ can be obtained by testing (12.126b) by ϑ_ε^-, exploiting that $\gamma(u,0)=0$ by definition of the nonlinearity γ; here we can, for a moment, assume $\gamma(u,\cdot)=0$ defined for negative arguments. □

Further, relying on $\vartheta_\varepsilon\geq 0$, physically motivated estimates are to be derived.

Lemma 12.48. *Let the assumptions of Lemma 12.47 be fulfilled and, in addition, also $\|\vartheta_{\mathrm{e},\varepsilon}\|_{L^1(\Sigma)}$ and $\|\vartheta_{0,\varepsilon}\|_{L^1(\Omega)}$ be bounded independently of ε. Then, with C and C_r independent of $\varepsilon>0$, it holds that*

$$\big\|u_\varepsilon\big\|_{L^\infty(I;W^{1,2}(\Omega;\mathbb{R}^3))\cap W^{1,2}(I;L^2(\Omega;\mathbb{R}^3))}\leq C, \tag{12.129a}$$

$$\big\|\vartheta_\varepsilon\big\|_{L^\infty(I;L^1(\Omega))\cap W^{1,1}(I;W^{3,2}(\Omega)^*)}\leq C, \tag{12.129b}$$

$$\big\|\nabla\vartheta_\varepsilon\big\|_{L^r(Q;\mathbb{R}^3)}\leq C_r,\qquad 1\leq r<5/4, \tag{12.129c}$$

$$\left\|\frac{\partial u_\varepsilon}{\partial t}\right\|_{L^p(Q;\mathbb{R}^3)}\leq C\sqrt[p]{1/\varepsilon}. \tag{12.129d}$$

Sketch of the proof. First, test (12.126b) by 1 and (12.126a) again by $\frac{\partial u_\varepsilon}{\partial t}$. We will see cancellation effects of $\pm\alpha|\frac{\partial u_\varepsilon}{\partial t}|^2$ and $\pm\gamma(u_\varepsilon,\vartheta_\varepsilon)\cdot\frac{\partial u_\varepsilon}{\partial t}$ and again the orthogonality $(\beta(|u_\varepsilon|)u_\varepsilon\times\frac{\partial u_\varepsilon}{\partial t})\cdot\frac{\partial u_\varepsilon}{\partial t}=0$, and obtain the energy balance:

$$\frac{\mathrm{d}}{\mathrm{d}t}\int_\Omega\frac{\mu}{2}\big|\nabla u_\varepsilon\big|^2+\varphi_0(u_\varepsilon)+\vartheta_\varepsilon\,\mathrm{d}x$$
$$+\int_\Gamma b\,\vartheta_\varepsilon\mathrm{d}S+\int_\Omega\varepsilon\left|\frac{\partial u_\varepsilon}{\partial t}\right|^p\mathrm{d}x=\int_\Gamma b\,\vartheta_\mathrm{e}\,\mathrm{d}S+\int_\Omega h\cdot\frac{\partial u_\varepsilon}{\partial t}\,\mathrm{d}x. \tag{12.130}$$

Then we integrate it over $[0,t]$ and make the by-part integration for the last term. Then (12.129a,d) and also the first part of (12.129b) follow by Gronwall's inequality.

Further, we test (12.126b) by $\chi(\vartheta_\varepsilon)=1-(1{+}\vartheta_\varepsilon)^\delta$ with $\delta>0$ and sum it with (12.126a) tested by $k\frac{\partial u_\varepsilon}{\partial t}$ with a sufficiently large k, and proceed like in (12.17)–(12.24) by using also (12.127b). Thus (12.129c) follows.

Eventually, testing (12.126b) by $v\in L^\infty(I;W^{3,2}(\Omega))$, we obtain the second part of (12.129b). □

Proposition 12.49 (Convergence with $\varepsilon\to 0$). *Let, in addition to the assumptions of Lemmas 12.47 and 12.48, also $\vartheta_{\mathrm{e},\varepsilon}\to\vartheta_\mathrm{e}$ in $L^1(\Sigma)$ and $\vartheta_{0,\varepsilon}\to\vartheta_0$ in $L^1(\Omega)$.*

There is a subsequence $\{(u_\varepsilon,\vartheta_\varepsilon)\}_{\varepsilon>0}$ *converging weakly* in the topology indicated in the estimates* (12.129a-c) *to some* (u,ϑ) *and any* (u,ϑ) *obtained by this way is a weak solution to the initial-boundary-value problem* (12.124)–(12.125).

Proof. The weakly* converging subsequence is to be selected by Banach's Theorem 1.7. By Aubin-Lions' theorem with interpolation, we have also strong convergence of $\vartheta_\varepsilon \to \vartheta$ in $L^{5/3-\epsilon}(Q)$. By the Sobolev embedding $W^{1,2}(Q) \Subset L^{4-\epsilon}(Q)$, we have $u_\varepsilon \to u$ in $L^{4-\epsilon}(Q;\mathbb{R}^3)$. The convergence in the "asymptotically semilinear" equation (12.126a) to (12.124a) is then by the weak convergence; more in detail, the only quasilinear term $\varepsilon\big|\frac{\partial u_\varepsilon}{\partial t}\big|^{p-2}\frac{\partial u_\varepsilon}{\partial t}$ is in (12.126a) but it disappears in the limit due to the estimate

$$\bigg|\int_Q \varepsilon\Big|\frac{\partial u_\varepsilon}{\partial t}\Big|^{p-2}\frac{\partial u_\varepsilon}{\partial t}\cdot v\,\mathrm{d}x\mathrm{d}t\bigg| \le \varepsilon\Big\|\frac{\partial u_\varepsilon}{\partial t}\Big\|_{L^p(Q;\mathbb{R}^3)}^{p-1}\big\|v\big\|_{L^p(Q;\mathbb{R}^3)} = \mathscr{O}(\varepsilon^{1/p}) \to 0$$

for any v smooth, cf. (12.129d).

Testing (12.124a) by $\frac{\partial u}{\partial t}$ gives the magnetic-energy balance

$$\int_\Omega \varphi_0\big(u(T)\big) + \frac{\mu}{2}|\nabla u(T)|^2\mathrm{d}x + \int_Q \alpha\Big|\frac{\partial u}{\partial t}\Big|^2\mathrm{d}x\mathrm{d}t = \int_\Omega \varphi_0\big(u_0\big) + \frac{\mu}{2}|\nabla u_0|^2\mathrm{d}x + \int_Q \big(h-\gamma(u,\vartheta)\big)\cdot\frac{\partial u}{\partial t}. \tag{12.131}$$

To execute rigorously the proof of (12.131), one needs the formulas

$$\int_Q \varphi_0'(u)\cdot\frac{\partial u}{\partial t}\,\mathrm{d}x\mathrm{d}t = \int_\Omega \varphi_0\big(u(T)\big) - \varphi_0\big(u_0\big)\,\mathrm{d}x, \quad \text{and} \tag{12.132a}$$

$$\int_Q \mu\Delta u\cdot\frac{\partial u}{\partial t}\,\mathrm{d}x\mathrm{d}t = \frac{\mu}{2}\int_\Omega |\nabla u_0|^2 - |\nabla u(T)|^2\,\mathrm{d}x. \tag{12.132b}$$

For (12.132a), we use that $\frac{\partial u}{\partial t} \in L^2(Q;\mathbb{R}^3)$ and $\varphi_0'(u) \in L^2(Q;\mathbb{R}^3)$ due to (12.127c) since $u \in L^4(Q;\mathbb{R}^3)$ has been proved. Since it has also been proved that $\frac{\partial u}{\partial t} \in L^2(Q;\mathbb{R}^3)$, it follows from (12.124a) that, in addition, $\Delta u \in L^2(Q;\mathbb{R}^3)$. Here we also use that we have $\vartheta \in L^{5/3-\epsilon}(Q)$ and, by (12.127b), we can see that $\gamma(u,\vartheta)\cdot\frac{\partial u}{\partial t} \in L^2(Q;\mathbb{R}^3)$. Moreover, $u: I \to W^{1,2}(\Omega;\mathbb{R}^3)$ is actually a weakly continuous function (although not necessarily strongly continuous). Thus the integration-by-parts formula (12.132b) has indeed a good sense. [61]

[61] More in detail, (12.132b) can be proved by mollifying u with respect to the spatial variables, not with respect to time as used in Lemma 7.3 through Lemma 7.2: Denoting by M_η the mollification operator, for $u_\eta := M_\eta u$ we have $\frac{\partial u_\eta}{\partial t} = M_\eta\frac{\partial u}{\partial t} \in L^2(I;C^1(\Omega;\mathbb{R}^3))$. By standard calculus, we obtain that (12.132b) holds for the mollified function u_η, namely: $\int_Q \Delta u_\eta\cdot\frac{\partial u_\eta}{\partial t}\,\mathrm{d}x\mathrm{d}t = \frac{1}{2}\int_\Omega |\nabla(M_\eta u_0)|^2 - |\nabla u_\eta(T)|^2\,\mathrm{d}x$. We then obtain (12.132b) by letting $\eta \to 0$, using the fact that $\Delta u_\eta \to \Delta u$ in $L^2(Q;\mathbb{R}^3)$, $\frac{\partial u_\eta}{\partial t} \to \frac{\partial u}{\partial t}$ in $L^2(Q;\mathbb{R}^3)$, and $\nabla u_\eta(T) \to \nabla u(T)$ in $L^2(\Omega;\mathbb{R}^{3\times 3})$.

Then the strong convergence $\frac{\partial u_\varepsilon}{\partial t} \to \frac{\partial u}{\partial t}$ in $L^2(Q)$ can be proved by

$$\begin{aligned}\int_Q \alpha\Big|\frac{\partial u}{\partial t}\Big|^2 \mathrm{d}x\mathrm{d}t &\le \liminf_{\varepsilon\to 0}\int_Q \alpha\Big|\frac{\partial u_\varepsilon}{\partial t}\Big|^2 \mathrm{d}x\mathrm{d}t \le \limsup_{\varepsilon\to 0}\int_Q \alpha\Big|\frac{\partial u_\varepsilon}{\partial t}\Big|^2 \mathrm{d}x\mathrm{d}t\\ &= \limsup_{\varepsilon\to 0}\int_\Omega \frac{\mu}{2}|\nabla u_0|^2 + \varphi_0(u_0) - \frac{\mu}{2}|\nabla u_\varepsilon(T)|^2 - \varphi_0(u_\varepsilon(T))\,\mathrm{d}x\\ &\qquad + \int_Q \big(h-\gamma(u_\varepsilon,\vartheta_\varepsilon)\big)\cdot\frac{\partial u_\varepsilon}{\partial t}\,\mathrm{d}x\mathrm{d}t\\ &\le \int_\Omega \frac{\mu}{2}|\nabla u_0|^2 + \varphi_0(u_0) - \frac{\mu}{2}|\nabla u(T)|^2 - \varphi_0(u(T))\,\mathrm{d}x\\ &\qquad + \int_Q \big(h-\gamma(u,\vartheta)\big)\cdot\frac{\partial u}{\partial t}\,\mathrm{d}x\mathrm{d}t = \int_Q \alpha\Big|\frac{\partial u}{\partial t}\Big|^2 \mathrm{d}x\mathrm{d}t, \end{aligned} \tag{12.133}$$

where the weak lower semicontinuity of the convex functionals $\dot u \mapsto \int_Q \alpha|\dot u|^2\,\mathrm{d}x\mathrm{d}t$ and $u \mapsto \int_\Omega \frac{\mu}{2}|\nabla u|^2\mathrm{d}x$, the equation (12.126a), and the strong convergence $u_\varepsilon(T) \to u(T)$ in $L^5(\Omega;\mathbb{R}^3)$ and $\gamma(u_\varepsilon,\vartheta_\varepsilon) \to \gamma(u,\vartheta)$ in $L^2(Q;\mathbb{R}^3)$ have successively been used, and eventually also (12.131) has been used for the last equality in (12.133). Thus, in fact, we have equalities everywhere in (12.133) and, in particular, $\lim_{\varepsilon\to 0}\int_Q \alpha|\frac{\partial u_\varepsilon}{\partial t}|^2\mathrm{d}x\mathrm{d}t = \int_Q \alpha|\frac{\partial u}{\partial t}|^2\mathrm{d}x\mathrm{d}t$. The weak convergence $\frac{\partial u_\varepsilon}{\partial t} \to \frac{\partial u}{\partial t}$ in $L^2(Q;\mathbb{R}^3)$ is thus converted to the strong convergence; cf. Exercise 2.52. Then the limit passage in the heat-transfer equation (12.126b) towards (12.124b) is easy. □

Remark 12.50 (Ferromagnets with eddy currents). By combination of (12.121) with (12.107), one obtains the system[62]

$$\alpha\frac{\partial u}{\partial t} - \beta(|u|)u\times\frac{\partial u}{\partial t} - \mu\,\Delta u + \varphi_0'(u) + \theta\varphi_1'(u) = \mu_0 h, \tag{12.134a}$$

$$\mathfrak{c}(\theta)\frac{\partial \theta}{\partial t} - \mathrm{div}\big(\kappa(\theta)\nabla\theta\big) = \alpha\Big|\frac{\partial u}{\partial t}\Big|^2 + \theta\varphi_1'(u)\cdot\frac{\partial u}{\partial t} + \sigma(\theta)|e|^2, \tag{12.134b}$$

$$\mu_0\frac{\partial h}{\partial t} + \mathrm{curl}\, e = -\mu_0\frac{\partial u}{\partial t}, \tag{12.134c}$$

$$\mathrm{curl}\, h = \sigma(\theta)e. \tag{12.134d}$$

Note that the right-hand sides of (12.134a) and of (12.134c) are now not given, in contrast to (12.121a) and (12.107b). This system describes thermodynamics of electrically conductive ferromagnets under a possible ferro/para magnetic transformation with Joule heating by the induced electric current governed by the Maxwell system considered in the *eddy-current approximation*.

Exercise 12.51. Apply Galerkin method to (12.126) and, using a-priori estimates of Lemma 12.47, show convergence of the Galerkin solutions.[63]

---

[62] For such a system even augmented by a mechanical part describing magnetostrictive materials under small strains see [373].

[63] Hint: Show strong convergence of $\frac{\partial}{\partial t}u_{\varepsilon k} \to \frac{\partial}{\partial t}u_\varepsilon$ in $L^p(Q;\mathbb{R}^3)$ for $k\to\infty$ (k indexing the Galerkin approximants) by using the d-monotonicity of the regularizing term $\varepsilon|\frac{\partial}{\partial t}u_{\varepsilon k}|^{p-2}\frac{\partial}{\partial t}u_{\varepsilon k}$.

Exercise 12.52. Considering an initial-boundary-value problem for the system (12.134), derive the a-priori estimates and prove existence of a weak solution.[64]

12.9 Thermo-visco-elasticity: fully nonlinear theory

The last example of coupled nonlinear systems refines the thermo-visco-elasticity from Section 12.1 where we used a partly linear ansatz (12.1) as $\psi'''_{e\theta\theta} \equiv 0$ and thus the heat capacity $\mathfrak{c} = -\theta\psi''_{\theta\theta}(e,\theta)$ was independent of θ. Now we consider the fully nonlinear specific Helmholtz free energy. This will require us also to use a gradient theory for e. Referring to the notation of Section 12.1, we use the free energy

$$\psi_1(e,\theta,\nabla e) := \psi(e,\theta) + \frac{\gamma}{2}|\nabla e|^2 \tag{12.135}$$

and the dissipation rate

$$\xi_1(\dot{e},\nabla\dot{e}) := \xi(\dot{e}) + \gamma_1|\nabla\dot{e}|^2. \tag{12.136}$$

Obviously, for $\gamma = 0 = \gamma_1$ and for ψ from (12.1) and ξ from (12.2), we obtain the model from Section 12.1.

For simplicity, like in (12.2), we will consider ξ quadratic so that ξ_1 is, up to the factor $\frac{1}{2}$, simultaneously the (pseudo)potential of the dissipative forces.

The force balance now reads as

$$\varrho\frac{\partial^2 u}{\partial t^2} - \mathrm{div}\big(\sigma - \mathrm{div}\,\mathfrak{h}\big) = g \tag{12.137}$$

with the stress $\sigma = \sigma(e,\nabla e,\dot{e},\nabla\dot{e},\theta) := [\psi_1]'_e(e,\theta,\nabla e) + \frac{1}{2}[\xi_1]'_{\dot{e}}(\dot{e},\nabla\dot{e})$ and the so-called hyperstress $\mathfrak{h} = \mathfrak{h}(e,\nabla e,\dot{e},\nabla\dot{e},\theta) := [\psi_1]'_{\nabla e}(e,\theta,\nabla e) + \frac{1}{2}[\xi_1]'_{\nabla\dot{e}}(\dot{e},\nabla\dot{e})$.[65] Thus, in view of the ansatz (12.135) and (12.136), we have simply $\sigma = \sigma(e,\dot{e},\theta) = \psi'_e(e,\theta) + \frac{1}{2}\xi'\dot{e}$ and $\mathfrak{h} = \mathfrak{h}(\nabla e,\nabla\dot{e}) = \gamma\nabla e + \gamma_1\nabla\dot{e}$, and thus we consider (12.137) in the specific form:

$$\varrho\frac{\partial^2 u}{\partial t^2} - \mathrm{div}\Big(\frac{1}{2}\xi' e\big(\frac{\partial u}{\partial t}\big) + \psi'_e(e(u),\theta) - \Delta e\big(\gamma_1\frac{\partial u}{\partial t} + \gamma u\big)\Big) = g. \tag{12.138a}$$

As in Section 12.1, the entropy is defined by $s = -[\psi_1]'_\theta(e,\theta,\nabla e) = -\psi'_\theta(e,\theta)$ and the entropy equation reads as $\theta\frac{\partial s}{\partial t} = \xi_1(e(\frac{\partial u}{\partial t}),\nabla e(\frac{\partial u}{\partial t})) + \mathrm{div}\,j$. Considering an anisotropic nonlinear Fourier law for the heat flux $j = \mathbb{K}(e(u),\theta)\nabla\theta$, we obtain the heat equation

$$\begin{aligned}\mathfrak{c}(e(u),\theta)\frac{\partial\theta}{\partial t} - \mathrm{div}(\mathbb{K}(e(u),\theta)\nabla\theta) &= \xi\Big(e\big(\frac{\partial u}{\partial t}\big)\Big) + \gamma_1\Big|\nabla e\big(\frac{\partial u}{\partial t}\big)\Big|^2\\ &+ \theta\psi''_{e\theta}\big(e(u),\theta\big){:}e\big(\frac{\partial u}{\partial t}\big) \qquad \text{with} \quad \mathfrak{c}(e,\theta) = -\theta\psi''_{\theta\theta}(e,\theta).\end{aligned} \tag{12.138b}$$

[64] Hint: Apply enthalpy transformation, and use the test from Section 12.7, noting also the cancellation of the terms $\pm\mu_0\frac{\partial u}{\partial t}\cdot h$.

[65] The concept of hyperstresses [338] is related with so-called 2nd-grade nonsimple materials (also called multipolar solids or complex materials), cf. e.g. [153, 385].

We consider again the initial conditions (12.6b). We will use natural Newton-type boundary conditions, which means[66]

$$\Big(\frac{1}{2}\xi'e\big(\frac{\partial u}{\partial t}\big)+\psi_e'(e(u),\theta)-\Delta e\big(\gamma_1\frac{\partial u}{\partial t}+\gamma u\big)\Big)\cdot\nu-\mathrm{div}_{\mathrm{S}}\Big(\nabla e\big(\gamma_1\frac{\partial u}{\partial t}+\gamma u\big)\cdot\nu\Big)=h, \tag{12.139a}$$

$$\nabla e(u){:}(\nu\otimes\nu)=0\,,\qquad\text{and} \tag{12.139b}$$

$$\mathbb{K}(e(u),\theta)\frac{\partial\theta}{\partial\nu}=f\qquad\text{on }\Sigma; \tag{12.139c}$$

cf. (2.107). In (12.139a), h stands for the prescribed true boundary force. This leads to the weak formulation of the mechanical part (12.138a)-(12.139a,b) as

$$\int_Q\Big(\frac{1}{2}\xi'e\big(\frac{\partial u}{\partial t}\big)+\psi_e'(e(u),\theta)\Big){:}e(v)+\nabla e\big(\gamma_1\frac{\partial u}{\partial t}+\gamma u\big)\vdots\nabla e(v)-\varrho\frac{\partial u}{\partial t}\cdot\frac{\partial v}{\partial t}\,\mathrm{d}x\mathrm{d}t$$
$$+\int_\Omega\varrho\frac{\partial u}{\partial t}(T)\cdot v(T)\,\mathrm{d}x=\int_\Omega\varrho v_0\cdot v(T)\,\mathrm{d}x+\int_Q g\cdot v\,\mathrm{d}x\mathrm{d}t+\int_\Sigma h\cdot v\,\mathrm{d}S\mathrm{d}t \tag{12.140}$$

for any $v\in W^{1,2}(I;L^2(\Omega;\mathbb{R}^n))\cap L^2(I;W^{2,2}(\Omega;\mathbb{R}^n))$; naturally, "$\vdots$" means summation over three indices.

The peculiarity is that the heat capacity $\mathfrak{c}$ depends also on the mechanical variable e and one cannot perform the conventional enthalpy transformation (8.198). Anyhow, one can make an *enhanced enthalpy transformation*, which requires gradient theory for e. This procedure is based on an elementary calculus:

$$\mathfrak{c}(e,\theta)\frac{\partial\theta}{\partial t}=\frac{\partial\widehat{\mathfrak{c}}(e,\theta)}{\partial t}-\widehat{\mathfrak{c}}_e'(e,\theta){:}\frac{\partial e}{\partial t} \tag{12.141}$$

where

$$\widehat{\mathfrak{c}}(e,\theta):=\int_0^\theta\mathfrak{c}(e,t)\,\mathrm{d}t\,,\qquad\text{and thus}\qquad\widehat{\mathfrak{c}}_e'(e,\theta)=\int_0^\theta\mathfrak{c}_e'(e,t)\,\mathrm{d}t. \tag{12.142}$$

We introduce the substitution

$$\vartheta=\widehat{\mathfrak{c}}(e(u),\theta). \tag{12.143}$$

It is physically natural to assume the heat capacity $\mathfrak{c}$ positive, which makes $\widehat{\mathfrak{c}}(e,\cdot)$ increasing and thus also invertible, which allows us to define

$$\Theta(e,\vartheta):=\big[\widehat{\mathfrak{c}}(e,\cdot)\big]^{-1}(\vartheta), \tag{12.144a}$$
$$\mathscr{K}_0(e,\vartheta):=\mathbb{K}(e,\Theta(e,\vartheta))\Theta_\vartheta'(e,\vartheta), \tag{12.144b}$$
$$\mathscr{K}_1(e,\vartheta):=\mathbb{K}(e,\Theta(e,\vartheta))\Theta_e'(e,\vartheta), \tag{12.144c}$$
$$\mathscr{F}(e,\vartheta):=\Theta(e,\vartheta)\psi_{\theta e}''(e,\Theta(e,\vartheta))+\widehat{\mathfrak{c}}_e'(e,\Theta(e,\vartheta)). \tag{12.144d}$$

---

[66] We use twice the Green formula $\int_\Omega\mathrm{div}\Delta e(u)\cdot z\,\mathrm{d}x=\int_\Gamma\Delta e(u){:}(z\otimes\nu)\,\mathrm{d}S-\int_\Omega\Delta e(u){:}\nabla z\,\mathrm{d}x=\int_\Omega\nabla e(u)\vdots\nabla^2z\,\mathrm{d}x+\int_\Gamma\Delta e(u){:}(z\otimes\nu)-\nabla e(u)\vdots(\nabla z\otimes\nu)\,\mathrm{d}S=\int_\Omega\nabla e(u)\vdots\nabla e(z)+\int_\Gamma\Delta e(u){:}(z\otimes\nu)-\nabla e(u)\vdots((\frac{\partial z}{\partial\nu}\nu+\nabla_{\mathrm{S}}z)\otimes\nu)\,\mathrm{d}S$ together with the formula (2.104). Note that, in (12.139a), one expects the term $(\mathrm{div}_{\mathrm{S}}\nu)\nabla e(\gamma_1\frac{\partial u}{\partial t}+\gamma u){:}(\nu\otimes\nu)$ which, however, vanishes because of (12.139b).

Realizing $\theta = \Theta(e(u), \vartheta)$, one has

$$\begin{aligned}\mathbb{K}(e(u),\theta)\nabla\theta &= \mathbb{K}(e(u),\Theta(e(u),\vartheta))\nabla\Theta(e(u),\vartheta)\\ &= \mathscr{K}_0(e(u),\vartheta)\nabla\vartheta + \mathscr{K}_1(e(u),\vartheta)\nabla e(u). \end{aligned} \tag{12.145}$$

Also, we have $\mathscr{F}(e,\vartheta) = \mathbb{F}(e,\theta) := \theta\psi''_{e\theta}(e,\theta) + \int_0^\theta \mathfrak{c}'_e(e,t)\,\mathrm{d}t$ for $\theta = \Theta(e,\vartheta)$ so that, since $\mathfrak{c}'_e(e,\theta) = -\theta\psi'''_{e\theta\theta}(e,\theta)$, we have

$$\big[\mathscr{F}-\psi'_e\circ\Theta\big](e,\vartheta) = \big[\mathbb{F}-\psi'_e\big](e,\theta) = \theta\psi''_{e\theta}(e,\theta) - \int_0^\theta t\psi'''_{e\theta\theta}(e,t)\,\mathrm{d}t - \psi'_e(e,\theta). \tag{12.146}$$

Furthermore,

$$\big[\mathbb{F}-\psi'_e\big]'_\theta(e,\theta) = \theta\psi'''_{e\theta\theta}(e,\theta) + \psi''_{e\theta}(e,\theta) - \theta\psi'''_{e\theta\theta}(e,\theta) - \psi''_{e\theta}(e,\theta) = 0$$

so that $\mathscr{F}-\psi'_e\circ\Theta$ is independent of ϑ. On the other hand, from (12.146), one can see that $[\mathbb{F}-\psi'_e](e,0) = -\psi'_e(e,0)$. Therefore, we can deduce that

$$\big[\psi'_e\circ\Theta - \mathscr{F}\big](e,\vartheta) = \psi'_e(e,0) = \varphi'(e), \tag{12.147}$$

where we again abbreviated $\varphi(\cdot) := \psi(\cdot,0)$ as in Section 12.1. This transforms the system (12.138) into the form

$$\varrho\frac{\partial^2 u}{\partial t^2} - \mathrm{div}\Big(\frac{1}{2}\xi'e(\frac{\partial u}{\partial t}) + \varphi'(e(u)) + \mathscr{F}(e(u),\vartheta) - \Delta e\big(\gamma_1\frac{\partial u}{\partial t}+\gamma u\big)\Big) = g, \tag{12.148a}$$

$$\begin{aligned}\frac{\partial\vartheta}{\partial t} - \mathrm{div}\big(\mathscr{K}_0(e(u),\vartheta)\nabla\vartheta\big) &= \xi\Big(e\big(\frac{\partial u}{\partial t}\big)\Big) + \gamma_1\Big|\nabla e\big(\frac{\partial u}{\partial t}\big)\Big|^2\\ &\quad + \mathscr{F}(e(u),\vartheta){:}e\big(\frac{\partial u}{\partial t}\big) + \mathrm{div}\big(\mathscr{K}_1(e(u),\vartheta)\nabla e(u)\big), \end{aligned}\tag{12.148b}$$

with the boundary conditions

$$\begin{aligned}&\Big(\frac{1}{2}\xi'e\big(\frac{\partial u}{\partial t}\big) + \varphi'(e(u)) + \mathscr{F}(e(u),\vartheta) - \Delta e\big(\gamma_1\frac{\partial u}{\partial t}+\gamma u\big)\Big)\cdot\nu\\ &\qquad\qquad - \mathrm{div}_{\mathrm{s}}\Big(\nabla e\big(\gamma_1\frac{\partial u}{\partial t}+\gamma u\big)\cdot\nu\Big) = h, \end{aligned}\tag{12.149a}$$

$$\nabla e(u){:}(\nu\otimes\nu) = 0, \qquad \text{and} \tag{12.149b}$$

$$\mathscr{K}_0(e(u),\vartheta)\frac{\partial\vartheta}{\partial\nu} + \mathscr{K}_1(e(u),\vartheta)\frac{\partial e(u)}{\partial\nu} = f \qquad \text{on } \Sigma, \tag{12.149c}$$

and the initial conditions

$$u(0,\cdot) = u_0, \quad \frac{\partial u}{\partial t}(0,\cdot) = v_0, \quad \vartheta(0,\cdot) = \widehat{\mathfrak{c}}(e(u_0),\theta_0) \qquad \text{on } \Omega. \tag{12.149d}$$

We analyse the system (12.148) by Rothe's method with a suitable regularization to compensate the growth of the non-monotone terms on the right-hand side of the heat equation (12.148b). More specifically, we consider

$$\varrho\frac{u_\tau^k-2u_\tau^{k-1}+u_\tau^{k-2}}{\tau^2}-\mathrm{div}\Big(\frac{1}{2}\xi' e\big(\frac{u_\tau^k-u_\tau^{k-1}}{\tau}\big)+\varphi'(e(u_\tau^k))$$
$$+\mathscr{F}(e(u_\tau^k),\vartheta_\tau^k)+\tau\big|e(u_\tau^k)\big|^{\eta-2}e(u_\tau^k)-\mathrm{div}\,\mathfrak{h}_\tau^k\Big)=g_\tau^k$$
$$\text{with }\ \mathfrak{h}_\tau^k=\nabla e\Big(\gamma_1\frac{u_\tau^k-u_\tau^{k-1}}{\tau}+\gamma u_\tau^k\Big)+\tau\big|\nabla e(u_\tau^k)\big|^{\eta-2}\nabla e(u_\tau^k), \tag{12.150a}$$

$$\frac{\vartheta_\tau^k-\vartheta_\tau^{k-1}}{\tau}-\mathrm{div}\big(\mathscr{K}_0(e(u_\tau^k),\vartheta_\tau^k)\nabla\vartheta_\tau^k\big)=\big(1-\frac{\sqrt{\tau}}{2}\big)\xi\Big(e\big(\frac{u_\tau^k-u_\tau^{k-1}}{\tau}\big)\Big)$$
$$+\gamma_1\Big|\nabla e\big(\frac{u_\tau^k-u_\tau^{k-1}}{\tau}\big)\Big|^2+\mathscr{F}\big(e(u_\tau^k),\vartheta_\tau^k\big){:}e\big(\frac{u_\tau^k-u_\tau^{k-1}}{\tau}\big)$$
$$+\mathrm{div}\big(\mathscr{K}_1(e(u_\tau^k),\vartheta_\tau^k)\nabla e(u_\tau^k)\big). \tag{12.150b}$$

We complete this system with the boundary conditions

$$\Big(\frac{1}{2}\xi' e\big(\frac{u_\tau^k-u_\tau^{k-1}}{\tau}\big)+\varphi'(e(u_\tau^k))+\mathscr{F}(e(u_\tau^k),\vartheta_\tau^k))$$
$$+\tau\big|e(u_\tau^k)\big|^{\eta-2}e(u_\tau^k)-\mathrm{div}\,\mathfrak{h}_\tau^k\Big){\cdot}\nu-\mathrm{div}_{\mathrm{S}}\big(\mathfrak{h}_\tau^k{\cdot}\nu\big)=h_\tau^k, \tag{12.151a}$$

$$\mathfrak{h}_\tau^k{:}(\nu\otimes\nu)=0\,,\qquad\text{and} \tag{12.151b}$$

$$\mathscr{K}_0(e(u_\tau^k),\vartheta_\tau^k)\frac{\partial\vartheta_\tau^k}{\partial\nu}+\mathscr{K}_1(e(u_\tau^k),\vartheta_\tau^k)\frac{\partial e(u_\tau^k)}{\partial\nu}=f_\tau^k\qquad\text{on }\Gamma. \tag{12.151c}$$

We start this recursive scheme for $k=1$ by considering the initial conditions

$$u_\tau^0=u_{0\tau},\quad u_\tau^{-1}=u_{0\tau}-\tau v_0,\quad \vartheta_\tau^0=\widehat{\mathfrak{c}}(e(u_{0\tau}),\theta_0)\qquad\text{on }\Omega \tag{12.151d}$$

involving a suitably regularized initial displacement $u_{0\tau}$.

The involved higher gradients allow for a general free energy $\psi(e,\theta)$. Recall that the function φ is called *semi-convex* if $\varphi(\cdot)+K|\cdot|^2$ is convex for some sufficiently large K. In terms of the transformed data, we assume:

$$\varphi(\cdot)=\psi(\cdot,0):\mathbb{R}^{n\times n}_{\mathrm{sym}}\to\mathbb{R}^+\text{ is semi-convex and} \tag{12.152a}$$
$$\exists C\in\mathbb{R}:\qquad \varphi(\cdot)\le C\big(1+|\cdot|^{2^*}\big),\quad \big|\varphi'(\cdot)\big|\le C\big(1+|\cdot|^{2^*-1-\epsilon}\big), \tag{12.152b}$$
$$\forall e\in\mathbb{R}^{n\times n}_{\mathrm{sym}}:\quad \varphi(e){:}e+C\ge 0, \tag{12.152c}$$
$$\exists C_{\mathscr{K}_0},C_{\mathscr{K}_1}\in\mathbb{R},\ \kappa_0,\epsilon>0\ \ \forall(e,\vartheta)\in\mathbb{R}^{n\times n}_{\mathrm{sym}}{\times}\mathbb{R},\ s\in\mathbb{R}^n:$$
$$\mathscr{K}_0(e,\vartheta)s{\cdot}s\ge\kappa_0|s|^2, \tag{12.152d}$$
$$\big|\mathscr{K}_0(e,\vartheta)\big|\le C_{\mathscr{K}_0}\big(1+|e|^{2^*/(n+2)-\epsilon}+|\vartheta|^{1/n-\epsilon}\big), \tag{12.152e}$$

$$\big|\mathscr{K}_1(e,\vartheta)\big| \le C_{\mathscr{K}_1}\sqrt{1+|\vartheta|}, \tag{12.152f}$$

$$\exists C_{\mathscr{F}}\in\mathbb{R},\ 0\le p_{\mathscr{F}}<2^*/2,\ 0\le q_{\mathscr{F}}<1\ \ \forall(e,\vartheta)\in\mathbb{R}^{n\times n}_{\rm sym}\times\mathbb{R}: \tag{12.152g}$$

$$\big|\mathscr{F}(e,\vartheta)\big| \le C_{\mathscr{F}}\big(1+|e|^{p_{\mathscr{F}}}+|\vartheta|^{q_{\mathscr{F}}}\big), \tag{12.152h}$$

$$\varrho,\gamma_1,\gamma>0,\qquad \xi:\mathbb{R}^{n\times n}_{\rm sym}\to\mathbb{R}\ \text{quadratic, coercive}, \tag{12.152i}$$

$$g\in L^2(I;L^{2^{*\prime}}(\Omega;\mathbb{R}^n)),\qquad h\in L^2(I;L^{2^{\#\prime}}(\Gamma;\mathbb{R}^n)), \tag{12.152j}$$

$$u_0\in W^{2,2}(\Omega;\mathbb{R}^n),\qquad v_0\in L^2(\Omega;\mathbb{R}^n), \tag{12.152k}$$

$$\theta_0\ge 0,\quad f\ge 0,\quad \mathfrak{c}(\theta_0,e(u_0))\in L^1(\Omega),\quad f\in L^1(\Sigma). \tag{12.152l}$$

In fact, we will need (12.152) only for $\vartheta\ge 0$ in what follows. The semiconvexity of φ assumed in (12.152a) is a rather technical assumption facilitating a simple usage of Rothe's method, otherwise the ψ'_e-term is in a position of the lower-order term and its monotonicity is not essentially needed. Note that (12.152e) ensures integrability of $\mathscr{K}_0(e(u),\vartheta)\nabla\vartheta$ if $e(u)\in L^{2^*}(Q;\mathbb{R}^{n\times n})$, $\vartheta\in L^{(n+2)/n-\epsilon}(Q)$, and $\nabla\vartheta\in L^{(n+2)/(n+1)-\epsilon}(Q;\mathbb{R}^n)$. Similarly, also (12.152f) ensures integrability of $\mathscr{K}_1(e(u),\vartheta)\nabla e(u)$.

Lemma 12.53 (Existence of Rothe's solution). *Let* (12.152) *hold and let* $u_{0\tau}\in W^{2,\eta}(\Omega;\mathbb{R}^n)$ *with* $\eta>\max(4,2p_{\mathscr{F}}+2,2/(1-q_{\mathscr{F}}))$. *Then the recursive boundary-value problem* (12.150)–(12.151) *possesses a weak solution* $(u_\tau^k,\vartheta_\tau^k)\in W^{2,\eta}(\Omega;\mathbb{R}^n)\times W^{1,2}(\Omega)$ *such that* $\vartheta_\tau^k\ge 0$ *for any* $k=1,...,T/\tau$.

Proof. To prove the coercivity of the underlying nonlinear operator, the particular equations in (12.150) are to be tested respectively by u_τ^k and ϑ_τ^k, and the non-monotone terms are to be estimated by Hölder's and Young's inequalities. Let us briefly discuss the estimation of the nonmonotone terms under this test. Using (12.152f), one can estimate

$$\begin{aligned}\bigg|\int_\Omega \mathscr{K}_1(e(u_\tau^k),\vartheta_\tau^k)\nabla e(u_\tau^k)\cdot\nabla\vartheta_\tau^k\,{\rm d}x\bigg| &\le \int_\Omega C_{\mathscr{K}_1}\sqrt{1+|\vartheta_\tau^k|}\,\big|\nabla e(u_\tau^k)\big|\,\big|\nabla\vartheta_\tau^k\big|\,{\rm d}x\\ &\le \int_\Omega C_{\mathscr{K}_1}\Big(1+\sqrt{|\vartheta_\tau^k|}\Big)\big|\nabla e(u_\tau^k)\big|\,\big|\nabla\vartheta_\tau^k\big|\,{\rm d}x\\ &\le C_{\delta,\eta}+\delta\Big\|\sqrt{|\vartheta_\tau^k|}\Big\|_{L^4(\Omega)}^4+\delta\big\|\nabla e(u_\tau^k)\big\|_{L^\eta(\Omega;\mathbb{R}^{n\times n\times n})}^\eta+\delta\big\|\nabla\vartheta_\tau^k\big\|_{L^2(\Omega;\mathbb{R}^n)}^2\end{aligned} \tag{12.153}$$

with $\delta>0$ arbitrarily small and some $C_{\delta,\eta}$ depending on η and δ; here we used that $\eta>4$. The heat sources with quadratic growth can be estimated as

$$\begin{aligned}\bigg|\int_\Omega\big(\xi(e(u_\tau^k))+\gamma_1|\nabla e(u_\tau^k)|^2\big)\vartheta_\tau^k\,{\rm d}x\bigg| &\le C_{\delta,\eta}+\delta\big\|e(u_\tau^k)\big\|_{L^\eta(\Omega;\mathbb{R}^{n\times n})}^\eta\\ &\quad+\delta\big\|\nabla e(u_\tau^k)\big\|_{L^\eta(\Omega;\mathbb{R}^{n\times n\times n})}^\eta+\delta\big\|\vartheta_\tau^k\big\|_{L^2(\Omega)}^2;\end{aligned} \tag{12.154}$$

here we again used that $\eta>4$. By (12.152h), the adiabatic term in (12.150b) can be estimated as

$$\left|\int_\Omega \mathscr{F}(e(u_\tau^k),\vartheta_\tau^k){:}e(u_\tau^k)\vartheta_\tau^k\,\mathrm{d}x\right| \le C_{\mathscr{F}}\int_\Omega\Big(1+|e(u_\tau^k)|^{p_{\mathscr{F}}}+|\vartheta_\tau^k|^{q_{\mathscr{F}}}\Big)|e(u_\tau^k)|\,|\vartheta_\tau^k|\,\mathrm{d}x$$
$$\le C_{\mathscr{F}}' + \big\|e(u_\tau^k)\big\|_{L^\eta(\Omega;\mathbb{R}^{n\times n})}^\eta + \delta\|\vartheta_\tau^k\|_{L^2(\Omega)}^2 \qquad (12.155)$$

with $C_{\mathscr{F}}'$ dependent on δ and on $C_{\mathscr{F}}$, $p_{\mathscr{F}}$, and $q_{\mathscr{F}}$ from (12.152h); here we used that $\eta > \max(2p_{\mathscr{F}}{+}2, 2/(1{-}q_{\mathscr{F}}))$.

The pseudomonotonicity of this operator is obvious because all non-monotone terms are either in a position of lower-order terms with the exception of the $\mathscr{K}_1$-term in (12.150b). This term is however linear in $\nabla e(u_\tau^k)$, hence weakly continuous. Then the claimed existence follows by using Brézis' Theorem 2.6.

Eventually, one can test (12.150b) by $(\vartheta_\tau^k)^-$, which reveals $\vartheta_\tau^k \ge 0$ when realizing that, for a moment, one can define $\mathscr{F}(e,\vartheta)=0$ for $\vartheta\le 0$, and then found that this possible re-definition is, in fact, irrelevant for this particular solution. □

Lemma 12.54 (Energy estimates). *Let the assumptions of Lemma* 12.53 *hold, and let also* $\|u_{0\tau}\|_{W^{2,\eta}(\Omega;\mathbb{R}^n)} = \mathscr{O}(\tau^{-1/\eta})$. *Then:*

$$\big\|u_\tau\big\|_{W^{1,\infty}(I;L^2(\Omega;\mathbb{R}^n))\cap L^\infty(I;W^{2,2}(\Omega;\mathbb{R}^n))} \le C, \qquad (12.156a)$$
$$\big\|\vartheta_\tau\big\|_{L^\infty(I;L^1(\Omega))} \le C, \qquad (12.156b)$$
$$\big\|e(u_\tau)\big\|_{L^\infty(I;W^{1,\eta}(\Omega;\mathbb{R}^{n\times n}))} \le C\tau^{-1/\eta}. \qquad (12.156c)$$

Sketch of the proof. First we test the mechanical part (12.150a) by $(u_\tau^k{-}u_\tau^{k-1})/\tau$. We use convexity of the kinetic energy $\frac{\varrho}{2}|\cdot|^2$, of $E\mapsto\frac{\gamma}{2}|E|^2+\frac{\tau}{\eta}|E|^\eta$, and of the regularizing functional $\frac{\tau}{\eta}|\cdot|^\eta$. We further use the semi-convexity of φ and the coercivity of the quadratic form ξ to estimate, as in (8.86),

$$\begin{aligned}
&\Big(\varphi'(e(u_\tau^k)) + \frac{1}{2}\xi'e\big(\frac{u_\tau^k-u_\tau^{k-1}}{\tau}\big)\Big){:}e\big(\frac{u_\tau^k-u_\tau^{k-1}}{\tau}\big)\\
&= \Big(\varphi'(e(u_\tau^k)) + \frac{1}{2\sqrt{\tau}}\xi'e(u_\tau^k)\Big){:}e\big(\frac{u_\tau^k-u_\tau^{k-1}}{\tau}\big)\\
&\qquad - \frac{\xi'e(u_\tau^{k-1})}{2\sqrt{\tau}}{:}e\big(\frac{u_\tau^k-u_\tau^{k-1}}{\tau}\big) + (1{-}\sqrt{\tau})\xi\Big(e\big(\frac{u_\tau^k{-}u_\tau^{k-1}}{\tau}\big)\Big)\\
&\ge \frac{1}{\tau}\Big(\varphi(e(u_\tau^k)) + \frac{\xi(e(u_\tau^k))}{2\sqrt{\tau}} - \varphi(e(u_\tau^{k-1})) - \frac{\xi(e(u_\tau^{k-1}))}{2\sqrt{\tau}}\Big)\\
&\qquad - \frac{\xi'e(u_\tau^{k-1})}{2\sqrt{\tau}}{:}e\big(\frac{u_\tau^k{-}u_\tau^{k-1}}{\tau}\big) + (1{-}\sqrt{\tau})\xi\Big(e\big(\frac{u_\tau^k{-}u_\tau^{k-1}}{\tau}\big)\Big)\\
&= \frac{\varphi(e(u_\tau^k)) - \varphi(e(u_\tau^{k-1}))}{\tau} + \Big(1{-}\frac{\sqrt{\tau}}{2}\Big)\xi\Big(e\big(\frac{u_\tau^k-u_\tau^{k-1}}{\tau}\big)\Big)
\end{aligned} \qquad (12.157)$$

provided τ is sufficiently small, namely so small that $\varphi(\cdot)+\xi(\cdot)/\sqrt{4\tau}:\mathbb{R}^{n\times n}\to\mathbb{R}$ is convex. Summing it for $k=1,...,l$ yields the discrete mechanical-energy balance

$$\begin{aligned}&\frac{\varrho}{2}\Big\|\frac{u_\tau^l-u_\tau^{l-1}}{\tau}\Big\|^2_{L^2(\Omega;\mathbb{R}^n)}+\tau\sum_{k=1}^{l}\Big(\Big(1-\frac{\sqrt{\tau}}{2}\Big)\xi\Big(e\Big(\frac{u_\tau^k-u_\tau^{k-1}}{\tau}\Big)\Big)+\gamma_1\Big|\nabla e\Big(\frac{u_\tau^k-u_\tau^{k-1}}{\tau}\Big)\Big|^2\Big)\\&\quad+\int_\Omega\varphi(e(u_\tau^l))+\frac{\gamma}{2}\big|\nabla e(u_\tau^l)\big|^2+\frac{\tau}{\eta}|e(u_\tau^l)|^\eta+\frac{\tau}{\eta}|\nabla e(u_\tau^l)|^\eta\,\mathrm{d}x\\&\quad\le\tau\sum_{k=1}^{l}\Big(\int_\Omega g_\tau^k\cdot\frac{u_\tau^k-u_\tau^{k-1}}{\tau}-\mathscr{F}\big(e(u_\tau^k),\vartheta_\tau^k\big):e\Big(\frac{u_\tau^k-u_\tau^{k-1}}{\tau}\Big)\,\mathrm{d}x\\&\quad+\int_\Gamma h_\tau^k\cdot\frac{u_\tau^k-u_\tau^{k-1}}{\tau}\,\mathrm{d}S\Big)+\frac{\varrho}{2}\big\|v_0\big\|^2_{L^2(\Omega;\mathbb{R}^n)}\\&\quad+\int_\Omega\varphi(u_{0\tau})+\frac{\gamma}{2}\big|\nabla e(u_{0\tau})\big|^2+\frac{\tau}{\eta}|e(u_{0\tau})|^\eta+\frac{\tau}{\eta}|\nabla e(u_{0\tau})|^\eta\,\mathrm{d}x.\end{aligned}\tag{12.158}$$

Further we test the heat part (12.150b) by 1 and add it to (12.158). Observing cancellation of dissipative terms and also of the adiabatic $\mathscr{F}$-terms, we arrive at the discrete total-energy balance:

$$\begin{aligned}&\frac{\varrho}{2}\Big\|\frac{u_\tau^l-u_\tau^{l-1}}{\tau}\Big\|^2_{L^2(\Omega;\mathbb{R}^n)}+\big\|\vartheta_\tau^l\big\|_{L^1(\Omega)}\\&\quad+\int_\Omega\varphi(e(u_\tau^l))+\frac{\gamma}{2}\big|\nabla e(u_\tau^l)\big|^2+\frac{\tau}{\eta}|e(u_\tau^l)|^\eta+\frac{\tau}{\eta}|\nabla e(u_\tau^l)|^\eta\,\mathrm{d}x\\&\quad\le\tau\sum_{k=1}^{l}\Big(\int_\Omega g_\tau^k\cdot\frac{u_\tau^k-u_\tau^{k-1}}{\tau}\,\mathrm{d}x+\int_\Gamma h_\tau^k\cdot\frac{u_\tau^k-u_\tau^{k-1}}{\tau}+f_\tau^k\,\mathrm{d}S\Big)\\&\quad+\frac{\varrho}{2}\big\|v_0\big\|^2_{L^2(\Omega;\mathbb{R}^n)}+\big\|C_\mathrm{v}(e(u_{0\tau}),\theta_0)\big\|_{L^1(\Omega)}\\&\quad+\int_\Omega\varphi(u_{0\tau})+\frac{\gamma}{2}\big|\nabla e(u_{0\tau})\big|^2+\frac{\tau}{\eta}|e(u_{0\tau})|^\eta+\frac{\tau}{\eta}|\nabla e(u_{0\tau})|^\eta\,\mathrm{d}x.\end{aligned}\tag{12.159}$$

Then we employ the discrete Gronwall inequality. □

Proposition 12.55 (Further estimates). *Under the assumptions of Lemma 12.54, it also holds that*

$$\big\|u_\tau\big\|_{W^{1,2}(I;W^{2,2}(\Omega;\mathbb{R}^n))}\le C,\tag{12.160a}$$

$$\Big\|\varrho\frac{\partial}{\partial t}\Big[\frac{\partial u_\tau}{\partial t}\Big]^{\mathrm{i}}\Big\|_{L^2(I;W^{2,2}(\Omega;\mathbb{R}^n)^*)+L^{\eta'}(I;W^{2,\eta}(\Omega;\mathbb{R}^n)^*)}\le C,\tag{12.160b}$$

$$\begin{aligned}&\Big\|\varrho\frac{\partial}{\partial t}\Big[\frac{\partial u_\tau}{\partial t}\Big]^{\mathrm{i}}-\tau\mathrm{div}\big(|e(\bar u_\tau)|^{\eta-2}e(\bar u_\tau)\big)\\&\qquad+\tau\mathrm{div}^2\big(|\nabla e(\bar u_\tau)|^{\eta-2}\nabla e(\bar u_\tau)\big)\Big\|_{L^2(I;W^{2,2}(\Omega;\mathbb{R}^n)^*)}\le C,\end{aligned}\tag{12.160c}$$

and the estimates (12.16b,c,f) for $\bar\vartheta_\tau$ and ϑ_τ.

Proof. The strategy (12.17) to get estimate (12.16b) for $\nabla\vartheta_\tau$ is to be modified by an additional term $\mathscr{K}_1(e(\bar{u}_\tau),\bar{\vartheta}_\tau)\nabla e(\bar{u}_\tau)\cdot\nabla\chi(\bar{\vartheta}_\tau)$ with $\chi(\vartheta) := 1 - 1/(1{+}\vartheta)^\varepsilon$. We proceed as follows:

$$\begin{aligned}\int_Q \mathscr{K}_1(e(\bar{u}_\tau),\bar{\vartheta}_\tau)\nabla e(\bar{u}_\tau)\cdot\nabla\chi(\bar{\vartheta}_\tau)\,\mathrm{d}x\mathrm{d}t &= \varepsilon\int_Q \nabla e(\bar{u}_\tau)\vdots\frac{\mathscr{K}_1(e(\bar{u}_\tau),\bar{\vartheta}_\tau)\otimes\nabla\bar{\vartheta}_\tau}{(1+\bar{\vartheta}_\tau)^{1+\varepsilon}}\,\mathrm{d}x\mathrm{d}t\\ &\le \varepsilon\int_Q \frac{1}{4\delta}|\nabla e(\bar{u}_\tau)|^2 + \delta\frac{|\mathscr{K}_1(e(\bar{u}_\tau),\bar{\vartheta}_\tau)|^2}{1+\bar{\vartheta}_\tau}\frac{|\nabla\bar{\vartheta}_\tau|^2}{(1{+}\bar{\vartheta}_\tau)^{1+\varepsilon}}\,\mathrm{d}x\mathrm{d}t\\ &\le \varepsilon\int_Q \frac{1}{4\delta}|\nabla e(\bar{u}_\tau)|^2 + \delta C_{\mathscr{K}_1}\frac{|\nabla\bar{\vartheta}_\tau|^2}{(1{+}\bar{\vartheta}_\tau)^{1+\varepsilon}}\,\mathrm{d}x\mathrm{d}t \end{aligned}\tag{12.161}$$

with $C_{\mathscr{K}_1}$ from (12.152f). Using (12.156a) and choosing $\delta > 0$ small, we can execute the interpolation strategy (12.17). Instead of (12.25), we now have

$$\begin{aligned}\Big|\int_Q \mathscr{F}(e(\bar{u}_\tau),\bar{\vartheta}_\tau){:}e\Big(\frac{\partial u_\tau}{\partial t}\Big)\,\mathrm{d}x\mathrm{d}t\Big| &\le C\big\|\mathscr{F}(e(\bar{u}_\tau),\bar{\vartheta}_\tau)\big\|^2_{L^2(Q)} + \delta\Big\|e\Big(\frac{\partial u_\tau}{\partial t}\Big)\Big\|^2_{L^2(Q;\mathbb{R}^{n\times n})}\\ &\le C_\delta + C_\delta\big\|e(\bar{u}_\tau)\big\|^{2^*}_{L^{2^*}(Q;\mathbb{R}^{n\times n})} + C_\delta\big\|\bar{\vartheta}_\tau\big\|^{2/\omega}_{L^{2/\omega}(Q)} + \delta\Big\|e\Big(\frac{\partial u_\tau}{\partial t}\Big)\Big\|^2_{L^2(Q;\mathbb{R}^{n\times n})}\end{aligned}$$

where we used the restriction $p_{\mathscr{F}} < 2^*/2$ in (12.152h). Note that, by (12.156a), we have $\|e(\bar{u}_\tau)\|_{L^{2^*}(Q;\mathbb{R}^{n\times n})} \le T^{1/2^*}\|e(\bar{u}_\tau)\|_{L^\infty(I;L^{2^*}(\Omega;\mathbb{R}^{n\times n}))}$ already estimated. Then we can finish the interpolation game as in (12.26)–(12.30). This yields both (12.16b,c) and (12.160a).

The dual estimates (12.16f) and (12.160b,c) can be obtained routinely; note that (12.160c) relies on the restriction $p_{\mathscr{F}} < 2^*/2 \le 2^* - 1$ used in (12.152h) so that $\mathscr{F}(e(\bar{u}_\tau),\bar{\vartheta}_\tau)$ is certainly bounded in $L^\infty(I;L^{2^{*\prime}}(\Omega;\mathbb{R}^{n\times n}))$. $\square$

Proposition 12.56 (Convergence for $\tau \to 0$). *Let* (12.152) *hold and* $\eta > \max(4, 2p_{\mathscr{F}}{+}2, 2/(1{-}q_{\mathscr{F}}))$, *and let also*

$$u_{0\tau} \to u_0 \quad \textit{in}\ \ W^{2,2}(\Omega;\mathbb{R}^n) \quad \textit{and} \quad \big\|u_{0\tau}\big\|_{W^{2,\eta}(\Omega;\mathbb{R}^n)} = o(\tau^{-1/\eta}). \tag{12.162}$$

Then there is a subsequence such that

$$u_\tau \to u \qquad \textit{strongly in}\ \ W^{1,2}(I;W^{2,2}(\Omega;\mathbb{R}^n)), \tag{12.163a}$$

$$\bar{\vartheta}_\tau \to \vartheta \qquad \textit{strongly in}\ \ L^q(Q)\ \ \textit{with any}\ \ q < (n{+}2)/n, \tag{12.163b}$$

and any (u,ϑ) *obtained by this way is a weak solution to the initial-boundary-value problem* (12.148)–(12.149). *In particular,* (12.148)–(12.149) *has a weak solution.*

Proof. Choose weakly* converging subsequence $\{(u_\tau,\bar{\vartheta}_\tau)\}_{\tau>0}$ in the topology of the estimates (12.156a)–(12.16b,c). By (12.160c), we can choose this subsequence so that also

$$\varrho\frac{\partial}{\partial t}\Big[\frac{\partial u_\tau}{\partial t}\Big]^{\mathrm{i}} - \tau\mathrm{div}\big(|e(\bar{u}_\tau)|^{\eta-2}e(\bar{u}_\tau)\big) + \tau\mathrm{div}^2\big(|\nabla e(\bar{u}_\tau)|^{\eta-2}\nabla e(\bar{u}_\tau)\big) \rightharpoonup \zeta \tag{12.164}$$

in $L^2(I;W^{2,2}(\Omega;\mathbb{R}^n)^*)$ for some ζ and, like in Exercise 8.85, we can show that $\zeta = \varrho\frac{\partial^2 u}{\partial t^2}$. By the interpolated Aubin-Lions' Lemma 7.8, combining (12.156b), (12.16b) and (12.16f), one obtains (12.163b).

Like in (12.32), we use the by-part summation (11.126) to obtain the identity

$$\int_\Omega \varrho\frac{\partial u_\tau}{\partial t}(T)\cdot v_\tau(T)\,\mathrm{d}x - \int_\tau^T\!\!\int_\Omega \varrho\frac{\partial u_\tau}{\partial t}(\cdot-\tau)\cdot\frac{\partial v_\tau}{\partial t}\,\mathrm{d}x\mathrm{d}t + \int_Q \Big(\frac12\xi'e\big(\frac{\partial u_\tau}{\partial t}\big)$$
$$+\varphi'(e(\bar u_\tau)) + \mathscr{F}(e(\bar u_\tau),\bar\vartheta_\tau) + \tau\big|e(\bar u_\tau)\big|^{\eta-2}e(\bar u_\tau)\Big):e(\bar v_\tau) + \bar{\mathfrak{h}}_\tau\vdots\nabla e(\bar v_\tau)\,\mathrm{d}x\mathrm{d}t$$
$$= \int_\Omega v_0\cdot v_\tau(\tau)\,\mathrm{d}x + \int_Q \bar g_\tau\cdot\bar v_\tau\,\mathrm{d}x\mathrm{d}t + \int_\Sigma \bar h_\tau\cdot\bar v_\tau\,\mathrm{d}x\mathrm{d}S \tag{12.165}$$

with $\bar{\mathfrak{h}}_\tau = \gamma_1\nabla e(\frac{\partial u_\tau}{\partial t})+\gamma\nabla e(\bar u_\tau)+\tau|\nabla e(\bar u_\tau)|^{\eta-2}\nabla e(\bar u_\tau)$ and $\bar v_\tau$ and v_τ as in (12.32).

For v smooth, by (12.156c), one has (12.33) and also

$$\Big|\int_Q \tau|\nabla e(\bar u_\tau)|^{\eta-2}\nabla e(\bar u_\tau)\vdots\nabla e(v)\,\mathrm{d}x\mathrm{d}t\Big|$$
$$\le \tau\big\|\nabla e(\bar u_\tau)\big\|^{\eta-1}_{L^\eta(Q;\mathbb{R}^{n\times n\times n})}\big\|\nabla e(v)\big\|_{L^\eta(Q;\mathbb{R}^{n\times n\times n})} = \mathscr{O}(\tau^{1/\eta}) \to 0, \tag{12.166}$$

and thus can see that the regularizing η-terms disappear in the limit for $\tau\to 0$. Altogether, we can pass to the limit in (12.165) directly (without using Minty's trick) to the weak formulation of the mechanical part (12.139).

Here it is again essential that $\varrho\frac{\partial^2 u}{\partial t^2}$ in duality with $\frac{\partial u}{\partial t}$ and also that both $\mathscr{F}(e(u),\vartheta)\in L^{q'}(Q;\mathbb{R}^{n\times n})$ and $\varphi'(e(u))\in L^{q'}(Q;\mathbb{R}^{n\times n})$ are in duality with $e(\frac{\partial u}{\partial t}) \in L^q(Q;\mathbb{R}^{n\times n})$, so that, by substituting $v = \frac{\partial u}{\partial t}$ into (12.139) with taking (12.147) into account, we obtain the mechanical-energy equality

$$\int_\Omega \frac{\varrho}{2}\Big|\frac{\partial u}{\partial t}(T)\Big|^2 + \varphi(e(u(T))) + \frac{\gamma}{2}\big|\nabla e(u(T))\big|^2\,\mathrm{d}x$$
$$+\int_Q \xi\Big(e\big(\frac{\partial u}{\partial t}\big)\Big) + \gamma_1\Big|\nabla e\big(\frac{\partial u}{\partial t}\big)\Big|^2\mathrm{d}x\mathrm{d}t = \int_\Omega \frac{\varrho}{2}|v_0|^2 + \varphi(e(u_0)) + \frac{\gamma}{2}\big|\nabla e(u_0)\big|^2\mathrm{d}x$$
$$+\int_Q g\cdot\frac{\partial u}{\partial t} - \mathscr{F}(e(u),\vartheta):e\big(\frac{\partial u}{\partial t}\big)\,\mathrm{d}x\mathrm{d}t + \int_\Sigma h\cdot\frac{\partial u}{\partial t}\,\mathrm{d}x\mathrm{d}S \tag{12.167}$$

with $\varphi(e) = \psi(e,0)$ as in (12.9). For the limit passage in the heat equation, we need to prove the strong L^2-convergence of $e(\frac{\partial}{\partial t}u_\tau)$ and of $\nabla e(\frac{\partial}{\partial t}u_\tau)$. We use

$$\int_Q \xi\Big(e\big(\frac{\partial u}{\partial t}\big)\Big) + \gamma_1\Big|\nabla e\big(\frac{\partial u}{\partial t}\big)\Big|^2\mathrm{d}x\mathrm{d}t \le \liminf_{\tau\to 0}\int_Q \xi\Big(e\big(\frac{\partial u_\tau}{\partial t}\big)\Big) + \gamma_1\Big|\nabla e\big(\frac{\partial u_\tau}{\partial t}\big)\Big|^2\mathrm{d}x\mathrm{d}t$$
$$\le \limsup_{\tau\to 0}\int_Q \big(1-\frac{\sqrt\tau}{2}\big)\xi\Big(e\big(\frac{\partial u_\tau}{\partial t}\big)\Big) + \gamma_1\Big|\nabla e\big(\frac{\partial u_\tau}{\partial t}\big)\Big|^2\,\mathrm{d}x\mathrm{d}t$$

$$
\begin{aligned}
&\leq \limsup_{\tau\to 0} \int_\Omega \frac{\varrho}{2}|v_0|^2 - \frac{\varrho}{2}\Big|\frac{\partial u_\tau}{\partial t}(T)\Big|^2 + \varphi(e(u_{0\tau})) + \frac{\gamma}{2}\big|\nabla e(u_{0\tau}))\big|^2 \\
&\quad - \varphi(e(u_\tau(T))) - \frac{\gamma}{2}\big|\nabla e(u_\tau(T))\big|^2 + \frac{\tau}{\eta}\big|e(u_{0\tau})\big|^\eta + \frac{\tau}{\eta}\big|\nabla e(u_{0\tau})\big|^\eta \,\mathrm{d}x \\
&\quad + \int_Q \bar g_\tau\cdot\frac{\partial u_\tau}{\partial t} - \mathscr{F}(e(\bar u_\tau),\bar\vartheta_\tau){:}e\Big(\frac{\partial u_\tau}{\partial t}\Big)\,\mathrm{d}x\mathrm{d}t + \int_\Sigma \bar h_\tau\cdot\frac{\partial u_\tau}{\partial t}\,\mathrm{d}x\mathrm{d}S \\
&\leq \int_\Omega \frac{\varrho}{2}|v_0|^2 + \varphi(e(u_0)) + \frac{\gamma}{2}\big|\nabla e(u_0)\big|^2 \\
&\quad - \frac{\varrho}{2}\Big|\frac{\partial u}{\partial t}(T)\Big|^2 - \varphi\big(e(u(T)\big) - \frac{\gamma}{2}\big|\nabla e(u(T))\big|^2\,\mathrm{d}x \\
&\quad + \int_Q g\cdot\frac{\partial u}{\partial t} - \mathscr{F}(e(u),\vartheta){:}e\Big(\frac{\partial u}{\partial t}\Big)\,\mathrm{d}x\mathrm{d}t + \int_\Sigma h\cdot\frac{\partial u}{\partial t}\,\mathrm{d}x\mathrm{d}S \\
&= \int_Q \xi\Big(e\Big(\frac{\partial u}{\partial t}\Big)\Big) + \gamma_1\Big|\nabla e\Big(\frac{\partial u_\tau}{\partial t}\Big)\Big|^2\,\mathrm{d}x\mathrm{d}t.
\end{aligned}
\tag{12.168}
$$

Note that the 3rd inequality in (12.168) is due to (12.158) used for $l=T/\tau$, while the 4rd inequality uses weak upper semicontinuity and (12.162), and eventually the last equality in (12.168) is exactly (12.167). Thus we obtain equalities in (12.168) and, in particular, we obtain $\lim_{\tau\to 0}\int_Q \xi(e(\frac{\partial u_\tau}{\partial t}))\,\mathrm{d}x\mathrm{d}t = \int_Q \xi(e(\frac{\partial u}{\partial t}))\,\mathrm{d}x\mathrm{d}t$ and also $\lim_{\tau\to 0}\int_Q |\nabla e(\frac{\partial u_\tau}{\partial t})|^2\,\mathrm{d}x\mathrm{d}t = \int_Q |\nabla e(\frac{\partial u}{\partial t})|^2\,\mathrm{d}x\mathrm{d}t$. By (12.152i), the quadratic form ξ is coercive and thus we obtain both $e(\frac{\partial u_\tau}{\partial t}) \to e(\frac{\partial u}{\partial t})$ strongly in $L^2(Q;\mathbb{R}^{n\times n})$, and also $\nabla e(\frac{\partial u_\tau}{\partial t}) \to \nabla e(\frac{\partial u}{\partial t})$ strongly in $L^2(Q;\mathbb{R}^{n\times n\times n})$. Then the limit passage in the semi-linear heat equation is simple.

In particular, using Lemma 12.53 and the above arguments, some weak solution to (12.148)–(12.149) indeed exists because, due to the qualification (12.152k) of u_0, the regularization $u_{0\tau}$ with the properties (12.162) always exist. □

Remark 12.57. In fact, even a bigger growth of $\mathscr{F}(\cdot,\vartheta)$ than (12.152h) can be admitted if further interpolation of this term would be done, cf. [369]. The existence for large data (for a very similar model) has been investigated by Pawłow and Zochowski [329] even allowing for $\gamma_1 = 0$, and by Pawłow and Zajączkowski [327] for a heat capacity depending on both θ and $e(u)$ by using regularity under assumption of a smooth domain with zero Dirichlet boundary conditions. The *enhanced enthalpy transformation* (12.143) is similar to [311] where dependence on space/time variables, leading to additional terms in the transformed system, has been treated in an analogous way. Nonconvexity of $\psi(\cdot,\theta)$ makes it possible to model *shape-memory alloys* and thermodynamics of so-called martensitic transformation occuring typically in such materials, as already mentioned in Exercise 11.43.

Remark 12.58 (Internal energy balance)**.** Like (12.3), we have now $\vartheta = w - \varphi(e) - \frac{1}{2}\mathbb{H}\nabla e\vdots\nabla e$, which suggests to subtract from the internal-energy balance $\frac{\partial w}{\partial t} = \psi_e'(e(u),\theta){:}e(\frac{\partial u}{\partial t}) + \mathbb{H}\nabla e(u)\vdots\nabla e(\frac{\partial u}{\partial t}) + \xi_1(e(\frac{\partial u}{\partial t}),\nabla e(\frac{\partial u}{\partial t})) + \mathrm{div}\,j$ the balance of the stored-energy rate versus the power of conservative parts of (hyper)stresses, i.e. $\frac{\partial}{\partial t}(\varphi(e(u))+\frac{1}{2}\mathbb{H}\nabla e(u)\vdots\nabla e(u) = \varphi'(e(u)){:}e(\frac{\partial u}{\partial t})+\mathbb{H}\nabla e(u)\vdots\nabla e(\frac{\partial u}{\partial t})$. In this way, we obtain $\frac{\partial\vartheta}{\partial t} + \mathrm{div}\, j = (\sigma-\varphi'(e(u))){:}e(\frac{\partial u}{\partial t}) + (\mathfrak{h}-\mathbb{H}\nabla e(u))\vdots\nabla e(\frac{\partial u}{\partial t})$ which, after eliminating temperature as we did by the substitution (12.143), would again result to the transformed heat equation (12.148b). This reveals the physical character of the transformed system (12.148) and a certain conceptual similarity with thermodynamics of fluid where internal energy is sometimes used instead of temperature for analysis, cf. e.g. [80, 82].

Bibliography

[1] ABEYARATNE, R., KNOWLES, J.K.: Implications of viscosity and strain-gradient effects for the kinetics of propagating phase boundaries in solids. *SIAM J. Appl. Math.* **51** (1991), 1205–1221.

[2] ACERBI, E., FUSCO, N.: Semicontinuity problems in the calculus of variations. *Archive Ration. Mech. Anal.* **86** (1984), 125–145.

[3] ADAMS, R.A.: *Sobolev Spaces.* Academic Press, New York, 1975.

[4] ADAMS, R.A., FOURNIER, J.J.F.: *Sobolev Spaces.* Elsevier, 2nd ed., Oxford, 2003.

[5] ADLER, G.: Sulla caratterizzabilita dell'equazione del calore dal punto di vista del calcolo delle variazioni. *Matematikai Kutató Intézenétek Közlemenyei* **2** (1957), 153–157.

[6] AGMON, S., DOUGLIS, A., NIRENBERG, L.: Estimates near the boundary for solutions of elliptic partial differential equations satisfying general boundary conditions. Parts I and II. *Comm. Pure Appl. Math.* **12** (1959), 623–727, and **17** (1964), 35–92.

[7] AIZICOVICI, S., HOKKANEN, V.-M.: Doubly nonlinear equations with unbounded operators. *Nonlinear Anal., Th. Meth. Appl.* **58** (2004), 591-607.

[8] AIZICOVICI, S., PAVEL, N.H.: Anti-periodic solutions to a class of nonlinear differential equations in Hilbert spaces. *J. Funct. Anal.* **99** (1991), 387-408.

[9] ALBER, H.-D.: *Materials with Memory.* Lect. Notes in Math. **1682**, Springer, Berlin, 1998.

[10] ALT, H.W.: *Lineare Funktional-analysis.* 3. Aufl., Springer, Berlin, 1999.

[11] ALT, H.W., LUCKHAUS, S.: Quasilinear Elliptic-Parabolic Differential Equations. *Math. Z.* **183** (1983), 311–341.

[12] ALT, H.W., PAWŁOW, I.: Existence of solutions for non-isothermal phase separation. *Adv. Math. Sci. Appl.* **1** (1992), 319–409.

[13] ALAOUGLU, L.: Weak topologies of normed linear spaces. *Ann. Math.* **41** (1940), 252–267.

[14] ALIKAKOS, N.D., BATES, P.W.: On the singular limit in a phase field model of phase transitions. *Ann. Inst. Henri Poincaré* **5** (1988), 141–178.

[15] ALLEN, S., CAHN, J.: A microscopic theory for antiphase boundary motion and its application to antiphase domain coarsening. *Acta Metall.* **27** (1979), 1084–1095.

[16] AMANN, H., QUITTNER, P.: Elliptic boundary value problems involving measures: existence, regularity, and multiplicity. *Adv. Differ. Equ.* **3** (1998), 753–813.

[17] AMBROSETTI, A., RABINOWITZ, P.H.: Dual variational methods in critical point theory and applications. *J. Funct. Anal.* **14** (1973), 349–380.

[18] AMMARI, H., BUFFA, A., NÉDÉLEC, J.-C.: A justification of eddy currents model for the Maxwell equations, *SIAM J. Appl. Math.* **60** (2000), 1805–1823.

[19] ARONSSON, G., EVANS, L.C., WU, Y: Fast/slow diffusion and growing sandpiles. *J. Diff. Eq.* **131** (1996), 304–335.

[20] ARTSTEIN, Z., SLEMROD, M.: Phase separation of the slightly viscous Cahn-Hilliard equation in the singular parturbation limit. *Indiana Univ. Math. J.* **47** (1998), 1147–1166.

[21] ASH, R.B.: *Real Analysis and Probablity.* Acad. Press, New York, 1972.

[22] ASPLUND, E.: Positivity of duality mappings. *Bull. A.M.S.* **73**, (1967), 200–203.

[23] ASPLUND, E.: Fréchet differentiability of convex functions. *Acta Math.* **121** (1968), 31–47.

[24] ATTOUCH, H., BOUCHITTÉ, G., MAMBROUK, H.: Variational formulation of semilinear elliptic equations involving measures. In: *Nonlinear Variat. Problems II*, Pitman Res. Notes in Math. **193**, Longman, Harlow, 1989.

[25] AUBIN, J.-P.: Un théorème de compacité. *C.R. Acad. Sci.* **256** (1963), 5042–5044.

[26] AUBIN, J.-P.: Variational principles for differential equations of elliptic, parabolic and hyperbolic type. In: *Math. Techniques of Optimization, Control and Decision* (Eds. J.-P.Aubin, A.Bensoussan, I.Ekeland) Birkhäuser, 1981, pp.31–45.

[27] AUBIN, J.-P.: *Optima and Equilibria.* 2nd ed. Springer, Berlin, 1998.

[28] AUBIN, J.P., CELLINA, A.: *Differential Inclusions.* J.Wiley, New York, 1984.

[29] BAIOCCHI, C.: Sur un problème à frontière libre traduisant le filtrage de liquides à travers des milieux poreaux. *C.R. Acad. Sci. Paris* **273** (1971), 1215–1217.

[30] BAIOCCHI, C., CAPELO, A.: *Variational and Quasivariational Inequalities.* J. Wiley, Chichester, 1984.

[31] BALL, J.M.: Convexity conditions and existence theorems in nonlinear elasticity. *Archive Ration. Mech. Anal.* **63** (1977), 337–403.

[32] BALL, J.M.: Some open problems in elasticity. In: *Geometry, Mechanics, and Dynamics.* (P.Newton, P.Holmes, A.Weinstein, eds.) Springer, New York, 2002.

[33] BANACH, S.: Sur les opérations dans les ensembles abstraits et leur applications aux équations intégrales. *Fund. Math.* **3** (1922), 133–181.

[34] BANACH, S.: Sur les fonctionelles linéaires. *Studia Math.* **1** (1929), 211–216, 223–239.

[35] BANACH, S.: *Théorie des Opérations Linéaires.* M.Garasiński, Warszawa, 1932 (Engl. transl. North-Holland, Amsterdam, 1987).

[36] BANACH, S., STEINHAUS, H.: Sur le principe de la condensation de singularités. *Fund. Math.* **9** (1927), 50–61.

[37] BARBU, V.: *Nonlinear Semigroups and Differential Equations in Banach Spaces.* Editura Academiei, Bucuresti, and Noordhoff, Leyden, 1976.

[38] BARBU, V.: *Analysis and Control of Nonlinear Infinite Dimensional Systems.* Acad. Press, Boston, 1993.

[39] BARBU, V., PRECUPANU, T.: *Convexity and Optimization in Banach spaces.* 3rd ed., D.Reidel, Dordrecht, 1986.

[40] BELLENI-MORANTE, A., MCBRIDE, A.C.: *Applied Nonlinear Semigroups.* J.Wiley, Chichester, 1998.

[41] BELLMAN, R.: The stability of solutions of linear differential equations. *Duke Math. J.* **10** (1943), 643–647.

[42] BELLOUT, H., NEUSTUPA, J., PENEL, P.: On the Navier-Stokes equations with boundary conditions based on vorticity. *Math. Nachrichten* **269–270** (2004), 59–72.

[43] BENEŠ, M.: Mathematical analysis of phase-field equations with numerically efficient coupling terms. *Interfaces & Free Boundaries* **3** (2001), 201–221.

[44] BENEŠ, M.: On a phase-field model with advection. In: *Num. Math. & Adv. Appl., ENUMATH 2003* (M.Feistauer at al., eds.) Springer, Berlin, 2004, pp.141–150.

[45] BÉNILAN, P.: Solutions intégrales d'évolution dans un espace de Banach. *C. R. Acad. Sci. Paris*, **A 274** (1972), 45–50.

[46] BÉNILAN, P.: Operateurs m-accretifs hemicontinus dans un espace de Banach quelconque. *C. R. Acad. Sci. Paris*, **A 278** (1974), 1029–1032.

[47] BÉNILAN, P., BOCCARDO, L., GALLOUËT, T., GARIEPY, R., PIERRE, M., VAZQUEQ, J.L.: An L^1-theory of existence and uniqueness of solutions of nonlinear elliptic equation. *Annali Scuola Norm. Sup. Pisa* **22** (1995), 241–273.

[48] BÉNILAN, P., BREZIS, H.: Solutions faibles d'équations d'évolution dans les espaces de Hilbert. *Ann. Inst. Fourier* **22** (1972), 311–329.

[49] BENILAN, P., CRANDALL, M.G., SACKS, P.: Some L^1 existence and dependence results for semilinear elliptic equations under nonlinear boundary conditions. *Appl. Math. Optim.* **17** (19988), 203–224.

[50] BENSOUSSAN, A., FREHSE, J.: *Regularity Results for Nonlinear Elliptic Systems and Applications.* Springer, Berlin, 2002.

[51] BEURLING, A., LIVINGSTON, A.E.: A theorem on duality mappings in Banach spaces. *Arkiv för Matematik* **4** (1954), 405–411.

[52] BIAZUTTI, A.G.: On a nonlinear evolution equation and its applications. *Nonlinear Analysis, Th. Meth. Appl.* **24** (1995), 1221–1234.

[53] BLANCHARD, P., BRÜNING, E.: *Variational Methods in Mathematical Physics.* Springer, Berlin, 1992.

[54] BLANCHARD, D., GUIBÉ, O.: Existence of a solution for a nonlinear system in thermoviscoelasticity. *Adv. Diff. Eq.* **5** (2000), 1221–1252.

[55] BOCCARDO, L., DACOROGNA, B.: A characterization of pseudomonotone differential operators in divergence form. *Comm. Partial Differential Equations* **9** (1984), 1107–1117.

[56] BOCCARDO, L., GALLOUËT, T.: Non-linear elliptic and parabolic equations involving measure data. *J. Funct. Anal.* **87** (1989), 149–169.

[57] BOCCARDO, L., GALLOUËT, T.: Non-linear elliptic equations with right hand side measures. *Commun. in Partial Diff. Equations* **17** (1992), 641–655.

[58] BOCCARDO, L., GALLOUËT, T., ORSINA, L.: Existence and uniqueness of entropy soutions for nonlinear elliptic equations with measure data. *Ann.Inst. H.Poincaré, Anal. Nonlinéaire* **13** (1996), 539–551.

[59] BOCHNER, S.: Integration von Funktionen, deren Werte die Elemente eines Vektorraumes sind. *Fund. Math.* **20** (1933), 262–276.

[60] BOLZANO, B.: *Schriften Bd.I: Funktionenlehre.* After a manuscript from 30ties of 19th century by K.Rychlík in: *Abh. Königl. Böhmischen Gesellschaft Wiss.* **XVI**+183+24+IV S (1930).

[61] BONETTI, E., BONFANTI, G.: Existence and uniqueness of the solution to a 3D thermoelastic system. *Electronic J. Diff. Eqs.* (2003), No.50, 1–15.

[62] BOSSAVIT, A.: *Computational Electromagnetism.* Acad. Press, San Diego, 1998.

[63] BOURBAKI, N.: Sur les espaces de Banach. *Comptes Rendus Acad. Sci. Paris* **206** (1938), 1701–1704.

[64] BRÉZIS, H.: Équations et inéquations non-linéaires dans les espaces vectoriel en dualité. *Ann. Inst. Fourier* **18** (1968), 115–176.

[65] BRÉZIS, H.: Problèmes unilatéraux. *J. Math. Pures Appl.* **51** (1972), 1–168.

[66] Brézis, H.: *Operateur Maximaux Monotones et Semi-groupes de Contractions dans les Espaces de Hilbert.* North-Holland, Amsterdam, 1973.

[67] Brezis, H.: Nonlinear elliptic equations involving measures. In: *Contributions to nonlinear partial differential equations.* (C.Bardos et al., eds.) Pitman Res. Notes Math. **89**, (1983), pp.82–89.

[68] Brézis, H., Ekeland, I.: Un principe variationnel associé à certaines équations paraboliques. *Compt. Rendus Acad. Sci. Paris* **282** (1976), 971–974 and 1197–1198.

[69] Brézis, H., Strauss, W.A.: Semi-linear second-order elliptic equation in L^1. *J. Math. Soc. Japan* **25** (1973), 565–590.

[70] Brokate, M., Krejčí, P., Schnabel, H.: On uniqueness in evolution quasivariational inequalities. *J. Convex Anal.* **11** (2004), 111-130.

[71] Brokate, M., Sprekels, J.: *Hysteresis and Phase Transitions.* Springer, New York, 1996.

[72] Brouwer, L.E.J.: Über Abbildungen von Mannigfaltigkeiten. *Math.Ann.* **71** (1912), 97–115.

[73] Browder, F.: Nonlinear elliptic boundary value problems. *Bull. A.M.S.* **69** (1963), 862–874.

[74] Browder, F.E.: Nonlinear accretive operators in Banach spaces. *Bull. A.M.S.* **73** (1967), 470–476.

[75] Browder, F.E.: Nonlinear equations of evolution and nonlinear accretive operators in Banach spaces. *Bull. A.M.S.* **73** (1967), 867–874.

[76] Browder, F.: *Nonlinear Operators and Nonlinear Equations of Evolution in Banach Spaces.* Proc.Symp.Pure Math. **18/2**, AMS, Providence, 1976.

[77] Browder, F., Hess, P.: Nonlinear mappings of monotone type in Banach spaces. *J. Funct. Anal.* **11** (1972), 251–294.

[78] Bulíček, M.: Navier's slip and evolutionary Navier-Stokes-Fourier-like systems with pressure, shear-rate and temperature dependent viscosity. PhD-thesis, Math.-Phys. Fac., Charles Uni., Prague, 2006.

[79] Bulíček, M., Consiglieri, L., Málek, J.: Slip boundary effects on unsteady flow of incompressible viscous heat conducting fluids with a non-linear internal energy-temperature relationships. In preparation.

[80] Bulíček, M., Feireisl, E., Málek, J.: A Navier-Stokes-Fourier system for incompressible fluids with temperature dependent material coefficients. *Nonlinear. Anal., Real World Appl.* **10** (2009), 992-1015.

[81] Bulíček, M., Gwiazda, P., Málek, J., Świerczewska-Gwiazda, A.: On unsteady flows of implicitly constituted incompressible fluids. *SIAM J. Math. Anal.* **44** (2012), 2756-2801.

[82] Bulíček, M., Kaplický, P., Málek, J.: An L^2-maximal regularity result for the evolutionary Stokes-Fourier system. *Applic. Anal.* **90** (2011), 31-45.

[83] Bulíček, M., Málek, J., Rajagopal, K.R.: Navier's slip and evolutionary Navier-Stokes-like system with pressure and shear-rate dependent viscosity. *Indiana Univ. Math. J.* **56** (2007), 51–85.

[84] Bulíček, M., Málek, J., Rajagopal, K.R.: On Kelvin-Voigt model and its generalizations. *Evolution Equations & Control Th.* **1** (2012), 17-42.

[85] Bunyakovskiĭ, V.Ya.: Sur quelques inequalités concernant les intégrales aux differences finis. *Mem. Acad. Sci. St. Peterbourg* (7) **1** (1859), 9.

[86] Caffarelli, L.A., Chabré, X.: *Fully Nonlinear Elliptic Equations.* AMS, Providence, 1995.

[87] Caginalp, G.: An analysis of a phase field model of a free boundary. *Archive Ration. Mech. Anal.* **92** (1986), 205–245.
[88] Cahn, J.W., Hilliard, J.E.: Free energy of a uniform system I. Interfacial free energy. *J. Chem. Phys.* **28** (1958), 258–267.
[89] Cahn, J.W., Hilliard, J.E.: Free Energy of a Nonuniform System III. Nucleation of a Two-Component Incompressible Fluid. *J. Chem. Phys.* **31** (1959), 688–699.
[90] Cazenave, T., Haraux, A.: *An introduction to Semilinear Evolution Equations.* Clarendon Press, Oxford, 1998.
[91] Chabrowski, J.: *Variational Methods for Potential Operator Equations.* W. de Gruyter, Berlin, 1997.
[92] Chen, Y.-Z., Wu, L.-C.: *Second Order Elliptic Equations and Elliptic Systems.* AMS, Providence, 1998.
[93] Ciarlet, P.G.: *Mathematical Elasticity. Vol.1.* North-Holland, Amsterdam, 1988.
[94] Ciarlet, P.G., Nečas, J.: Injectivity and self-contact in nonlinear elasticity. *Archive Ration. Mech. Anal.* **19** (1987), 171–188.
[95] Cioranescu, I.: *Geometry of Banach Spaces, Duality Mappings and Nonlinear Problems.* Kluwer, Dordrecht, 1990.
[96] Clarke, F.: *Optimization and Nonsmooth Analysis.* J.Wiley, New York, 1983.
[97] Clarkson, J.A.: Uniformly convex spaces. *Trans. Amer. Math. Soc.* **40** (1936), 396–414.
[98] Clement, P.: Approximation by finite element function using local regularization. *Rev. Francais Autom. Informat. Recherche Opérationnelle Sér. Anal. Numér.* R-2 (1975), 77–84.
[99] Chipot, M.: *Variational Inequalities and Flow in Porous Media.* Springer, Berlin, 1984.
[100] Chow, S.-N., Hale, J.K.: *Methods of Bifurcation Theory.* Springer, New York, 1982.
[101] Colli, P.: On some doubly nonlinear evolution equations in Banach spaces. *Japan J. Indust. Appl. Math.* **9** (1992), 181–203.
[102] Colli, P., Sprekels, J: Positivity of temperature in the general Frémond model for shape memory alloys. *Continuum Mech. Thermodyn.* **5** (1993), 255–264.
[103] Colli, P., Sprekels, J: Stefan problems and the Penrose-Fife phase field model. *Adv. Math. Sci. Appl.* **7** (1997), 911–934.
[104] Colli, P., Visintin, A.: On a class of doubly nonlinear evolution equations. *Comm. P.D.E.* **15** (1990), 737–756.
[105] Conca, C., Murat, F., Pironneau, O.: The Stokes and Navier-Stokes equations with boundary conditions involving the pressure. *Japan J. Math.* **20** (1994), 279–318.
[106] Constantin, P., Foias, C.: *Navier-Stokes Equations.* The Chicago Univ. Press, Chicago, 1988.
[107] Crandall, M.G., Liggett, T.: Generations of semi-groups of nonlinear transformations on general Banach spaces. *Amer. J. Math.* **93** (1971), 265–298.
[108] Crandall, M.G., Liggett, T.: A theorem and a counterexample in the theory of semigroups of nonlinear transformations. *Trans. Amer. Math. Soc.* **160** (1971), 263–278.
[109] Crandall, M.G., Pazy, A.: Semigroups of nonlinear contractions and dissipative sets. *J. Funct. Anal.* **3** (1969), 376–418.
[110] Crandall, M.G., Pazy, A.: Nonlinear evolution equations in Banach spaces. *Israel J. Math.* **11** (1972), 57–94.

[111] CRANK, J.: *Free and Moving Boundary Problems.* Clarendon Press, Oxford, 1984.

[112] DACOROGNA, B.: *Direct Methods in the Calculus of Variations.* Springer, Berlin, 1989.

[113] DAFERMOS, C.M.: Global smooth solutions to the initial boundary value problem for the equations of one-dimensional thermoviscoelasticity. *SIAM J. Math. Anal.* **13** (1982), 397–408.

[114] DAFERMOS, C.M.: *Hyperbolic Conservation Laws in Continuum Physics.* Springer, Berlin, 2000.

[115] DAFERMOS, C.M., HSIAO, L.: Global smooth thermomechanical processes in one-dimensional nonlinear thermoviscoelasticity. *Nonlinear Anal.* **6** (1982), 435–454.

[116] DAL MASO, G., FRANCFORT, G.A., TOADER, R.: Quasistatic crack growth in nonlinear elasticity. *Archive Ration. Mech. Anal.* **176** (2005), 165–225.

[117] DAL MASO, G., MURAT, F., ORSINA, L., PRIGNET, A.: Renormalization solutions of elliptic equations with general measure data. *Ann. Sc. Norm. Super. Pisa, Cl. Sci.*, IV.Ser., **28**, 741–808 (1999).

[118] DEIMLING, K.: *Nonlinear Functional Analysis.* Springer, Berlin, 1985.

[119] DIAZ, J.I., GALIANO, G., JÜNGEL, A.: On a quasilinear degenerate system arising in semiconductor theory. *Nonlinear Anal., Real World Appl.* **2** (2001), 305–336.

[120] DIBENEDETTO, E.: *Degenerate Parabolic Equations.* Springer, New York, 1993.

[121] DIBENEDETTO, E., SHOWALTER, R.E.: Implicit degenerate evolution equations and applications. *SIAM J. Math. Anal.* **12** (1981), 731–751.

[122] DIENING, L., MÁLEK, J., STEINHAUER, M.: On Lipschitz truncations of Sobolev functions (with variable exponent) and their selected applications. *ESAIM Control Optim. Calc. Var.* **14** (2008), 211232.

[123] DIENING, L., RŮŽIČKA, M., WOLF, J.: Existence of weak solutions for unsteady motions of generalized Newtonian fluids. *Ann. Scuola Norm. Sup. Pisa Cl. Sci.* **9** (2010), 1–46.

[124] DIPERNA, R.J.: Measure-valued solutions to conservation laws. *Archive Ration. Mech. Anal.* **88** (1985), 223–270.

[125] DOLZMANN, G., HUNGERBÜHLER, N., MÜLLER, S.: Non-linear elliptic systems of with measure-valued right hand side. *Math. Z.* **226** (1997), 545–574.

[126] DONG, G.: *Nonlinear Partial Differential Equations of Second Order.* Amer. Math. Soc., Providence, 1991.

[127] DUBINSKIĬ, YU.A.: Weak convergence in nonlinear elliptic and parabolic equations. (In Russian) *Mat. Sbornik* **67 (109)** (1965), 609–642.

[128] DUNFORD, N., PETTIS, J.T.: Linear operators on summable functions. *Trans. Amer. Math. Soc.* **47** (1940), 323–392.

[129] DUVAUT, G.: Resolution d'un probleme de Stefan (fusion d'un bloc de glace a zero degre). *C.R. Acad. Sci. Paris* **276A** (1973), 1461–1463.

[130] DUVAUT, G., LIONS, J.L.: *Les Inéquations en Mécanique et en Physique.* Dunod, Paris, 1972 (Engl. transl. Springer, Berlin, 1976).

[131] ECK, CH., JARUŠEK, J.: Existence of solutions for the dynamic frictional contact problems of isotropic viscoelastic bodies. *Nonlinear Anal., Th. Meth. Appl.* **53A** (2003), 157–181.

[132] ECK, CH., JARUŠEK, J., KRBEC, M.: *Unilateral Contact Problems; Variational Methods and Existence Theorems.* Chapman & Hall, CRC, Boca Raton, 2005.

[133] ECKART, C.: The thermodynamics of irreversible processes. II.Fluid mixtures. *Physical Rev.* **58** (1940), 269–275.

[134] EHRLING, G.: On a type of Eigenvalue problems for certain elliptic differential operators. *Math. Scand.* **2** (1954), 267–285.

[135] ELLIOTT, C., OCKENDON, J.: *Weak and Variational Methods for Moving Boundary Problems.* Pitman, Boston, 1982.

[136] ELLIOTT, C., ZHENG, S.: On the Cahn-Hilliard equation. *Archive Ration. Mech. Anal.* **96** (1986), 339–357.

[137] ELLIOTT, C., ZHENG, S.: Global existence and stability of solutions to the phase field equations. In: *Free Boundary Problems.* (K.-H.Hoffmann, J.Sprekels, eds.) ISNM **95**, Birkhäuser, Basel, 1990, pp.47–58.

[138] EVANS, L.C.: *Partial differential equations.* AMS, Providence, 1998.

[139] FAN, K.: Fixed-point and minimax theorems in locally convex topological linear spaces. *Proc. Nat. Acad. Sci. U.S.A.* **38** (1952), 121–126.

[140] FAN, K., GLICKSBERG, I.L.: Some geometric properties of the spheres in a normed linear space. *Duke Math. J.* **25** (1958), 553–568.

[141] FAEDO, S.: Un nuovo metodo per l'analysi esistenziale e qualitativa dei problemi di propagazione. *Ann. Sc. Norm. Sup. Pisa* Sér.III, **1** (1949), 1–40.

[142] FARAGÓ, I.: Splitting methods and their application to the abstract Cauchy problems. Lect. Notes Comp.Sci. **3401**, Springer Verlag, Berlin, 2005, pp.35–45.

[143] FATOU, P.: Séries trigonométriques et séries de Taylor. *Acta Math.* **30** (1906), 335–400.

[144] FATTORINI, H.O.: *Infinite Dimensional Optimization and Control Theory.* Cambridge Univ. Press, Cambridge, 1999.

[145] FEIREISL, E., MÁLEK, J.: On the Navier-Stokes equations with temperature-dependent transport coefficients. *Diff. Equations Nonlin. Mech.* (2006), 14pp. (electronic), Art.ID 90616.

[146] FEIREISL, E., PETZELTOVÁ, H., ROCCA, E.: Existence of solutions to a phase transition model with microscopic movements. *Math. Methods Appl. Sci.* **32** (2009), 1345–1369.

[147] FEISTAUER, M.: *Mathematical Methods in Fluid Dynamics.* Longman, Harlow, 1993.

[148] FENCHEL, W: *Convex Cones, Sets, and Functions.* Princeton Univ., 1953.

[149] FRANCŮ, J.: Weakly continuous operators. Applications to differential equations. *Appl. Mat.* **39** (1994), 45–56.

[150] FREHSE, J., MÁLEK, J.: Problems due to the no-slip boundary in incompressible fluid dynamics. In: *Geometric Anal. and Nonlin. P.D.E.s.* (Eds. S.Hildebrant, H.Karcher), Springer, Berlin, 2003, pp.559–571.

[151] FREHSE, J., MÁLEK, J., STEINHAUER, M.: On analysis of steady flow of fluids with shear dependent viscosity based on the Lipschitz truncation method. *SIAM J. Math. Anal.* **34** (2003), 1064–1083.

[152] FREHSE, J., NAUMANN, J.: An existence theorem for weak solutions of the basic stationary semiconductor equations. *Applicable Anal.* **48** (1993), 157–172.

[153] FRIED, E., GURTIN, M.E.: Tractions, balances, and boundary conditions for non-simple materials with application to liquid flow at small-lenght scales. *Archive Ration. Mech. Anal.* **182** (2006), 513–554.

[154] FRÉMOND, M.: *Nonsmooth Thermomechanics.* Springer, Berlin, 2002.

[155] FRIEDMAN, A.: *Variational Principles and Free-Boundary Problems.* J.Wiley, New York, 1982.

[156] FRIEDMAN, A., NEČAS, J.: Systems of nonlinear wave equations with nonlinear viscosity. *Pacific J. Math.* **135** (1988), 29–55.

[157] FRIEDRICHS, K.: Spektraltheorie halbbeschränkter Operatoren I, II. *Math. Ann.* **106** (1934), 465–487, 685–713.

[158] FUČÍK, S., NEČAS, J., SOUČEK, V.: *Einführung in die Variationsrechnung.* Teubner, Leipzig, 1977.

[159] FUČÍK, S., KUFNER, A.: *Nelineární diferenciální rovnice.* SNTL, Praha, 1978 (Engl. Transl.: *Nonlinear Differential Equations.* Elsevier, Amsterdam, 1980.)

[160] FUBINI, G.: Sugli integrali multipli. *Rend. Accad. Lincei Roma* **16** (1907), 608–614.

[161] GAGLIARDO, E.: Ulteriori proprietà di alcune classi di funzioni in piu variabli. *Ricerche Mat.* **8** (1959), 102–137.

[162] GAJEWSKI, H.: On the existence of steady-state carrier distributions in semiconductors. In: *Probleme und Methoden der Math. Physik.* Teubner, Lepzig, 1984, pp.76–82.

[163] GAJEWSKI, H.: On existence, uniqueness and asymptotic behaviour of solutions of the basic equations for carrier transport in semiconductors. *ZAMM* **65**, 101–108.

[164] GAJEWSKI, H.: On a variant of monotonicity and its application to differential equations. *Nonlin. Anal., Th. Meth. Appl.* **22** (1994), 73–80.

[165] GAJEWSKI, H.: The drift-diffusion model as an evolution equation of special structure. In: *Math. Problems in Semiconductor Physics.* (P.Marcati, P.A.Markowich, R.Natalini, eds.), Pitman Res. Notes in Math. **340**, Longman, Harlow, 1995, pp.132–142.

[166] GAJEWSKI, H., GRÖGER, K.: On the basic equations for carrier transport in semiconductors. *J. Math. Anal. Appl.* **113** (1986), 12–35.

[167] GAJEWSKI, H., GRÖGER, K.: Semiconductor equations for variable mobilities based on Boltzmann statistics or Fermi-Dirac statistics. *Math. Nachr.* **140** (1989), 7–36.

[168] GAJEWSKI, H., GRÖGER, K., ZACHARIAS, K.: *Nichtlineare Operatorgleichungen und Operatordifferentialgleichungen.* Akademie-Verlag, Berlin, 1974.

[169] GALAKTIONOV, V.A.: *Geometric Sturmian Theory of Nonlinear Parabolic Equations and Applications.* Chapman & Hall / CRC, Boca Raton, 2004.

[170] GALDI, P.G.: *An Introduction to the Mathematical Theory of the Navier-Stokes equations II.* Springer, New York, 1994.

[171] GALERKIN, B.G.: Series development for some cases of equilibrium of plates and beams. (In Russian). *Vestnik Inzhinierov Teknik.* **19** (1915), 897–908.

[172] GÅRDING, L.: Dirichlet's problem for linear elliptic differential equations. *Math. Scand.* **1** (1953), 55–72.

[173] GÂTEAUX, R.: Sur les fonctionnelles continues at les fonctionnelles analytiques. *C.R. Acad. Sci. Paris* Sér I Math. **157** (1913), 325–327.

[174] GEAR, C.W.: *Numerical Initial Value Problems in Ordinary Differential Equations.* Prentice–Hall, Englewood Cliffs, 1971.

[175] GIAQUINTA, M.: *Introduction into Regularity of Nonlinear Elliptic Systems.* Birkhäuser, Basel, 1993.

[176] GIAQUINTA, M., HILDEBRANDT, S.: *Calculus of Variations I,II.* Springer, Berlin, 1996 (2nd ed. 2004).

[177] GIAQUINTA, M., MODICA, G., SOUČEK, J.: *Cartesian Currents in the Calculus of Variations I,II.* Springer, Berlin, 1998.

[178] GILBARG, D., TRUDINGER, N.S.: *Elliptic Partial differential Equations of Second Order.* Springer, Berlin, 2nd ed., 1983; revised printing 2001.

[179] GILBERT, T.L.: A Lagrangian formulation of the gyromagnetic equation of the magnetization field. *Phys. Rev.* **100** (1955), 1243.

[180] Giusti, E.: *Direct Methods in Calculus of Variations.* World Scientific, Singapore, 2003.

[181] Glicksberg, I.L.: A further generalization of the Kakutani fixed point theorem, with application to Nash equilibrium points. *Proc. Amer. Math. Soc.* **3** (1952), 170–174.

[182] Glowinski, R, Lions, J.L., Trémolières, R.: *Analyse Numérique des Inéquations Variationnelles.* Dunod, Paris, 1976, Engl. transl. North-Holland, Amsterdam, 1981.

[183] Goeleven, D., Motreanu, D.: *Variational and Hemivariational Inequalities. Theory, Methods and Applications I, II.* Kluwer, Boston, 2003.

[184] Gossez, J.-P., Mustonen, V.: Pseudomonotonicity and the Leray-Lions condition. *Diff. Integral Equations* **6** (1993), 37–45.

[185] Ghoussoub, N.: *Self-dual Partial Differential Systems and Their Variational Principles.* Springer, New York, 2009.

[186] Ghoussoub, N., Tzou, L.: A variational principle for gradient flows. *Mathematische Annalen* **330** (2004), 519–549.

[187] Grange, O., Mignot, F.: Sur le résolution d'une equation et d'une inequation paraboliques non-linéaires. *J. Funct. Anal.* **11** (1972), 77–92.

[188] Green, G.: An essay on the application of mathematical analysis to the theories on electricity and magnetism. Nottingham, 1828.

[189] Grinfeld, M., Novick-Cohen, A.: The viscous Cahn-Hilliard equation: Morse decomposition and structure of the global attractor. *Trans. Amer. Math. Soc.* **351** (1999), 2375–2406.

[190] Grisvard, P.: *Singularities in Boundary Value Problems.* Masson/Springer, Paris/Berlin, 1992.

[191] Gröger, K.: On steady-state carrier distributions in semiconductor devices. *Apl. Mat.* **32** (1987), 49–56.

[192] Gröger, K., Nečas, J.: On a class of nonlinear initial-value problems in Hilbert spaces. *Math. Nachrichten* **93** (1979), 21–31.

[193] Gronwall, T.H.: Note on the derivatives with respect to a parameter of the solution of a system of differential equations. *Ann. Math.* **20** (1919), 292–296.

[194] Hackl, K: Generalized standard media and variational principles in classical and finite strain elastoplasticity. *J. Mech. Phys. Solids* **45** (1997), 667-688.

[195] Hahn, H.: Über Annäherung an Lebesgue'sche Integrale durch Riemann'sche Summen. *Sitzungber. Math. Phys. Kl. K. Akad. Wiss. Wien* **123** (1914), 713–743.

[196] Hahn, H.: Über lineare Gleichungsysteme in linearen Räume. *J. Reine Angew. Math.* **157** (1927), 214–229.

[197] Halphen, B, Nguyen, Q.S.: Sur les matériaux standards généralisés. *J. Mécanique* **14** (1975), 39–63.

[198] Haraux, A.: Anti-periodic solutions of some nonlinear evolution equations. *Manuscripta Math.* **63** (1989), 479–505.

[199] Haslinger, J., Mietinen, M., Panagiotopoulos, P.D.: *Finite Element Method for Hemivariational Inequalities.* Kluwer, Dordrecht, 1999.

[200] Henri, D. *Geometric Theory of Semilinear Parabolic Equations.* Springer, Berlin, 1981.

[201] Hess, P.: On nonlinear problems of monotone type with respect to two Banach spaces. *J. Math. Pures Appl.* **52** (1973), 13–26.

[202] HILBERT, D.: Mathematatische probleme. *Archiv d. Math. u. Physik* **1** (1901), 44–63, 213–237. French transl.: *Comp. Rendu du Deuxième Cong. Int. Math.*, Gauthier-Villars, Paris, 1902, pp.58–114. Engl. transl.: *Bull. Amer. Math. Soc.* **8** (1902), 437–479.

[203] HLAVÁČEK, I.: Variational principle for parabolic equations. *Apl. Mat.* **14** (1969), 278–297.

[204] HLAVÁČEK, I., HASLINGER, J., NEČAS, J., LOVÍŠEK, J.: *Solution of Variational Inequalities in Mechanics.* Springer, New York, 1988.

[205] HOFFMANN, K.-H., TANG, Q.: *Ginzburg-Landau Phase Transition Theory and Superconductivity.* Birkhäuser, Basel, 2001.

[206] HOFFMANN, K.-H., ZOCHOWSKI, A.: Existence of solutions to some non-linear thermoelastic systems with viscosity. *Math. Methods in the Applied Sciences* **15** (1992), 187–204.

[207] HOKKANEN, V.-M., MOROSANU, G.: *Functional Methods in Differential Equations.* Chapman & Hall/CRC, Boca Raton, 2002.

[208] HÖLDER, O.: Ueber einen Mittelwerthsatz. *Nachr. Ges. Wiss. Göttingen* (1889), 38–47.

[209] HU, S., PAPAGEORGIOU, N.S.: *Handbook of Multivalued Analysis I,II.* Kluwer, Dordrecht, Part I: 1997, Part II: 2000.

[210] ILLNER, R., WICK, J.: On statistical and measure-valued solutions of differential equations. *J. Math. Anal. Appl.* **157** (1991), 351–365.

[211] IOFFE, A.D., TIKHOMIROV, V.M.: *Theory of Extremal Problems.* (In Russian.) Nauka, Moscow, 1974. Engl. transl.: North-Holland, Amsterdam, 1979.

[212] ITO, K., KAPPEL, F.: *Evolution Equations and Approximation.* World Scientific, New Jersey, 2002.

[213] JARUSEK, J., MÁLEK, J., NEČAS, J., ŠVERÁK, V.: Variational inequality for a viscous drum vibrating in the presence of an obstacle. *Rend. di Matematica* **12** (1992), 943–958.

[214] JÄGER, W., KAČUR, J.: Solutions of porous medium type systems by linear approximation schemes. *Numer. Math.* **60** (1991), 407-427.

[215] JEROME, J.W.: The method of lines and the nonlinear Klein-Gordon equation. *J. Diff. Eq.* **30** N-1 (1978), 20–31.

[216] JEROME, J.W.: Consistency of semiconductor modeling: an existence/stability analysis for the stationary Van Roosbroeck system. *SIAM J. Appl. Math.* **45** (1985), 565–590.

[217] JIANG, S., RACKE, R.: *Evolution Equations in Thermoelasticity.* Chapman & Hall / CRC, Boca Raton, 2000.

[218] JOST, J., LI-JOST, X.: *Calculus of Variations.* Cambridge Univ. Press, Cambridge, 1998.

[219] KAČUR, J.: *Method of Rothe in Evolution Equations.* Teubner, Lepzig, 1985.

[220] KAČUR, J.: Solution of degenerate parabolic problems by relaxation schemes. In: *Recent Advances in Problems of Flow and Transport in Porous Media* (J.M.Crolet and M.E.Hatri, eds.), Kluwer Acad. Publ., 1998, pp.89-98.

[221] KAČUR, J.: Solution to strongly nonlinear parabolic problems by a linear approximation scheme. *IMA J. Numer. Anal.* **19** (1999), 119–145.

[222] KAČUR, J., LUCKHAUS, S.: Approximation of degenerate parabolic systems by nondegenerate elliptic and parabolic systems. *Applied Numerical Mathematics* **26** (1998), 307-326.

[223] Kagei, Y., Růžička, M., Thäter, G.: Natural Convection with Dissipative Heating. *Comm. Math. Physics* **214** (2000), 287–313.

[224] Kakutani, S.: A generalization of Brouwer's fixed-point theorem. *Duke Math. J.* **8** (1941), 457–459.

[225] Kato, T.: Nonlinear semigroups and evolution equations. *J. Math. Soc. Japan* **19** (1967), 508–520.

[226] Kato, T.: Accretive operators and nonlinear evolution equations in Banach spaces. In: *Nonlinear Funct. Anal.* (Ed.: F.E.Browder) Proc.Symp.Pure Math. XVIII, Part I, 1968, 138–161.

[227] Kenmochi, N.: Nonlinear operators of monotone type in reflexive Banach spaces and nonlinear perturbations. *Hiroshima Math. J.* **4** (1974), 229–263.

[228] Kenmochi, N.: Systems of nonlinear PDEs arising from dynamical phase transition. In: *Phase Transitions and Hysteresis.* (A.Visintin, ed.) L.N. in Math. 1584, Springer, Berlin, 1994, pp.39–86.

[229] Kenmochi, N., Niezgódka, M.: Evolution systems of nonlinear variational inequalities arising from phase change problems. *Nonlinear Anal., Th. Meth. Appl.* **23** (1994), 1163–1180.

[230] Kenmochi, N., Niezgódka, M.: Non-linear system for non-isothermal diffusive phase separation. *J. Math. Anal. Appl.* **188** (1994), 651–679.

[231] Kikuchi, N., Oden, J.T.: *Contact Problems in Elasticity.* SIAM, Philadelphia, 1988.

[232] Kinderlehrer, D., Stampacchia, G.: *An Introduction to Variational Inequalities and their Applications.* Academic Press, New York, 1980.

[233] Klei, H.-A., Miyara, M.: Une extension du lemme de Fatou. *Bull. Sci. Math.* 2nd *serie* **115** (1991), 211–221.

[234] Kobayashi, Y., Oharu, S.: Semigroup of locally Lipschitzian operators and applications. In: *Funct. Anal. and Related Topics.* (Ed. H.Komatsu.) Springer, Berlin, 1991.

[235] Komura, Y.: Nonlinear semigroups in Hilbert spaces. *J. Math. Soc. Japan* **19** (1967), 493–507.

[236] Kondrachov, V.I.: Sur certaines propriétés fonctions dans l'espace L^p. *C.R. (Doklady) Acad. Sci. USSR (N.S.)* **48** (1945), 535–538.

[237] Korn, A.: Sur les équations d'élasticité. *Ann. École Norm.* **24** (1907), 9–75.

[238] Krasnoselskiĭ, M.A., Zabreĭko, P.P., Pustylnik, E.I., Sobolevskiĭ, P.E.: *Integral Operators in Spaces of Summable Functions.* Nauka, Moscow, Russia, 1966, in Russian. Engl. transl.: Noordhoff, Leyden, 1976.

[239] Krejčí, P.: Evolution Variational Inequalities and Multidimensional Hysteresis Operators. In: Drábek, P., Krejčí, P., Takáč, P.: *Nonlinear Differential Equations.* Chapman & Hall / CRC, Boca Raton, 1999.

[240] Krejčí, P., Laurencot, Ph.: Generalized variational inequalities. *J. Convex Anal.* **9** (2002), 159–183.

[241] Krejčí, P., Roche, T.: Lipschitz continuous data dependence of sweeping processes in BV spaces. *Disc. Cont. Dynam. Systems, Sec.B* **15** (2011), 637–650.

[242] Kristensen, J.: On the non-locality of quasiconvexity. *Ann. Inst. Henri Poincaré, Anal. Non Linéaire* **16** (1999), 1-13.

[243] Křížek, M., Liu, L.: On a comparison principle for a quasilinear elliptic boundary value problem of a nonmonotone type. *Applicationes Mathematicae* **24** (1996), 97–107.

[244] Kružík, M.: Variational methods in multipolar elasticity. MS thesis, Math.-Phys. Fac., Charles Univ., Prague, 1993.

[245] Kufner, A., John, O., Fučík, S.: *Function Spaces.* Academia, Praha, and Nordhoff Int. Publ., Leyden, 1977.

[246] Kunze, M., Monteiro Marques, M.D.P.: Existence of solutions for degenerate sweeping processes. *J. Convex Anal.* **4** (1997), 165–176.

[247] Kuzin, L., Pohozaev, S.: *Entire Solutions of Semilinear Elliptic Equations.* Birkhäuser, Basel, 1997.

[248] Ladyzhenskaya, O.A.: *The Mathematical Theory of Viscous Incompressible Flow.* Gordon and Beach, New York, 1969.

[249] Ladyzhenskaya, O.A., Solonnikov, V.A., Ural'tseva, N.N.: *Linear and Quasilinear Equations of Parabolic Type.* Nauka, Moscow, 1967. (Engl. Transl.: AMS, Providence, 1968.)

[250] Ladyzhenskaya, O.A., Ural'tseva, N.N.: *Linear and Quasilinear Equations of Elliptic Type.* Nauka, Moscow, 1964. (Engl. Transl.: Acad. Press, New York, 1968.)

[251] Landau, L., Lifshitz, E.: On the theory of dispersion of magnetic permeability in ferromagnetic bodies. *Phys. Z. Sowjet.* **8** (1935), 153–169.

[252] Larsen, C.J., Ortner, C., Süli, E.: Existence of solution to a regularized model of dynamic fracture. *Math. Models Meth. Appl. Sci.*, **20** (2010), 1021-1048.

[253] Lax, P., Milgram, N.: Parabolic equations. *Annals of Math. Studies* **33** (1954), 167–190, Univ. Press, Princeton, NJ.

[254] Lebesgue, H.: Sur les intégrales singuliéres. *Ann. Fac. Sci Univ. Toulouse, Math.-Phys.* **1** (1909), 25–117.

[255] Lees, M.: Apriori estimates for the solutions of difference approximations to parabolic differential equations. *Duke Math. J.* **27** (1960), 297–311.

[256] Leray, J.: Sur le mouvement d'un liquide visqueux emplissant l'espace. *Acta Mathematica* **63** (1934), 193–248.

[257] Leray, J., Lions, J.L.: Quelques résultats de Višik sur les problèmes elliptiques non linéaires par les méthodes de Minty-Browder. *Bull. Soc. Math. France* **93** (1965), 97–107.

[258] Levi, B.: Sul prinzipio di Dirichlet. *Rend. Circ. Mat. Palermo* **22** (1906), 293–359.

[259] Liebermann, G.M.: *Second Order Parabolic Differential Equations.* World Scientific, Singapore, 1996.

[260] Lions, J.L.: Sur certaines équations paraboliques non linéaires. *Bull Soc. math. France* **93** (1965), 155–175.

[261] Lions, J.L.: *Quelques Méthodes de Résolution des Problémes aux Limites non linéaires.* Dunod, Paris, 1969.

[262] Lions, J.L., Magenes, E.: *Problèmes aux Limites non homoegènes et Applications.* Dunod, Paris, 1968.

[263] Ljusternik, L., Schnirelman, L.: *Méthodes topologiques dans les problémes variationnels.* Hermann, Paris, 1934.

[264] Lotka, A.: Undamped oscillations derived from the law of mass action. *J. Am. Chem. Soc.* **42** (1920), 1595–1599.

[265] Lucchetti, R., Patrone, F.: On Nemytskii's operator and its application to the lower semicontinuity of integral functionals. *Indiana Univ. Math. J.* **29** (1980), 703–713.

[266] Lumer, G., Phillips, R.S.: Dissipative operators in a Banach space. *Pacific J. Math.* **11** (1961), 679–698.

[267] MAGENES, E, VERDI, C, VISINTIN, A.: Semigroup approach to the Stefan problem with non-linear flux. *Atti Acc. Lincei Rend. fis.* - S.VIII, **75** (1983), 24–33.

[268] MÁLEK, J., NEČAS, J., ROKYTA, M., RŮŽIČKA, M.: *Weak and Measure-Valued Solutions to Evolution Partial Differential Equations.* Chapman & Hall, London, 1996.

[269] MÁLEK, J., PRAŽÁK, D., STEINHAUER, M.: On existence of solutions for a class of degenerate power-law fluids. *Diff. Integral Equations* **19** (2006), 449–462.

[270] MÁLEK, J., RŮŽIČKA, M., THÄTER, G.: Fractal dimension, attractors and Boussinesq approximation in three dimensions. *Act. Appl. Math.* **37** (1994), 83–98.

[271] MALÝ, J., ZIEMER, P.: *Fine Regularity of Solutions of Elliptic Partial Differential Equations.* Amer. Math. Soc., Providence, 1997.

[272] MARCHUK, G.I.: Splitting and alternating direction methods. In: *Handbook of Numerical Analysis I* (Eds: P.G.Ciarlet, J.L.Lions), Elsevier, Amsterdam, 3rd. ed. 2003, pp.197–462.

[273] MARCUS, M., MIZEL, V.J.: Nemitsky operators on Sobolev spaces. *Archive Ration. Mech. Anal.* **51** (1973), 347–370.

[274] MARCUS, M., MIZEL, V.J.: Every superposition operator mapping one Sobolev space into another is continuous. *J. Funct. Anal.* **33** (1979), 217–229.

[275] MARKOWICH, P.A.: *The Stationary Semiconductor Device Equations.* Springer, Wien, 1986.

[276] MARKOWICH, P.A., RINGHOFER, C.A., SCHMEISER, C.: *Semiconductor Equations.* Springer, Wien, 1990.

[277] MAZ'YA, V.G.: *Sobolev spaces.* Springer, Berlin, 1985.

[278] MIELKE, A.: Evolution of rate-independent systems. In: *Handbook of Differential Equations: Evolutionary Diff. Eqs., vol. 2* (Eds. C.Dafermos, E.Feireisl), Elsevier B.V., Amsterdam, 2005, pp. 461–559.

[279] MIELKE, A.: Differential, energetic and metric formulations for rate-independent processes. In: *Nonlinear PDEs and Applications* (Eds: L.Ambrosio, G.Savaré), Springer, 2010, pp.87–170.

[280] MIELKE, A., ROSSI, R., SAVARÉ, G.: Nonsmooth analysis of doubly nonlinear evolution equations *Calc. Var., P.D.E.* (2012), in print; DOI: 10.1007/s00526-011-0482-z.

[281] MIELKE, A., ROUBÍČEK, T., STEFANELLI, U.: Γ-limits and relaxations for rate-independent evolutionary problems. *Calc. Var., P.D.E.* **31** (2008), 387–416.

[282] MIELKE, A., THEIL, F.: A mathematical model for rate-independent phase transformations with hysteresis. In: *Models of continuum mechanics in analysis and engineering.* (Eds.: H.-D.Alber, R.Balean, R.Farwig), Shaker Verlag, Aachen, 1999, pp.117–129.

[283] MIELKE, A., THEIL, F.: On rate-independent hysteresis models. *Nonlinear Diff. Equations Appl.* **11** (2004), 151–189.

[284] MIKLAVČIČ, M.: *Applied Functional Analysis and Partial Differential Equations.*, World Scientific, Singapore, 1998.

[285] MILMAN, D.: On some criteria for the regularity of spaces of the type (B). *C. R. Acad. Sci. URSS (Doklady Akad. Nauk SSSR)* **20** (1938), 243–246.

[286] MINTY, G.: On a monotonicity method for the solution of non-linear equations in Banach spaces. *Proc. Nat. Acad. Sci. USA* **50** (1963), 1038–1041.

[287] MIYADERA, I.: *Nonlinear Semigroups.* AMS, Providence, R.I., 1992.

[288] MOCK, M.S.: On equations describing steady-state carrier distributions in semiconductor devices. *Comm. Pure Appl. Math.* **25** (1972), 781–792.

[289] MOCK, M.S.: *Analysis of Mathematical Models of Semiconductor Devices.* Boole Press, Dublin, 1983.

[290] MORREY, JR., C.B.: Quasi-convexity and the lower semicontinuity of multiple integrals. *Pacific J. Math.* **2** (1952), 25–53.

[291] MORREY, JR., C.B.: *Multiple Integrals in the Calculus of Variations.* Springer, Berlin, 1966.

[292] MOSCO, U.: A remark on a theorem of F.E.Browder. *J. Math. Anal. Appl.* **20** (1967), 90–93.

[293] MOSCO, U.: Convergence of convex sets and of solutions of variational inequalities. *Adv. in Math.* **3** (1969), 510–585.

[294] MOSCO, U.: Implicit variational problems and quasivariational inequalities. In: *Nonlinear Oper. Calc. Var.*, Lect. Notes in Math. **543**, Springer, Berlin, 1976, pp.83–156.

[295] MOSER, J. A new proof of De Giorgi's theorem concerning the regularity problem for elliptic differential equations. *Comm. Pure Appl. Math.* **13** (1960), 457–468.

[296] MÜLLER, S.: Higher integrability of determinants and weak convergence in L^1. *J. reine angew. Math.* **412** (1990), 20–34.

[297] MÜLLER, S.: Variational models for microstructure and phase transitions. In: *Calculus of Variations and Geometric Evolution Problems.* (Eds.: S.Hildebrandt et al.) Lect. Notes in Math. **1713** (1999), Springer, Berlin, pp.85–210.

[298] NANIEWICZ, Z., PANAGIOTOPOULOS, P.D.: *Mathematical Theory of Hemivariational Inequalities.* Marcel Dekker, 1995.

[299] NAUMANN, J.: *Einführung in die Theorie parabolischer Variationsungleichungen.* Teubner, Leipzig, 1984.

[300] NAUMANN, J., POKORNÝ, M., WOLF, J.: On the existence of weak solutions to the equations of steady flow of heat-conducting fluids with dissipative heating. *Nonlinear Anal.-Real World Appl.* **13** (2012), 1600-1620.

[301] NAYROLES, B.: Deux théorèmes de minimum pour certains systèmes dissipatifs. *C.R. Acad. Sci. Paris Sér. A-B* **282** (1976), A1035–A1038.

[302] NEČAS, J.: *Les Méthodes Directes en la Théorie des Equations Elliptiques.* Academia, Praha & Masson, Paris, 1967.

[303] NEČAS, J.: Les équations elliptiques non linéaires. *Czech. Math. J.* **19** (1969), 252–274.

[304] NEČAS, J.: Application of Rothe's method to abstract parabolic equations. *Czech. Math. J.* **24** (1974), 496–500.

[305] NEČAS, J.: *Introduction to the Theory of Nonlinear Elliptic Equations.* Teubner, Leipzig, 1983 & J.Wiley, Chichester, 1986.

[306] NEČAS, J.: Dynamic in the nonlinear thermo-visco-elasticity.In: Symposium *Partial Differential Equations* Holzhau 1988 (Eds.: B.-W.Schulze, H.Triebel.), Teubner-Texte zur Mathematik **112**, pp. 197–203. Teubner, Leipzig, 1989.

[307] NEČAS, J.: Sur les normes équivalentes dans $W_p^k(\Omega)$ et sur la coercivité des formes formellement positives. In: *Séminaire Equations aux Dérivées Partielles.* Montréal (1996), 102–128.

[308] NEČAS, J., HLAVÁČEK, I.: *Mathematical Theory of Elastic and Elasto-Plastic Bodies: An Introduction.* Elsevier, Amsterdam, 1981.

[309] NEČAS, J., NOVOTNÝ, A., ŠVERÁK, V.: On the uniqueness of solution to the nonlinear thermo-visco-elasticity. *Math. Nachr.* **149** (1990), 319–324.

[310] NEČAS, J., ROUBÍČEK, T.: Buoyancy-driven viscous flow with L^1-data. *Nonlinear Anal., Th. Meth. Appl.* **46** (2001), 737–755.

[311] NIEZGÓDKA, M., PAWŁOW, I.: A generalize Stefan problem in several space variables. *Appl. Math. Optim.* **9** (1983), 193–224.

[312] NIRENBERG, L.: On elliptic partial differential equations. *Ann. Scuola Norm. Sup. Pisa Cl. Sci.* **13** (1959), 115–162.

[313] NIRENBERG, L.: An extended interpolation inequality. *Ann. Scuola Norm. Sup. Pisa Cl. Sci.* **20** (1966), 733–737.

[314] NOCHETTO, R.H., SAVARÉ, G., VERDI, C.: A posteriori error estimates for variable time-step discretization of nonlinear evolution equations. *Comm. Pure Appl. Math.* **53** (2000), 525–589.

[315] NOVICK-COHEN, A.: On Cahn-Hilliard type equations. *Nonlinear Anal., Theory Methods Appl.* **15** (1990), 797–814.

[316] NOVICK-COHEN, A.: The Cahn-Hilliard equation: Mathematical and modeling perspectives. *Adv. Math. Sci. Appl.* **8** (1998), 965–985.

[317] OHTA, T., MIMURA, M., KOBAYASHI, R.: Higher-dimensional localized patterns in excitable media. *Physica D* **34** (1989),115–144.

[318] ORNSTEIN, D.: A non-inequality for differential operators in the L^1-norm. *Archive Ration. Mech. Anal.* **11** (1962), 40–49.

[319] OTTO, F.: The geometry of dissipative evolution equations: the porous medium equations. *Comm. P.D.E.* **26** (2001), 101–174.

[320] OTTO, F.: L^1-contraction and uniqueness for unstationary saturated-unsaturated porous media flow. *Adv. Math. Sci. Appl.* **7** (1997), 537–553.

[321] OUTRATA, J.V., KOČVARA, M., ZOWE, J.: *Nonsmooth Approaches to Optimization Problems with Equilibrium Constraints.* Kluwer, Dordrecht, 1998.

[322] PANAGIOTOPOULOS P.D.: *Inequality Problems in Mechanics and Applications: Convex and Nonconvex Energy Functions.* Boston, Birkhäuser, 1985.

[323] PAPAGEORGIOU, N.S.: On the existence of solutions for nonlinear parabolic problems with nonmonotone discontinuities. *J. Math. Anal. Appl.* **205** (1997), 434–453.

[324] PAO, V.C.: *Nonlinear Parabolic and Elliptic Equations.* Plenum Press, New York, 1992.

[325] PASCALI, D., SBURLAN, S.: *Nonlinear Mappings of Monotone Type.* Editura Academiei, Bucuresti, 1978.

[326] PAVEL, N.H.: *Nonlinear Evolution Operators and Semigroups.* Lect. Notes in Math. **1260**, Springer, Berlin, 1987.

[327] PAWŁOW, I., ZAJĄCZKOWSKI, W.M.: Global existence to a three-dimensional nonlinear thermoelasticity system arising in shape memory materials. *Math. Methods Appl. Sci.* **28** (2005), 407–442.

[328] PAWŁOW, I., ZAJĄCZKOWSKI, W.M.: Global regular solutions to a Kelvin-Voigt type thermoviscoelastic system. A preprint.

[329] PAWŁOW, I., ZOCHOWSKI, A.: Existence and uniqueness of solutions for a three-dimensional thermoelastic system. *Dissertationes Mathematicae* **406**, IM PAN, Warszawa, 2002.

[330] PAZY, A.: *Semigroups of Linear Operators and Applications to Partial Differential Equations.* Springer, New York, 1983.

[331] PEDREGAL, P.: *Parametrized Measures and Variational Principles.* Birkhäuser, Basel, 1997.

[332] PEDREGAL, P.: *Variational Methods in Nonlinear Elasticity.* SIAM, Philadelphia, 2000.

[333] PENROSE, O., FIFE, P.C.: Theromdynamically consistent models of phase-field type for kinetics of phase transitions. *Physica D* **43** (1990), 44–62.

[334] PETTIS, B.J.: On integration in vector spaces. *Trans. Amer. Math. Soc.* **44** (1938), 277–304.

[335] PETTIS, B.J.: A proof that every uniformly convex space is reflexive. *Duke Math. J.* **5** (1939), 249–253.

[336] PODIO-GUIDUGLI, P., ROUBÍČEK, T., TOMASSETTI, G.: A thermodynamically-consistent theory of the ferro/paramagnetic transition. *Archive Ration. Mech. Anal.* **198** (2010), 1057–1094.

[337] PODIO-GUIDUGLI, P., VERGARA CAFFARELLI, G.: Surface interaction potentials in elasticity. *Archive Ration. Mech. Anal.* **109** (1990), 343-381.

[338] PODIO-GUIDUGLI, P., VIANELLO, M.: Hypertractions and hyperstresses convey the same mechanical information. *Cont. Mech. Thermodynam.* **22** (2010), 163–176.

[339] POINCARÉ, H.: Sur les équations aux dérivées partialles de la physique mathématique. *Amer. J. Math.* **12** (1890), 211–294.

[340] PRIGOGINE, I.: *Étude Thermodynamique des Processes Irreversibles.* Desoer, Lieg, 1947.

[341] QUARTERONI, A., VALLI, A.: *Numerical Approximation of Partial Differential Equations.* Springer, Berlin, 1994.

[342] RADEMACHER, H.: Über partielle und totale Differenzierbarkeit I. *Math. Ann.* **79** (1919), 340–359.

[343] RAJAGOPAL, K.R., ROUBÍČEK, T.: On the effect of dissipation in shape-memory alloys. *Nonlinear Anal., Real World Appl.* **4** (2003), 581–597.

[344] RAJAGOPAL, K. R., RŮŽIČKA, M., SRINIVASA, A. R.: On the Oberbeck-Boussinesq Approximation. *Math. Models Methods Appl. Sci.* **6** (1996), 1157–1167.

[345] RAKOTOSON, J.M.: Generalized solutions in a new type of sets for problems with measures as data. *Diff. Integral Eq.* **6** (1993), 27–36.

[346] RAMOS, A.M., ROUBÍČEK, T.: Noncooperative game with a predator-prey system. *Appl. Math. Optim.*, **56** (2007), 211–241.

[347] REKTORYS, K.: *The Method of Discretization in Time.* D.Reidel, Dordrecht, 1982.

[348] RELLICH, F.: Ein Satz über mittlere Konvergenz. *Nachr. Akad. Wiss. Göttingen*, 1930, pp.30–35.

[349] RENARDY, M., ROGERS, R.C.: *An Introduction to Partial Differential Equations.* Springer, New York, 1993.

[350] RITZ, W.: Über eine neue Methode zur Lösung gewisser Variationsprobleme der mathematischen Physik. *J. für reine u. angew. Math.* **135** (1908), 1–61.

[351] ROBINSON, J.C.: *Infinite-dimensional dynamical systems.* Cambridge Univ. Press, Cambridge, 2001.

[352] ROCKAFELLAR, R.T.: On maximal monotonicity of subdifferential mappings. *Pacific J. Math.* **33** (1970), 209–216.

[353] ROCKAFELLAR, R.T., WETTS, R.J.-B.: *Variational analysis.* Springer, Berlin, 1998.

[354] RODRIGUES, J.-F.: *Obstacle Problems in Mathematical Physics.* North-Holland, Amsterdam, 1987.

[355] ROOSBROECK, W. VAN: Theory of flow of electrons and holes in germanium and other semiconductors. *Bell System Tech. J.* **29** (1950), 560–607.

[356] ROTHE, E.: Zweidimensionale parabolische Randwertaufgaben als Grenzfall eindimensionaler Randwertaufgaben. *Math. Ann.* **102** (1930), 650–670.

[357] ROUBÍČEK, T.: Unconditional stability of difference formulas. *Apl. Mat.* **28** (1983), 81–90.

[358] ROUBÍČEK, T.: A model and optimal control of multidimensional thermoelastic processes within a heating of large bodies. *Probl. Control Inf. Theory* **16** (1987), 283–301.

[359] ROUBÍČEK, T.: A generalization of the Lions-Temam compact imbedding theorem. *Časopis pěst. mat.* **115** (1990), 338–342.

[360] ROUBÍČEK, T.: *Relaxation in Optimization Theory and Variational Calculus*, W. de Gruyter, Berlin, 1997.

[361] ROUBÍČEK, T.: Nonlinear heat equation with L^1-data. *Nonlinear Diff. Eq. Appl.* **5** (1998), 517–527.

[362] ROUBÍČEK, T.: Direct method for parabolic problems. *Adv. Math. Sci. Appl.* **10** (2000), 57–65.

[363] ROUBÍČEK, T.: Steady-state buoyancy-driven viscous flow with measure data. *Mathematica Bohemica* **126** (2001), 493–504.

[364] ROUBÍČEK, T.: Incompressible fluid mixtures of ionized constituents. In: Proc. *STAMM 2004* (Y.Wang, K.Hutter, eds.) Shaker Ver., Aachen., 2005, pp. 429–440.

[365] ROUBÍČEK, T.: Incompressible ionized fluid mixtures. *Cont. Mech. Thermodyn.* **17** (2006), 493–509.

[366] ROUBÍČEK, T.: Incompressible ionized non-Newtonean fluid mixtures. *SIAM J. Math. Anal.* **39** (2007), 863–890.

[367] ROUBÍČEK, T.: On non-Newtonian fluids with energy transfer. *J. Math. Fluid Mech.* **11** (2009), 110–125.

[368] ROUBÍČEK, T.: Thermo-visco-elasticity at small strains with L^1-data. *Quarterly Appl. Math.* **67** (2009), 47–71.

[369] ROUBÍČEK, T.: Nonlinearly coupled thermo-visco-elasticity. *Nonlin. Diff. Eq. Appl.*, to appear.

[370] ROUBÍČEK, T., HOFFMANN, K.-H.: About the concept of measure-valued solutions to distributed parameter systems. *Math. Methods in the Applied Sciences* **18** (1995), 671–685.

[371] ROUBÍČEK, T., TOMASSETTI, G.: Thermodynamics of shape-memory alloys under electric current. *Zeit. angew. Math. Phys.* **61** (2010), 1–20.

[372] ROUBÍČEK, T, TOMASSETTI, G.: Ferromagnets with eddy currents and pinning effects: their thermodynamics and analysis. *Math. Models Meth. Appl. Sci.* **21** (2011), 29–55.

[373] ROUBÍČEK, T, TOMASSETTI, G.: Phase transformations in electrically conductive ferromagnetic shape-memory alloys, their thermodynamics and analysis. *Archive Ration. Mech. Anal.*, to appear.

[374] ROUBÍČEK, T, TOMASSETTI, G., ZANINI, C.: The Gilbert equation with dry-friction-type damping. *J. Math. Anal. Appl.* **355** (2009), 453–468

[375] RULLA, J.: Weak solutions to Stefan problems with prescribed convection. *SIAM J. Math. Anal.* **18** (1987), 1784–1800.

[376] RŮŽIČKA, M.: *Nichtlineare Funktionalanalysis.* Springer, Berlin, 2004.

[377] SAADOUNE, M., VALADIER, M.: Extraction of a "good" sequence from a bounded sequence of integrable functions. *J. Convex Anal.* **2** (1994), 345–357.

[378] SAMOHÝL, I.: Application of Truesdell's model of mixture to ionic liquid mixture. *Computers Math. Appl.* **53** (2007), 182–197.

[379] SCHAUDER J.: Der Fixpunktsatz in Fuktionalräumen. *Studia Math.* **2** (1930), 171–180.

[380] SELBERHERR, S.: *Analysis and Simulation of Semiconductor Devices.* Springer, Wien, 1984.

[381] SERRIN, J.: Pathological solutions of elliptic differential equations. *Ann. Scuola Norm. Sup. Pisa Cl. Sci.* **18** (1964), 385–387.

[382] SHOWALTER, R.E.: Nonlinear degenerate evolution equations and partial differential equations of mixed type. *SIAM J. Math. Anal.* **6** (1975), 25–42.

[383] SHOWALTER, R.E.: *Monotone Operators in Banach Space and Nonlinear Partial Differential Equations.* AMS Math. Surveys and Monographs **49**, 1997.

[384] SHOWALTER, R.E., SHI, P.: Plasticity models and nonlinear semigroups. *J. Math. Anal. Appl.* **216** (1997), 218–245

[385] ŠILHAVÝ, M.: Phase transitions in non-simple bodies. *Archive Ration. Mech. Anal.* **88** (1985), 135–161.

[386] SIMON, J.: Compact sets in the space $L^p(0,T;B)$. *Annali di Mat. Pura Applic.* **146** (1987), 65–96.

[387] SKRYPNIK, I.V.: *Methods for Analysis of Nonlinear Elliptic Boundary Value Problems.* Nauka, Moskva, 1990; Engl. Transl. AMS, Providence, 1994.

[388] SLEMROD, M.: Global existence, uniqueness and asymptotic stability of classical smooth solutions in one-dimensional non-linear thermoelasticity. *Archive Ration. Mech. Anal.* **76** (1981), 97–133.

[389] SOBOLEV, S.L.: On some estimates relating to families of functions having derivatives that are square integrable. (In Russian.) *Dokl. Akad. Nauk SSSR* **1** (1936), 267–270.

[390] SOBOLEV, S.L.: *Applications of Functional Analysis to Mathematical Physics.* (In Russian) Izdat. LGU, Leningrad, 1950; Engl. transl.: AMS Transl. **7**, 1963.

[391] SOHR, H.: *The Navier-Stokes equations.* Birkhäuser, Basel, 2001.

[392] SRAUGHAN, B.: *The Energy Method, Stability, and Nonlinear Convection.* 2nd ed., Springer, New York, 2004.

[393] STAMPACCHIA, G.: Le problème de Dirichlet pour les équations elliptiques du second ordre à coefficients discontinus. *Ann. Inst. Fourier (Grenoble)* **15** (1965), 189–258.

[394] STARÁ, J., JOHN, O.: *Funkcionální analýza. Nelineární úlohy.* Skripta, MFF UK, SPN, Praha, 1986.

[395] STEFANELLI, U.: On some nonlocal evolution equations in Banach spaces. *J. Evol. Equ.* **4** (2004), 1-26.

[396] STEFANELLI, U.: The Brezis-Ekeland principle for doubly nonlinear equations. *SIAM J. Control Optim.* **47** (2008), 1615-1642.

[397] STEFANELLI, U.: A variational principle for hardening elastoplasticity. *SIAM J. Math. Anal.* **40** (2008), 623–652.

[398] STEFANELLI, U.: A variational characterization of rate-independent evolution. *Math. Nachr.* **282** (2009), 1492–1512.

[399] STRAŠKRABA, I., VEJVODA, O.: Periodic solutions to abstract differential equations. *Czech Math. J.* **23** (1973), 635–669.

[400] STRUWE, M.: *Variational Methods.* Springer, Berlin, 1990.

[401] ŠVERÁK, V.: Rank-one convexity does not imply quasiconvexity. *Proc. Royal Soc. Edinburgh* **120** (1992), 185–189.

[402] TAYLOR, M.E.: *Partial Differential Equations* III. *Nonlinear Equations.* Springer, 1996.

[403] TEMAM, R.: *Navier-Stokes Equations.* North-Holland, Amsterdam, 1979.

[404] TIBA, D.: *Optimal Control of Nonsmooth Distributed Parameter Systems.* Lect. Notes in Math. **1459**, Springer, Berlin, 1990.

[405] THOMÉE, V.: *Galerkin Finite Element Methods for Parabolic Problems.* Springer, Berlin, 1997.
[406] TONELLI, L.: Sula quadratura delle superficie. *Atti Reale Accad. Lincei* **6** (1926), 633–638.
[407] TOUPIN, R.A.: Elastic materials with couple stresses. *Archive Ration. Mech. Anal.* **11** (1962), 385–414.
[408] TRIEBEL, H.: *Theory of Function Spaces.* Birkhäuser, Basel, 1983.
[409] TROIANIELLO, G.M.: *Elliptic Differential Equations and Obstacle Problems.* Plenum Press, New York, 1987.
[410] TRÖLTZSCH, F.: *Optimality Conditions for Parabolic Control Problems and Applications.* Teubner, Leipzig, 1984.
[411] TROTTER, H.F.: On the product of semigroups of operators. *Trans. Amer. Math. Soc.* **10** (1959), 545-551.
[412] TROYANSKI, S.: On locally uniformly convex and differentiable norms in certain non-separable spaces. *Studia Math.* **37** (1971), 173–180.
[413] TYCHONOFF, A.: Ein Fixpunktsatz. *Math. Anal.* **111** (1935), 767–776.
[414] VAINBERG, M.M.: *Variational Methods and Method of Monotone Operators in the Theory of Nonlinear Equations.* J.Wiley, New York, 1973.
[415] VEJVODA, O. ET AL.: *Partial Differential Equations.* Noordhoff, Alphen aan den Rijn, 1981.
[416] VISHIK, M.I.: Quasilinear strongly elliptic systems of differential equations in divergence form. *Trans. Moscow Math. Soc.* (1963), 140–208.
[417] VISINTIN, A.: Strong convergence results related to strict convexity. *Comm. Partial Diff. Equations* **9** (1984), 439–466.
[418] VISINTIN, A.: *Models of Phase Transitions.* Birkhäuser, Boston, 1996.
[419] VISINTIN, A.: Extension of the Brezis-Ekeland-Nayroles principle to monotone operators. *Adv. Math. Sci. Appl.* **18** (2008), 633–650.
[420] VISINTIN, A.: Variational formulation and structural stability of monotone equations. *Calc. Var.* (2012), in print: DOI 10.1007/s00526-012-0519-y
[421] VITALI, G.: Sull'integrazione par serie. *Rend. del Circolo Mat. di Palermo* **23** (1907), 137–155.
[422] VON WAHL, W.: On the Cahn-Hilliard equation $u' + \Delta^2 u - \Delta f(u) = 0$. *Delft Prog. Res.* **10** (1985), 291–310.
[423] VOLTERRA, V.: Variazionie fluttuazioni del numero d'individui in specie animali convienti. *Mem. Acad. Lincei* **2** (1926), 31–113.
[424] WLOKA, J.: *Partial Differential Equations.* Cambridge Univ. Press, Cambridge, 1987. (German orig.: Teubner, Stuttgart, 1982).
[425] YOSIDA, K.: *Functional Analysis.* 6nd ed., Springer, Berlin (1980).
[426] YOSIDA, K., HEWITT, E.: Finitely additive measures. *Trans. Amer. Math. Soc.* **72** (1952), 46–66.
[427] ZEIDLER, E.: *Nonlinear Functional Analysis and its Applications, I. Fixed-Point Theorems, II. Monotone Operators, III. Variational Methods and Optimization, IV. Applications to Mathematical Physics.* Springer, New York, 1985–1990.
[428] ZHENG, SONGMU: *Nonlinear Parabolic Equations and Hyperbolic-Parabolic Coupled Systems.* Longman, Harlow, 1995.
[429] ZHENG, SONGMU: *Nonlinear Evolution Equations.* Chapman & Hall / CRC, Boca Raton, 2004.
[430] ZIEMER, W.P.: *Weakly Differentiable Functions.* Springer, New York, 1989.

Index

MIX
Papier aus verantwortungsvollen Quellen
Paper from responsible sources
FSC® C105338

If you have any concerns about our products,
you can contact us on
ProductSafety@springernature.com

In case Publisher is established outside the EU,
the EU authorized representative is:
Springer Nature Customer Service Center GmbH
Europaplatz 3, 69115 Heidelberg, Germany

Printed by Libri Plureos GmbH
in Hamburg, Germany